GENERA

EUPHORBIACEARUM

1. *Acalypha hispida* Burm. f.
2. *A. wilkesiana* Müll. Arg.
3. *Breynia disticha* J. R. & G. Forst. var. *disticha* f. *nivosa* (W. Bull) Croizat
4. *Codiaeum variegatum* (L.) A. Juss. f. *appendiculatum* Čelak.
5. *C. variegatum* (L.) A. Juss. f. *lobatum* Pax
6. *C. variegatum* (L.) A. Juss. f. *taeniosum* Müll. Arg.
7. *Dalechampia scandens* L.
8. *Euphorbia marginata* Pursh
9. *E.* (*Poinsettia*) *pulcherrima* Willd. ex Klotzsch
10. *Excoecaria cochinchinensis* Lour.
11. *Jatropha integerrima* Jacq.
12. *J. multifida* L.
13. *Ricinus communis* L. var. *purpurascens* (Bertol.) Müll. Arg.

GENERA EUPHORBIACEARUM

Alan Radcliffe-Smith

Tribes Hippomaneae,

Pachystromateae & Hureae

by H.-J. Esser

Illustrations

by Camilla Speight

Cover illustration & Frontispiece

by Christabel King

Royal Botanic Gardens, Kew

2001

Production Editor: R. Linklater

Cover and page make-up by Media Resources,
Information Services Department,
Royal Botanic Gardens, Kew

ISBN 1 84246 022 6

Printed in Great Britain by
The Cromwell Press Ltd

CONTENTS

INTRODUCTION

This account of the genera of the *Euphorbiaceae* is based primarily on the unpublished MSS of the family housed in the Kew Archives, which was to have been included by Dr John Hutchinson in Volume 3 of his "The Genera of Flowering Plants", only the first two volumes of which were published at the time of his death in 1972. However, whereas his generic descriptions were largely drawn upon for this account, his system of classification of the family, outlined in the American Journal of Botany in 1969, in which the genera are grouped in forty tribes but not in subfamilies, has not been followed. It was superseded by the system of Professor Grady L. Webster, first outlined in "Taxon" in 1975, and subsequently extended and considerably revised by the time the much fuller treatment of that system appeared in the Annals of the Missouri Botanic Garden in 1994. Apart from the addition of three new tribes, the segregation of two new subtribes and the reassignment, resurrection or subsumption of a handful of genera, this treatment closely follows the system of Webster. No attempt has been made to take account of recent findings based on DNA studies.

The Websterian system, revolutionary at the time of its first appearance, but now fairly generally accepted by most workers on the family, recognizes five subfamilies, primarily on a palynological basis, two of which, namely the *Phyllanthoideae* s. str. and the *Oldfieldioideae*, more or less correspond to the biovulate subfamily *Phyllanthoideae* s. l. in the system of Pax & Hoffmann, the last version of which appeared in 1931, whilst the other three, namely the *Acalyphoideae*, the *Crotonoideae* s. str. and the *Euphorbioideae*, more or less correspond to their uniovulate subfamily *Crotonoideae* s. l.

Pax & Hoffmann (1931) also had another category between the family and the subfamily, to which they gave no hierarchical designation, but which corresponded to Mueller's (1864, 1866) "Tribuum Series" (a category having no counterpart in modern classifications) and which comprised two taxa, namely the *Platylobeae* and the much smaller Australasian *Stenolobeae*, which depended on the width of the cotyledons. The former was divided into the two subfamilies *Phyllanthoideae* s. l. and *Crotonoideae* s. l., whilst the latter was also divided into two subfamilies, namely the biovulate *Porantheroideae* and the uniovulate *Ricinocarpoideae*.

Webster held the cotyledonar character to be of no significance at such a high rank, and so in his system he split the various stenolobe tribes among four of the platylobe subfamilies, the porantheroid tribe *Poranthereae* being assigned to the *Phyllanthoideae* as the subtribe *Porantherinae* of the tribe *Antidesmeae*, whilst the tribe *Caletieae* was assigned to the *Oldfieldioideae* and considerably expanded by the addition of several related platylobe genera. Of the ricinocarpoid tribes, the tribe *Ampereae* was assigned to the *Acalyphoideae* whilst the tribe *Ricinocarpeae* was assigned to the *Crotonoideae* with the addition of the genera *Alphandia, Myricanthe, Cocconerion & Borneodendron.* No stenolobe tribes or genera were assigned to the *Euphorbioideae.*

Of the taxa included here which are additional to the system of Webster, the three tribes mentioned above are the *Centroplaceae* mihi, the *Martretieae* Eg. Köhler ex J. Léonard (1989), dismissed by Webster because of a conflicting suggestion of Meeuse (1990), and the *Sphyranthereae* mihi, the two former belonging to the *Phyllanthoideae*, and the latter to the *Acalyphoideae*. Webster included the genus *Sphyranthera* in the *Crotonoideae*, placing it in the tribe *Codiaeae*. However, a recent investigation of the pollen carried out at Kew has shown that it is not possible to retain the genus in that subfamily, as the pollen is aperturate and lacks the characteristic *Croton*-pattern. It is clearly an acalyphoid genus.

The two new subtribes mentioned above are the acalyphoid *Mareyinae* (Hutch.) mihi, based on the tribe *Mareyeae* Hutch., which Webster had amalgamated with the subtribe *Claoxylinae* Hurus., and the crotonoid *Benoistiinae* (mihi) mihi, based on my tribe *Benoistieae*, which Webster had amalgamated with the subtribe *Neoboutoniinae* (Hutch.) G. L. Webster.

This treatment differs primarily from that of Webster (1994) in providing generic descriptions in addition to those of subfamilies, tribes and subtribes. Pollen morphological descriptions have been largely based on those provided by Punt (1962). Some attempt has been made to modify Webster's keys in order to take account of exceptions to the chief diagnostic characters of the genera, which has resulted in the blurring of the edges of the apparently sharp distinctions between taxa in certain instances.

I have attempted to be rather more comprehensive than Webster in the matter of literature-citations, but I also have tended to ignore non-taxonomic references. Typifications mostly follow ING (Farr *et al.*, 1979) and Webster's additions or neotypifications; my own neotypifications are indicated as such. I have also given a more extensive synonymy than is to be found in Webster's treatment, but as in that treatment, most of the names proposed by Otto Kuntze (1891) have been omitted.

As in Webster's treatment, generic concepts are relatively conservative, that is to say, traditional usage is followed where possible, and in borderline cases, genera are enumerated as distinct even though their claim to generic status may be somewhat questionable. The greatest departure from Webster's treatment will be found in the tribe *Hippomaneae*, the account of which, and of 2 small related tribes, has been very kindly provided by Dr Hans-Joachim Esser of Hamburg, for whom this group has constituted a special topic of research over many years.

The following abbreviations are used in the work:–

B. Basionym
f. figure
LT. Lectotype-species
RS. Referred synonym
S. Synonym
t. tabula
T. Type-genus or species
[] Square brackets used to enclose accepted names and their basionyms when citing types of generic synonyms

TAXONOMY

Family **EUPHORBIACEAE** *Juss.*, Gen. Pl.: 384 (1789). T.: *Euphorbia* L.

Monoecious or dioecious prostrate, erect, scandent or twining, annual, biennial or perennial herbs, shrubs or trees, succulent or not, spiny or unarmed, perulate or not, sometimes with phylloclades, with or without a milky latex or coloured sap, sometimes poisonous. Indumentum simple, sometimes urticating, malpighiaceous, dendritic, stellate, lepidote or 0. Leaves alternate, opposite or verticillate or any two or all three, green or scarious and squamiform, petiolate or sessile, stipulate or exstipulate; stipules free or connate, sometimes spathaceous, membranaceous, capilliform, glandular or spiny, subpersistent, readily deciduous or 0; lamina simple, lobate or compound, peltate or not, entire or variously toothed, palminerved or penninerved, often with 1–more pairs of basal glands at or near the base, sometimes minutely pellucid-punctate beneath. Inflorescences terminal, axillary, lateral or leaf-opposed, cymose, thyrsiform, paniculate, racemose or spicate, or with the flowers aggregated into 2-bracteate terminal capitate pseudanthia (*Dalechampia*) or glanduliferous involucrate radially or bilaterally symmetrical pseudanthia with connate bracts forming a cyathium, the glands often with petaloid appendages, and the pseudanthia themselves often aggregated into pseudoplciochasial hyperflorescences (tribe *Euphorbieae*), or else the flowers fasciculate or solitary, ramiflorous or cauliflorous; bracts 1–2-glandular, 1–more-flowered. Flowers unisexual, usually actinomorphic, small to minute, mostly pedicellate; calyx in both sexes usually of (1)3–6(8) valvate, imbricate or open, equal or unequal lobes, or of free sepals, often dissimilar as between the two sexes, the female calyx sometimes accrescent, rarely spathaceous, or minute or 0; corolla in one or both sexes of (1)3–6(8) free or rarely coherent or united, subvalvate or imbricate petals, sometimes brightly coloured (e.g. *Jatropha*), or petals minute or 0; disc in the male flowers often of 5–6, occasionally more, free extrastaminal and/or interstaminal glands, less often the disc annular or cupular and extrastaminal or intrastaminal and lobate or lobulate with the stamens enfolded by the lobes (e.g. *Drypetes*), or else the disc receptacular, domed and aperturate with the stamens arising through the apertures, or disc 0; disc in the female flowers hypogynous, usually annular and flat or cupular, entire or lobate, rarely of separate glands, usually persistent, or 0; stamens 1 (tribe *Euphorbieae*) 3–100(1000) (*Ricinus*), the filaments free or variously connate, simple or rarely branched (*Ricinus, Lasiococcinae*), anthers erect or inflexed (*Croton*) in bud, introrse, latrorse or extrorse, 2–4-thecate with the thecae usually parallel and adnate to the connective throughout, sometimes free at the base, occasionally completely free and erect (*Claoxylinae*), divaricate, horizontal or pendulous (*Mareyinae*), rarely with the thecae superposed (*Ptychopyxis*) or laterally fused to form an annulus (e.g. *Omphalea*), dehiscing longitudinally, obliquely or laterally by slits, less often apically by pores (*Porantherinae*); pollen tectate or semitectate, often reticulate, sometimes with the "*Croton*-pattern" (*Crotonoideae*), sometimes spinose (*Oldfieldioideae*), mostly 3-colporate, less often inaperturate or polyporate; pistillode often present, variously-shaped; staminodes occasionally +, subulate, mostly 0; gynoecium syncarpous, ovary superior, sessile or occasionally stipitate, (1)2–5(20)-locular, but most frequently 3-locular, placentation axile, ovules 1–2 per locule, collateral, pendulous, anatropous or hemitropous, rarely orthotropous (*Panda*), usually inserted beneath an obturator, crassinucellate, bitegumental, nucellus often beaked, embryo sac mostly 8-nucleate, hilum between the ventral raphe and micropylar caruncle, when present, funicle often thickened; styles (1)3–4(20), free or connate, erect or spreading, entire, 2-lobate, -fid or -partite to multifid or laciniate, the adaxial surface usually stigmatic throughout, smooth, granulate, papillose, plumose or fimbriate, sometimes reddish- or purplish-tinged (e.g. *Acalypha*). Fruit smooth, rugulose, tuberculate, verrucose, cornute or alate, rarely inflated, often schizocarpic, dehiscing septifragally and septicidally into 1–4(20), but mostly 3, 2-valved cocci or mericarps leaving a persistent angulate or alate columella, the cocci then dehiscing loculicidally, or the fruit dehiscing loculicidally into 3 valves, or breaking up irregularly (e.g. *Margaritaria*), or else indehiscent, baccate, drupaceous or subdrupaceous, epicarp thin, mesocarp crustaceous or fleshy, endocarp paleaceous, crustaceous, ligneous or osseous. Seeds 1–2 per locule, or by abortion 1 per fruit, carunculate or not, sometimes myrmecochorous; sarcotesta sometimes +; sclerotesta thin and crustaceous to thick and osseous; endosperm copious, often fleshy, or 0; embryo straight, curved or plicate, extending for most of the length of the seed; radicle superior; cotyledons usually broader than the radicle, mostly flat, less often thick and fleshy or folded.

The family *Euphorbiaceae* is here construed as having 334 genera grouped in 52 tribes and 5 subfamilies, and it is the sixth largest family of *Angiospermae*, after the *Compositae*, *Gramineae*, *Leguminosae*, *Orchidaceae* and *Rubiaceae*. The increase in the number of genera from 317 in Webster's (1994) account is partly accounted for by the incorporation of most of the genera which he placed in his "Incertae Sedis" and both of the ones placed in his "Addendum" into the main generic sequence, *Chlamydojatropha* Pax & K. Hoffm. and *Tacarcuna* Huft still being left unaccounted for, as is also *Myladenia* Airy Shaw, which Webster overlooked, and partly by the segregate genera now being recognized in the tribe *Hippomaneae* by Dr. Hans-Joachim Esser.

As to the segregate families from the *Euphorbiaceae*, those of Hurusawa (1954), namely the *Antidesmataceae*, the *Porantheraceae* and the *Ricinocarpaceae* simply represented the inflation of Pax & Hoffmann's (1931) four subfamilies, and they have not received general acceptance. Those of Airy Shaw (1965, 1966) however, namely the *Androstachydaceae*, the *Bischofiaceae*, the *Hymenocardiaceae*, the *Peraceae*, the *Picrodendraceae*, the *Stilaginaceae* and the *Uapacaceae* were accepted as a whole or in part for a time, but have now been shown to be better regarded as capable of reabsorption into the parent family. The *Pandaceae*, recognized as distinct by, among others, Takhtajan (1980),

Cronquist (1981) and Meeuse (1990) has, however, been incorporated into the *Euphorbiaceae* in this account, following Webster (1994), whilst realizing that it has probably more claim to recognition than the rest.

KEY TO THE SUBFAMILIES

1. Ovary-locules 2-ovulate, except for 75. *Scagea*; milky latex, intraxylary phloem and stinging hairs 0; indumentum simple, rarely lepidote (52. *Hieronyma*) or dendritic; embedded foliar glands rare; pollen 2-nucleate · · · · · · · · 2
1. Ovary-locules 1-ovulate, except for 98. *Dicoelia*; latex + or 0; indumentum various; pollen 2- or 3-nucleate · 3
2. Leaves alternate, very rarely opposite, stipulate, lamina simple, elobate, but 3-foliolate in 57. *Bischofia*; petals + or 0; pollen 3-colporate or -porate, not conspicuously spiny, but irregularly spiny in 12. *Amanoa*; seeds ecarunculate. · I. *PHYLLANTHOIDEAE*
2. Leaves alternate, opposite or verticillate, stipulate or exstipulate, lamina simple or 3-foliolate; petals 0, except in 61. *Croizatia*; pollen colpoidorate or porate, spiny; seeds carunculate or ecarunculate; endosperm copious, except in 65. *Hyaenanche* and 81. *Picrodendron* · · II. *OLDFIELDIOIDEAE*
3. Milky latex 0; laticifers, if +, inarticulate; leaves elobate or lobate; indumentum simple or stellate; petals + or 0; pollen 2-nucleate, mostly 3-colporate or 3-porate · III. *ACALYPHOIDEAE*
3. Latex milky or coloured, rarely 0; laticifers articulate or inarticulate; leaves elobate to compound; pollen 2- or 3- nucleate · · · · · · · · · · · · · · · · · · · 4
4. Latex reddish or yellowish to milky, sometimes scanty or 0; laticifers articulate or inarticulate; leaves elobate to lobate or compound; indumentum simple, stellate, dendritic or lepidote; bracts usually not 2-glandular at the base; sepals imbricate or valvate, usually covering the anthers in bud; petals mostly +, at least in the staminate flowers; pollen 3-colporate, porate or inaperturate, with "*Croton*"-pattern of processes · · · IV. *CROTONOIDEAE*
4. Latex whitish, often caustic or toxic; laticifers inarticulate; leaves usually elobate; indumentum simple or 0, rarely dendritic (309. *Mabea*), never stellate or lepidote; bracts often 2-glandular at the base; sepals imbricate or open, anthers mostly not covered in bud; petals 0; pollen 3-colporate, mostly perforate-reticulate, never with "*Croton*"-pattern · V. *EUPHORBIOIDEAE*

Subfamily I. **PHYLLANTHOIDEAE** *Asch.*, Fl. Brandenburg 1: 59 (1864); G. L. Webster, Ann. Missouri Bot. Gard. 81: 34 (1994). T.: *Phyllanthus* L.

Euphorbiaceae subordo *Dispermae* Zoll., Natuur-Geneesk. Arch. Ned.-Indië 2: 17 (1845).

Phyllanthaceae Klotzsch & Garcke, Monatsber. Königl. Preuss. Akad. Wiss. Berlin 1859: 246 (1859).

Trees, shrubs, or herbs; indumentum simple (rarely branched; lepidote in *Hieronyma* and *Uapaca*); leaves alternate (very rarely opposite), spiral to distichous, usually stipulate; leaf blade simple, entire [except in *Drypetes* and *Phyllanoa*], without embedded foliar glands, but sometimes superficially glandular-granulose. Inflorescences axillary (rarely terminal), racemiform, spiciform, glomerulate or flowers solitary; bracts eglandular, mostly inconspicuous. Sepals mostly 4–6, imbricate or (*Bridelia*) valvate, sometimes connate; petals and disc + or 0; stamens (2)4–8(50), filaments free or united; pollen grains mostly 3–4-colporate (rarely porate; periporate in *Phyllanthus*, zonoporate in *Hymenocardia*), exine semitectate, rarely echinate (tribe *Amanoeae*); male gametophyte binucleate; pistillode + or 0; ovary (1)2–5(15)-locular, ovules 2 per locule, anatropous or hemitropous, nucellar beak sometimes prominent; styles mostly free except at base, usually 2-fid. Fruit dehiscent, baccate or drupaceous; seeds ecarunculate or with a small caruncle; endosperm + or 0; cotyledons flat or folded. Base chromosome numbers mostly 12 or 13.

Key to the tribes

1. Ovary 2-locular above, 4-locular below; fruits 4-gonous; pollen with atria · · · · · · · · 10. *Martretieae*
1. Ovary loculi not as above; fruits commonly 1–3-gonous; pollen exatriate · 2
2. Leaves pinnately 3–5-foliate; inflorescences paniculate or racemose; petals and disc 0; styles entire; fruits drupaceous · · · · · · · · 8. *Bischofieae*
2. Leaves simple, mostly entire; styles usually bifid · · · · · · · · 3
3. Flowers in axillary or terminal spicate, racemose or paniculate inflorescences (axis usually over 1 cm long; capitate in *Uapaca* and *Poranthera*), or flowers solitary (*Oreoporanthera*); petals + or 0; leaves with paracytic or anomocytic stomata; endosperm + · · · · · · · · 4
3. Flowers glomerulate (inflorescence axes less than 1 cm long except in some *Phyllantheae*); leaves mostly with paracytic stomata, lacking tanniniferous epidermal cells; endosperm + or 0 · · · · · · · · 6
4. Fruits winged, samaroid or doubly scrotiform; leaves granulose-glandular; tanniniferous epidermal cells 0; floral disc 0; pollen zonoporate; styles simple · · · · · · · · 7. *Hymenocardieae*
4. Fruits not winged; leaves not granulose-glandular; tanniniferous epidermal cells +; floral disc + or 0; pollen 3-colporate · · · · · · · · 5
5. Pollen usually >20μ; styles simple or 2-fid · · · · · · · · 6. *Antidesmeae*
5. Pollen <20μ; styles simple · · · · · · · · 9. *Centroplaceae*
6. Male disc intrastaminal or 0 (*Putranjiva*); leaves often dentate; wood fibres thick-walled, not septate; ovary 1–4-locular; fruit dehiscent (*Lingelsheimia*) or drupaceous · · · · · · · · 5. *Drypeteae*
6. Male disc extrastaminal or 0; leaves entire; ovary 2–3(6)-locular; fruit mostly dehiscent or baccate, less often drupaceous · · · · · · · · 7
7. Petals usually +, at least in male flowers; sepals imbricate; wood parenchyma conspicuous, fibres mostly thick-walled; vessel element perforations scalariform or simple; ovules anatropous; endosperm + or 0 · · · · · · · · 8

7. Petals usually 0, or if + then sepals valvate, or wood and ovule characters otherwise; endosperm usually + · 9
8. Pollen exine finely reticulate, not coarsely reticulate or spinose; stipules deciduous · 1. *Wielandieae*
8. Pollen exine coarsely and deeply reticulate or spinose; stipules persistent · 2. *Amanoeae*
9. Sepals valvate; petals +, usually shorter than sepals · · · · · · · · · 3. *Bridelieae*
9. Sepals imbricate; petals 0, or if + then often as long as sepals · 4. *Phyllantheae*

Tribe 1. **WIELANDIEAE** *Baill. ex Hurus.*, J. Fac. Sci. Univ. Tokyo, Sect. 3, Bot. 6: 339 (1954); G. L. Webster, Ann. Missouri Bot. Gard. 81: 35 (1994). T.: *Wielandia* Baill.

Wielandiiées Baill., Étude Euphorb. 568 (1858).
Phyllantheae subtribe *Wielandiinae* Pax & K. Hoffm., Pflanzenr. 81: 180 (1922).

Monoecious or dioecious trees or shrubs; indumentum simple. Leaves alternate, entire, penninerved, eglandular; stipules deciduous. Inflorescences axillary. Flowers in glomerules or the female solitary; bracts usually inconspicuous; sepals mostly 4–6, imbricate; petals usually 5, sometimes 0; disc usually annular; stamens 5–12, free or filaments connate at base; anthers dehiscing longitudinally, connective not enlarged; pollen-grains 3-colporate, tectate, semitectate, or reticulate; pistillode +; ovary 3–5-locular, ovules anatropous, usually with a single obturator; styles 2-fid. Fruit a regma. Seeds 1 or 2 per locule; endosperm + or absent; embryo with cotyledons flat or folded, much longer than the radicle.

The taxa with the greatest numbers of unspecialized morphological features are gathered together in this tribe, which is unified on the basis of pollen morphology (Köhler, 1965) and leaf venation (Levin, 1986a, b, c). Webster (1994) considers that it is probably paraphyletic in cladistic terms, which he feels to be inevitable for the basal tribe in a large family such as the *Euphorbiaceae.* Relationships among the 10 genera appear reticulate, and it does not seem feasible to group them into subtribes. It is notable that all of the genera occur in South America, Africa, and the Indian Ocean islands, and only *Savia* extends into the northern hemisphere. Stuppy (1995), however, considers *Savia* and *Actephila* to be closer to *Andrachne* and *Leptopus* than to the other genera in tribe on the basis of their seed-morphology.

Key to the genera

1. Pistillode large, hypocrateriform; male petals 0 · · · · · · · · 9. *Chonocentrum*
1. Pistillode usually small, 3–5-fid, -lobate or -partite; male petals usually + · · 2
2. Seeds with copious endosperm (albuminous) · · · · · · · · · · · · · · · · · · · 3

2. Seeds without endosperm (exalbuminous), or endosperm thin or meagre · 5
3. Stamens 8–12; sepals 3; petals 5, larger, otherwise ± similar; ovary 4–5-locular · 1. *Heywoodia*
3. Stamens mostly 5, sepals and petals mostly 5, ovary mostly 3-locular · · · · 4
4. Female disc entire, not glandular · 2. *Savia*
4. Female disc multiglandular-lobate · · · · · · · · · · · · · · · · · · · 3. *Gonatogyne*
5. Male disc dissected, the segments petaloid, sometimes in 2 whorls · 4. *Petalodiscus*
5. Male disc annular or cupular, sometimes lobate · · · · · · · · · · · · · · · · · · 6
6. Pistillode quinquelobate; ovary 5-locular · · · · · · · · · · · · · · · · 10. *Wielandia*
6. Pistillode 3(4)-lobate, -fid or -partite; ovary (2)3–4-locular · · · · · · · · · · 7
7. Monoecious or rarely subdioecious · 8
7. Dioecious; leaves coriaceous · 9
8. Leaves coriaceous; petals mostly longer than sepals; stamens free except at base · 5. *Blotia*
8. Leaves chartaceous; petals scale-like, shorter than sepals or 0; stamens free or connate into a column · 6. *Actephila*
9. Male disc glabrous; styles dilated · 7. *Discocarpus*
9. Male disc villous; styles slender · 8. *Lachnostylis*

1. **Heywoodia** *Sim*, Forest Fl. Cape: 326, t.140/1 (1907); Hutch. in Fl. Cap. 5(2): 384 (1915), and Kew Bull. 1922: 115 (1922), descr. emend. et ampl.; Pax & K. Hoffm., Pflanzenr. 81: 280 (1922), and Nat. Pflanzenfam. ed. 2, 19C: 74 (1931); Milne-Redh., Bull. Jard. Bot. État 27: 327, t. 10 (1957); Radcl.-Sm. in Fl. Trop. E. Afr., Euph. 1: 85, t. 12 (1987); G. L. Webster, Ann. Missouri Bot. Gard. 81: 36 (1994); Radcl.-Sm. in Fl. Zamb. 9(4): 6, t. 1 (1996). T.: *H. lucens* Sim.

Dioecious large glabrous evergreen tree. Leaves of seedlings and sucker-shoots peltate, other leaves alternate, stipulate, simple, entire, symmetrical, coriaceous, shiny. Flowers axillary, the male flowers small, in dense glomerules, bracts 4, free, imbricate, broad, concave, unequal, the females in 3–4-flowered fascicles, bracts ± as in male. Male flowers sessile; sepals 3, free, imbricate, unequal; petals 5, free, twice as large as the sepals, strongly imbricate, subequal; disc peripheral, invaginated amongst the filaments of the outer staminal whorl; stamens 8–12, in two whorls, filaments free, anthers dorsifixed, introrse, thecae parallel; pollen prolate-spheroidal, with ecostate colpi; pistillode minute, 3-fid. Female flowers shortly pedicellate; bracteoles 1–2; sepals and petals more or less as in male; staminodes 6–8, filiform; ovary 4–5-locular; ovules 2 per locule; stigmas 4–5, sessile. Fruit slightly subglobose, 4–5-locular, separating into 4–5 2-valved cocci, at length loculicidally dehiscent; exocarp wrinkled when dry. Seeds 1 per locule by abortion, ecarunculate, smooth, closely longitudinally striate; endosperm papyraceous; embryo minute. Fig.: See Forest Fl. Cape: t. 140/1 (1907); Bull. Jard. Bot. État. 27: 331, t. 10 (1957); Fl. Trop. E. Afr., Euph. 1: 87, t. 12 (1987); Fl. Zamb. 9(4): 7, t. 1 (1996).

A monotypic African genus (S. Kenya, NW. Tanzania and W. Uganda; S. Mozambique, Swaziland and S. Africa: Natal and Transkei). *Heywoodia* has a marked distributional discontinuity — some 2850 km separate the Tanzanian and Mozambican localities.

On the basis of foliar venation characters, Levin (1986b) suggests a relationship between *Heywoodia* and *Astrocasia*, and regards both genera as closer to the tribe *Phyllanthae*. There are indeed some resemblances between *Heywoodia* and *Astrocasia* but Webster (1994) considers the former genus to be the more primitive, and nearer in morphological characters to the hypothetical ancestor of the family. This would not preclude an affinity with *Astrocasia*, which may connect the tribes *Weilandieae* and *Phyllantheae*.

2. **Savia** *Willd.*, Sp. Pl. 4: 771 (1805); A. Juss., Euphorb. Gen.: 15 (1824); Baill., Étude Euphorb.: 569 (1858); Müll. Arg. in DC., Prodr. 15(2): 228 (1866); Benth. & Hook. f., Gen. Pl. 3: 270 (1880); Fawc. & Rendle, Fl. Jamaica 4: 271 (1920); Pax & K. Hoffm., Pflanzenr. 81: 181 (1922); Urb., Repert. Spec. Nov. Regni Veg. 28: 209 (1928); Pax & K. Hoffm., Nat. Pflanzenfam. ed. 2, 19C: 67 (1931); Alain, Fl. Cuba 3: 41 (1953); G. L. Webster, J. Arnold Arbor. 48: 325 (1967); Radcl.-Sm., Kew Bull. 27: 508 (1972); Correll, Fl. Bah. Arch.: 842 (1982); G. L. Webster, Taxon 31: 535 (1982); Proctor, Kew Bull. Addit. Ser. 11: 520 (1984); Alain, Fl. Español. 4: 211 (1986); Radcl.-Sm. in Fl. Trop. E. Afr., Euph. 1: 83, t. 11 (1987); Alain, Descr. Fl. Puerto Rico 2: 425 (1988); G. L. Webster, Ann. Missouri Bot. Gard. 81: 36 (1994); Gillespie in Acev.-Rodr., ed., Fl. St. John: 222 (1996). T.: *Savia sessiliflora* (Sw.) Willd. (B.: *Croton sessiliflorus* Sw.)

Maschalanthus Nutt., Trans. Amer. Philos. Soc. 5: 137 (1837). T.: *M. obovatus* Nutt. [=*S. phyllanthoides* (Nutt.) Pax & K. Hoffm.]

Savia Sect. *Maschalanthus* (Nutt.) Pax in Nat. Pflanzenfam. ed. 1, 3(5): 15 (1890), Pflanzenr. 81: 183 (1922), and Nat Pflanzenfam. ed. 2, 19C: 67 (1931).

Kleinodendron L. B. Sm. & Downs, Sellowia 16: 177 (1964). T.: *K. riosulense* L. B. Sm. & Downs [=*S. dictyocarpa* Müll. Arg.]

Monoecious or dioecious shrubs or small trees. Leaves alternate, shortly petiolate, stipulate, entire, penninerved. Flowers axillary, males densely clustered, female solitary or few together, or mixed. Male flowers shortly to long-pedicellate; sepals (4)5(6), imbricate, subequal; petals (4)5(6), small or exceeding the sepals; disc annular, entire, lobate or crenate; stamens 5, alternate with the petals, filaments ± free, anthers erect, loculi parallel, contiguous, introrse, dehiscing longitudinally; pollen suboblate, 3-colporate, narrow-aperturate, lumina reticuli $<3\mu$; pistillode short, 3–5-fid. Female flowers: pedicels, sepals and petals as in the males; disc annular; ovary 3-locular, ovules 2 per locule; styles free, spreading, 2-fid. Fruit dehiscing into 2-valved cocci; endocarp crustaceous or slightly indurated. Seeds often 1 per locule by abortion, ovoid or 3-sided; endosperm fleshy; cotyledons broad, flat or slightly folded. Fig.: See Euphorb. Gen.: t. 2, f. 5.1–6 (1824); Étude Euphorb.: t. 26, f. 20–23 (1858); Fl. Jamaica 4: 272, t. 89 (1920); Fl. Cuba 3: 43, t. 7 (1953); Fl.

FIG. 1. *Gonatogyne brasiliensis* Müll. Arg. (S.: *Savia brasiliensis* (Müll. Arg.) Pax & K. Hoffm.). **A** Habit, old fruiting specimen × 1; **B** Fruit-valve × 6; **C** Columella × 5. All from *Sellow* s.n. Drawn by Camilla Speight.

Madag. 111(1): 117–127, tt. XIX.7–11, XX.1–5 and XXI.1–7 (1958); Fl. Bah. Arch.: 843, t. 350 (1982); Fl. Español. 4: 214, t. 116-34 (1986); Fl. Trop. E. Afr., Euph. 1: 84, t. 11 (1987); Descr. Fl. Puerto Rico 2: 428, t. 59-27 (1988); Bol. Mus. Paraense Emilio Goeldi, N.S., Bot. 7: 29, t. 1 (1991); Fl. St. John: 223, t. 101A–D (1996).

About 25 species in tropical America, E. and S. Africa and Madagascar.

Three sections have been recognized: sect. *Savia*, 2 spp., the type and *S. dictyocarpa* of Brazil; sect. *Heterosavia* Urb., c. 20 spp. of the West Indies, S. Africa and Madagascar; and the recently described sect. *Afrosavia* Radcl.-Sm. of Kenya (which P. Hoffmann, pers. comm., thinks may be generically distinct). A number of other genera of tribe *Wielandieae* have been included in *Savia* by Müller and other workers.

3. **Gonatogyne** *Klotzsch ex Müll. Arg.* in Mart., Fl. Bras. 11(2): 13 (1873); Pax & K. Hoffm., Repert. Spec. Nov. Regni Veg 31: 190 (1933); G. L. Webster, Ann. Missouri Bot. Gard. 81: 36 (1994). T.: *Gonatogyne brasiliensis* (Baill.) Müll. Arg. (B.: *Amanoa ? brasiliensis* Baill.; S.: *Savia brasiliensis* (Baill.) Pax & K. Hoffm.)

Savia sect. *Gonatogyne* (Klotzsch ex Müll. Arg.) Pax & K. Hoffm., Pflanzenr. 81: 187 (1922), and Nat. Pflanzenfam. ed. 2, 19C: 68 (1931).

Very like 2. *Savia* s. str., differing only in the thickly and strongly multi-glandular female disc. Fig.: See Fig. 1, p. 8.

A monotypic genus from S. Brazil.

4. **Petalodiscus** *(Baill.) Pax,* Nat. Pflanzenfam. ed. 1, 3(5): 15 (1890); G. L. Webster, Ann. Missouri Bot. Gard. 81: 36 (1994). LT.: *P. laureola* (Baill.) Pax (B.: *Savia laureola* Baill.) (lectotype, selected here).

Savia sect. *Petalodiscus* Baill., Étude Euphorb.: 571 (1858); Pax & K. Hoffm., Pflanzenr. 81: 182 (1922), and Nat. Pflanzenfam. ed. 2, 19C: 67 (1931).

Very close to 2. *Savia* s. str., but monoecious, disc-glands petaloid, sometimes in 2 whorls, and seeds exalbuminous. Fig.: See Étude Euphorb.: t. XXII.11–14 (1858); Pflanzenr. 81: 174, t. 14B (1922); Nat. Pflanzenfam. ed. 2, 19C: 67, t. 31B (1931); Fl. Madag. 111(1): 117, t. XIX.7,8 and 123, t. XX.6–8 (1958).

A genus of 5 species in Madagascar, recombined with *Savia* by Pax & Hoffman (1922, 1931), Leandri (1958), and Hutchinson (ined.).

5. **Blotia** *Leandri,* Mém. Inst. Sci. Madagascar, Sér. B., Biol. Vég. 8: 240 (1957); Fl. Madag. 111(1): 126 (1958); G. L. Webster, Ann. Missouri Bot. Gard. 81: 36 (1994); Petra Hoffm. & McPherson, Adansonia, Sér. 3, 20(2): 248 (1998). T.: *Blotia oblongifolia* (Baill.) Leandri (B.: *Savia oblongifolia* Baill.).

Savia sect. *Charidia* Baill., Étude Euphorb.: 572 (1858); Müll. Arg. in DC., Prodr. 15(2): 229 (1866).

Petalodiscus sect. *Charidia* (Baill.) Pax, Nat. Pflanzenfam. ed. 1, 3(5): 15 (1890).

Monoecious or subdioecious trees. Indumentum simple, weakly-developed. Leaves alternate, mostly very shortly petiolate, stipulate, simple, entire, penninerved, eglandular. Stipules persistent. Flowers axillary, the males and hermaphrodites in several-flowered fascicles, the females few or solitary. Male flowers pedicellate; pedicel inarticulate; sepals (4)5–6, quincuncial when 5-merous, slightly imbricate, stout, involute; petals 1–3 times as long as the sepals, membranous, striate, inserted on the disc; disc extrastaminal, thick, fleshy, saucer-shaped, lobate between the petals; stamens 5–6, episepalous, filaments united at the base with the pistillode, free above, anthers introrse, thecae ovoid or globose, longitudinally laterally dehiscent; pollen as in *Savia*; pistillode stout, obconical, 3(4)-lobate at the apex. Female flowers pedicellate; sepals, petals and disc similar to those of the male; ovary 3(4)-locular, ovules 2 per locule, pendulous beneath a common obturator; styles 3(4), united at the base, free parts each 2-lobate. Fruit ovoid-subglobose, 3(4)-lobate, dehiscing into 6 valves; columella trigonous, persistent. Seeds 1–2 per locule, ovoid-subglobose, apiculate, bitegmic, ecarunculate, with chalazal ring; endosperm very meagre; cotyledons large, thin, several times folded; radicle c. 3 times longer than wide. Fig.: See Fl. Madag. 111(1): 127, t. XXI.8–16 (1958); Adansonia, Sér. 3, 20(2): 249–255, tt. 1–3 (1998).

An endemic Madagascan genus of 5 species.

6. **Actephila** *Blume*, Bijdr. Fl. Ned. Ind.: 581 (1825); Müll. Arg. in DC., Prodr. 15(2): 221 (1866); Baill., Adansonia I. 6: 231 (1866); Benth. & Hook. f. Gen. Pl. 3: 269 (1880); Hook. f., Fl. Brit. India 5: 282 (1887); J. J. Sm., Add. Cogn. Fl. Arb. Jav. 12: 45 (1910); Pax & K. Hoffm., Pflanzenr. 81: 191 (1922); Gagnep., Fl. Indo-Chine 5: 530 (1927); Pax & K. Hoffm., Nat. Pflanzenfam. ed. 2, 19C: 68 (1931); Backer & Bakh. f., Fl. Java 1: 470 (1963); Airy Shaw, Kew Bull. 25: 496 (1971), and 26: 208 (1972); Whitmore, Tree Fl. Mal. 2: 51 (1973); Airy Shaw, Kew Bull. Addit. Ser. 4: 25 (1975) and 8: 21 (1980), Kew Bull. 35: 586 (1980), 36: 248 (1981), 37: 4 (1982) and Euph. Philipp.: 3 (1983); G. L. Webster, Ann. Missouri Bot. Gard. 81: 37 (1994); Philcox in Dassanayake, ed., Fl. Ceyl. XI: 82 (clav. tant.) (1997). T.: *A. javanica* Miq.

Lithoxylon Endl., Gen.: 1122 (1840); Baill., Étude Euphorb.: 590 (1858). T.: *L. lindleyi* Steud. (RS.: *Securinega nitidum* Willd. sensu Lindl. [=*A. nitida* (Lindl.) Benth. & Hook. f. ex Drake])
Anomospermum Dalzell, Hooker's J. Bot. Kew Gard. Misc. 3: 228 (1851), non Miers (1851). T.: *A. excelsum* Dalzell [= *Actephila excelsa* (Dalzell) Müll. Arg.]
Savia sect. *Actephila* (Blume) Baill., Étude Euphorb.: 571 (1858)

Monoecious trees or shrubs. Leaves alternate or subverticillate, entire, penninerved. Flowers fasciculate, axillary, extra-axillary or rarely on short densely minutely bracteate brachyblasts, or solitary. Male flowers pedicellate; sepals 5–6, imbricate; petals 2–6, scale-like below the disc, or 0; disc entire or

5-lobate; stamens 3–6, filaments free or shortly connate around the pistillode, anther-thecae parallel, longitudinally dehiscent; pollen as in 24. *Andrachne*, q.v.; pistillode 3-fid. Female flowers long-pedicellate; calyx and petals as in the male but calyx often enlarged and persistent in fruit; disc flat or cupular; ovary sessile on the disc, 3-locular, ovules 2 per locule; styles short, free or connate at base, undivided or 2-fid. Fruit hard, loculicidally 3-valved or separating into 2-valved cocci. Seeds exarillate, often solitary; endosperm scanty or 0; cotyledons thick, fleshy, sometimes plicate. Fig.: See Étude Euphorb.: t. 26, f. 24 (1858); Pflanzenr. 81: 174, t. 14E,F (1922); Fl. Indo-Chine 5: 511, t. 65.21, 22 and 538, t. 66.2, 3 (1926); Nat. Pflanzenfam. ed. 2, 19C: 67, t. 31E,F (1931); Kew Bull. 35: 590, t. 1A1, B1,2 (1980).

An Asiatic and Australasian genus of c. 20 described species, still poorly understood taxonomically.

The position of the genus has been variously interpreted. Pax & Hoffman (1922) associated it with *Amanoa*, which has similar flowers but very different pollen. Köhler (1965) pointed out a palynological similarity to *Andrachne*, and Levin (1986a) found a resemblance in leaf venation to that genus. However, the exalbuminous seeds with plicate embryo of *Actephila* would appear to rule out a close association with *Andrachne*. It is worth noting that Mennega (1987) recommended assignment of *Actephila* to the *Wielandieae*, and so for the moment, it seems best to include it here.

7. **Discocarpus** *Klotzsch*, Arch. Naturgesch. 7: 201 (1841); Baill., Étude Euphorb.: 585 (1858); Müll. Arg. in DC., Prodr. 15(2): 223 (1866), and in Mart., Fl. Bras. 11(2): 11, t. 2 (1873); Benth. & Hook. f., Gen. Pl. 3: 269 (1880); Pax & K. Hoffm., Pflanzenr. 81: 202 (1922), and Nat. Pflanzenfam. ed. 2, 19C: 70 (1931); Jabl., Mem. New York Bot. Gard. 17(1): 84 (1967); A. H. Gentry, Woody Pl. NW. S. Amer.: 419 (1993); G. L. Webster, Ann. Missouri Bot. Gard. 81: 37 (1994); Hayden & Hayden, Ann. Missouri Bot. Gard. 83: 154 (1996). T.: *D. essequiboensis* Klotzsch.

Dioecious divaricately-branched trees or shrubs. Leaves alternate, stipulate, shortly petiolate, entire, coriaceous, shiny; perulae covering the terminal bud acuminate, spinulose. Flowers axillary, sessile, males numerous, densely glomerate, females fewer or solitary. Male flowers: sepals 4–5, imbricate; petals (0)4–5; disc membranous, cyathiform or cupular, 5-lobate; stamens 4–5, filaments connate into a column, free and spreading above, anther-thecae parallel, longitudinally dehiscent; pollen oblate-spheroidal, stephanocolporate, colpi ecostate, columellae small, reticulum 0; pistillode 3-partite, terminating the column. Female flowers: sepals more rigid than in male; petals smaller than the sepals; disc thicker, spreading, subentire; ovary 3-locular, ovules 2 per locule; styles 3, free, spreading, 2-fid or dilated into a broad crenulate limb. Fruit hard, globose, smooth or verruculose, pubescent, 3-coccous or by abortion 1–2-coccous, cocci 2-valved; columella short, squat, 3-alate, persistent. Seed solitary in the cocci; testa thick; endosperm thin or 0; cotyledons very broad and folded. Fig.: See Étude Euphorb.: t. XXII.1 (1858); Fl. Bras. 11(2): t. 2 (1873); Pflanzenr. 81: 174, t. 14G, H (1922); Nat. Pflanzenfam. ed. 2, 19C: 67, t. 31G, H (1931); Bol. Mus. Paraense Emilio

Goeldi, N.S., Bot. 7: 30, t. 2 (1991); Woody Pl. NW. S. Amer.: 416, t. 119.5 (1993); Ann. Missouri Bot. Gard. 83: 158–162, tt. 10–12 (1996).

A tritypic genus in the Amazon and Orinoco River basins of Brazil, Colombia, Peru and Venezuela, as well as the smaller rivers of Guyana and Surinam.

The taxonomic position of *Discocarpus* is somewhat uncertain. Pax & Hoffmann (1922) assigned it to a separate subtribe (including *Lachnostylis*); Köhler (1965) suggested it was close to *Amanoeae* and *Bridelieae*; Mennega (1987) regards it as out of place in *Wielandieae* and Hayden & Hayden (1996) prefer to place it in the *Amanoeae* on the grounds of similarity in wood, foliar anatomy, inflorescence-architecture, flower structure and pollen.

8. **Lachnostylis** *Turcz.*, Bull. Soc. Imp. Naturalistes Moscou 19: 503 (1846); Baill., Étude Euphorb.: 663 (1858); Müll. Arg. in DC., Prodr. 15(2): 224 (1866); Benth. in Hooker's Icon. Pl. 13: 61, t. 1279 (1879) and in Gen. Pl. 3: 269 (1880); Hutch. in Fl. Cap. 5(2): 383 (1915); Dyer, Gen. S. Afr. Fl. Pl. 1: 309 (1975); G. L. Webster, Ann. Missouri Bot. Gard. 81: 37 (1994). T.: *Lachnostylis capensis* Turcz. [=*L. hirta* (L.f.) Müll. Arg. (B.: *Cluytia hirta* L. f.)]

Discocarpus sensu Pax & K. Hoffm., non Klotzsch.

Dioecious shrubs. Leaves small, alternate, coriaceous, dull, entire. Flowers axillary, males fasciculate, numerous, females fewer, often solitary. Male flowers shortly pedicellate to subsessile; sepals 5, imbricate, subequal; petals 5, larger than the sepals; disc glands thick, villous; stamens 5, filaments connate for $^1/_2$ their length into a column in the middle of the disc, spreading above around the pistillode; anther-thecae parallel, longitudinally dehiscent; pollen as in *Savia*; pistillode 3-partite, villous. Female flowers long-pedicellate; sepals and petals as in the male; disc annular, thick, villous; ovary 2–3-locular, tomentose, ovules 2 per locule; styles 2–3, short, flat, recurved-patent, shortly 2-fid. Fruit dehiscing into 3 2-valved cocci; pericarp thick, crustaceous. Seeds subglobose; testa smooth; endosperm thin, scanty; cotyledons broad, much-folded. Fig.: See Hooker's Icon. Pl. 13: t. 1279 (1879).

A South African endemic genus with a single variable species (or 2 according to some, e.g. Dyer, 1975).

Although combined with *Discocarpus* by Pax & Hoffmann (1922), it differs in a number of respects, the most obvious being the small size and dull texture of the leaves.

9. **Chonocentrum** *Pierre ex Pax & K. Hoffm.*, Pflanzenr. 81: 205 (1922), and Nat. Pflanzenfam. ed. 2, 19C: 70 (1931); Jabl., Mem. New York Bot. Gard. 17: 121 (1967); G. L. Webster, Ann. Missouri Bot. Gard. 81: 37 (1994). T.: *Chonocentrum cyathophorum* (Müll. Arg.) Pax & K. Hoffm. (B.: *Drypetes cyathophora* Müll. Arg.)

Dioecious tree or large shrub. Leaves alternate, shortly petiolate, stipulate, oblong, acuminate, entire, penninerved, lateral nerves few; axillary buds sharply pointed. Male flowers axillary, glomerate, small; calyx cupular, 4–5-toothed, valvate, soon open; petals 0; disc shortly cupular, thin, extrastaminal;

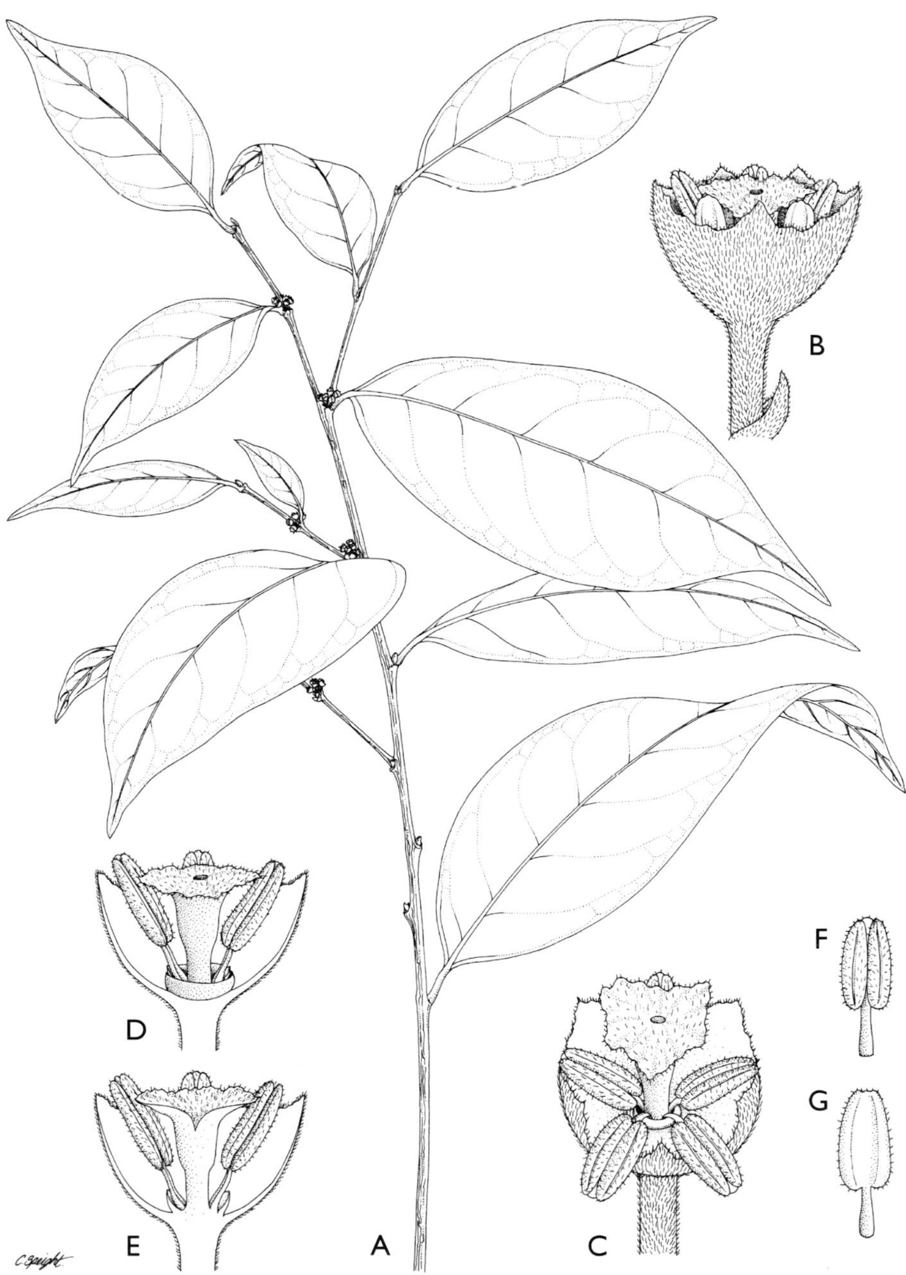

FIG. 2. *Chonocentrum cyathophorum* (Müll. Arg.) Pierre ex Pax & K. Hoffm. (B.: *Drypetes cyathophora* Müll. Arg.). **A** Habit, male × 1; **B** Male flower × 40; **C** Male flower, viewed from above × 40; **D** Male flower, part of calyx removed × 40; **E** LS male flower × 40; **F** Stamen, ventral view × 40; **G** Stamen, dorsal view × 40. **A** from *Spruce* 3781; **B–G** based on Delpy's drawings of *Spruce* 3781. Drawn by Camilla Speight.

stamens 4–6, episepalous if isomerous, filaments free, short, anthers broadly oblong, introrse, thecae parallel, contiguous, puberulous, longitudinally dehiscent; pollen prolate, tetracolporate, the colpi costate, the transverse colpi rounded at the end; pistillode very large, a little exceeding the calyx, cyathiform, with a broad spreading undulate-4–5-lobulate rim. Female flowers, fruits and seeds not yet known. Fig.: See Delpy's unpublished drawings in K and P based on *Spruce* 3781, the only known collection of the genus; Bol. Mus. Paraense Emilio Goeldi, N.S., Bot. 7: 44, t. 16 (1991). See also Fig. 2, p. 13.

A monotypic genus of Amazonian South America.

Pax & Hoffmann regarded it as related to *Discocarpus*; however, the pollen is more like that of certain *Phyllanthus* spp.

10. **Wielandia** *Baill.*, Étude Euphorb.: 568 (1858); Benth. & Hook. f., Gen. Pl. 3: 270 (1880); Hemsl., Hooker's Icon. Pl. 29: t. 2813 (1906); Pax & K. Hoffm., Pflanzenr. 81: 181 (1922), and Nat. Pflanzenfam. ed. 2, 19C: 67 (1931); Leandri, Notul. Syst. (Paris) 7: 190 (1939), and Fl. Madag. 111 (1): 135 (1958); G. L. Webster, Ann. Missouri Bot. Gard. 81: 37 (1994); Petra Hoffm., Adansonia, sér. 3, 20(2): 334 (1998). T.: *W. elegans* Baill.

Savia sect. *Wielandia* Müll. Arg., Linnaea 32: 78 (1863), and in DC., Prodr. 15 (2): 228 (1866).

Monoecious shrub. Young branches flattened. Indumentum 0 or very scarce, simple. Leaves alternate, petiolate, stipulate, simple, entire, coriaceous, finely penninerved, reticulate; petiole canaliculate; stipules peltate, deciduous. Flowers axillary, fasciculate, fascicles few-flowered, flowering branchlets sometimes resembling small racemes, on account of the minute leaves. Male flowers pedicellate; calyx-lobes 5, imbricate; petals broad, imbricate, emarginate, longer than the calyx-lobes; disc extrastaminal, cupular, 5-angled, the angles oppositisepalous; stamens 5, alternipetalous, filaments connate into a thick column, free and recurved at the apex, anthers basifixed, at first erect, then decurved, thecae contiguous, longitudinally dehiscent; pollen ± as in *Savia*, but 3–4-colporate; pistillode terminating the staminal column, apically flattened, stellately 5-lobate, the lobes 3-dentate, extending between the stamens. Female flowers long-pedicellate; calyx and petals as in the male; disc deeply cupular; ovary 5–6-locular; ovules 2 per locule; styles somewhat recurved or reflexed, apex truncate or shortly 2-lobate. Fruit 4–6-locular, each locule 1-seeded by abortion; mericarps 2-valved; exocarp thinly coriaceous, separating; endocarp woody, hard; columella persistent, cylindric. Seeds mostly aborting, mature ones roundly reniform, bitegmic; exotesta thinly crustaceous; endotesta membranous; endosperm 0; cotyledons large, lobate, oleiferous; radicle minute. Fig.: See Étude Euphorb.: t. XXII.6–10 (1858); Hooker's Icon. Pl. 29: t. 2813 (1906); Fl. Madag. 111(1): 117, t. XIX. 1–6 (1958); Adansonia, sér. 3, 20(2): 337, t. 1 (1998).

A monotypic genus restricted to Madagascar and the Seychelles.

Although combined with *Savia* by Müller, *Wielandia* appears closer to *Petalodiscus* by virtue of its monoecism and exalbuminous seeds.

Tribe 2. **AMANOEAE** *(Pax & K. Hoffm.) G. L. Webster*, Taxon 24: 594 (1975), and Ann. Missouri Bot. Gard. 81: 37 (1994). T.: *Amanoa* Aubl.

Phyllantheae subtribe *Amanoinae* Pax & K. Hoffm., Pflanzenr. 81: 190 (1922).

Monoecious or dioecious shrubs or trees; indumentum, when +, simple. Leaves alternate, simple, entire, penninerved, eglandular; stipules persistent, sometimes adnate to petiole. Inflorescences glomerulate, spicate, racemose or paniculate; bracts conspicuous or not; sepals and petals (4)5(6), imbricate; male disc annular or dissected; stamens 5(7), free or united at the base; anthers introrse; pollen grains 3-colporate, very coarsely reticulate or irregularly spinose; pistillode prominent; female disc cupular; ovary 3-locular; ovules anatropus; styles twice 2-fid or dilated, emarginate. Fruit tardily dehiscent. Seeds 1 or 2 per locule; endosperm +, scanty or 0; cotyledons massive, not plicate, much longer than radicle.

Following Webster (1994), this tribe includes the genera *Amanoa* and *Pentabrachion*. In the original circumscription of the tribe (Webster, 1975), *Actephila* was also included. However, evidence from pollen morphology (Köhler, 1965), leaf venation (Levin, 1986a, b, c), and wood anatomy (Mennega, 1987) argues against this.

Key to the genera

1. Stipules not intrapetiolar; floral bracts inconspicuous, not indurate; styles slender, twice 2-fid; fruit thin-walled, seeds 2 per locule, endosperm +, fleshy · 11. *Pentabrachion*
1. Stipules more or less intrapetiolar; floral bracts conspicuous, indurate; styles dilated, emarginate; fruit thick-walled, seeds 1 per locule, endosperm thin or 0 · 12. *Amanoa*

11. **Pentabrachion** Müll. Arg., Flora 47: 532 (1864) and in DC., Prodr. 15 (2): 223 (1866); Pax & K. Hoffm., Pflanzenr. 81: 188 (1922), and Nat. Pflanzenfam. ed. 2, 19C: 68 (1931) (as *Pentabrachium*); Keay (ed.), Fl. W. Trop. Afr. ed. 2, 1(2): 371 (1958); G. L. Webster, Ann. Missouri Bot. Gard. 81: 38 (1994). T.: *P. reticulatum* Müll. Arg.

Monoecious or dioecious subglabrous shrub or small tree. Leaves alternate, shortly petiolate, stipulate, oblong or elliptic, long-caudate-acuminate, entire, penninerved, tertiary nerves closely subparallel; stipules subulate from a triangular base. Flowers in axillary fascicles, or arising below the leaves, males numerous, females few. Male flowers pedicellate; calyx 4–6-partite, segments imbricate; petals 4–6, small; disc urceolate, elobate; stamens 5–7, episepalous if isomerous; filaments united at the base to a short column; anthers basifixed, introrse, longitudinally and laterally dehiscent; pollen ± as in *Amanoa* (q.v. infra); pistillode 3-fid, the lobes subpetaloid. Female flowers long-pedicellate; sepals, petals and disc as in the male; ovary ovoid, 3-locular, ovules 2 per locule; styles (fide Pax) free, twice 2-fid, erecto-patent. Fruit deeply 3-lobate, dehiscing into 2-valved cocci; exocarp very thin, separating; endocarp thin, bony. Seeds 3-gonous-subglobose, ecarunculate, shining; endosperm firmly fleshy; cotyledons thin, broad, flat; radicle cylindric. Fig.: See Pflanzenr. 81: 189, t. 15 (1922); Nat. Pflanzenfam. ed. 2, 19C: 68, t. 32 (1931).

A monotypic W. African genus, confined to Cameroon and Gabon.

Levin (1986a) suggests a possible relationship to *Dicoelia* on the basis of foliar venation characters, but this is not supported by pollen characters (Punt, 1962; Köhler, 1965). Indeed, Köhler associated *Pentabrachion* with the *Amanoeae* and *Bridelieae* because of its coarsely reticulate pollen grains. Webster (1994) considers that this palynological similarity may be a better guide to affinity, and so associates *Pentabrachion* with *Amanoa* in the same tribe, despite the obvious differences between the 2 genera. Possibly the best solution would be to place each in their own tribe as sole representatives, although this would tend to help to devalue the rank of tribe.

12. **Amanoa** *Aubl.*, Hist. Pl. Guiane 256, pl. 101 (1775); A. Juss., Euphorb. Gen.: 15 (1824); Planch., Hooker's Icon. Pl. 8: t. 797 (1848); Baill., Étude Euphorb.: 579 (1858); Müll. Arg. in DC., Prodr. 15 (2): 219 (1866), and Mart., Fl. Bras. 11(2): 10 (1873); Benth. & Hook. f., Gen. Pl. 3: 268 (1880); Baill. in Grandid., Atlas Hist. Phys. Madagascar tt. 210, 211 (1892); Hutch. in Fl. Trop. Afr. 6(1): 630 (1912); Pax & K. Hoffm., Pflanzenr. 81: 195 (1922), and Nat. Pflanzenfam. ed. 2, 19C: 68 (1931); Keay (ed.), Fl. W. Trop. Afr. ed. 2, 1(2): 371 (1958); Jabl., Mem. New York Bot. Gard. 17: 82 (1967); G. L. Webster, Ann. Missouri Bot. Gard. 54: 215 (1968); J.-G. Adam, Mém. Mus. Natl. Hist. Nat., B, Bot. 20: 455 (1971);); R. A. Howard, Fl. Lesser Antilles 5: 13 (1989); W. J. Hayden, Brittonia 42 (4): 260 (1990), and in IAWA J. 14: 205 (1993); A. H. Gentry, Woody Pl. NW. S. Amer.: 419 (1993); G. L. Webster, Ann. Missouri Bot. Gard. 81: 38 (1994); Murillo-Aldana & Franco-Rosselli, Euf. Reg. Ararac.: 44 (1995). T.: *A. guianensis* Aubl.

Monoecious glabrous trees or shrubs. Leaves alternate, petiolate, stipulate, simple, entire, coriaceous, penninerved with numerous interstitials; stipules intrapetiolar, sometimes connate. Inflorescences axillary, closely bracteate, strobiliform, the strobili densely imbricate, single or tightly pseudospicate, -racemose or -paniculate; male flowers glomerate-fasciculate or in dichasia, female solitary or few; bracts large, foliaceous; bracteoles minute. Male flowers sessile; sepals 5, imbricate; petals 5, short, scale-like, clawed; disc short, extrastaminal, lobate, lobes oppositisepalous; stamens 5, inserted on the receptacle, oppositisepalous, filaments free, commonly short, anthers thick, basifixed, introrse, ovoid, loculi parallel, longitudinally dehiscent; pollen oblate, 3–5-colporate, coarsely reticulate, baculate or echinate; pistillode columnar, 3-lobate at apex; receptacle short, thick, prominent. Female flowers pedicellate; sepals and petals narrower than in male, petals sometimes 0; hypogynous disc small, lobate, lobes oppositisepalous; ovary subglobose, 3-locular, ovules 2 per locule; styles connate, stigmas 3, sessile, thick, fleshy, 2-fid, reflexed, or disciform, crenate. Fruit massive, (1)3-seeded, tardily dehiscing into 2-valved cocci; endocarp thick-walled; columella massive, tapering from the base. Seeds 1 per coccus, ecarunculate, emarginate; testa crustaceous, shining; endosperm thin or 0; cotyledons thick, fleshy, flat within, angular, radicle short. Fig.: See Euphorb. Gen.: t. 2, f. 6.1–5 (1824); Hooker's Icon. Pl. 8: t. 797 (1848); Étude Euphorb.: t. 26, f. 48–50; 27, f. 1–4 (1858); Fl. Bras. 11(2): t. 1 (1873); Pflanzenr. 81: 199, t. 16 (1922); Nat. Pflanzenfam. ed. 2, 19C: 69, t. 33 (1931); Ann. Missouri Bot. Gard. 54: 216, t. 1 (1968); Mém. Mus. Natl. Hist. Nat., B, Bot. 20: 456, t. 158 (1971); Fl. Lesser Antilles 5: 11, t. 2 (1989); Bol. Mus. Paraense Emilio Goeldi, N.S., Bot. 7: 31, t. 3 (1991); Woody Pl. NW. S. Amer.: 416, t. 119.3 (1993); Euf. Reg. Ararac.: 46, t. 3 (1995).

A neotropical and W. and C. African genus of 16 species (13 of which are neotropical, mostly in northern S. America), following Hayden (1990).

Tribe 3. **BRIDELIEAE** *Müll. Arg.*, Bot. Zeitung (Berlin) 22: 324 (1864); G. L. Webster, Ann. Missouri Bot. Gard. 81: 38 (1994). T.: *Bridelia* Willd.

Tribe *Biovulatae* sect. *Bridelieae* Thwaites, Enum. Pl. Zeyl.: 279 (1861). *Antidesmataceae* subfam. *Bridelioideae* Hurus., J. Fac. Sci. Univ. Tokyo, Sect. 3, Bot. 6: 321 (1954).

Monoecious or rarely dioecious trees or shrubs. Leaves alternate, stipulate, entire, penninerved, eglandular; stipules usually deciduous. Flowers in bracteate glomerules, these axillary or in pseudospikes; sepals mostly 5, valvate; petals 5, imbricate or valvate, usually smaller than the sepals, sometimes dentate or reduced; disc annular or dissected; stamens mostly 5, filaments connate into a column; anthers introrse; pollen grains 3-colporate, reticulate or striate; pistillode +; female sepals 5, valvate or rarely apically imbricate, mostly persistent, very rarely caducous; disc double, annular and cupular; ovary 2–4-locular; ovules anatropous; styles 2–4-fid. Fruit dehiscent or drupaceous, 1–3(4)-celled. Seeds with endosperm; cotyledons much longer and broader than the radicle.

The tribe has traditionally been considered to comprise two closely related genera, *Cleistanthus* and *Bridelia*. According to Dressler (pers. comm.), the tribe is a natural one despite differences in the leaf-venation and fruit-type. Baillon (1858) regarded *Bridelia* as being close to *Amanoa*, and Köhler (1965) saw resemblances between pollen of *Amanoa* and the African species of *Cleistanthus*; following the latter's suggestion, it might be possible to associate the *Amanoeae* and *Bridelieae* as subtribes of a single tribe. Nevertheless, neither the floral nor pollen characters of the *Bridelieae* appear sufficiently close to *Amanoa* to warrant this.

Key to the genera

1. Ovary (2)3(4)-locular; fruits dehiscent, lobate, mainly 3-locular with 2 seeds per locule; leaf-venation brochidodromous · · · · · · · · · 13. *Cleistanthus*
1. Ovary 2(3)-locular; fruit drupaceous or rarely dehiscent, unlobate, 1–2-locular with 1 seed per locule; leaf-venation mainly craspedodromous · 14. *Bridelia*

13. **Cleistanthus** *Hook. f. ex Planch.*, Hooker's Icon. Pl. 8: t. 779 (1848); Müll. Arg. in DC., Prodr. 15(2): 503 (1866); Benth. & Hook. f., Gen. Pl. 3(1): 268 (1880); Hook. f., Fl. Brit. India 5: 274 (1887); Jabl., Pflanzenr. 65: 8 (1915); Gagnep., Fl. Indo-Chine 5: 482 (1926); Pax & K. Hoffm., Nat. Pflanzenfam. ed. 2, 19C: 80 (1931); Croizat, J. Arnold Arbor. 23: 38 (1942); Leandri, Fl. Madag. 111(1): 181 (1958); Keay (ed.), Fl. W. Trop. Afr. ed. 2, 1: 371 (1958); J. Léonard, Bull. Jard. Bot. État. 30: 421 (1960), and in Fl. Cong. Rwa.-Bur. 8 (1): 5 (1962); Backer & Bakh. f., Fl. Java 1: 474 (1963); Airy Shaw, Kew Bull. 21: 362 (1968), and 26: 235 (1972); Whitmore, Tree Fl. Mal. 2: 79 (1973); Airy Shaw, Hooker's Icon. Pl. 38: t. 3711 (1974), Kew Bull. Addit. Ser. 4: 75 (1975), and 8: 58 (1980), Kew Bull. 35: 608 (1980), 36: 279 (1981), 37: 13 (1982), and Euph. Philipp.: 14 (1983); Radcl.-Sm., Fl. Trop. E. Afr., Euph. 1: 130 (1987); Lebrun & Stork, Enum. Pl. Afr. Trop. 1: 207 (1991); McPherson, Fl. Nouv.-Caléd. 17: 26 (1991); G. L. Webster, Ann. Missouri Bot. Gard. 81: 38 (1994); Radcl.-Sm., Fl. Zamb., 9(4): 8 (1996); Philcox in Dassanayake, ed., Fl. Ceyl. XI: 82 (clav. tant.) (1997). T.: *Cl. polystachyus* Hook. f. ex Planch.

Nanopetalum Hassk.,Verslagen Meded. Afd. Natuurk. Kon. Akad. Wetensch. 4: 140 (1855). T.: *N. myrianthum* Hassk. [=*Cl. myrianthus* (Hassk.) Kurz]

Lebidiera Baill., Étude Euphorb.: 50, t. XXVII.1, 3–4 (1858). T.: *L. ferruginea* Baill. [=*Cl. ferrugineus* (Thwaites) Müll. Arg. (B.: *Amanoa ferruginea* Thwaites)]

Stenonia Baill., Étude Euphorb.: 578, t. XXII.2–5 (1858), non Didr. nec Endl. T.: *S. boiviniana* Baill. [=*Cl. stenonia* (Baill.) Jabl. (B.: *Amanoa stenonia* Baill.)]

Leiopyxis Miq., Fl. Ned. Ind., Eerste Bijv. 445 (1861). T.: *L. sumatrana* Miq. [=*Cl. sumatranus* (Miq.) Müll. Arg.)

Lebidieropsis Müll. Arg., Linnaea 32: 79 (1863). T.: *L. collina* (Roxb.) Müll. Arg. (B.: *Cluytia collina* Roxb.) [=*Cleist. collinus* (Roxb.) Benth.]

Kaluhaburunghos Kuntze, Revis. Gen. Pl. 2: 607 (1891). T.: (as for *Cleistanthus*)

Stenoniella Kuntze in T. Post & Kuntze, Lex. Gen. Phan.: 535 (1903), nom. nov. for *Stenonia* Baill.

Schistostigma Lauterb., Fl. Schutzgeb. Südsee, Nachtr. 299 (1905). T.: *S. papuanum* Lauterb. [=*Cl. papuanus* (Lauterb.) Jabl.]

Zenkerodendron Gilg ex Jabl., Pflanzenr. 65: 48 (1915), nom. nud. T.: *Z. bipindensis* Gilg ex Jabl., pro syn. [=*Cl. bipindensis* Pax]

Godefroya Gagnep., Bull. Soc. Bot. France 70: 135 (1923), and Fl. Indo-Chine 5: 481 (1926). T.: *G. rotundata* (Jabl.) Gagnep. [=*Cl. rotundatus* Jabl.]

Paracleisthus Gagnep., Bull. Soc. Bot. France 70: 499 (1923), and Fl. Indo-Chine 5: 496 (1926). LT.: *P. subgracilis* Gagnep. [=*Cl. saichikii* Merr. & Chun] (designated by Wheeler, 1975).

Monoecious or more rarely dioecious trees or shrubs. Indumentum simple. Leaves alternate, petiolate, stipulate, simple, entire, penninerved, secondary nerves anastomosing within the margin, tertiary reticulate. Inflorescences axillary, rarely terminal, fasciculate, or pseudospicate to pseudoracemose, uni- or 2-sexual; bracts soon falling. Flowers proterogynous, fragrant. Male flowers usually sessile; calyx-lobes (4)5(6), valvate; petals (4)5(6), scale-like; disc annular or cupular, entire or sinuate-lobate; stamens (4)5(6), filaments connate at the base, free and spreading above, anthers basifixed, introrse, thecae parallel, longitudinally dehiscent; pollen generally resembling that of *Amanoa*, q.v.; pistillode at the apex of the column, style-like, 3-lobate to 3-partite. Female flowers sessile or pedicellate; sepals and petals ± perigynous, otherwise as in the male; disc double, outer annular, inner annular or cupuliform, closely investing the ovary; ovary (2)3(4)-locular, ovules 2 per locule; styles (2)3(4), free or united at the base, 1–4 times 2-lobate or 2-fid. Fruit (2)3(4)-lobate, subglobose or depressed, dehiscing into 2-valved cocci; endocarp woody, separating from the pericarp; columella persistent, apically dilated. Seeds (1)2 per coccus, ecarunculate; endosperm scanty or copious; cotyledons broad, thin or thick and fleshy, flat or folded. Fig.: See Hooker's Icon. Pl. 8: t. 779 (1848); Pflanzenr. 65: 4–49, tt. 1–10 (1915); Fl. Indo-Chine 5: 480, t. 61.9, 10 and 486, t. 62.1–11 (partly as *Godefroya*), 495, t. 63.9–18 and 502, t. 64.1–10 (as *Paracleisthus*) (1926); Nat. Pflanzenfam. ed. 2, 19C: 80, t. 39 (1931); Fl. Madag. 111(1): 183–189, tt. XXXI–XXXIII.1–5 (1958); Fl. Cong. Rwa.-Bur. 8 (1): 9–19, tt. 1, I–II (1962); Hooker's Icon. Pl. 38: t. 3711 (1974); Kew Bull. Addit. Ser. 8: 225, t. 3.2 (1980); Kew Bull. 35: 590, t. 1C1,2 (1980); Fl. Trop. E. Afr., Euphorb. 1: 132, t. 24 (1987); Fl. Nouv.-Caléd. 17: 27, t. 5 (1991); Fl. Zamb., 9(4): 10, t. 2 (1996).

A large and rather diverse genus of c. 140 species, of which 23 occur in Africa, 6 in Madagascar, and the rest in tropical Asia, from India to northern Australia and Melanesia.

Both Köhler (1965) and Levin (1986a) found *Cleistanthus* to be heterogeneous on the basis of pollen and leaf characters respectively; clearly, generic limits here need to be critically reviewed.

14. **Bridelia** *Willd.*, Sp. Pl. 4(2): 978 (1806) (as *Briedelia*); Baill., Étude Euphorb.: 582 (1858); A. Juss., Euphorb. Gen.: 26 (1824) (as *Briedelia*); Müll. Arg. in DC., Prodr. 15(2): 492 (1866); Benth. & Hook. f., Gen. Pl. 3(1): 267 (1880); Hook. f., Fl. Brit. India 5: 267 (1887); Gehrm., Bot. Jahrb. Syst. 41 (Beibl. 95): 1 (1908); Hutch. in Fl. Trop. Afr. 6(1): 611 (1912); Jabl., Pflanzenr. 65: 54 (1915); Gagnep., Fl. Indo-Chine 5: 485 (1926); Pax & K. Hoffm., Nat. Pflanzenfam. ed. 2, 19C: 82 (1931); Robyns, Fl. Parc Nat. Alb. 1: 446 (1948); Hurus., J. Fac. Sci. Univ. Tokyo, Sect. 3, Bot. 6: 321 (1954); Leandri, Fl. Madag. 111(1): 192 (1958); Keay (ed.), Fl. W. Trop. Afr. ed. 2, 1: 368 (1958); Léonard, Fl. Cong. Rwa.-Bur. 8(1): 27 (1962); Backer & Bakh. f., Fl. Java 1: 475 (1963); Airy Shaw, Kew Bull. 23: 65 (1969); J.-G. Adam, Mém. Mus. Natl. Hist. Nat., B, Bot. 20: 457 (1971); Airy Shaw, Kew Bull. 26: 227 (1972); Whitmore, Tree Fl.

Mal. 2: 74 (1973); Berhaut, Fl. Ill. Sénég. 3: 391 (1975); Airy Shaw, Kew Bull. Addit. Ser. 4: 63 (1975), and 8: 43 (1980), Kew Bull. 35: 601 (1980), 36: 272 (1981), 37: 10 (1982), and Euph. Philipp.: 11 (1983); Radcl.-Sm. in Fl. Pak. 172. Euph.: 4 (1986), and in Fl. Trop. E. Afr., Euph. 1: 120 (1987); Grierson & D. G. Long, Fl. Bhut. 1: 767 (1987); Lebrun & Stork, Enum. Pl. Afr. Trop. 1: 207 (1991); G. L. Webster, Ann. Missouri Bot. Gard. 81: 39 (1994); S. Dressler, Kew Bull. 51: 601(1996), and Blumea 41: 263 (1996); Radcl.-Sm. in Fl. Zamb., 9(4): 12 (1996); Philcox in Dassanayake, ed., Fl. Ceyl. XI: 82 (clav. tant.) (1997). LT.: *B. scandens* (Roxb.) Willd. (B.: *Cluytia scandens* Roxb.) (chosen by Webster, 1994)

Candelabria Hochst., Flora 26: 79 (1843). T.: *C. micrantha* Hochst. [=*B. micrantha* (Hochst.) Baill.]
Pentameria Klotzsch ex Baill., Étude Euphorb.: 584 (1858) T.: *P. melanthesoides* Klotzsch ex Baill. [=*B. cathartica* Bertol.]
Neogoetzea Pax, Bot. Jahrb. Syst. 28: 419 (1900). T.: *N. brideliifolia* Pax [=*B. brideliifolia* (Pax) Fedde]
Gentilia Beille, Compt. Rend. Hebd. Séances Acad. Sci. 145: 1294 (1907). T.: *G. hygrophila* Beille [=*B. ndellensis* Beille]
Tzellemtinia Chiov., Ann. Bot. (Rome) 9: 55 (1911). T.: *T. nervosa* Chiov. [=*B. scleroneura* Müll. Arg.]

Monoecious or rarely dioecious trees or shrubs; trunk and branches often armed with blunt woody thorns. Twigs lenticellate. Indumentum simple. Leaves alternate, shortly petiolate, stipulate, simple, usually entire, penninerved, secondary nerves looped or continued to the margin, tertiary usually transverse, parallel. Flowers small, axillary, glomerulate or fasciculate, or in spikes or panicles of fascicles, males numerous, female fewer or solitary, sometimes fascicles 2-sexual; bracts small, scale-like. Male flowers sessile or subsessile; calyx-segments 5, valvate; petals 5, small, erect or inflexed, imbricate or separate, scale-like, stipitate or spathulate, subdentate; disc annular or cupular, entire or sinuate-lobate; stamens 5, filaments connate at the base into a short column, free and spreading above, anthers horizontal, basifixed, thecae parallel, longitudinally dehiscent; pollen oblate-spheroidal, 3-colporate, often pseudostriatoreticulate; pistillode at the top of the column, ampulliform, entire or 2–4-lobate. Female flowers mostly pedicellate; sepals and petals perigynous, otherwise ± as in the male; disc double, outer annular, inner lobate or cupular, closely enveloping the ovary; ovary 2(3)-locular, ovules 2 per locule; styles 2(3), free or shortly connate at the base, 2-fid, 2-lobate or subentire. Fruit a berry or drupe, small, smooth, ovoid or subglobose, usually indehiscent; exocarp fleshy or pulpy; endocarp crustaceous or slightly indurated, of 2 pyrenes (or 1 by abortion). Seeds often solitary per pyrene by abortion, plano-convex and adaxially grooved (2 pyrenes) or upsiloid in TS and adaxially excavate (1 pyrene); endosperm copious, fleshy, often deeply excavated on the inner face; embryo curved; cotyledons broad, foliaceous, thin. Fig.: See Euphorb. Gen.: t. 7, f. 22.1–6 (1824); Étude Euphorb.: t. 25, f. 25–34 (1858); Pflanzenr. 65: 59–84, tt. 11–15 (1915); Fl. Indo-Chine 5: 486, t. 62.12–16 and 495, t. 63.1–8 (1926);

Nat. Pflanzenfam. ed. 2, 19C: 82, t. 40 (1931); Formos. Trees: 333, t. 287 and 334, t. 288 (1936); Fl. Madag. 111(1): 189, t. XXXIII.6–9 and 195, t. XXXIV (1958); Fl. W. Trop. Afr. ed. 2, 1: 370, t. 131 (1958); Fl. Cong. Rwa.-Bur. 8 (1): 33–47, tt. 2, III–IV (1962); Amer. J. Bot. 56: 748, t. 6A–H (1969); Mém. Mus. Natl. Hist. Nat., B, Bot. 20: 461, t. 161 and 463, t. 162 (1971); Fl. Ill. Sénég. 3: 390 (1975); Kew Bull. Addit. Ser. 8: 223, t. 1.2 (1980); Fl. Pak. 172. Euph.: 7, t. 1A–D (1986); Fl. Trop. E. Afr., Euphorb. 1: 124, t. 23 (1987); Fl. Bhut. 1: 763, t. 47.e–h (1987); Kew Bull. 51: 603, t. 1(1996); Blumea 41: 263–331, tt. 1–5 (1996); Fl. Zamb., 9(4): 14, t. 3 (1996).

A Palaeotropical genus of more than 60 species, c. 20 in Africa, 2 in Madagascar and the Mascarenes, the rest Asiatic.

Tribe 4. **PHYLLANTHEAE** *Dumort.*, Anal. Fam. Pl.: 45 (1829); G. L. Webster, Ann. Missouri Bot. Gard. 81: 39 (1994). T.: *Phyllanthus* L.

Euphorbiaceae sect. *Phyllantheae* Blume, Bijdr. Fl. Ned. Ind. 578 (1826) (nom illeg.).

Ordnung *Phyllanthaceae* Klotzsch & Garcke, Monatsber. Königl. Preuss. Akad. Wiss. Berlin 1859: 246 (1859).

Monoecious or dioecious trees, shrubs or herbs. Leaves alternate, stipulate, entire, penninerved, eglandular; stipules persistent or deciduous. Flowers in axillary glomerules or racemes; bracts inconspicuous; male sepals 4–6, free or connate, imbricate; petals + or 0; disc extrastaminal, dissected or entire (rarely 0); stamens (2)3–5(15), filaments free or connate; anthers mostly extrorse; pollen grains prolate to oblate, mostly 3-colporate (sometimes stephanocolporate or porate); pistillode + or 0; female sepals mostly 5, imbricate, persistent or deciduous; petals usually 0; disc annular, cupular to lobate (less commonly dissected or 0), rarely double; ovary usually 3–5-locular; ovules anatropous or hemitropous; styles 2-fid or entire (rarely multifid). Fruit usually dehiscent (rarely baccate or drupaceous). Seeds 1 or 2 per locule; endosperm +; embryo with cotyledons usually broader than, and longer than or as long as the radicle.

As Webster (1994) interprets it, the tribe *Phyllantheae* is considerably narrower in scope than Pax & Hoffmann's (1922) circumscription, but broader than his earlier (Webster, 1975) one. The relationships of some of the taxa (e.g. *Astrocasia, Leptopus*) point towards the *Wielandieae*, but *Phyllantheae* are anatomically more advanced in such characters as septate fibres and scarcity of scalariform perforations. The inclusion of petaliferous and apetalous genera together seems justified by similarities in pollen (Punt, 1962; Köhler, 1965) and some of the leaf venation characters pointed out by Levin (1986a).

Subdivision of this heterogeneous tribe into subtribes is still controversial; the present classification with 6 subtribes must be considered provisional. The distinction between anatropous and hemitropous ovules, pointed out by Baillon (1858), appears systematically significant even though not all genera have been carefully studied, and there may prove to be transitional stages.

KEY TO THE SUBTRIBES

1. Petals +; ovules anatropous; seeds smooth ········· 2
1. Petals 0 (+ in some *Andrachninae*); ovules anatropous, hemitropous or amphitropous; seeds smooth or sculptured ········· 3
2. Stipules and female sepals deciduous; staminal column adnate to peltate pistillode ········· 4a. *Astrocasiinae*
2. Stipules and female sepals persistent; stamens free or connate below ········· 4b. *Leptopinae*
3. Ovules anatropous; pistillode + ········· 4
3. Ovules hemitropous or amphitropous; male disc annular or dissected, or 0; pistillode + or 0; seeds smooth or sculptured ········· 5
4. Male disc usually ± annular; pollen grains (where known) reticulate, rarely verrucate; seed coat dull, smooth or sculptured ········· 4c. *Pseudolachnostylidinae*
4. Male disc dissected; pollen grains verruculose; seed coat blackish, smooth ········· 4d. *Securineginae*
5. Male petals +; stamens 5(6); male disc dissected; pistillode + ········· 4e. *Andrachinae*
5. Male petals 0; stamens 2–15 or more; male disc annular or dissected, or 0; pistillode usually 0 ········· 4f. *Flueggeinae*

Subtribe 4a. **Astrocasiinae** *G. L. Webster*, Syst. Bot. 17: 315 (1992), and Ann. Missouri Bot. Gard. 81: 40 (1994). T.: *Astrocasia* B. L. Rob. & Millsp.

Dioecious or subdioecious trees or shrubs; stipules deciduous; flowers in axillary glomerules; male sepals 5; disc annular; stamens 3–5, filaments connate into a column; pollen grains 3-colporate, reticulate; pistillode peltate, flat; female sepals 5, caducous; disc annular; ovary 3-locular; ovules anatropous; styles 2-fid; fruits dehiscent; seeds 1 or 2 per locule, smooth; cotyledons much longer than radicle.

This monogeneric subtribe includes only the type genus *Astrocasia*, which was included in tribe *Wielandieae* by Webster (1975), following Pax & Hoffmann (1922). However Pax & Hoffmann referred *Astrocasia* to the *Wielandieae* on the basis of a mistaken interpretation of the floral morphology. Pollen studies of Köhler (1965) and the leaf venation studies of Levin (1986a, b, c) suggest on the contrary that *Astrocasia* shows more similarity to the tribe *Phyllantheae*, particularly to genera of subtribes *Leptopinae* and *Pseudolachnostylidinae*.

15. **Astrocasia** *B. L. Rob. & Millsp.*, Bot. Jahrb. Syst. 36 (Beibl. 80): 19 (1905); Pax & K. Hoffm., Pflanzenr. 81: 189 (1922), and Nat. Pflanzenfam. ed. 2, 19C: 68 (1931); Proctor, Kew Bull. Addit. Ser. 11: 521 (1984); G. L. Webster & Huft, Ann. Missouri Bot. Gard. 75: 1091 (1988); G. L. Webster, Syst. Bot. 17: 311 (1992), and Ann. Missouri Bot. Gard. 81: 40 (1994). T.: *A. phyllanthoides* B. L. Rob. & Millsp. [=*A. tremula* (Griseb.) G. L. Webster (B.: *Phyllanthus tremulus* Griseb.)]

Dioecious or monoecious slender glabrous shrubs or small trees. Leaves alternate, often long-petiolate, stipulate, sometimes stipellate, thin, more or less rhomboid, basifixed or peltate, entire, penninerved or palminerved, deciduous or evergreen; stipules linear, costate, deciduous. Flowers small, axillary, fasciculate-glomerulate; male fascicles several to many-flowered, female 1–3-flowered. Male flowers pedicellate; sepals 5, unequal, imbricate, patent; petals 5, trullate, ascending, almost twice as long as the sepals, venose, conspicuous; disc annular to shallowly patelliform or cupular, 5-lobate, the lobes episepalous; stamens 3–5, filaments connate into a slender column, anthers sessile or stipitate, extrorse, horizontal or deflexed, 2-thecate, thecae ellipsoid, transversely dehiscent; pollen oblate-spheroidal, 3-colporate, colpi long, ora well-defined, exine reticulate; pistillode peltate, flat, sessile or stipitate atop the column. Female flowers long-pedicellate; sepals and petals ± as in the male, soon caducous; disc annular to cupular; ovary 3(4)-locular, ovules 2 per locule, anatropous, obturator massive; styles free, 2-fid, spreading or deflexed. Fruit 3-coccous, septicidally dehiscent into 3 2-valved cocci; exocarp separating; endocarp thinly woody. Seeds 1 or 2 per locule, irregularly 3-gonous-subglobose, ecarunculate; testa dry, smooth or roughened, pale dull brownish; raphe conspicuous; endosperm copious, firmly fleshy; embryo straight, cotyledons thin, flat, white, much longer and broader than the radicle. Fig.: See Syst. Bot. 17: 312–319, tt. 1–7 (1992).

A neotropical genus of 5 species distributed from Mexico and Cuba south to Bolivia and eastern Brazil.

The pollen-morphology indicates a relationship with *Securinega* (Köhler, 1965), which is at variance with the nature of the gross morphology, e.g. the presence of well-developed petals.

Subtribe 4b. **Leptopinae** *G. L. Webster*, Ann. Missouri Bot. Gard. 81: 40 (1994). T.: *Leptopus* Decne. ex Jacquem.

Dioecious or monoecious shrubs or herbs; stipules persistent; flowers in axillary glomerules; male sepals and petals 5; disc annular or dissected; stamens 5, filaments free or connate below; pollen grains 3-colporate, reticulate; pistillode 3-fid; female sepals 5, persistent; petals minute; disc annular or dissected; ovary 3-locular; ovules anatropous; styles 2-fid; fruit dehiscent; seeds 2 per locule, testa smooth; cotyledons much longer than radicle.

This subtribe includes only the single genus *Leptopus*, which was submerged in *Andrachne* by Müller (1866), Bentham (1880), and Pax & Hoffmann (1922). However, *Leptopus* differs from *Andrachne* (s.str.) in its anatropous ovules, and appears to represent a connecting link between *Astrocasia* and *Andrachne.* Webster therefore assigned it to a separate subtribe, but Stuppy (1995) feels that this does not best reflect the undoubted close relationship that exists between *Leptopus* and *Andrachne.*

16. **Leptopus** *Decne. in Jacquem.*, Voy. Inde 155 (1836); Pojark., Bot. Mater. Gerb. Bot. Inst. Komarova Akad. Nauk SSSR 20: 269 (1960); Airy Shaw, Kew Bull. 26: 285 (1972); Whitmore, Tree Fl. Mal. 2: 105 (1973); Airy Shaw, Kew Bull. 35: 645 (1980), 37: 26 (1982), and Euph. Philipp.: 33 (1983); Li Ping-T'ao, Notes Roy. Bot. Gard. Edinburgh 40: 467 (1983); G. L. Webster, Ann. Missouri Bot. Gard. 81: 40 (1994). T.: *L. cordifolius* Decne.

Lepidanthus Nutt., Trans. Amer. Philos. Soc. 5: 175 (1837), non Nees (1830). T.: *L. phyllanthoides* Nutt. [= *Leptopus phyllanthoides* (Nutt.) G. L. Webster]

Hexakistra Hook. f., Fl. Brit. India 5: 283 (1887). T.: (none designated)

Thelypetalum Gagnep., Bull. Soc. Bot. France 71: 876 (1924). T.: *T. pierrei* Gagnep. [=*L. australis* (Zoll. & Moritzi) Pojark. (B.: *Andrachne australis* Zoll. & Moritzi)]

Chorisandrachne Airy Shaw, Kew Bull. 23: 40 (1969), 26: 232 (1972), and Hooker's Icon. Pl. 38: t. 3707 (1974). T.: *C. diplosperma* Airy Shaw [=*L. diplospermus* (Airy Shaw) G. L. Webster]

Very close to 24. *Andrachne*, from which it differs primarily in having anatropous ovules. Fig.: See Hooker's Icon. Pl. 38: t. 3707 (1974) (as *Chorisandrachne*); Fl. Pak., 172 Euph.: 40, t. 8A–C (1986) (as *Andrachne cordifolia*).

A genus of c. 10 species of widely scattered distribution in the Old World (from the Caucasus through India to China, Indonesia, and tropical Australia) and North America (Mexico, southern U.S., and Greater Antilles).

The genus *Archileptopus* P. T. Li, J. S. China Agric. Univ. 12 (3): 38 (1991) (T.: *A. fangdingianus* P. T. Li), described from Guangxi, S.E. China, is stated to differ from *Leptopus* in having extrorse as opposed to introrse anthers, and 4–5-locular ovaries and fruit. It may be, however, that these features are not generically significant, and that the taxon should only be accorded specific rank under *Leptopus*.

Subtribe 4c. **Pseudolachnostylidinae** *Pax & K. Hoffm.*, Pflanzenr. 81: 206 (1922), and Nat. Pflanzenfam. ed. 2, 19C: 70 (1931); G. L. Webster, Ann. Missouri Bot. Gard. 81: 40 (1994). T.: *Pseudolachnostylis* Pax.

Monoecious or dioecious trees, shrubs, or undershrubs; stipules persistent or deciduous. Flowers in axillary glomerules or cymes, bracts inconspicuous; male sepals usually 5, petals 0; disc ± annular; stamens usually 5, filaments free or connate; pollen, where known, 3-colporate, reticulate, rarely verrucate; pistillode +; female sepals 5, persistent or deciduous; petals 0; disc usually annular, single or double; ovary 3-locular; ovules anatropous; styles 2-lobate, 2-fid or 2-partite, sometimes undivided. Fruit usually dehiscent, rarely drupaceous. Seeds 1 or 2 per locule, testa smooth or sculptured; cotyledons often much longer than radicle.

This subtribe of 6 genera is heterogeneous and possibly unnatural, as it is quite possible that *Keayodendron* and *Pseudolachnostylis* may not be closely related to the rest.

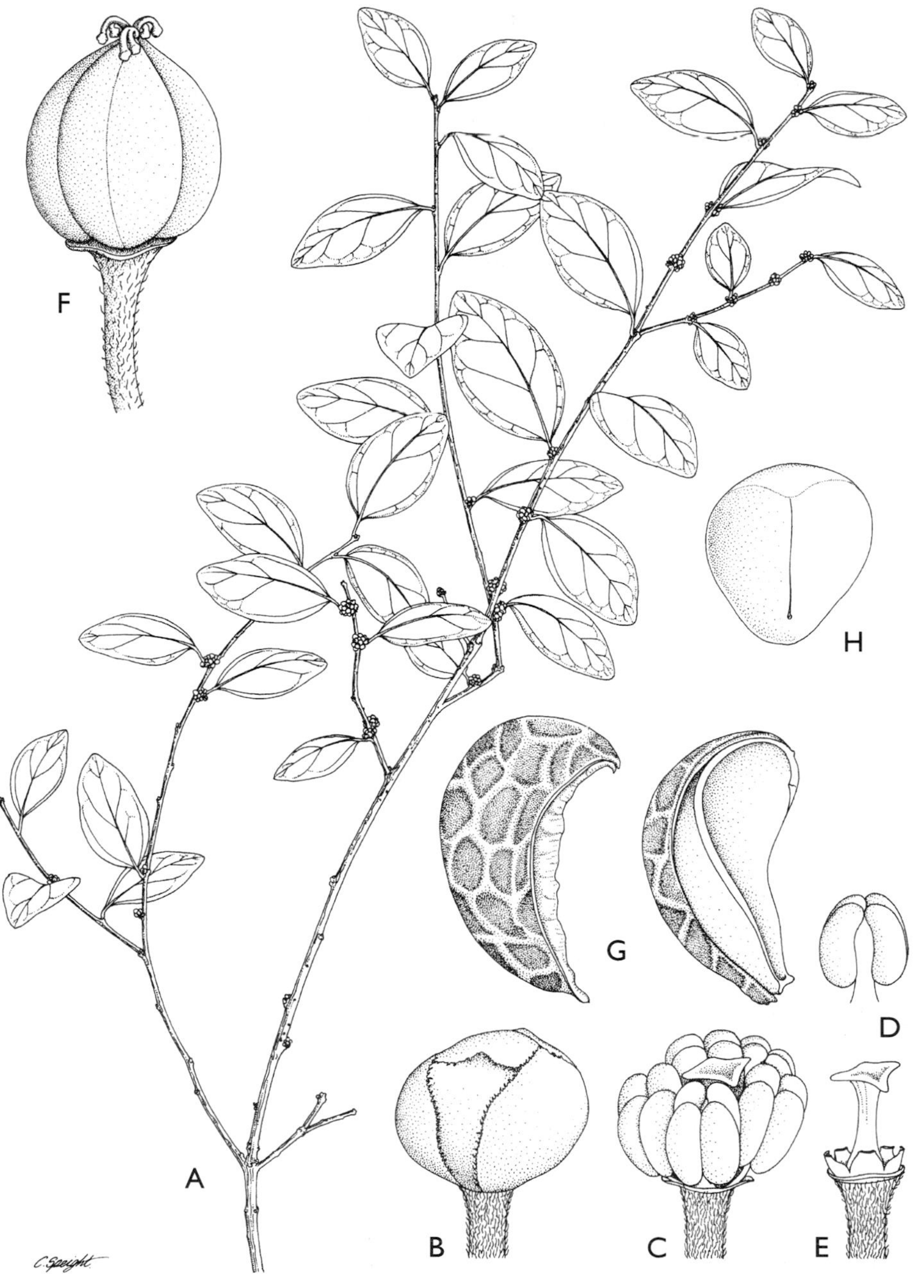

FIG. 3. *Chascotheca neopeltandra* Urb. **A** Habit × 1; **B** Male flower-bud × 25; **C** Male flower-bud, calyx removed × 25; **D** Stamen × 25; **E** Pistillode × 25; **F** Ovary × 25; **G** Coccus-valves × 15; **H** Seed × 15. **A–E** from *Shafer* 12080; **F–H** from *Rugel* 281. Drawn by Camilla Speight.

KEY TO THE GENERA

1. Fruit drupaceous, 2-locular, one-seeded; male disc dissected, female double · 22. *Keayodendron*
1. Fruit dehiscent, 3-locular, 3–6-seeded; male disc annular, female single · · 2
2. Fruit thick-walled; female cymes pedunculate; seeds with smooth testa; female sepals deciduous; filaments connate · · · · · 21. *Pseudolachnostylis*
2. Fruit thin-walled; cymes (glomerules) not pedunculate; seeds with sculptured testa; female sepals persistent; filaments free or connate · · · · · · · · · · 3
3. Chalaza dorsal on seed; male flowers subsessile · · · · · · · · 17. *Chascotheca*
3. Chalaza ventral; male flowers pedicellate · 4
4. Filaments adnate to pistillode; stipules narrow; styles 2-lobate, 2-fid or 2-partite ·20. *Meineckia*
4. Filaments free from pistillode; stipules foliaceous; styles ± undivided · · · 5
5. Tepals broad; styles 3; stipules persistent · · · · · · · · · · · 18. *Zimmermannia*
5. Tepals narrow; styles 6; stipules deciduous · · · · · · · · · 19. *Zimmermanniopsis*

17. **Chascotheca** *Urb.*, Symb. Antill. 5: 14 (1904); Alain, Fl. Cuba 3: 44 (1953); Proctor, Kew Bull. Addit. Ser. 11: 521 (1984); Alain, Fl. Español. 4: 106 (1986); G. L. Webster, Ann. Missouri Bot. Gard. 81: 41 (1994). T.: *Ch. neopeltandra* (Griseb.) Urb. (B.: *Phyllanthus neopeltandrus* Griseb.)

Chaenotheca Urb., Symb. Antill. 3: 284 (1902), non Th. Fr. (1860). T.: *Ch. domingensis* Urb. [=*Chascotheca domingensis* (Urb.) Urb.]

Dioecious trees or shrubs. Leaves alternate, distichous, stipulate, entire, 3–5-nerved at the base, otherwise penninerved. Flowers clustered among numerous bracts; pedicels clothed with jointed hairs. Male flowers subsessile; sepals 5, imbricate, orbicular, concave, ± reflexed during flowering; petals 0; disc adnate to the bottom of the calyx, free margin crenulate; stamens 5, opposite the sepals, filaments connate at the base; anthers suborbicular-reniform, dorsifixed, thecae parallel, extrorse, longitudinally dehiscent, valves gaping; pollen prolate, 3-colpate; pistillode columnar, linear-triangular, 3-fid, united with the base of the filaments. Female flowers: sepals as in the male; petals 0; ovary 3-locular, ovules 2 per locule; styles 3, very short, free, 2-fid, spreading or reflexed, lobes linear, thickened at the apex. Fruit dehiscing into 2-valved cocci; exocarp nervose, separating from the endocarp; endocarp hardened; columella linear-triangular, persistent. Seed 1 per locule by abortion, ovoid, ecarunculate; testa reticulate; endosperm thick and fleshy; cotyledons ovate, thin, flat; radicle curved. Fig.: See Fl. Cuba 3: 44, t. 8 (1953). See also Fig. 3, p. 25.

Two species have been described in this Greater Antillean genus (Cuba and Hispaniola), but *Chascotheca domingensis* (Urb.) Urb. is probably conspecific with the type species.

18. **Zimmermannia** *Pax*, Bot. Jahrb. Syst. 45: 235 (1910); Hutch. in Fl. Trop. Afr. 6(1): 739 (1912); Pax & K. Hoffm., Nat. Pflanzenfam. ed. 2, 19C: 59 (1931); Verdc., Kew Bull. 9: 38 (1954); M. M. Poole, Kew Bull. 36: 129 (1981);

Radcl.-Sm., Kew Bull. 36: 127 (1981), and in Fl. Trop. E. Afr. Euph. 1: 71 (1987); G. L. Webster, Ann. Missouri Bot. Gard. 81: 41 (1994). T.: *Z. capillipes* Pax (S.: *Meineckia paxii* Brunel ex Radcl.-Sm.)

Meineckia sect. *Decaryana* Brunel ex Radcl.-Sm., Kew Bull. 52: 173 (1997).
Meineckia sect. *Zimmermannia* (Pax) Brunel ex Radcl.-Sm., Kew Bull. 52: 174 (1997).

Monoecious or occasionally dioecious glabrous trees or shrubs with flexuous branches. Leaves alternate, shortly petiolate, stipulate, simple, entire, thin, penninerved; stipules large, foliaceous, green, persistent. Flowers axillary, the males solitary, geminate or fasciculate, females solitary, usually in the distal axils. Male flowers often long-pedicellate; sepals 5, imbricate, venose; petals 0; disc large, flat, shallowly 5-lobate; stamens 5, filaments free, anthers introrse, medifixed, thecae parallel, longitudinally dehiscent; pollen spheriodal to prolate, 3-colporate or 3-porate, reticulate or verrucate; pistillode small to minute, 3-fid or 3-lobate. Female flowers long-pedicellate; sepals and disc as in the male but larger, especially in fruit; ovary 3-locular, ovules 2 per locule; styles 3, ± free or connate at the base, ascending or spreading, recurved, swollen at the apex, ± entire. Fruit dehiscing into 3 2-valved cocci, or 6 valves; exocarp rugulose, separating from the endocarp; endocarp thinly woody; columella persistent. Seeds 2 per locule, roundly 3-quetrous-subreniform, foveolate-lineate or rugulose, ecarunculate. Fig.: See Fl. Trop. E. Afr. Euph., 1: 72, t. 8 (1987).

As interpreted by Verdcourt (1954) and by Radcliffe-Smith (1987), *Zimmermania* includes 6 east African species, plus a dubious Madagascan species transitional in some respects to *Meineckia.*

Brunel (1987) reduced *Zimmermannia* and *Zimmermanniopsis* to sectional status under *Meineckia,* a course adopted by Radcliffe-Smith (1997), but as both Webster (1994) and Stuppy (1995) maintain them, he here reverts to his former position.

19. **Zimmermanniopsis** *Radcl.-Sm.*, Kew Bull. 45: 152 (1990); G. L. Webster, Ann. Missouri Bot. Gard. 81: 41 (1994). T.: *Z. uzungwaënsis* Radcl.-Sm. (S.: *Meineckia uzungwaënsis* (Radcl.-Sm.) Radcl.-Sm.).

Meineckia sect. *Zimmermanniopsis* (Radcl.-Sm.) Radcl.-Sm., Kew Bull. 52: 175 (1997).

Very close to 18. *Zimmermannia,* but differing in the membranous, readily decidous stipules, the narrow non-overlapping sepals and the more numerous styles. Fig.: See Kew Bull. 45: 153, t. 2 (1990).

A monotypic genus from S. Tanzania (Iringa District).

20. **Meineckia** *Baill.*, Étude Euphorb. 586 (1858); G. L. Webster, Acta Bot. Neerl. 14: 323 (1965); Radcl.-Sm., Hooker's Icon. Pl. 37: t. 3698 (1971); Brunel & J. P. Roux, Bull. Mus. Natl. Hist. Nat., Sér. 4, 4: 79 (1982); Radcl.-Sm., Fl. Trop. E. Afr. Euph., 1: 76 (1987); G. L. Webster, Ann. Missouri Bot. Gard. 81:

41 (1994); Philcox in Dassan., ed., Fl. Ceyl. XI: 82 (clav. tant.) (1997). T.: *M. phyllanthoides* Baill.

Cluytiandra Müll. Arg., J. Bot. 2: 328 (1864), and in DC., Prodr. 15(2): 225 (1866); Benth. & Hook. f., Gen. Pl. 3(1): 272 (1880); Hutch. in Fl. Trop. Afr. 6(1): 738 (1912); Pax & K. Hoffm., Pflanzenr. 81: 209 (1922), and Nat. Pflanzenfam. ed. 2, 19C: 70 (1931); Robyns, Fl. Parc Nat. Alb. 1: 446 (1948); Leandri, Fl. Madag. 111(1): 138 (1958). T.: *C. trichopoda* Müll. Arg. [=*M. trichopoda* (Müll. Arg.) G. L. Webster]

Peltandra Wight, Icon. Pl. Ind. Orient. 5(2): 24 (1852) (non *Peltandra* Raf., 1819). LT.: *P. parvifolia* Wight [=*M. parvifolia* (Wight) G. L. Webster] (chosen by Webster, 1994).

Neopeltandra Gamble, Fl. Madras 2: 1285 (1925). LT.: *N. longipes* (Wight) Gamble (B.: *Peltandra longipes* Wight) [=*M. longipes* (Wight) G. L. Webster] (chosen here).

Monoecious or dioecious shrubs or shrublets, sometimes subherbaceous. Indumentum simple, hairs multicellular. Leaves alternate, distichous, shortly petiolate, stipulate, simple, entire, penninerved, deciduous or not. Flowers axillary, fasciculate, the fascicles unisexual or 2-sexual — if unisexual, male flowers several, females usually solitary. Male flowers long-pedicellate; sepals 5(6), imbricate; petals 0; disc annular or saucer-shaped, slightly lobate, the lobes oppositisepalous, partly adnate to the calyx; stamens 5, filaments partly united into a column, anthers extrorse, 2-thecate, dehiscing laterally or longitudinally; pollen prolate-spheroidal, indistinctly reticulate; pistillode at the top of the column, 3-fid, 3-lobate or subentire. Female flowers: pedicels long and slender, jointed above the base; calyx as in the male, becoming more or less scarious in fruit; disc as in the male; ovary 3-locular, ovules 2 per locule; styles 3, slightly connate at the base, 2-lobate, 2-fid or 2-partite, stigmas capitate or clavate. Fruit 3-lobate, dehiscing into 3 venose 2-valved cocci; endocarp thinly woody; columella persistent, slender. Seed usually 1 per locule by abortion, reniform, pitted, ecarunculate; endosperm copious; cotyledons broad, thin, flat, radicle over half their length, bent. Fig.: See Fl. Madag. 111(1): 139, t. XXII (1958) (as *Cluytiandra*); Hooker's Icon. Pl. 37: t. 3698 (1971); Fl. Trop. E. Afr. Euph., 1: 77, t. 9 (1987).

A genus of 20 species with a disjunct distribution in the New World and Old World, from Mexico to Colombia and Brazil, and from central Africa to Madagascar, S. Arabia, Socotra, southern India, Sri Lanka and Assam, following Webster (1965).

21. **Pseudolachnostylis** *Pax*, Bot. Jahrb. Syst. 23: 19 (1899); Hutch. in Fl. Trop. Afr. 6(1): 671 (1912), and in Hooker's Icon. Pl. 31: t. 3011 (1915); Pax & K. Hoffm., Pflanzenr. 81: 206 (1922), and Nat. Pflanzenfam. ed. 2, 19C: 70 (1931); E. Phillips, Gen. S. Afr. Fl. Pl. ed 2, 457 (1951); Radcl.-Sm. in Fl. Trop. E. Afr. Euph., 1: 80 (1987), and in Proc. XIIIth. Plenary Meeting, AETFAT, Malawi, 1: 385 (1994); G. L. Webster, Ann. Missouri Bot. Gard. 81: 41 (1994); Radcl.-Sm. in Fl. Zamb. 9(4): 22 (1996). T.: *P. maprouneifolia* Pax.

Dioecious or rarely monoecious trees or shrubs; buds perulate. Indumentum simple. Leaves alternate, petiolate, stipulate, simple, entire, somewhat coriaceous, penninerved; stipules small. Flowers axillary, or arising below the leaves, males in few-flowered pedunculate or subsessile bracteate cymes, females solitary or subsolitary. Male flowers: buds ovoid; sepals 5(6), imbricate; petals 0; disc extrastaminal, annular, ± lobate, the lobes alternisepalous; stamens (4)5(7), filaments partly connate into a column, anthers basifixed, introrse, thecae parallel, longitudinally dehiscent; pollen suboblate, 2–3-colporate, lumina up to 4.5μ; pistillode columnar, often 3-fid. Female flowers: pedicels 2-bracteolate; buds and sepals as in the male; disc annular to cupular, often irregularly toothed; ovary 3-locular, ovules 2 per locule; styles 3, very slightly connate at the base, 2-lobate or 2-fid, recoiled. Fruit globose or depressed-globose, subdrupaceous, tardily septicidally dehiscent; exocarp smooth, shiny, becoming wrinkled on drying; mesocarp somewhat fleshy to spongy; endocarp thick and bony or woody. Seeds 1 per locule by abortion, ellipsoid, ecarunculate; endosperm fleshy; cotyledons broad and flat. Fig.: See Hooker's Icon. Pl. 31: t. 3011 (1915); Pflanzenr. 81: 208, t. 17 (1922); Fl. Trop. E. Afr. Euph., 1: 82, t. 10 (1987); Fl. Zamb. 9(4): 25, t. 4 (1996).

A monotypic African genus.

The phylogenetic position of *Pseudolachnostylis* remains uncertain, although the disposition of Pax & Hoffmann (1922) in locating it adjacent to *Meineckia*, followed by Webster (1994), is also followed here. However, the pedunculate inflorescence and massive fruits of *Pseudolachnostylis* are quite different from the preceding genera, and it is not clear that it really belongs in the same subtribe. Köhler (1965) has proposed that *Pseudolachnostylis* be classified with *Amanoa* because of its coarsely reticulate pollen grains and large fruits. However, the flowers differ from those of *Amanoa* in lacking petals, in lacking conspicuous bracts, and in having bifid styles; furthermore, the plants are dioecious, and the stipules are deciduous. Because of all these differences, it seems that the similarity in pollen ornamentation is of no taxonomic significance. Stuppy (1995) argues the case for its placement near *Cleistanthus*.

22. **Keayodendron** *Leandri*, Bull. Soc. Bot. France 105: 517 (1959); Breteler, Novit. Gabon. 10, Bull. Jard. Bot. Belg. 62: 187 (1993); G. L. Webster, Ann. Missouri Bot. Gard. 81: 42 (1994). T.: *K. bridelioides* Leandri (S.: *Casearia bridelioides* Gilg. ex Engl.; *C. bridelioides* Mildbr. ex Hutch. & Dalziel).

Large dioecious tree. Leaves alternate, entire, elliptic-obovate. Flowers axillary, fasciculate, males several, shortly pedicellate or subsessile, females fewer or solitary, shortly pedicellate; bracts acute. Male flowers: sepals 5(7), imbricate, unequal, ciliate; petals 0; disc-lobes 3–5, ± triangular; stamens 5–6, filaments adnate to the pistillode at the base; anthers dorsally sub-basifixed, introrse, thecae ovoid-oblong, longitudinally dehiscent; pollen undescribed; pistillode 2–4-fid. Female flowers: sepals thicker than in the male, otherwise similar; disc double, outer similar to the male, inner saucer-shaped, strigillose; ovary 2-locular, with 2 ovules per locule; styles 4, recurved. Fruit drupaceous, ovoid-subglobose, apiculate, with persistent calyx, 1-locular by abortion; exocarp soft, endocarp woody. Seed 1 by abortion, C-shaped in TS; testa

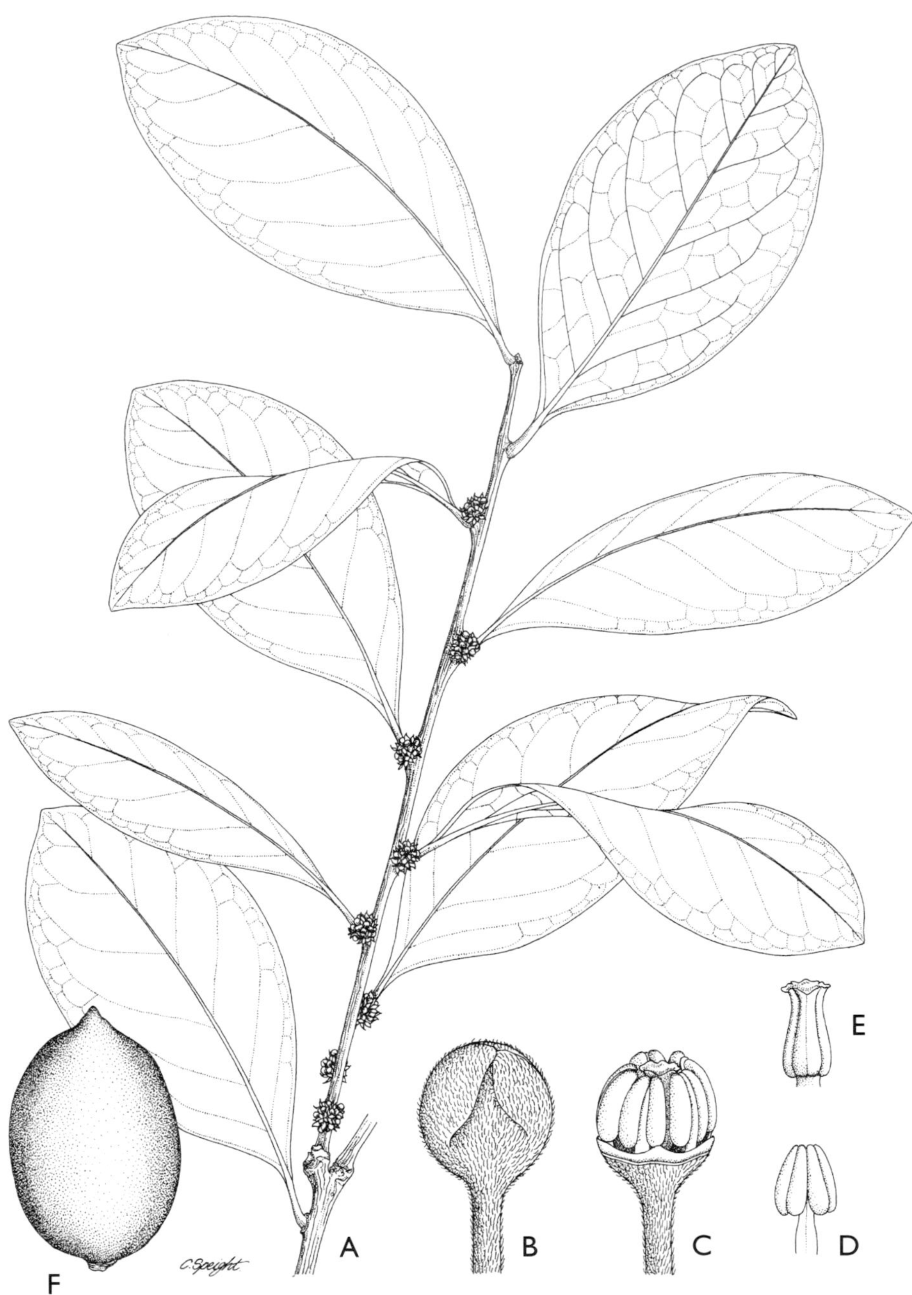

FIG. 4. *Keayodendron bridelioides* (Mildbr. ex Hutch. & Dalziel) Leandri (B.: *Casearia bridelioides* Mildbr. ex Hutch. & Dalziel). **A** Habit × 1; **B** Male flower-bud × 15; **C** Male flower-bud, calyx removed × 15; **D** Stamen × 15; **E** Pistillode × 15; **F** Fruit × 2. **A–E** from *Mildbraed* 8768; **F** from *Olorunfemi* in FHI 35662. Drawn by Camilla Speight.

papery; endosperm thin; cotyledons broad, rounded, folded transversely C-wise and longitudinally S-wise, apically undulate; radicle superior. Fig.: See Bull. Soc. Bot. France 105: 515, t. 1 (1959). See also Fig. 4, p. 30.

A monotypic genus from Cote d'Ivoire to Cameroun, still incompletely known.

As the specific epithet suggests, the fruits and aspect suggest *Bridelia*, but the flowers are apetalous, with imbricate sepals. *Keayodendron* is somewhat tentatively referred to this subtribe for lack of a better alternative.

Subtribe 4d. **Securineginae** *Müll. Arg. in DC.*, Prodr. 15(2): 446 (1866); G. L. Webster, Ann. Missouri Bot. Gard. 81: 42 (1994). T.: *Securinega* Comm. ex Juss.

Monoecious or dioecious shrubs or small trees; stipules deciduous. Flowers in axillary glomerules, bracts inconspicuous; male sepals 5; petals 0; disc dissected; stamens (4)5(10), free; pollen 3-colporate, sexine verruculose; pistillode +; female sepals 5, persistent; disc angled or 0; ovary 3-locular; ovules anatropous; styles 2-fid. Fruit dehiscent. Seeds 1 or 2 per locule, testa dark and smooth; cotyledons about equalling the radicle.

Although Müller's name is applied to this subtribe, its present circumscription is very different, since Müller (1866) included *Hymenocardia* and genera of tribe *Aporoseae*. The distinctive pollen and seeds of *Securinega* set it apart from other taxa in tribe *Phyllantheae*, and suggest a possible affinity with subfam. *Oldfieldioideae*.

23. **Securinega** *Comm. ex Juss.*, Gen. Pl.: 388 (1789) (nom. cons.); Baill., Étude Euphorb.: 588 (1858); Pax & K. Hoffm., Nat. Pflanzenfam. ed. 2, 19C: 60 (1931), sens. ampliss.; Leandri, Fl. Madag. 111(1): 107 (1958); Coode in Fl. Masc. 160. Euph.: 32 (1982), p.p.; G. L. Webster, Ann. Missouri Bot. Gard. 81: 42 (1994). T.: *S. durissima* J. F. Gmel. (typ. cons.).

Monoecious or dioecious shrubs. Leaves alternate or spiral, entire, often small. Flowers axillary, in all axils, glomerate, males numerous, small, females fewer or solitary. Male flowers subsessile to shortly pedicellate; sepals 5, free, imbricate; petals 0; disc-glands 5, alternisepalous; stamens (4)5(10); filaments free, opposite the sepals; anthers erect, dorsifixed, introrse, thecae separate, parallel, longitudinally dehiscent; pollen 3-colporate, verruculose; pistillode small or elongate, 2–3-fid. Female flowers pedicellate; sepals persistent, otherwise as in the male; petals 0 or vestigial; disc annular, subentire; staminodes 0; ovary 3(4)-locular, ovules 2 per locule; styles 3(4), free, sessile, recurved, 2-fid. Fruit dry or slightly fleshy, dehiscing into 2-valved cocci. Seed 1 per locule by abortion, testa thin or thinly coriaceous, smooth, blackish; endosperm fleshy; embryo straight; cotyledons broad, flat. Fig.: See Étude Euphorb.: t. 26, f. 32–38 (1858); Notul. Syst. (Paris) 6: 32, t. 3 (1937); Fl. Madag. 111(1): 109, t. XVII.9–15 and 113, t. XVIII.6–16 (1958); Fl. Masc. 160. Euph.: 34, t. 5 (1982).

A Mascarene genus of 5 or 6 closely related species, the type from Réunion, Mauritius and Rodriguez, the others from Madagascar.

The majority of workers have confounded the genus with *Flueggea*, and only Leandri (1958) among modern writers has presented the genus as accepted by Webster (1994).

Subtribe 4e. **Andrachninae** *Müll. Arg.*, Linnaea 34: 64 (1865) (as *Andrachneae*); Pax, Nat. Pflanzenfam. ed. 1, 3(5): 15 (1890); Pax & K. Hoffm., Pflanzenr. 81: 169 (1922), and Nat. Pflanzenfam. ed. 2, 19C: 66 (1931); G. L. Webster, Ann. Missouri Bot. Gard. 81: 42 (1994). T.: *Andrachne* L.

Monoecious herbs or subshrubs; stipules persistent. Flowers in axillary glomerules, bracts inconspicuous; male sepals and petals 5; disc dissected; stamens 5(6), filaments free or connate; pollen 3-colporate, sexine reticulate or striate; pistillode +; female sepals 5, persistent; petals 5 or 0; disc annular or dissected; ovary 3-locular; ovules hemitropus; styles 2-fid. Fruit dehiscent. Seeds 2 per locule, testa smooth or sculptured; cotyledons much longer than radicle.

The subtribe includes only the genus *Andrachne* (s.str.); Müller (1866) and most later workers included *Leptopus* within *Andrachne.* Except for the presence of petals, species of *Andrachne* resemble herbaceous species of *Phyllanthus*; however, it is not clear whether this indicates close affinity. Earlier Webster (1975) associated the Australian genus *Poranthera* with *Andrachne* because of similarity in the pollen pointed out by Köhler (1965). However, other characters of *Poranthera* now seem to him to point towards affinity with tribe *Antidesmeae.*

24. **Andrachne** *L.*, Sp. Pl.: 1014 (1753); Gen. Pl. ed. 5: 444 (1754); A. Juss., Euphorb. Gen.: 24 (1824); Baill., Étude Euphorb.: 575 (1858); Müll. Arg. in DC., Prodr. 15(2): 232 (1866); Benth. & Hook. f. Gen. Pl. 3: 270 (1880); Hook. f., Hooker's Icon. Pl. 18: t. 1704 (1887), and Fl. Brit. India 5: 283 (1887), p.p.; Kossinky, Bot. Mater. Gerb. Glavn. Bot. Sada RSFSR 2: 77 (1921); Pax & K. Hoffm., Pflanzenr. 81: 169 (1922); Beille, Fl. Indo-Chine 5: 537 (1927); Pax & K. Hoffm., Nat. Pflanzenfam. ed. 2, 19C: 66 (1931); Croizat, J. Wash. Acad. Sci. 33: 11 (1942); Pojark. in Fl. USSR 14: 269 (1949); Vindt, Monogr. Euph. Maroc: 4 (1953); Alain, Fl. Cuba 3: 40 (1953); Pojark., Bot. Mater. Gerb. Bot. Inst. Komarova Akad. Nauk SSSR 20: 256 (1960); Backer & Bakh. f., Fl. Java 1: 469 (1963); Rech. f. & Schiman-Czeika, Fl. Iran. 6: 1 (1964); G. L. Webster, J. Arnold Arbor. 48: 327 (1967); Tutin in Fl. Eur. 2: 211 (1968); Radcl.-Sm., Hooker's Icon. Pl. 37: t. 3697 (1971), in Fl. Iraq 4(1): 311 (1980), in Fl. Turk. 7: 566 (1982), in Fl. Cyp. 2: 1448 (1985), and in Fl. Pak., 172 Euph.: 36 (1986); Alain, Fl. Español. 4: 80 (1986); M. G. Gilbert, Kew Bull. 42: 351 (1987); Radcl.-Sm. in Fl. Trop. E. Afr., Euph. 1: 7 (1987); M. G. Gilbert & Thulin, Nordic J. Bot. 8: 159 (1988); G. L. Webster, Ann. Missouri Bot. Gard. 81: 42 (1994); Radcl.-Sm. in Fl. Zamb. 9(4): 27 (1996). LT.: *A. telephioides* L. (designated by Small, 1913).

Telephioides Ortega, Tab. Bot.: 15 (1773). T.: *T. procumbens* Moench [=*A. telephioides* L.]

Eraclissa Forssk., Fl. Aegypt.-Arab.: 208 (1775). T.: *E. hexagyna* Forssk. [=*A. telephioides* L.]

Arachne Neck., Elem. Bot. II: 348 (1790); Pojark. in Komarov, Fl. U.S.S.R. 14: 269 (1949); Hurus., J. Fac. Sci. Univ. Tokyo, Sect. 3, Bot. 6: 338 (1954). T.: (none designated).

Phyllanthidea Didr., Vidensk. Meddel. Naturhist. Foren. Kjøbenhavn 1857: 150 (1857). T.: *P. microphylla* (Lam.) Didr. (B.: *Croton microphyllus* Lam.) [=*A. microphylla* (Lam.) Baill.]

Monoecious shrublets or perennial herbs from a woody taproot. Indumentum simple, glandular or 0. Leaves alternate, petiolate, stipulate, simple, entire, penninerved, often small. Flowers axillary, males often fasciculate, female solitary. Male flowers pedicellate; sepals 5(6), free or ± so, imbricate; petals 5(6), shorter than or equalling the sepals; disc-glands 5(6), oppositipetalous, often 2-lobate, or 10(12), or disc cupular, dentate; stamens 5(6), oppositisepalous, filaments free or connate to $^{1}/_{2}$-way, anthers erect, dorsifixed, introrse, thecae parallel, distinct, longitudinally dehiscent; pollen prolate, 3-colporate with broad transverse colpi, sexine reticulate-striate; pistillode small, 3-lobate, 3-fid or 3-partite, the segments capitate. Female flowers pedicellate; calyx segments often much larger than in the male; petals minute or 0; disc-glands free, oppositipetalous, or united; ovary 3-locular, ovules 2 per locule, hemitropous; styles 3, short, 2-fid or 2-partite, stigmas capitate. Fruit 3-lobate to subglobose, dehiscing into 3 2-valved cocci; endocarp thinly woody; columella small, persistent. Seeds 2 per locule, curved, segmentiform, 3-quetrous or 3-gonous, smooth or rugose, ecarunculate; endosperm fleshy; embryo curved, cotyledons broad and flat, radicle long. Fig.: See Euphorb. Gen.: t. 6, f. 20.1–6 (1824); Étude Euphorb.: t. 27, f. 18 (1858); Hooker's Icon. Pl. 18: t. 1704 (1887); Pflanzenr. 81: 174, t. 14A (1922); Fl. Indo-Chine 5: 538, t. 66.4–11 (1926); Nat. Pflanzenfam. ed. 2, 19C: 67, t. 31A (1931); Fl. USSR 14: 275, t. XV.1,2 and 285, t. XVI.1–3 (1949); Monogr. Euph. Maroc: 7, t. 5 (1953); Fl. Iran. 6: t. 1 (1964); Hooker's Icon. Pl. 37: t. 3697 (1971); Fl. Iraq 4(1): 313, t. 57 (1980); Fl. Pak., 172 Euph.: 40, t. 8D–L (1986); Kew Bull. 42: 352, t. 1 (1987); Fl. Trop. E. Afr., Euph. 1: 8, t. 1 (1987); Fl. Zamb. 9(4): 28, t. 5 (1996).

In the restricted circumscription adopted by Webster (1994), *Andrachne* includes about 15 species with a primarily Tethyan distribution from Persia through the Mediterranean to the West Indies, with one species (*A. microphylla*) disjunct from Baja California to Pacific South America.

The majority of species listed under *Andrachne* by most authors are better classified in the separate genus *Leptopus*.

Subtribe 4f. **Flueggeinae** *Müll. Arg.*, Linnaea 34: 64 (1865) (as *Fluggeae*); G. L. Webster, Ann. Missouri Bot. Gard. 81: 43 (1994). T.: *Flueggea* Willd.

Tribe *Phyllantheae* subtribe *Euphyllantheae* Müll. Arg., Linnaea 34: 64 (1865). T.: *Phyllanthus* L.

Tribe *Phyllantheae* subtribe *Sauropodeae* Müll. Arg., Linnaea 34: 64 (1865). T.: *Sauropus* Blume.

Tribe *Phyllantheae* subtribe *Phyllanthinae* Pax, Nat. Pflanzenfam. ed. 1, 3(5): 17 (1890). T.: *Phyllanthus* L.

Monoecious or dioecious trees, shrubs, or herbs; branchlets deciduous in some taxa. Flowers in axillary glomerules (except *Richeriella*); male sepals mostly 4–6, free or connate; petals 0; stamens 2–15, free or connate; disc extrastaminal (intrastaminal in *Reverchonia*) and usually dissected (less commonly cupular or 0); anthers introrse or extrorse; pistillode + or 0; female sepals mostly 5 or 6, persistent in fruit; disc usually cupular, often lobate (sometimes dissected or 0); ovary (2)3(15)-locular; styles free or connate, 2-fid or entire; ovules hemitropous, occasionally amphitropous. Fruit usually dehiscent (sometimes baccate or drupaceous). Seeds usually 2 per locule (except *Richeriella*); testa smooth or sculptured; endosperm copious; cotyledons broader than and 1–2 times longer than the radicle (except *Reverchonia*).

A rather homogeneous tribe of 8 genera with over 1,000 species.

The vast majority of the species in genera lacking pistillodes (except *Margaritaria* and some *Phyllanthus* spp.) share a distinctive vegetative specialization known as "phyllanthoid branching" (Webster, 1956), where the leaves are 2-morphic: small, scarious and scale-like on the main axes, and larger, green and membranous to chartaceous on often also floriferous and sometimes deciduous lateral axes.

Key to the genera

1. Pistillode + in male flower; usually dioecious · 2
1. Pistillode 0; monoecious or dioecious · 3
2. Flowers in axillary clusters; male flowers distinctly pedicellate; seeds mostly 2 per locule · 25. *Flueggea*
2. Flowers in axillary or ramiflorous spikes, racemes or panicles; male flowers subsessile; seeds mostly 1 per locule · · · · · · · · · · · · · · · · · 26. *Richeriella*
3. Seed coat with bony endotesta; fruits dry, irregularly dehiscent; disc annular in both male and female flowers; dioecious shrubs or trees · 27. *Margaritaria*
3. Seed coat usually lacking bony endotesta; fruits, when dehiscent, usually regularly septicidal; disc usually dissected or 0 in male flower; monoecious or dioecious trees, shrubs, or herbs · · · · · · · · · · · · · · · 4
4. Floral disc usually +; seeds dry, not usually ventrally invaginated; branching phyllanthoid or not · 5
4. Floral disc 0; pollen grains stephanocolporate; styles 2-fid or entire; seeds usually with thickened ± fleshy exotesta, and often ventrally invaginated; branching phyllanthoid · 6

5. Male disc extrastaminal; stamens various, but rarely 2 and introrse; cotyledons much broader than radicle · · · · · · · · · · · · · 28. *Phyllanthus*
5. Male disc lobate, intrastaminal; stamens 2, anthers introrse; cotyledons scarcely broader than radicle · · · · · · · · · · · · · · · · · · · 29. *Reverchonia*
6. Styles usually entire; anthers apiculate; seed-coat usually fleshy; ovary 3–15-locular · 32. *Glochidion*
6. Styles 2-fid or emarginate; anthers not apiculate; ovary 3-locular · · · · · · 7
7. Male calyx ± discoid, usually not turbinate · · · · · · · · · · · · · · 30. *Sauropus*
7. Male calyx turbinate · 31. *Breynia*

25. **Flueggea** *Willd.*, Sp. Pl. 4:637 (1806); A. Juss., Euphorb. Gen.: 16 (1824); Baill., Étude Euphorb.: 590 (1858); Benth. & Hook. f., Gen. Pl. 3: 276 (1880); Hook. f., Fl. Brit. India 5: 328 (1887); Beille, Fl. Indo-Chine 5: 528 (1927); Berhaut, Fl. Ill. Sénég. 3: 595 (1975) (as *Securinega*); G. L. Webster, Allertonia 3: 273 (1984); Radcl.-Sm. in Fl. Pak., 172 Euph.: 18 (1986), and in Fl. Trop. E. Afr. Euph. 1: 68 (1987); W. J. Hayden, Brittonia 39: 268 (1987); Grierson & D. G. Long, Fl. Bhut. 1: 775 (1987); Alain, Descr. Fl. Puerto Rico 2: 395 (1988); G. L. Webster, Ann. Missouri Bot. Gard. 81: 43 (1994); Radcl.-Sm. in Fl. Zamb. 9(4): 29 (1996); Gillespie in Acev.-Rodr., ed., Fl. St. John: 213 (1996); Philcox in Dassanayake, ed., Fl. Ceyl. XI: 235 (1997). T.: *F. leucopyrus* Willd.

Acidoton P. Browne, Civ. Nat. Hist. Jamaica 335 (1756), nom. rej., non Sw. (1788). T.: *Ac. frutescens aculeatum et diffusum* P. Browne, nom. rej. [=*F. acidoton* (L.) G. L. Webster (B.: *Adelia acidoton* L.)]

Bessera Spreng., Pl. Min. Cogn. Pug. 2: 90 (1815). T.: *B. inermis* Spreng. [prob. =*F. virosa* (Roxb. ex Willd.) Voigt (B.: *Phyllanthus virosus* Roxb. ex Willd.)]

Geblera Fisch. & C. A. Mey., Index Sem. Hort. Petrop. 1: 28 (1835). T.: *G. suffruticosa* (Pall.) Fisch. & C. A. Mey. (B.: *Pharnaceum ? suffruticosum* Pall.) [=*F. suffruticosa* (Pall.) Baill.]

Colmeiroa Reut., Biblioth. Universelle Genève 38: 215 (1842). T.: *C. buxifolia* Reut. [=*F. tinctoria* (L. in Loefl.) G. L. Webster (B.: *Rhamnus tinctoria* L. in Loefl.)]

Pleiostemon Sond., Linnaea 23: 135 (1850); Baill., Étude Euphorb.: 615 (1858). T.: *P. verrucosum* (Thunb.) Sond. (B.: *Phyllanthus verrucosus* Thunb.) [=*F. verrucosa* (Thunb.) G. L. Webster]

Securinega sect. *Flueggea* (Willd.) Müll. Arg. in DC., Prodr. 15(2): 448 (1866); Pax & K. Hoffm., Nat. Pflanzenfam. ed. 2, 19C: 60 (1931). T.: as above

Neowawraea Rock, Indig. Trees Haw. Isl.: 243 (1913). T.: *N. phyllanthoides* Rock [=*F. neowawraea* W. J. Hayden]

Dioecious or rarely monoecious often glabrous mostly deciduous trees or shrubs. Leaves alternate, spiral or distichous, shortly petiolate, stipulate, simple, entire, penninerved. Flowers small, axillary, males often numerous, fasciculate, females few or solitary; bracts inconspicuous. Male flowers: pedicels capillary, not jointed; sepals (4)5(7), imbricate, unequal, subpetaloid to scarious, entire or denticulate; petals 0; disc-glands (4)5(7), free or united, alternisepalous, interstaminal; stamens exserted, (4)5(7), filaments free, anthers erect, dorsifixed, extrorse, thecae distinct, parallel, longitudinally dehiscent; pollen subglobose, 3-colporate, semitectate; pistillode usually 2–3-lobate or -partite, the divisions often appendiculate or 2-lobate, rarely 0. Female flowers: pedicels and calyx ± as in the male; petals 0; disc flat, annular, entire, or shallowly 5–6-lobate; ovary (2)3(4)-locular, ovules 2 per locule, hemitropous; styles free or connate at the base, recurved or spreading, 2-lobate or 2-fid, the arms sometimes dilated. Fruit 3-locular or rarely 2- or 1-locular by abortion, slightly fleshy or dry, indehiscent and baccate or loculicidally dehiscent into 3 cocci; exocarp thin; endocarp crustaceous; columella persistent. Seeds 2 per locule, trigonous, dorsally convex, ventrally often invaginated at the hilum; testa thin, crustaceous, smooth, reticulate or verruculose, double on the ventral side, missing between the folds; endosperm thin or copious, incurved around the cavity; embryo straight to distinctly curved, lacking chlorophyll, cotyledons flat, broad, larger than the radicle. Fig.: See Euphorb. Gen.: t. 2, f. 7A.1–10, 7B. 1–5 (1824); Étude Euphorb.: t. 26, f. 39–47 (1858); Fl. Bras. 11(2): t. 11.II (1873) (as *Securinega*); Fl. Jamaica 4: 267, t. 86 (1920); Fl. Indo-Chine 5: 538, t. 66.1 (1927); Nat. Pflanzenfam. ed. 2, 19C: 60, t. 27A, B (1931) (as *Securinega*); Fl. USSR 14: 285, t. XVI.4,5 (1949); J. Fac. Sci. Univ. Tokyo, Sect. 3, Bot. 6: 330–6, tt. 51–55.L–V (1954) (mostly as *Securinega*); Fl. W. Trop. Afr. ed. 2, 1: 388, t. 135 (1958) (as *Securinega*); Fl. Ill. Sénég. 3: 594 (1975) (as *Securinega*); Fl. Bah. Arch.: 845, t. 351 (1982) (as *Securinega*); Fl. Pak., 172 Euph.: 20, t. 4C–E (1986); Brittonia 39: 269, t. 1 (1987); Fl. Trop. E. Afr. Euph. 1: 69, t. 7 (1987); Fl. Zamb. 9(4): 31, t. 6 (1996); Fl. St. John: 214, t. 96H–K (1996).

A genus of 15 species, widespread in the tropics and temperate Eurasia, but with a disjunct relict distribution: 1 from the Bahamas, Greater Antilles and Virgin Is., 1 in Ecuador, 1 in NE. Brazil, 1 from Spain and Portugal, 1 from Turkey, 1 widespread in the Palaeotropics, 1 from E. South Africa, 1 in E. Asia, 1 in SW. China, 1 in C. China, 1 from S. India and Sri Lanka, 1 from Indo-China, 1 from Laos, 1 from the Philippines, New Guinea, Solomon Is. and SW. Pacific and 1 from Hawai'i (Webster, 1984a).

The genus was merged with *Securinega* by Müller (1866), who was followed by many subsequent workers. However, Baillon (1858), Bentham (1880), and Hooker (1887) maintained *Flueggea* as distinct; and indeed it is not at all closely related to *Securinega*, as indicated by both pollen and seed characters.

26. **Richeriella** *Pax & K. Hoffm.*, Pflanzenr. 81: 30 (1922), and Nat. Pflanzenfam. ed. 2, 19C: 48 (1931); M. R. Hend., Gard. Bull. Straits Settlem. 7: 122, t. 32 (1933); Airy Shaw, Kew Bull. 25: 489 (1971), and 26: 328 (1972); Whitmore, Tree Fl. Mal. 2: 51 (1973); Airy Shaw, Hooker's Icon. Pl. 38: t. 3703 (1974), Kew Bull. Addit. Ser. 4: 190 (1975), and Euph. Philipp.: 43 (1983); G. L. Webster, Ann. Missouri Bot. Gard. 81: 44 (1994). T.: *R. gracilis* (Merr.) Pax & K. Hoffm. (B.: *Baccaurea gracilis* Merr.)

Dioecious glabrous trees. Leaves large, alternate, entire, penninerved, laurel-like; stipules caducous. Inflorescences axillary or ramiflorous, simply spicate or racemose, or branched, slender, lax, wide-spreading, unisexual. Flowers very small. Male flowers subsessile; sepals 5, free, imbricate; petals 0; disc-glands 5, small; stamens 5, opposite the sepals; filaments free, at length long-exserted; anthers dorsifixed, extrorse, longitudinally dehiscent; pollen subprolate, colpi transversales ± circular; pistillode 2–3-lobate or -fid. Female flowers pedicellate, pedicels elongating in fruit; sepals 5, imbricate; hypogynous disc annular; ovary 3-locular, ovules 2 per locule; styles 3, free, adpressed to the ovary, shortly 2-fid. Fruit (2)3-lobate, dehiscing into 2-valved cocci; endocarp corneo-crustaceous, separating from the pericarp. Seeds mostly 1 per locule, ecarunculate, dorsally convex; testa chartaceous, ventrally excavated, with a thickened raphe; endosperm thin, cotyledons broad, concave, thin. Fig.: See Gard. Bull. Straits Settlem. 7: 122, t. 32 (1933); Hooker's Icon. Pl. 38: t. 3703 (1974).

A genus of 2 species native to southern China, Philippines, Malaya and Borneo.

It is close to *Flueggea*, but the inflorescences especially set it apart.

27. **Margaritaria** *L. f.*, Suppl. Pl. 66 (1781); A. Juss., Euphorb. Gen.: 59 (1824); Airy Shaw, Kew Bull. 20: 386 (1966); Jabl., Mem. New York Bot. Gard. 17: 118 (1967); G. L. Webster, Ann. Missouri Bot. Gard. 54: 217 (1968); Airy Shaw, Kew Bull. 26: 308 (1972); Whitmore, Tree Fl. Mal. 2: 117 (1973); Airy Shaw, Kew Bull. Addit. Ser. 4: 175 (1975); Berhaut, Fl. Ill. Sénég. 3: 557 (1975) (as *Phyllanthus*); Fosberg, Kew Bull. 33: 184 (1978); Philcox, Fl. Trin. Tob. 2(10): 637 (1979); G. L. Webster, J. Arnold Arbor. 60: 407 (1979); Airy Shaw, Kew Bull. 35: 657 (1980), Addit. Ser. 8: 173 (1980), and Kew Bull. 36: 330 (1981); Radcl.-Sm., Kew Bull. 36: 219 (1981); Coode in Fl. Masc. 160 Euph.: 29 (1982); Correll, Fl. Bah. Arch.: 828 (1982); Alain, Fl. Español. 4: 173 (1986); Alain, Descr. Fl. Puerto Rico 2: 407 (1988); Airy Shaw, Kew Bull. 37: 30 (1982), and Euph. Philipp.: 38 (1983); Radcl.-Sm. in Fl. Trop. E. Afr. Euph., 1: 63 (1987); R. A. Howard, Fl. Lesser Antilles 5: 66 (1989); A. H. Gentry, Woody Pl. NW. S. Amer.: 417 (1993); G. L. Webster, Ann. Missouri Bot. Gard. 81: 44 (1994); Murillo-Aldana & Franco-Rosselli, Euf. Reg. Ararac.: 108 (1995); Radcl.-Sm. in Fl. Zamb. 9(4): 32 (1996); Gillespie in Acev.-Rodr., ed., Fl. St. John: 216 (1996). T.: *M. nobilis* L. f.

Prosorus Dalzell, Hooker's J. Bot. Kew Gard. Misc. 4: 345 (1852). T.: *P. indicus* Dalzell [=*M. indica* (Dalzell) Airy Shaw]

Zygospermum Thwaites ex Baill., Étude Euphorb. 620 (1858). T.: *Z. zeylanicum* Thwaites ex Baill. [=*M. cyanosperma* (Gaertn.) Airy Shaw (B.: *Croton* ? *cyanospermus* Gaertn.)]

Wurtzia Baill., Adansonia I. 1: 186 (1861). T.: *W. tetracocca* Baill. [=*M. tetracocca* (Baill.) G. L. Webster]

Calococcus Kurz ex Teijsm. & Binn., Natuurk Tijdschr. Ned.-Indië 27: 48 (1864). T.: *C. sundaicus* Kurz ex Teisjm. & Binn. [=*M. indica* (Dalzell) Airy Shaw (B.: *Prosorus indicus* Dalzell)]

Dioecious or rarely monoecious trees or shrubs, rarely scandent; bark lenticellate. Indumentum simple. Leaves alternate, distichous, shortly petiolate, stipulate, simple, entire, penninerved, sometimes deciduous. Flowers in the proximal axils of young leafy branches or short shoots, the males several, fasciculate, the females few or solitary. Male flowers: pedicels capillary; calyx-lobes 4, imbricate, unequal; petals 0; disc annular, ± entire; stamens 4, filaments free, anthers muticous, extrorse, longitudinally dehiscent; pollen subglobose, 3-colporate, semitectate; pistillode 0. Female flowers: pedicels stouter than in the males; calyx and disc as in the males; ovary (2)3–4(6)-locular, with 2 ovules per locule; styles free or united to up to $^1/_2$ their length, 2(3)-fid or -partite. Fruit (2)3–4(6)-lobate, irregularly dehiscent or subindehiscent; exocarp green, usually separating from the thinly paleacous-crustaceous endocarp. Seeds usually 2 per locule, sometimes 1 by abortion, hemispherical, ecarunculate; sarcotesta fleshy, bluish; sclerotesta thickly woody or bony, smooth or rugulose, invaginated at the chalazal end; endosperm copious; embryo ± straight; cotyledons thin, flat, much longer than the radicle. Fig.: See Fl. Madag. 111(1): 95, t. XV (1958) (as *Phyllanthus* spp.); Ann. Missouri Bot. Gard. 54: 218, t. 2 (1968); Fl. Ill. Sénég. 3: 556 (1975) (as *Phyllanthus*); Kew Bull. 31: 358, t. 1 (1976); J. Arnold Arbor. 60: 421, tt. 11–13 (1979); Fl. Masc. 160 Euph.: 31, t. 4 (1982); Fl. Bah. Arch.: 829, t. 344 (1982); Fl. Español. 4: 177, t. 116-25 (1986); Fl. Trop. E. Afr. Euph., 1: 64, t. 6 (1987); Descr. Fl. Puerto Rico 2: 409, t. 59-22 (1988); Fl. Lesser Antilles 5: 68, t. 24 (1989); Bol. Mus. Paraense Emilio Goeldi, N.S., Bot. 7: 32, t. 4 (1991); Woody Pl. NW. S. Amer.: 416, t. 119.7 (1993); Euf. Reg. Ararac.: 110, t. 28 (1995); Fl. Zamb. 9(4): 34, t. 7 (1996); Fl. St. John: 218, t. 99A–G (1996); Árvores Brasileiras 2: 96 (1998).

A genus of 14 species ocurring in tropical America (4 species, of which 3 only in Greater Antilles and Bahamas), Africa (2 species), Madagascar and the Mascarenes (4 species), tropical Asia (3 species, of which 1 endemic to Sri Lanka and 1 to Luzon) and N Australia (1 species).

28. **Phyllanthus** *L.*, Sp. Pl. 981 (1753), and Gen. Pl. ed. 5, 422 (1754); A. Juss., Euphorb. Gen.: 21 (1824); Baill., Étude Euphorb.: 621 (1858), and Adansonia I. 1: 24 (1860); Müll. Arg. in DC., Prodr. 15(2): 274 (1866); Benth. & Hook. f., Gen. Pl. 3(1): 272 (1880); Hook. f., Fl. Brit. India 5: 285 (1887), and Hooker's Icon. Pl. 16: t. 1569 (1887); Oliv., *op. cit.* 24: t. 2372 (1895); Hutch., Fl. Trop. Afr. 6(1): 692 (1912); Fawc. & Rendle, Fl. Jamaica 4: 251 (1920); Beille, Fl. Indo-Chine 5: 571 (1927); Pax & K. Hoffm., Nat. Pflanzenfam. ed. 2, 19C: 60 (1931); Robyns, Fl. Parc Nat. Alb. 1: 439 (1948); Alain, Fl. Cuba 3: 44 (1953); Hurus., J. Fac. Sci. Univ. Tokyo, Sect. 3, Bot. 6: 329 (1954); Leandri, Fl. Madag. 111(1): 30 (1958); Keay (ed.), Fl. W. Trop. Afr. ed. 2, 1: 384 (1958); Paul G. Wilson, Hooker's Icon. Pl. 36: t. 3589 (1962); Backer & Bakh. f., Fl. Java 1: 466 (1963); Rech. f. & Schiman-Czeika, Fl. Iran. 6: 5 (1964); G. L. Webster, J. Arnold Arbor. 37: 91 (1956), and 48: 332 (1967); Jabl., Mem. New York Bot. Gard. 17: 85 (1967); G. L. Webster, Ann. Missouri Bot. Gard. 54: 194, 220 (1967/8), and Brittonia 22: 44 (1970); J.-G. Adam, Mém. Mus. Natl. Hist. Nat., B, Bot. 20: 501

(1971); G. L. Webster & Airy Shaw, Kew Bull. 26: 85 (1971); Airy Shaw, Kew Bull. 26: 312 (1972); Whitmore, Tree Fl. Mal. 2: 121 (1973); Airy Shaw, Hooker's Icon. Pl. 38: tt. 3704–3706 (1974), and Kew Bull. Addit. Ser. 4: 181 (1975); Berhaut, Fl. Ill. Sénég. 3: 547 (1975); Philcox, Fl. Trin. Tob. 2(10): 624 (1979); Airy Shaw, Kew Bull. Addit. Ser. 8: 179 (1980); Brunel & J. P. Roux, Willdenowia 11: 69 (1981), and Bull. Mus. Natl. Hist. Nat. Sér. 4, 3B: 185 (1981); Radcl.-Sm., Kew Bull. 35: 765 (1981); Airy Shaw, Kew Bull. 36: 335 (1981), and 37: 31 (1982); Coode, Fl. Masc. 160 Euph.: 8 (1982); Correll, Fl. Bah. Arch.: 834 (1982); Airy Shaw, Euph. Philipp.: 40 (1983); Brunel & J. P. Roux, Nordic J. Bot. 4: 469 (1984); Proctor, Kew Bull. Addit. Ser. 11: 523 (1984); Alain, Fl. Español. 4: 187 (1986); Radcl.-Sm. in Fl. Pak., 172 Euph.: 22 (1986), and in Fl. Trop. E. Afr. Euph., 1: 9 (1987); M. G. Gilbert, Kew Bull. 42: 354 (1987); Rossignol *et al.*, Amer. J. Bot. 74: 1853 (1987); Brunel, Phyll. Afr. Intertrop. Mad.: 261 (1987) (ineff. publ.); Grierson & D. G. Long, Fl. Bhut. 1: 770 (1987); Alain, Descr. Fl. Puerto Rico 2: 414 (1988); Santiago, Bradea 5(2): 44 (1988); G. L. Webster & Huft, Ann. Missouri Bot. Gard. 75: 1096 (1988); R. A. Howard, Fl. Lesser Antilles 5: 70 (1989); Lebrun & Stork, Enum. Pl. Afr. Trop. 1: 232 (1991); Schmid, Fl. Nouv.-Caléd. 17: 31 (1991); G. L. Webster, Ann. Missouri Bot. Gard. 81: 44 (1994); Murillo-Aldana & Franco-Rosselli, Euf. Reg. Ararac.: 129 (1995); Radcl.-Sm., Kew Bull. 51: 305 (1996), and Fl. Zamb. 9(4): 36 (1996); Gillespie in Acev.-Rodr., ed., Fl. St. John: 219 (1996); Wheeler & G. L. Webster in Dassanayake, ed., Fl. Ceyl. XI: 206 (1997). LT.: *Ph. niruri* L. (designated by Small, 1913)

Niruri Adans., Fam. Pl. 2: 356 (1763). LT.: *Ph. niruri* L. (selected by Webster, 1994)

Cicca L., Syst. Nat. ed. 12, 2: 621 (1767); A. Juss., Euphorb. Gen.: 20 (1824); Baill., Étude Euphorb.: 617 (1858). T.: *C. disticha* L. [=*Ph. acidus* (L.) Skeels (B.: *Averrhoa acida* L.)]

Xylophylla L., Mant. Alt. 147, 221 (1771); A. Juss., Euphorb. Gen.: 23 (1824). LT.: *X. latifolia* L. [=*Ph. epiphyllanthus* L.]

Conami Aubl., Hist. Pl. Guiane 2: 926 (1775). T.: *C. brasiliensis* Aubl. [=*Ph. brasiliensis* (Aubl.) Müll. Arg.]

Meborea Aubl., Hist. Pl. Guiane 2: 825 (1775); Baill., Étude Euphorb.: 656 (1858); Lanj., Euphorb. Surinam: 9 (1931). T.: *M. guianensis* Aubl., non *Ph. guianensis* Klotzsch, nec Müll. Arg. [=*Ph. aubletianus* Radcl.-Sm., **nom. nov.**]

Genesiphylla L'Hér., Sert. Angl. 29 (1778). T.: *G. asplenifolia* L'Hér. [=*Ph. latifolius* Sw.]

Cathetus Lour., Fl. Cochinch. 607 (1790). T.: *C. fasciculata* Lour. [=*Ph. cochinchinensis* Spreng. (1826), non (Lour.) Müll. Arg. (1866) =*Ph. cathetus* Radcl.-Sm., **nom. nov.**]

Nymphanthus Lour., Fl. Cochinch. 543 (1790). T.: *N. ruber* Lour. [=*Ph. ruber* (Lour.) Spreng.]

Emblica Gaertn., Fruct. Sem. Pl. 2: 122 (1790); A. Juss., Euphorb. Gen.: 20 (1824). T.: *E. officinalis* Gaertn. [=*Ph. emblica* L.]

Kirganelia Juss., Gen. Pl.: 387 (1789); A. Juss., Euphorb. Gen.: 21 (1824); Baill., Étude Euphorb.: 612 (1858). T.: *K. virginea* J. F. Gmel. [=*Ph. virgineus* (J. F. Gmel.) Pers. = *P. casticum* P. Willemet]

Tricarium Lour., Fl. Cochinch.: 557 (1790). T.: *T. cochinchinense* Lour. [=*Ph. acidus* (L.) Skeels (B.: *Averrhoa acida* L.)]

Epistylium Sw., Fl. Ind. Occid. 2: 1095 (1800); A. Juss., Euphorb. Gen.: 17 (1824); Baill., Étude Euphorb.: 646 (1858). LT.: *E. axillare* (Sw.) Sw. (B.: *Omphalea axillaris* Sw.) [=*Ph. axillaris* (Sw.) Müll. Arg.] (chosen by Webster, 1994).

Geminaria Raf., West. Minerva: 42 (1821). T.: *G. obovata* Raf. [=*Ph. caroliniensis* Walter]

Menarda Comm. ex A. Juss., Euphorb. Gen.: 23 (1824); Baill., Étude Euphorb.: 608 (1858). T.: *M. cryptophila* Comm. ex A. Juss. [=*Ph. cryptophilus* (Comm. ex A. Juss.) Müll. Arg.]

Anisonema A. Juss., Euphorb. Gen.: 19 (1824) (nom. rej.). T.: *A. reticulatum* (Poir.) A. Juss. [=*Ph. reticulatus* Poir.]

Scepasma Blume, Bijdr. 582 (1825); Baill., Étude Euphorb.: 648 (1858). T.: *S. buxifolia* Blume [=*Ph. buxifolius* (Blume) Müll. Arg.]

Synexemia Raf., Neogenyton: 2 (1825). T.: *S. caroliniana* Raf. [=*Ph. caroliniensis* Walter]

Hexadena Raf., Sylva Tellur.: 92 (1838). T.: *H. angustifolia* (Sw.) Raf. (B.: *Xylophylla angustifolia* Sw.) [=*Ph. angustifolius* (Sw.) Sw.]

Moeroris Raf., Sylva Tellur.: 91 (1838). T.: *M. stipulata* Raf. [=*Ph. stipulatus* (Raf.) G. L. Webster]

Nellica Raf., Sylva Tellur.: 92 (1838). T.: *N. maderaspatana* (L.) Raf. [=*Ph. maderaspatensis* L.]

Asterandra Klotzsch, Arch. Naturgesch. 7(1): 200 (1841); Baill., Étude Euphorb.: 610 (1858). T.: *A. cornifolia* (Humb., Bonpl. & Kunth) Klotzsch [=*Ph. cornifolius* Humb., Bonpl. & Kunth]

Eriococcus Hassk., Tijdschr. Natuurl. Gesch. Physiol. 10: 143 (1843). T.: *E. gracilis* Hassk. [=*Ph. gracilipes* Müll. Arg.]

Ceramanthus Hassk., Cat. Hort. Bot. Bogor 240 (1844). T.: *C. gracilis* Hassk. [=*Ph. albidiscus* (Ridl.) Airy Shaw (B.: *Cleistanthus albidiscus* Ridl.)]

Macraea Wight, Icon. Pl. Ind. Orient. 5(2): 27, t. 1901, 1902 (1852) (non *Macraea* Lindley, 1828). LT.: *M. oblongifolia* Wight [=*Ph. simplex* Retz. = *Ph. virgatus* G. Forst.] (chosen by Webster, 1994)

Reidia Wight, Icon. Pl. Ind. Orient. 5(2): 27, t. 1903, 1904 (1852). LT.: *R. floribunda* Wight [=*Ph. wightianus* Müll. Arg.] (chosen by Webster, 1994)

Chorisandra Wight, Icon. Pl. Ind. Orient. 6: 13, t. 1994 (1853). T.: *C. pinnata* Wight [=*Ph. pinnatus* (Wight) G. L. Webster]

Dichelactina Hance, Ann. Bot. Syst. 3: 375 (1852). T.: *D. nodicaulis* Hance [=*Ph. acidus* (L.) Skeels (B.: *Averrhoa acida* L.)]

Staurothylax Griff., Not. Pl. Asiat. 4: 476 (1854). T.: (none given) [=*Ph. acidus* (L.) Skeels (B.: *Averrhoa acida* L.), ex descr., *fide* Webster, 1994]

Hemicicca Baill., Étude Euphorb.: 645 (1858). T.: *H. japonica* Baill. [=*Ph. flexuosus* (Siebold & Zucc.) Müll. Arg. (B.: *Cicca flexuosa* Siebold & Zucc.)]

Williamia Baill., Étude Euphorb.: 559 (1858). T.: *W. pruinosa* Baill. [=*Ph. discolor* Poepp. ex Spreng.]

Orbicularia Baill., Étude Euphorb.: 616 (1858). T.: *O. phyllanthoides* Baill. [nom. illeg., =*Ph. orbicularis* Humb., Bonpl. & Kunth]

Phyllanthodendron Hemsl., Hooker's Icon. Pl. 26: t. 2563, 2564 (1898); Craib, *op. cit.* 30: t. 2935 (1911). T.: *Ph. mirabile* (Müll. Arg.) Hemsl. [=*Phyllanthus mirabilis* Müll. Arg.]

Aporosella Chodat, Bull. Herb. Boissier, II. 5: 488 (1905). T.: *A. hassleriana* Chodat [=*Ph. chacoensis* Morong]

Flueggeopsis (Müll. Arg.) K. Schum. in K. Schum & Lauterb., Fl. Schutzgeb. Südsee Nachtr.: 289 (1905). T.: *Fl. microspermus* K. Schum. [=*Ph. ciccoides* Müll. Arg.]

Nymania K. Schum. in K. Schum & Lauterb., Fl. Schutzgeb. Südsee Nachtr.: 291 (1905). T.: *N. insignis* K. Schum. [=*Ph. schumannianus* L. S. Sm.]

Uranthera Pax & K. Hoffm., Pflanzenr. 47: 95 (1911). T.: *U. siamensis* Pax & K. Hoffm. [=*Ph. roseus* (Craib & Hutch.) Beille (B.: *Phyllanthodendron roseum* Craib & Hutch.)]

Dimorphocladium Britton, Mem. Torrey Bot. Club 16: 74 (1920). T.: *D. formosum* (Urb.) Britton [=*Ph. formosus* Urb.]

Ramsdenia Britton, Mem. Torrey Bot. Club 16: 72 (1920). T.: *R. incrustata* (Urb.) Britton [=*Ph. incrustatus* Urb.]

Roigia Britton, Mem. Torrey Bot. Club 16: 73 (1920). T.: *R. comosa* (Urb.) Britton [=*Ph. comosus* Urb.]

Dendrophyllanthus S. Moore, J. Linn. Soc., Bot. 45: 395 (1921). T.: *D. comptonii* S. Moore [=*Ph. moorei* M. Schmid]

Pseudoglochidion Gamble, Kew Bull. 1925: 329 (1925). T.: *Ps. anamalayanum* Gamble [=*Ph. anamalayanus* (Gamble) G. L. Webster]

Hexaspermum Domin, Biblioth. Bot. 89: 315 (1927). T.: *H. paniculatum* Domin [=*Ph. clamboides* (F. Muell.) Diels (B.: *Leichardtia clamboides* F. Muell.)]

Arachnodes Gagnep., Notul. Syst. (Paris) 14: 32 (1950); Airy Shaw, Kew Bull. 14: 469 (1960). T.: *A. chevalieri* Gagnep. [=*Ph. arachnodes* Gov. & Radcl.-Sm.], non *Ph. chevalieri* Beille

Monoecious or rarely dioecious trees, shrubs or herbs of diverse habit, often with the stems and branches differentiated into 2 or 3 types: orthotropic long shoots of unlimited growth; orthotropic short shoots (brachyblasts) and plagiotropic leafy and/or floriferous shoots of limited growth, resembling pinnate leaves or pseudoracemose inflorescences. Indumentum simple, rarely dendritic. Leaves often of 2 types: scale-like cataphylls on the orthotropic shoots, and foliage-leaves usually only on the plagiotropic shoots. Foliage-leaves alternate, often distichous, shortly petiolate, stipulate, simple, entire, penninerved, the nerves brochidodromous. Flowers small, axillary, the males usually geminate or fasciculate in the lower axils, females solitary in the upper, rarely the fascicles 2-sexual. Male flowers pedicellate, pedicels often capillary; sepals (4)5–6, subequal, imbricate; petals 0; disc glands (4)5–6, free, alternisepalous, or rarely disc annular; stamens 2–6(15), filaments free or some or all partially or completely connate, anthers basifixed, extrorse, vertically, obliquely or horizontally-held, the thecae parallel or convergent, free or united, longitudinally, obliquely, laterally, apically or circumferentially dehiscent; pollen prolate to globose, variously ornamented; pistillode 0. Female flowers: pedicels more robust than in the male; calyx ± as in the male but often larger; petals 0; disc thin or thick, flat, annular, saucer-shaped or cupular, entire or variously-lobate or toothed, rarely of distinct glands or 0; staminodes rarely +; ovary sessile or stipitate, 3- or more locular, ovules 2 per locule; styles 3, rarely more, free or connate, variously-held, usually 2-lobate or 2-fid, the stigmas slender or swollen, usually recurved. Fruit 3-more-lobate, dry and septicidally and locudicidally dehiscent into 3 or more 2-valved cocci or separate valves, or fleshy and subindehiscent; endocarp usually crustaceous. Seeds 2 per locule, segmentiform, 3-quetrous or 3-gonous, dorsally convex, rarely ovoid, verruculose, tuberculate, ridged, lineate or smooth, ecarunculate; testa usually thinly crustaceous; endosperm fleshy; embryo straight or slightly curved; cotyledons flat and straight, rarely flexuose, much broader than the radicle. Fig.: See Euphorb. Gen.: t. 3, f. 8.1–6 (as *Epistylium*), t. 4, f. 11.1–7 (as *Anisonema*), t. 4, f. 13A.1–4, B.1–7 (as *Cicca*), t. 4, f. 14.1–8 (as *Kirganelia*), t. 5, f. 15.1–6 (as *Emblica*), t. 5, f. 16A.1–8, t. 5, f. 17.1–9 (as *Xylophylla*), and t. 6, f. 18.1–9 (as *Menarda*) (1824); Étude Euphorb.: t. 22, f. 15–36; 23, f. 1–21; 24, 15–33; 25, 10–15 and 22–24; 27, 5–6 and 9–10 (1858); Fl. Bras. 11(2): tt. 3.II–10 (1873); Hooker's Icon. Pl. 16: t. 1569 (1887); *op. cit.* 24: t. 2372 (1895); *op. cit.* 26: t. 2563, 2564 (1898) (as *Phyllanthodendron*); *op. cit.* 30: t. 2935 (1911) (as *Phyllanthodendron*); Fl. Jamaica 4: 251, t. 85 (1920); Pflanzenr. 81: 106, t. 11 (1922) (as *Aporosella*); Fl. Indo-Chine 5: 567, t. 70.4–15, 582, t. 71, 595, t. 72, 603, t. 73.1–13 and 613, t. 74.1, 2 (1927); Nat. Pflanzenfam. ed. 2, 19C: 61–65, tt. 28–30 (1931); Formos. Trees: 356, t. 311 and 357, t. 312 (1936); Notul. Syst. (Paris) 6: 192–195, tt. 4, 5 (1938); Fl. Parc Nat. Alb. 1: 443, t. 43 (1948); Fl. Cuba 3: 52, t. 9 (1953); J. Fac. Sci. Univ. Tokyo, Sect. 3, Bot. 6: 336, t. 55.H–K (1954); J. Arnold Arbor. 37–39: tt. I–XXXII (1956–8); Fl. Madag. 111(1): 41–87, t. VII–XIV and 103, t. XVI (1958); Fl. W. Trop. Afr. ed. 2, 1: 386, t. 134 (1958); Hooker's Icon. Pl. 36: t. 3589 (1962); Mem. New York Bot. Gard. 17: 94–109, tt. 17–22 (1967); Ann. Missouri Bot. Gard. 54: 230, t. 3 (1968); Amer. J. Bot. 56: 745, t. 1A–F (1969); Mém. Mus. Natl. Hist. Nat., B, Bot. 20: 502–6, tt. 186–8 (1971); Tree Fl. Mal. 2: 124, t. 11 (1973); Hooker's Icon. Pl. 38: tt. 3704–3706 (1974); Fl. Ill. Sénég. 3: 546–554, 560–566, 572–580 (1975); Fl. Bah. Arch.: 839, t. 348 (1982); Fl. Masc. 160 Euph.: 13–23, tt. 1–3 (1982); Kew Bull. Addit. Ser. 11: 527, t. 159 (1984); Fl. Español. 4: 193, t. 116-29 and 204, t. 116-30 (1986); Fl. Pak., 172 Euph.: 26–34, tt. 5–7 (1986); Kew Bull. 42: 355, t. 2 and 358, t. 3 (1987); Fl. Trop. E. Afr. Euph., 1: 11–40, tt. 2–5 (1987); Fl. Bhut. 1: 763, t. 47.i–o (1987); Descr. Fl. Puerto Rico 2: 417, t. 59.24 (1988); Ann. Missouri Bot. Gard. 75: 1097, t. 1 (1988); Fl. Lesser Antilles 5: 88, t. 34 (1989); Fl. Nouv.-Caléd. 17: 45–319, tt. 7–68 (1991); Bol. Mus. Paraense Emilio Goeldi, N.S., Bot. 7: 33–39, tt. 5–11 (1991); Euf. Reg. Ararac.: 131–6, tt. 35–8 (1995); Fl. Zamb. 9(4): 37–81, tt. 8–12 (1996); Fl. St. John: 218, t. 99L–P (1996).

A large and very diverse genus of 750–800 species, about 200 of which are American, 100 African, 70 Madagascan, and the remainder Asiatic and Australasian.

Many attempts have been made to subdivide the genus, which is very possibly an unnatural one. The broad concept of Müller (1866) has been modified by Webster (1979, 1994) and others by the recognition of *Glochidion* and *Margaritaria* as distinct. From his findings in seed-morphology, Stuppy (1995), upholds the view that *Phyllanthus* as currently conceived represents several genera, and would resurrect some that have been sunk into it, e.g. *Phyllanthodendron* Hemsl.

29. **Reverchonia** *A. Gray*, Proc. Amer. Acad. Arts 16: 107 (1880); Pax & K. Hoffm., Nat. Pflanzenfam. ed. 2, 19C: 66 (1931); G. L. Webster & K. I. Mill., Rhodora 65: 200 (1963); G. L. Webster, Ann. Missouri Bot. Gard. 81: 45 (1994). T.: *R. arenaria* A. Gray

Monoecious annual herb. Leaves spirally alternate, petiolate, stipulate, simple, narrowly oblanceolate, entire; stipules very small. Flowers pedicellate, in axillary androgynous cymules, each with one central female and several lateral male flowers. Male flowers: sepals 4, biseriate, spathulate-oblong, slightly imbricate, inflated, sigmoid; disc central, deeply 4-lobate, H-shaped, partially enveloping the filaments; stamens 2, opposite the outer sepals; filaments short, free; anthers 2-locular, thecae contiguous, parallel, introrse, longitudinally and vertically dehiscent; pollen prolate, 3-colporate, tectate, psilate; pistillode 0. Female flowers: sepals 6, not inflated; disc saucer-shaped, subentire or 6-angled; ovary 3-locular, ovules 2 per locule, collateral, amphitropous; styles short, united at the base, shortly 2-lobate; stigmas swollen. Fruit berry-like when young, depressed-globose, dehiscing into 2-valved cocci; columella not persistent. Seeds 2 per locule, 3-gonous, with a conspicuous subchalazal invagination, ± smooth; cotyledons very narrow, only slightly broader than the radicle. Fig.: See Rhodora 65: 194, t. 1 (1963). See also Fig. 5, p. 44.

A monotypic genus confined to sand dunes in the southwestern United States and adjacent Mexico.

Although it is very close to *Phyllanthus*, Webster & Miller (1963) state that it would unwarrantably extend the boundaries of that genus if placed in it, and so recommend its recognition.

30. **Sauropus** *Blume*, Bijdr. 595 (1826); Baill., Étude Euphorb.: 634 (1858); Müll. Arg. in DC., Prodr. 15(2): 239 (1866); Benth. & Hook. f., Gen. Pl. 3: 271 (1880); Hook. f., Fl. Brit. India 5: 332 (1887); Pax & K. Hoffm., Pflanzenr. 81: 215 (1922); Beille, Fl. Indo-Chine 5: 643 (1927); Pax & K. Hoffm., Nat. Pflanzenfam. ed. 2, 19C: 71 (1931); Backer & Bakh. f., Fl. Java 1: 471 (1963); Airy Shaw, Kew Bull. 23: 42 (1969), and 26: 330 (1972); Whitmore, Tree Fl. Mal. 2: 130 (1973); Airy Shaw, Hooker's Icon. Pl. 38: tt. 3708, 3709 (1974), Kew Bull. Addit. Ser. 4: 192 (1975), 8: 199, 221 (1980), Kew Bull. 35: 669 (1980), 36: 342 (1981), 37: 34 (1982), and Euph. Philipp.: 44 (1983); Grierson & D. G. Long, Fl. Bhut. 1: 782 (1987); G. L. Webster, Ann. Missouri Bot. Gard. 81: 46 (1994); Philcox in Dassanayake, ed., Fl. Ceyl. XI: 82 (clav. tant.) (1997). LT.: *S. albicans* Blume [=*S. androgynus* (L.) Merr. (B.: *Cluytia androgyna* L.)] (chosen by Webster, 1994).

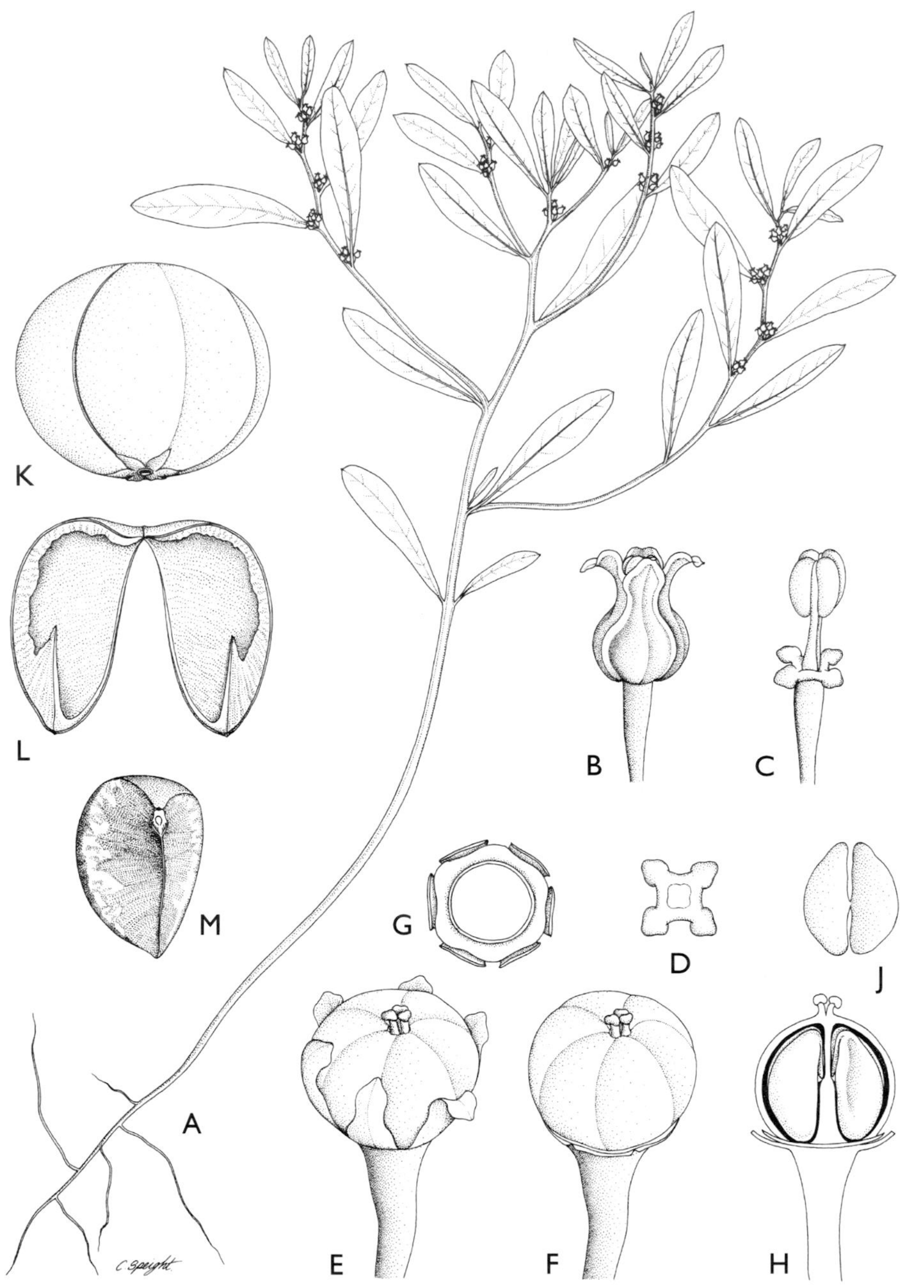

FIG. 5. *Reverchonia arenaria* A. Gray. **A** Habit × 1; **B** Male flower × 12; **C** Stamens × 12; **D** Male disc × 12; **E** Female flower × 12; **F** Female flower, perianth removed to show ovary × 12; **G** Female disc × 12; **H** VS ovary × 12; **J** Ovules × 12; **K** Fruit × 6; **L** Coccus-valves × 6; **M** Seed × 6. **A–J** from *McVaugh* 10707; **K–M** from *Pringle* 792. Drawn by Camilla Speight.

Agyneia Vent., Descr. Pl. Nouv.: t. 23 (1800), non L.; A. Juss., Euphorb. Gen.: 24 (1824); Hook. f., Fl. Brit. India 5: 285 (1887). T.: *A. impubes* Vent., non L. [=*S. bacciformis* (L.) Airy Shaw (B.: *Phyllanthus bacciformis* L.)]

Ceratogynum Wight, Icon. Pl. Ind. Orient. 5(2): 26 (1852). T.: *C. rhamnoides* Wight [=*S. quadrangularis* (Willd.) Müll. Arg. (B.: *Phyllanthus quadrangularis* Willd.)]

Diplomorpha Griff., Not. Pl. Asiat. 4: 479 (1854). T.: *D. herbacea* Griff. [=*S. bacciformis* (L.) Airy Shaw (B.: *Phyllanthus bacciformis* L.)]

Synostemon F. Muell., Fragm. 1: 32 (1858); Backer & Bakh. f., Fl. Java 1: 470 (1963); Airy Shaw, Kew Bull. 26: 343 (1972); Whitmore, Tree Fl. Mal. 2: 133 (1973); Airy Shaw, Kew Bull. Addit. Ser. 4: 199 (1975), and 8: 205 (1980). LT.: *S. ramosissimus* F. Muell. [=*Sauropus ramosissimus* (F. Muell.) Airy Shaw] (chosen by Wheeler, 1975)

Breyniopsis Beille, Bull. Soc. Bot. France 72: 157 (1925), and Fl. Indo-Chine 5: 630 (1927). T.: *B. pierrei* Beille [=*S. pierrei* (Beille) G. L. Webster]

Heterocalymnantha Domin, Biblioth. Bot. 89: 313 (1927). T.: *H. minutifolia* Domin [=*S. rigens* (F. Muell.) Airy Shaw (B.: *Synostemon rigens* F. Muell.)]

Monoecious or apparently dioecious shrubs, scramblers or woody herbs. Leaves alternate, shortly petiolate, minutely stipulate, distichous, thin, entire, penninerved; stipules paired, subulate. Flowers axillary, fasciculate or pseudoracemose, males minute, female solitary in the same or in separate axils, or rarely paired. Male flowers often long-pedicellate; sepals 6, sub-2-seriate, free or connate into a tube or inflexed towards the androecium; petals 0; disc 0; stamens 3, opposite the outer sepals, filaments connate into a short thick column; anthers sessile or subsessile, horizontally or vertically-held, extrorse, longitudinally dehiscent; pollen ± as in *Breynia*, q.v. infra; pistillode 0. Female flowers often long-pedicellate; sepals 6, biseriate, connate at the base, larger than the male and accrescent, or smaller; petals 0; disc 0; ovary 3-locular, truncate or concave at the top, ovules 2 per locule; styles very short, broad, spreading, free or connate at the base, or 2-lobate or -partite, with linear coiled or recurved segments. Fruit depressed-globose or ovoid, ± crustaceous, inconspicuously-lobate, occasionally berry-like, dehiscing into 2-valved cocci. Seeds usually 2 per locule, 3-quetrous, ecarunculate; endosperm fleshy; embryo straight or nearly so; cotyledons broad and flat. Fig.: See Euphorb. Gen.: t. 6, f. 19.1–6 (1824) (as *Agyneia*); Étude Euphorb.: t. 27, f. 19–22 (1858); Pflanzenr. 81: 213, t. 18 (as *Agyneia*) and 217, t. 19 (1922); Fl. Indo-Chine 5: 629, t. 75.1–9 (as *Breyniopsis*), 75.16–18 and 639, t. 76.2–5 (as *Agyneia*) and 639, t. 76.6–14 (1927); Nat. Pflanzenfam. ed. 2, 19C: 57, t. 26F and 71, t. 34 (1931); Hooker's Icon. Pl. 38: tt. 3708, 3709 (1974); Kew Bull. 35: 678, t. 7 (1980); 36: 356, t. 12A1–4 (1981).

A genus of about 80 species distributed from India and Ceylon to southern China, the Philippines, Borneo, New Guinea and Australia.

Airy Shaw (1980c) has transferred a large number of Australian species of *Synostemon* to *Sauropus*, and it appears he may have good reason to do so, although the demarcation of *Sauropus* from *Breynia* is almost as problematical as its separation from *Synostemon*. Stuppy (1995) in fact considers that, on the basis of seed-morphology, *Sauropus bacciformis* and *S. huntii* differ more from other *Sauropus* species than the latter do from *Breynia*.

31. **Breynia** *J. R. & G. Forst.*, Char. Gen. Pl. 145, t. 73 (1776) (nom. cons.); A. Juss., Euphorb. Gen.: 22 (1824), *in adnot.*; Müll. Arg. in DC., Prodr. 15(2): 438 (1866); Benth. & Hook. f., Gen. Pl. 3: 276 (1880); Hook. f., Fl. Brit. India 5: 329 (1887); Beille, Fl. Indo-Chine 5: 631 (1927); Pax & K. Hoffm., Nat. Pflanzenfam. ed. 2, 19C: 59 (1931); Alain, Fl. Cuba 3: 59 (1953); Backer & Bakh. f., Fl. Java 1: 464 (1963); Airy Shaw, Kew Bull. 26: 224 (1972); Whitmore, Tree Fl. Mal. 2: 73 (1973); Berhaut, Fl. Ill. Séneg. 3: 571 (1975) (as *Phyllanthus*); Airy Shaw, Kew Bull. Addit. Ser. 4: 61 (1975), 8: 38 (1980), Kew Bull. 35: 599 (1980), 36: 271 (1981), 37: 9 (1982), and Euph. Philipp.: 10 (1983); Radcl.-Sm., Kew Bull. 35: 498 (1980), and 37: 612 (1983); Coode in Fl. Masc. 160. Euph.: 35 (1982); Proctor, Kew Bull. Addit. Ser. 11: 526 (1984); Radcl.-Sm. in Fl. Pak. 172: 12 (1986) and in Fl. Trop. E. Afr., Euph. 1: 103 (1987); Grierson & D. G Long, Fl. Bhut. 1: 781 (1987); R. A. Howard, Fl. Lesser Antilles 5: 19 (1989); McPherson in Fl. Nouv.-Caléd. 17: 14 (1991); G. L. Webster, Ann. Missouri Bot. Gard. 81: 46 (1994); Radcl.-Sm. in Fl. Zamb. 9(4): 85 (1996); Philcox in Dassanayake, ed., Fl. Ceyl. XI: 238 (1997). T.: *B. disticha* J. R. & G. Forst.

Foersteria Scop., Intr. Hist. Nat. 98 (1777). T.: (none designated)

Melanthesa Blume, Bijdr. 590 (1826); Baill., Étude Euphorb.: 632 (1858). LT.: *M. racemosa* Blume [=*B. racemosa* (Blume) Müll. Arg.] (chosen by Webster, 1994)

Melanthesopsis Müll. Arg., Linnaea 32: 74 (1863). LT.: *M. lucens* (Poir.) Müll. Arg. (B.: *Phyllanthus lucens* Poir.) [=*B. fruticosa* (L.) Hook. f. (B.: *Andrachne fruticosa* L.)] (designated by Wheeler, 1975)

Monoecious or apparently dioecious shrubs or small trees; cataphylls + on orthotropic shoots. Indumentum, when present, simple. Leaves alternate, distichous on plagiotropic shoots, shortly petiolate, stipulate, simple, entire, penninerved, blackening when dried. Flowers axillary, small, males solitary or fasciculate, rarely shortly racemose, in the proximal axils, females solitary in the distal. Male flowers: pedicels often capillary; calyx obconic or shortly turbinate, lobes 6, sharply inflexed, imbricate; petals 0; disc 0; stamens 3, connate into a short column, anthers sessile, extrorse, elongate, thecae parallel, linear, longitudinally dehiscent; pollen stephanocolporate, oblate, usually with 2 circular colpi transversales per colpus, intectate, reticulate; pistillode 0. Female flowers: pedicels capillary or not; calyx turbinate or campanulate, sometimes enlarged and flat in fruit, lobes 6, imbricate, sometimes short or reduced to minute teeth, or sometimes accrescent, not inflexed, usually larger than the male; petals 0; disc 0; ovary sessile or shortly stipitate, 3-locular, ovules 2 per locule; styles 3, free, short, erect or spreading, simple, 2-lobate or 2-fid. Fruit globose or depressed, indehiscent or tardily or often incompletely loculicidally dehiscent; exocarp sometimes somewhat fleshy; endocarp crustaceous. Seeds 3-quetrous or 3-gonous, rounded on the back, ecarunculate; testa membranous to crustaceous, ventrally duplicated, hollow between the lobes; endosperm not copious, incurved around the cavity, fleshy; embryo curved, cotyledons broad, flat, radicle long. Fig.: See

Char. Gen. Pl. 145, t. 73 (1776); Euphorb. Gen.: t. 5, f. 16B.1–8 (1824) (as *Phyllanthus*); Fl. Indo-Chine 5: 629, t. 75.10–15 and 639, t. 76.1 (1927); Nat. Pflanzenfam. ed. 2, 19C: 57, t. 26 E (1931); Fl. Ill. Sénég. 3: 570 (1975) (as *Phyllanthus*); Fl. Pak. 172: 15, t. 3 (1986); Fl. Bhut. 1: 789, t. 48.a–d (1987); Fl. Lesser Antilles 5: 16, t. 6 (1989); Fl. Nouv.-Caléd. 17: 15, t. 3.6–10 (1991).

A difficult genus of 10–15 (Webster, 1994), 25 (Airy Shaw, 1966) c. 40 (Hutchinson, ined.) or 50 (IK) highly variable species, found in tropical Asia, Indonesia and the Pacific islands.

The genus is barely separable from *Sauropus* on the one hand, and from *Glochidion* on the other.

Some species, varieties and forms widely cultivated, e.g. the "Snow Bush" (*B. disticha* var. *disticha* f. *nivosa*) (See Frontispiece).

32. **Glochidion** *J. R. & G. Forst.*, Char. Gen. Pl. 113, t. 57 (1776) (nom. cons.); A. Juss., Euphorb. Gen.: 18 (1824); Baill., Étude Euphorb.: 636 (1858); Müll. Arg., Flora 1865: 369 (1865); Hook. f., Fl. Brit. India 5: 305 (1887); Beille, Fl. Indo-Chine 5: 608 (1927); Alston, Ann. Roy. Bot. Gard. (Peradeniya) 11: 1 (1928); Pax & K. Hoffm., Nat. Pflanzenfam. ed. 2, 19C: 56 (1931); Hurus., J. Fac. Sci. Univ. Tokyo, Sect. 3, Bot. 6: 329 (1954); Leandri, Fl. Madag. 111(1): 21 (1958); Backer & Bakh. f., Fl. Java 1: 460 (1963); Airy Shaw, Kew Bull. 23: 6 (1969), 25: 481 (1971), 26: 271 (1972) and 27: 6 (1972); Whitmore, Tree Fl. Mal. 2: 98 (1973); Airy Shaw, Kew Bull. Addit. Ser. 4: 115 (1975), 8: 92 (1980), Kew Bull. 35: 633 (1980), 36: 298 (1981), 37: 21 (1982), and Euph. Philipp.: 27 (1983); A. C. Sm., Fl. Vit. Nov. 2: 467 (1981); Radcl.-Sm. in Fl. Pak., 172: 9 (1986); Grierson & D. G. Long, Fl. Bhut. 1: 776 (1987); McPherson in Fl. Nouv.-Caléd. 17: 18 (1991); G. L. Webster, Ann. Missouri Bot. Gard. 81: 46 (1994); Philcox in Dassanayake, ed., Fl. Ceyl. XI: 242 (1997). T.: *Gl. ramiflorum* J. R. & G. Forst.

Agyneia L., Mant. Pl. 2: 161 (1771); Baill., Étude Euphorb.: 630 (1858). LT.: *A. pubera* L. [=*Gl. puberum* (L.) Müll. Arg.] (chosen by Webster, 1994)

Bradleja Banks ex Gaertn., Fruct. Sem. Pl. 2: 127 (1790). T.: *B. sinica* Gaertn. [=*Gl. sinicum* (Gaertn.) Hook. & Arn.]

Gynoon A. Juss., Mém. Mus. Hist. Nat. 10: 335 (1823), and Euphorb. Gen.: 17 (1824). T.: *G. rigidum* A. Juss. [=*Gl. rigidum* (A. Juss.) Müll. Arg.]

Glochidionopsis Blume, Bijdr.: 588 (1826); Baill., Étude Euphorb.: 639 (1858). T.: *G. sericea* Blume [=*Glochidion sericeum* (Blume) Hook. f.]

Lobocarpus Wight & Arn., Prodr. Fl. Ind. Orient. 1: 7 (1834). T.: *L. candolleanus* Wight & Arn. [=*Gl. neilgherrense* Wight]

Glochisandra Wight, Icon. Pl. Ind. Orient. 5: 26 (1852). T.: *G. acuminata* Wight [=*Glochidion lanceolarium* (Roxb.) Voigt (B.: *Bradleia lanceolaria* Roxb.)]

Zarcoa Llanos, Bot. Zeitung (Berlin) 15: 423 (1857). T.: *Z. philippica* Llanos [=*Gl. album* (Blanco) Boerl. (B.: *Kirganelia alba* Blanco)]

Coccoglochidion K. Schum. in K. Schum. & Lauterb., Fl. Schutzgeb. Südsee Nachtr.: 292 (1905). T.: *C. erythrococcus* K. Schum. [=*Gl. philippicum* (Cav.) C. B. Rob. (B.: *Bradleia philippica* Cav.)]

Hemiglochidion (Müll. Arg.) K. Schum. in K. Schum. & Lauterb., Fl. Schutzgeb. Südsee Nachtr.: 289 (1905). NT.: *Gl. ramiflorum* J. R. & G. Forst. (designated by Wheeler (1975); the choice of *H. hylodendron* K. Schum. as lectotype by Farr *et al.* (ING, 1979) appears invalid, as the species was not included in *Glochidion* sect. *Hemiglochidion* by Müller, and is in fact a *Phyllanthus*).

Tetraglochidion K. Schum. in K. Schum. & Lauterb., Fl. Schutzgeb. Südsee Nachtr.: 291 (1905). T.: *T. gimi* K. Schum. [=*Gl. gimi* (K. Schum.) Pax & K. Hoffm.]

Monoecious or dioecious evergreen trees or shrubs. Indumentum simple. Leaves alternate, 2-farious, petiolate, stipulate, simple, entire, penninerved. Flowers small, axillary, fasciculate. Male flowers pedicellate; sepals 5–6, 2-seriate, imbricate; petals 0; disc 0; stamens 3–8, connate into a column, anthers extrorse, thecae linear, longitudinally dehiscent, connectives produced separately or connate in an umbonate head; pollen as in *Breynia* but with only 1 colpus transversalis per colpus; pistillode 0 or very minute. Female flowers pedicellate; sepals usually 6, imbricate, or more or less united; petals 0; disc 0; ovary 3–15-locular, ovules 2 per locule; styles connate in a globose, columnar, conical or subclavate column, lobate or toothed at the tip, usually lengthening during or after flowering, rarely styles 3, slender, free. Fruit 3 or more-lobate, often with twice as many lobes as loculi, globose or apically or basally depressed, dehiscing into 3 or more 2-valved cocci; epicarp separable or not; endocarp crustaceous or coriaceous. Seeds hemispherical or laterally compressed; exotesta often fleshy; endotesta crustaceous; endosperm fleshy; cotyledons flat. Fig.: See Char. Gen. Pl. 113, t. 57 (1776); Euphorb. Gen.: t. 3, f. 9.1–6 (as *Gynoon*), and t. 3, f. 10A.1–11, B1–8 (1824); Étude Euphorb.: t. 24, f. 1–14; 27, f. 12–17 (1858); Fl. Indo-Chine 5: 603, t. 73.14–17 and 613, t. 74.5–23 (1927); Nat. Pflanzenfam. ed. 2, 19C: 57, t. 26A–D (1931); Formos. Trees: 342–347, tt. 296–302 (1936); Notul. Syst. (Paris) 6: 30, t. 2 (1937); Fl. Madag. 111(1): 23, t. V and 25, t. VI (1958); Kew Bull. Addit. Ser. 8: 227, t. 5.1,2 (1980); Kew Bull. 35: 639, t. 3A1,2 and D1–3 (1980); 36: 305, t. 6 (1981); Fl. Pak. 172: 11, t. 2A–C (1986); Fl. Bhut. 1: 763, t. 47.p–s (1987); Fl. Nouv.-Caléd. 17: 21 t. 4 (1991).

A large predominantly Asiatic, Pacific and Australasian genus of c. 300 species, a few also in Madagascar.

The genus was combined with *Phyllanthus* by Müller (1866) and Bentham (1880), but most subsequent workers have treated it as distinct, even though it is an undoubted *Phyllanthus*-vicariad.

Tribe 5. **DRYPETEAE** *(Griseb.) Hurus.*, J. Fac. Sci. Univ. Tokyo, Sect. 3, Bot. 6: 334 (1954) (as tribe in *Antidesmataceae*); G. L. Webster, Ann. Missouri Bot. Gard. 81: 46 (1994). T.: *Drypetes* Vahl

Phyllantheae subtribe *Drypeteae* Griseb., Fl. Brit. W. I.: 31 (1859).
Antidesmeae * *Putranjiveae* Endl., Gen. Pl. 287 (1837).
Ordo *Putranjiveae* Endl., Gen. Pl. 287 (1837). T.: *Putranjiva* Wall.

Putranjivaceae Endl. ex A. Meeuse, Euph. Auct. Pl.: 30 (1990).
Cyclostemonées Baill., Étude Euphorb.: 561 (1858).
Tribe *Biovulatae* sect. *Pierardieae* Thwaites, Enum. Pl. Zeyl. 286 (1861). T.: *Pierardia* Roxb. ex Jack [=*Baccaurea* Lour.]
Phyllantheae subtribe *Cyclostemoneae* Baill. ex Müll. Arg., Linnaea 34: 64 (1865). T.: *Cyclostemon* Blume [=*Drypetes* Vahl]

Monoecious or dioecious trees or shrubs; leaves entire or dentate, stipulate. Flowers in axillary clusters, or cauliflorous; sepals mostly 4–5, imbricate; petals 0; male disc intrastaminal or 0; stamens (2)3–20(50), usually free, anthers introrse or extrorse; pollen 3-colporate, often reticulate; pistillode minute or 0; female sepals usually deciduous; disc annular or 0; ovary 1–4-locular; ovules anatropous, with massive obturator; styles usually entire and dilated. Fruit dehiscent or drupaceous. Seeds 1 per locule or fruit; endosperm copious; cotyledons flat, broader and somewhat longer than the radicle.

As circumscribed by Webster (1994), tribe *Drypeteae* includes 4 genera, with the majority of species in the large genus *Drypetes.* The recent study by Hayden (1987) has demonstrated that *Neowawraea,* included in the *Drypeteae* by Webster (1975), is really a species of *Flueggea.*

KEY TO THE GENERA

1. Flowers usually monoecious; ovary 3-locular; fruit dehiscent; styles 2-fid or 2-partite, style-branches slender · · · · · · · · · · · · · · · · · 33. *Lingelsheimia*
1. Flowers usually dioecious; ovary 1–2-locular; fruit drupaceous; styles stigmatiform or petaloid · 2
2. Disc 0; stamens 2–3(4), anthers extrorse; styles petaloid-dilated · 36. *Putranjiva*
2. Disc usually +; stamens mostly 4 or more, anthers usually introrse; styles stigmatiform · 3
3. Female sepals imbricate in bud, deciduous in fruit · · · · · · · · 34. *Drypetes*
3. Female sepals open in bud, persistent in fruit · · · · · · · · · · · · 35. *Sibangea*

33. **Lingelsheimia** *Pax,* Bot. Jahrb. Syst. 43: 317 (1909); Mildbr., Wiss. Erg. Deut. Zentr.-Afr. Exped., Bot. 2: t. 54, 55 (1912); Hutch. in Fl. Trop. Afr. 6 (1): 690 (1912), pro min. parte; Pax & K. Hoffm., Pflanzenr. 81: 279 (1922), and Nat. Pflanzenfam. ed. 2, 19C: 74 (1931); J. Léonard, Bull. Soc. Roy. Bot. Belg. 84: 49 (1951), and in Bull. Jard. Bot. État. 32: 513 (1962); G. L. Webster, Ann. Missouri Bot. Gard. 81: 47 (1994); Radcl.-Sm., Kew Bull. 52: 172 (1997). LT.: *L. frutescens* Pax (chosen by Wheeler, 1975).

Danguyodrypetes Leandri, Bull. Soc. Bot. France 85: 524 (1938). T.: *D. manongarivensis* Leandri [=*L. manongarivensis* (Leandri) G. L. Webster]
Aerisilvaea Radcl.-Sm., Kew Bull. 45: 149 (1990). T.: *A. sylvestris* Radcl.-Sm. [=*L. sylvestris* (Radcl.-Sm.) Radcl.-Sm.]

Monoecious or sometimes dioecious glabrous shrubs; branches angled.

Leaves alternate, shortly petiolate, stipulate, simple, entire, penninerved, not oblique at the base. Flowers axillary, fasciculate, males and females sometimes mixed. Male flowers: pedicels short or long; sepals 4–6(7), imbricate; petals 0; stamens 15–20(35), filaments free, slender, inserted in the disc, anthers basifixed, introrse, thecae parallel, longitudinally dehiscent; pollen prolate-spheroidal, 3-colporate, often almost syncolpate, tectate, psilate; disc central, irregularly 6–7-lobate; pistillode 0 or very minute. Female flowers long-pedicellate; sepals 5–6, imbricate, often accrescent; petals 0; disc annular, shallowly 6-lobate, lobes alternisepalous; ovary 2–3-locular, ovules 2 per locule; styles 3, united at the base, reflexed or patent, 2-lobate, -fid or -partite, often exceeding the ovary. Fruit subglobose, dehiscing septicidally into 3 2-valved cocci or 6 valves; exocarp thin, finely reticulate when dried, separating; endocarp woody. Seeds 1 per locule, broadly ovoid-3-gonous, ecarunculate, with an invaginated hilum. Fig.: See Wiss. Erg. Deut. Zentr.-Afr. Exped., Bot. 2: t. 54, 55 (1912); Bull. Soc. Bot. France 85: 525, t. 1.1–3 (1938) (as *Danguyodrypetes*); Fl. Madag. 111(1): 159, t. XXVI (1958); Kew Bull. 45: 150, t. 1 (1990).

An Afro-Malagasy genus of 7 species.

34. **Drypetes** *Vahl*, Eclog. Amer. 3: 49 (1810); Baill., Étude Euphorb.: 606 (1858); Müll. Arg. in DC., Prodr. 15(2): 453 (1866), and in Mart., Fl. Bras. 11(2): 78 (1873); Benth. & Hook. f., Gen. Pl. 3: 278 (1880); Hutch. in Fl. Trop. Afr. 6 (1): 674 (1912); Fawc. & Rendle, Fl. Jamaica 4: 267 (1920); Pax & K. Hoffm., Pflanzenr. 81: 229 (1922); Gagnep., Fl. Indo-Chine 5: 563 (1927); Pax & K. Hoffm., Nat. Pflanzenfam. ed. 2, 19C: 72 (1931); Brenan, Kew Bull. 7: 444 (1953); Alain, Fl. Cuba 3: 59 (1953); Hurus., J. Fac. Sci. Univ. Tokyo, Sect. 3, Bot. 6: 334 (1954); Leandri, Fl. Madag. 111(1): 144 (1958); Keay (ed.), Fl. W. Trop. Afr. ed. 2, 1: 377 (1958); van Steenis, Blumea 10: 140 (1960); J. Léonard, Bull. Jard. Bot. État. 32: 513 (1962); Backer & Bakh. f., Fl. Java 1: 471 (1963); A. C. Sm. & Ayensu, Brittonia 16: 220 (1964); Airy Shaw, Kew Bull. 18: 272 (1965); G. L. Webster, J. Arnold Arbor. 48: 329 (1967); Jabl., Mem. New York Bot. Gard. 17: 118 (1967); J.-G. Adam, Mém. Mus. Natl. Hist. Nat., B, Bot. 20: 471 (1971); Airy Shaw, Kew Bull. 26: 252 (1972), and 27: 309 (1972); Whitmore, Tree Fl. Mal. 2: 87 (1973); Airy Shaw, Hooker's Icon. Pl. 38: t. 3710 (1974), Kew Bull. Addit. Ser. 4: 97 (1975), 8: 74 (1980), Kew Bull. 35, 627 (1980), and 36: 286 (1981); Berhaut, Fl. Ill. Sénég. 3: 429 (1975); A. C. Sm., Fl. Vit. Nov. 2: 455 (1981); Airy Shaw, Kew Bull. 37: 16 (1982); Coode in Fl. Masc. 160. Euph.: 38 (1982); Correll, Fl. Bah. Arch.: 795 (1982); Airy Shaw, Euph. Philipp.: 21 (1983); Alain, Fl. Español. 4: 137 (1986); Radcl.-Sm. in Fl. Trop. E. Afr., Euph. 1: 88 (1987); Grierson & D. G. Long, Fl. Bhut. 1: 784 (1987); Alain, Descr. Fl. Puerto Rico 2: 386 (1988); G. L. Webster & Huft, Ann. Missouri Bot. Gard. 75: 1096 (1988); R. A. Howard, Fl. Lesser Antilles 5: 43 (1989); Lebrun & Stork, Enum. Pl. Afr. Trop. 1: 211 (1991); McPherson in Fl. Nouv.-Caléd. 17: 12 (1991); G. L. Webster, Ann. Missouri Bot. Gard. 81: 47 (1994); Murillo-Aldana & Franco-Rosselli, Euf. Reg. Ararac.: 74 (1995); Radcl.-Sm. in Fl. Zamb. 9(4): 87 (1996); Gillespie in Acev.-Rodr., ed., Fl. St. John: 210 (1996); Philcox in Dassanayake, ed., Fl. Ceyl. XI: 260 (1997). T.: *D. glauca* Vahl.

Koelera Willd., Sp. Pl., ed. 4: 750 (1806), nom. illeg., non *Koeleria* Pers. (1805). T.: *K. laurifolia* Willd. [=*Dr. lateriflora* (Sw.) Krug & Urb. (B.: *Schaefferia lateriflora* Sw.)]

Limacia F. Dietr., Nachtr. Vollst. Lex. Gärtn. 4: 334 (1818), nom. illeg., non *Limacia* Lour. (1790). T.: *L. laurifolia* F. Dietr. [=*Dr. lateriflora* (Sw.) Krug & Urb. (B.: *Schaefferia lateriflora* Sw.)]

Liparena Poit. ex Léman, Dict. Sci. Nat. 27: 6 (1823). LT.: *L. crocea* Poit. ex Baill. [=*Dr. lateriflora* (Sw.) Krug & Urb. (B.: *Schaefferia lateriflora* Sw.)] (chosen here)

Cyclostemon Blume, Bijdr. 597 (1826); Baill., Étude Euphorb.: 561 (1858); Hook. f., Fl. Brit. India 5: 339 (1887). LT.: *C. macrophyllum* Blume [=*Dr. macrophylla* (Blume) Pax & K. Hoffm.] (chosen by Webster, 1994)

Hemicyclia Wight & Arn., Edinb. New Philos. J. 14: 297 (1833); Baill., Étude Euphorb.: 562 (1858); Hook. f., Fl. Brit. India 5: 337 (1887); Gamble, Hooker's Icon. Pl. 28: t. 2701 (1901). T.: *H. sepiaria* Wight & Arn. [=*Dr. sepiaria* (Wight & Arn.) Pax & K. Hoffm.]

Astylis Wight, Icon. Pl. Ind. Orient. 6: t. 1992 (1853). T.: *A. venusta* Wight [=*Dr. venusta* (Wight) Pax & K. Hoffm.]

Sphragidia Thwaites, Hooker's J. Bot. Kew Gard. Misc. 7: 269 (1855). T.: *S. zeylanica* Thwaites [=*Dr. longifolia* (Blume) Pax & K. Hoffm. (B.: *Cyclostemon longifolius* Blume)]

Dodecastemon Hassk., Versl. Meded. Afd. Natuurk. Kon. Akad. Wetensch. 4: 141 (1856). T.: *D. teysmannii* Hassk. [=*Dr. teysmannii* (Hassk.) Bakh. f. & van Steenis]

Pycnosandra Blume, Mus. Bot. 2: 191 (1856). T.: *P. serrata* (Blume) Blume (B.: *Cyclostemon serratus* Blume) [=*Dr. serrata* (Blume) Pax & K. Hoffm., nom. illeg. (non *Dr. serrata* (Maycock) Krug & Urb.) =*Dr. teysmannii* (Hassk.) Bakh. f. & van Steenis (B.: *Dodecastemon teysmannii* Hassk.)]

Cometia Thouars ex Baill., Étude Euphorb.: 642 (1858) (female only). T.: *C. thouarsii* Baill. [=*Dr. thouarsii* (Baill.) Leandri]

Anaua Miq., Fl. Ned. Ind., Suppl. 1: 410 (1861). T.: *A. sumatrana* Miq. [=*Dr. sumatrana* (Miq.) Pax & K. Hoffm.]

Laneasagum Bedd., Madr. J. Lit. Sci., Ser. 2, 22: 71 (1861). T.: *L. oblongifolium* Bedd. [=*Dr. longifolia* (Blume) Pax & K. Hoffm. (B.: *Cyclostemon longifolius* Blume)]

Stelechanteria Thouars ex Baill., Adansonia I. 4: 147 (1864). T.: *S. thouarsiana* Baill. [=*Dr. thouarsiana* (Baill.) Capuron]

Freireodendron Müll. Arg. in DC., Prodr. 15(2): 245 (1866). T.: *F. sessiliflorum* (Allemão.) Müll. Arg. [=*Dr. sessiliflora* Allemão]

Humblotia Baill., Bull. Mens. Soc. Linn. Paris 1: 593 (1886). T.: *H. comorensis* Baill. [=*Dr. comorensis* (Baill.) Pax & K. Hoffm.]

Guya Frapp. ex Cordem., Fl. Réunion: 350 (1895). T.: *G. caustica* Frapp. ex Cordem. [=*Dr. caustica* (Frapp. ex Cordem.) Airy Shaw]

Riseleya Hemsl., J. Bot. 55: 286 (1917). T.: *R. griffithii* Hemsl. [=*Dr. riseleyi* Airy Shaw, non *Cyclostemon griffithii* Hook. f.]

Calyptosepalum S. Moore, J. Bot. 63 (Suppl.): 91 (1925). T.: *C. sumatranum* S. Moore [=*Dr. calyptosepala* Airy Shaw]

Brexiopsis H. Perrier, Notul. Syst. (Paris) 10: 192 (1942). T.: *B. aquifolia* H. Perrier [=*Dr. bathiei* Capuron & Leandri]

Dioecious, polygamo-dioecious or rarely monoecious trees or shrubs; buds sometimes perulate. Indumentum simple. Leaves alternate, sometimes subdistichous, petiolate, stipulate, simple, entire or spiny-dentate, often coriaceous, penninerved, usually asymmetrical at the base. Flowers fasciculate, axillary among or below the leaves or cauliflorous on the main branches or trunk. Male flowers usually pedicellate; sepals 4–5(7), strongly imbricate, broad, concave, often unequal; petals 0; stamens 3–20(50), filaments free, anthers erect, basifixed, commonly introrse, rarely latrorse or extrorse, thecae parallel, longitudinally dehiscent; pollen prolate, 3-colporate, colpi transversales broad, elongate, tectate, psilate; disc central, intrastaminal, flat, cupular, entire or often lobulate, laciniate, plicate or convoluted and then the lobes enfolding the filaments; pistillode minute or 0. Female flowers often pedicellate; sepals deciduous, otherwise ± as in the male; disc annular, cupular, or rarely 0; ovary 1–4-locular, ovules 2 per locule, covered with an obturator; styles usually short or 0, stigmas thick, flattened, dilated, sometimes subpeltate, 2-fid, reniform or discoid. Fruit drupaceous, 1–4-locular, globose, ovoid, ellipsoid or rarely angular; exocarp somewhat fleshy, becoming hardened on drying; endocarp coriaceous, chartaceous or osseous. Seeds solitary per locule or per fruit by abortion, ecarunculate; sarcotesta, when present, thin; endosperm copious, fleshy; embryo straight; cotyledons broad, flat. Fig.: See Étude Euphorb.: t. 23, f. 22–25; 24, f. 34–40; 27, f. 7–8 (1858); Fl. Bras. 11(2): t. 3.I (as *Freireodendron*) and 12 (1873); Hooker's Icon. Pl. 28: t. 2701 (1901) (as *Hemicyclia*); Fl. Jamaica 4: 268, t. 87 (1920); Pflanzenr. 81: 239, t. 20 and 257, t. 21 (1922); Fl. Indo-Chine 5: 556, t. 69.5–12 and 567, t. 70.1–3 (1927); Nat. Pflanzenfam. ed. 2, 19C: 73, t. 35 (1931); Formos. Trees: 338, t. 292 and 339, t. 293 (1936); Fl. Cuba 3: 60, t. 10 (1953); J. Fac. Sci. Univ. Tokyo, Sect. 3, Bot. 6: 336, t. 55A–G (1954); Fl. Madag. 111(1): 109, t. XVII.6–8, 145–155, tt. XXIII–XXV and 167, t. XXVII.1–4 (1958); Fl. W. Trop. Afr. ed. 2, 1: 383, t. 133 (1958); Amer. J. Bot. 56: 748, t. 2A–M (1969); Mém. Mus. Natl. Hist. Nat., B, Bot. 20: 472–477, tt. 167–171 (1971); Hooker's Icon. Pl. 38: t. 3710 (1974); Fl. Ill. Séneg. 3: 428, 430 (1975); Kew Bull. 35, 619, t. 2B1 (1980), and 36: 290, t. 5 (1981); Fl. Masc. 160. Euph.: 38, t. 6 (1982); Fl. Bah. Arch.: 796, t. 333 and 797, t. 334 (1982); Fl. Español. 4: 140, t. 116-14 and 141, t. 116-15 (1986); Fl. Trop. E. Afr., Euph. 1: 90–100, tt. 13–15 (1987); Descr. Fl. Puerto Rico 2: 388, t. 59-14 (1988); Fl. Lesser Antilles 5: 51, t. 17 (1989); Fl. Nouv.-Caléd. 17: 11, t. 2.6–9 (1991); Fl. Zamb. 9(4): 90, t. 13 (1996); Fl. St. John: 211, t. 95F–I (1996).

A large and variable genus of c. 200 species, poorly represented in the New World, with c. 10 species, 70 species in Africa and Madagascar, and the remainder in Asia and Australasia.

Drypetes exhibits considerable anatomical and palynological variation, and the generic limits require further study. Cronquist (1981), Stuppy (1995) and others also note a divergence from other phyllanthoids and suggest an affinity to *Erythroxylaceae*.

35. **Sibangea** *Oliv.*, Hooker's Icon. Pl. 15(1): 9, t. 1411 (1883); Benth. & Hook. f., Gen Pl. 3: 1223 (1883); Radcl.-Sm., Kew Bull. 32: 480 (1978), and in Fl. Trop. E. Afr., Euph. 1: 101 (1987); G. L. Webster, Ann. Missouri Bot. Gard. 81: 48 (1994). T.: *S. arborescens* Oliv.

Very close to *Drypetes*, differing chiefly in having the female sepals open in bud and persistent in fruit. Fig.: See Hooker's Icon. Pl. 15(1): t. 1411 (1883); Fl. Trop. E. Afr., Euph. 1: 102, t. 16 (1987).

A small Tropical African genus of 3 species, 2 from W. Africa and one from E. Africa

36. **Putranjiva** *Wall.*, Tent. Fl. Napal. 2: 61 (1826); Baill., Étude Euphorb.: 641 (1858); Müll. Arg. in DC., Prodr. 15(2): 443 (1866); Benth. & Hook. f., Gen. Pl. 3: 277 (1880); Hook. f., Fl. Brit. India 5: 336 (1887); Pax & K. Hoffm., Nat. Pflanzenfam. ed. 2, 19C: 59 (1931); Backer & Bakh. f., Fl. Java. 1: 465 (1963); Radcl.-Sm. in Fl. Pak., 172: 16 (1986); G. L. Webster, Ann. Missouri Bot. Gard. 81: 48 (1994); Philcox in Dassanayake, ed., Fl. Ceyl. XI: 258 (1997). T.: *P. roxburghii* Wall.

Nageia Roxb., Hort. Bengal: 71 (1814). T.: *N. putranjiva* Roxb.[=*P. roxburghii* Wall.]

Palenga Thwaites, Hooker's J. Bot. Kew Gard. Misc. 8: 270 (1856); Baill., Étude Euphorb.: 649 (1858). T.: *P. zeylanica* Thwaites [=*Putranjiva zeylanica* (Thwaites) Müll. Arg.]

Liodendron Keng, J. Wash. Acad. Sci. 41: 201 (1951) T.: *L. matsumurae* (Koidz.) Keng [=*P. matsumurae* Koidz.]

Drypetes subgen. *Putranjiva* (Wall.) Hurus., J. Fac. Sci. Univ. Tokyo, Sect. 3, Bot. 6(6): 335 (1954).

Dioecious or monoecious evergreen trees. Indumentum simple. Leaves alternate, petiolate, stipulate, simple, entire or toothed, penninerved, reticulate. Flowers axillary, males in dense many-flowered clusters, females solitary or up to 3 per axil. Male flowers shortly pedicellate; calyx deeply 3–6-lobate, lobes imbricate, unequal; petals 0; disc 0; stamens 2–3(4), filaments free or subconnate, anthers erect, basifixed, extrorse, thecae separate, parallel, longitudinally dehiscent; pollen spheroidal, with a thick endexine, otherwise resembling that of *Drypetes*; pistillode 0. Female flowers long-pedicellate, pedicels extending in fruit; calyx ± as in the male; petals 0; disc 0; ovary ovoid, 2–3-locular, ovules 2 per locule; styles 2–3, connate at the base, short, spreading, stigmas dilated, fleshy, papillose. Fruit drupaceous, elobate, ovoid or globose, 1-locular and 1-seeded by abortion; endocarp hard or osseous. Seed ovoid; testa crustaceous; endosperm fleshy; embryo straight; cotyledons broad, flat, palminerved. Fig.: See Fl. Pak. 172: 11, t. 2D–F (1986). See also Fig. 6, p. 54.

A small genus of 3 Asiatic species (India, Ceylon, Indonesia, Taiwan and the Ryukyu Islands).

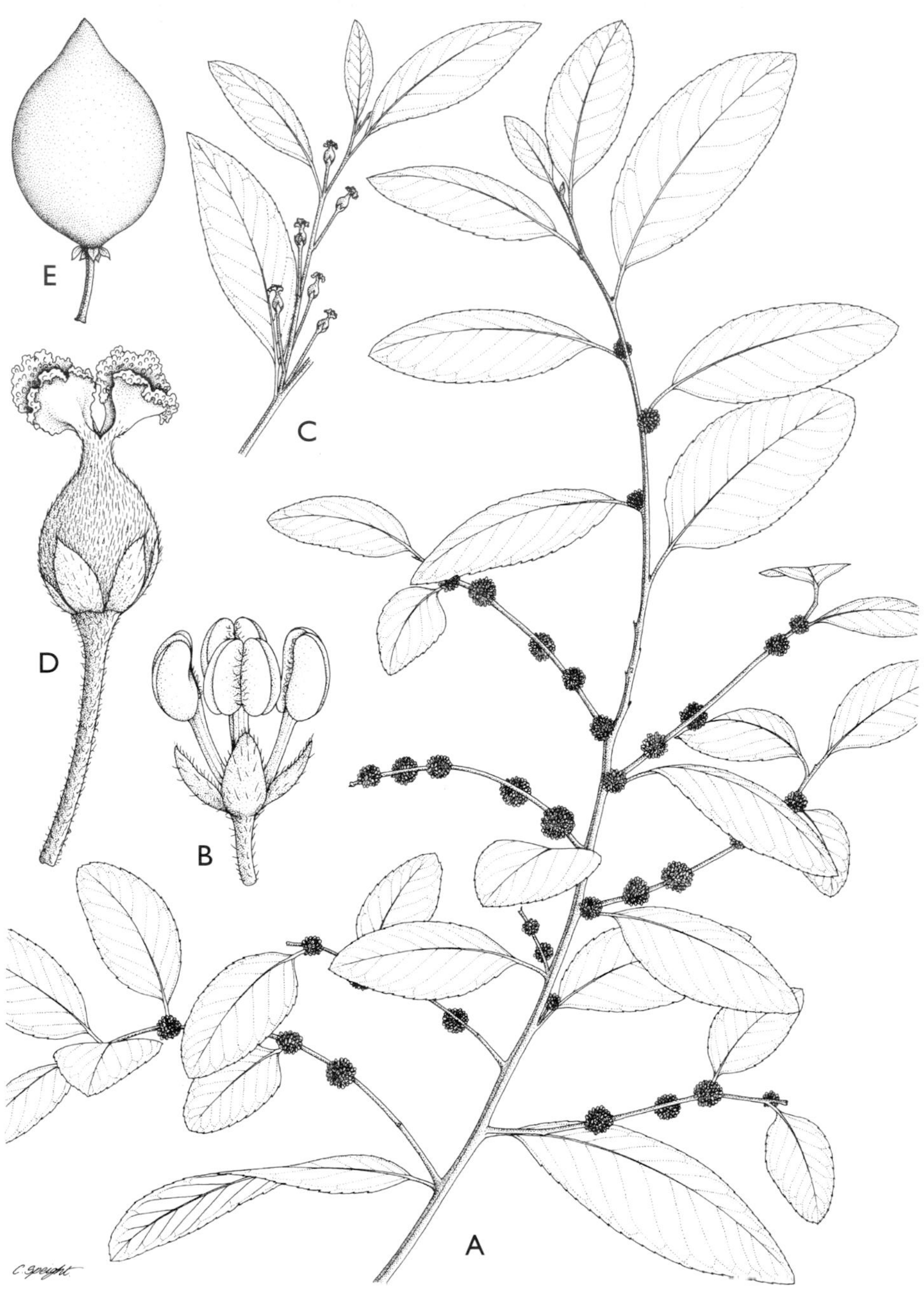

FIG. 6. *Putranjiva roxburghii* Wall. (S.: *Drypetes roxburghii* (Wall.) Hurus.). **A** Habit, male × 1; **B** Male flower × 18; **C** Female flowering shoot × 1; **D** Female flower × 7.5; **E** Fruit × 3. **A, B** from *Dawkins* 678; **C, D** from *Nasir* 5321; **E** from *Broadway* 9249. Drawn by Camilla Speight.

It is very close to *Drypetes* in many respects, but differs chiefly in the absence of a disc, and in the extrorse anthers. Meeuse (1990) supports familial rank and placement near *Capparidales* owing to phytochemical similarities, but the latter are undoubtedly not significant.

Tribe 6. **ANTIDESMEAE** *(Sweet) Hurus.*, J. Fac. Sci. Univ. Tokyo, Sect. 3, Bot. 6: 322 (1954) [as tribe of *Antidesmataceae*]; G. L. Webster, Ann. Missouri Bot. Gard. 81: 48 (1994). T.: *Antidesma* L.

Ordo *Stilagineae* C. Agardh, Aphor. Bot. 199, Class. Pl.: 9 (1825).
Ordo *Antidesmeae* Sweet, Hort. Brit. ed. 2: 460 (1830).
Tribe *Biovulatae* sect. *Antidesmeae* Thwaites, Enum. Pl. Zeyl.: 289 (1861).

Dioecious or rarely monoecious trees or shrubs, rarely herbs; leaves alternate, stipulate, entire or rarely dentate, penninerved, glandular or eglandular. Inflorescences various but usually axillary, amentiform; sepals 3–6(8), free or connate, imbricate; petals reduced or 0 (except *Spondianthus*); disc annular, lobate or dissected; stamens 2–8, free or nearly so; anthers extrorse or introrse; pollen grains mostly prolate, 3-colporate, reticulate; pistillode + or 0; ovary (1)2–3(5)-locular; ovules anatropous; styles 2-fid or 2-partite, rarely entire, sometimes dilated. Fruit dehiscent or drupaceous. Seeds mostly 1 per locule, or 1 per fruit; seed-coat dry or fleshy; endosperm thin to copious; cotyledons mostly broader than radicle.

KEY TO THE SUBTRIBES

1. Stems with resinous secretion; inflorescence paniculate or capitate with large involucrate bracts · 2
1. Stems lacking resin; inflorescence not paniculate or capitate; petals mostly 0 · 3
2. Petals +; inflorescence terminal, paniculate, bracts small; pollen sexine not spinulose; fruit dehiscent; stomata paracytic · · · · · · · 6a. *Spondianthinae*
2. Petals 0; inflorescences axillary, capitate, bracts large and involucrate; pollen sexine spinulose; fruit drupaceous; stomata anisocytic · · · 6b. *Uapacinae*
3. Inflorescences terminal, corymbiform, or if capitate, then bracts not large; herbs or shrublets; anthers opening by pores · · · · · · · 6e. *Porantherinae*
3. Inflorescences axillary, often amentiform; mostly trees or shrubs; anthers opening by slits · 4
4. Pollen grains spheroidal or slightly prolate; anther-thecae not pendulous; leaves sometimes glandular; stomata anisocytic · · · · · · · · · 6c. *Scepinae*
4. Pollen grains distinctly prolate; anther-thecae ± pendulous; leaves not glandular; stomata paracytic · · · · · · · · · · · · · · · · · · · 6d. *Antidesminae*

Subtribe 6a. **Spondianthinae** *(G. L. Webster) G. L. Webster*, Ann. Missouri Bot. Gard. 81: 49 (1994). T.: *Spondianthus* Engl.

Tribe *Spondiantheae* G. L. Webster, Taxon 24: 594 (1975).

Dioecious trees; leaves entire, penninerved, stipules caducous. Inflorescences terminal, at least in part, paniculate, bracts small; petals +, at least in the male flowers; male disc dissected; stamens 5, filaments free, anthers introrse; pistillode +; female disc shallowly lobate; styles ± free. Fruit loculicidally dehiscent; columella persistent. Seeds 1 per locule; testa dry; endosperm scanty; embryo with cotyledons much broader than radicle.

A monotypic subtribe, including only the African genus *Spondianthus.* Pax & Hoffmann (1922) placed it adjacent to *Thecacoris,* a genus of Subtribe *Antidesminae.* Levin (1986c), on the basis of foliar characters, suggests that it is the primitive sister group of the other woody tribes of *Antidesmeae.* The observations of Köhler (1965) on pollen suggest a similar position.

37. **Spondianthus** *Engl.*, Bot. Jahrb. Syst. 36: 215 (1905); Hutch., Hooker's Icon. Pl. 30: t. 2986 (1911), and Fl. Trop. Afr. 6(1): 1044 (1913), in adnot.; Pax & K. Hoffm., Pflanzenr. 81: 13 (1922), and Nat. Pflanzenfam. ed. 2, 19C: 46 (1931); Keay (ed.), Fl. W. Trop. Afr. ed. 2, 1: 372 (1958); Radcl.-Sm. in Fl. Trop. E. Afr., Euph. 1: 104 (1987); J. Léonard & Nkounkou, Bull. Jard. Bot. Belg. 59: 133 (1989); G. L. Webster, Ann. Missouri Bot. Gard. 81: 49 (1994); J. Léonard, Fl. Afr. Cent. Euph. 2: 40 (Nov. 1995). T.: *S. preussii* Engl.

Megabaria De Wild., Études Fl. Bas-Moyen-Congo. 2: 284 (1908). T.: *M. trillesii* De Wild. [=*S. preussii* Engl.]

Dioecious trees. Indumentum, when present, simple. Leaves alternate or subverticillate, crowded at the ends of the shoots, long- and short-petiolate, stipulate, simple, entire, penninerved, varying greatly in size; stipules readily deciduous. Inflorescences terminal, subterminal or axillary, paniculate, shortly pedunculate; bracts small; male flowers glomerulate, female solitary. Male flowers subsessile; sepals (4)5, imbricate; petals (4)5, smaller than the sepals; disc-glands 5, epipetalous; stamens 5, episepalous, filaments free, anthers medifixed, introrse, connective glandular and pigmented at the apex, thecae parallel, longitudinally dehiscent; pollen prolate, 3-colporate, colpae transversales elongate; pistillode obconic-cylindric, truncate and slightly 3-lobate at the apex. Female flowers shortly pedicellate; sepals and petals ± as in the male, or petals 0; disc shallowly cupular, 5-lobate, the lobes alternisepalous; ovary 3-locular, ovules 2 per locule; styles 3, ± free, short, reflexed, shortly 2-lobate, stigmas papillose. Fruit ± ellipsoid, entire, dehiscing loculicidally into 3 valves; exocarp crustaceous; endocarp thin and horny; columella persistent, 3-partite. Seeds 1 per locule by abortion, compressed, ovoid-ellipsoid; exotesta papyraceous, red or reddish brown, shiny; mesotesta spongy or chartaceous, white; endotesta thin, crustaceous; endosperm scanty, forming thin strata; embryo straight; cotyledons broad, flat. Fig.: See Hooker's Icon. Pl. 30: t. 2986 (1911); Pflanzenr. 81: 14, t. 2 (1922); Fl. Trop. E. Afr., Euph. 1: 106, t. 17 (1987); Fl. Afr. Cent. Euph. 2: 45, t. 6 (Nov. 1995).

A monotypic genus. The single variable species is widespread in tropical Africa from Guinée to Angola, Uganda and Tanzania.

Although data from anatomical and palynological studies support affinity to other *Antidesmeae*, seed morphology does not, according to Stuppy (1995), but rather shows similarities to that of uniovulates.

Subtribe 6b. **Uapacinae** *Müll. Arg.*, Linnaea 34: 64 (1865) (as *Phyllantheae* subtr. *Uapaceae*); G. L. Webster, Ann. Missouri Bot. Gard. 81: 49 (1994). T.: *Uapaca* Baill.

Tribe *Uapaceae* (Müll. Arg.) Hutch., Amer. J. Bot. 56: 747 (1969).
Family *Uapacaceae* (Müll. Arg.) Airy Shaw, Kew Bull. 18: 270 (1965)

Dioecious trees or shrubs; leaves entire, penninerved, stipules caducous. Inflorescences axillary, capitate, involucrate, pedunculate; male calyx gamosepalous; petals 0; disc 0; stamens 5, filaments free, anthers introrse; pistillode apically dilated; female calyx gamosepalous; petals 0; disc 0; ovary mostly 3-locular; styles free, multifid. Fruit drupaceous. Seeds 1 per locule; endosperm +; cotyledons much longer than radicle.

This subtribe includes only the single genus *Uapaca*, which has always appeared isolated because of its remarkable involucrate inflorescences; Airy Shaw (1965) and Meeuse (1990) have assigned it to a separate family. However, Bentham (1880) juxtaposed it with *Aporosa*, and Pax (1890) referred it to his subtribe *Antidesminae*. These intuitions appear confirmed by the foliar anatomical studies of Levin (1986a, b, c), who has found that the tanniniferous epidermal cells and anisocytic stomata indicate an affinity with subtribe *Scepinae*. However, the distinctive pollen and resiniferous stems of *Uapaca* are divergent characters, and it appears best referred to a separate subtribe.

38. **Uapaca** *Baill.*, Étude Euphorb. 595 (1858); Müll. Arg. in DC., Prodr. 15(2): 489 (1866); Benth., Hooker's Icon. Pl. 13: t. 1287 (1879); Benth. & Hook. f., Gen. Pl. 3: 282 (1880); Hutch. in Fl. Trop. Afr. 6 (1): 634 (1912); Pax & K. Hoffm., Pflanzenr. 81: 298 (1922), and Nat. Pflanzenfam. ed. 2, 19C: 76 (1931); De Wild., Mém. Inst. Roy. Colon. Belge, Sect. Sci. Nat.: 71 (1936); Leandri, Fl. Madag. 111(1): 163 (1958); Keay (ed.), Fl. W. Trop. Afr. ed. 2, 1: 389 (1958); J.-G. Adam, Mém. Mus. Natl. Hist. Nat., Sér. B, Bot. 20: 513 (1971); Berhaut, Fl. Ill. Sénég. 3: 603 (1975); Radcl.-Sm. in Fl. Trop. E. Afr., Euph. 2: 566 (1988); Lebrun & Stork, Enum. Pl. Afr. Trop. 1: 239 (1991); Radcl.-Sm., Kew Bull. 48: 611 (1993); G. L. Webster, Ann. Missouri Bot. Gard. 81: 49 (1994); Radcl.-Sm. in Fl. Zamb. 9(4): 93 (1996). LT.: *U. thouarsii* Baill. (selected by Airy Shaw, Kew Bull. 18: 271 (1965)).

Dioecious pachycaul vernicifluous trees or shrubs, superficially resembling large-leaved *Rhododendra* in habit; trunks often stilt-rooted; twigs often cicatricose. Indumentum simple, sometimes minutely pseudolepidote. Leaves alternate, crowded at the shoot-apices, petiolate or subsessile, stipulate or exstipulate, simple, often obovate-oblong, entire, penninerved; stipules deciduous, sometimes large, or 0. Inflorescences axillary, hornotine or annotine, solitary or fasciculate, pedunculate, involucrate, the males capituliform, shortly stipitate, globose, dense, many-flowered, the females 1-flowered; bracts 5–12, subverticillate, tepaloid, imbricate, at length reflexed. Male flowers sessile; calyx very small, campanulate or turbinate, truncate, dentate or 4–6-lobate, the lobes imbricate; petals 0; disc 0; stamens 4–6, filaments free, episepalous, anthers erect, subbasifixed, introrse, thecae parallel, longitudinally dehiscent; pollen resembling that of *Savia* (q.v. supra); pistillode cylindric-obconic, infundibuliform, hypocrateriform, pileiform or sometimes lobate. Female flowers sessile; calyx minute, truncate, sinuate or lobate, disciform; petals 0; disc 0; ovary (2)3(5)-locular, ovules 2 per locule; styles (2)3(5), free, thick, recurved, covering the ovary, flattened, plurilobate, the lobes laciniate. Fruit drupaceous; mesocarp spongy; endocarp of (2)3(4) pyrenes; pyrenes usually dorsally carinate, bisulcate, indurate, tardily loculicidally dehiscent into 2 valves each. Seeds often 1 per pyrene, compressed, ecarunculate; endosperm fleshy; embryo straight; cotyledons flat, longer and broader than the radicle, green. Fig.: See Hooker's Icon. Pl. 13: t. 1287 (1879); Pflanzenr. 81: 309, t. 25 (1922); Nat. Pflanzenfam. ed. 2, 19C: 77, t. 37 (1931); Fl. Madag. 111(1): 167, t. XXVII.5–11 and 169–179, tt. XXVIII–XXX (1958); Fl. W. Trop. Afr. ed. 2, 1: 391, t. 136 (1958); Amer. J. Bot. 56: 748, t. 5A–K (1969); Mém. Mus. Natl. Hist. Nat., Sér. B, Bot. 20: 516–520, tt. 194–7 (1971); Fl. Ill. Sénég. 3: 602–606 (1975); Bot. J. Linn. Soc. 94: 55, tt. 5,6 (1987); Fl. Trop. E. Afr., Euph. 2: 569, t. 105 (1988); Kew Bull. 48: 613, t. 1 (1993); Fl. Zamb. 9(4): 99, t. 14 (1996).

An Afro-Malagasy genus of 61 species, 49 in tropical Africa and 12 in Madagascar.

Subtribe 6c. **Scepinae** *(Lindl.) G. L. Webster*, Ann. Missouri Bot. Gard. 81: 49 (1994). T.: *Aporosa* Blume.

Ordo *Scepaceae* Lindl., Intr. Nat. Syst. Bot. ed. 2, 171 (1836). T.: *Scepa* Lindl.
Ordo *Aporuseae* Lindl. ex Miq., Fl. Ned. Ind. 1(2): 430 (1858).
Tribe *Aporuseae* (Lindl. ex Miq.) Airy Shaw, Hooker's Icon. Pl. 38: t. 3701 (1974); G. L. Webster, Taxon 24: 594 (1975).

Dioecious or occasionally monoecious trees or shrubs; stems without resinous secretion; leaves often glandular; stomata mostly anisocytic; stipules sometimes persistent. Inflorescences axillary, sometimes cauliflorous, thyrsiform, racemose, spicate or amentiform; male bracts often several-flowered; sepals 3–5(6), free or connate; petals 0; disc-glands free, alternisepalous, rarely 0; stamens 2–8, episepalous; filaments free, rarely anthers sessile; connectives usually not enlarged; thecae often contiguous and not pendulous; pollen prolate to spheroidal, 3-colporate, colpi transversales parallel, open-ended, elongate or not, sometimes reticulate; pistillode usually +; ovary (1)3(5)-locular, ovules anatropous; styles 2-partite, 2-fid, 2-lobate or entire. Fruit regularly or irregularly dehiscent, sometimes drupaceous. Seeds (where known) 1 or 2 per locule or fruit; exotesta often fleshy; endosperm thin or copious; cotyledons mostly broader than but scarcely exceeding the radicle, green or white.

This taxon, although early designated by Lindley (1836) as an "ordo", was confounded by Müller (1866) and later authors with the *Antidesminae*, and was first recognized in the 20th century by Airy Shaw (1974) at the tribal rank. Since Lindley published the *Scepaceae* over 20 years before the Ordo *Aporuseae* was proposed, it seems preferable to preserve his name. The 8 genera of *Scepinae* are scattered through the tropics in America, Africa, and Asia, with only *Baccaurea* reaching the Pacific islands.

Key to the genera

1. Bracts and bracteoles in 3 series, the middle cupular; female disc lobate; styles 2-lobate or 2-fid · 39. *Protomegabaria*
1. Bracts uniform, or if in 2 series, never cupular; female disc cupular or 0 · · 2
2. Male disc 0, or, if +, exceedingly minute; dioecious · · · · · · · · · · · · · · 3
2. Male disc usually + · 4
3. Pistillode relatively large, peltate; stamens 5–6; ovary (3)4(5)-locular; styles stigmatiform · 45. *Ashtonia*
3. Pistillode minute or 0; stamens 2–3(5); ovary 2–3(4)-locular; styles usually 2-lobate, 2-fid, or 2-partite · 46. *Aporosa*
4. Female disc 0; female sepals caducous · 5
4. Female disc +; female sepals persistent in fruit · · · · · · · · · · · · · · · · · · 6
5. Dioecious; styles 2-fid; cotyledons flat · · · · · · · · · · · · · · · · · 43. *Baccaurea*
5. Monoecious; styles elobate; cotyledons thick · · · · · · · · · · · · 44. *Apodiscus*
6. Monoecious; flowers in axillary glomerules; fruit irregularly dehiscent · 42. *Jablonskia*
6. Dioecious; flowers racemose · 7
7. Fruit loculicidal; male flowers mostly 1 per bract · · · · · · · 40. *Maesobotrya*
7. Fruit septicidal; male flowers several per bract · · · · · · · · · · · · 41. *Richeria*

39. **Protomegabaria** *Hutch.*, Hooker's Icon. Pl. 30: t. 2929 (1911) and Fl. Trop. Afr. 6(1): 656 (1912); Pax & K. Hoffm., Pflanzenr. 81: 43 (1922), and Nat. Pflanzenfam. ed. 2, 19C: 49 (1931); Keay (ed.), Fl. W. Trop. Afr. ed. 2, 1: 373 (1958); Aubrév., Fl. For. Côte d'Ivoire ed. 2, 2: 68, t. 147 (1959); J.-G. Adam, Mém. Mus. Natl. Hist. Nat., Sér. B, Bot. 20: 507 (1971); G. L. Webster, Ann. Missouri Bot. Gard. 81: 50 (1994); J. Léonard, Bull. Jard. Bot. Belg. 64: 53 (Jun. 1995), and Fl. Afr. Cent. Euph. 2: 79 (Nov. 1995). LT.: *P. stapfiana* (Beille) Hutch. (B.: *Maesobotrya stapfiana* Beille) (chosen by Webster, 1994).

Dioecious, semipachycaul trees; trunk buttressed or stilt-rooted. Indumentum, when present, simple. Leaves alternate, petiolate, stipulate, simple, large, narrowed to the base, entire, penninerved; stipules very fugacious. Inflorescences axillary or extra-axillary, racemose or subspicate, solitary, geminate, ternate or fasciculate; flowers subtended by bracts and bracteoles in 3 series: outer bract ovate, middle bract cupular and enclosing 3(4) flowers in bud, inner bracteoles 4, obovate. Male flowers pedicellate, pedicels articulate, soon caducous; sepals (4)5(6), imbricate, cucullate, later reflexed; petals 0; disc-glands (4)5(6), large, fleshy, free or ± contiguous, alternisepalous; stamens (4)5, oppositisepalous, filaments free, anthers basifixed, introrse, thecae parallel, contiguous, longitudinally dehiscent; pollen ± as in *Spondianthus*; pistillode columnar, entire or bipartite, fleshy. Female flowers pedicellate, pedicels articulate; calyx persistent, otherwise ± as in the male; petals 0; disc annular, shallowly lobate, adnate to the sepals, lobes oppositisepalous; ovary 3(4)-locular, ovules 2 per locule, collateral, obturator large; styles 3(4), 2-lobate or 2-fid, erect at first, later spreading, recurved, persistent. Fruit large, subglobose-ellipsoid to rounded-3-gonous, 6(8)-ribbed, septicidally dehiscent into 3(4) 2-valved cocci; exocarp thin; endocarp woody; columella often persistent. Seeds 1–2 per locule, small, funiculate, ellipsoid; testa thin, smooth, shiny. Fig.: See Hooker's Icon. Pl. 30: t. 2929 (1911); Mém. Mus. Natl. Hist. Nat., Sér. B, Bot. 20: 508–9, tt. 189, 190 (1971); Bull. Jard. Bot. Belg. 64: 59, t. 1 (Jun. 1995); Fl. Afr. Cent. Euph. 2: 81, t. 13 (Nov. 1995).

A tropical African genus with 3 species, 2 in W. Africa and 1 in Congo.

40. **Maesobotrya** *Benth.*, Hooker's Icon. Pl. 13: 75, t. 1296 (1879); Benth. & Hook. f., Gen. Pl. 3: 284 (1880); Hutch., Fl. Trop. Afr. 6(1): 663 (1912); Pax & K. Hoffm., Pflanzenr. 81: 17 (1922), and Nat. Pflanzenfam. ed. 2, 19C: 46 (1931); J. Léonard, Bull. Jard. Bot. État. 17: 256 (1945); Keay (ed.), Fl. W. Trop. Afr., ed. 2, 1: 373 (1958); Aubrév., Fl. For. Côte d'Ivoire ed. 2, 2: 74 (1959); J.-G. Adam, Mém. Mus. Natl. Hist. Nat., Sér. B, Bot. 20: 493 (1971); Radcl.-Sm. in Fl. Trop. E. Afr., Euph. 1: 112 (1987); G. L. Webster, Ann. Missouri Bot. Gard. 81: 50 (1994); J. Léonard, Bull. Jard. Bot. Belg. 63: 4 (1994), and Fl. Afr. Cent. Euph. 2: 46 (Nov. 1995); Radcl.-Sm. in Fl. Zamb. 9(4): 103 (1996). T.: *M. floribunda* Benth.

Staphysora Pierre, Bull. Mens. Soc. Linn. Paris 1233 (1896); Pax in Engl., Bot. Jahrb. Syst. 23: 521 (1897). T.: *S. dusenii* Pax [=*M. dusenii* (Pax) Hutch.]

Dioecious shrubs or small trees. Indumentum simple. Leaves alternate, often long-petiolate, stipulate, simple, entire or toothed, penninerved; petioles bipulvinate; stipules minute and deciduous or foliaceous and persistent. Inflorescences axillary or cauliflorous, solitary or fasciculate, subspicate or racemose; bracts mostly 1-flowered. Male flowers shortly pedicellate; calyx (4)5-lobate, lobes imbricate; petals 0; disc-glands (4)5, alternistemonous, fleshy, contiguous; stamens (4)5(6), oppositisepalous, filaments free, anthers erect, dorsifixed, introrse, thecae ovoid, parallel, longitudinally dehiscent, connective small; pollen smaller than, but otherwise resembling that of *Spondianthus* and *Protomegabaria*; pistillode thick, cylindric, elobate. Female flowers shortly pedicellate; calyx as in the male; petals 0; disc hypogynous, cupular, entire; ovary (1)2(4)-locular, ovules 2 per locule; styles short, stigmas 2-lobate to 2-fid, slightly recurved. Fruit subglobose or ellipsoid, subdrupaceous, tardily loculicidally dehiscent; exocarp membranous to thinly coriaceous; endocarp 1-locular by supression. Seeds solitary by abortion, ellipsoid, ecarunculate; testa thin; endosperm copious; cotyledons broad, flat, green. Fig.: See Hooker's Icon. Pl. 13: 75, t. 1296 (1879); Pflanzenr. 81: 22, t. 4 (1922); Nat. Pflanzenfam. ed. 2, 19C: 47, t. 18 (1931); Mém. Mus. Natl. Hist. Nat., Sér. B, Bot. 20: 494, t. 181 (1971); Fl. Trop. E. Afr., Euph. 1: 113, t. 20 (1987); Bull. Jard. Bot. Belg. 63: 7–53, tt. 1–7 (1994); Fl. Afr. Cent. Euph. 2: 55–75, tt. 7–12 (Nov. 1995); Fl. Zamb. 9(4): 104, t. 15 (1996).

An African genus of c. 20 species, mostly in W. and C. Africa, but with 1 in Uganda and another extending S. to Zambia.

Maesobotrya is extremely close to the neotropical *Richeria.*

41. **Richeria** *Vahl,* Eclog. Amer. 1: 30, t. 4 (1797); Baill., Étude Euphorb.: 597 (1858); Müll. Arg. in DC., Prodr. 15(2): 466 (1866) and in Mart., Fl. Bras. 11(2): 13 (1873); Benth. & Hook. f., Gen. Pl. 3: 286 (1880); Pax & K. Hoffm., Pflanzenr. 81:26 (1922), and Nat. Pflanzenfam. ed. 2, 19C: 47 (1931), p.p.; Jabl., Mem. New York Bot. Gard. 17: 124 (1967); Philcox, Fl. Trin. Tob. 2(10): 638 (1979); R. A. Howard, Fl. Lesser Antilles 5: 82 (1989); G. L. Webster & Huft, Ann. Missouri Bot. Gard. 75: 1093 (1989); Secco & G. L. Webster, Bol. Mus. Paraense Emilio Goeldi, N.S., Bot. 6: 143 (1990); A. H. Gentry, Woody Pl. NW. S. Amer.: 420 (1993); G. L. Webster, Ann. Missouri Bot. Gard. 81: 50 (1994). T.: *R. grandis* Vahl.

Guarania Wedd. ex Baill., Étude Euphorb.: 598 (1858). LT.: *G. gardneriana* Baill. [=*R. gardneriana* (Baill.) Baill.] (chosen by Webster, 1994).

Dioecious trees or shrubs. Indumentum simple. Leaves alternate, large, petiolate, stipulate, simple, entire or ± so, ± papillose beneath, penninerved, sometimes with basal glands. Stipules readily deciduous. Inflorescences interruptedly racemose or spicate, axillary or lateral below the leaves; males slender, bracts many-flowered, flowers glomerulate; females short, dense, bracts 1-flowered. Male flowers very small; sepals 3–5, imbricate; petals 0; disc-glands 3–5, separate, protruding between the filaments; stamens 3–6, episepalous; filaments free, exsert; anthers ± versatile, introrse, small, oblong, thecae longitudinally dehiscent; pollen prolate, 3-colporate, colpus transversalis narrow, elongate, semitectate, coarsely reticulate; pistillode short, often villous. Female flowers shortly pedicellate; calyx as in the male; disc annular or shallowly cupulate, sometimes slightly lobulate; ovary 3-locular, ovules 2 per locule, anatropous; styles short, spreading, thick, broad, entire or 2-fid. Fruit ovoid to ellipsoid, somewhat fleshy, tardily dehiscing septicidally into (2)3 2-valved cocci; mesocarp corky; columella slender, apically dilated, with papery wings. Seeds usually 1 per locule or 1 per fruit by suppression and abortion, ecarunculate; exotesta fleshy; endotesta woody; endosperm fleshy; embryo straight; cotyledons broad, flat, basally cordate. Fig.: See Nat. Pflanzenfam. ed. 2, 19C: 48, t. 19A–C (1931); Fl. Lesser Antilles 5: 84, t. 28 (1989); Bol. Mus. Paraense Emilio Goeldi, N.S., Bot. 6: 156–158, tt. 2–4 (1990); Woody Pl. NW. S. Amer.: 416, t. 119.2 (1993).

A neotropical genus formerly considered to comprise 5–6 closely-related species, reduced to 2 by Secco & Webster (1990), which extends in lowland and montane tropical rain forest from Costa Rica and the Lesser Antilles to Peru and Brazil.

42. **Jablonskia** *G. L. Webster*, Syst. Bot. 9: 232 (1984), and Ann. Missouri Bot. Gard. 81: 50 (1994); A. H. Gentry, Woody Pl. NW. S. Amer.: 417 (1993); Murillo-Aldana & Franco-Rosselli, Euf. Reg. Ararac.: 87 (1995). T.: *J. congesta* (Benth. ex Müll. Arg) G. L. Webster (B.: *Phyllanthus congestus* Benth. ex Müll. Arg.; S.: *Securinega congesta* (Benth. ex Müll. Arg.) Müll. Arg.).

Monoecious glabrous trees or shrubs. Leaves alternate, distichous, shortly petiolate, stipulate, simple, entire, penninerved, with small laminar glands near the base; stipules elongate, deciduous. Flowers axillary, glomerulate; bracts fairly conspicuous. Male flowers sessile or subsessile; sepals 5, imbricate; petals 0; disc-glands 5, free, alternisepalous; stamens 5; filaments free, equalling or exceeding the sepals; anthers versatile, introrse, thecae parallel, contiguous, longitudinally dehiscent; pollen prolate, 3-colporate, colpus transversalis narrow, elongate, exine tectate-perforate; pistillode cylindric, shortly lobate, striate. Female flowers sessile or subsessile; sepals recurved at the apex, persistent, otherwise as in the male; petals 0; disc patelliform; ovary 3-locular, ovules 2 per locule, anatropous; styles free, erect, 2-fid. Fruit thin-walled, baccate, irregularly dehiscent; columella subpersistent. Seeds 2 per locule, ecarunculate; exotesta thin, fleshy, smooth; endotesta crustaceous; endosperm copious; embryo not green; cotyledons broader than but ± as long as the radicle. Fig.: See Fl. Bras. 11(2): t. 11.I (1873) (as *Securinega*); Nat. Pflanzenfam. ed. 2, 19C: 60, t. 27C (1931) (as *Securinega*); Syst. Bot. 9: 231–2, tt. 7–10 (1984); Woody Pl. NW. S. Amer.: 416, t. 119.4 (1993); Euf. Reg. Ararac.: 89, t. 20 (1995).

Monotypic, the single species confined to Amazonian South America.

The genus is isolated in subtribe *Scepinae* on account of its pollen-type and paracytic stomata, although it shows some similarities to *Celianella* Jabl. in the next subtribe, *fide* Webster (1984b).

43. **Baccaurea** *Lour.*, Fl. Cochinch. 661 (1790); Müll. Arg. in DC., Prodr. 15(2): 456 (1866); Benth. & Hook. f., Gen. Pl. 3: 283 (1880); Hook. f., Fl. Brit. India 5: 367 (1887); Pax & K. Hoffm., Pflanzenr. 81: 45 (1922); Gagnep., Fl. Indo-Chine 5: 547 (1927); Pax & K. Hoffm., Nat. Pflanzenfam. ed. 2, 19C: 50 (1931); Airy Shaw, Kew Bull. 14: 353 (1960), 26: 219 (1972), Kew Bull. Addit. Ser. 4: 43 (1975), 8: 34 (1980), Kew Bull. 36: 258 (1981), 37: 8 (1982) and Euph. Philipp.: 9 (1983); Backer & Bakh. f., Fl. Java 1: 453 (1963); Whitmore, Tree Fl. Mal. 2: 63 (1973); A. C. Sm., Fl. Vit. Nov. 2: 450 (1981); Thin, Công Trình Nghiên Cú'u Khoa Hoc 10: 78 (1986); Grierson & D. G. Long, Fl. Bhut. 1: 788 (1987); G. L. Webster, Ann. Missouri Bot. Gard. 81: 51 (1994); Haegens, Blumea Suppl. 12: 80 (2000). LT.: *B. ramiflora* Lour. (chosen by Merrill, 1935).

Pierardia Roxb. ex Jack, Trans. Linn. Soc. London 14: 119 (1823); Baill., Étude Euphorb.: 557 (1858). T.: *P. dulcis* Jack [=*B. dulcis* (Jack) Müll. Arg.]
Adenocrepis Blume, Bijdr. 579 (1825); Baill., Étude Euphorb.: 600 (1858). T.: *A. javanica* Blume [=*B. javanica* (Blume) Müll. Arg.]
Calyptroon Miq., Fl. Ned. Ind., Erste Bijv. 471 (1861). T.: *C. sumatranum* Miq. [=*B. sumatrana* (Miq.) Müll. Arg.]
Microsepala Miq., Fl. Ned. Ind., Erste Bijv. 444 (1861). T.: *M. acuminata* Miq. [=*B. javanica* (Blume) Müll. Arg. (B.: *Adenocrepis javanica* Blume)]
Everettiodendron Merr., Philipp. J. Sci. 4: 279 (1909). T.: *E. philippinense* Merr. [=*B. philippinensis* (Merr.) Merr.]
Gatnaia Gagnep., Bull. Soc. Bot. France 71: 870 (1924), and Fl. Indo-Chine 5: 540 (1927). T.: *G. annamica* Gagnep. [=*B. oxycarpa* Gagnep.]

Dioecious trees or shrubs. Indumentum simple, branched or minutely stellate. Leaves alternate, large, often long-petiolate, stipulate, usually crowded at the branch-apices, entire or subentire, penninerved; petioles apically and basally pulvinate; stipules subulate, readily deciduous. Inflorescences axillary or cauliflorous, often fasciculate, pendulous or erect; males narrowly thyrsiform-pseudoracemose, females racemose. Male flowers: sepals 4–5, free, imbricate, sometimes papillose; petals 0; disc-glands 4–8, mostly small, interstaminal, free or connate, sometimes 0; stamens 4–8; filaments free; anthers basifixed, introrse, lateral or extrorse, subquadrithecate, thecae contiguous or separated by a connective, longitudinally dehiscent; pollen subprolate, 3-colporate, colpus transversalis small, parallel, open-ended; pistillode squat, crowned with a peltate stigmatode. Female flowers: sepals larger than the male, vestite, caducous; petals 0; disc 0; ovary 2–5-locular, ovoid, ovules 2 per locule, obturator thick; styles very short or 0, stigmas 2–5, very small, 2-fid, papillose or plumose, reflexed. Fruit fleshy, indehiscent or dry, loculicidally dehiscent into 3 valves, sometimes pilose within. Seeds few or solitary, large; sarcotesta often +, brightly-coloured; endosperm fleshy; embryo very often curved; cotyledons broad. Fig.: See Pflanzenr. 81: 57, t. 7 (1922); Fl. Indo-Chine 5: 538, t. 66.12–14 and 541, t. 67.1–5 (as *Gatnaia*), 11–13, 545, t. 68.2–3 and 556, t. 69.1(1927); Nat. Pflanzenfam. ed. 2, 19C: 51, t. 21 (1931); Kew Bull. 36: 262, t. 2 (1981); Fl. Bhut. 1: 789, t. 48.k–m (1987); Blumea Suppl. 12: 92–189, tt. 1–27 & pl. 1–12 (2000).

A genus of 43 species (Haegens, 2000) extending from India to the Philippines, New Guinea & the SW Pacific Islands.

Haegens (2000) has segregated two new genera from *Baccaurea*: *Distichirhops* (sic — should be *Distichorrhops*) with 3 species from Borneo and New Guinea, and *Nothobaccaurea* with 2 species from the SW Pacific islands.

44. **Apodiscus** *Hutch.*, Bull. Soc. Bot. France 58, Mém. 8: 205 (1912), Fl. Trop. Afr. 6(1): 1045 (1913) and Hooker's Icon. Pl. 31: t. 3032 (1915); Pax & K. Hoffm., Pflanzenr. 81: 45 (1922), and Nat. Pflanzenfam. ed. 2, 19C: 50 (1931); Keay, Fl. W. Trop. Afr., ed. 2, 1: 373 (1958); G. L. Webster, Ann. Missouri Bot. Gard. 81: 51 (1994). T.: *A. chevalieri* Hutch.

Small monoecious tree. Indumentum simple, confined to the inflorescences. Leaves alternate, shortly petiolate, stipulate, large, coriaceous, entire, penninerved; stipules lanceolate, deciduous. Inflorescences spicate, axillary, binate or fasciculate, male sometimes with a solitary female towards the base; bracts small, 1-flowered. Male flowers: sepals 5, imbricate; petals 0; disc-glands 5, alternisepalous; stamens 5, oppositisepalous; filaments free; anthers introrse, thecae subglobose, distinct, diverging from the apex, pendulous, longitudinally dehiscent, connective slightly produced; pollen prolate, 3-colporate, colpus transversalis elongate, parallel, open-ended; pistillode depressed-globose. Female flowers pedicellate; sepals ± as in the male; disc 0; ovary (3)4(5)-locular, ovules 2 per locule; styles (3)4(5), short, thick, incurved, elobate. Fruit strongly (3)4(5)-lobate, the lobes humerate, somewhat inflated, septicidally dehiscent into 3 cocci; exocarp thin; endocarp crustaceous. Seeds 2 per locule, spheroidal, venose; embryo large; cotyledons hemispherical, endospermic. Fig.: See Hooker's Icon. Pl. 31: t. 3032 (1915).

An imperfectly-known monotypic West African genus from Guinée, Liberia and Sierra Leone.

Pax and Hoffmann (1922) consider it to be related to *Maesobotrya*, but it has quite a different fruit. Stuppy (1995) points out that the storage embryo distinguishes *Apodiscus* from all other *Antidesmeae*.

45. **Ashtonia** *Airy Shaw* , Kew Bull. 21: 357 (1968); Whitmore, Tree Fl. Malaya 2: 62 (1973); Airy Shaw, Kew Bull. 27: 4 (1972), Hooker's Icon. Pl. 38: t. 3702 (1974), and Kew Bull. Addit. Ser. 4: 42 (1975); G. L. Webster, Ann. Missouri Bot. Gard. 81: 51 (1994). T.: *A. excelsa* Airy Shaw.

Dioecious, glabrous large trees. Leaves alternate, often long-petiolate, stipulate, simple, entire, penninerved, ± aggregated towards the branch-apices, biglandular-pitted at the base, drying yellow-green; petiole slender, apically pulvinate; stipules peruliform, soon deciduous. Male inflorescences axillary among and below the leaves, solitary, interruptedly spicate; flowers 3 or more in small sessile glomerules; bracts and bracteoles small, numerous, dense; females laxly racemose, few-flowered. Male flowers sessile; sepals 3–4, connate at the base, strongly imbricate; petals 0; disc 0 or very minute; stamens 5–6, sessile or filaments short, free; anthers basifixed, didymous, latrorse, thecae subglobose, discrete, longitudinally dehiscent; pollen spheroidal, 3-colporate, colpi subpolar, lumina coarsely reticulate; pistillode relatively large with a large pileiform stigmatode. Female flowers shortly pedicellate; sepals 4, imbricate, caducous; petals 0; disc 0; ovary 3–5 locular, ovules 2 per locule; styles very short or 0; stigmas 3–5, short, 2-lobate, recurved, coarsely papillose. Fruit long-pedicellate, globose, (3)4(5)-locular, loculicidally dehiscent into (3)4(5) valves; exocarp thick, somewhat fleshy, separating; endocarp woody or corneous; columella persistent, 4-gonous. Seeds obovoid, minutely carunculate; sarcotesta coarsely reticulate, orange, greyish when dry. Fig.: See Tree Fl. Malaya 2: 62, t. 1 (1973); Hooker's Icon. Pl. 38: t. 3702 (1974).

An Old World genus of 2 species in Borneo and the Malay peninsula.

It comes very close to *Aporosa, Baccaurea,* and *Richeria.*

46. **Aporosa** *Blume,* Bijdr. 514 (1826), and Fl. Javae 1: viii (1828); Baill., Étude Euphorb.: 643 (1858); Müll. Arg. in DC., Prodr. 15 (2): 469 (1866); Benth. & Hook. f., Gen. Pl. 3: 282 (1880); Hook. f., Hooker's Icon. Pl. 16: t. 1583 (1887), Fl. Brit. India 5: 345 (1887), and Hooker's Icon. Pl. 23: t. 2204 (1892); Pax & K. Hoffm., Pflanzenr. 81: 80 (1922); Gagnep., Fl. Indo-Chine 5: 552 (1927); Pax & K. Hoffm., Nat. Pflanzenfam. ed. 2, 19C: 52 (1931); Backer & Bakh.f., Fl. Java 1: 455 (1963); Airy Shaw, Kew Bull. 20: 380 (1966), 23: 2 (1969), 25: 474 (1971), 26: 213 (1972); Whitmore, Tree Fl. Malaya 2: 58 (1973); Airy Shaw, Hooker's Icon. Pl. 38: t. 3701 (1974), Kew Bull. Addit. Ser. 4: 30 (1975), 8: 28 (1980), Kew Bull. 36: 252 (1981), 37: 7 (1982), and Euph. Philipp.: 8 (1983); Thin, Công Trình Nghiên Cú'u Khoa Hoc 10: 82 (1986); Grierson & D. G. Long, Fl. Bhut. 1: 785 (1987); G. L. Webster, Ann. Missouri Bot. Gard. 81: 51 (1994); Schot, Blumea 40: 449 (1995); Philcox in Dassanayake, ed., Fl. Ceyl. XI: 269 (1997). T.: *A. frutescens* Blume.

Leiocarpus Blume, Bijdr.: 581 (1826); Baill., Étude Euphorb.: 655 (1858). LT.: *L. fruticosus* Blume [=*A. fruticosa* (Blume) Müll. Arg.] (chosen by Wheeler, 1975).

Lepidostachys Wall. ex Lindl., Intr. Nat. Syst. Bot. ed. 2: 441 (1836). T.: *L. roxburghii* Wall. ex Lindl. [nom. illeg.; =*A. dioica* (Roxb.) Müll. Arg. (B.: *Alnus dioica* Roxb.)]

Scepa Lindl., Intr. Nat. Syst. Bot. ed. 2: 441 (1836). T.: *S. stipulacea* Lindl. [=*A. dioica* (Roxb.) Müll. Arg. (B.: *Alnus dioica* Roxb.)]

Tetractinostigma Hassk., Hort. Bogor. Descr.: 55 (1858). T.: *T. microcalyx* Hassk. [=*A. microcalyx* (Hassk.) Hassk.]

Dioecious trees or shrubs. Leaves alternate, shortly petiolate, stipulate, simple, entire, penninerved, glanduliferous; petioles apically biglandular; stipules soon deciduous or persistent and often oblique. Male inflorescences axillary, often fasciculate, usually densely spicate, amentiform; bracts broad, concave, often subimbricate, subtending dense glomerules; flowers minute. Females shorter, fewer-flowered and laxer; bracts smaller, the lowest barren. Male flowers ± sessile; sepals 3–6, imbricate; petals 0; disc 0; stamens 2–5, filaments free, equalling or exceeding the sepals, anthers small, basifixed, extrorse, subglobose, 2-thecate, thecae parallel, distinct, longitudinally dehiscent; pollen ± as in *Baccaurea*; pistillode very minute or 0. Female flowers ± sessile; sepals 2–5, larger than male, persistent; petals 0; disc 0; ovary 2–3(4)-locular, ovules 2 per locule; styles short, 2-lobate, 2-fid or 2-partite, rarely undivided; stigmas papillose, fimbriate or laciniate. Fruit subdrupaceous, tardily irregularly dehiscent or sub-2–4-valved; exocarp thick or thin, spongy or fleshy; endocarp crustaceous, hard, or rarely papyraceous, septa often pilose. Seeds usually 1–2 per fruit by abortion, obovoid, ecarunculate; exotesta somewhat fleshy; endosperm fleshy; cotyledons broad, flat. Fig.: See Étude Euphorb.: t. 27, f. 23 (1858) (as *Scepa*); Hooker's Icon. Pl. 16: t. 1583 (1887); *op. cit.* 23: t. 2204 (1892); Pflanzenr. 81: 89, t. 10 (1922); Fl. Indo-Chine 5: 541, t. 67.14, 545, t.

68.4–12 and 556, t. 69.2–4 (1927); Nat. Pflanzenfam. ed. 2, 19C: 53, t. 23 (1931); Hooker's Icon. Pl. 38: t. 3701 (1974); Kew Bull. Addit. Ser. 8: 224, t. 2.1 (1980); Kew Bull. 36: 250, t. 1C,D (1981); Fl. Bhut. 1: 789, t. 48.e–g (1987).

A genus of c. 80 species of the Old World tropics, ranging from India and Sri Lanka to Indonesia and Melanesia (Solomon Is.), *fide* Schot (1995).

Subtribe 6d. **Antidesminae** *Müll. Arg.*, Linnaea 34: 64 (1865) (as *Antidesmeae*); Pax, Nat. Pflanzenfam. ed. 1, 3(5): 26 (1890); Pax & K. Hoffm., Pflanzenr. 81: 3 (1922); G. L. Webster, Ann. Missouri Bot. Gard. 81: 51 (1994). T.: *Antidesma* L.

Ordo *Stilaginae* C. Agardh, Aphor. Bot. 14: 199 (1824). T.: *Stilago* L. [=*Antidesma* L.]
Subtribe *Hieronymeae* Müll. Arg., Linnaea 34: 64 (1865). T.: *Hieronyma* Allemão.

Dioecious or rarely monoecious trees or shrubs, rarely herbs; leaves eglandular, stomata paracytic; stipules caducous. Inflorescences axillary, usually ± amentiform; male bracts mostly 1-flowered; sepals 3–6(8), free or connate; petals 0 or reduced; disc annular, lobate or dissected; stamens 3–6, free or nearly so, anthers with ± enlarged glandular connectives, thecae usually pendulous; pollen perprolate to prolate, 3-colporate, reticulate; pistillode +, rarely 0; ovary 1–3(5)-locular, ovules anatropous; styles bifid, rarely entire. Fruit dehiscent or drupaceous. Seeds mostly 1 per locule or fruit; testa dry; endosperm thin to copious; cotyledons broader than but scarcely exceeding the radicle.

The *Stilagineae* (*Stilaginaceae*), kept up by Airy Shaw in most of his publications, has been shown not to be separable from the *Euphorbiaceae* on the basis of evidence from palynology and foliar anatomy. Two genera in this subtribe, *Phyllanoa* and *Leptonema*, are not well understood, and their inclusion here is provisional.

Key to the genera

1. Indumentum lepidote; ovary 2(3)-locular; fruit indehiscent · · 52. *Hieronyma*
1. Indumentum 0 or of simple or stellate hairs · · · · · · · · · · · · · · · · · · 2
2. Ovary 1-locular; fruit indehiscent; endocarp reticulate, foveolate · 51. *Antidesma*
2. Ovary 2–3(5)-locular; fruit dehiscent; endocarp ± smooth · · · · · · · · · · · 3
3. Disc 0; ovary 4–5-locular · 50. *Leptonema*
3. Disc +; ovary (2)3-locular · 4
4. Pistillode 0; female sepals accrescent; female flowers lacking petals or staminodes · 49. *Celianella*
4. Pistillode (where known) +; female sepals not accrescent in fruit; female flower sometimes with petals and staminodes · · · · · · · · · · · · · · · · · · 5

5. Leaves entire; female flowers not bracteolate; fruits pedicellate · 47. *Thecacoris*
5. Leaves dentate; female flowers bracteolate; fruits subsessile · · 48. *Phyllanoa*

47. **Thecacoris** *A. Juss.*, Euphorb. Gen.: 12 (1824); Baill., Étude Euphorb.: 605 (1858); Benth. & Hook. f., Gen. Pl. 3: 286 (1880); Hutch. in Fl. Trop. Afr. 6(1): 658 (1912); Pax & K. Hoffm., Pflanzenr. 81: 8 (1922), and Nat. Pflanzenfam. ed. 2, 19C: 45 (1931); Keay (ed.), Fl. W. Trop. Afr. ed. 2, 1: 371 (1958); Leandri, Fl. Madag. 111(1): 4 (1958); Radcl.-Sm., Fl. Trop. E. Afr. Euph., 1: 107 (1987); G. L. Webster, Ann. Missouri Bot. Gard. 81: 52 (1994); J. Léonard, Bull. Jard. Bot. Belg. 64: 26 (1995), and Fl. Afr. Cent. Euph. 2: 82 (Nov. 1995); Radcl.-Sm., Fl. Zamb. 9(4): 110 (1996). T.: *Th. madagascariensis* A. Juss.

Cometia Thouars ex Baill., Étude Euphorb. 642 (1858) (male only). T.: *C. lucida* Baill. [=*Th. cometia* Leandri, non *Th. lucida* (Pax) Hutch. (B.: *Baccaureopsis lucida* Pax)]
Cyathogyne Müll. Arg., Flora 47: 536 (1864); Benth., Hooker's Icon. Pl. 13: t. 1278 (1879). T.: *C. viridis* Müll. Arg. [=*Th. viridis* (Müll. Arg.) Leandri]
Baccaureopsis Pax, Bot. Jahrb. Syst. 43: 319 (1909). T.: *B. lucida* Pax [=*Th. lucida* (Pax) Hutch.]

Dioecious or rarely monoecious trees, shrubs, subshrubs or perennial herbs. Indumentum simple. Leaves alternate or sometimes subfasciculate, shortly petiolate, stipulate, simple, entire or repand, penninerved. Inflorescences axillary, solitary, geminate or fasciculate, pedunculate, spicate, racemose or subpaniculate, few- to many-flowered; bracts small, 1-flowered. Male flowers shortly pedicellate; sepals 5(6), membranous, imbricate; petals 5, small, or 0; disc-glands 5, free, alternisepalous, interstaminal, thick, pubescent; stamens 5, oppositisepalous, filaments free, anthers apicifixed, introrse, later extrorse, thecae separate, subglobose, parallel and pendulous at first, later divaricate and erect, longitudinally dehiscent, connective thickened, apiculate; pollen perprolate, 3-colporate, colpi transversales long, parallel, open-ended; pistillode large, cylindric, obconic, turbinate or cupuliform, truncate or 3–5-lobate. Female flowers: pedicels patent, elongating and often geniculate in fruit; sepals ± as in male; petals 5, minute or 0; staminodes sometimes +; disc hypogynous, annular, sometimes crenulate; ovary 3-locular, ovules 2 per locule; styles 3, free or connate at the base, recurved, 2-fid, stigmas ± smooth. Fruit apically depressed, 3-dymous, dehiscent into 3 2-valved cocci; exocarp thin, separating; endocarp thinly woody; columella persistent. Seeds obovoid to pyriform-cypraeiform, subecarunculate; testa thin, striate to smooth, shiny; endosperm copious, fleshy; cotyledons greenish. Fig.: See Euphorb. Gen.: t. 1, f. 1.1–6 (1824); Hooker's Icon. Pl. 13: t. 1278 (1879) (as *Cyathogyne*); Pflanzenr. 81: 10, t. 1 and 42, t. 6 (1922), and Nat. Pflanzenfam. ed. 2, 19C: 45, t. 17 and 49, t. 20 (1931); Notul. Syst. (Paris) 6: 20, t. 1 (1937); Fl. Madag. 111(1): 5–15, tt. I–III.1–7 (1958); Fl. Trop. E. Afr. Euph., 1: 109, t. 18 and 111, t. 19 (1987); Bull. Jard. Bot. Belg. 64: 17–35, tt. 1–5 and 49, t. 7 (1995) (t. 7 as *Cyathogyne*); Fl. Afr. Cent. Euph. 2: 85–91, tt. 14–16 and 99, t. 17 (Nov. 1995) (t. 17 as *Cyathogyne*); Fl. Zamb. 9(4): 112, t. 17 (1996).

As delimited by Webster (1994), *Thecacoris* includes about 25 species of Africa and Madagascar.

The herbaceous species were segregated as a separate genus *Cyathogyne* by Müller (1866) and Pax & Hoffmann (1922, 1931), but Radcliffe-Smith (1987) stated that *Th. usambarensis* bridged the gap between the two generic concepts. Léonard (1995), however, does not agree, and again recognizes *Cyathogyne* as distinct. Stuppy (1995) considers the seeds of *Cyathogyne* to be more specialized than those of *Thecacoris* s. str., somewhat resembling those of *Andrachne* and *Phyllanthus*.

48. **Phyllanoa** *Croizat*, Caldasia 2: 123 (1943); G. L. Webster, Ann. Missouri Bot. Gard. 81: 52 (1994). T.: *P. colombiana* Croizat.

Dioecious tree. Indumentum simple, scabrid, adpressed. Leaves alternate, very rarely subopposite, shortly petiolate, stipulate, simple, crenate-dentate, penninerved. Male inflorescences and flowers not yet known. Female inflorescences racemose, fasciculate, lateral or cauliflorous. Female flowers bracteolate; calyx-lobes 5, persistent, toothed; petals 5, minute, alternisepalous; staminodes 5, petaloid, oppositisepalous; disc thick, hypogynous, undulate to 5–10-lobate; ovary 2–3-locular, ovules 2 per locule, borne halfway up the columella; style very short, stigmas forming a flat 6-lobate disc. Fruit obovoid, subsessile, dehiscent into 2–3 cocci; exocarp thin; endocarp forming imperfect dissepiments; columella short. Seed (immature) 1 per locule by abortion, ovoid, ecarunculate, apically deflexed. Fig.: See Caldasia 2: 125, t. A–G (1943).

A monotypic genus from the Colombian Andes, known only from the imperfect type specimen.

The floral details of the female flower, especially the staminodes, suggest a possible affinity with *Thecacoris*, although there is also a general resemblence to *Richeria*, in subtribe *Scepinae*.

49. **Celianella** *Jabl.*, Mem. New York Bot. Gard. 12(3): 176 (1965); G. L. Webster, Ann. Missouri Bot. Gard. 81: 52 (1994). T.: *C. montana* Jabl.

Dioecious glabrous shrub. Leaves alternate, sessile, stipulate, simple, entire, obovate, semisucculent, penninerved, the nerves numerous, close; stipules soon falling, leaving prominent scars. Male inflorescences subterminal, pseudo-racemosely narrowly paniculate; flowers in pedunculate bracteate glomerules, each one 4-more flowered, bracteolate; females axillary, 1–3-flowered. Male flowers: sepals 5, imbricate; petals 0; stamens 5, oppositisepalous; filaments free, anthers introrse, 2-thecate, thecae pendulous, longitudinally dehiscent; pollen prolate, 3-colporate, colpi elongate, finely reticulate; disc central, 5-lobate, the lobes protruding between the filaments; pistillode 0. Female flowers: sepals 5, large, reticulate-nerved, accrescent and reddish-purple in fruit; petals 0; disc annular; ovary 3-locular, ovules 2 per locule, pendulous beneath an obturator; styles 3, connate in the lower half, 2-lobate. Fruit dehiscing septicidally into 3 2-valved cocci; exocarp thin; endocarp thickly woody; columella persistent. Seeds elongated, fusiform, shiny, minutely caruncuIate; endosperm fleshy, copious; embryo straight, central; radicle superior; cotyledons much broader than the radicle, flat. Fig.: See Mem. New York Bot. Gard. 12(3): 177, t. 28 (1965).

A monotypic genus of the Venezuela highlands.

The plant is distinctive on account of the semisucculent leaves and accrescent female sepals. Levin (1986c) allies it with *Hieronyma* on the basis of foliar characters.

50. **Leptonema** *A. Juss.*, Euphorb. Gen.: 19 (1824); Baill., Étude Euphorb.: 609 (1858); Benth. & Hook. f., Gen. Pl. 3: 275 (1880); Pax & K. Hoffm., Nat. Pflanzenfam. ed. 2, 19C: 59 (1931); Leandri, Notul. Syst. (Paris) 6: 22 (1937), Mém. Inst. Sci. Madagascar, Sér. B, Biol. Vég. 8: 212 (1957) and Fl. Madag. 111(1): 12 (1958); G. L. Webster, Ann. Missouri Bot. Gard. 81: 52 (1994). T.: *L. venosum* (Poir.) A. Juss. (B.: *Acalypha venosa* Poir.)

Dioecious shrub. Indumentum simple. Leaves alternate, long-petiolate, stipulate, ovate-orbicular, cordate, entire, penninerved, softly hairy below. Male inflorescences axillary, umbellate, pedunculate; bracts 1-flowered; female flowers solitary. Male flowers: calyx 5-partite, segments imbricate; petals 0; disc 0; stamens 5, oppositisepalous, filaments free, anthers apicifixed, recurved or pendulous and extrorse in bud, becoming introrse at anthesis, thecae separate, longitudinally dehiscent, connective globose, glandular; pollen subprolate, colpi transversales large, costate, colpi narrow, ecostate, columellae imperceptible; pistillode 0. Female flowers long-pedicellate; calyx ± as in the male; petals 0; disc 0; ovary 4–5-locular, ovules 2 per locule, collateral, funicles distinct; styles 4–5, 2-partite. Fruit globose, depressed, splitting into 4–5 2-valved cocci. Seeds 2 per locule, small, ovoid, asperulous. Fig.: See Euphorb. Gen.: t. 4, f. 12.1–6 (1824); Fl. Madag. 111(1): 19, t. IV. 1–6 (1958).

A little-known genus of 2 species endemic to Madagascar.

51. **Antidesma** *L.*, Sp. Pl.: 1027 (1753) and Gen. Pl.: 451 (1754); Baill., Étude Euphorb.: 601 (1858); Müll. Arg. in DC., Prodr. 15(2): 247 (1866); Benth. & Hook. f., Gen. Pl. 3: 284 (1880); Hook. f., Fl. Brit. India 5: 354 (1887); Hutch. in Fl. Trop. Afr. 6(1): 642 (1912); Pax & K. Hoffm., Pflanzenr. 81: 107 (1922); Gagnep., Fl. Indo-Chine 5: 501 (1926); Pax & K. Hoffm., Nat. Pflanzenfam. ed. 2, 19C: 54 (1931); Hurus., J. Fac. Sci. Univ. Tokyo, Sect. 3, Bot. 6: 324 (1954); Keay (ed.), Fl. W. Trop. Afr. ed. 2, 1(2): 374 (1958); Leandri, Fl. Madag. 111(1): 14 (1958); Backer & Bakh.f., Fl. Java 1: 457 (1963); Airy Shaw, Kew Bull. 23: 277 (1969); J.-G. Adam, Mém. Mus. Natl. Hist. Nat., N.S. Bot. 20: 455 (1971); Airy Shaw, Kew Bull. 26: 351, 457 (1972), and 28: 269 (1973); Whitmore, Tree Fl. Malaya 2: 54 (1973); Berhaut, Fl. Ill. Sénég. 3: 383 (1975); Airy Shaw, Kew Bull. Addit. Ser. 4: 207 (1975), Kew Bull. 33: 15, 423 (1979), Kew Bull. Addit. Ser. 8: 208 (1980), Kew Bull. 35: 692 (1980), 36: 358, 635 (1981), 37: 5 (1982), and Euph. Philipp.: 4 (1983); Grierson & D. G. Long, Fl. Bhut. 1: 786 (1987); Radcl.-Sm. in Fl. Trop. E. Afr. Euph. 2: 572 (Apr. 1988); J.

Léonard, Bull. Jard. Bot. Belg. 58: 4 (Jun. 1988); G. L. Webster, Ann. Missouri Bot. Gard. 81: 52 (1994); J. Léonard, Fl. Afr. Cent. Euph. 2: 16 (Nov. 1995); Radcl.-Sm., Fl. Zamb. 9(4): 105 (1996); Philcox in Dassanayake, ed., Fl. Ceyl. XI: 275 (1997). T.: *A. alexiteria* L.

Bestram Adans., Fam. Pl. 354 (1763). T.: (none given).
Stilago L., Mant. Pl.: 16 (1767). T.: *S. bunius* L. [=*A. bunius* (L.) Spreng.]
Rhytis Lour., Fl. Cochinch.: 660 (1790). T.: *R. fruticosa* Lour. [=*A. fruticosum* (Lour.) Müll. Arg.]

Dioecious trees or shrubs. Indumentum simple. Leaves alternate, shortly petiolate, stipulate, simple, entire, penninerved, with acarodomatia; stipules entire and deciduous or laciniate and persistent. Inflorescences leaf-opposed, axillary, terminal or cauliflorous, shortly pedunculate, spicate or racemose, sometimes subpaniculate, solitary or paucifasciculate, dense-flowered; bracts small, 1-flowered; flowers very small. Male flowers sessile or subsessile; calyx ± cupular, deeply 3–5(8)-lobate or -partite, segments imbricate; petals 0; disc-glands variable, free or connate, extrastaminal or covering the receptacle; stamens 2–5(10), oppositisepalous, filaments free, exserted, anthers inflexed in bud, later erect, apicifixed, 2-lobate, thecae distinct, divergent, basally and longitudinally dehiscent, connective thick; pollen perprolate to prolate, 3-colporate, colpi transversales narrow, parallel, elongate, open; pistillode small, ± cylindric, or 0. Female flowers sessile or subsessile; calyx as in the male; petals 0; disc hypogynous, annular or cupular; ovary 1(2)-locular, ovules 2 per locule, pendulous; styles 2–3(5), very short, united at the base, terminal or lateral, usually 2-lobate, often persistent. Fruit drupaceous, 1-locular, small, often oblique, laterally compressed, indehiscent, often red or black; mesocarp fleshy; endocarp indurated, reticulate to foveolate. Seeds 1(2) per fruit by abortion, ecarunculate; endosperm fleshy, not copious; cotyledons broad, flat. Fig.: See Pflanzenr. 81: 112, t. 12 and 146, t. 13 (1922); Fl. Indo-Chine 5: 502, t. 64.11–19 and 511, t. 65.1–14 (1926); Nat. Pflanzenfam. ed. 2, 19C: 54–5, tt. 24–5 (1931); Formos. Trees: 330, t. 284 and 331, t. 285 (1936); J. Fac. Sci. Univ. Tokyo, Sect. 3, Bot. 6: 323–327, tt. 48–50 (1954); Fl. Madag. 111(1): 15, t. III.8–14 and 19, t. IV.7–19 (1958); Mém. Mus. Natl. Hist. Nat., N.S., Bot. 20: 458, t. 159 and 459, t. 160 (1971); Fl. Ill. Sénég. 3: 382, 386 (1975); Kew Bull. Addit. Ser. 8: 224, t. 2.4 (1980); Kew Bull. 36: 356, t. 12.C,D (1981); Bot. J. Linn. Soc. 94: 51, tt. 2,3 (1987); Fl. Bhut. 1: 789, t. 48.h–j (1987); Fl. Trop. E. Afr. Euph. 2: 575, t. 106 (1988); Bull. Jard. Bot. Belg. 58: 7–31, tt. 1–5 and 41, t. 7 (1988); Fl. Afr. Cent. Euph. 2: 21–37, tt. 2–5 (Nov. 1995); Fl. Zamb. 9(4): 109, t. 16 (1996).

A large Old World genus of c. 170 species, only 7 of which occur in Africa and 2–3 in Madagascar. The remainder of the species are Asiatic except for a few in Australia and the Pacific islands.

From time to time this genus has been considered as constituting a monotypic family, the *Stilaginaceae* (e.g. by C.A. Agardh, 1825, and by Airy Shaw in numerous publications), on account of the rather anomalous fruit characters. However, from the palynological, cytological, embryological, serological and anatomical standpoints, *Antidesma* is typical of the *Euphorbiaceae.*

52. **Hieronyma** *Allemão*, Pl. Novas Brasil: 22 (1848); Müll. Arg. in DC., Prodr. 15(2): 268 (1866), and in Mart., Fl. Bras. 11(2): 19 (1873); Benth. & Hook. f., Gen. Pl. 3: 284 (1880); Fawc. & Rendle, Fl. Jamaica 4: 270 (1920); Pax & K. Hoffm., Pflanzenr. 81: 31 (1922), and Nat. Pflanzenfam. ed. 2, 19C: 48 (1931); Alain, Fl. Cuba 3: 61 (1953); Jabl., Mem. New York Bot. Gard. 17: 122 (1967); G. L. Webster, Ann. Missouri Bot. Gard. 54: 231 (1968); Philcox, Fl. Trin. Tob. 2(10): 639 (1979); Alain, Fl. Español. 4: 162 (1986), and Descr. Fl. Puerto Rico 2: 401 (1988); G. L. Webster, Ann. Missouri Bot. Gard. 75: 1094 (1988); R. A. Howard, Fl. Lesser Antilles 5: 59 (1989); R. P. Franco, Bot. Jahrb. Syst. 111(3): 299 (1990); A. H. Gentry, Woody Pl. NW. S. Amer.: 411 (1993); G. L. Webster, Ann. Missouri Bot. Gard. 81: 52 (1994); Murillo-Aldana & Franco-Rosselli, Euf. Reg. Ararac.: 84 (1995). T.: *H. alchorneoides* Allemão.

Stilaginella Tul., Ann. Sci. Nat. Bot. III, 15: 240 (1851); Baill., Étude Euphorb.: 603 (1858). T.: *S. laxiflora* Tul. [=*H. laxiflora* (Tul.) Müll. Arg.]

Dioecious trees or shrubs. Indumentum thinly lepidote, or rarely simply ferrugineous-tomentose. Leaves alternate, spirally arranged, often large, petiolate, stipulate, simple, entire, penninerved; stipules usually small, cochleate, often soon deciduous. Inflorescences racemose or paniculate, pedunculate, axillary, the males large, the females smaller and simpler; bracts very small. Male flowers sessile or shortly pedicellate; calyx campanulate, lepidote, 4–6-dentate, the lobes scarcely imbricate; petals 0; disc annular or lobate, or disc-glands usually 5, free or connate at the base, alternisepalous, thick; stamens (3)4–6, oppositisepalous, outside or within the disc, filaments free, anthers apicifixed, extrorse and inflexed in bud, at length apparently basifixed, introrse and erect, thecae distinct, longitudinally dehiscent, connective thick; pollen perprolate, 3-colporate, colpi elongate, colpi transversales parallel, open, microreticulate or tectum perforate; pistillode cylindric-obconic, usually lepidote. Female flowers: pedicels & calyx as in the male; petals 0; disc annular; ovary 2(3)-locular, ovules 2 per locule; styles very short or 0, stigmas 2–3, 2-partite, reflexed. Fruits drupaceous, often small, 1–2-locular by suppression; exocarp fleshy, blackish; endocarp hard and bony or fibrous, irregularly rugulose. Seed often 1 per fruit by abortion, ecarunculate; endosperm fleshy; cotyledons broad and flat. Fig.: See Fl. Jamaica 4: 271, t. 88 (1920); Pflanzenr. 81: 35, t. 5 (1922); Fl. Cuba 3: 62, t. 11 (1953); Ann. Missouri Bot. Gard. 54: 232, t. 4 (1968); Fl. Español. 4: 165, t. 116-21 (1986); Descr. Fl. Puerto Rico 2: 402, t. 59-19 (1988); Fl. Lesser Antilles 5: 60, t. 22 (1989); Bot. Jahrb. Syst. 111(3): 301–341, tt. 1–16 (1990); Árvores Brasileiras: 104 (1992); Woody Pl. NW. S. Amer.: 412, t. 117.2 (1993); Euf. Reg. Ararac.: 86, t. 19 (1995).

A neotropical genus of 40 species ranging from S. Mexico & Cuba south to S. Brazil, *fide* Franco (1990); many of these are, however, very difficult to distinguish.

Subtribe 6e. **Porantherinae** *(Müll. Arg.) Eg. Köhler,* Grana Palynol. 6: 99 (1965); G. L. Webster, Ann. Missouri Bot. Gard. 81: 53 (1994). T.: *Poranthera* Rudge.

Tribe *Poranthereae* Müll. Arg., Bot. Zeitung (Berlin) 2: 324 (1864).

Monoecious or dioecious herbs or subshrubs; leaves sometimes ± ericoid. Flowers in dense terminal capitate racemes, often aggregated into hypercorymbs, or solitary; sepals (3)5; petals + or 0; stamens (3)5; disc +; anthers pendulous, opening by pores; pistillode small or 0; female disc annular; ovary 3-locular; ovules anatropous; styles 2-fid. Fruits dehiscent. Seeds 2 per locule; endosperm copious; cotyledons about as broad as radicle.

Two closely-related Australasian genera. Pax & Hoffmann (1931) followed Müller (1866) in placing the tribe *Poranthereae* in the group *"Stenolobeae"* on account of the narrow cotyledons. Köhler (1965) demonstrated the similarity of the pollen of the *Porantherinae* to that of the *Antidesmineae.* Stuppy (1995) considers that the seed-morphology supports a relationship with *Andrachne.*

KEY TO THE GENERA

1. Monoecious; leaves alternate or subopposite; flowers in dense racemes; petals +; pistillode 3-partite · 53. *Poranthera*
1. Dioecious; leaves opposite; flowers solitary; petals 0; pistillode minute or 0 · 54. *Oreoporanthera*

53. **Poranthera** *Rudge,* Trans. Linn. Soc. London 10: 302 (1811); Baill., Étude Euphorb.: 573 (1858); Benth., Fl. Austral. 6: 54 (1873); Benth. & Hook. f., Gen. Pl. 3: 262 (1880); Grüning, Pflanzenr. 58: 13 (1913); Pax & K. Hoffm., Nat. Pflanzenfam. ed. 2, 19C: 224 (1931); G. L. Webster, Ann. Missouri Bot. Gard. 81: 53 (1994). T.: *P. ericifolia* Rudge.

Monoecious annual herbs or much-branched shrublets. Leaves alternate or subopposite, ± sessile, stipulate, simple, entire, small, often narrow, flat or revolute; stipules small, acuminate or laciniate. Inflorescences densely racemose; racemes capitate, corymbiform or umbelliform, solitary and terminal or more often aggregated into a leafy hypercorymb; bracts 1-flowered; male flowers numerous, females few at the base of the racemes. Male flowers pedicellate; calyx-lobes (3)5, petaloid, imbricate; petals (3)5, small; disc-glands (3)5, free, oppositisepalous, 2-fid; stamens (3)5, oppositisepalous, filaments free, anthers 4-thecate, thecae opening by separate pores at the apex or by two, confluent in pairs; pollen subprolate, 3-colporate, colpus transversalis broadly elliptic, tectate, perforate or sexine reticulate; pistillode 3-partite, the segments clavate, membranous. Female flowers pedicellate; calyx and petals as in the males; disc hypogynous, annular, 10-lobate; ovary depressed-globose, 6-lobate, 3-locular, ovules 2 per locule; styles 3, 2-partite or 2-crurate, the segments subulate. Fruit depressed-globose, rugulose, loculicidally dehiscent into 3 valves or septicidally into 3 2-valved cocci or 6 valves; exocarp thin, reticulate-rugulose when dry; endocarp thinly woody; columella small, persistent. Seeds 6, 3-gonous, reticulate, foveolate, rugulose, byssaceous or ± smooth, often whitish; endosperm copious; embryo slender, terete, incurved; cotyledons semicylindric, not broader than the radicle. Fig.: See Étude Euphorb.: t. 25, f. 1–9 (1858) Pflanzenr. 58: 17, t. 4 (1913); Nat. Pflanzenfam. ed. 2, 19C: 225, t. 121 (1931).

A genus of 8 species in Australia, Tasmania and New Zealand.

54. **Oreoporanthera** *(Grüning) Hutch.*, Amer. J. Bot. 56: 747 (1969); G. L. Webster, Ann. Missouri Bot. Gard. 81: 53 (1994). T.: *O. alpina* (Cheeseman) Hutch. (B.: *P. alpina* Cheesman).

Poranthera sect. *Oreoporanthera* Grüning, Pflanzenr. 58: 21 (1913).

Very small dioecious shrublet, much branched from the base; branches densely leafy. Leaves opposite, sessile, stipulate, simple, entire, small, narrow, revolute; stipules interfoliar, obtuse, entire, ± coriaceous. Flowers minute, solitary in the upper leaf-axils. Male flowers shortly pedicellate; sepals 5, imbricate; petals 0; disc-glands 5, alternisepalous, small, circular, flat; stamens 5, oppositisepalous; filaments free; anthers 4-thecate, opening by pores; pollen not described; pistillode minute or 0. Female flowers: pedicels, calyx and disc ± as in the males; petals 0; ovary 3-locular, ovules 2 per locule; styles 3, free, 2-lobate or 2-fid. Fruit subglobose, 3-lobate. Seeds not known. Fig.: See Hooker's Icon. Pl. 14: t. 1366B (1881) (as *Poranthera*).

A monotypic genus endemic to New Zealand.

Although Hutchinson (1969) described *Oreoporanthera* as a new genus, it is clear that he was raising Grüning's section of *Poranthera* to generic rank.

Tribe 7. **HYMENOCARDIEAE** *(Müll. Arg.) Hutch.*, Amer. J. Bot. 56: 746 (1969); G. L. Webster, Ann. Missouri Bot. Gard. 81: 53 (1994). T.: *Hymenocardia* Wall. ex Lindl.

Phyllantheae subtribe *Hymenocardieae* Müll. Arg., Linnaea 34: 64 (1865).
Hymenocardiaceae Airy Shaw, Kew Bull. 18: 261 (1965); J. Léonard & Mosango, Fl. Afr. Cent., *Hymenocardiaceae*: 1 (1985).

Dioecious trees or shrubs; indumentum of simple hairs and small sessile glands; leaves alternate, stipulate, entire, penninerved, granulose-glandular. Inflorescences terminal or axillary, amentiform, spicate, racemose or paniculate; male sepals 4–6, free or connate; petals 0; disc 0; stamens 4–6, filaments free or basally connate; anthers longitudinally dehiscent; pollen (where known) zonoporate; pistillode +; female flowers subsessile or pedicellate; sepals 4–8, free or connate, sometimes caducous; ovary 2-locular; styles 2, short or elongated, entire or lacerate, free or connate at the base; ovules anatropous, 2 per locule. Fruit winged, samaroid or scrotiform. Seeds 1(2) per locule; endosperm scanty; cotyledons broader and longer than the radicle.

A "tribe" of two rather diverse genera, possibly not really closely related. Although the type-genus has been considered to constitute a distinct family by Airy Shaw (1965), Léonard & Mosango (1985) and Meeuse (1990), most anatomical and floral details indicate that it is better retained within the *Euphorbiaceae*. The genus *Didymocistus* could perhaps be placed in a separate tribe or subtribe.

Key to the genera

1. Fruit samaroid or cymbiform; inflorescence racemose, spicate, subpaniculate or amentiform; female sepals caducous; stigmas smooth or papillose · 55. *Hymenocardia*
1. Fruit doubly scrotiform; inflorescence paniculate; female calyx persistent; stigmas fimbriate-lacerate · 56. *Didymocistus*

55. **Hymenocardia** *Wall. ex Lindl.*, Intr. Nat. Syst. Bot. ed. 2: 441 (1836); Baill., Étude Euphorb.: 599 (1858); Müll. Arg. in DC., Prodr. 15(2): 476 (1866); Oliv., Hooker's Icon. Pl. 12: 29, t. 1131 (1873); Benth. & Hook. f., Gen. Pl. 3: 284 (1880); Hook. f., Fl. Brit. India 5: 376 (1887); Hutch. in Fl. Trop. Afr. 6(1): 648 (1912); Pax & K. Hoffm., Pflanzenr. 81: 72 (1922); Gagnep., Fl. Indo-Chine 5: 543 (1927); Pax & K. Hoffm., Nat. Pflanzenfam. ed. 2, 19C: 51 (1931); Keay (ed.), Fl. W. Trop. Afr. ed. 2, 1(2): 375 (1958); J.-G. Adam, Mém. Mus. Natl. Hist. Nat., N.S. Bot. 20: 481 (1971); Airy Shaw, Kew Bull. 26: 363 (1972); Whitmore, Tree Fl. Mal. 2: 103 (1973); Berhaut, Fl. Ill. Sénég. 3: 493 (1975); J. Léonard & Mosango, Fl. Afr. Cent., Hym.: 2 (1985); Radcl.-Sm. in Fl. Trop. E. Afr., Euph. 2: 577 (1988); G. L. Webster, Ann. Missouri Bot. Gard. 81: 53 (1994); Radcl.-Sm., Fl. Zamb. 9(4): 113 (1996). T.: *H. punctata* Wall. ex Lindl.

Samaropyxis Miq., Fl. Ned. Ind., Eerste Bijv.: 464 (1860). T.: *S. elliptica* Miq. [=*H. punctata* Wall. ex Lindl.]

Dioecious deciduous trees or shrubs. Indumentum simple and glandular. Leaves alternate or rarely opposite, petiolate, stipulate, simple, entire, penninerved or subtriplinerved, glandular-punctate beneath. Male inflorescences axillary, densely spicate or subpaniculate, amentiform, precocious; females terminal, racemose, lax, few-flowered or flowers axillary, solitary, coaëtaneous. Male flowers sessile; calyx cupular, (4)5(6)-lobate, the lobes short, imbricate or almost valvate; petals 0; disc 0; stamens (4)5(6), episepalous, filaments short, spreading, free or connate at the base, anthers large, dorsifixed, reflexed and introrse in bud, becoming horizontal and extrorse at anthesis, thecae distinct, parallel, longitudinally dehiscent, opening until almost flat, connective with a dorsal gland; pollen oblate, 3-porate, the pori costate, tectate, psilate, brevicolumellate; pistillode minute, cylindric, entire or shortly 2-lobate. Female flowers pedicellate; sepals (4)5(8), ± free, narrow, soon caducous; petals 0; disc 0; ovary 2-locular, compressed at right-angles to the septal plane, ovules 2 per locule, apical, pendulous, anatropous; styles 2, free, elongate, simple, entire, stigmas smooth or papillose. Fruit compressed, 2-lobate, the lobes (cocci) samaroid or cymbiform, winged or not, separating from the axis, adaxially

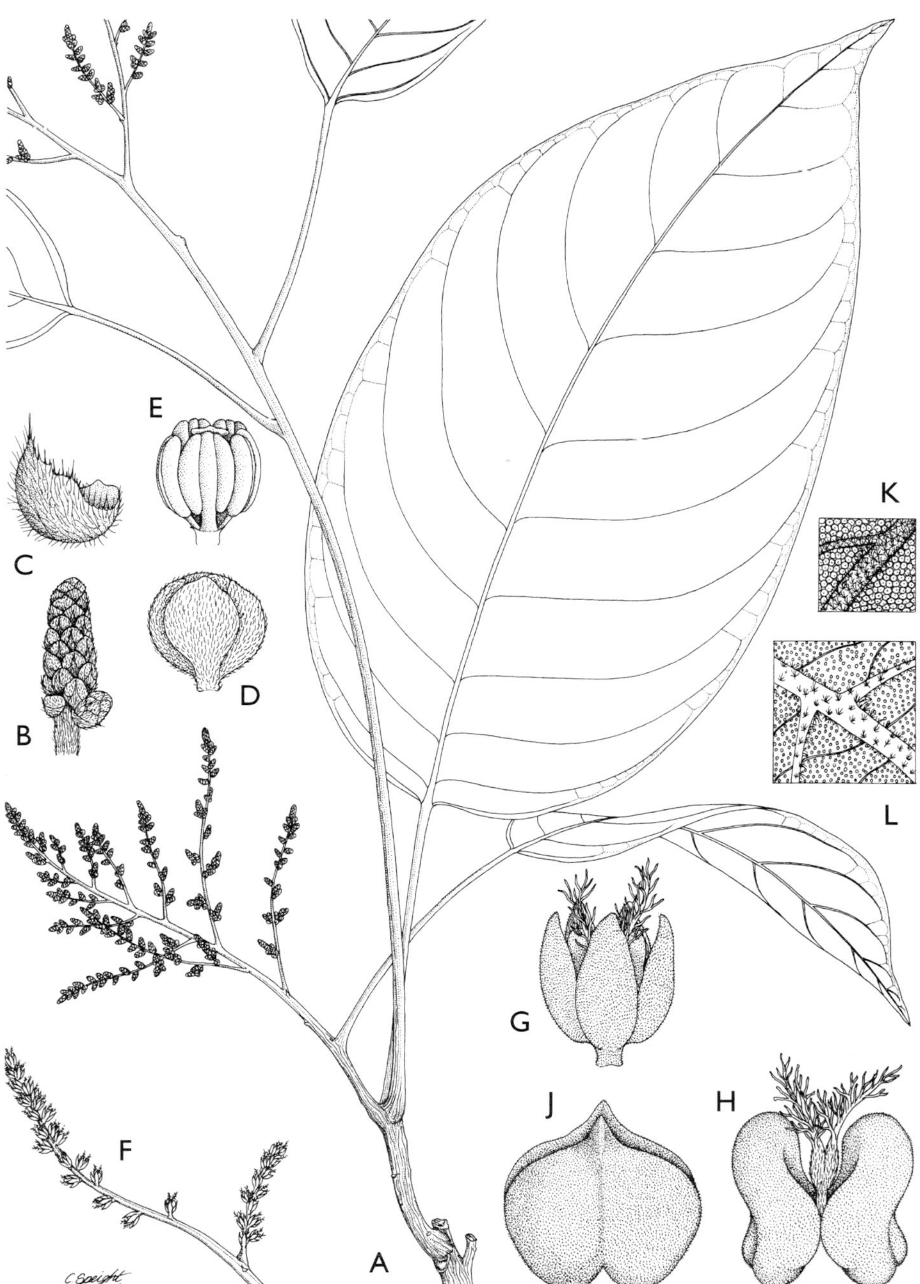

FIG. 7. *Didymocistus chrysadenius* Kuhlm. **A** Habit, male × 1; **B** Portion of male inflorescence × 6; **C** Male bract × 75; **D** Male flower-bud × 75; **E** Male flower-bud, calyx removed × 75; **F** Portion of female inflorescence × 1; **G** Female flower × 10; **H** Young fruit × 10; **J** Fruit × 10; **K** Young leaf-glands × 7; **L** Mature leaf-glands × 4. **A–E** from *Krukoff* 6579; **F–H** from *Ducke* 1830; **J–L** from *Croat* 19711. Drawn by Camilla Speight.

dehiscent; exocarp subcrustaceous; endocarp thin, membranous; columella persistent. Seeds usually 1 per coccus, flat, sometimes alate; testa thin, striate; endosperm sparse; cotyledons broad, flat, very thin; radicle elongate. Fig.: See Étude Euphorb.: t. 27, f. 24–25 (1858); Hooker's Icon. Pl. 12: 29, t. 1131 (1873); Pflanzenr. 81: 73, t. 8 and 76, t. 9 (1922); Fl. Indo-Chine 5: 545, t. 68.1 (1927); Nat. Pflanzenfam. ed. 2, 19C: 52, t. 22 (1931); Fl. W. Trop. Afr. ed. 2, 1(2): 376, t. 132 (1958); Amer. J. Bot. 56: 745, t. 3A–K (1969); Mém. Mus. Natl. Hist. Nat., N.S. Bot. 20: 484, t. 174 (1971); Fl. Ill. Sénég. 3: 492–498 (1975); Fl. Afr. Cent., Hym.: 4–9, tt. 1,2 and pl. I (1985); Bot. J. Linn. Soc. 94: 53, t. 4 (1987); Fl. Trop. E. Afr., Euph. 2: 578, t. 107 (1988); Fl. Zamb. 9(4): 115, t. 18 (1996).

An Old World genus of 9 species, of which 8 are African, the other in Asiatic; absent from Madagascar.

Considered by several recent workers as constituting a separate family, but shown by Levin (1986a, b, c) to be best retained in the *Euphorbiaceae*.

56. **Didymocistus** *Kuhlm.*, Anais Reunião Sul-Amer. Bot. 3: 82 (1938); A. H. Gentry, Woody Pl. NW. S. Amer.: 424 (1993); G. L. Webster, Ann. Missouri Bot. Gard. 81: 53 (1994); Murillo-Aldana & Franco-Rosselli, Euf. Reg. Ararac.: 68 (1995). T.: *D. chrysadenius* Kuhlm.

Dioecious small tree. Indumentum simple and glandular. Leaves large, alternate, long-petiolate, stipulate, entire, penninerved, minutely glandular-punctate beneath; stipules readily deciduous. Inflorescences terminal, pyramidal, paniculate, males copiously branched, ultimate branchlets amentiform, females few-branched; bracts minute, triangular, soon caducous; male flowers extremely small, females larger. Male flowers sessile; sepals 5, free, imbricate; petals 0; disc 0; stamens 5, erect; filaments free; anthers dorsifixed, introrse, obcordate, 2-thecate, thecae longitudinally dehiscent; pollen undescribed; pistillode short, 2- or 3-partite. Female flowers subsessile; calyx-lobes 5, imbricate, connate to halfway, persistent; petals 0; disc 0; ovary 2-locular, ovules 2 per locule; style short, 2-partite; stigmas fimbriate-lacerate, recurved at the apex, persistent. Fruit doubly scrotiform-obcordate, the lobes separating, falling, loculicidally dehiscent; exocarp thin; endocarp crustaceous; columella persistent. Seeds carunculate, subpersistent on the placentas. Fig.: See Woody Pl. NW. S. Amer.: 416, t. 119.8 (1993); Euf. Reg. Ararac.: 71, t. 14 (1995). See also Fig. 7, p. 75.

A monotypic genus of Amazonian Brazil, Colombia and Peru.

Kuhlmann related it to *Aporosa*, but the anatomical studies of Levin (1986a, b, c) show that it is probably closer to *Hymenocardia*. It appears that *Didymocistus* may link the tribes *Antidesmeae* and *Hymenocardieae*, but its curious fruit-structure is not readily assignable to either tribe.

Tribe 8. **BISCHOFIEAE** *(Müll. Arg.) Hurus.*, J. Fac. Sci. Univ. Tokyo, Sect. 3, Bot. 6: 339 (1954); G. L. Webster, Ann. Missouri Bot. Gard. 81: 54 (1994). T.: *Bischofia* Blume.

Phyllantheae subtribe *Bischoffieae* Müll. Arg., Linnaea 34: 64 (1865).
Family *Bischofiaceae* (Müll. Arg.) Airy Shaw, Kew Bull. 18: 252 (1965).

Description: see generic description.

A monogeneric tribe of 2 Asiatic species, very distinct from other *Phyllanthoideae* by virtue of the pinnately 3–5-foliolate leaves. Although Airy Shaw (1965) suggested a relationship with the *Staphyleaceae*, evidence from embryology (Bhatnagar & Kapil, 1974) and leaf anatomy (Levin, 1986a, b, c) supports retention in the *Euphorbiaceae*.

57. **Bischofia** *Blume*, Bijdr. 17: 1168 (1826/7); Hook., Icon. Pl. 9: t. 844 (1852); Baill., Étude Euphorb.: 594 (1858); Müll. Arg. in DC., Prodr. 15(2): 478 (1866) ("*Bischoffia*"); Benth. & Hook. f., Gen. Pl. 3: 280 (1880); Hook. f., Fl. Brit. India 5: 344 (1887); Pax & K. Hoffm. in Engl., Pflanzenr. 81: 312 (1922); Gagnep., Fl. Indo-Chine 5: 542 (1927); Pax & K. Hoffm. in Nat. Pflanzenfam. ed. 2, 19C: 78 (1931); Hurus., J. Fac. Sci. Univ. Tokyo, Sect. 3, Bot. 6: 339 (1954); W. C. Cheng & C. D. Chu, Sci. Sylvae 8: 13 (1963); Backer & Bakh.f., Fl. Java 1: 473 (1963); Airy Shaw, Kew Bull. 18: 253 (1965), 21: 327 (1967), and 27: 271 (1972); A. C. Sm., Fl. Vit. Nov. 2: 494 (1981); Radcl.-Sm. in Fl. Trop. E. Afr., Euph. 2: 565 (1988); G. L. Webster, Ann. Missouri Bot. Gard. 81: 54 (1994). T.: *B. javanica* Blume.

Microelus Wight & Arn., Edinb. New Philos. J. 14: 298 (1833). T.: *M. roeperianus* Wight & Arn. [=*B. javanica* Blume]

Stylodiscus Benn., Pl. Jav. Rar. 133 (1840). T.: *S. trifoliatus* (Roxb.) Benn. (B.: *Andrachne trifoliata* Roxb.) [=*B. javanica* Blume]

Dioecious or rarely monoecious trees. Indumentum simple, localized. Leaves alternate, long-petiolate, stipulate, pinnately 3(5)-foliolate, the leaflets petiolulate, crenate-serrate, penninerved; stipules soon deciduous. Inflorescences axillary, paniculate or racemose, solitary, floribund; bracts 1-flowered, the female ones readily caducous; flowers very small. Male flowers shortly pedicellate; sepals 5, imbricate, concave, cucullate, later reflexed; petals 0; disc 0; stamens 5, free, episepalous; filaments very short, anthers large, subbasifixed, introrse, oblong, 2-thecate, thecae parallel, longitudinally and laterally dehiscent; pollen spheroidal, 3-colporate, colpi narrow, elongate, ecostate, tectate, psilate, brevicolumellate; pistillode shortly stipitate, infundibuliform, peltate, pentagonal. Female flowers pedicellate; sepals 5, imbricate, flat, soon caducous; petals 0; disc 0; staminodes very minute or 0; ovary 3(4)-locular, ovules 2 per locule, pendulous; styles 3(4), slightly connate at the base, elongate, simple, entire, filiform-subulate, spreading or reflexed, completely stigmatic adaxially. Fruit drupaceous, indehiscent, globose, often 3-locular; epicarp thin; mesocarp fleshy; endocarp corneo-pergamaceous. Seeds 1 or 2 per locule, oblong, ecarunculate; testa corneo-pergamaceous; endosperm fleshy; cotyledons broad, flat. Fig.: See Icon. Pl. 9: t. 844 (1852); Étude Euphorb.: t. 26, f. 25–32 (1858) (as *Microelus*); Pflanzenr. 81: 314, t. 26 (1922); Fl. Indo-Chine 5: 538, t. 66.15–18 and 541, t. 67.6–10 (1927); Nat.

Pflanzenfam. ed. 2, 19C: 79, t. 38 (1931); Formos. Trees: 332, t. 286 (1936); Bot. J. Linn. Soc. 94: 56, t. 7 (1987).

A ditypic E. Asiatic and Indopacific genus.

Bischofia is anomalous in the *Euphorbiaceae* on account of its pinnate leaves.

Tribe 9. **CENTROPLACEAE** *Radcl.-Sm.*, **trib. nov.**

Arbor dioica. Inflorescentiae axillares, paniculatae vel subpaniculatae; petala mascula adsunt, foeminea desunt; discus staminalis cupularis, carnosus; stamina sepalis opposita; filamentis perbrevibus; antherae introrsae; pollinis grana <20μ ea *Microdesmidis* similia; pistillodium columnare; disci pistillati lobi sepalis alternantes; staminodia sepalis opposita; ovarium 3-loculare; ovula anatropa; obturator deëst. Fructus loculicidus; columella deëst. Semina nigra, nitentia, ea *Clutiae* revocantia; endospermium copiosum; cotyledones ellipticae. T.: *Centroplacus* Pierre.

Tribus monogenericus occidentali-africanus.

Description as for genus q.v. infra.

58. **Centroplacus** *Pierre*, Bull. Mens. Soc. Linn. Paris. n.s. 14: 144 (1899); Gilg, Bot. Jahrb. Syst. 40: 516 (1908); Hutch., Fl. Trop. Afr. 6(1): 629 (1912); Pax & K. Hoffm., Nat. Pflanzenfam. ed. 2, 19C: 46 (1931); G. L. Webster, Ann. Missouri Bot. Gard. 81: 54 (1994). T.: *C. glaucinus* Pierre.

Dioecious tree. Indumentum simple, confined to the inflorescences. Leaves alternate, shortly petiolate, stipulate, simple, obscurely serrate, penninerved, tertiary nerves subparallel; stipules small. Inflorescences axillary, solitary or subfasciculate, males paniculate, females racemiform-subpaniculate; bracts minute, 1-flowered. Male flowers shortly pedicellate; sepals 5, imbricate; petals 5, imbricate, larger than the sepals; disc cupular, fleshy; stamens 5, oppositisepalous, filaments very short, anthers basifixed, introrse, thecae distinct, ellipsoid, longitudinally dehiscent; pollen oblate-spheroidal, 3-colporate, convex-triangular in polar view, colpus transversalis isodiametric, costae colpi 0, tectate, psilate, brevicolumellate, <20μ; pistillode columnar, entire or 3-fid, densely villous. Female flowers shortly pedicellate; sepals as in the male; petals 0; disc acetabuliform, 5-lobate, lobes alternisepalous; staminodes small, oppositisepalous; ovary 3-locular, ovules 2 per locule, anatropous, exobturatorate; styles 3, short, subulate, entire, slightly recurved. Fruit small, obpyriform-6-lobate, apically depressed, loculicidally dehiscent; exocarp subosseous; endocarp thin; columella 0. Seeds ovoid-ellipsoid; exotesta black, shining; caruncle long, narrow; endosperm copious, fleshy; embryo minute; cotyledons elliptic, 2× the radicle; radicle superior. Fig.: See Fig. 8, p. 79.

A monotypic genus from Cameroon, Equatorial Guinea and Gabon.

Centroplacus was referred to a flacourtiaceous affinity by Pierre, but this was doubted by Gilg (1908). Hallier (1910) placed it in the *Celastraceae*, but Hutchinson (1912) referred it to

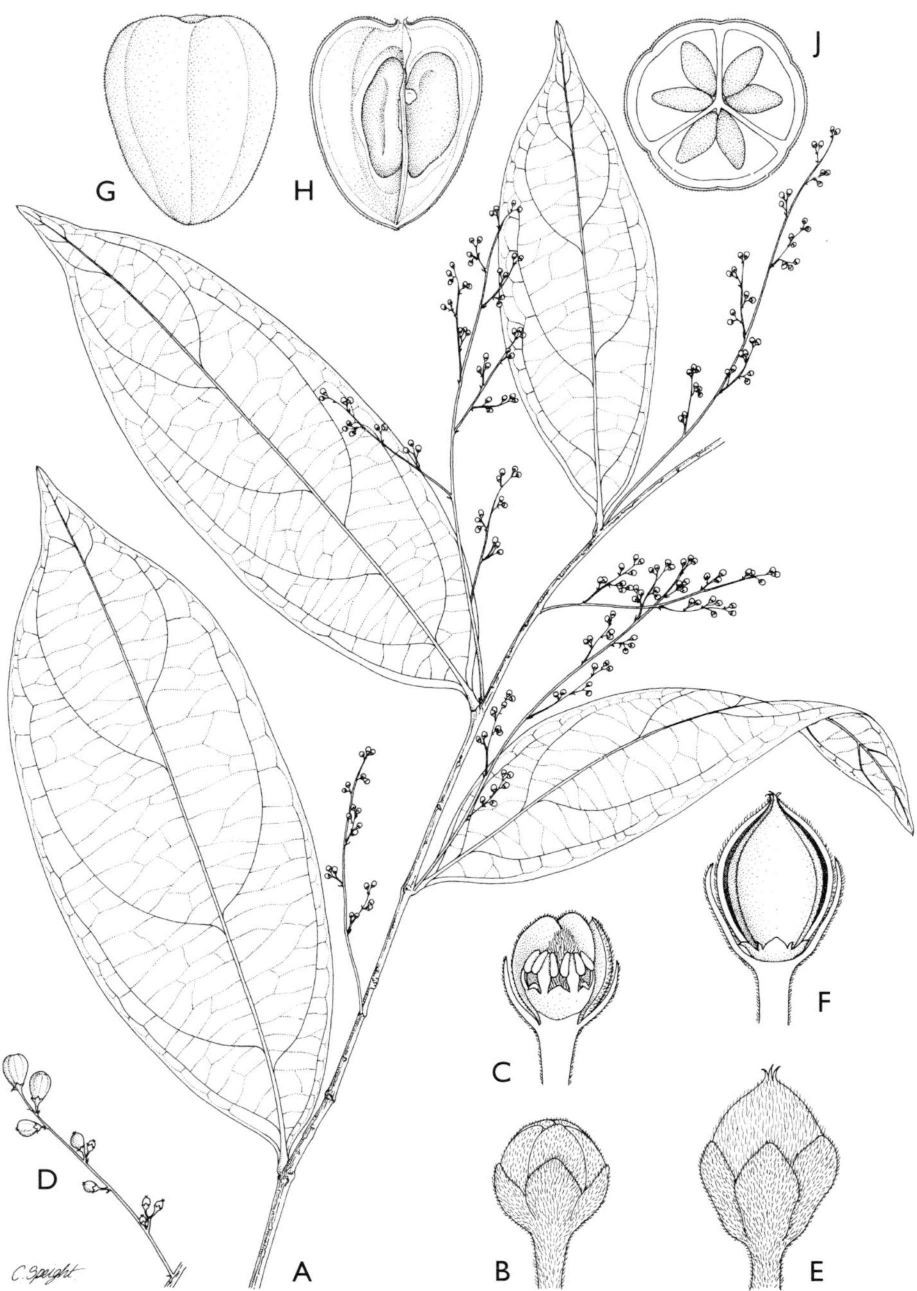

FIG. 8. *Centroplacus glaucinus* Pierre. **A** Habit, male × 1; **B** Male flower-bud × 27; **C** Male flower, part of perianth removed to show disc, stamens & pistillode × 27; **D** Infructescence × 1; **E** Female flower × 27; × **F** Female flower, part of perianth removed to show ovary × 27; **G** Fruit × 7.5; **H** Fruit, vertical section × 7.5; **J** Fruit, transverse section × 7.5. **A–C** from *Zenker* 2030; **D, G–J** from *de Wilde* 8251; **E, F** from *Zenker* 4041. Drawn by Camilla Speight.

the *Euphorbiaceae.* Pax & Hoffmann (1931) placed it in the *Antidesminae* next to *Spondianthus,* but Punt (1962) showed that despite the 2-ovulate loculi and the dehiscent fruit, palynologically the strongest affinity of *Centroplacus* is with *Microdesmis* and *Galearia* of the *Acalyphoïdeae.* Airy Shaw (1966) placed it with these in the *Pandaceae,* a family which has now been re-absorbed into the *Acalyphoïdeae* where it has been accorded tribal status (q.v.), but it is rather anomalous there. Stuppy (1995) places it among his 'genera excludenda' on the basis of seed morphology. In view of its linkages in different directions, it has been deemed best here to constitute separate tribal status for it, near the end of the *Phyllanthoideae.*

Tribe 10. **MARTRETIEAE** *Eg. Köhler ex J. Léonard,* Bull. Jard. Bot. Belg. 59: 326 (Dec. 1989). T.: *Martretia* Beille.

Mostly dioecious trees or shrubs; laticifers 0; leaves alternate, simple, entire, eglandular, penninerved; stipules deciduous. Inflorescences subspiciform-racemose or sometimes paniculate; male sepals 4–5, slightly imbricate; petals 0; disc 0; stamens 4–8, filaments slightly connate at the base, anthers 2-thecate, connective slightly produced; pollen oblate-spheroidal, 3-colporate, transverse costae rounded at ends, costae colpi 0, atrium +; pistillode 2-fid; female sepals 4–6, imbricate; disc glands 5–6, minute, free; ovary 2-locular above, 4-locular below, divided by a pseudophragm, ovules 2 per locule above, 1 below; styles 2, connate at the base, elongate, entire, resembling *Alchornea.* Fruit dehiscing into 4 cocci. Seeds ecarunculate; testa shiny.

A monogeneric W. and C. African tribe. The distinctive pollen, ovary and fruit separates this tribe from all others in the subfamily.

59. **Martretia** *Beille,* Compt. Rend. Hebd. Séances Acad. Sci. 145: 1294 (1907), and Bull. Soc. Bot. France 55, Mém. 2, 8: 64 (1908); Hutch., Fl. Trop. Afr. 6(1): 655 (1912); Pax & K. Hoffm., Pflanzenr. 81: 79 (1922), and Nat. Pflanzenfam. ed. 2, 19C: 52 (1931); Keay (ed.), Fl. W. Trop. Afr. ed. 2, 1: 372 (1958); J. Léonard, Bull. Jard. Bot. Belg. 59: 319 (Dec. 1989); G. L. Webster, Ann. Missouri Bot. Gard. 81: 130 (1994); J. Léonard, Fl. Afr. Cent. Euph. 2: 12 (Nov. 1995). T.: *M. quadricornis* Beille.

Dioecious or rarely monoecious tree or shrub. Latex 0. Indumentum simple, confined to the buds, ovaries and fruits. Leaves alternate, shortly petiolate, stipulate, simple, entire, penninerved, eglandular; stipules soon deciduous. Inflorescences subspiciform-racemose or sometimes paniculate, axillary or arising below the leaves, solitary or subfasciculate; bracts small, 1-flowered; flowers small. Male flowers pedicellate, pedicels bracteolate; sepals 4–5, slightly imbricate; petals 0; disc 0; stamens 4–8, filaments slightly connate at the base, oppositisepalous, anthers subbasifixed, extrorse or latrorse, oblong, 2-thecate, thecae parallel, contiguous, longitudinally dehiscent, connective slightly produced, glandular at the apex; pollen oblate-spheroidal, 3-colporate, colpi transversales costate, costae rounded at ends, atrium +, colpi narrow, costae 0, tectate, psilate; pistillode 2-fid, variable in magnitude. Female flowers shortly pedicellate; sepals 4–6, imbricate; petals 0; disc glands 5–6, minute, free, variable in position; ovary 2-carpellate, 2-locular above, 4-locular below, divided by a pseudophragm, ovules 2 per locule above, appearing as if 1 per locule below; styles 2, connate at the base, divergent, elongate, subulate, entire, slightly twisted, parallel to the pseudophragm, somewhat resembling those of *Alchornea* spp. Fruit obselliform, dehiscing septicidally into 4 1-seeded cocci; cocci each with a narrow transverse cleft. Seeds subellipsoid, ecarunculate; testa shiny, bright orange; cotyledons median, broad, herbaceous. Fig.: See Bull. Jard. Bot. Belg. 59: 327, t. 1 (Dec. 1989); Fl. Afr. Cent. Euph. 2: 15, t. 1 (Nov. 1995).

A monotypic Guineo-Congolean genus ranging from Sierra Leone to C.A.R., Congo and Gabon.

Pax & Hoffmann (1922, 1931) placed *Martretia* between *Cometia* and *Aporosa* in their subtribe *Antidesminae*, but on account of its several aberrant features, Köhler's suggestion, taken up by Léonard (1989), that it should constitute a distinct tribe, although not followed by Webster (1994), seems to be the best solution. Meeuse (1990) considered it to be an acalyphoid genus, but 'the whole cut of the jib' (to use an expression favoured by the late H. K. Airy Shaw) militates against this.

INCERTAE SEDIS SUBFAMILIAE

60. **Tacarcuna** *Huft*, Ann. Missouri Bot. Gard. 76: 1080 (1989); A. H. Gentry, Woody Pl. NW. S. Amer.: 419 (1993); G. L. Webster, Ann. Missouri Bot. Gard. 81: 130 (1994). T.: *T. gentryi* Huft.

A 2-ovulate genus of 3 species from Panama and S. America.

From the floral structure an affinity to 2. *Savia* is indicated, but pollen studies (Nowicke & Webster, ined.) point rather to the *Antidesmeae* or the *Bridelieae*. When it is better known, separate tribal status may be right for this genus also.

Subfamily II. **OLDFIELDIOIDEAE** *Eg. Köhler & G. L. Webster*, J. Arnold Arbor. 48: 308 (1967); G. L. Webster, Taxon 24: 595 (1975), and Ann. Missouri Bot. Gard. 81: 54 (1994). T.: *Oldfieldia* Benth.

Euphorbiaceae-Phyllanthoideae-Phyllantheae-Petalostigmatinae, Toxicodendrinae, Dissiliariinae and *Paivaeusinae* Pax & K. Hoffm., Pflanzenr. 81: 281–297 (1922). TT.: *Petalostigma* F. Muell., *Toxicodendron* Thunb. *(= Hyaenanche* Benth.*), Dissiliaria* F. Muell. and *Paivaeusa* Welw. *(= Oldfieldia* Benth.*)*

Monoecious or dioecious trees, shrubs or subshrubs; leaves alternate, opposite or verticillate, simple or palmately compound, entire or toothed; stipules persistent, caducous or 0. Inflorescences axillary, glomerulate, spicate, racemose, capitulate or paniculate; flowers usually apetalous; male sepals (3)4–8(12), imbricate, usually free; disc extrastaminal, intrastaminal, interstaminal or 0; stamens (2)3–55, filaments free or rarely connate; pollen grains ellipsoid to spheroidal, tectate, exine mostly with conspicuous spines, brevicolporate or porate; pistillode present or 0; female sepals (3)4–8(13), free; disc annular to lobed or dissected, or 0; ovary 2–4(5)-locular, ovules 2 per locule (except in *Scagea*); styles entire or stigmatoid (rarely bifid or twice bifid). Fruit dehiscent (rarely drupaceous). Seeds 1 or 2 per locule, carunculate or ecarunculate, testa often smooth and shiny; endosperm usually copious (rarely 0).

This most recently recognized subfamily includes 26 genera and c. 100 species, found predominantly in the Southern Hemisphere.

Key to the tribes

1. Petals present; male disc extrastaminal, annular; pollen apertures 3; styles twice bifid; ovules hemitropous · · · · · · · · · · · · · · · · · · · 11. *Croizatieae*
1. Petals 0; male disc dissected, intrastaminal, interstaminal or 0; styles entire or rarely once bifid; ovules anatropous · 2
2. Pollen apertures four; vessel elements at least in part with scalariform perforation plates (except in *Tetracoccus*); leaves simple · 12. *Podocalyceae*
2. Pollen with more than 4 apertures (rarely inaperturate); vessel elements with simple perforation plates; leaves simple, unifoliolate, or palmately compound · 3
3. Leaves simple; monoecious or dioecious trees, shrubs, or herbs; seeds mostly carunculate · 13. *Caletieae*
3. Leaves compound, or if simple then lamina stipellate or stipules adnate to petiole; usually dioecious trees or shrubs; seeds mostly carunculate · 14. *Picrodendreae*

Tribe 11. **CROIZATIEAE** *G. L. Webster*, Ann. Missouri Bot. Gard. 81: 55 (1994). T.: *Croizatia* Steyerm.

Dioecious trees or shrubs; leaves alternate, entire; stipules deciduous or persistent. Flowers in axillary clusters; male sepals and petals 5, imbricate; disc extrastaminal, annular; stamens 5, anthers introrse or latrorse; pollen grains globose, 3-colporate, exinous spines conspicuous; pistillode present; female sepals and petals 5, imbricate; disc annular, glabrous; ovary pubescent, 3-locular; ovules hemitropous; styles free, twice 2-fid. Fruit dehiscent; columella persistent, dilated. Seeds 1–2 per locule, ecarunculate; endosperm 0; cotyledons broader than and about equalling radicle.

Represented by the single genus *Croizatia*. This taxon shows clear affinities with the tribe *Amanoeae* in subfamily *Phyllanthoideae* and it is more or less arbitrary where it is placed, but in this treatment it is placed in the *Oldfieldioideae* solely on the basis of pollen characters.

61. **Croizatia** *Steyerm.*, Fieldiana, Bot. 28: 308, t. 57 (1952); G. L. Webster, L. Gillespie & Steyerm., Syst. Bot. 12: 6 (1987); G. L. Webster & Huft, Ann. Missouri Bot. Gard. 75: 1092 (1988); A. H. Gentry, Woody Pl. NW. S. Amer.: 419 (1993); G. L. Webster, Ann. Missouri Bot. Gard. 81: 55 (1994). T.: *C. neotropica* Steyerm.

Dioecious trees or shrubs. Leaves alternate or subopposite, shortly petiolate, stipulate, simple, entire, penninerved, eglandular; stipules persistent or deciduous. Flowers pedicellate, axillary, fasciculate, fascicles few-flowered, females sometimes solitary. Male flowers: sepals 5, imbricate, entire; petals 5, small, subentire, pubescent; disc extrastaminal, annular; stamens 5, filaments free or connate, anthers introrse or latrorse; thecae ± contiguous, longitudinally dehiscent; pollen spheroidal, 3-brevicolporate, exine tectate-perforate, echinate; pistillode 3-fid. Female flowers: sepals 5, imbricate, persistent and reflexed in fruit; petals ± as in the male; disc annular; staminodes present; ovary 3-locular, ovules 2 per locule, hemitropous; styles 3, free, twice 2-fid. Fruit hard, dehiscing into 3 2-valved cocci; exocarp crustaceous, separating from the woody endocarp; columella persistent, prominently 3-winged in the upper part, wings broad and recurved. Seeds 1–2 per locule, smooth, ecarunculate; endosperm 0; cotyledons contorted, broader than and as long as the radicle, greenish. Fig.: See Fieldiana, Bot. 28: 310, t. 57A and 311, t. 57B–E (1952); Syst. Bot. 12: 2, t.1 (1987).

A neotropical genus of 4 species in Panama, Colombia, Venezuela and Ecuador, and possibly also in Peru, *fide* Webster *et al.* (1987).

Tribe 12. **PODOCALYCEAE** *G. L. Webster*, Ann. Missouri Bot. Gard. 81: 55 (1994). T.: *Podocalyx* Klotzsch.

Dioecious trees or shrubs; leaves alternate or opposite, simple, entire or serrate; stipules caducous. Inflorescences in spikes, racemes, thyrses or ebracteate cymes; male sepals (3)4–5, free or connate, imbricate; petals 0; disc intrastaminal or 0; stamens 4–13, free; anthers extrorse or introrse; pollen grains (where known) 4-brevicolporate, echinate, tectum smooth; pistillode + or 0; female sepals 4–5, imbricate, persistent; petals 0; disc + or 0; ovary 3-locular, ovules 2 per locule, anatropous; styles entire, dilated, stigmatiform. Fruit dehiscent or drupaceous. Seeds ecarunculate; endosperm +; cotyledons plane or plicate, much broader than radicle.

This tribe includes three Tropical American genera that are very different in many respects, but are classified together on the basis of pollen characters (pollen grains with 4 shortened colpi); however, the differences between the genera are so striking that it has been necessary to assign them to different subtribes:

Key to the subtribes

1. Peduncle partially adnate to the petiole; male sepals free; anthers introrse; pistillode 0; fruit drupaceous · · · · · · · · · · · · · · · · · 12c. *Paradrypetinae*
1. Peduncle free from petiole; anthers extrorse; fruit dehiscent · · · · · · · · 2
2. Male sepals connate; pistillode +; male and female flowers in spikes or racemes; seed ecarunculate; leaves alternate, entire · 12a. *Podocalycinae*
2. Male sepals free; pistillode 0; female flowers solitary or clustered; seeds carunculate; leaves alternate, opposite or whorled, entire or dentate · 12b. *Tetracoccinae*

Subtribe 12a. **Podocalycinae** *G. L. Webster*, Ann. Missouri Bot. Gard. 81: 55 (1994). T.: *Podocalyx* Klotzsch.

Dioecious trees; leaves alternate; stipules caducous. Inflorescences axillary, spicate or racemose; male sepals 4, connate below; stamens 4, anthers extrorse; pollen 4-colporate, echinate; disc intrastaminal; pistillode +; female flowers pedicellate; sepals 5–6, imbricate, persistent; disc cupular. Fruit dehiscent; columella slender, persistent. Seeds ecarunculate; cotyledons flat.

This subtribe includes only the single genus *Podocalyx.*

62. **Podocalyx** *Klotzsch*, Arch. Naturgesch. 7: 202 (1841); G. L. Webster, Ann. Missouri Bot. Gard. 81: 55 (1994); Murillo-Aldana & Franco-Rosselli, Euf. Reg. Ararac.: 141 (1995). T.: *P. loranthoides* Klotzsch.

Richeria sect. *Podocalyx* (Klotzsch) Müll. Arg. in DC., Prodr. 15(2): 469 (1866), and in Fl. Bras. 11(2): 18, t. 13 (1873); Pax & K. Hoffm., Pflanzenr. 81: 29 (1922), and Nat. Pflanzenfam. ed. 2, 19C: 48 (1931).

Dioecious shrub or small tree. Leaves alternate, petiolate, stipulate, simple, entire, penninerved, thinly coriaceous, eglandular; stipules minute, readily caducous. Inflorescences axillary, solitary or fasciculate, sericeous, the males elongate, interruptedly pseudospicate or in lax panicles of pseudospikes, the females abbreviate, simply spicate, later racemose. Male flowers: pedicels short; calyx-lobes 4, imbricate; petals 0; stamens 4, oppositisepalous, filaments free, long-exserted, anthers medifixed, extrorse, 2-thecate, thecae distinct, not contiguous, longitudinally dehiscent, valves hiant; pollen prolate-spheroïdal, 4-brevicolporate, echinate; disc intrastaminal, 4-lobed, the lobes extruded between the filaments; pistillode minute, conical, bilobate. Female flowers: pedicels short, extending considerably and thickening in fruit; calyx lobes 5–6, otherwise as in male; petals 0; disc cupular, sericeous; ovary 3-locular, sericeous, ovules 2 per locule; styles 3, very short; stigmas entire, gongyloid. Fruit dehiscing into 3 2-valved cocci; exocarp thinly crustaceous, separating from the woody endocarp; columella persistent, much narrower than the pedicel, capitate. Seeds 1 per locule, ellipsoid, black, shiny, ecarunculate; testa subcrustaceous; albumen fleshy; embryo straight, sometimes multiple; cotyledons broad, flat. Fig.: See Fl. Bras. 11(2): t. 13 (1873); Woody Pl. NW. S. Amer.: 416, t. 119.1 (1993); Euf. Reg. Ararac.: 142, t. 41 (1995).

A monotypic genus restricted to Amazonian South America.

Although Pax & Hoffmann (1922, 1931) and more recent workers (e.g., Jablonski, 1967 and Hutchinson, 1969) have followed Müller in treating *Podocalyx* as a section of *Richeria*, its pollen is completely different and characteristic of the subfamily *Oldfieldioideae.* Within the subfamily, *Podocalyx* has an isolated position, but is perhaps very distantly related to *Hyaenanche* and *Mischodon.*

Subtribe 12b. **Tetracoccinae** *G. A. Levin ex G. L. Webster*, Ann. Missouri Bot. Gard. 81: 56 (1994). T.: *Tetracoccus* Engelm. ex Parry.

Dioecious shrubs; leaves alternate, opposite or whorled, simple, margins toothed or entire; stipules 0. Male flowers in axillary racemes or panicles, or fasciculate; female flowers solitary or fasciculate, axillary; male sepals 4–10, free; disc intrastaminal, lobate; stamens 4–10, anthers extrorse; pollen grains 4-porate, echinate, verruculose between spines; pistillode 0; female sepals 5–13; disc glabrous, lobate; ovary (2)3–4(5)-locular; styles elongate. Fruit dehiscent. Seeds carunculate; endosperm copious; cotyledons plicate.

This monogeneric subtribe is very difficult to place and perhaps should constitute a distinct tribe. Vegetatively, *Tetracoccus* is intermediate between *Podocalyx* and *Paradrypetes* in phyllotaxy and leaf margins. In some characters, such as the intrastaminal male disc and verruculose pollen sexine, there is more resemblance between *Tetracoccus* and other tribes of *Oldfieldioideae*. The anthers, however, are very like those of *Podocalyx*. Levin & Simpson (1994) would refer the *Tetracoccinae* to the tribe *Picrodendreae*.

63. **Tetracoccus** *Engelm. ex Parry*, W. Amer. Sci. 1: 13 (1885); Pax & K. Hoffm., Nat. Pflanzenfam. ed. 2, 19C: 74 (1931); Croizat, Bull. Torrey Bot. Club 69: 456 (1942); Dressler, Rhodora 56: 49 (1954); G. L. Webster, Ann. Missouri Bot. Gard. 81: 56 (1994). T.: *T. dioicus* Parry.

Halliophytum I. M. Johnst., Contr. Gray Herb. 68: 88 (1923). LT.: *H. fasciculatum* (S. Watson) I. M. Johnst. (B.: *Bernardia ? fasciculata* S. Watson) [=*T. fasciculatum* (S. Watson) Croizat] (designated by Croizat, 1942b; but according to Wheeler (1975), *H. hallii* (Brandegee) I. M. Johnst. would have been a better choice).

Dioecious shrubs. Leaves alternate, opposite, ternate or whorled, sessile or shortly petiolate, exstipulate, often densely fasciculate on very short shoots, simple, entire or serrate. Male inflorescences axillary, racemose or paniculate or fasciculate on brachyblasts. Females axillary, fasciculate or flowers solitary. Male flowers pedicellate; sepals 4–10, free; petals 0; stamens 4–10, filaments free, anthers medifixed, extrorse, 2-thecate, thecae distinct, not contiguous, longitudinally dehiscent, valves hiant; pollen oblate-spheroïdal, 4-brevicolporate, tectate, echinate and verruculose; disc intrastaminal, variously irregularly lobed, the lobes usually extruded between the filaments, often crested; pistillode 0. Female flowers pedicellate; sepals 5–13, caducous; disc (2)3–4(5)-lobate; ovary (2)3–4(5)-locular, ovules 2 per locule; styles free, simple. Fruit (2)3–4(5)-lobate-subglobose, septicidally dehiscent into (2)3–4(5) 2-valved cocci; exocarp thinly crustaceous; endocarp thinly woody; columella persistent, capitate. Seeds 1 or 2 per locule, carunculate, smooth, shining; endosperm copious; cotyledons plicate. Fig.: See Rhodora 56: tt. 1199, 1200 (habit photos.) (1954). See also Fig. 9, p. 86.

As interpreted by Dressler (1954), a genus of 4 species endemic to southwestern North America (southern California and adjacent Arizona,

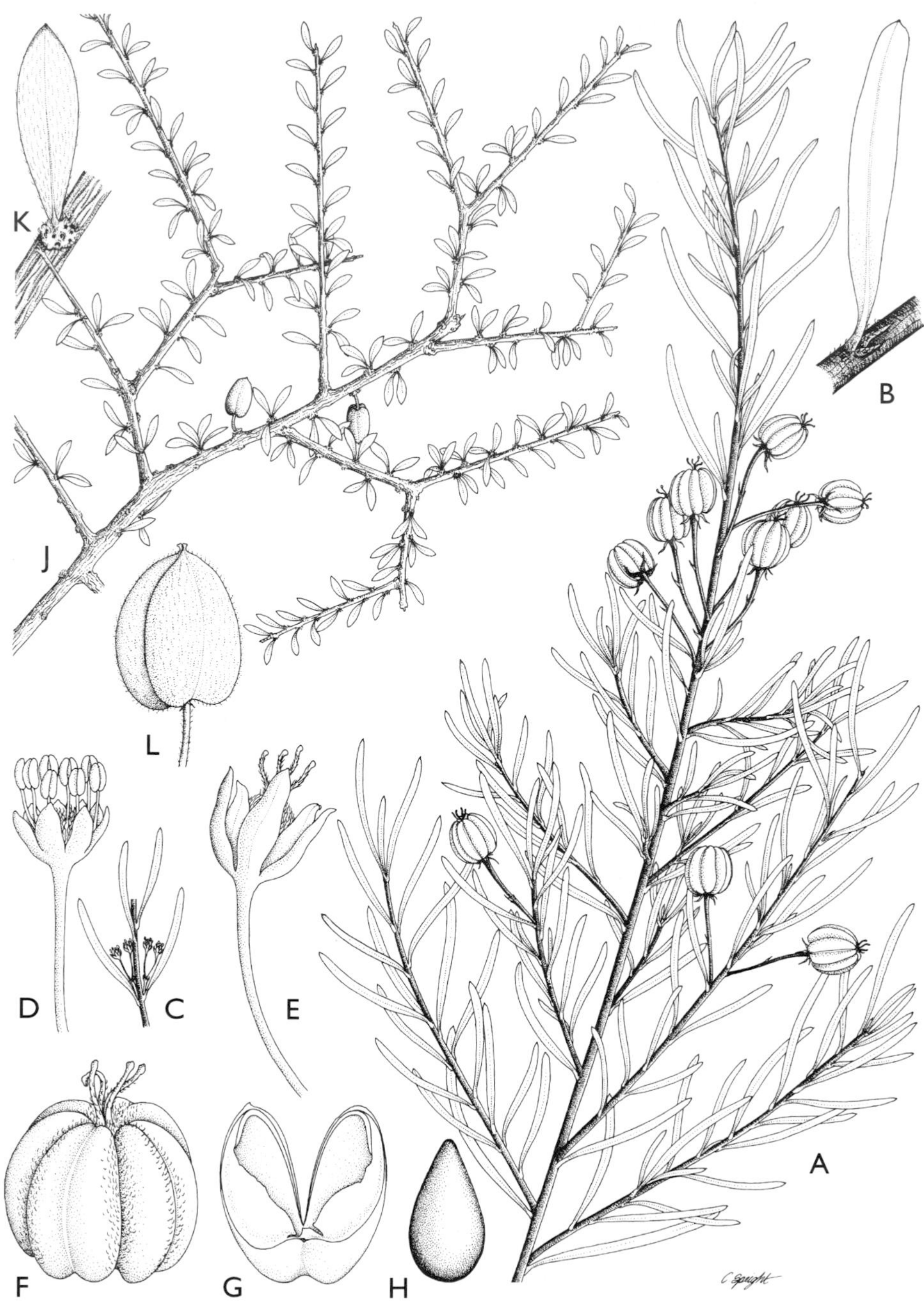

FIG. 9. *Tetracoccus dioicus* Parry. **A** Habit, fruiting specimen × 1; **B** Leaf × 3; **C** Male inflorescences × 1; **D** Male flower × 7.5; **E** Female flower × 7.5; **F** Fruit × 4.5; **G** Fruit-coccus × 4.5; **H** Seed × 4.5. *T. fasciculatus* **J** Habit, fruiting specimen × 1; **K** Leaf × 6; **L** Fruit × 4.5. **A, B, F–H** from *Reid Moran* 13170; **C–E** from *Orcutt* s.n.; **J–L** from *Koune* 679. Drawn by Camilla Speight.

southern Baja California, and northern Mexico). Croizat (1942b) had recognized 5 species disposed in 3 sections.

The ring-porous wood is unique in the subfamily (Hayden 1994). The caruncle recalls that of some genera of the *Pseudanthinae*, q.v. (Stuppy, 1995).

Subtribe 12c. **Paradrypetinae** *G. A. Levin*, Syst. Bot. 17: 78 (1992); G. L. Webster, Ann. Missouri Bot. Gard. 81: 56 (1994). T.: *Paradrypetes* Kuhlm.

Leaves opposite, margin ± spinose; stipules enclosing apical bud. Male flowers in ebracteate cymes, female flowers solitary, the peduncle in both sexes adnate to the subtending petiole; male sepals 3–4, free; disc 0; stamens 9–13, anthers introrse; pistillode 0; female sepals 4; ovary 3-locular; style 0. Fruit drupaceous, 1-seeded; embryo with plicate cotyledons.

A monogeneric subtribe.

64. **Paradrypetes** *Kuhlm.*, Arq. Inst. Biol. Veg. 2: 84 (1935); G. A. Levin, Syst. Bot. 17: 79 (1992); G. L. Webster, Ann. Missouri Bot. Gard. 81: 56 (1994). T.: *P. ilicifolia* Kuhlm.

Dioecious glabrous tree; branching monopodial. Leaves opposite, petiolate, stipulate, simple, spinulose-dentate or subentire, penninerved, strongly reticulate; stipules large, interpetiolar, enclosing the apical bud, deciduous, leaving a prominent scar. Inflorescences axillary, cymose, pedunculate, partially adnate to the petioles; males 2-partite, glomerate, many flowered; females 1–3-flowered. Male flowers: sepals 3–4, free, closely imbricate; petals 0; disc 0; stamens (9)10(13), erect, anthers introrse; thecae longitudinally dehiscent; pollen undescribed; pistillode 0. Female flowers: sepals 4, caducous; petals 0; disc 0; staminodes 4, alternisepalous, antherodes empty; ovary 3-locular, ovules 2 per locule, anatropous; stigmas sessile, broad, dilated, 3-lobate. Fruit drupaceous, indehiscent, ellipsoid, 1-seeded by abortion; exocarp fleshy, sparse, sweet; endocarp crustaceous, fibrous-reticulate. Seeds thick, ellipsoid, unilaterally sulcate, caruncalate; testa venose-reticulate; endosperm starchy; cotyledons multiplicate; radicle fairly thick. Fig.: See Syst. Bot. 17: 81, t. 8 (1992).

A ditypic genus from Brazil, with 1 species from the coast region, the other from the Amazon Basin.

Tribe 13. **CALETIEAE** *Müll. Arg.*, Bot. Zeitung (Berlin) 22: 34 (1865) and in DC., Prodr. 12(2): 190 (1866); Grüning, Pflanzenr. 58: 21 (1913); G. L. Webster, Ann. Missouri Bot. Gard. 81: 56 (1994). T.: *Caletia* Baill. [=*Micrantheum* Desf.]

Dioecious or monoecious trees or shrubs; leaves alternate, opposite or verticillate, simple, entire, or dentate, stipules obscure or 0. Flowers apetalous in axillary clusters or panicles; male sepals 4–8(12), free; disc + or 0; stamens

4–30(55), filaments usually free; anthers extrorse or latrorse; pollen grains ± spheroidal, with exinous spines, colporate, zonoporate or pantoporate; pistillode + or 0; female sepals 3–8(13), free; disc + or 0; ovary (2)3(5)-locular, ovules (1)2 per locule; styles short, linear or dilated. Fruit dehiscent, dry or fleshy. Seeds carunculate or not; endosperm often copious, fleshy; cotyledons broader and longer than the radicle, or not.

Key to the subtribes

1. Leaves whorled (at least in part); filaments shorter than anthers; pollen tectum granulose, exinous spines long · · · · · · · · · · 13a. *Hyaenanchinae*
1. Leaves alternate or opposite; filaments longer than anthers; pollen tectum smooth · 2
2. Pollen grains pantoporate; stipules deciduous or persistent; leaves mostly alternate; fruit dry; ovules (1)2 per locule · · · · · · · 13d. *Pseudanthinae*
2. Pollen grains zonoaperturate; stipules deciduous or reduced; leaves opposite or if alternate then fruit fleshy; ovules 2 per locule · · · · · · · 3
3. Leaves opposite; stamens free; anthers not apiculate; female sepals persistent; styles short, stigmas obcordate or linear, often sulcate, persistent; fruit dry · 13b. *Dissiliariinae*
3. Leaves alternate; stamens connate; anthers apiculate; female sepals caducous; styles large, stigmas flabelliform or petaloid, fleshy, caducous; fruit fleshy · 13c. *Petalostigmatinae*

Subtribe 13a. **Hyaenanchinae** *Baill. ex Müll. Arg.*, Linnaea 34: 64 (1865) (as *Hyaenancheae*) and DC., Prodr. 15(2): 479 (1866); G. L. Webster, Ann. Missouri Bot. Gard. 81: 57 (1994). T.: *Hyaenanche* Lamb. & Vahl.

Tribe *Hyaenancheae* (Baill. ex Müll. Arg.) Hutch., Amer. J Bot. 56: 746 (1969).

Toxicodendrinae Pax, Nat. Pflanzenfam. ed. 1, 3(5): 31 (1890); Pax & K. Hoffm., Pflanzenr. IV, 147, xv (Heft 81): 284 (1922). T.: *Toxicodendrum* Thunb. (non Mill.) [= *Hyaenanche* Lamb.]

Dioecious trees or shrubs; leaves opposite or verticillate, simple, coriaceous, stipules 0. Flowers in axillary cymes or solitary; male sepals 4–8(12), more or less connate; disc 0; stamens 8–30, filaments free, shorter than anthers; anthers latrorse; pollen grains 6- or 7-zonoporate, spinose, tectum granulose; pistillode 0; female sepals 3–8, free, caducous in fruit; disc 0; ovary 3–4-locular, sericeous; styles elongated, more or less dilated. Fruit dehiscent. Seeds carunculate; endosperm sparse.

This subtribe is here circumscribed to include only the type genus *Hyaenanche*, as treated by Müller and Pax & Hoffmann. Clearly *Hyaenanche* is a very isolated genus whose relationships are still uncertain. It is isolated in the tribe *Caletieae* because of the verruculose microsculpturing of the pollen sexine, which recalls genera of tribe *Picrodendreae*. There is some resemblance to the S. Indian and Sri Lankan genus *Mischodon* in habit and pollen morphology, but it is still not clear whether the two genera should be grouped together. Also it has similar pollen and fruits to *Tetracoccus* (q.v. supra).

65. **Hyaenanche** *Lamb. & Vahl*, Descr. *Cinchona* 52, t. 10 (1797); A. Juss., Euphorb. Gen.: 40 (1824); Baill., Étude Euphorb.: 565 (1858); Müll. Arg. in DC., Prodr. 15(2): 479 (1866); Dyer, Gen. S. Afr. Fl. Pl. 1: 311 (1975); G. L. Webster, Ann. Missouri Bot. Gard. 81: 57 (1994). T.: *H. globosa* (Gaertn.) Lamb. & Vahl (B.: *Jatropha globosa* Gaertn.)

Toxicodendrum Thunb., Kongl. Vetensk. Acad. Handl. 17: 188 (1796) (non *Toxicodendron* Mill., 1754); Benth. & Hook. f., Gen. Pl. 3: 289 (1880); Hutch. in Fl. Cap. 5(2): 408 (1920); Pax & K. Hoffm., Pflanzenr. 81: 284 (1922), and Nat. Pflanzenfam. ed. 2, 19C: 74 (1931). T.: *T. capense* Thunb. [=*H. globosa* (Gaertn.) Lamb. & Vahl (B.: *Jatropha globosa* Gaertn.)]

Dioecious shrub or small tree. Leaves opposite or verticillate, in whorls of 4, shortly petiolate, exstipulate, simple, entire, obscurely penninerved, thickly and rigidly coriaceous. Inflorescences axillary, males short, densely cymulose, female flowers solitary. Male flowers: sepals 4–12, short, irregularly imbricate; petals 0; disc 0; stamens 8–30, crowded, filaments very short, free; anthers oblong, erect, latrorse, thecae parallel, longitudinally dehiscent; valves hiant; pollen suboblate, stephanocolporate, colpi 6–7, echinate, tectate, granulate; pistillode 0; receptacle broad, undulate or subplicate. Female flowers: sepals 3–8, in two whorls, the inner larger than the outer, caducous; ovary villous, 3–4 locular; ovules 2 per locule; styles 3–4, connate at the base, thick, undivided, stigmas densely papillose. Fruits 6–8-lobed, subglobose, thick, hard, septicidally dehiscent into 3–4 2-valved cocci; exocarp suberose, becoming crustaceous, separating; endocarp woody; columella triquetrous, persistent. Seeds oblong, smooth, black, shining, carunculate; embryo straight; endosperm scanty, fleshy; cotyledons broad, flat, green. Fig.: See Étude Euphorb.: t. 23, f. 29–39 (1858); Fl. Pl. S. Afr. 21: t. 837 (1941); Amer. J. Bot. 56: 745, t. 4A–L (1969); Bot. J. Linn. Soc. 94 (1/2): 63, t. 11 (1987).

A monotypic genus of the Cape region of South Africa.

The seeds of *Hyaenanche* are unusual within the *Oldfieldioideae* in the scanty endosperm, green cotyledons, and production of a unique toxin (Hyaenanchin), but Stuppy (1995) postulates an affinity to *Whyanbeelia* (q.v.) on the basis of seed-morphology and anatomy.

Subtribe 13b. **Dissiliariinae** *Pax & K. Hoffm.*, Pflanzenr. 81: 288 (1922); G. L. Webster, Ann. Missouri Bot. Gard. 81: 57 (1994). T.: *Dissiliaria* F. Muell.

Monoecious or dioecious trees or shrubs; leaves opposite, simple, entire; stipules mostly reduced or 0, rarely large. Flowers in axillary panicles or clusters; male sepals 4–6, imbricate, free; disc 0; stamens (2)4–55, free; pollen grains longispinous or brevispinous, tectum smooth; pistillode usually 0; female sepals

(3)4–6, free; disc cupulate or 0; ovary 2–3(5)-locular, glabrous; styles mostly short. Fruit dehiscent or drupaceous. Seeds carunculate or ecarunculate.

This Australasian subtribe includes 7 genera with c. 25 species. Generic limits require further critical study.

Key to the genera

1. Pollen exine with reduced spines; pistillode + or 0 · · · · · · · · · · · · · · · · 2
1. Pollen exine with elongated spines; pistillode 0 · · · · · · · · · · · · · · · · · · 4
2. Stipules large; pistillode 0; female sepals 3; fruit ovoid · · · · 69. *Sankowskia*
2. Stipules small or 0; pistillode +; female sepals 6; fruit globose or subglobose · 3
3. Stamens 4–6; styles separated at base, linear, recurved; seeds ecarunculate · 71. *Choriceras*
3. Stamens 9–17; styles proximate, ovate-cordate, not recurved; seeds carunculate · 72. *Longetia*
4. Monoecious; stamens 50–55; female disc of subulate segments · 70. *Whyanbeelia*
4. Dioecious; stamens 8–27; female disc cupular or 0 · · · · · · · · · · · · · · · · 5
5. Ovary 2-locular; styles 2, stigmas ovate-cordate; seeds ecarunculate · 67. *Canaca*
5. Ovary 3–4(5)-locular; seeds carunculate · 6
6. Caruncle arilloid, fimbriate; ovary 3(5)-locular; styles short, erect, stigmas ovate, subsessile · 66. *Austrobuxus*
6. Caruncle entire; ovary 3(4)-locular; styles elongate, linear, recurved · 68. *Dissiliaria*

66. **Austrobuxus** *Miq.*, Fl. Ned. Ind. Suppl. 444 (1861); Müll. Arg. in DC., Prodr. 15(2): 1254 (1866); van Steenis, Blumea 12: 362 (1964); Airy Shaw, Kew Bull. 25: 506 (1971), 29: 303 (1974) and 35: 597 (1980); A. C. Sm., Fl. Vit. Nov. 2: 495 (1981); McPherson & Tirel, Fl. Nouv.-Caléd. 14(1): 193 (1987); G. L. Webster, Ann. Missouri Bot. Gard. 81: 57 (1994). T.: *A. nitidus* Miq.

Buraeavia Baill., Adansonia 11: 83 (1873). T.: *B. carunculata* (Baill.) Baill. (B.: *Baloghia carunculata* Baill.) [=*A. carunculatus* (Baill.) Airy Shaw]
Choriophyllum Benth, Hooker's Icon. Pl. 13: 62, t. 1280 (1879); Hook. f., Fl. Brit. India 5: 343 (1887). T.: *Ch. malayanum* Benth. [= *A. nitidus* Miq.]

Dioecious trees or shrubs. Leaves opposite, subsessile or shortly petiolate, stipulate or exstipulate, simple, usually entire, penninerved, usually coriaceous; stipules minute or 0. Inflorescences axillary, cymose, the males several-flowered, the females few-flowered or flowers solitary. Male flowers: sepals 4–6, free, strongly imbricate, suborbicular; petals 0; disc 0; stamens (2)8–27, filaments free, arising from receptacular pits, anthers medifixed, extrorse, thecae distinct, longitudinally dehiscent; pollen oblate-spheroidal, stephanoporate, pori 5–6, echinate, tectate, psilate; pistillode 0; receptacle convex. Female flowers: sepals

4–6, free, imbricate, ovate; petals 0; disc shortly cupular or 0; ovary 3(5)-locular, ovules 2 per locule; styles very short; stigmas 3(5), subsessile, cordate, sulcate, ± smooth. Fruit (2)3–4(5)-locular, dehiscing into (2)3–4(5) 2-valved cocci; exocarp crustaceous, separating; endocarp woody; columella narrow, persistent. Seeds ovoid-ellipsoid, smooth, shiny, with or without an intricately byssaceous caruncle; albumen fleshy; cotyledons broad, flat or shallowly concave. Fig.: See Hooker's Icon. Pl. 13: t. 1280 (1879) (as *Choriophyllum*); Fl. Nouv.-Caléd. 14(1): 189, t. 38.1,2; 191, t. 39.7,8 and 197–217, tt. 40–43, excl. 42.2–6 (1987).

In the circumscription of Airy Shaw (1974) and McPherson & Tirel (1987), *Austrobuxus* is the largest genus of subfamily *Oldfieldioideae*, with c. 20 species. The distribution is disjunct, namely Malaya and Western Indonesia, Queensland, New Caledonia, and Fiji. New Caledonia is the main centre, with 14 endemic species.

67. **Canaca** *Guillaumin*, Arch. Bot. Bull. Mens. 1: 74 (1927); Airy Shaw, Kew Bull. 25: 508 (1971); G. L. Webster, Ann. Missouri Bot. Gard. 81: 58 (1994). T.: *C. vieillardii* Guillaumin (S.: *Austrobuxus vieillardii* (Guillaumin) Airy Shaw).

Similar to 66. *Austrobuxus*, but differing in having very shortly petiolate exstipulate leaves, with a large lamina strongly cordate at the base; a 2-locular ovary and 2 styles; a 2-locular, drupaceous fruit, and ecarunculate seeds. Fig.: See Fl. Nouv.-Caléd. 14(1): 213, t. 42.2–6 (1987).

A monotypic genus, endemic to N. New Caledonia.

68. **Dissiliaria** *F. Muell. ex Baill.*, Adansonia I. 7: 366 (1867); Benth., Fl. Austral. 6: 90 (1873); Pax & K. Hoffm., Pflanzenr. 81: 291 (1922), and Nat. Pflanzenfam. ed. 2, 19C: 76 (1931); Airy Shaw, Muelleria 4: 220 (1980) and Kew Bull. 35: 625 (1980); G. L. Webster, Ann. Missouri Bot. Gard. 81: 58 (1994). LT.: *D. baloghioides* F. Muell. ex Baill. (chosen by Webster, 1994).

Dioecious trees or shrubs. Leaves opposite or 3-nate, shortly petiolate, stipulate, simple, entire or crenulate, penninerved, reticulate; stipules interpetiolar, soon caducous. Inflorescences axillary or subterminal, cymose, the males few-flowered, the females fewer-flowered or solitary. Male flowers: sepals 4–6, free, imbricate, inner a little larger and semipetaloid; petals 0; disc of small interstaminal glands or 0; stamens 15–20, filaments free, anthers dorsifixed, extrorse, thecae parallel, longitudinally dehiscent; pollen undescribed; pistillode 0; receptacle pubescent. Female flowers: sepals 6, in 2 whorls of 3, imbricate, inner sometimes smaller; petals 0; disc annular or shallowly cupular, often denticulate; ovary short, 3(4)-locular, ovules 2 per locule; obturator thick; styles 3, linear, simple, recurved, ± smooth, shortly connate at the base. Fruit subglobose, 3(4)-locular, 6–8-lobed, dehiscing into 3(4) 2-valved cocci or 6 valves; exocarp thick, indurated, separating; endocarp thinly woody or crustaceous; columella stout, persistent. Seeds 1–2 per locule, carunculate; caruncle entire; endosperm fleshy; embryo straight; cotyledons broad, flat. Fig.: See Fig. 10, p. 92.

FIG. 10. *Dissiliaria baloghioides* F. Muell. **A** Habit, male × 1; **B** Male flower × 7.5; **C** Female flower × 7.5; **D** Young fruit × 1.5; **E** Mature fruit × 1.5; **F** Fruit-valve, showing separating exocarp × 1.5; **G** Seed × 1.5. **A, B** from *Guymer* 1857; **C** from *Forster* 13126A; **D** from *McDonald* & *Stanton* 2347; **E** from *Smith* s.n.; **F, G** from *Hubbard* 4172. Drawn by Camilla Speight.

According to Airy Shaw (1980c), *Dissiliaria* includes 3 species endemic to Australia (Queensland).

Stuppy (1995) places *Dissiliaria* and *Whyanbeelia* with *Hyaenanche* and the *Mischodontinae* rather than with *Austrobuxus, Choriceras* and *Longetia* on the basis of seed morphology.

69. **Sankowskia** *P. I. Forst.*, Austrobaileya 4(3): 329 (1995). T.: *S. stipularis* P. I. Forst.

Small monoecious evergreen tree. Indumentum simple, hairs multicellular. Leaves opposite, petiolate, stipulate, simple, crenulate, penninerved; stipules interpetiolar, large, linear- to oblong-lanceolate, not persistent. Inflorescences axillary, cymose, bracteate. Male flowers pedicellate; sepals 4, 2-seriate, imbricate; petals 0; disc 0; stamens 12–15, filaments free, filiform, anthers dorsifixed, extrorse, 2-thecate, thecae distinct, oblong, longitudinally dehiscent; pollen smooth; pistillode 0; receptacle convex. Female flowers pedicellate; sepals 3, imbricate; petals 0; disc 0; ovary 3-locular, ovules 2 per locule; styles 3, free, simple, linear, erect or spreading, later recurved, stigmas papillose-rugulose. Fruit 3-lobate-ovoid, dehiscing into 3 2-valved cocci, style-bases persistent; exocarp not separating; endocarp thinly woody. Seeds oblong-ovoid, carunculate; caruncle entire. Fig.: See Austrobaileya 4(3): 334, t. 1 (1995).

A monotypic genus of tropical Australia (NE. Queensland).

70. **Whyanbeelia** *Airy Shaw & B. Hyland*, Kew Bull. 31: 375, t. 2 (1976) and Kew Bull. 35: 691 (1980); G. L. Webster, Ann. Missouri Bot. Gard. 81: 58 (1994). T.: *W. terra-reginae* Airy Shaw & B. Hyland.

Monoecious tree. Leaves opposite, petiolate, exstipulate, simple, entire, penninerved. Inflorescences axillary or subterminal, laxly cymose, usually mostly male with a few females admixed, sometimes all male; bracts subulate. Male flowers long-pedicellate; buds 3-lobate; sepals 6, 2-seriate, imbricate, 2-morphic; petals 0; stamens 50–55, filaments free, filiform, anthers oblong-subglobose, dorsifixed, extrorse, longitudinally dehiscent; pollen spinose; disc central, consisting of several small angular glands in 3 radiating rows among the filaments; pistillode 0; receptacle hemispherical, densely pilose. Female flowers long-pedicellate; sepals 6, biseriate, imbricate, narrowly lanceolate, acute; petals 0; disc-segments several, subulate, staminodiiform; ovary 3-locular, ovules 2 per locule; styles 3, shortly connate at the base, simple, linear, spreading and recurved, stigmas sulcate. Fruit globose, 6-sulcate, dehiscing into 3 2-valved cocci; exocarp not separating; endocarp woody; columella long, slender, capitate, persistent. Seeds laterally-compressed, carunculate; caruncle entire. Fig.: See Kew Bull. 31: 377, t. 2 (1976), and 35: 668, t. 6D1–5 (1980).

A monotypic genus of tropical Australia (Queensland).

Stuppy (1995) places *Whyanbeelia* and *Dissiliaria* with *Hyaenanche* and the *Mischodontinae* rather than with *Austrobuxus, Choriceras* and *Longetia* on the basis of seed morphology.

71. **Choriceras** *Baill.*, Adansonia I. 11: 119 (1873); Airy Shaw, Kew Bull. 14: 356 (1960), Muelleria 4: 220 (1980), Kew Bull. Addit. Ser. 8: 46 (1980) and Kew Bull 35: 604 (1980); G. L. Webster, Ann. Missouri Bot. Gard. 81: 58 (1994). T.: *C. australiana* Baill. [=*Ch. tricorne* (Benth.) Airy Shaw (B.: *Dissiliaria tricornis* Benth.)]

Dioecious tree; branchlets opposite. Leaves opposite, shortly petiolate, stipulate, simple, finely crenulate or subentire, penninerved; stipules subulate, soon falling. Inflorescences axillary, cymose, uni- or bisexual, the males dense, many-flowered, pseudofasciculate, the females lax, 1–3-flowered. Male flowers: sepals 4–6, 2-seriate, imbricate; petals 0; disc 0; stamens 4–6, filaments free, filiform, exserted; anthers small, globose, extrorse, thecae adnate to the connective, longitudinally dehiscent; pollen suboblate, stephanoporate, pori 6, tectum brevispinous, psilate; pistillode conical-cylindrical, fulvous-pilose; receptacle fulvous-pilose. Female flowers: sepals 6, 2-seriate; disc glands 3, hypogynous; ovary sessile, 3(4)-locular, loculi shortly connate, ovules 2 per locule, collateral; styles 3, free, divaricate, undivided, recurved. Fruit 3-lobate-subglobose, 3-coccous, 3-horned due to the erect conical style-bases, dehiscing into 3 2-valved cocci; exocarp thinly crustaceous; endocarp woody; columella slender, short. Seeds ovoid, compressed, smooth, ecarunculate. Fig.: See Kew Bull. Addit. Ser. 8: 223, t. 1.3 (1980) (as *Dissiliaria tricornis*).

A genus of 2 species, both in tropical Australia and 1 in New Guinea.

Although Pax & Hoffmann (1922) reduced *Choriceras* to a synonym of *Dissiliaria*, Airy Shaw (1960) pointed out its distinguishing characters.

72. **Longetia** *Baill. ex Müll. Arg.* in DC., Prodr. 15(2): 244 (Aug. 1866); Baill., Adansonia I. 6: 352, t. 9 (Oct. 1866); Benth. & Hook. f., Gen. Pl. 3: 279 (1880); Pax & K. Hoffm., Pflanzenr. 81: 289 (1922), and Nat. Pflanzenfam. ed. 2, 19C: 75 (1931); McPherson & Tirel, Fl. Nouv.-Caléd. 14(1): 188 (1987); G. L. Webster, Ann. Missouri Bot. Gard. 81: 58 (1994). T.: *L. buxoides* Baill.

Monoecious shrub. Leaves opposite, petiolate, exstipulate, simple, entire, coriaceous, penninerved. Inflorescences axillary or terminal, cymose, the males several-flowered, dense, the females few-flowered, lax, or flowers solitary, or inflorescences bisexual with a terminal female and lateral male flowers. Male flowers: sepals 6, biseriate, imbricate; petals 0; disc 0; stamens 9–17, filaments free; anthers dorsifixed, thecae extrorse, parallel, longitudinally dehiscent; pollen suboblate, stephanocolporate, colpi (5)6(7), small, short, crassicostate, tectum brevispinous, psilate; pistillode small, subconic, pubescent; receptacle slightly elevated. Female flowers: sepals ± as in male; ovary 3-locular, ovules 2 per locule; styles 3, very short, stigmas obcordate, spreading. Fruit ovoid, subglobose, subtriquctrous, loculicidally and septicidally dehiscent into 3 2-valved cocci: exocarp crustaceous; endocarp woody; columella persistent. Seeds ovoid-ellipsoid, carunculate; embryo straight; endosperm fleshy; cotyledons broad and flat. Fig.: See Adansonia I. 6: 352, t. 9 (Oct. 1866); Fl. Nouv.-Caléd. 14(1): 189, t. 38. 3–5 and 191, t. 39.1–6 (1987).

As interpreted by McPherson & Tirel (1987), *Longetia* is a monotypic genus endemic to New Caledonia.

Pax & Hoffmann (1922) construed *Longetia* in a broad sense, including species now placed in *Austrobuxus*. Airy Shaw (1972, 1974) held a similar broad sense, except that *Longetia* species were now transferred to *Austrobuxus*. In view of the distinct difference in pollen, McPherson & Tirel's (1987) return to the classical concept of Baillon and Müller seems fully justified.

Subtribe 13c. **Petalostigmatinae** *Pax & K. Hoffm.*, Pflanzenr. 81: 281 (1922); G. L. Webster, Ann. Missouri Bot. Gard. 81: 58 (1994). T.: *Petalostigma* F. Muell.

Tribe *Petalostigmateae* (Pax & K. Hoffm.) G. L. Webster, Taxon 24: 595 (1975).

Dioecious trees or shrubs; leaves alternate, simple, mostly entire; stipules caducous. Flowers on abbreviated axillary cymes; male sepals 4–6, imbricate; disc 0; stamens 20–40, filaments connate below into a column; anther connective typically pilose; pollen grains with short spines; pistillode obscure or 0; female sepals 4–8, imbricate; disc 0; ovary 3–4-locular, pubescent; styles dilated. Fruit fleshy. Seeds usually carunculate.

A monogeneric tribe restricted to Australia and Papua New Guinea.

73. **Petalostigma** *F. Muell.*, Hook. J. Bot. Kew Gard. Misc. 9: 17 (1857); Baill., Étude Euphorb.: 657 (1858); Müll. Arg. in DC., Prodr. 15 (2): 273 (1866); Baill., Adansonia I. 7: 352 (1867); Benth., Fl. Austral. 6: 92 (1873), and Gen. Pl. 3: 277 (1880); Pax & K. Hoffm., Pflanzenr. 81: 281, t. 22 (1922), and Nat. Pflanzenfam. ed. 2, 19C: 74 (1931); Airy Shaw, Kew Bull. 31: 366 (1976), Kew Bull. Addit. Ser. 8: 178 (1980), and Kew Bull. 35: 661 (1980); G. L. Webster, Ann. Missouri Bot. Gard. 81: 58 (1994). T.: *P. quadriloculare* F. Muell.

Dioecious often densely pubescent trees or shrubs. Leaves alternate, shortly petiolate, stipulate, simple, entire or shallowly crenate, penninerved, nerves often impressed above. Male inflorescences axillary, cymose, dense, several-flowered, abbreviated, pseudofasciculate; female flowers usually solitary or paired per axil. Male flowers: sepals 4–6, suborbicular, closely imbricate; petals 0; disc 0; stamens 20–40, filaments connate below into a short column, free above; anthers erect, elongate, extrorse, 2-thecate, thecae connate, connective produced, apiculate, often pilose, longitudinally dehiscent; pollen oblate-spheroidal, substephanoporate, pori 4–6, brevispinous; pistillode irregularly 2-lobate, or more usually 0. Female flowers: sepals 4–8, narrower than in the male, imbricate, soon caducous; petals 0; disc 0; ovary 3–4-locular, ovules 2 per locule; styles 3–4, large, spreading, stigmas flabelliform or petaloid, undulate or lobulate, fleshy, caducous. Fruit subdrupaceous, tardily dehiscent into 3–4 2-valved cocci; exocarp fleshy, bright orange; endocarp woody; columella narrowly alate, persistent. Seeds oblong-ellipsoid, 1 per locule by abortion, usually carunculate; testa smooth, shiny, black or dark brown; endosperm fleshy; embryo straight; cotyledons broad, flat. Fig.: See Pflanzenr. 81: 282, t. 22 (1922); Kew Bull. Addit. Ser. 8: 226, t. 4.4 (1980); Kew Bull 35: 660, t. 5C1–3 (1980).

As treated by Airy Shaw (1980c), *Petalostigma* is a genus of 6 variable Australian species, one of which extends to Papua New Guinea.

Subtribe 13d. **Pseudanthinae** *Müll. Arg.*, Linnaea 34: 55 (1865) and in DC., Prodr. 15(2): 195 (1866); G. L. Webster, Ann. Missouri Bot. Gard. 81: 58 (1994). T.: *Pseudanthus* Spreng.

Monoecious or rarely dioecious trees, shrubs, or herbs; leaves mostly alternate, entire; stipules deciduous or persistent. Flowers in axillary clusters; male sepals 3–6, imbricate, free; disc intrastaminal, central or 0; stamens 3–20, rarely more, filaments free or connate; pollen grains pantoporate, with long or short spines, tectum smooth; pistillode + or 0; female sepals 4–6, free, usually persistent; disc + or 0; ovary 2–3-locular; styles simple, dilated or elongated. Fruit dehiscent. Seeds usually carunculate.

The subtribe *Pseudanthinae* is here circumscribed to include 5 Australasian genera. The referral of *Kairothamnus* and *Scagea* to the *Pseudanthinae* needs confirmation, but appears compatible with the anatomical and palynological evidence (Webster, 1994).

Key to the genera

1. Cotyledons scarcely broader than radicle; herbs or undershrubs · · · · · · 2
1. Cotyledons broader than radicle; trees and shrubs · · · · · · · · · · · · · · · · 3
2. Stipules foliose; fruits 3–6-seeded; pollen grains with long exinous spines · 77. *Micrantheum*
2. Stipules not foliose; fruits 1-seeded; pollen grains with reduced spines · 78. *Pseudanthus*
3. Male sepals 4; stamens 10–12 · 74. *Kairothamnus*
3. Male sepals 5(6); stamens 4–6 · 4
4. Ovules 1 per locule; pistillode +; male inflorescence pseudoracemose, with umbels or fascicles along the axes; male disc 0; stipules conspicuous but quickly deciduous · 75. *Scagea*
4. Ovules 2 per locule; pistillode 0; male flowers in axillary fascicles; male disc central; stipules minute or almost obsolete · · · · · · · · · · · 76. *Neoroepera*

74. **Kairothamnus** *Airy Shaw*, Kew Bull. 34: 596 (1980), and Kew Bull. Addit. Ser. 8: 121, t. 1 fig. 1 (1980); G. L. Webster, Ann. Missouri Bot. Gard. 81: 59 (1994). T.: *K. phyllanthoides* (Airy Shaw) Airy Shaw (B.: *Austrobuxus phyllanthoides* Airy Shaw).

Monoecious shrub or small tree. Leaves alternate, petiolate, stipulate, simple, entire, long-caudate, penninerved; stipules readily deciduous. Flowers axillary, often in the axils of fallen leaves, fasciculate; fascicles each usually with 10–12 male flowers and 1(2) female flowers. Male flowers: pedicels capillary; sepals 4, orbicular, strongly imbricate; petals 0; disc covering the receptacle, fenestrate; stamens 10–12, filaments free, short, arising through the pits in the disc, anthers oblong, extrorse, longitudinally dehiscent; pollen spheroidal, pantoporate, echinate, tectum smooth; pistillode 0; receptacle convex. Female flowers: pedicels long, elongating to over 5 cm in fruit; sepals 6, 2-seriate, narrower than the male, inner subcucullate; petals 0; disc 0; ovary 3-locular, ovules 2 per locule; stigmas 3, suberect, sessile, distinct, triangular or soleiform, apically connivent, smooth. Fruit ovoid-subglobose, elobate, dehiscing septicidally into 3 2-valved cocci; exocarp thinly crustaceous, separating; endocarp thinly woody; columella slender, persistent. Seeds 1 per locule by abortion, fusiform-ellipsoid; testa smooth, shiny, brown; ecarunculate; cotyledons broader than the radicle. Fig.: See Kew Bull. Addit. Ser. 8: 121, t. 1 fig. 1 (1980).

A monotypic genus of eastern New Guinea (Morobe district).

The alternate caudate-acuminate leaves of this plant indeed give it an aspect suggestive of *Phyllanthus*, as noted by Airy Shaw (1980a,b), but its pollen is typical of the *Oldfieldioideae*, and the genus appears very close to *Austrobuxus*.

75. **Scagea** *McPherson*, Bull. Mus. Natl. Hist. Nat., Sér. 4, B, Adansonia 7 (3): 247 (1985); McPherson & Tirel, Fl. Nouv.-Caléd. 14 (1): 90 (1987); G. L. Webster, Ann. Missouri Bot. Gard. 81: 59 (1994). T.: *S. depauperata* (Baill.) McPherson (B.: *Longetia depauperata* Baill.) (S.: *Austobuxus depauperatus* (Baill.) Airy Shaw).

Monoecious shrubs or small trees. Leaves alternate, shortly petiolate, stipulate, simple, entire, penninerved, coriaceous; stipules fugacious. Inflorescences axillary or terminal, uni- or bisexual, the male racemose or pseudoracemose with fasciculate or umbelliform clusters on the axes, the female racemose and few-flowered, or flowers solitary, the bisexual with 1–2 terminal female flowers, the rest male. Male flowers: pedicels capillary; sepals (5)6(7), 2-seriate, imbricate; petals 0; disc 0; stamens (4)6, filaments free, anthers dorsifixed, extrorse, oblong, thecae with a narrow connective, longitudinally dehiscent; pollen spheroidal, pantoporate, pori c. 16, echinate; pistillode small, conic-cylindric, vertically 3-sulcate. Female flowers: pedicels shorter and stouter than male; sepals 6, imbricate; petals 0; disc-glands 3, ovate, flattened, alternilocular; staminodes 0; ovary 3-locular, with only 1 ovule per locule; styles 3, free, erect, short, thick; stigmas triangular, introrse, entire, smooth. Fruit 3-lobate, humerate, style-bases widely separate, dehiscing septicidally into 3 2-valved cocci; exocarp thinly crustaceous, separating; endocarp corneous; columella capitate, persistent. Seeds 1 per locule, ovoid, minutely carunculate or not; funicle-scar large, ovate; testa ± smooth, dull, brownish; endosperm copious; embryo almost straight; cotyledons broad, thin. Fig.: See Bull. Mus. Natl. Hist. Nat., Sér. 4, B, Adansonia 7 (3): 249, t. 1 (1985); Fl. Nouv.-Caléd. 14 (1): 93, t. 18 (1987).

A genus of 2 species endemic to New Caledonia.

The genus was referred to subfamily *Crotonoideae* by McPherson (1985, 1987) because of the uniovulate carpels, but the pollen clearly suggests placement in the *Oldfieldioideae*. It is the only uniovulate member of the subfamily. The genus appears very similar to *Choriceras*, but differs in ovule number and pollen morphology.

76. **Neoroepera** *Müll. Arg. & F. Muell.* in DC., Prodr. 15(2): 488 (1866); Benth., Fl. Austral. 6: 116 (1873); Benth. & Hook. f., Gen. Pl. 3: 276 (1880); Pax & K. Hoffm., Nat. Pflanzenfam., ed. 2, 19C: 73 (1931); Airy Shaw, Muellera 4: 217 (1980) and Kew Bull. 35: 658 (1980); G. L. Webster, Ann. Missouri Bot. Gard. 81: 59 (1994). T.: *N. buxifolia* Müll. Arg. & F. Muell.

Dioecious or less often monoecious shrubs or small trees. Leaves alternate, small, very shortly petiolate, minutely stipulate, simple, entire, indistinctly penninerved. Male flowers axillary, fasciculate or solitary, the fascicles few- to several-flowered; females usually solitary; Male flowers: pedicels capillary; sepals 5–6, 2-seriate, the 3 inner larger, subpetaloid, imbricate; petals 0; stamens 5–6, filaments free, pilose; anthers erect, dorsifixed, introrse, thecae parallel, distinct, minutely granulate, longitudinally dehiscent; pollen spheroidal, pantoporate, pori c. 30, echinate; disc central, flat, 3-lobate; pistillode 0. Female flowers: pedicels much longer than in the male, but not extending in fruit; sepals longer and narrower than in the male and persistent, otherwise similar; disc 6-lobed, the lobes opposite the inner sepals larger than the others; ovary 3-locular, ovules 2 per locule; styles 3, united at the base, spreading; stigmas sessile or stipitate, entire, clavate or suborbicular-flabellate, introrse, ± smooth. Fruit 6-lobate-subglobose, dehiscing septicidally into 3 2-valved cocci; exocarp crustaceous, venose-rugulose, separating; endocarp corneous; columella somewhat irregularly triquetrous. Seeds ovoid-ellipsoid, carunculate; testa smooth, shiny; endosperm copious; embryo almost straight; cotyledons broad, flat. Fig.: See Kew Bull. 35: 660, t. 5A1–4 (1980).

A genus of two species endemic to Australia (Queensland).

77. **Micrantheum** *Desf.*, Mém. Mus. Hist. Nat. 4: 253, t. 14 (1818); Baill., Étude Euphorb.: 555 (1858) (as *Micranthea*); Müll. Arg. in DC., Prodr. 15 (2): 195 (1866); Benth., Fl. Austral. 6: 58 (1873); Benth. & Hook. f., Gen. Pl. 3: 262 (1880); Grüning in Engl., Pflanzenr. 58: 21 (1913); Pax & K. Hoffm., Nat. Pflanzenfam. ed. 2, 19C: 224 (1931); G. L. Webster, Ann. Missouri Bot. Gard. 81: 59 (1994). T.: *M. ericoides* Desf.

Caletia Baill., Étude Euphorb. 553 (1858). T.: *C. micrantheoides* Baill., nom illeg. [=*M. hexandrum* Hook. f.]

Allenia Ewart, Proc. Roy. Soc. Victoria II, 22: 7 (1909). T.: *A. blackiana* Ewart, nom. illeg. [=*M. demissum* F. Muell.]

Monoecious much-branched ericoid shrubs or shrublets. Leaves alternate, small, narrow, very shortly petiolate, stipulate, simple, entire, uninerved; stipules foliaceous, appearing as if 2 further leaves per node. Flowers small, axillary, solitary or in few-flowered fascicles, usually in the upper axils, shortly pedicellate. Male flowers: sepals 4–6, 2-seriate, the inner 2–3 larger, imbricate; petals 0; stamens 3, 4 or 6(9), opposite the outer or all the sepals; filaments free; anthers dorsifixed, extrorse, thecae parallel, distinct, minutely papillose, longitudinally dehiscent, valves hiant; pollen prolate-spheroidal, pantoporate, pori c. 25–50, echinate, spines c. 3μ long; disc central, flat or ± hemispherical, 3-, 4- or 6-lobate; pistillode 0. Female flowers: sepals 6, longer and narrower than in the male, persistent; hypogynous disc annular, shallowly 6-lobed, or 0; ovary 2–3-locular, ovules 2 per locule; styles 2–3, ± united at the base, suberect or spreading and recurved, stigmas entire, linear, ± smooth. Fruit shallowly 2–3-lobate, ovoid-ellipsoid, septicidally dehiscent into 2–3 2-valved cocci or 4–6 valves; exocarp thin, crustaceous, separating; endocarp thinly corneous; columella slender, capitate. Seeds 2 per locule (or 1 by abortion), oblong, smooth, carunculate; endosperm present; embryo linear, straight; cotyledons longer but scarcely broader than the radicle. Fig.: See Étude Euphorb.: t. 26, f. 1–19 (1858); Pflanzenr. 58: 24, t. 5 (1913).

A genus of 4 Australian species, 1 of which reaches Tasmania.

78. **Pseudanthus** *Sieber ex Spreng.*, Syst. Veg. 4(2): 22, 25 (1827); Baill., Étude Euphorb.: 556 (1858); Müll. Arg. in DC., Prodr. 15(2): 196 (1866); Benth., Fl. Austral. 6: 58 (1873); Benth. & Hook. f., Gen. Pl. 3: 262 (1880); Grüning in Engl., Pflanzenr. 58: 26 (1913); Pax & K. Hoffm., Nat. Pflanzenfam. ed. 2, 19C: 224 (1931); G. L. Webster, Ann. Missouri Bot. Gard. 81: 59 (1994). T.: *P. pimeleoides* Sieber ex Spreng.

Stachystemon Planch., London J. Bot. 4: 471, t. 15 (1845); Baill., Étude Euphorb.: 560 (1858); G. L. Webster, Ann. Missouri Bot. Gard. 81: 59 (1994). T.: *St. vermicularis* Planch. [=*Ps. vermicularis* (Planch.) F. Muell.]
Chrysostemon Klotzsch in Lehm., Pl. Preiss. 2: 232 (1848); Baill., Étude Euphorb.: 654 (1858). T.: *Chr. virgatus* Klotzsch [=*Ps. virgatus* (Klotzsch) Müll. Arg.]
Chorizotheca Müll. Arg., Linnaea 32: 76 (1863). T.: *Ch. micrantheoides* Müll. Arg. [=*Ps. virgatus* (Klotzsch) Müll. Arg. (B.: *Chrysostemon virgatus* Klotzsch)]

Monoecious or rarely dioecious much-branched ericoid shrubs or shrublets. Leaves opposite or alternate, small or narrow, subsessile or very shortly petiolate, stipulate, simple, entire, costate, coriaceous; stipules small, free or ± connate, subulate or triangular, often scarious, persistent. Flowers usually small, axillary or subterminal, solitary or in few-flowered fascicles, the males often in the upper axils, the females often at the end of the previous season's growth, sessile or shortly pedicellate. Male flowers: sepals (3–5)6, ± equal or rarely 1 nematomorphous, imbricate, erect, entire, rigid or often petaloid; petals 0; stamens 3, 6, 9–25 or ∞, episepalous when few; filaments free or some or all united, sometimes into a long amentiform column, anthers dorsifixed, basifixed or sessile on the column, extrorse or apicitrorse, thecae convergent, disjunct, minutely papillose or ± smooth, connective often shortly bifid, longitudinally (laterally or apically) dehiscent; pollen spheroidal, pantoporate, pori c. 7–12, brevispinous; disc central, ± flat, 4–5-lobate, or disc-glands c. 3, free, interstaminal, vestigial or 0; pistillode 0. Female flowers: sepals sometimes paleaceous and denticulate or fimbriate, but otherwise similar to the male; petals 0; hypogynous disc minute or 0; staminode 0(1); ovary 2–3-locular, often conical, ovules 2 per locule; styles 2–3, connate at the base or to about half-way, short, thick, erect or divergent or elongate, slender, recurved, stigmas entire, ± smooth. Fruit ± conical, triquetrous, 1-locular by suppression, dehiscing loculicidally into 3–6 valves; exocarp thin, chartaceous; endocarp thin, coriaceous; columella 0. Seeds 1 per fruit by abortion, anatropous, ovoid, smooth, carunculate; caruncle terminal or eccentric; endosperm thin; embryo linear, straight; cotyledons longer but scarcely broader than the radicle. Fig.: See Étude Euphorb.: t. 25, f. 16–21 (1858); Pflanzenr. 58: 29, t. 6 and 34, t. 7 (1913).

An Australian genus of 11 species.

The alleged distinctions between *Pseudanthus* and *Stachystemon* seem to me to be of degree rather than kind. In terms of the androecium (numbers of stamens, and degree of fusion of filaments), for example, a more or less continuous series from few to many and from free to fused exists as follows:– *Pseudanthus* sect. *Microcaletia* → *P.* sect. *P.* → *P.* sect. *Chrysostemon* → *Stachystemon axillaris* → *S. polyandrus* → *S. brachyphyllus* → *S. vermicularis.* This also has a geographical basis, since *P.* sect. *Chrysostemon,* like *Stachystemon,* is also W. Australian. Both genera share the same unusual type of fruit, where the septa disappear during development, whilst the dehiscence into 6 valves is as in many genera where the septa are retained. Both also share the single ovoid seed (Radcliffe-Smith, 1993). However, Stuppy (1995) disagrees with this, stating that seed-characters support their recognition as separate genera: whilst the seeds of *Stachystemon* are strictly anatropous, those of *Pseudanthus* s. str. are better defined as anacampylotropous.

Tribe 14. **PICRODENDREAE** *(Small) G. L. Webster,* Taxon 24: 595 (1975); G. L. Webster, Ann. Missouri Bot. Gard. 81: 60 (1994). T.: *Picrodendron* Planch.

Family *Picrodendraceae* Small, J. New York Bot. Gard. 18: 184 (1917).

Dioecious or rarely monoecious trees or shrubs; leaves alternate, opposite or whorled, 3–8-(rarely unifoliolate), entire; stipules caducous, fused to petiole, or 0. Inflorescences axillary, cymose, racemose or glomerulate, or flowers solitary; male sepals 4–9 (or obsolete), ± free; disc intrastaminal, interstaminal, or 0; stamens 3–54, free; anthers extrorse or introrse; pollen grains ellipsoid to sphaeroidal, commonly brevicolporate or porate, with exinous spines, tectum more or less granulose; pistillode small, minute or 0; female sepals 4–9, persistent or deciduous; disc + or 0; ovary 2–3(5) locular; styles elongate or stigmatiform. Fruit dehiscent or drupaceous. Seeds carunculate or ecarunculate; endosperm usually copious, rarely 0; cotyledons broader and longer than radicle.

This tribe is generally well characterized by its compound leaves, although the leaf blade is simple or unifoliolate in *Mischodon, Androstachys* and *Parodiodendron.*

Key to the subtribes

1. Stipules, where present, minute, free from petiole · · · · · · · · · · · · · · · · 2
1. Stipules where present, persistent or ± so, more or less adnate to petiole · 14c. *Mischodontinae*

2. Stipules +; fruit septicidally dehiscent or drupaceous · · 14a. *Picrodendrinae*
2. Stipules 0; fruit loculicidally dehiscent · · · · · · · · · · · · · · 14b. *Paiveusinae*

Subtribe 14a. **Picrodendrinae** *(Small) G. L. Webster*, Ann. Missouri Bot. Gard. 81(1): 60 (1994). T.: *Picrodendron* Planch.

Family *Picrodendraceae* Small, J. New York Bot. Gard. 18: 184 (1917); Alain, Fl. Cuba 2: 38 (1951); Correll, Fl. Bah. Arch.: 410 (1982).

Dioecious or rarely monoecious trees; leaves alternate, 1–3(5)-foliolate, margins entire; stipules small and caducous. Inflorescences axillary, often from the axils of fallen leaves, cymose, racemose or spicate, sometimes reduced to catkins or glomerules; male sepals 4–9, ± free, or obsolete; disc intrastaminal, interstaminal or 0; stamens 3–54, free; pollen grains 5–8-stephanoporate or brevicolporate, echinate, tectum verruculose or vermiculate; pistillode 0; female sepals 4–9, deciduous or subpersistent; disc annular or 0; ovary 2- or 3-locular; styles elongate or stigmatiform. Fruit dehiscent or drupaceous. Seeds 1–2 per locule, ecarunculate or scarcely carunculate.

Key to the genera

1. Leaves unifoliolate · 80. *Parodiodendron*
1. Leaves 3-5-foliolate · 2
2. Male sepals developed; disc +; ovary 3-locular; fruit dehiscent; endosperm +; cotyledons flat · 79. *Piranhea*
2. Male sepals reduced; disc 0 in both sexes; ovary 2-locular; fruit indehiscent; endosperm 0; cotyledons plicate · · · · · · · · · · · · · · · · · 81. *Picrodendron*

79. **Piranhea** *Baill.*, Adansonia I. 6: 235, t. 6 (1866); Müll. Arg. in Mart., Fl. Bras. 11(2): 79 (1873); Benth., Gen. Pl. 3: 281 (1880); Pax & K. Hoffm., Pflanzenr. 81: 295 (1922), and Nat. Pflanzenfam. ed. 2, 19C: 77 (1931); Jabl., Mem. New York Bot. Gard. 17: 121 (1967); G. L. Webster, Ann. Missouri Bot. Gard. 81: 60 (1994); Radcl.-Sm. & Ratter, Kew Bull. 51: 543 (1996). T.: *P. trifoliolata* Baill.

Celaenodendron Standl., Contr. Dudley Herb. 1: 76, t.1.5 (1927); G. L. Webster, Ann. Missouri Bot. Gard. 81: 60 (1994). T.: *C. mexicanum* Standl. [= *P. mexicanum* (Standl.) Radcl.-Sm.]

Dioecious evergreen or deciduous tree; bark smooth, striate, or flaking off in chips; wood hard. Indumentum simple. Leaves alternate, stipulate, petiolate, digitately 3–5-foliolate, leaflets shortly petiolulate or sessile, entire, penninerved; stipules usually small, oblong or filiform at apex, readily deciduous. Male inflorescences axillary, often arising from the axils of fallen leaves, slender, interruptedly pseudoracemose; bracts minute or 0. Male flowers: pedicels slender; calyx irregularly 4–6-lobate, the lobes small, basally imbricate, apically open in bud; petals 0; receptacle slightly domed; disc-glands interstaminal, myloid, some central, aggregated and pseudopistillodoid; stamens 8–16, filaments free, slender, anthers baso-medifixed, small, introrse, latrorse or extrorse, 2-thecate, thecae parallel, free at apex and base, papillose, longitudinally dehiscent; pollen suboblate-spheroidal, stephanoporate, pori 6–8, echinate, interspinally verrucate; pistillode 0. Female inflorescences axillary, usually arising from the axils of fallen leaves, racemose or flowers solitary. Female flowers: pedicels often long, and extending in fruit; calyx lobes 6–9, unequal, imbricate, subpersistent or caducous; petals 0; staminodes sometimes present, several, subulate; disc annular, shallowly hexalobate; ovary 3-locular, ovules 2 per locule, covered by a large strophiole; styles (1)3(6), simple, free or united, linear, acute or bifid, strongly recurved, stigmas papillose or minutely puberulous. Fruit hexangular-subglobose, acuminate or apically-depressed, hard, 3-locular, dehiscing septicidally into 3 2-valved cocci, the cocci elastically loculicidally dehiscent; exocarp crustaceous, partially separating from the thick woody or bony endocarp; septa complete, strongly venose; columella persistent. Seeds solitary per locule by abortion, anatropous, ecarunculate or ± so, ovoid or oblong; testa crustaceous or coriaceous, shiny, black or dark brown; endosperm fleshy; cotyledons broad and flat. Fig.: See Adansonia I. 6: 236, t. 6 (1866); Bol. Mus. Paraense Emilio Goeldi, N.S., Bot. 7: 43, t. 15 (1991); Kew Bull. 51: 545, t. 1 (1996).

A tetratypic genus from W Mexico (Sinaloa, Nayarit, Jalisco and Colima), Venezuela, Guyana, N and SE Brazil (Minas Gerais), *fide* Radcliffe-Smith & Ratter (1996).

80. **Parodiodendron** *Hunz.*, Kurtziana 5: 331, tt. 1–3 (1969); G. L. Webster, Ann. Missouri Bot. Gard. 81: 60 (1994). T.: *P. marginivillosum* (Speg.) Hunz. (B.: *Phyllanthus marginivillosus* Speg.)

Dioecious tree; bark smooth. Indumentum simple. Leaves small, alternate, clustered at the ends of the branchlets, unifoliolate, stipulate, shortly petiolate, often bistipellate, elliptic, emarginate, entire, penninerved; stipules caducous. Male flowers in axillary cymose fasciculate glomerules, pedicellate, surrounded at the base by rounded bracts; sepals (5)6(9) in 2 whorls, imbricate, closed in bud; petals 0; stamens (13)15(19), arising from pits in the convex disc covering the receptacle, filaments free, anthers dorsifixed, extrorse, 2-thecate, free at base, papillose, especially at the apex, longitudinally dehiscent; pollen suboblate-spheroidal, stephanoporate, pori 7, echinate, verrucate between the spines; pistillode 0. Female flowers axillary, solitary, long-pedicellate; sepals 6, imbricate, 2-seriate, foliaceous, caducous; petals 0; hypogynous disc annular, 6-lobed; ovary 3-locular, ovules 2 per locule, epitropous, with a large fleshy obturator; styles 3, very short, connate at the base, entire, revolute. Fruit trilobate, subglobose, dehiscing septicidally into 3 2-valved cocci; exocarp thinly crustaceous, tardily separating; endocarp bony; septa ± complete, venose; columella persistent, conoidal. Seeds 1–2 per locule, ecarunculate, ovoid; testa smooth, shiny, dark greyish-brown; endosperm abundant; cotyledons broad and flat. Fig.: See Kurtziana 5: 331, tt. 1–3 (1969).

A monotypic genus of northern Argentina (Salta and Jujuy Provs.) and S. Bolivia (Chuquisaca Prov.)

Stuppy (1995) states that as the seed-morphology of *Parodiodendron* and *Piranhea* is almost identical, he doubts whether they are generically distinct.

81. **Picrodendron** *Planch.*, London J. Bot. 5: 579 (1846); nom. cons. prop.; Krug & Urb., Bot. Jahrb. Syst. 15: 308 (1893); Fawc. & Rendle, J. Bot. 55: 268 (1917), and Fl. Jamaica 4: 273 (1920); Pax & K. Hoffm., Nat. Pflanzenfam. ed. 2, 19C: 232 (1931); Alain, Fl. Cuba 2: 38 (1951); W. J. Hayden *et al.*, J. Arnold Arbor. 58: 257 (1977), and 65: 109 (1984); Correll, Fl. Bah. Arch.: 410 (1982); Proctor, Kew Bull. Addit. Ser. 11: 526 (1984); Alain, Fl. Español. 4: 206 (1986); G. L. Webster, Ann. Missouri Bot. Gard. 81: 60 (1994). T. (typ. cons.): *P. baccatum* (L.) Krug & Urb. (B.: *Juglans baccata* L.).

Dioecious or rarely monoecious deciduous tree; bark bitter, peeling off in long strips. Indumentum simple. Leaves alternate, stipulate, long-petiolate, digitately 3-foliolate, leaflets petiolulate, minutely and irregularly stipellate, entire, penninerved; stipules minute, subulate, deciduous. Male inflorescences arising from the axils of fallen leaves, coaëtaneously with the new leaves, interruptedly spicate or subpaniculate; bracts (1)3(7) per flower. Male flowers: sepals 0; petals 0; disc 0; stamens 3–54, borne on a hemispherical receptacle, filaments free, short, anthers basifixed, oval, slightly extrorse, apiculate, puberulous at apex, 2-thecate, thecae free at base, longitudinally dehiscent; pollen oblate-spheroidal, stephanobrevicolporate, apertures 5–8, echinate, vermiculate; pistillode 0. Female flowers axillary, solitary, borne on the hornotinous shoots; pedicels elongate; sepals (or bracts (Hayden, 1984)) 4(5), valvate, unequal, ascending-recurved, glanduliferous, subentire; disc and staminodes 0; ovary 2-locular, narrowing into the style, ovules 2 per locule, anatropous, obturator pulviniform, withering in fruit; styles 2, connate, stigmas 2, divergent, elongate, subulate, revolute. Fruit globose, indehiscent, drupaceous, 2-locular, yellowish, drying black; exocarp thin, fleshy, bitter, adherent; endocarp woody to bony, 4-lineate, brittle. Seed usually 1 by abortion (rarely 2, and then 1 per locule), irregularly subglobose; testa membranaceous, invaginated; endosperm 0; cotyledons plicate, corrugated. Fig.: See Fl. Jamaica 4: 274, t. 90 (1920); Fl. Bah. Arch.: 411, t. 165 (1982); J. Arnold Arbor. 65: 107, t. 1 and 112, t. 2 (1984); Kew Bull. Addit. Ser. 11: 528, t. 160 (1984); Fl. Español. 4: 207, t. 116-31 (1986).

A monotypic genus endemic to the West Indies (Greater Antilles, Bahamas, Cayman Is., Swan Is.).

The anomalous structure of the fruit and seed, as well as the male inflorescence, seems to indicate that, despite pollen and other similarities, *Picrodendron* does not really sit comfortably in the *Oldfieldioideae*, nor indeed in the *Euphorbiaceae*. Various authors have in the past assigned it to a number of families in different orders, and in an attempt to resolve this unsatisfactory situation Small (1917) accorded it separate familial status. Thus, for the moment, the dispositions of Hayden (1977), Hayden *et al.* (1984) and Webster (1994) are followed here pending further investigation and assessment. Stuppy (1995) would exclude it from the *Euphorbiaceae* on the grounds of seed morphology, favouring separate family status, but with the proviso that many of its features do corroborate association with the *Oldfieldioideae*.

Subtribe 14b. **Paivaeusinae** *Pax & K. Hoffm.*, Pflanzenr. IV, 147, xv (Heft 81): 294 (1922); G. L. Webster, Ann. Missouri Bot. Gard. 81: 60 (1994). T.: *Paivaeusa* Welw. ex Benth. [=*Oldfieldia* Benth. & Hook. f.]

Dioecious trees or shrubs; leaves alternate or opposite, 3–8-foliolate; stipules 0. Male inflorescences in pedunculate more or less congested cymes, the female reduced to 1–3 flowers; male sepals 5–8, basally connate; disc intrastaminal, lobed; stamens 4–12, free, anthers extrorse; pistillode minute or 0; female sepals 5–8, persistent; disc annular; ovary 2–3 locular; stigmas dilated, lobed. Fruit tardily loculicidally dehiscent. Seeds with a carunculoid funicle.

As circumscribed here, in agreement with the studies of Levin, this subtribe includes only the single African genus *Oldfieldia.*

82. **Oldfieldia** *Benth. & Hook. f.*, Hooker's J. Bot. Kew Gard. Misc. 2: 184, t. 6 (1850); Baill., Étude Euphorb.: 657 (1858); Müll. Arg. in DC., Prodr. 15(2): 1259 (1866); Benth., Gen. Pl. 3: 281 (1880); Hutch. in Fl. Trop. Afr. 6(1): 625 (1912); Pax & K. Hoffm., Pflanzenr. 81: 297 (1922), and Nat. Pflanzenfam. ed. 2, 19C: 78 (1931); Milne-Redh., Kew Bull. 3: 456 (1949); J. Léonard, Bull. Jard. Bot. État. 26: 338 (1956); Keay (ed.), Fl. W. Trop. Afr. ed. 2, 1: 368 (1958); Radcl.-Sm., Fl. Trop. E. Afr., Euph. 1: 114 (1987); G. L. Webster, Ann. Missouri Bot. Gard. 81: 61 (1994); Radcl.-Sm., Fl. Zamb. 9(4): 117 (1996). T.: *O. africana* Benth. & Hook. f.

Paivaeusa Welw. ex Benth., Gen. Pl. 1: 993 (1867); Hutch. in Fl. Trop. Afr. 6(1): 626 (1912); Pax & K. Hoffm., Pflanzenr. 81: 296 (1922), and Nat. Pflanzenfam. ed. 2, 19C: 78 (1931). T.: *P. dactylophylla* Welw. ex Oliv. [=*O. dactylophylla* (Welw. ex Oliv.) J. Léonard]
Cecchia Chiov., Fl. Somala 2: 397, fig. 227 (1932). T.: *C. somalensis* Chiov. [=*O. somalensis* (Chiov.) Milne-Redh.]

Dioecious pachycaul trees or shrubs. Twigs cicatricose. Leaves alternate, opposite or verticillate, long-petiolate, exstipulate, digitately 3–8-foliolate; leaflets subsessile or petiolulate, entire, penninerved, lateral nerves brochidodromous. Inflorescences axillary, solitary or geminate, cymose, the males subsessile or pedunculate, many-flowered, lax or capitate, the females pedunculate, the peduncles extending in fruit, 1–3-flowered. Male flowers small; calyx 5–8-fid; petals 0; stamens 4–12, free, inserted between the lobes of the disc; filaments unequal; anthers dorsifixed, extrorse, finely papillose, longitudinally widely dehiscent; pollen suboblate to oblate-spheroidal, stephanocolpate, colpi small, short, costate, echinate; disc central, thick, fleshy, lobate-sinuate, pubescent; pistillode minute, filiform, entire, truncate or 0. Female flowers: pedicels short; sepals longer, narrower and thicker than in the male, persistent in fruit; disc annular, fleshy, crenellate; ovary 2–3-locular, ovules 2 per locule; styles 2–3, short, connate at the base; stigmas reniform, 2–4-lobulate, caducous. Fruit 2–3-locular, depressed-ovoid globose, marked by 1–3 circular vertical lines, tardily loculicidally dehiscent into 3 valves each with an attached septum; pericarp coriaceous, mesocarp crustaceous, endocarp chartaceous; columella persistent, often with the seeds attached. Seeds 1–2 per locule, usually 3 per fruit, slightly compressed; exotesta fleshy; endotesta crustaceous; funicle thickened, carunculoid; albumen fleshy; embryo green; cotyledons broad, flat. Fig.: See Hooker's J. Bot. Kew Gard. Misc. 2: 184, t. 6 (1850); Fl. Somala 2: 397, fig. 227 (1932); Fl. W. Trop. Afr., ed. 2, 1(2): 369, t. 130 (1958); Bot. J. Linn. Soc. 94 (1/2): 60, t. 8 (1987); Fl. Trop. E. Afr., Euph. 1: 116, t. 21 (1987); Fl. Zamb. 9(4): 119, t. 19 (1996).

A tetratypic Tropical African genus, *fide* Milne-Redhead (1949), Léonard (1956) and Radcliffe-Smith (1987).

The distinction in phyllotaxy (alternate vs. opposite leaves) used by Pax & Hoffmann (1922, 1931) to distinguish *Paivaeusa* is clearly unworkable, as alternate and opposite leaves occur in both *Paivaeusa* and *Oldfieldia*. The features of wood-anatomy of *Oldfieldia* are unique in the family (Metcalfe & Chalk, 1950), and this, taken in conjunction with distinctive features of seed-morphology, suggests to Stuppy (1995) that its affiliation with the *Euphorbiaceae* should be critically reviewed. It has some seed-characters in common with *Meliaceae* and *Rutaceae*, which is of interest, since Müller (1866) excluded it from the *Euphorbiaceae* and suggested an affinity with the *Sapindaceae*, whilst Bentham (1867) originally placed *Paivaeusa* with the *Burseraceae*!

Subtribe 14c. **Mischodontinae** *Müll. Arg.*, Linnaea 34: 202 (1865) and in DC., Prodr. 15(2): 1124 (1866); G. L. Webster, Taxon 24: 595 (1975), and Ann. Missouri Bot. Gard. 81: 61 (1994). T.: *Mischodon* Thwaites.

Androstachydaceae Airy Shaw, Kew Bull. 18: 250 (1965). T.: *Androstachys* Prain.

Dioecious trees or shrubs; leaves alternate or opposite, simple or 3–7-foliolate, entire; stipules usually persistent, adnate to petiole or 0; flowers in axillary or subterminal cymes or clusters, or solitary; male sepals (2) 5–8, free; disc intrastaminal, interstaminal or 0; stamens 5–50, free; anthers extrorse; pistillode small or 0; female sepals 5–7, deciduous or persistent; disc annular or 0; ovary 3–5-locular, glabrous or pubescent; styles slightly to distinctly connate below; fruit dehiscent; seeds carunculate or ecarunculate.

Key to the genera

1. Stipules large, small, epipetiolar or 0; fruiting pedicel thicker than the columella 2
1. Stipules large, intrapetiolar; fruiting pedicel not thicker than the columella 3

2. Leaves mostly alternate; flowers in fascicles below the leaves · 83. *Aristogeitonia*
2. Leaves verticillate; flowers in axillary or subterminal panicles · 84. *Mischodon*
3. Stamens free, filaments long; male disc + · · · · · · · · · · · · · · · 85. *Voatamalo*
3. Stamens with filaments united into a long column, anthers subsessile; male disc 0 · 4
4. Stipules free from the petiole, soon caducous; leaves simple, palminerved, often peltate; fruit apically depressed · · · · · · · · · · · · · · 86. *Androstachys*
4. Stipules adnate to the petiole, long-persistent; leaves 3–7-foliolate; leaflets penninerved; fruit apiculate · 87. *Stachyandra*

83. **Aristogeitonia** *Prain*, Bull. Misc. Inform., Kew 1908: 439 (1908), Hooker's Icon. Pl. 30: t. 2926 (1911) and Fl. Trop. Afr. 6(1): 625 (1912); Pax & K. Hoffm., Pflanzenr. 81: 296 (1922), and Nat. Pflanzenfam. ed. 2, 19C: 78 (1931); Airy Shaw, Kew Bull. 26: 495 (1972); Radcl.-Sm., Fl. Trop. E. Afr. Euph. 1: 118 (1987) and Kew Bull. 43: 627 (1988); G. L. Webster, Ann. Missouri Bot. Gard. 81: 61 (1994); Radcl.-Sm., Kew Bull. 51: 799 (1996), and 53: 977 (1998). T.: *A. limoniifolia* Prain.

Paragelonium Leandri, Bull. Soc. Bot. France 85: 531 (1939). T.: *P. perrieri* Leandri [= *A. perrieri* (Leandri) Radcl.-Sm.]

Dioecious small trees or shrubs, not or little-branched. Leaves alternate or occasionally whorled, 1-foliolate or digitately 2–3-foliolate, stipulate, long-petiolate; leaflets sessile, entire, penninerved; stipules filiform-subulate, epipetiolar, sometimes 0; petiole not or obscurely biglandular at the apex; flowers in small bracteate fascicles on old wood in axils of fallen leaves. Male flowers: sepals 6, 2-seriate, imbricate; petals 0; stamens 9–15(20), filaments free, arising from convolutions in the disc; anthers dorsifixed or basifixed, extrorse, oblong or ovoid, thecae 2, parallel, longitudinally dehiscent; pollen suboblate, stephanocolpate, echinate; disc central, 6–8-lobed; pistillode small, 2–3-lobed. Female flowers: pedicels elongating and thickening in fruit; sepals 6, 2-seriate, imbricate, soon falling; petals 0; disc conspicuous, annular, crenate-lobulate; ovary 3-locular, ovules 2 per locule; styles short, flat, spreading, obcordate, slightly connate at the base. Fruit trilobate or trigonous-subglobose; exocarp thin, separating and dehiscing loculicidally, endocarp dehiscing septicidally into 3 2-valved cocci; endocarp thinly woody, crustaceous; seed often 1 per locule by abortion, ovoid with a small apical hook, ecarunculate; testa crustaceous, shining, black; endosperm copious; cotyledons flat, wide-cordate, palminerved; radicle superior. Fig.: See Hooker's Icon. Pl. 30: t. 2926 (1911); Bull. Soc. Bot. France 85: 525, t. 1.8–12 (1938) (as *Paragelonium*); Bot. J. Linn. Soc. 94 (1/2): 62, t. 9 (1987); Fl. Trop. E. Afr. Euph. 1: 119, t. 22 (1987); Kew Bull. 43: 626, t. 1 and 628, t. 2 (1988); 51: 800, t. 1 (1996), and 53: 979, t. 1 (1998). See also Fig. 11, p. 107.

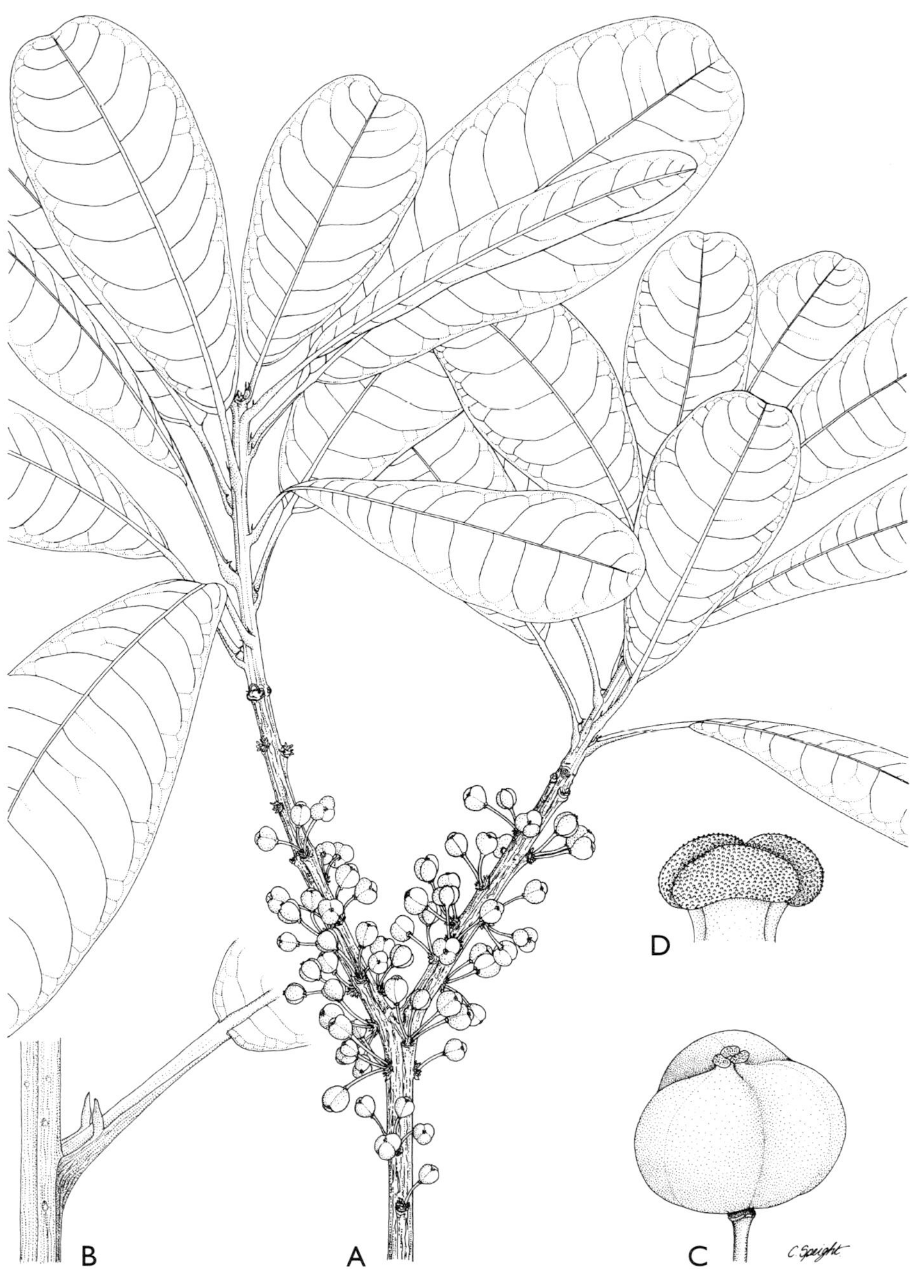

FIG. 11. *Aristogeitonia uapacifolia* Radcl.-Sm. **A** Habit × $^2/_3$ (after reduction); **B** Petiole showing epipetiolar stipules × 3 (after reduction); **C** Young fruit × 4 (after reduction); **D** Styles × 20 (after reduction). All from *McPherson* 14400. Drawn by Camilla Speight.

A heptatypic Afro-Malagasy genus with 1 species in Angola, 2 in Kenya and Tanzania, and 4 in Madagascar, *fide* Radcliffe-Smith (1988, 1996a, 1998 and ined.).

84. **Mischodon** *Thwaites*, Hooker's J. Bot. Kew Gard. Misc. 6: 299 (1854); Baill., Étude Euphorb.: 335 (1858); Müll. Arg. in DC., Prodr. 15(2): 1124 (1866); Benth., Gen. Pl. 3: 280 (1880); Hook. f., Fl. Brit. India 5: 344 (1887); Pax & K. Hoffm., Pflanzenr. 81: 292 (1922), and Nat. Pflanzenfam. ed. 2, 19C: 76 (1931); V. S. Raju, J. Econ. Taxon. Bot. 5: 165 (1984); G. L. Webster, Ann. Missouri Bot. Gard. 81: 61 (1994); Philcox in Dassanayake, ed., Fl. Ceyl. XI: 267 (1997). T.: *M. zeylanicus* Thwaites.

Dioecious evergreen tree; twigs quadrangular. Leaves 3–4-verticillate, stipulate or not, petiolate or subsessile, simple, long, narrow, entire, penninerved; stipules minute, epipetiolar, or 0; Inflorescences axillary or subterminal, paniculate, males dense, subcapitate, females fewer-flowered and laxer. Male flowers: sepals 5–8, imbricate; petals 0; stamens 5–6 (10), episepalous, filaments free, elongate, hirsute; anthers ovate, dorsifixed, extrorse, connective pubescent, thecae parallel, free at apex and base, longitudinally dehiscent; pollen oblate-spheroidal, stephanocolpate, colpi 5–6, echinate, tectate; disc central, 5–6(10) lobed; pistillode 0. Female flowers: pedicels elongating and thickening in fruit; sepals 6, imbricate, caducous; petals 0; staminodes 0; disc annular, lobulate; ovary 3–4-locular, ovules 2 per locule; styles 3(4), united at the base, simple, short, spreading, stigmas flat, thick, obovate, papillose. Fruit deeply 3-lobate, lobes obtusely keeled, dehiscing septicidally (endocarp) and loculicidally (exocarp), separating; endocarp thinly woody; columella persistent, much narrower than the fruiting pedicels. Seeds often solitary per locule by abortion, ovoid with a small apical hook; testa crustaceous, shining, black; endosperm fleshy; cotyledons broad, flat; radicle short. Fig.: See Pflanzenr. 81: 293, t. 23 (1922); Nat. Pflanzenfam. ed. 2, 19C: 75, t. 36 (1931); Bot. J. Linn. Soc. 94 (1/2): 62, t. 10 (1987).

A monotypic genus of southern India and Ceylon; very rare in the former.

85. **Voatamalo** *Capuron ex Bosser*, Adansonia II. 15 (3): 333 (1976); G. L. Webster, Ann. Missouri Bot. Gard. 81: 61 (1994). T.: *V. eugenioides* Capuron ex Bosser.

Dioecious trees or shrubs; branching monopodial. Leaves opposite, decussate, stipulate, petiolate, simple, entire, penninerved, the lateral nerves closely parallel, glabrous; stipules intrapetiolar, each pair connate, the pairs contiguous, covering the terminal bud, partially caducous. Inflorescences axillary or subterminal, the males umbellate, with the peduncles simple or bifid, the females dichasial. Male flowers: pedicels long, capillary; sepals 6, in 2 whorls, imbricate; petals 0; disc-glands interstaminal, arising from a fleshy receptacle; stamens 9–14(27), in 2 whorls, the outer alternate with the inner sepals, anthers medifixed, extrorse, connective not produced, thecae longitudinally dehiscent;

pollen globose, zoni-aperturate, echinate, tectum baculate/microperforate; pistillode 0. Female flowers: sepals 6(7), in two whorls, imbricate, the outer much larger than the inner, foliaceous, persistent; petals 0; disc annular; staminodes 0; ovary 3–5-locular, ovules 2 per locule, attenuate into the stylar column; styles 3, free in upper part, simple, divaricate, stigmas lingulate, minutely papillose. Fruit 3–5-locular, dehiscing into bivalved cocci; pericarp lineate; epicarp and mesocarp thin; endocarp corky; columella persistent. Seeds carunculate. Fig.: See Adansonia II. 15 (3): 335, t. 1 (1976).

A genus of 2 species in N. Madagascar.

86. **Androstachys** *Prain*, Bull. Misc. Inform., Kew 1908: 438 (1908); Hutch. in Fl. Trop. Afr. 6(1): 740 (1912); Pax & K. Hoffm. Pflanzenr. 81: 287 (1922), and Nat. Pflanzenfam. ed. 2, 19C: 75 (1931); Leandri, Fl. Madag. 111: 197 (1958); Airy Shaw, Kew Bull. 18: 251 (1965), and Adansonia II. 10: 519 (1970); Dyer, Gen. S. Afr. Fl. Pl. ed. 3: 312 (1975); J.-F. Leroy, Compt. Rend.-Hebd. Séances Acad. Sci., Sér. D 283: 147 (1976); Alvin, Ann. Bot. (London) 59: 579 (1987); G. L. Webster, Ann. Missouri Bot. Gard. 81: 61 (1994); Radcl.-Sm., Fl. Zamb. 9(4): 120 (1996). T.: *A. johnsonii* Prain.

Weihea Sim, Forest Fl. Port. E. Afr.: 66 (1909), non Spreng. T.: *W. subpeltata* Sim (S.: *A. subpeltatus* (Sim) E. Phillips) [=*A. johnsonii* Prain]

Dioecious trees; branchlets closely nodose; branching monopodial. Leaves opposite, decussate, stipulate, petiolate, sometimes peltate, simple, entire, palminerved, coriaceous, villous below; stipules large, intrapetiolar and free from the petiole, connate and covering the terminal bud, soon deciduous. Flowers axillary; males in triads, pendent, females solitary, long-pedicellate. Male flowers: sepals 2–5, free, spirally arranged, or ± whorled, bracteiform; petals 0; disc 0; stamens numerous, spirally inserted on an elongate receptacle; lower filaments short and recurved, upper 0; anthers dorsifixed, elongated, extrorse, 2-thecate, thecae free at apex and base, connective setose at first, longitudinally dehiscent; pollen ellipsoïdal, inaperturate or pantoporate, brevispinous; pistillode 0. Female flowers: sepals 5–6, imbricate, caducous; petals 0; disc 0; ovary 3(5)-locular, silky, ovules 2 per locule; styles connate into an elongate column; stigmas 3, simple, reflexed. Fruit depressed-globose, 3-lobate, the keels subcarinate, smooth, dehiscing septicidally and loculicidally into 3 2-valved cocci or 6 valves; indumentum detersible; endocarp crustaceous; columella narrow, persistent. Seeds 6, ovate, laterally compressed, shiny, ecarunculate; endosperm fleshy; cotyledons flat, much broader than the radicle. Fig.: See Fl. Madag. 111: 198, t. XXXV (1958); Fl. Zamb. 9(4): 122, t. 20 (1996).

A monotypic genus native to south-east Africa and Madagascar.

87. **Stachyandra** *J. F. Leroy ex Radcl.-Sm.*, Kew Bull. 45(3): 562 (1990); G. L. Webster, Ann. Missouri Bot. Gard. 81: 62 (1994). T.: *S. merana* (Airy Shaw) J.-F. Leroy ex Radcl.-Sm. (B.: *Androstachys merana* Airy Shaw).

Androstachys sensu Airy Shaw, Adansonia II. 10(4): 519–524 (1970), non Prain.
Androstachys Prain subgen. *Archandrostachys* J.-F. Leroy, Compt. Rend. Hebd. Séances Acad. Sci., Sér. D, 283 (2): 150 (1976). T.: *A. merana* Airy Shaw [=*Stachyandra merana* (Airy Shaw) J.-F. Leroy ex Radcl.-Sm.]

Large dioecious trees; branching monopodial. Leaves opposite, decussate, stipulate, sessile or petiolate, digitately 3-, 5- or 7-foliate, the leaflets petiolulate or sessile, involute at first, later flat or slightly revolute, entire, penninerved, coriaceous, variously indumented; stipules large, connate, completely enclosing the terminal bud, persistent, with the petioles or petiolules partially or completely adnate to them. Male inflorescence axillary, cymose, 3-flowered. Male flowers: sepals 3–6, free, irregularly-arranged, bract-like; petals 0; disc 0; stamens ∞, spirally arranged on a spirally elongate pendent receptacle; filaments very short; anthers dorsifixed, extrorse, elongate, 2-thecate, longitudinally dehiscent, connective produced, densely ferrugineous-setiferous; pollen globose, zoni-aperturate, echinate, tectum paucimicroperforate; pistillode 0. Female flowers solitary; pedicels extending in fruit; sepals in 2 whorls of 3, imbricate; petals 0; disc 0; staminodes 0; ovary 3-locular, ovules 2 per locule; styles connate to about half-way, stigmas 3, simple, contorted. Fruit rounded-3-lobate, apiculate, often rugulose, dehiscing septicidally and loculicidally into 3 2-valved cocci; indumentum persistent; exocarp separating; endocarp woody; columella stout, persistent. Seeds 6, ovoid, shiny, carunculate. Fig.: See Kew Bull. 45(3): 563–566, tt. 1–4 (1990).

A genus of 4 species endemic to Madagascar, *fide* Radcliffe-Smith (1990).

Subfamily III. **ACALYPHOIDEAE** *Asch.*, Fl. Brandenburg 1: 58 (1864); G. L. Webster, Ann. Missouri Bot. Gard. 81: 62 (1994). Type: *Acalypha* L.

Acalyphaceae Klotzsch & Garcke, Monatsber. Königl. Preuss. Akad. Wiss. Berlin 1859: 246 (1859).

Trees, shrubs or herbs; milky latex mostly absent; indumentum simple, stellate or lepidote; leaves usually alternate, stipulate, blade simple or palmatilobate, often glandular. Inflorescences mostly axillary or terminal, racemose, spicate or glomerulate, or occasionally flowers solitary; bracts sometimes glandular. Male sepals imbricate or valvate; petals and disc + or 0; stamens 2–1000, free or connate; pollen 2-nucleate, mostly 3–4-colporate, semitectate, rarely echinate; pistillode+ or 0; female sepals (2)3–6(12), imbricate or apert, sometimes connate; petals and disc + or 0; ovary mostly 2–4-locular, ovules 1 per locule [except *Dicoelia*], anatropous; styles entire, 2-fid or ∞-fid. Fruit usually dehiscent. Seeds carunculate or not, exotesta sometimes fleshy; endosperm usually copious; cotyledons usually longer and broader than the radicle; n mostly = 9, 10 or 11.

With 119 genera disposed in 20 tribes, or c. $^3/_8$ of the genera in $^2/_5$ of the tribes, this subfamily is the largest and most complex in the family. Through *Clutia* and *Dicoelia* it approaches the *Phyllanthoideae* in different ways, and through *Omphalea* both the *Crotonoideae* and the *Euphorbioideae*.

Key to the tribes

1. Male sepals and petals imbricate; stamens 5; pollen tectate-perforate; fruit dehiscent · · · · · · · · · · 2
1. Male sepals, petals and stamens not as above, or, if so, then pollen or fruit not as above · · · · · · · · · · 3
2. Petals not adaxially barbate; filaments connate; disc-glands distinct; seeds carunculate; indumentum simple · · · · · · · · · · 15. *Clutieae*
2. Petals adaxially barbate; filaments free; disc urceolate; seeds ecarunculate; indumentum simple or malpighiaceous · · · · · · · · · 16. *Pogonophoreae*
3. Male sepals distinctly imbricate, or else flowers enclosed in a 2-bracteolate involucre; petals 0 · · · · · · · · · · 4
3. Male sepals valvate or slightly imbricate; flowers not in a 2-bracteolate involucre; petals + or 0 · · · · · · · · · · 7
4. Leaves opposite; pollen exine reticulate; seeds ecarunculate, testa dry · · · · · · · · · · · 20. *Erismantheae*
4. Leaves alternate; pollen exine rugulose or micropunctate · · · · · · · · · · 5
5. Filaments free; pollen-tectum spinulose; seeds ecarunculate; testa fleshy · · · · · · · · · · 19. *Cheiloseae*
5. Filaments connate; pollen-tectum not spinulose; seeds carunculate; testa dry · · · · · · · · · · 6
6. Flowers not involucrate; stamens 5–20; styles elongate, bifid; leaves stipulate · · · · · · · · · · 17. *Chætocarpeae*
6. Flowers involucrate; stamens 2–6(8); styles abbreviate, stigmatiform; leaves usually exstipulate · · · · · · · · · · 18. *Pereae*
7. Male petals with oppositistemonous depressions, or fruit drupaceous with a channelled endocarp · · · · · · · · · · 8
7. Male petals lacking oppositistemonous depressions; fruit not drupaceous with a channelled endocarp · · · · · · · · · · 9
8. Monoecious; ovules 2 per locule; styles simple; fruit dehiscent 21. *Dicoelieae*
8. Dioecious; ovules 1 per locule; styles 2-fid; fruit drupaceous · · 22. *Galearieae*
9. Cotyledons scarcely broader than the radicle; foliage ericoid · 23. *Ampereae*
9. Cotyledons distinctly broader than the radicle; foliage not ericoid · · · 10
10. Petals usually +, at least in the male flowers · · · · · · · · · · 11
10. Petals 0 · · · · · · · · · · 13
11. Indumentum, at least in part, malpighiaceous, stellate or lepidote; anthers not as below; pollen often heterobrochous · · · · · · · 26. *Chrozophoreae*
11. Indumentum simple; anther-connective enlarged, thecae pendulous; pollen reticulate · · · · · · · · · · 12
12. Leaves stipulate; inflorescences paniculate, racemose or spicate, bracts not crowded; pollen spheroidal; female petals usually +; styles bifid · · · · · · · · · · · 24. *Agrostistachydeae*

12. Leaves exstipulate; inflorescences subumbellately capitellate, bracts crowded; pollen prolate; female petals 0; stigmas multifid, fimbriate or laciniate · 25. *Sphyranthereae*
13. Styles simple, mostly connate into a distinct column; often scandent or twining, sometimes with urticating hairs · 14
13. Styles free or basally connate (if long-connate, then styles 2-fid or indumentum stellate); rarely scandent; urticating hairs 0 · · · · · · · · 15
14. Male sepals valvate; latex 0; inflorescences capitate, racemose or spicate; bracts eglandular; stamens up to 100, filaments free or connate · 34. *Plukenetieae*
14. Male sepals imbricate; latex clear or reddish; inflorescences paniculate; bracts ± foliaceous, 2-glandular; stamens 2–3, filaments connate · 35. *Omphaleae*
15. Male disc +; pollen coarsely reticulate, colpi mostly inoperculate · · · · 16
15. Male disc 0 (or if +, colpi often operculate) · · · · · · · · · · · · · · · · · · · 18
16. Leaves eglandular; indumentum simple; inflorescence axillary; stamens ∞ · 29. *Pycnocomeae*
16. Leaves usually glandular; indumentum simple or stellate; inflorescence axillary or terminal; stamens (2)20–60(90) · · · · · · · · · · · · · · · · · · 17
17. Male disc massive, pubescent; stamens 4–15, anther-connective enlarged; colpi non-marginate · 27. *Caryodendreae*
17. Male disc tenuous or 0; stamens (2)20–60(90), anther-connective usually not enlarged; colpi marginate · · · · · · · · · · · · · · · · · · 28. *Bernardieae*
18. Pollen coarsely reticulate or perforate-tectate; indumentum stellate; stamens usually inflexed in bud; female calyx sometimes accrescent · 30. *Epiprineae*
18. Pollen finely perforate-tectate to rugulose; indumentum simple or stellate; stamens not inflexed in bud · 19
19. Pollen colpi inoperculate, often reduced; seeds carunculate or ecarunculate, testa dry or fleshy · · · · · · · · · · · · · · · · · 33. *Acalypheae*
19. Pollen colpi operculate; male disc + or 0; seeds mostly ecarunculate, testa not fleshy; stipules deciduous or 0 · 20
20. Leaves usually eglandular; male disk extrastaminal, interstaminal, annular, dissected or 0; pollen perforate-tectate, polygonally nanospinulose; styles usually laciniate · 31. *Adelieae*
20. Leaves usually with embedded laminar glands; male disk, if +, intrastaminal; pollen rugulose to striate, irregularly nanospinulose; styles entire, or if divided, inflorescences terminal · · · · 32. *Alchorneae*

Tribe 15. **CLUTIEAE** *(Müll. Arg.) Pax,* Nat. Pflanzenfam. ed. 1, 3(5): 81 (1890) (as *Cluytieae*); G. L. Webster, Ann. Missouri Bot. Gard. 81: 63 (1994). Type: *Clutia* L.

Hippomaneae subtribe *Cluytieae* Müll. Arg., Linnaea 34: 202 (1865).
Cluytieae subtribe *Cluytiinae* Pax, Pflanzenr. 47: 49 (1911).

Dioecious or rarely monoecious perennial herbs or shrubs; indumentum simple or 0; leaves alternate, simple, entire, penninerved, often pellucid-punctate, eglandular; stipules small or 0. Flowers in axillary glomerules, females often solitary; sepals 5, imbricate; petals 5, free; disc in 1–3 series; stamens 5, filaments connate; anthers muticous, extrorse; pollen prolate, 3-colporate, colpae inoperculate, marginate, exine tectate-perforate; pistillode +; ovary 3(4)-locular; styles free or nearly so, 2-fid. Fruit dehiscent. Seeds carunculate; testa dry, smooth or rarely punctate; endosperm copious; cotyledons flat, broader and somewhat longer than the radicle.

As here interpreted, this tribe contains only the type-genus *Clutia,* in accordance with the original concept of Müller. For further information on the history of the classification of the genus, see Radcliffe-Smith (1992).

88. **Clutia** *L.*, Sp. Pl. 1042 (1753); Gen. Pl. ed. 5, 464 (1754); A. Juss., Euphorb. Gen.: 25 (1824); Baill., Étude Euphorb.: 328 (1858); Müll. Arg. in DC., Prodr. 15(2): 1043 (1866) (as *Cluytia*); Benth., Gen. Pl. 3: 302 (1880); Pax & K. Hoffm., Pflanzenr. 47: 50 (1911); Hutch., Fl. Trop. Afr. 6(1): 801 (1912); Prain in Harv., Fl. Cap. 5(2): 427 (1920); Pax & K. Hoffm., Nat. Pflanzenfam., ed. 2, 19C: 167 (1931); Robyns, Fl. Parc Nat. Alb. 1: 471 (1948); Keay (ed.), Fl. W. Trop. Afr., ed. 2, 1: 396 (1958); J. Léonard, Fl. Cong. Rwa.-Bur. 8(1): 93 (1962); Dyer, Gen. S. Afr. Fl. Pl. 1: 321 (1975); Radcl.-Sm., Fl. Trop. E. Afr. Euph. 1: 331 (1987); J.-P. Lebrun & Stork, Enum. Pl. Afr. Trop. 1: 208 (1991); Radcl.-Sm., Kew Bull. 47: 111 (1992); G. L. Webster, Ann. Missouri Bot. Gard. 81: 64 (1994); Radcl.-Sm., Fl. Zamb. 9(4): 123 (1996). T.: *C. pulchella* L.

Altora Adans., Fam. Pl. 2: 356 (1763). T.: none designated.
Cratochwilia Neck., Elem. Bot.: 339 (1791). T.: none designated.

Dioecious or rarely monoecious shrubs, shrublets or woody herbs. Indumentum simple. Leaves alternate, shortly petiolate or subsessile, often exstipulate, simple, entire, penninerved, often pellucid-punctate, eglandular, often small, ericoid (S. Afr.). Inflorescences axillary, male flowers fasciculate, females often solitary. Male flowers pedicellate; sepals 5, ± free, imbricate, spreading; petals 5, free, unguiculate, often longer; disc of ∞ glands in 1–3 series at the base of the sepals, petals and staminal column; stamens 5, oppositisepalous, filaments united into a column but free above and spreading verticillately below the pistillode, anthers short, dorsifixed, muticous, thecae parallel, contiguous, extrorse, longitudinally dehiscent; pollen subprolate, triangular in polar view, 3-colporate, colpae narrow, inoperculate, marginate, colpus transversalis small, exine perforate-tectate; pistillode cylindric, truncate or slightly dilated and hollow at the apex. Female flowers often long-pedicellate; calyx and petals ± as in the male, but persistent and becoming indurated in fruit; disc usually only in one series at the base of the sepals; ovary 3(4)-locular, ovule 1 per locule; styles free or nearly so, spreading, 2-fid. Fruit small, ovoid or subglobose, septicidally dehiscent into 3 2-valved partially loculicidal cocci; septa thin, detaching or adnate to the columella; endocarp woody. Seeds ovoid, strophiolate; testa smooth or rarely punctate, dry, black, shining; endosperm fleshy, copious; cotyledons flat, broader and somewhat longer than the radicle. Fig.: See Euphorb. Gen.: t. 6, f. 21.1–10 (1824); Étude Euphorb.: t. 16, f. 1–21 (1858); Pflanzenr. 47: 55–82, tt. 19–26 (1911); Fl. Trop. E. Afr., Euph. 1: 335, t. 63 (1987); Kew Bull. 47: 113–118, tt. 1–3 (1992); Fl. Zamb. 9(4): 126, t. 21 (1996).

A large African genus of c. 75 species, with 2 extending into tropical Arabia. The greatest diversity is in S. Africa.

Tribe 16. **POGONOPHOREAE** *(Müll. Arg.) G. L. Webster*, Taxon 24: 595 (1975), and Ann. Missouri Bot. Gard. 81(1): 64 (1994). Type: *Pogonophora* Miers ex Benth.

Hippomaneae subtribe *Pogonophoreae* Müll. Arg., Linnaea 34: 202 (1865).

Dioecious trees or shrubs; indumentum simple or malpighiaceous; leaves alternate, simple, entire, penninerved, eglandular; stipules very small or 0. Inflorescences axillary, thyrsiform or glomerulate; male sepals 5, broadly imbricate, ± biseriate; petals 5, imbricate, adaxially barbate; stamens 5, filaments free, anthers basifixed, elongate; pollen prolate, 3-colporate, colpi inoperculate, not marginate, perforate-tectate; disc intrastaminal, urceolate, 5-lobate; pistillode +; female disc cupular; ovary 3-locular; styles bilobed. Fruit dehiscent. Seeds carunculate, smooth, black, shining; hilum large; endosperm copious; cotyledons much longer and broader than the radicle.

Another monogeneric group, as circumscribed by Müller. Various affinities have been suggested, but as the pollen, fruit and seed are similar to those of *Clutia*, this is here considered its closest relative.

89. **Pogonophora** *Miers ex Benth.*, Hooker's J. Bot. & Kew Gard. Misc. 6: 372 (1854); Baill., Étude Euphorb.: 332 (1858); Müll. Arg. in DC., Prodr. 15(2): 1040 (1866), and Fl. Bras. 11(2): 435 (1874); Benth., Gen. Pl. 3: 288 (1880); Pax & K. Hoffm., Pflanzenr. 47: 108, t. 35 (1911), and Nat. Pflanzenfam., ed. 2, 19C: 172 (1931); Jabl., Mem. New York Bot. Gard. 17: 153 (1967); Letouzey, Adansonia II. 9: 275 (1969); Secco, Rev. Gen. Euph.-Crot. Amer. Sul: 88 (1990); Feuillet, Novon 3(1): 23 (1993); G. L. Webster, Ann. Missouri Bot. Gard. 81: 64 (1994); Murillo-Aldana & Franco-Rosselli, Euf. Reg. Ararac.: 143 (1995). T.: *P. schomburgkiana* Miers ex Benth.

Dioecious trees or shrubs. Indumentum simple or (in inflorescence) malpighiaceous. Leaves alternate, petiolate, ± exstipulate, simple, entire, penninerved, eglandular. Inflorescences axillary, paniculate, thyrsiform or glomerulate, shorter than the leaves. Male flowers sessile; buds cylindric; sepals 5, strongly imbricate, ± biseriate, coriaceous; petals 5, longer than the calyx, imbricate, bearded in the middle within; stamens 5, episepalous, inserted below the disc, filaments free, anthers basifixed, introrse, linear, apiculate, thecae adnate, parallel, dehiscing longitudinally; pollen prolate, triangular in polar view, 3-colporate, colpi inoperculate, not marginate, perforate-tectate; disc of 5 thick glands fused into a cup enclosing the pistillode; pistillode deeply

2–3-lobed, hirsute. Female flowers shortly pedicellate; calyx and petals as in the male; disc cupular, tenuous; ovary 3-locular, ovule 1 per locule; styles 3, shortly connate at the base, bilobed or bifid, stigmas fimbriate. Fruit ovoid, dehiscing septicidally into 3 2-valved cocci; endocarp hard; septum incomplete; columella thin, straight, unwinged. Seeds ovoid, carunculate; testa smooth, shining, black; hilum large; caruncle large, yellow; endosperm copious, fleshy; cotyledons flat, much longer and broader than the radicle. Fig.: See Étude Euphorb.: t. 19, f. 21–23 (1858); Fl. Bras. 11(2): t.62 (1874); Pflanzenr. 47: 109, t. 35 (1911); Nat. Pflanzenfam., ed. 2, 19C: 172, t. 89 (1931); Adansonia II. 9: 274, t. 1 (1969); Euf. Reg. Ararac.: 144, t. 42 (1995).

A genus of 3 species, 2 neotropical (1 Venezuela, Amazonian and coastal Brazil and the Guianas, and 1 Colombia), the other West African (Gabon).

Tribe 17. **CHÆTOCARPEAE** *(Müll. Arg.) G. L. Webster*, Taxon 24: 595 (1975), and Ann. Missouri Bot. Gard. 81(1): 64 (1994). Type: *Chætocarpus* Thwaites.

Hippomaneae subtribe *Chætocarpeae* Müll. Arg., Linnaea 34: 202 (1865).

Dioecious trees or shrubs; indumentum simple or 0; leaves alternate, simple, entire, penninerved, eglandular; stipules deciduous. Flowers in axillary bracteate glomerules, male sepals 4–6, imbricate, free or connate; petals 0 or +; disc-glands 4–10, free; stamens 5–20, filaments connate, hirsute; anthers extrorse, oblong, longitudinally dehiscent; pollen spheroidal, 3-colporate, margins irregular, micropunctate-rugulose; pistillode prominent, hirsute; female sepals 4–8, imbricate; disc urceolate; ovary 3-locular; styles 2-partite, papillose. Fruit dehiscent. Seeds carunculate; testa smooth, blackish; cotyledons broader and longer than the radicle.

Webster (1975) augmented this tribe by the inclusion therein of *Trigonopleura* on palynological grounds. Pax & Hoffmann (1911) and Hutchinson (1969) had associated *Trigonopleura* with *Clutia.*

Key to the genera

Petals +; anthers ± sessile; ovary smooth · · · · · · · · · · · · · · 90. *Trigonopleura*
Petals 0; filaments free above; ovary echinate · · · · · · · · · · · 91. *Chætocarpus*

90. **Trigonopleura** *Hook. f.*, Fl. Brit. India 5: 399 (1887), and Hooker's Icon. Pl. 18, t. 1753 (1888); Pax & K. Hoffm., Pflanzenr. 47: 95 (1911); Merr., Philipp. J. Sci. 11C: 76 (1916); Ridl., Fl. Malay. Penins. 3: 263 (1924); Pax & K. Hoffm., Nat. Pflanzenfam., ed. 2, 19C: 170 (1931); Whitmore, Tree Fl. Mal. 2: 134 (1973); Airy Shaw, Kew Bull. Addit. Ser. 4: 201 (1975), Kew Bull. 36: 350 (1981), and Euph. Philipp.: 46 (1983); G. L. Webster, Ann. Missouri Bot. Gard. 81: 65 (1994); van Welzen *et al.*, Blumea 40: 368 (1995). T.: *Trigonopleura malayana* Hook. f.

Dioecious trees. Indumentum simple and quasi-stellate. Leaves alternate, distichous, shortly petiolate, stipulate, simple, entire, penninerved, punctate; stipules large, oblanceolate-falcate, deciduous. Flowers sessile or subsessile, axillary, clustered. Male flowers: pedicels articulate; sepals 5, broadly imbricate; petals 5, slightly longer than the sepals; disc-glands 5, free, antheriform; stamens 8, slightly exsert, filaments completely connate into a column, anthers biverticillate, 5 below, 3 above, sessile, oblong, extrorse, longitudinally dehiscent, connective apiculate; pollen oblate-spheroidal, otherwise as in 91. *Chætocarpus* (q.v. infra); pistillode at the top of the staminal column, deeply 3-lobed. Female flowers : sepals and petals ± as in the male; ovary 3-locular, smooth, ovules 1 per cell, styles 3, erect, stigmas papillose. Fruit small, ellipsoid-3-gonous, dehiscing septicidally into 3 2-valved cocci or loculicidally into 6 valves; epicarp smooth, rugulose-reticulate when dried; endocarp bony, fairly thick; columella broadly 3-winged, wings papyraceous, hyaline. Seeds solitary per loculus, broadly dorsiventrally-compressed-ovoid, carunculate-arillate, aril extending $^{3}/_{4}$ the length of the seed; exotesta smooth, black, shiny, endotesta grey, dull; endosperm 0; cotyledons flat, broader and longer than the radicle. Fig.: See Hooker's Icon. Pl. 18, t. 1753 (1888); Pflanzenr. 47: 10, t. 1.B–D and 96, t. 30 (1911); Nat. Pflanzenfam. ed. 2, 19C: 155, t. 79.B–D (1931); Blumea 40: 370, t. 2 (1995).

A genus of 3 species, 1 widespread from Malaya to Celebes (Sulawesi), 1 in the C. Philippines and the other in W. Sarawak, according to van Welzen *et al.* (1995).

Considered by Airy Shaw (1981) perhaps to be scarcely different from the following genus.

91. **Chætocarpus** *Thwaites*, Hooker's J. Bot. Kew Gard. Misc. 6: 300, t. 10a (1854) (nom. cons.); Baill., Étude Euphorb.: 323 (1858); Müll. Arg. in DC., Prodr. 15(2): 1121 (1866), and in Mart., Fl. Bras. 11(2): 505 (1874); Benth., Gen. Pl. 3: 323 (1880); Hook. f., Fl. Brit. India 5: 460 (1887); Prain in Fl. Trop. Afr. 6(1): 946 (1912); Pax & K. Hoffm., Pflanzenr. 52.IV: 7 (1912); Fawc. & Rendle, Fl. Jamaica 4: 318 (1920); Gagnep., Fl. Indo-Chine 5: 471 (1926); Pax & K. Hoffm., Nat. Pflanzenfam. ed. 2, 19C: 179 (1931); J. Léonard, Fl. Cong. Rwa.-Bur. 8(1): 127 (1962); Jabl., Mem. New York Bot. Gard. 17: 161 (1967); Capuron, Adansonia II. 12: 209 (1972); Airy Shaw, Kew Bull. 26: 231 (1972); Whitmore, Tree Fl. Mal. 2: 76 (1973); Airy Shaw, Kew Bull. Addit. Ser. 4: 67 (1975), and Kew Bull. 36: 275 (1981); Alain, Fl. Español. 4: 93 (1986); G. L. Webster, Ann. Missouri Bot. Gard. 81: 65 (1994); van Welzen, Rheedea 4(2): 94 (1994); Radcl.-Sm., Fl. Zamb. 9(4): 134 (1996); Philcox in Dassanayake, ed., Fl. Ceyl. XI: 177 (1997). T.: *Ch. pungens* Thwaites [= *Ch. castanocarpus* (Roxb.) Thwaites (B.: *Adelia castanocarpa* Roxb.)]

Mettenia Griseb., Fl. Brit. W. I.: 43 (1859); Alain, Fl. Cuba 3: 124 (1953). T.: *M. globosa* Griseb. [= *Ch. globosus* (Griseb.) Fawc. & Rendle]

Regnaldia Baill., Adansonia I. 1: 87 (1861). T.: *R. cluytioides* Baill. [= *Ch. castanocarpus* (Roxb.) Thwaites (B.: *Adelia castanocarpa* Roxb.)]

Gaedawakka [L., Fl. Zeyl.: 203 (1747)] ex Kuntze, Revis. Gen. Pl. 2: 606 (1891). T.: *G. castanocarpa* (Roxb.) Kuntze (B.: *Adelia castanocarpa* Roxb.) [= *Ch. castanocarpus* (Roxb.) Thwaites]

Neochevaliera Beille, Compt. Rend. Hebd. Séances Acad. Sci. 145: 1295 (1907). T.: *N. brazzavillensis* Beille [= *Ch. africanus* Pax]

Dioecious trees or shrubs. Indumentum simple. Leaves alternate, shortly petiolate, stipulate, simple, entire, penninerved, eglandular; stipules leafy, lateral, asymmetric, mostly early caducous. Flowers in small, dense axillary clusters, males many, females few; bracts minute. Male flowers shortly pedicellate; sepals 4–6, free, strongly imbricate; petals 0; disc-glands 4–10, extrastaminal, free; stamens 5–20, filaments connate at the base into a column, free above, anthers dorsifixed, extrorse, oblong, thecae parallel, longitudinally dehiscent, connective broad; pollen spheroidal, 3-colporate, convex-triangular in polar view, colpus transversalis isodiametric with large ora, colpi narrow, tectate, psilate; pistillode at the apex of the column, 3-lobed, villous. Female flowers sometimes pedicellate; sepals ± as in the male; petals 0; disc entire to crenellate; ovary 3-locular, ovules 1 per locule; styles 3, ± free, incurved-erect, 2-partite, branches introrsely papillose-fimbriate. Fruit subglobose, dehiscing into 3 2-valved cocci, epicarp densely echinate-setose or tuberculate; endocarp hard, woody; columella persistent. Seeds ovoid-subglobose, caruncuIate; testa smooth, black, shining; caruncle large, thin, bipartite; endosperm fleshy; cotyledons flat, broader and longer than the radicle. Fig.: See Fl. Bras. 11(2): t. 100 (1874); Pflanzenr. 52.IV: 9, t. 1 and 11, t. 2 (1912); Fl. Jamaica 4: 319, t. 106 (1920); Fl. Indo-Chine 5: 472, t. 59.12–16 (1926); Nat. Pflanzenfam. ed. 2, 19C: 180, t. 94 (1931); Fl. Cuba 3: 125 t. 44 (1953); Adansonia II. 12: 210, t. 2 (1972); Rheedea 4(2): 95, t. 1 (1994); Kew Bull. 50: 122, t. 2 (1995); Fl. Zamb. 9(4): 135, t. 23 (1996).

A genus of 13 species with a disjunct distribution: West Indies (Greater Antilles), South America (Bolivia, Venezuela, Brazil, Guyana, Surinam), Africa (Gabon, Cabinda, Congo (Brazzaville), Congo (Kinshasa), Zambia, Angola), Madagascar and Tropical Asia (Sri Lanka, Bangladesh, Bhutan, Myanmar {Burma}, Thailand, Cambodia, S. Vietnam, Malay Peninsula, Sumatra, Bangka, Billiton, Borneo).

The wood of *Ch. castanocarpus* (Roxb.) Thwaites is close-grained, hard, durable and reddish-brown, and is used for building.

Tribe 18. **PEREAE** *(Baill.) Pax & K. Hoffm.*, Pflanzenr. 68. XIII: 1 (1919); G. L. Webster, Ann. Missouri Bot. Gard. 81: 65 (1994). Type: *Pera* Mutis.

Gp. *Perideae* Baill., Étude Euphorb.: 433 (1858).

Peraceae (Baill.) Klotzsch & Garcke, Monatsber. Königl. Preuss. Akad. Wiss. Berlin 1859: 246 (1859).

Prosopidoclineae Klotzsch, Arch. Naturgesch. 7: 176 (1841).

Monoecious or dioecious trees or shrubs; indumentum stellate, lepidote or simple; leaves alternate, rarely opposite, minutely stipulate or exstipulate, entire, penninerved. Flowers in axillary involucrate glomerules; involucre usually of 2 valvate bracts completely enclosing the flowers in bud; male calyx 2–4(6)-fid or 0; petals 0; disc 0; stamens 2–6(8), filaments connate; pollen 3-colporate, rugulose; "pistillodes" (reduced female flowers) + or 0, surrounding the male flower; female flowers achlamydeous; ovary 3-locular; style stigmatiform. Fruit dehiscent; valves non-elastic, eseptate, persistent. Seeds caruncluate; testa smooth, dark; endosperm copious; cotyledons much longer and broader than the radicle.

This tribe, by unanimous consensus of all from Baillon onwards, is monogeneric. The relationships of *Pera* are obscure, on account of the reduced flowers and the peculiar inflorescences unlike anything else in the family. Webster (1994) thinks the *Chætocarpeae* to be the closest group on account of similar pollen, fruit and seeds.

92. **Pera** *Mutis*, Kongl. Vetensk. Acad. Nya Handl. 5: 299 (1784); Baill., Étude Euphorb.: 433 (1858); Müll. Arg. in DC., Prodr. 15(2): 1025 (1866), and Fl. Bras. 11(2): 421 (1874); Benth., Gen. Pl. 3: 340 (1880); Pax & K. Hoffm., Pflanzenr. 68.XIII: 2 (1919), and Nat. Pflanzenfam., ed. 2, 19C: 153 (1931); Alain, Fl. Cuba 3: 104 (1953); Jabl., Mem. New York Bot. Gard. 17: 147 (1967); G. L. Webster, Ann. Missouri Bot. Gard. 54: 316 (1968); Philcox, Fl. Trin. Tob. II(X): 640 (1979); Correll, Fl. Bah. Arch.: 834 (1983); Alain Fl. Español. 4: 184 (1986), and Descr. Fl. Puerto Rico 2: 412 (1988); G. L. Webster, Ann. Missouri Bot. Gard. 81: 65 (1994); Murillo-Aldana & Franco-Rosselli, Euf. Reg. Ararac.: 128 (1995); L J. Gillespie, Smithsonian Contr. Bot. 86: 8 (1997). T.: *P. arborea* Mutis.

Perula Schreb., Gen. Pl. 2: 703 (1791). T.: *Perula arborea* (Mutis) Willd. [=*P. arborea* Mutis]

Spixia Leandro, Denkschr. Königl. Akad. Wiss. München 7: 231 (1821), non Schrank. T.: *Sp. heteranthera* Schrank [=*P. heteranthera* (Schrank) I. M. Johnst.]

Peridium Schott in Spreng., Syst. Veg., ed. 16, 4(2), Cur. Post. App.: 410 (1827). T.: *Peridium glabratum* Schott [=*P. glabrata* (Schott) Baill.]

Schismatopera Klotzsch, Arch. Naturgesch. 7: 178 (1841). T.: *Sch. martiana* Klotzsch [=*P. distichophylla* (Mart.) Baill. (B.: *Spixia distichophylla* Mart.)]

Monoecious or dioecious trees or shrubs. Indumentum stellate, lepidote or occasionally simple. Leaves alternate or very rarely opposite, shortly petiolate, minutely stipulate or exstipulate, simple, entire, penninerved. Flowers axillary, in 3–4-flowered unisexual or bisexual involucrate capitula; involucre of 1–2 small free outer and 2 larger valvate variously connate spathaceous inner bracts completely enclosing the flowers in bud. Male flowers sessile; calyx 2–4(6)-lobed or -toothed, cupuliform or 0; petals 0; disc 0; stamens 2–6(8), filaments connate below into a column, anthers basifixed, introrse or extrorse, connective apiculate, thecae 2, longitudinally dehiscent; pollen prolate-spheroidal, 3-colporate, colpi narrow, tectate, rugulose; "pistillodes" 3–4 or 0, surrounding the male flower, never central, thus representing reduced female flowers and not solely a suppressed gynæcium. Female flowers subsessile or very shortly pedicellate, achlamydeous; ovary 3-locular, ovule 1 per locule; style very short, stigma peltate, umbraculiform, discoid or 3-lobed. Fruit trigonous-ellipsoid to subglobose, subdrupaceous-regmatiform, tardily dehiscent into 6 non-elastic eseptate valves which remain attached to the base of the columella; epicarp smooth at first, much-wrinkled when dried; mesocarp spongy or fleshy; endocarp thinly woody, readily separating from the epicarp; columella slender, splitting longitudinally into 3 strands. Seeds compressed, carunculate; testa smooth, shiny, dark; endosperm copious, fleshy; cotyledons broad, flat, shallowly cordate, much longer than the radicle. Fig.: See Étude Euphorb.: t. 2, f. 25–27 (1858); Fl. Bras. 11(2): t. 61 (1874); Pflanzenr. 68.XIII: 5, t. 1 and 12, t. 2 (1919); Nat. Pflanzenfam., ed. 2, 19C: 154, t. 78 (1931); Fl. Cuba 3: 106, t. 31 (1953); Ann. Missouri Bot. Gard. 54: 317, t. 19 (1968); Amer. J. Bot. 56: 754, t. 16A–K (1969); Fl. Bah. Arch.: 835, t. 347 (1983); Fl. Español. 4: 186, t. 116-28 (1986); Descr. Fl. Puerto Rico 2: 413, t. 59-23 (1988); Árvores Brasileiras: 109 (1992); Smithsonian Contr. Bot. 86: 10, t. 2 and12, t. 3 (1997).

A neotropical genus of c. 30–40 species, extending from Cuba and Central America south to Brazil, with the greatest concentration in the Amazon basin.

Tribe 19. **CHEILOSEAE** *(Müll. Arg.) Airy Shaw & G. L. Webster,* Taxon 24: 595 (1975); G. L. Webster, Ann. Missouri Bot. Gard. 81(1): 65 (1994). Type: *Cheilosa* Blume.

Hippomaneae subtribe *Cheiloseae* Müll. Arg., Linnaea 34: 202 (1865).
Gelonieae subtribe *Chætocarpinae* ser. *Cheilosiformes* Pax & K. Hoffm., Pflanzenr. 68, Add. 6: 50 (1919).

Dioecious trees; indumentum simple and/or stellate; leaves alternate, simple, entire or toothed, penninerved, usually glandular; petiole bipulvinate; stipules caducous. Inflorescences pseudoterminal or axillary, thyrsiform; flowers apetalous; sepals 4–5, imbricate; male disc dissected, or glands free, intra- or extrastaminal; stamens 5–12, free; pollen 3-colporate, colpi narrow, inoperculate, tectate, echinate; pistillode +; female disc + or 0; ovary 2–4-locular; stigmas 2-fid. Fruit tardily loculicidally dehiscent; seeds ecarunculate; testa fleshy; cotyledons longer and broader than the radicle.

This tribe is differently circumscribed from Pax & Hoffmann's (1919) *Cheilosiformes,* since *Elateriospermum* is removed to the *Crotonoideae,* following Webster (1975), and *Neoscortechinia* is added, following Airy Shaw (1972).

Key to the genera

Indumentum simple; stamens 8–10(12), anthers apiculate, pollen echinate; female disc +; fruit thin- or thick-walled, with up to 3 seeds · · 93. *Cheilosa*
Indumentum simple and/or stellate; stamens 5–9, anthers muticous, pollen scabrate; female disc 0; fruit thin-walled, with 1 seed · · 94. *Neoscortechinia*

93. **Cheilosa** *Blume*, Bijdr. 613 (1826); Baill., Étude Euphorb.: 420 (1858); Müll. Arg. in DC., Prodr. 15(2): 1122 (1866); Benth., Gen. Pl. 3: 322 (1880); J. J. Sm., Meded. Dept. Landb. Ned.-Indië 10: 604 (1910); Pax & K. Hoffm., Pflanzenr. 52. IV: 12 (1912), and Nat. Pflanzenfam., ed. 2, 19C: 180 (1931); Airy Shaw, Kew Bull. 16: 364 (1963); Backer & Bakh. f., Fl. Java 1: 496 (1963); Whitmore, Tree Fl. Mal. 2: 77 (1973); Airy Shaw, Kew Bull. Addit. Ser. 4: 68 (1975), Kew Bull. 36: 276 (1981), and Euph. Philipp.: 11 (1983); van Welzen *et al.*, Blumea 38: 162 (1993); G. L. Webster, Ann. Missouri Bot. Gard. 81: 66 (1994). T.: *Ch. montana* Blume.

Tall dioecious tree. Indumentum simple. Leaves alternate, petiolate, stipulate, simple, obovate, entire, shallowly crenulate or denticulate, teeth glandular beneath, penninerved; petiole bipulvinate; stipules small, deciduous. Inflorescences ramiflorous, axillary or subterminal, male thyrsiform, female racemose, male flowers fasciculate, females single to each bract. Male flowers: sepals 5(6), free, slightly imbricate; petals 0; disc 8–10(12)-lobed, lobes enfolding inner whorl of 5(6) stamens; stamens 8–10(12) in 2 whorls, filaments free, anthers short, basifixed, latrorse, connective slightly produced, thecae longitudinally dehiscent; pollen suboblate, 3-colporate, convex-triangular in polar view, colpi narrow, inoperculate, tectate, echinate; pistillode (2)3(4)-lobed, pubescent. Female flowers: sepals unequal; petals 0; disc denticulate; ovary 3(4)-locular, ovule 1 per locule; styles very shortly connate at the base, stigmas recurved, 2-lobed at apex, papillose. Fruit ellipsoid-subglobose, 6-grooved, tardily dehiscing loculicidally into 3 valves; endocarp suberiform, often massive; columella slender. Seeds 1–3 per fruit, ecarunculate, large, keeled; endosperm +; embryo central; cotyledons broad, flat. Fig.: See Pflanzenr. 52.IV: 13, t. 3 (1912); Blumea 38: 164, t. 1 (1995).

A monotypic genus from the Malay Peninsula, S. Sumatra, W. Java, Borneo and the Philippines.

94. **Neoscortechinia** *Hook. f. ex Pax*, Nat. Pflanzenfam. Nachtr. 1: 213 (1897); Pax & K. Hoffm., Pflanzenr. 68. XIV(Addit. VI): 52 (1919), and Nat. Pflanzenfam., ed. 2, 19C: 183 (1931); Backer & Bakh. f., Fl. Java 1: 497 (1963); Airy Shaw, Kew Bull. 16: 368 (1963), and 26: 310 (1971); Whitmore, Tree Fl. Mal. 2:119 (1973); Airy Shaw, Kew Bull. Addit. Ser. 4: 177 (1975), 8: 175 (1980), Kew Bull. 36: 333 (1981), 37: 31 (1982), and Euph. Philipp.: 38 (1983); van Welzen, Blumea 39: 307 (1994); G. L. Webster, Ann. Missouri Bot. Gard. 81: 66 (1994). T.: *N. kingii* (Hook. f.) Pax & K. Hoffm.

Scortechinia Hook. f., Hooker's Icon. Pl. Ser. 3, 8: t. 1706 (1887) (inaccurate as to male fls.), and Fl. Brit. India 5: 366 (1887), non Saccardo (1885). T.: *S. kingii* Hook. f. [=*N. kingii* (Hook. f.) Pax & K. Hoffm.]

Alcinaeanthus Merr., Philipp. J. Sci. 7: 379 (1912). T.: *A. philippinensis* Merr. [=*N. philippinensis* (Merr.) van Welzen]

Dioecious trees. Indumentum simple and/or stellate. Leaves alternate, petiolate, stipulate, simple, entire or serrulate, penninerved, often biglandular at the base; petiole bipulvinate; stipules minute, early caducous. Inflorescences pseudoterminal or axillary, paniculate-thyrsiform, unisexual, male with smaller and more flowers than the female. Male flowers: sepals 4–5, rounded, imbricate; petals 0; disc-glands 4–7, squamiform, pubescent; stamens 5–9, filaments short, free, anthers basifixed, latrorse-introrse, ovoid, thecae separated by a connective, longitudinally dehiscent; pollen oblate-spheroidal, 3-colporate, convex-triangular in polar view, colpi narrow, inoperculate, tectate, microechinate to scabrate; pistillode short, broad, truncate, 3-lobed, pubescent. Female flowers: sepals ± as in the male; petals 0; disc 0; ovary 2-locular, ovule 1 per locule; style 0; stigmas 4, minute, forming a sessile cone. Fruit obovoid-ellipsoid, slightly 3–4-ribbed, tardily loculicidally dehiscent into 3–4 valves, 1-seeded by abortion; endocarp thinly woody; columella elongate, fibrous, split, not rigid or winged. Seed 1 per fruit, ecarunculate, pendulous from the apex of the column, oblong-ellipsoid, smooth, black, aril-membrane red, aril yellow; endosperm +; embryo flat; cotyledons amygdaline, radicle superior, minute. Fig.: See Hooker's Icon. Pl. Ser. 3, 8: t. 1706 (1887) (as *Scortechinia*); Blumea 39: 306, t. 3 (1995).

A tropical Asiatic genus of 6 species, according to van Welzen (1994), ranging from Myanmar and the Nicobars to the Philippines and the Solomon Is.

Tribe 20. **ERISMANTHEAE** *G. L. Webster*, Taxon 24: 595(1975), and Ann. Missouri Bot. Gard. 81(1): 66 (1994); Van Welzen, Blumea 40: 375(1995). Type: *Erismanthus* Wall. ex Müll. Arg.

Monecious trees or shrubs; indumentum simple; leaves opposite, mostly subsessile, stipulate, simple, toothed, basally oblique, penninerved; stipules interpetiolar. Inflorescences axillary, pseudospicate or racemose to subpaniculate; male calyx-lobes 4–5, imbricate; petals 2–7; disc minute or 0; stamens 4–15, filaments free; anthers latrorse; pollen 3-colpate or -colporate, colpi inoperculate, not marginate, reticulate; pistillode +; female sepals 4–6, imbricate; petals + or 0; disc 0; ovary 3-locular, ovules anatropous; styles 2-fid. Fruit dehiscent. Seeds ecarunculate; testa smooth, dry.

This tribe is characterized by the opposite leaves with interpetiolar stipules, monoecy, male flowers with a torus and exarilloid seeds. The position of the axillary buds is distinctive; recent research on seedlings of *Moultonianthus* has shown that the leaves become opposite, after an initial alternate seedling stage, by the drastic reduction of internode-length, and the reduction to nothing of the stipule of the distal node of each conjoined pair of nodes. There is also the retention of an 'unused' terminal bud of the lower of the two conjoined nodes, the distal node appearing to arise from an axillary bud of the proximal node (Cam Webb, pers. comm. 1996).

Key to the genera

1. Male inflorescences amentiform, covered with closely-imbricate bracts when young; flowers long-pedicellate; petals shorter than the calyx; stamens 12–15; pistillode elongate, styliform, clavate; female inflorescences uniflorous; flowers apetalous · · · · · · · · · · · · · · · · · · · 95. *Erismanthus*
1. Male inflorescences thyrsiform, not as above; flowers short-pedicellate; stamens 4–11; pistillode not elongate, stigmatiform, tripartite; female inflorescences pluriflorous, or mixed inflorescences with several female flowers · 2
2. Stipules large, foliaceous, cordate, amplexicaul, persistent; inflorescences unisexual; male petals longer than the calyx; female flowers and fruits long-pedicellate, female flowers petaliferous · · · · · · 96. *Moultonianthus*
2. Stipules small, deciduous; inflorescences often bisexual; male petals shorter than the calyx; female flowers subsessile, apetalous; fruits shortly pedicellate · 97. *Syndyophyllum*

95. **Erismanthus** *Wall. ex Müll. Arg.* in DC., Prodr. 15(2): 1138 (1866); Benth., Gen. Pl. 3: 325 (1880); Hook. f., Fl. Brit. India 5: 405 (1887); Pax & K. Hoffm., Pflanzenr. 47: 33, t. 9 (1911); Gagnep., Fl. Indo-Chine 5: 461 (1926); Pax & K. Hoffm., Nat. Pflanzenfam., ed. 2, 19C: 158 (1931); Airy Shaw, Kew Bull. 26: 260 (1972); Whitmore, Tree Fl. Mal. 2: 95(1973); Airy Shaw, Kew Bull. Addit. Ser. 4: 110 (1975), and Kew Bull. 36: 294 (1981); G. L. Webster, Ann. Missouri Bot. Gard. 81: 66 (1994); van Welzen, Blumea 40: 379 (1995). T.: *E. obliquus* Müll. Arg.

Monecious trees or shrubs, sometimes scrambling; branching sympodial. Indumentum simple. Leaves opposite, distichous, subsessile or shortly petiolate, stipulate, simple, shallowly toothed, obliquely rounded-cordate at the base, penninerved, punctate; buds alternately axillary to leaves and stipules; stipules interpetiolar, subpersistent. Male inflorescences axillary, in alternating axils of successive leaf-pairs, racemose, short, densely bracteate; female flowers solitary, in same axils. Male flowers long-filiform-pedicellate, pedicels bracteolate at the base; calyx-lobes (4)5, imbricate, at length reflexed; petals shorter than the calyx; disc 0; stamens 12–15, filaments free, very short, inserted on a pilose receptacle, anthers small, basifixed in cleft, latrorse, triangular, connective exappendiculate, thecae longitudinally dehiscent; pollen oblate-spheroidal, 3-colpate or -colporate, circular in polar view, colpi not marginate, tectate; pistillode filiform, much-produced, bifid or trifid. Female flowers long-pedicellate, pedicels bracteolate in middle; sepals 5, large, unequal, foliaceous, imbricate, accrescent and persistent in fruit; petals 0; disc 0; ovary 3-locular, ovule 1 per locule; styles 3, connate at the base, slender, erect, deeply 2-fid; stigmas densely papillose-pilosulous. Fruit depressed-trilobate, cocci shouldered, dehiscing septicidally and loculicidally into 3 2-valved cocci; endocarp thinly woody, septum 1-veined; columella apically obtriangular. Seeds 1–3 per fruit, rounded, ecarunculate; testa smooth, glossy brown-mottled, micropyle visible. Embryo not known. Fig.: See Pflanzenr. 47: 34, t. 9 (1911); Fl. Indo-Chine 5: 463, t. 57.6 (1926); Nat.

Pflanzenfam., ed. 2, 19C: 159, t. 82 (1931); Amer. J. Bot. 56: 748, t. 7A–J (1969); Blumea 40: 382, t. 2 (1995).

An Asiatic genus of 2 species, the type-species occurring from peninsular Thailand to Sumatra and Borneo, the other species (*E. sinensis* Oliv.) in E. Thailand, Indochina and Hainan.

96. **Moultonianthus** *Merr.*, Philipp. J. Sci. 11: 70 (1916); Pax & K. Hoffm., Pflanzenr. 68. XIV(Addit. VI): 41 (1919), and Nat. Pflanzenfam. ed. 2, 19C: 170 (1931); van Steenis, Bull. Jard. Bot. Buitenzorg III. 17: 404 (1948); Airy Shaw, Kew Bull. Addit. Ser. 4: 176 (1975), and Kew Bull. 36: 332 (1981); G. L. Webster, Ann. Missouri Bot. Gard. 81: 67 (1994); van Welzen, Blumea 40: 384 (1995). T.: *M. borneensis* Merr. [= *M. leembruggianus* (Boerl. & Koord.) van Steenis (B.: *Erismanthus leembruggianus* Boerl. & Koord.)]

Monecious shrub or small tree. Indumentum simple or rarely stellate. Leaves opposite, distichous, subsessile or shortly petiolate, stipulate, simple, shallowly crenate-serrate, oblique at the base, penninerved, not punctate; buds between leaves and stipules; stipules interpetiolar, large, foliaceous, ovate, deeply cordate, persistent. Inflorescences axillary, 1–2 per axil in alternating axils of successive leaf-pairs, pseudoracemose-subpaniculate, unisexual, elongated, male flowers numerous, fasciculate on brachyblasts along the rhachis, female flowers few, solitary, very long-pedicellate, pedicels thickened upwards, flattened, articulated below the middle. Male flowers: pedicels long, capillary; sepals 5, imbricate, free, 3 smaller, 2 larger; petals 5(7), much longer than the sepals, imbricate; disc-glands 5, minute, alternipetalous; stamens 9–11, on a short torus, in 2 whorls, filaments free, anthers basifixed, latrorse, elliptic, connective appendiculate, thecae longitudinally dehiscent; pollen spheroidal, 3-colporate, circular in polar view, colpi costate, intectate, finely reticulate; pistillode of 3 free slender "styles". Female flowers: sepals 5, imbricate, free, unequal; petals 5(7), elongate, imbricate; disc 0; ovary 3-locular, ovule 1 per locule; styles 3, free or shortly connate, 2-partite, recurved; stigmas dendritic-papillose. Fruit depressed-trilobate-subglobose, dehiscing septicidally and loculicidally into 3 2-valved cocci; endocarp thinly woody, septa 2-veined; columella narrowly obtriangular. Seeds 1–3 per fruit, globose, ecarunculate; testa smooth, glossy, brown, ± concolorous, micropyle visible. Embryo not known. Fig.: See Blumea 40: 386, t. 3 (1995).

A monotypic genus known only from C. Sumatra and Borneo.

97. **Syndyophyllum** *Lauterb. & K. Schum.*, Fl. Schutzgeb. Südsee Nachtr.: 403 (1901); Pax & K. Hoffm., Pflanzenr. 47: 104, t. 33 (1911), and Nat. Pflanzenfam. ed. 2, 19C: 172 (1931); Airy Shaw, Kew Bull. 14: 392 (1960), Hooker's Icon. Pl. 38: t. 3722 (1974), Kew Bull. Addit. Ser. 4: 199 (1975), 8: 204 (1980), and Kew Bull. 36: 349, t. 10 (1981); G. L. Webster, Ann. Missouri Bot. Gard. 81: 67 (1994); van Welzen, Blumea 40: 389 (1995). LT.: *S. excelsum* Lauterb. & K. Schum. (chosen by Wheeler, 1975).

Monecious trees with sympodial branching. Indumentum simple. Leaves opposite, distichous, petiolate, stipulate, simple, shallowly serrate to crenate-serrate, slightly unequal at the base, penninerved, tertiary venation closely parallel, punctate; buds axillary to the leaf, or each alternate one of a pair epipetiolar; stipules interpetiolar, deciduous. Inflorescences axillary, 1–2 per axil in alternating axils of successive leaf-pairs, interruptedly pseudospicate-pseudoracemose, elongate, pendulous, uni- or bisexual, if the latter usually with one female and several male flowers per cymule; bracts small, bracteoles minute. Male flowers sessile; calyx-lobes 4–5, imbricate, one-third united; petals 2–4(5), shorter than the sepals, imbricate, erose; disc-glands 0; stamens 4–5 or 8–10, on a short torus, filaments free, in 2 whorls, the outer epipetalous, anthers deeply basifixed, latrointrorse, triangular, subsaggitate, connective apiculate, thecae longitudinally dehiscent; pollen unknown; pistillode deeply tripartite. Female flowers 1(2) per glomerule, subsessile to shortly pedicellate, pedicel extending a little in fruit; calyx shortly 4–6-lobed, lobes imbricate, one-third united; petals 0; disc 0; ovary 3-locular, ovule 1 per locule; styles 3, elongate, united in the lower $^1/_2$, 2-fid, suberect or recurved; stigmas fimbriate-papillose. Fruit strongly trilobate, depressed, dehiscing septicidally and loculicidally into 3 2-valved cocci; endocarp thinly woody. Seeds 1–3 per fruit, spherical, ecarunculate; testa smooth, glossy, dark brown, micropyle visible, endosperm 0. Fig.: See Pflanzenr. 47: 104, t. 33 (1911); Hooker's Icon. Pl. 38: t. 3722 (1974); Kew Bull. 36: 351, t. 10 (1981); Blumea 40: 390, t. 4 (1995).

A genus of 2 species according to van Welzen (1995), one in N. Sumatra and Borneo, the other in N. New Guinea.

Tribe 21. **DICOELIEAE** *Hurus.*, J. Fac. Sci. Univ. Tokyo, Sect. 3, Bot. 6: 322 (1954); G. L. Webster, Bot. J. Linn. Soc. 94: 6 (1987), and Ann. Missouri Bot, Gard. 81: 67 (1994). Type: *Dicoelia* L.

Monoecious tree or shrub; indumentum simple; leaves simple, entire, penninerved, eglandular; stipules soon falling. Inflorescences pseudo-racemose, bisexual; male sepals 5, open in bud; petals 5, valvate, oppositistemonously 2-concave; disc 0; stamens 5, filaments adnate to the base of the pistillode; pollen 3-colporate, tectate; pistillode 3–5-fid; female sepals 5, valvate; petals 5, slightly imbricate; disc 0; ovary 3-locular, ovules 2 per locule, anatropous; styles free or ± so, simple, slender. Fruit dehiscent. Seeds ecarunculate.

A monogeneric tribe, anomalous in the *Acalyphoideae* on account of the biovulate carpels, but with a strong affinity to the next tribe. Stuppy (1995) however considers that *Dicoelia* belongs in the *Phyllanthoideae* and would affiliate it to the typical tribe *Phyllantheae*.

98. **Dicoelia** *Benth.*, Hooker's Icon. Pl. 13: 70, t. 1289 (1879), and Gen. Pl. 3: 286 (1880); J. J. Sm., Bull. Jard. Bot. Buitenzorg III. 1: 392 (1920); Pax & K. Hoffm., Pflanzenr. 81: 15 (1922), and Nat. Pflanzenfam. ed. 2, 19C: 46 (1931); Croizat, J. Arnold Arbor. 23: 38 (1942); Airy Shaw, Kew Bull. 27: 3 (1972), Kew

Bull. Addit. Ser. 4: 95 (1975), and Kew Bull. 36: 285 (1981); G. L. Webster, Ann. Missouri Bot. Gard. 81: 67 (1994). T.: *D. beccariana* Benth.

Monoecious shrub or small tree; branchlets angled. Indumentum simple. Leaves alternate, long-petiolate, stipulate, large, simple, obovate or elliptic, entire, penninerved, coriaceous, eglandular; petioles pulvinate at apex; stipules conspicuous, but soon falling. Inflorescences axillary, interruptedly pseudo-racemose, elongate, ascending, rigid, bisexual, the flowers fasciculate on the axis, the central flower of the clusters female, the others male, or sometimes all male. Male flowers: pedicels slender; buds pentagonal; sepals 5, small, open in bud; petals 5, valvate, convex outside, thick and fleshy at the apex, concave in the lower $^1/_2$ within, the concavity divided by a vertical median partition, reflexed at anthesis; disc 0; stamens 5, alternipetalous, filaments short, united at the base into a short column, and adnate to the base of the pistillode, anthers medifixed, latrorse, thecae ovoid, parallel, separate, resting in the cavities of adjacent petals, longitudinally dehiscent; pollen prolate, 3-colporate, colpi transversales broad, elongate, costae +, colpi narrow, costae +, tectate, psilate, indistinctly reticulate; pistillode turbinate, pentagonal, produced beyond the petals into 3–5 subulate hooked appendages. Female flowers: pedicels stout, longer than in the male, but not elongating in fruit; calyx ± as in the male; petals 5, ovate, concave, not much thickened, slightly imbricate; disc 0; ovary subglobose, 3-locular, ovules 2 per locule beneath a common obturator, anatropous; styles 3, free or ± so, rigid, filiform, undivided, divaricate, tapered into small capitate stigmas. Fruit subglobose, dehiscing septicidally into 3 2-valved cocci or loculicidally into 6 valves; exocarp thinly crustaceous, tomentellous; endocarp thinly woody, separating from the exocarp; septa incomplete; columella slender. Seeds 1 per locule, ovoid, plano-convex; testa thin, smooth, light brown, ecarunculate; endosperm fleshy; cotyledons orbicular, cordate, flat. Fig.: See Hooker's Icon. Pl. 13: 70, t. 1289 (1879); Pflanzenr. 81: 16, t. 3 (1922).

A monotypic genus from Malaya, Sumatra and Borneo according to Airy Shaw (1981), who did not accept *D. similis* J. J. Sm. as distinct.

Tribe 22. **GALEARIEAE** *Benth.*, Gen. Pl. 3: 247, 287 (1880); G. L. Webster, Ann. Missouri Bot. Gard. 81: 67 (1994). T.: *Galearia* Zoll. & Moritzi.

Cluytieae subtribe *Galeariinae* Pax, Nat. Pflanzenfam., ed. 1, 3(5): 81 (1890).

Bennettieaceae R. Br. in Benn., Pl. Jav. Rar.: 250 (1850) (nom. prov.).

Bennettieae R. Br. ex Schnizl., Iconogr. Fam. Nat. Regn. Veg. 3: 172 (1860). T.: *Bennettia* R. Br. [non *Bennettia* Gray (1821)] [= *Galearia* Zoll. & Moritzi]

Pandaceae Pierre, Bull. Mens. Soc. Linn. Paris: 1255 (1896); Engl. & Gilg, Syllabus ed. 7: 223 (1912); Forman, Kew Bull. 20: 309 (1966); G. L. Webster, Bot. J. Linn. Soc. 94: 6 (1987). T.: *Panda* Pierre.

Dioecious trees or shrubs; indumentum simple; leaves alternate, stipulate, simple, entire or toothed, penninerved, eglandular; stipules small, generally persistent. Inflorescences terminal or cauliflorous and thyrsiform or axillary and fasciculate; bracts minute; male sepals free or connate; petals valvate, contorted or imbricate, sometimes oppositistemonously 2-concave; disc 0; stamens 5–15, free or ± so, anthers usually introrse; pollen small, subglobose, 3-colporate, often finely reticulate; pistillode columnar, non-lobate, sometimes peltate; female calyx ± as in the male; petals mostly valvate; disc 0; ovary 2–5-locular, ovules 1 per locule, anatropous or orthotropous; styles simple, 2-fid or ∞-fid. Fruit drupaceous; exocarp fleshy, tartareous or woody; endocarp bony, thin and entire or thick and much-perforated and channelled. Seeds ecarunculate; endosperm copious.

The tribe *Galearieae* as treated here corresponds exactly with the family *Pandaceae* as circumscribed by Forman (1966), and the key to the 3 genera follows his treatment:

KEY TO THE GENERA

1. Inflorescences axillary, flowers solitary or in glomerules; male petals imbricate; leaves usually pellucid-punctate ·········· 99. *Microdesmis*
1. Inflorescences terminal or cauliflorous, pseudoracemose-thyrsiform; male petals valvate or ± imbricate; leaves not pellucid-punctate ········· 2
2. Endocarp mostly thin-walled; ovules usually anatropous; petals valvate; inflorescences mostly terminal ····················· 100. *Galearia*
2. Endocarp thick-walled; ovules orthotropous; petals ± imbricate; inflorescences cauliflorous ························ 101. *Panda*

99. **Microdesmis** *Planch.*, Hooker's Icon. Pl. 8: t. 758 (1848); Müll. Arg. in DC., Prodr. 15(2): 1041 (1866); Benth., Gen. Pl. 3: 287 (1880); Hook. f., Fl. Brit. India 5: 380 (1887); Pax & K. Hoffm., Pflanzenr.47: 105 (1911); Gagnep. Fl. Indo-Chine 5: 458 (1926); Pax & K. Hoffm., Nat. Pflanzenfam. ed. 2, 19C: 172 (1931); Robyns, Fl. Parc Nat. Alb. 1: 472 (1948); Keay (ed.), Fl. W. Trop. Afr., ed. 2, 1: 392 (1958); J. Léonard, Bull. Jard. Bot. État. 31: 159 (1961), and Fl. Cong. 8(1): 102 (1962); J.-G. Adam, Mém. Mus. Natl. Hist. Nat., B, Bot. 20: 499 (1971); Airy Shaw, Kew Bull. 26: 362 (1972); Whitmore, Tree Fl. Mal. 2:118 (1973); Berhaut, Fl. Ill. Sénég. 3: 539 (1975); Airy Shaw, Kew Bull. Addit. Ser. 4: 222 (1975), and Kew Bull. 36: 367 (1981); Radcl.-Sm., Fl. Trop. E. Afr. Euph. 2: 581 (1988); G. L. Webster, Ann. Missouri Bot. Gard. 81: 68 (1994). T.: *M. puberula* Hook. f. ex Planch.

Dioecious trees or shrubs. Indumentum simple. Leaves alternate, distichous, shortly petiolate, stipulate, simple, entire or toothed, mucronulate, often unequal at the base, penninerved, eglandular, finely pellucid-punctate; stipules small, generally persistent. Inflorescences axillary or supraaxillary, fasciculate or pseudoracemose-fasciculiform, males many-flowered, females fewer-flowered; bracts minute. Male flowers shortly pedicellate; calyx 5-lobed or -partite, the lobes slightly imbricate; petals 5, larger than the calyx-lobes, contorted or imbricate in bud, fleshy; disc 0; stamens 5–10, uniseriate, oppositisepalous (Africa) or biseriate, inner oppositipetalous (Asia), filaments short, the lower part broad or thick, alternating with and free from or adnate to the lobes of the pistillode, anthers basifixed, introrse or latrorse, longitudinally dehiscent, connective produced or not; pollen small, oblate-spheroidal, 3-colporate, convex-triangular in polar view, colpus transversalis isodiametric, ora large, colpi narrow, tectate; pistillode columnar (Asia), or pentagonal in the lower $1/2$ and cylindric at the apex (Africa). Female flowers: pedicels shorter and stouter than in the male, extending a little in fruit; calyx as in the male, but with glandular hairs at base, ± persistent; petals as in the male, caducous; disc 0; ovary 2–5-locular, fleshy above, with 1 ovule per locule, exobturatorate; styles 2–5, short, deeply bipartite, spreading, laciniate-papillose. Fruit drupaceous, ovoid-subglobose, not or scarcely lobed and smooth when fresh, verrucose-muricate when dry; mesocarp fleshy, endocarp woody, tubercled or muricate, dark, (1) 2–5-locular. Seeds compressed-ovoid, curved or not, ecarunculate; testa crustaceous, shiny; endosperm fleshy; cotyledons broad, flat. Fig.: See Hooker's Icon. Pl. 8: t. 758 (1848); Pflanzenr. 47: 107, t. 34 (1911); Fl. Indo-Chine 5: 459, t. 56.7–13 and 463, t. 57.1–5 (1926); Fl. Cong. Rwa.-Bur. 8(1): 113, t. 8 (1962); Mém. Mus. Natl. Hist. Nat., B, Bot. 20: 500, t. 185 (1971); Fl. Ill. Sénég. 3: 538 (1975); Bot. J. Linn. Soc. 94: 50, t. 1 (1987); Fl. Trop. E. Afr. Euph. 2: 582, t. 108 (1988).

A palaeotropical genus of 10 species, 8 African and 2 Asiatic.

100. **Galearia** *Zoll. & Moritzi*, Syst. Verz.: 19 (1846) (nom. cons. prop.); Müll. Arg. in DC., Prodr. 15(2): 1036 (1866); Benth., Gen. Pl. 3: 287 (1880); Hook. f., Fl. Brit. India 5: 377 (1887); Pax & K. Hoffm., Pflanzenr. 47: 97 (1911); Gagnep. Fl. Indo-Chine 5: 456 (1926); Pax & K. Hoffm., Nat. Pflanzenfam. ed. 2, 19C: 171, t. 88 (1931); Backer & Bakh. f., Fl. Java 1: 495 (1963); Forman, Kew Bull. 26: 155 (1971); Airy Shaw, Kew Bull. 26: 362 (1972); Whitmore, Tree Fl. Mal. 2: 97 (1973); Airy Shaw, Kew Bull. Addit. Ser. 4: 220 (1975), 8: 220 (1980), Kew Bull. 36: 365 (1981), and 37: 36 (1982); G. L. Webster, Ann. Missouri Bot. Gard. 81: 68 (1994). LT.: *G. sessilis* Zoll. & Moritzi [= *G. filiformis* (Blume) Boerl. (B.: *Antidesma filiforme* Blume)] (chosen by Webster, 1994).

Cremostachys Tul., Ann. Sci. Nat. Bot. III. 15: 259 (1851). LT.: *C. filiformis* (Blume) Tul. (B.: *Antidesma filiforme* Blume) [= *G. filiformis* (Blume) Boerl.] (chosen by Webster, 1994).

Bennettia R. Br. in Benn., Pl. Jav. Rar.: 249(1852) (non Gray, 1821); Baill., Étude Euphorb.: 311 (1858). T.: *B. javanica* R. Br. .[= *G. filiformis* (Blume) Boerl. (B.: *Antidesma filiforme* Blume)]

Dioecious trees; buds in shoot-axils, not leaf-axils. Indumentum simple. Leaves alternate, distichous, shortly petiolate, stipulate, simple, usually entire, often unequal at the base, penninerved, eglandular, often coriaceous; one stipule of a pair higher than the other. Inflorescences terminal or cauliflorous, racemose, elongate and ± pendulous or short and ± erect; bracts minute, rarely elongate, subtending either a cymose fascicle of male flowers or a solitary female flower. Male flowers shortly pedicellate; calyx 5-lobed or -toothed and cupular; petals 5, valvate, shallowly to deeply concave and cucullate, with a thickened median ridge, or straight and flat, reflexed or not; disc 0 or rarely toroidal; stamens 10–15, in 1 or 2 whorls (the outer episepalous), equal or unequal, filaments free, variously thickened, long or short, glabrous or variously indumented, anther-thecae basifixed, introrse or latrorse, parallel, distinct, longitudinally dehiscent, connective broad or not, prolonged or not; pollen sometimes suboblate, otherwise ± as in 99. *Microdesmis*; pistillode columnar, broadly ovoid, turbinate or peltate, not lobed, glabrous or indumented. Female flowers shortly pedicellate to subsessile, pedicels not or only slightly extending in fruit; calyx as in the male; petals often smaller than in the male, otherwise similar; disc 0; ovary 2–5-locular, with 1 pendulous anatropous ovule per locule; stigma sessile, variously lobed or laciniate. Fruit a drupe, small, bilaterally compressed and transversely elongate, or larger, depressed-subglobose, -trigonous or -pentagonal; exocarp fleshy or woody; endocarp bony, thin and entire or thick and much-perforated and -channelled, 1–5-celled. Seed compressed, transversely oblong or broadly cuneate, ecarunculate; testa membranaceous; endosperm fleshy; cotyledons broad, flat. Fig.: See Pflanzenr. 47: 99, t. 31 and 103, t. 32 (1911); Fl. Indo-Chine 5: 459, t. 56.1–6 (1926); Nat. Pflanzenfam. ed. 2, 19C: 171, t. 88 (1931); Kew Bull. 26: 154, t. 1 and 156, t. 2 (1971).

An Asiatic genus of 6 species ranging from Myanmar (Burma) to the Solomon Is., according to Forman (1971).

101. **Panda** *Pierre*, Bull. Mens. Soc. Linn. Paris : 1255 (1896); Aubrév., For. Fl. Cote d'Iv. 1: 300 (1959); Keay (ed.), Fl. W. Trop. Afr., ed. 2, 1: 634 (1958); Forman, Kew Bull. 20: 309 (1966); G. L. Webster, Ann. Missouri Bot. Gard. 81: 68 (1994). T.: *P. oleosa* Pierre.

Porphyranthus Engl., Bot. Jahrb. Syst. 26: 367 (1899). T.: *P. zenkeri* Engl. [=*Panda oleosa* Pierre]

Small dioecious trees; buds in shoot-axils, not leaf-axils. Indumentum simple. Leaves alternate, distichous, shortly petiolate, stipulate, simple, subentire to shallowly toothed, slightly unequal at the base, penninerved, eglandular; stipules minute, soon deciduous. Inflorescences cauliflorous, less often ramiflorous, racemose, the male racemes fasciculate on bosses on the trunk and branches, the female solitary or paired, sometimes in the axil of a shoot; bracts minute, subtending either a few-flowered fascicle of male flowers or a solitary female flower. Male flowers shortly pedicellate; calyx cupular, truncate, open in bud; petals 5, valvate below, weakly imbricate above, minutely cucullate; disc 0; stamens 10, in 2 whorls of 5, the 5 outer episepalous with long filaments, the 5 inner epipetalous with short ones, filaments slender, anthers basifixed, introrse, parallel, distinct, thecae longitudinally dehiscent, connective broader dorsally, not prolonged; pollen ± as in 99. *Microdesmis* and

100. *Galearia* as to shape, size and apertures, but surface reticulate; pistillode narrowly columnar. Female flowers: pedicels shorter and stouter than in the male, not extending in fruit; calyx as in the male; petals 5, imbricate; disc 0; staminodes 0; ovary (2)3–4-locular, ovule 1 per locule, pendulous, orthotropous; styles very short, united at the base, stigmas (2)3–4, otioid, undivided, smooth, erect at first, later reflexed. Fruit drupaceous, subglobose, (2)3–4-seeded; exocarp thick, firmly tartareous when cut and dried; endocarp massive, apiculate, bony, pitted, (2)3–4-sutured, very tardily dehiscent along these sutures into (2)3–4 valves, leaving a large capped tri- or tetrapterygoid columella. Seeds obtriangular, flattened, curved, lower angle obtuse, upper ones somewhat acute; testa thinly coriaceous, greyish; endosperm copious, oleaginous; cotyledons cordate, truncate. Fig.: See Fam. Fl. Pl. ed. 2, 1: 311, t. 169 (1959).

A monotypic genus of W. Tropical Africa.

Although diverging from nearly all other *Euphorbiaceae* in its distinctive characters, e.g. orthotropous ovules, Webster (1994) considers it best to regard it as the terminus of a morphological sequence within the *Galearieae*. Stuppy (1995), however, feels that from the ovule- and seed-morphology, the *Pandaceae* should be re-instated.

Tribe 23. **AMPEREAE** *Müll. Arg.*, Bot. Zeitung (Berlin) 22: 324 (1864), and in DC., Prodr. 15(2): 211 (1866); G. L. Webster, Ann. Missouri Bot. Gard. 81: 68 (1994). T.: *Amperea* A. Juss.

Monoecious or dioecious glabrous herbs or subshrubs; leaves alternate, often ericoid, stipulate. Flowers in terminal or axillary clusters; male calyx 3–5(6)-lobed, lobes imbricate or valvate; petals + or 0; disc 0, entire or dissected; stamens (2)3–10, 2-seriate, filaments free or ± so, anther-thecae disjunct; pollen 2–3-colporate, colpi inoperculate, marginate, perforate-tectate; pistillode 0 or +; female sepals 4–5, imbricate or valvate; petals 0 or +; disc 0 or +; ovary 3-locular, style-branches fimbriate or entire. Fruit toothed or smooth, dehiscent. Seeds smooth or ± so , carunculate; endosperm copious; embryo cylindric, cotyledons scarcely broader than the radicle.

This tribe of 2 disparate genera was upheld by Pax (1890), Grüning (1913) & Webster (1994). Despite the many differences between the genera, both the pollen and seeds have points in common.

KEY TO THE GENERA

1. Cymes terminal; sepals imbricate; male petals +; anther-thecae disjunct; pollen 3-angled, sexine reticulate; style-branches fimbriate; fruit unadorned · 102. *Monotaxis*
1. Flowers axillary; sepals valvate; male petals 0; anther-thecae pendulous from a glandular connective; pollen 3-lobed, sexine perforate-foveolate; style-branches entire; fruit toothed · 103. *Amperea*

102. **Monotaxis** *Brongn.* in Duperrey, Voy. Monde: 223 (1829); Baill., Étude Euphorb.: 307 (1858); Müll. Arg. in DC., Prodr. 15(2): 212 (1866); Benth., Fl. Austral. 6: 78 (1873), and Gen. Pl. 3: 264 (1880); Grüning, Pflanzenr. 58: 75 (1913); Pax & K. Hoffm., Nat. Pflanzenfam. ed. 2, 19C: 227 (1931); Airy Shaw, Muelleria 4: 239 (1980); G. L. Webster, Ann. Missouri Bot. Gard. 81: 69 (1994). LT.: *M. linifolia* Brongn. (chosen by Webster 1994).

Hippocrepandra Müll. Arg., Linnaea 34: 61 (1865). LT.: *H. gracilis* Müll. Arg. [= *M. gracilis* (Müll. Arg.) Baill.] (chosen by Wheeler, 1975).

Monoecious glabrous annual herbs or shrublets. Leaves alternate, subopposite or subverticillate, simple, entire or subentire, rarely toothed, subsessile or petiolate, stipulate, penninerved or lateral nerves invisible; stipules minute, commonly subulate. Inflorescences small, terminal, densely cymulose, glomerulate or subcapitate, usually surrounded by leafy lateral shoots; males numerous, females few or solitary in the centre of the glomerule; bracts small, numerous, scale-like. Male flowers: pedicels slender; calyx-segments 4–5, imbricate, subpetaloid; petals 4–5, hastate or auriculate at the base; disc-glands 4–6, alternipetalous; stamens 8–10, filaments free or shortly connate, attenuate to the apex, anthers apicifixed, pendulous, introrse at first, later extrorse, connective thick, curved or hippocrepiform, thecae separate, subglobose, longitudinally dehiscent; pollen 3-colporate, oblate, colpi elongate, colpi transversales long and narrow, coarsely reticulate and crumpet-like; pistillode 0, filiform or 3-partite. Female flowers: pedicels stouter than in the males, but calyx, petals and disc ± as in the males; ovary sessile, 3-locular, ovule 1 per locule; styles free, divided into 2 fimbriate-papillose branches. Fruit globose or didymous, exappendiculate, dehiscing septicidally and loculicidally into 3 2-valved cocci or 6 valves. Seeds ovoid or oblong, smooth, carunculate; endosperm copious; embryo linear, straight or curved; cotyledons narrow, much longer than the radicle. Fig.: See Étude Euphorb.: t. 16, f. 22–25 (1858); Pflanzenr. 58: 80 t. 13 and 83, t. 14 (1913).

An Australian genus of c. 10 species, 2 in the East, the rest in the West.

Brongniart in his original description suggested an affinity to *Chiropetalum* and *Ditaxis* in the tribe *Chrozophoreae*. Punt (1962) related the pollen of *Monotaxis* to his "Sumbavia" type, which includes genera of this tribe. Therefore *Monotaxis* may well prove to be an aberrant member of the *Chrozophoreae*.

103. **Amperea** *A. Juss.*, Euphorb. Gen. 35 (1824); Baill., Étude Euphorb.: 454 (1858); Müll. Arg. in DC., Prodr. 15(2): 213 (1866); Benth., Fl. Austral. 6: 81 (1873), and Gen. Pl. 3: 265 (1880); Grüning, Pflanzenr. 58: 86 (1913); Pax & K. Hoffm., Nat. Pflanzenfam. ed. 2, 19C: 228 (1931); R. J. F. Henderson, Austral. Syst. Bot. 5: 10 (1992); G. L. Webster, Ann. Missouri Bot. Gard. 81: 69 (1994). T.: *A. ericoides* A. Juss.

Monoecious or dioecious glabrous perennial herbs or shrublets, with stems arising from a hard woody stock; stems erect, ascending, decumbent or twining, sometimes subaphyllous and angular. Leaves when developed alternate, often linear, sessile or shortly petiolate, stipulate, simple, entire or subdentate, margins revolute or rarely flat, persistent or deciduous; stipules entire, lacerate or fimbriate. Flowers small, clustered at the nodes or in the axils, rarely aggregated into pseudospikes, males often numerous, females often solitary; bracts small, numerous, scarious. Male flowers sessile or shortly pedicellate; calyx campanulate, 3–5(6)-lobed, lobes valvate, subpetaloid; petals 0; disc 0 or indistinct; stamens (2)3–10, exscrted, in 2 whorls, rarely partly aborted, filaments free or shortly connate, anthers apicifixed, the outer introrse, the inner extrorse, 2-thecate or the outer 1-thecate, thecae separate, globose, ellipsoid or ovoid, parallel, longitudinally dehiscent, yellow, but with the dehiscence-slits reddish, connective often tipped by a small gland; pollen 2-colporate and oblate-spheroidal or 3-colporate and prolate-spheroidal, colpi long and narrow, subsyncolpate, exine thick, coarsely reticulate to perforate-tectate; pistillode 0. Female flowers shortly pedicellate; calyx 4–5-lobed, persistent in fruit; petals 0; disc 0; ovary sessile, 3-locular, ovule 1 per locule; styles 3, free, deeply 2-lobed, entire. Fruit ovoid, crowned by a ring of 6 or 12 tooth-like appendages, dehiscing septicidally into 3 2-valved cocci or 6 valves; valves thinly crustaceous; columella slender, persistent. Seeds 3, obovoid-oblong, smooth or ± so, carunculate; endosperm copious; embryo linear, straight or very slightly curved; cotyledons narrow, longer than the radicle. Fig.: See Euphorb. Gen. t. 10, f. 32.1–6 (1824); Étude Euphorb.: t. 14, f. 1–9 (1858); Pflanzenr. 58: 88 t. 15 and 90, t. 16 (1913); Austral. Syst. Bot. 5: 5, t. 1 and 6, t. 2 (1992).

An Australian genus of 8 species, 6 of which are confined to the West, according to Henderson (1992).

Punt (1962) related the pollen of *Amperea* to that of genera in the tribe *Epiprineae.*

Tribe 24. **AGROSTISTACHYDEAE** *(Müll. Arg.) G. L. Webster,* Taxon 24: 596 (1975), and Ann. Missouri Bot. Gard. 81(1): 69(1994). Type: *Agrostistachys* Dalzell.

Acalypheae subtribe *Agrostistachydeae* Müll. Arg., Linnaea 34: 143 (1865).

Dioecious or occasionally monoecious trees or shrubs; indumentum simple or 0; leaves alternate, simple, entire or denticulate, glandular or eglandular; stipules deciduous. Inflorescences axillary, paniculate, racemose or spicate, bracts sometimes scarious, imbricate, rigid; male calyx closed in bud, splitting into 2–5 valvate lobes; petals (0)3–8; disc receptacular or extrastaminal and dissected; stamens (8)10–55, filaments free, at least in part, anthers usually extrorse, thecae often pendulous from an enlarged connective; pollen subglobose, 3-colporate, colpi inoperculate, weakly marginate, usually coarsely reticulate; pistillode + or 0; female sepals 4–5, open or valvate; petals (0)5, caducous; disc annular, lobed, pitted or dissected; ovary 3-locular; styles 2-fid. Fruit dehiscent. Seeds mostly smooth, ecarunculate; cotyledons broad, flat.

A tribe of 4 genera akin to the *Ampereae* but with broad cotyledons and foliage, and to the *Chrozophoreae* but mostly dioecious, with simple indumentum and enlarged anther-connectives.

KEY TO THE GENERA

1. Inflorescences paniculate; petals 0; monoecious; leaves glandular at base · 107. *Chondrostylis*
1. Inflorescences spicate or racemose · 2
2. Monoecious; male petals not imbricate in bud; petals 0 in the female flower; leaves glandular at base · 106. *Cyttaranthus*
2. Dioecious; male petals imbricate in bud; petals + in the female flower · · 3
3. Stipules free; leaves eglandular; male disc dissected; stamens 8–13; pistillode + · 104. *Agrostistachys*
3. Stipules connate, leaving an annular scar; leaves glandular at base; male disc receptacular; stamens 20–55; pistillode 0 or much-reduced · 105. *Pseudagrostistachys*

104. **Agrostistachys** *Dalzell*, Hooker's J. Bot. Kew Gard. Misc. 2: 41 (1850); Baill., Étude Euphorb.: 310 (1858); Müll. Arg. in DC., Prodr. 15(2): 725 (1866); Benth., Gen. Pl. 3: 302 (1880); Hook. f., Fl. Brit. India 5: 405 (1887); Pax & K. Hoffm., Pflanzenr. 57: 98 (1912); Gagnep. Fl. Indo-Chine 5: 465 (1926); Pax & K. Hoffm., Nat. Pflanzenfam. ed. 2, 19C: 96 (1931); Airy Shaw, Kew Bull. 14: 472 (1971), and 26: 210 (1972); Whitmore, Tree Fl. Mal. 2: 52 (1973); Airy Shaw, Kew Bull. Addit. Ser. 4: 26 (1975), 8: 23 (1980), Kew Bull. 36: 248 (1981), and Euph. Philipp.: 3 (1983); G. L. Webster, Ann. Missouri Bot. Gard. 81: 69 (1994); Philcox in Dassanayake, ed., Fl. Ceyl. XI: 123 (1997). T.: *A. indica* Dalzell.

Sarcoclinium Wight, Icon. Pl. Ind. Orient. v. II: 24, t. 1887–8 (1852). T.: *S. longifolium* Wight [= *A. longifolia* (Wight) Benth.]
Heterocalyx Gagnep., Notul. Syst. (Paris) 14: 33 (1950). T.: *H. laoticus* Gagnep. [= *A. indica* Dalzell]

Dioecious shrubs or small trees; young parts often gummiferous. Indumentum, when present, of small, simple hairs. Leaves alternate, shortly petiolate or subsessile, stipulate, large, elongate, simple, entire or denticulate, penninerved, eglandular; stipules 2, free, small, soon falling, leaving conspicuous transverse scars. Inflorescences axillary, unisexual, spicate, solitary or clustered; bracts broad, concave, striate, rigid; male flowers 3–5 per bract, rarely solitary, female solitary, sometimes only 1 per spike. Male flowers subsessile or shortly pedicellate; calyx globose in bud, valvately splitting into 2–5 segments; petals 5–8, shorter than calyx-lobes, delicate, hyaline, denticulate; disc-glands 5, episepalous, thick, fleshy; stamens (8)10–13, rarely fewer, 2-seriate, outer epipetalous, filaments free or more or less shortly connate, anther-thecae versatile, pendulous from an often somewhat thickened glanduliform connective, extrorse, longitudinally dehiscent; pollen subprolate to oblate-spheroidal, 3-colporate, convex-triangular in polar view, colpi transversales large, colpi narrow, tectate, intrareticulate; pistillode 2–3-lobed or undivided. Female flowers: pedicels longer than in the males; calyx (4)5-lobed, lobes narrow, more rigid than in the male; petals exceeding the calyx, caducous; disc-glands 5, thick, ± confluent; ovary 3-locular, ovule 1 per locule; styles short, thick, entire, 2-lobed or 2-fid. Fruit tricoccous, ± smooth, dry or fleshy, dehiscing into 3 2-valved cocci; endocarp thinly crustaceous; columella persistent, 3-alate at the apex. Seeds subglobose, smooth, shining, ecarunculate; cotyledons broad, flat. Fig.: See Pflanzenr. 57: 101, t. 19 and 104, t. 20 (1912); Fl. Indo-Chine 5: 463, t. 57. 7–11 and 58.1, 2 (1926); Nat. Pflanzenfam. ed. 2, 19C: 89, t. 44D,E (1931); Kew Bull. 36: 250, t. 1 A1–6 (1981), and 50: 120, t. 1 (1995).

An Asiatic genus of c. 10 species, some difficult to distinguish, from India and Sri Lanka to New Guinea.

105. **Pseudagrostistachys** *Pax & K. Hoffm.*, Pflanzenr. 57: 96 (1912), and Nat. Pflanzenfam. ed. 2, 19C: 96 (1931); Lebrun, Bull. Soc. Roy. Bot. Belgique 67: 97 (1934); Keay (ed.), Fl. W. Trop. Afr., ed. 2, 1: 399 (1958); J. Léonard, Fl. Cong. Rwa.-Bur. 8(1): 183 (1962); Radcl.-Sm., Fl. Trop. E. Afr. Euph. 1: 166 (1987); G. L. Webster, Ann. Missouri Bot. Gard. 81: 70 (1994); Radcl.-Sm., Fl. Zamb. 9(4): 136 (1996). T.: *P. africana* (Müll. Arg.) Pax & K. Hoffm. (B.: *Agrostistachys africana* Müll. Arg.)

Agrostistachys sensu auctt. plur., non Dalzell.

Small dioecious trees or shrubs. Indumentum simple, mostly confined to the inflorescences. Leaves large, alternate, petiolate, stipulate, stipellate or not, simple, elliptic, entire or shallowly denticulate, penninerved with numerous lateral nerves looped within the margin and very many parallel tertiary nerves, 2–6-glandular at the base; stipules intrapetiolar, fused into a very long sheath enveloping the terminal bud, soon deciduous, leaving a circular scar. Inflorescences axillary or cauliflorous, solitary or fasciculate, racemose, at first condensed and catkin-like, short, later slender and lax; bracts broad, concave, overlapping when young, males 1-more-flowered, females usually 1-flowered; bracteoles 2. Male flowers: pedicels articulate near the base; calyx ovoid, closed in bud, splitting into 2–5 membranous valvate segments; petals (5)6(8), free, imbricate, longer than the calyx; receptacle convex, alveolate-glandular, pubescent; stamens 20–55, filaments free, inserted in the receptacular alveoli, anther-thecae apicifixed, extrorse, ± free, pendulous from a thickened apiculate glandular connective, longitudinally dehiscent; pollen subprolate, otherwise as in 104. *Agrostistachys*; pistillode 0. Female flowers: pedicels stouter than in the males; calyx and petals as in the males; disc thick, annular, alveolate; staminodes subulate-filiform, arising from the alveoli of the disc; ovary 3-locular, ovule 1 per locule; styles 3, free, 2-partite, stigmas densely papillose. Fruit depressed-trilobate-subglobose, dehiscing into 3 2-valved cocci; endocarp thinly woody; columella persistent. Seeds subglobose, large; testa crustaceous, smooth, shiny, mottled, ecarunculate; endosperm thick, fleshy; embryo straight or slightly curved; cotyledons broad, flat. Fig.: See Pflanzenr. 57: 97, t. 18 (1912); Fl. Cong. Rwa.-Bur. 8(1): 185, t. 15 (1962); Fl. Trop. E. Afr. Euph. 1: 168, t. 30 (1987); Fl. Zamb. 9(4): 137, t. 24 (1996).

An African genus of 2 species ranging from Nigeria and São Tomé to Uganda and Zambia.

106. **Cyttaranthus** *J. Léonard*, Bull. Jard. Bot. État. 25: 286 (1955), and Fl. Cong. Rwa.-Bur. 8(1): 180 (1962); G. L. Webster, Ann. Missouri Bot. Gard. 81: 70 (1994). T.: *C. congolensis* J. Léonard.

Small monoecious shrub. Indumentum simple. Leaves alternate, long-petiolate, stipulate, simple, obovate, subentire to dentate, 3-nerved at the base; petiole with 2 elongate erect glands at the apex; stipules small, soon deciduous. Inflorescences axillary, cauliflorous or pseudoterminal, racemose, usually unisexual, very slender; bracts adnate to the rhachis, embracing 2, rarely more, bracteoles, all ciliolate, and 1(2) buds. Male flowers: pedicels articulate; calyx membranous, closed in bud, splitting into 2–3 reflexed valvate lobes; petals 3, rarely 6–8, not imbricate; receptacle strongly convex, entirely covered between the filaments with thick, free, fleshy glands; stamens 25–40, filaments free, erect in bud, inserted among the glands, anther-thecae apicifixed, extrorse, subfree, pendulous, subequal, connective apiculate, ± longitudinally dehiscent; pollen prolate-spheroidal, not reticulate, otherwise ± as in 104. *Agrostistachys*; pistillode 0. Female flowers: pedicels articulate near the base, extending in fruit; calyx open in bud, 3-toothed; petals 0; disc annular, hypogynous; ovary 3-locular, ovule 1 per locule; styles 3, shortly connate at the base, 2-partite. Fruit 3-lobed, dehiscing into 3 2-valved cocci; columella persistent, obtriangular. Seeds ovoid, minutely rugulose-verruculose, ecarunculate; endosperm copious; cotyledons broad, flat. Fig.: See Bull. Jard. Bot. État. 25: 288, t. VII (1955); Fl. Cong. Rwa.-Bur. 8(1): 181, t. XII (1962).

A monotypic genus confined to Congo (Kinshasa) and Cabinda.

Very close to the next genus.

107. **Chondrostylis** *Boerl.*, Icon. Bogor. 1: t. 23 (1897); Koord., Ann. Jard. Bot. Buitenzorg 19: 45 (1904); Pax & K. Hoffm., Pflanzenr. 63: 15 (1914), and Nat. Pflanzenfam. ed. 2, 19C: 104 (1931); Airy Shaw, Kew Bull. 14: 358 (1960), and 26: 231 (1972); Whitmore, Tree Fl. Mal. 2: 77 (1973); Airy Shaw, Kew Bull. Addit. Ser. 4: 69 (1975), and Kew Bull. 36: 276 (1981); G. L. Webster, Ann. Missouri Bot. Gard. 81: 70 (1994) T.: *Ch. bancana* Boerl.

Kunstlerodendron Ridl., Fl. Malay Penins. 3: 283 (1924). LT.: *K. sublanceolata* Ridl., nom. illeg. [=*Ch. kunstleri* (King ex Hook. f.) Airy Shaw (B.: *Mallotus* ? *kunstleri* King ex Hook. f.)] (S.: *Kunstlera glumacea* King ex Hook. f.) (chosen by Wheeler, 1975).

Monoecious glabrous shrubs, branches densely leafy at the apex. Leaves alternate, large, narrow, very short- (*bancana*) or long-petiolate (*kunstleri*), bistipulate, simple, shallowly and remotely serrate, penninerved, coriaceous, sometimes glandular at the base or on the petiole; stipules soon falling. Male and female inflorescences in the axils of fallen leaves, paniculate, male elongated, female short; bracts glumaceous; flowers bibracteolate. Male flowers: calyx ovoid, acute, closed, valvately 3-partite; petals 0; stamens c. 50, filaments free, elongated, anthers basifixed, introrse, connective dilated, thecae half-free, divergent, longitudinally dehiscent; disc-glands numerous, juxtastaminal, pilose; pollen spheroidal, 3-colporate, convex-triangular in polar view, colpi transversales elongate, colpi narrow, tectate, psilate; pistillode 0. Female flowers: calyx campanulate, shortly 5-lobed, lobes imbricate; petals 0; disc annular, tomentose; ovary 3-locular, ovule 1 per locule; styles 3, almost free, 2-lobed; stigmas fimbriate. Fruit 3-coccous, ± smooth, dehiscing into 3 2-valved cocci; exocarp thin; endocarp thinly woody; columella persistent, narrowly 3-winged, slightly expanded at the apex. Seeds ± spherical, smooth, slightly shiny, ecarunculate; testa thinly crustaceous; endosperm hard; cotyledons broad, flat. Fig.: See Kew Bull. 36: 268, t. 3 C1–6 (1981).

A ditypic genus ranging from Thailand to Sumatra and Borneo.

Tribe 25. **SPHYRANTHEREAE** *Radcl.-Sm.*, **trib. nov.**, a tribu *Agrostistachydeis* foliis exstipulatis, inflorescentiis subumbellatim capitellatis confertim bracteatis, filamentis malleiformibus, connectivis 2-fidis, polline prolato, 3-colpato, petalis foemineis deficientibus, stigmatibus multifidis, fimbriatis vel laciniatis, fructibus pendulis differt. Genus solum typicum: *Sphyranthera* Hook. f.

Sphyranthera was placed by Webster (1994) in the *Crotonoïdeae–Codiaeae* before it was discovered that the pollen is not of the crotonoid type. Although there is some connection with the *Acalypheae–Claoxylinae* and *–Mareyinae* in androecial and stigmatic characters, the presence of petals in the male flowers *inter alia* would seem to suggest that the tribe is best placed next to the *Agrostistachydeae.*

108. **Sphyranthera** *Hook. f.*, Hooker's Icon. Pl. 18: t. 1702 (1887); Hook. f., Fl. Brit. India 5: 477 (1888); Pax in Engl. & Prantl, Nat. Pflanzenfam. ed. 1, 3(5): 118 (1890); Pax & K. Hoffm., Nat. Pflanzenfam. ed. 2, 19C: 231 (1931); Airy Shaw, Kew Bull. 39: 807 (1984); Chakrab. & Vasudeva Rao, J. Econ. Taxon. Bot. 5: 959 (1984), and 6: 429 (1985). T.: *S. capitellata* Hook. f., nom. illeg. [=*S. lutescens* (Kurz) Pax & K. Hoffm. (B.: *Codiaeum lutescens* Kurz)]

Dioecious shrubs. Indumentum simple, mostly confined to innovations and inflorescences. Leaves alternate, petiolate, exstipulate, simple, entire, penninerved, eglandular. Inflorescences axillary or arising below the leaves,

FIG. 12. *Sphyranthera lutescens* (Kurz) Pax & K. Hoffm. (B.: *Codiaeum lutescens* Kurz). **A** Habit, female × 1; **B** Male inflorescence × 4.5; **C** Male flower × 18; **D** Bifid petaloid disc-segment × 22.5; **E** Stamen × 22.5; **F** Pollen-grain × 650; **G** Reticulate exine × 6000; **H** Female flower × 18; **J** Female flower, vertical section × 18; **K** Fruit × 6; **L** Columella × 6. **A, H–L** from *Chakrabarty* 10254; **B–G** from *Chakrabarty* 10253. Drawn by Camilla Speight.

subumbellately capitellate, pedunculate, solitary or fasciculate, 1-sexual, male many-flowered, female 2–4-flowered or flowers solitary; bracts crowded at apex of peduncle, deltate. Male flowers pedicellate, pedicels slender, articulate near the base; buds ovoid-subglobose; sepals (3)4(5), concave, valvate; petals 4, small, entire, 2-lobate, -partite or -crurate; disc-glands 4, transversely oblong to obcordate, 2-lobate at the apex; stamens (8)12–20, central, filaments free, hammer-shaped, anther-thecae transversely oblong to subglobose, distant, divaricate, longitudinally dehiscent, connective broad, produced and minutely 2-fid at the apex; pollen prolate, 3-colporate, exine thickly reticulate, lumina small; pistillode 0. Female flowers long-pedicellate, pedicels slender, articulate, extending in fruit; sepals 3–4(5), valvate or open in bud, persistent in fruit; petals 0; disc-glands 3–4, free, oppositisepalous or alternisepalous, variously-shaped, thin; ovary 3-locular, globose, ovules 1 per locule; styles 3, short, free, deflexed, 2-fid, stigmas multifid or fimbriate-laciniate. Fruit pendulous, 3-lobate, depressed at the apex, lobes shouldered, septicidally dehiscent into 3 2-valved cocci; endocarp crustaceous; columella subpersistent, furcate at the apex. Seeds globose, ecarunculate; testa smooth, ochraceous. Fig.: See Hooker's Icon. Pl. 18: t. 1702 (1887); Fig. 12, p. 136.

A ditypic genus endemic to the Andaman and Nicobar Is.

Tribe 26. **CHROZOPHOREAE** *(Müll. Arg.) Pax & K. Hoffm.*, Pflanzenr. 68: 5 (1919); G. L. Webster, Ann. Missouri Bot. Gard. 81: 70 (1994). Type: *Chrozophora* Neck. ex A. Juss.

Chrozophoreae subtribe *Regulares* (Pax & K. Hoffm.) Pax & K. Hoffm., Naturl. Pflanzenfam. ed. 2, 19C: 89 (1931).

Monoecious or dioecious herbs, shrubs or trees; indumentum simple, malpighiaceous, stellate or lepidote; leaves alternate, simple or lobate, entire or dentate, penni- or palminerved, with or without basal laminar glands; stipules persistent, deciduous or 0. Inflorescences terminal or axillary, spicate or racemose; male sepals 3–5, valvate; petals (0)4–5(10); disc dissected or 0; stamens 3–15 or 30–200+, filaments free or connate, anthers introrse, muticous, thecae not pendulous; pollen subglobose to oblate, 3–6-colporate, colpi sometimes operculate, weakly marginate, reticulate, often heterobrochate; pistillode + or 0; female sepals 5–6(10), imbricate or valvate; petals 5(6) or 0; disc annular, dissected or 0; ovary 2–3-locular, smooth or muricate; styles simple to 2-fid or 2 × 2-fid, sometimes lacerate. Fruit dehiscent. Seeds fleshy or not, ecarunculate; endotesta smooth, roughened or weakly pitted.

A tribe of 4 subtribes possibly related to the tribe *Epiprineae*, following Webster (1975, 1994), and corresponding very closely to Pax & K. Hoffmann's subtribe *Regulares* (1931), apart from the inclusion of 118. *Melanolepis* Rchb. f. & Zoll., and the exclusion of 225. *Oligoceras* Gagnep. and 226. *Deutzianthus* Gagnep. (q.v.).

Key to the subtribes

1. Indumentum simple, malpighiaceous or stellate; leaves lacking basal laminar glands; pollen not distinctly heterobrochate · · · · · · · · · · · · 2
1. Indumentum stellate; leaves with basal laminar glands; pollen distinctly heterobrochate · 3
2. Indumentum simple; inflorescences terminal; stamens 10–15, free; female disc annular; ovary muricate · · · · · · · · · · · · · · · · · · 26a. *Speranskiinae*
2. Indumentum malpighiaceous or stellate; inflorescences axillary; stamens, if more than 6, connate; female disc dissected or obsolete · · 26b. *Ditaxinae*
3. Shrubs or trees; stamens free or connate, 30–200+; pollen 3-colporate · 26c. *Doryxylinae*
3. Herbs; stamens connate, 3–15; pollen stephanocolporate · 26d. *Chrozophorinae*

Subtribe 26a. **Speranskiinae** *G. L.Webster*, Taxon 24: 596 (1975), and Ann. Missouri Bot. Gard. 81(1): 70 (1994). Type: *Speranskia* Baill.

Monoecious herbs; indumentum simple; leaves alternate, glandular-dentate or -lobulate, but lacking basal laminar glands. Inflorescences terminal, spicate, bisexual, female flowers proximal; male sepals and petals (4)5; disk-glands free; stamens 10–15, filaments free; pollen 3-colporate, evenly reticulate; pistillode 0; female sepals 5; petals 5, minute or 0; disc annular; ovary 3-locular, verrucose; styles free, 2-fid, arms lacerate. Fruit dehiscent. Seeds ecarunculate; testa dry, foveolate-asperate.

This monogeneric subtribe is simlar to the next, where Müller (1866) included it, but differs primarily in the terminal inflorescence and free stamens.

109. **Speranskia** *Baill.*, Étude Euphorb. : 388 (1858); Benth., Gen. Pl. 3: 305 (1880); Oliv., Hooker's Icon. Pl. 16: t. 1577 (1887); Pax & K. Hoffm., Pflanzenr. 57: 14 (1912), and Nat. Pflanzenfam. ed. 2, 19C: 90 (1931); Hurus., J. Fac. Sci. Univ. Tokyo, Sect. 3, Bot. 6: 310 (1954); G. L. Webster, Ann. Missouri Bot. Gard. 81: 71 (1994). T.: *S. tuberculata* (Bunge) Baill. (B.: *Croton tuberculatus* Bunge).

Monoecious plants with 1 or more simple or little-branched herbaceous stems arising from a woody root. Indumentum simple. Leaves alternate, petiolate or sessile, stipulate, simple, glandular-dentate or -lobulate, palminerved. Inflorescences terminal, racemose, flowers often 3 per bract, male in the upper $^1/_2$, in the lower $^1/_2$ with 1 female or 1 female and 2 males. Male flowers sessile or subsessile; calyx globose, closed in bud, at length splitting valvately into 5 lobes; petals 4–5, short, rounded, clawed, thin and scale-like; glands 5, free, alternipetalous; stamens 10–15, filaments free, in 2 whorls on a convex receptacle, outer whorl epipetalous, anthers medifixed, extrorse, thecae subglobose, subpendulous and separated by the connective, longitudinally dehiscent; pollen subprolate, 3-colporate, convex-triangular in polar view, colpi transversales large, colpi narrow, tectate, intrareticulate; pistillode 0. Female flowers: calyx-segments 5, narrow; petals 5, minute or 0; disc urceolate; ovary 3-locular, ovule 1 per locule, verruculose or tuberculate; styles 3, free, 2-partite, stigmas papillose-lacerate. Fruit broadly tricoccous, dehiscing septicidally into 3 2-valved cocci; epicarp tubercled or warty; endocarp thinly crustaceous; columella slender, slightly expanded at the apex, readily frangent. Seeds ecarunculate, globose; testa shallowly foveolate or rugulose, blackish or brownish; endosperm fleshy; cotyledons broad, flat. Fig.: See Hooker's Icon. Pl. 16: t. 1577 (1887); Pflanzenr. 57: 16, t. 3 (1912).

A genus of 2 (Hurusawa, 1954) or 3 (Pax & Hoffmann, 1912b) species restricted to China and Upper Burma (Myanmar).

Subtribe 26b. **Ditaxinae** *Griseb.*, Fl. Brit. W. I.: 43 (1859), and Abh. Königl. Ges. Wiss. Göttingen 9: 15 (1861); G. L. Webster, Ann. Missouri Bot. Gard. 81: 71 (1994). T.: *Ditaxis* Vahl ex A. Juss.

Acalypheae subtribe *Caperonieae* Müll. Arg., Linnaea 34: 143 (1865). T.: *Caperonia* A. St. Hil.

Monoecious or rarely dioecious herbs, shrubs or trees, spiny or not, often aromatic when dried; indumentum simple, malpighiaceous, stellate or glandular; leaves alternate, entire or serrate, eglandular. Inflorescences axillary, racemose or glomerulate, usually bisexual; male sepals 4–5, scarious or not; petals 4–5, entire, toothed or lacerate; disc 0 or +, at least in the male; stamens 4–15, filaments connate, rarely ± free, anthers in 1 or 2 whorls, introrse; pollen 3–6-colporate, bilaterally or radially symmetrical, not distinctly heterobrochate; pistillode +, scarcely-developed or 0; female sepals 5–6(10), imbricate to valvate; petals 5(6), sometimes 0; disc 0 or +; ovary sometimes muricate; styles once or twice 2-fid to laciniate. Fruit dehiscent. Seeds ecarunculate; testa dry, smooth, foveolate, asperate or reticulate.

According to Webster (1975), this subtribe of 5 genera is a mostly New world taxon except for a few *Caperonia* spp., and approaches the *Speranskiinae* via *Caperonia,* which resembles *Speranskia* in its muricate ovary and laciniate style-branches, but differs primarily in its axillary inflorescences and stephanocolporate pollen. Generic delimitation within the subtribe is still controversial, but pollen characters support the narrow circumscription of Pax & Hoffmann rather than the broad delimitation of Müller (1866) and Ingram (1967, 1980a and b).

Key to the genera

1. Leaves finely serrate with straight parallel lateral nerves; indumentum simple, sometimes glandular; disc 0; pollen 6-colporate; pistillode +; styles dissected . . . . 110. *Caperonia*

1. Leaves not as above; disc +, at least in the male; pollen 3- or 4-colporate; pistillode 0 or scarcely-developed; styles bifid to dissected · · · · · · · · · 2
2. Dioecious shrubs or trees, often spiny; whole plant smelling strongly of fenugreek when dried; indumentum simple; sepals scarious · 111. *Philyra*
2. Monoecious herbs or undershrubs, not spiny; dried plants not as strongly odoriferous; indumentum malpighiaceous, at least in part; sepals not scarious · 3
3. Stellate hairs sometimes +; petals toothed or lacerate; pollen tricolpate; styles once bifid · 114. *Chiropetalum*
3. Stellate hairs 0; petals entire; styles twice bifid · · · · · · · · · · · · · · · · · · 4
4. Stamens 8–10, anthers in 2 whorls; pollen 3- or rarely 4-colpate, bilaterally symmetrical · 112. *Ditaxis*
4. Stamens 4–6, anthers in 1 whorl; pollen stephanocolpate, radially symmetrical · 113. *Argythamnia*

110. **Caperonia** *A. St.-Hil.*, Hist. Pl. Remarq. Brésil : 244 (1826); Baill., Étude Euphorb.: 299 (1858); Müll. Arg. in DC., Prodr. 15(2): 751 (1866), and Fl. Bras. 11(2): 315 (1874); Benth., Gen. Pl. 3: 304 (1880); Pax & K. Hoffm., Pflanzenr. 57: 27 (1912); Prain in Fl. Trop. Afr. 6(1): 829 (1912); Fawc. & Rendle, Fl. Jamaica 4: 288 (1920); Pax & K. Hoffm., Nat. Pflanzenfam. ed. 2, 19C: 92 (1931); Alain, Fl. Cuba 3: 82 (1953); J. Léonard, Bull. Jard. Bot. État. 26: 313 (1956); Keay (ed.), Fl. W. Trop. Afr., ed. 2, 1: 397 (1958); J. Léonard, Fl. Cong. Rwa.-Bur.: 8(1): 166 (1962); G. L. Webster, J. Arnold Arbor. 48: 363 (1967); G. L. Webster, Ann. Missouri Bot. Gard. 54: 265 (1968); Berhaut, Fl. Ill. Sénég. 3: 395 (1975); Dyer, Gen S. Afr. Fl. Pl. 1: 314 (1975); Philcox, Fl. Trin. Tob. II(X): 643 (1979); Alain, Fl. Español. 4: 91 (1986); Radcl.-Sm., Fl. Trop. E. Afr. Euph. 1: 163 (1987); Alain, Descr. Fl. Puerto Rico 2: 361 (1988); R. A. Howard, Fl. Lesser Antilles 5: 20 (1989); G. L. Webster, Ann. Missouri Bot. Gard. 81: 71 (1994); Radcl.-Sm., Fl. Zamb. 9(4): 138 (1996). LT.: *C. castaneifolia* (L.) A. St.-Hil. (B.: *Croton castaneifolius* L.) (designated by Britton & Wilson, 1925).

Cavanilla Vell., Fl. Flumin.: 226 (1825). T.: *Cav. spinosa* Vell. [=*Cap. vellosiana* Müll. Arg.]

Lepidococca Turcz., Bull. Soc. Imp. Naturalistes Moscou 21(1): 588 (1848). LT.: *L. sieberi* Turcz. [=*C. palustris* (L.) A. St.-Hil. (B.: *Croton palustris* L.)] (designated by Pfeiffer, Nomencl. Bot. 1, 2: 1334 (1874)).

Monoecious or rarely dioecious erect annual or perennial generally paludicolous herbs, rarely shrubby at the base. Indumentum simple, hispid, hairs rigid, often glandular at the apex, rarely glabrate. Leaves alternate, shortly petiolate, stipulate, simple, often serrate, narrow or rarely ovate, penni- or palminerved. Inflorescences axillary, spicate or racemose, pedunculate, lax, flowers solitary to each bract, upper male, lower female, few-flowered. Male flowers shortly pedicellate to subsessile; sepals 5, valvate; petals 5, free, inserted on the staminal column, subequal or unequal, imbricate; disc 0; stamens 10 or fewer, filaments fused at the base, 2-verticillate around the column, free portion of filaments spreading, anthers medifixed or subpendulous, introrse, ovoid, thecae separate on a short connective, longitudinally dehiscent; pollen oblate-spheroidal or spheroidal, stephanocolporate, colpi long and narrow, colpi transversales broad, intectate, reticulate with small lumina; pistillode cylindric or clavate, entire or tridenticulate, inserted at the top of the column. Female flowers shortly pedicellate to subsessile; sepals 5–6(10), equal or unequal, accrescent; petals 5(6), free, imbricate, often obsolete; disc 0; ovary 3-locular, ovule 1 per locule; styles 3, ± free or slightly connate at the base, lacerate-multilobate. Fruit tridymous, often hispid or echinate, dehiscing into 3 2-valved cocci; endocarp thinly crustaceous; columella expanded and minutely denticulate at the apex, often not persistent. Seeds subglobose, ecarunculate; testa smooth or weakly rugulose, thinly arillate; endosperm fleshy; cotyledons broad, flat. Fig.: See Euphorb. Gen.: t. 8, f. 26B.1–6 (1824) (as *Croton*); Pflanzenr. 57: 32–46, tt. 6–9 (1912); Fl. Jamaica 4: 289, t. 93 (1920); Nat. Pflanzenfam. ed. 2, 19C: 89, t. 44A–C (1931); Fl. Cuba 3: 82, t. 19 (1953); Fl. Cong. Rwa.-Bur.: 8(1): 169, t. XI and 170, fig. 12 (1962); Ann. Missouri Bot. Gard. 54: 270, t. 10 (1968); Fl. Ill. Sénég. 3: 394 (1975); Fl. Trop. E. Afr. Euph. 1: 165, t. 29 (1987); Fl. Lesser Antilles 5: 22, t. 9 (1989); Fl. Zamb. 9(4): 139, t. 25 (1996).

A genus of 34 species, mostly tropical American, with 5 in Africa and 1 in Madagascar.

Caperonia is quite distinct from the other members of the subtribe, and Müller placed it in a subtribe of its own. Hans (1973) reported a different base-number for it (x=11) from the other genera (x=13).

111. **Philyra** *Klotzsch,* Arch. Naturgesch. 7(1): 199 (1841); Baill., Étude Euphorb.: 297 (1858); Pax & K. Hoffm., Pflanzenr. 57: 49 (1912), and Nat. Pflanzenfam. ed. 2, 19C: 92 (1931); O'Donell & Lourteig, Lilloa 8: 60 (1942); G. L. Webster, Ann. Missouri Bot. Gard. 81: 71 (1994). T.: *Ph. brasiliensis* Klotzsch.

Argyrothamnia sect. *Philyra* (Klotzsch) Müll. Arg., Linnaea 34: 144 (1865), in DC., Prodr. 15(2): 733 (1866) and in Mart., Fl. Bras. 11(2): 308 (1874).
Argithamnia sect. *Philyra* (Klotzsch) Benth., Gen. Pl. 3: 303 (1880).

Dioecious mostly glabrous shrub or small tree, dried specimens smelling strongly of crushed *Trigonella foenum-graecum* seeds, fresh ones of curry-powder. Indumentum simple, mostly on innovations and inflorescence. Leaves alternate, shortly petiolate, stipulate, simple, obovate-oblanceolate, entire, subcoriaceous, penninerved; stipules chaffy, rigid, often with paired, straight spines arising below them. Inflorescences axillary, racemose, males congested, dense-flowered, females lax, few-flowered; bracts chaffy, sharply acute, persistent. Male flowers: pedicels articulate below the middle; calyx closed in bud, later valvately 5-partite, sepals narrow; petals 5, equalling the calyx, entire, free; disc-glands 5, oppositisepalous, adnate to the base of the androphore; stamens usually 15, in 3 whorls on the androphore, lower and middle normal, upper all or partly staminodal, free parts of filaments short, anthers basifixed, introrse, ovoid, longitudinally dehiscent, connective narrow; pollen prolate-spheroidal, 3-colporate, colpi long, narrow, colpus transversalis large, circular, intectate, reticulate, lumina small; pistillode 0 or scarcely-developed. Female flowers: pedicels elongate, bibracteolate, thickened above, articulate above the middle; sepals 5; petals 5, longer than the sepals; disc 0; ovary stipitate, 3-locular, ovule 1 per locule; styles 3, connate at the base, 2-fid almost to the middle, the arms themselves 2- to 3-fid, the lobes incised-lobulate. Fruit tridymous, dehiscing into 3 2-valved cocci; epicarp reticulate-rugulose, pubescent; endocarp thinly woody; columella 3-alate and 3-furcate, persistent. Seeds globose, ecarunculate; testa smooth, greyish-brown; endosperm fleshy; cotyledons broad, flat. Fig.: See Étude Euphorb.: t. 12, f. 16–22 (1858); Pflanzenr. 57: 50, t. 10 (1912).

A monotypic genus from Paraguay and S. Brazil.

112. **Ditaxis** *Vahl ex A. Juss.*, Euphorb. Gen.: 27 (1824); Baill., Étude Euphorb.: 298 (1858); Pax, Nat. Pflanzenfam. ed. 1, 3(5): 44 (1890); Pax & K. Hoffm., Pflanzenr. 57: 51 (1912), and Nat. Pflanzenfam. ed. 2, 19C: 93 (1931); Philcox, Fl. Trin. Tob. II(X): 641 (1979); G. L. Webster, Ann. Missouri Bot. Gard. 81: 72 (1994). T.: *D. fasciculata* Vahl ex A. Juss.

Aphora Nutt., Trans. Amer. Philos. Soc. II, 5: 174 (1837). T.: *A. mercurialina* Nutt. [=*D. mercurialina* (Nutt.) J. M. Coult.]

Serophyton Benth., Bot. Voy. Sulphur: 52 (1844). T.: *S. lanceolatum* Benth. [=*D. lanceolata* (Benth.) Pax & K. Hoffm.]

Stenonia Didr., Vidensk. Meddel. Dansk. Naturhist. Foren. Kjøbenhavn 1857f: 146 (1857)(non Endl., 1847). T.: *S. montevidensis* Didr. [=*D. montevidensis* (Didr.) Pax]

Paxia Herter, Fl. Urug. Pl. Vasc.: 80 (1931)(non Gilg, 1892). T.: *P. acaulis* (Herter ex Arechav.) Herter [=*D. acaulis* Herter]

Paxiuscula Herter, Revista Sudamer. Bot. 6: 92 (1939). T.: *P. acaulis* (Herter ex Arechav.) Herter [=*D. acaulis* Herter]

Argyrothamnia sect. *Ditaxis* (Vahl ex A. Juss.) Müll. Arg., Linnaea 34: 145 (1865), in DC., Prodr. 15(2): 734 (1866) and in Mart., Fl. Bras. 11(2): 309 (1874).

Argyrothamnia sect. *Aphora* (Nutt.) Müll. Arg., Linnaea 34: 147 (1865), in DC., Prodr. 15(2): 738 (1866) and in Mart., Fl. Bras. 11(2): 312 (1874).

Argithamnia sect. *Aphora* (Nutt.) Benth., Gen. Pl. 3: 303 (1880).

Argithamnia sect. *Ditaxis* (Vahl ex A. Juss.) Benth., Gen. Pl. 3: 303 (1880).

Argythamnia subg. *Ditaxis* (Vahl ex A. Juss.) Croizat, J. Arnold Arbor. 26(2): 191 (1945); G. L. Webster, J. Arnold Arbor. 48: 364 (1967), and Ann. Missouri Bot. Gard. 54: 271 (1968).

Monoecious or rarely dioecious shrubs, shrublets, perennial herbs with subsimple stems arising from a stout rootstock, or sometimes annuals. Indumentum malpighiaceous, sometimes also simple. Leaves alternate, shortly petiolate, stipulate, simple, entire or dentate, membranaceous, 3-nerved from the base; stipules small. Inflorescences axillary, racemose, mostly short and dense, rarely elongated, usually bisexual, male flowers distal, female proximal; bracts small, 1-flowered. Flowers imparting an intense purple coloration to water. Male flowers usually very shortly pedicellate; calyx closed in bud, ovoid, later valvately 5-partite; petals 5, free, equalling or exceeding the calyx, entire, clawed, sometimes the claw ± adnate to the androphore; disc-glands 5, oppositisepalous, ± free but sometimes ± adnate to the androphore; stamens 8–10, filaments united into a column, with a short free apical portion, fertile stamens 2-verticillate, 4–7 in the lower whorl, 3–4 in the upper, with sometimes also a third whorl of staminodes above, anthers basifixed, introrse, ovoid, longitudinally dehiscent; pollen oblate to suboblate, 3- or 4-colporate, bilaterally symmetrical, with 1 lobe smaller than the other 2, colpi long, not narrow, tectate, psilate, not reticulate; pistillode 0. Female flowers: pedicels short, often reflexed in fruit; sepals 5(6); petals 5(6), equalling or shorter than the calyx, entire; disc-glands 5(6), oppositisepalous, quadrate, cylindric or filamentiform; ovary ± sessile, 3-locular, ovule 1 per locule; styles 3, free or connate at the base, 2-fid, the style-arms cylindric or dilated and capitate or 2-lobate or multifid at the apex. Fruit tridymous, dehiscing into 3 2-valved cocci; columella persistent. Seeds subglobose, ecarunculate; testa often reticulate or foveolate; endosperm fleshy; cotyledons broad, flat. Fig.: See Euphorb. Gen.: t. 7, f. 24.1–10 (1824); Étude Euphorb.: t. 15, f. 23–29 (1858); Fl. Bras. 11(2): t. 46 (1874) (as *Argyrothamnia simoniana*); Pflanzenr. 57: 57–76, tt. 11–15 (1912); Nat. Pflanzenfam. ed. 2, 19C: 89, t. 44H–K and 93, t. 47 (1931); J. Arnold Arbor. 48: 365, t. 3 (1967) (as *Argythamnia blodgettii*); Gentes Herb. 11(7): 432, t. 3A, B and 433, t. 4A–C (1980); Fl. Lesser Antilles 5: 16, t. 5 (1989) (as *Argythamnia fasciculata*).

A neotropical genus of c. 60 species in drier areas from the USA south to Argentina.

It is very close to 113. *Argythamnia*, differing chiefly in the androecium, with 8–10 stamens disposed in 2 whorls, and in the bilaterally symmetrical pollen.

113. **Argythamnia** *P. Browne*, Civ. Nat. Hist. Jamaica: 338 (1756); Sw., Fl. Ind. Occid. 1: 335 (1797) ("*Argithamnia*"); A. Juss., Euphorb. Gen.: 26 (1824) (as *Argithamnia*); Baill., Étude Euphorb.: 337 (1858); Müll. Arg. in DC., Prodr. 15(2): 732, 741 (1866) ("*Argyrothamnia* sect. *Euargyrothamnia*"); Benth., Gen. Pl. 3: 304 (1880) ("*Argithamnia* sect. *Euargithamnium*"); Pax & K. Hoffm., Pflanzenr. 57: 78 (1912) ("*Argithamnia*"); Fawc. & Rendle, Fl. Jamaica 4: 286 (1920); Pax & K. Hoffm., Nat. Pflanzenfam. ed. 2, 19C: 94 (1931) ("*Argithamnia*"); Alain, Fl. Cuba 3: 83 (1953); J. W. Ingram, Gentes Herb. 10(1): 7 (1967), and 11(7): 435 (1980) (*Argythamnia* subg. *Argythamnia*); Correll, Fl. Bah. Arch.: 777 (1982); Proctor, Kew Bull. Addit. Ser. 11: 539 (1984); Alain, Fl. Español. 4: 82 (1986), and Descr. Fl. Puerto Rico 2: 356 (1988); R. A. Howard, Fl. Lesser Antilles 5: 14 (1989); G. L. Webster, Ann. Missouri Bot. Gard. 81: 72 (1994); L. J. Gillespie in Acev.-Rodr., ed., Fl. St. John: 203 (1996). T.: *A. candicans* Sw.

Monoecious or rarely dioecious perennial herbs, subshrubs, shrubs or rarely small trees. Indumentum malpighiaceous. Leaves alternate, petiolate to subsessile, stipulate, simple, entire or serrulate, 3-nerved from the base; stipules small. Inflorescences axillary, racemose to subspicate, bracteate, usually short, usually bisexual, male above, 1 or more females below; bracts small, 1-flowered. Flowers imparting a purple tinge to water. Male flowers shortly pedicellate to subsessile; calyx closed in bud, valvately 4–5-partite; petals (4)5, free, equalling or shorter than the calyx, mostly entire or sometimes shallowly 3-lobed, imbricate or not; disc-glands 4–5, oppositisepalous, adnate to the base of the column, usually glabrous; stamens 4–6, oppositipetalous, filaments united at the base, sometimes forming a short column, anthers in a single whorl, medifixed, introrse, ovoid, thecae parallel, contiguous, longitudinally dehiscent; pollen oblate-spheroidal, stephanocolpate, radially symmetrical, colpi long, fairly wide, inoperculate, tectate with a superimposed reticulum; pistillode represented by a shallowly convex prolongation of the staminal column. Female flowers shortly pedicellate; sepals 5, valvate; petals 0–5, shorter than the calyx; disc-glands 5, free, oppositisepalous; ovary 3-locular, ovule 1 per locule; styles 3, free or ± so, once or twice 2-partite, the segments shortly 2-lobate, stigmas sometimes flattened. Fruit small, tridymous, septicidally dehiscent into 3 2-valved 1-seeded cocci; epicarp thin, pubescent; endocarp crustaceous; columella narrow, brittle. Seeds ± spheroidal, ecarunculate; testa irregularly asperous, lineate, verruculose, tuberculate, rugulose, reticulate or rarely alveolate; endosperm fleshy; cotyledons broad, flat. Fig.: See Euphorb. Gen.: t. 7, f. 23.1–11 (1824); Étude Euphorb.: t. 15, f. 30–36 (1858); Pflanzenr. 57: 81, t. 16 (1912); Fl. Jamaica 4: 287, t. 92 (1920); Nat. Pflanzenfam. ed. 2, 19C: 89, t. 44L (1931); Fl. Cuba 3: 84, t. 20 (1953); Gentes Herb. 11(7): 432, t. 3C–F and 433, t. 4D–F (1980); Fl. Bah. Arch.: 779, t. 325 (1982); Fl. Español. 4: 84, t. 116-6 (1986); Descr. Fl. Puerto Rico 2: 358, t. 59-6 (1988); Fl. St. John: 204, t. 93A–G (1996).

A circum-Caribbean genus of 18 species, 10 of which occur in the West Indies, the other 8 occurring in Central America. Ingram (1967, 1980a) treats it as the typical subgenus in a trisubgeneric genus also including 112. *Ditaxis* and 114. *Chiropetalum.*

114. **Chiropetalum** *A. Juss.*, Ann. Sci. Nat. (Paris) 25: 21 (1832); Baill., Étude Euphorb.: 336 (1858); Pax & K. Hoffm., Pflanzenr. 57: 86 (1912), and Nat. Pflanzenfam. ed. 2, 19C: 95 (1931); G. L. Webster, Ann. Missouri Bot. Gard. 81: 72 (1994). T.: *Ch. tricuspidatum* (Lam.) A. Juss. (B.: *Croton tricuspidatus* Lam.)

Desfontaena Vell., Fl. Flumin.: 95 (1825). T.: *D. tricocca* Vell. [=*Ch. tricoccum* (Vell.) Chodat & Hassl.]

Chlorocaulon Klotzsch ex Endl., Gen. Pl. Suppl. 4(3): 89 (1850). T.: *Chl. schiedianum* Klotzsch, nomen. [=*Ch. schiedianum* (Klotzsch ex Müll. Arg.) Pax]

Argyrothamnia sect. *Chiropetalum* (A. Juss.) Müll. Arg., Linnaea 34: 148 (1865), in DC., Prodr. 15(2): 742 (1866) and Fl. Bras. 11(2): 313 (1874).

Argithamnia sect. *Chiropetalum* (A. Juss.) Benth., Gen. Pl. 3: 303 (1880).
Aonikena Speg., Anales Mus. Nac. Hist. Nat. Buenos Aires, Ser. 2, 7: 162 (1902).
T.: *A. patagonica* Speg. [=*Ch. patagonicum* (Speg.) O'Donell & Lourteig]
Argythamnia subg. *Chiropetalum* (A. Juss.) J. W. Ingram, Gentes Herb. 11(7): 435 (1980).

Monoecious shrubs or subshrubs, smelling faintly of fenugreek. Indumentum malpighiaceous, sometimes stellate and/or simple hairs admixed. Leaves alternate, petiolate or subsessile, stipulate, simple, entire to serrulate or serrate, gland-tipped or not, membranaceous, 3–5-nerved from the base. Inflorescences axillary, racemose to subspicate, pedunculate, usually short, unisexual (all male) or bisexual (females at the base); bracts small, 1-flowered. Flowers imparting a pink tinge to water. Male flowers shortly pedicellate or subsessile; calyx closed in bud, ovoid or globose, acute, valvately 5-partite; petals 5, free, equalling or shorter than the calyx-lobes, unguiculate, 3–7-lobed or -partite, imbricate or not; disc-glands 5, free, oppositisepalous, glabrous or pubescent; stamens 3–5, oppositipetalous, filaments $^{2}/_{3}$ united into a column, anthers in a single whorl, basifixed or subbasifixed, introrse, ovoid, thecae parallel, contiguous, longitudinally dehiscent; pollen spheroidal to prolate-spheroidal, 3-colpate, radially symmetrical, colpi wide, operculate, intectate, finely reticulate; pistillode represented by a shallowly convex prolongation of the staminal column. Female flowers: pedicels short, becoming reflexed in fruit; sepals 5, valvate; petals 0–5, much shorter than the calyx, entire or rarely lobed; disc-glands 5, free, oppositisepalous; ovary 3-locular, ovule 1 per locule; styles 3, free or connate only at the base, 2-fid to midway, stigmas terete. Fruit small, tridymous, septicidally dehiscent into 3 2-valved 1-seeded cocci, or loculicidally into 6 valves; epicarp thin, pubescent; endocarp crustaceous; columella narrow, expanded apically, subpersistent. Seeds spheroidal, ecarunculate; testa irregularly rugulose and pitted or not; cotyledons broad, flat. Fig.: See Euphorb. Gen.: t. 8, f. 26C.1–4 (1824) (as *Croton*); Étude Euphorb.: t. 15, f. 37–41 (1858); Pflanzenr. 57: 88, t. 17 (1912); Nat. Pflanzenfam. ed. 2, 19C: 95, t. 48 (1931); Gentes Herb. 11(7): 432, t. 3G–M, 433, t. 4G–K and 442–448, tt. 4–9 (1980).

A neotropical genus of 20 species, 2 in Mexico, the rest ranging from Peru and S. Brazil south to Chile and Argentina.

Although reduced to sectional and subgeneric rank under *Argy(ro)thamnia* by Müller (1865, 1866) and Ingram (1980b) respectively, *Chiropetalum* is sufficiently distinct in the combination of male petal, androphore, pollen and style characters to merit generic rank.

Subtribe 26c. **Doryxylinae** *G. L. Webster*, Taxon 24: 596 (1975), and Ann. Missouri Bot. Gard. 81(1): 72 (1994). T.: *Doryxylon* Zoll.

Monoecious shrubs or trees; indumentum stellate; leaves alternate, entire or dentate, with basal laminar glands; stipules small. Inflorescences terminal or axillary, usually bisexual; male sepals 3–5; petals 5–10 or 0; disc reduced or

0; stamens 30–250, free or connate; pollen 3-colporate, angulaperturate, coarsely reticulate, heterobrochate along colpi; pistillode + or 0; female sepals 5–6, imbricate or valvate; petals reduced or 0; disc annular or 0; ovary 2–3-locular; styles ± free or basally connate, 2-fid or simple, entire. Seeds: exotesta fleshy; endotesta foveolate.

This subtribe of 4 Asiatic genera stands apart from other *Chrozophoreae* by the combination of woody habit, stellate pubescence and polystemonous male flowers.

KEY TO THE GENERA

1. Petals +; stamens c. 30–100; ovary 3-locular · 2
1. Petals 0; stamens 200+; ovary usually 2-locular; inflorescences terminal; styles entire · 118. *Melanolepis*
2. Inflorescences axillary; stamens free; styles basally connate, simple · 115. *Doryxylon*
2. Inflorescences terminal; styles ± free, 2-fid · 3
3. Male disc 0; stamens free; female sepals imbricate · · · · · 116. *Sumbaviopsis*
3. Male disc +; stamens connate; female sepals valvate · · · · 117. *Thyrsanthera*

115. **Doryxylon** *Zoll.*, Tijdschr. Ned.-Indië 14: 172 (1857), and Linnaea 29: 469 (1859); N. P. Balakr., Bull. Bot. Surv. India 9: 56 (1967); Airy Shaw, Kew Bull. 37: 16 (1982), and Euph. Philipp.: 21 (1983); G. L. Webster, Ann. Missouri Bot. Gard. 81: 72 (1994). T.: *D. spinosum* Zoll.

Sumbavia Baill., Étude Euphorb. : 390 (1858); Müll. Arg. in DC., Prodr. 15(2): 727 (1866); Benth., Gen. Pl. 3: 304 (1880); Hook. f., Fl. Brit. India 5: 408 (1887); Pax & K. Hoffm., Pflanzenr. 57: 11 (1912), and Nat. Pflanzenfam. ed. 2, 19C: 89 (1931). T.: *S. rottleroides* Baill. [=*D. spinosum* Zoll.]
Mercadoa Náves in Blanco, Fl. Filip., ed. 3: t. 463 (1880). T.: *M. mandalojonensis* Náves [=*D. spinosum* Zoll.]

Monoecious trees or treelets. Branches sometimes spiny. Indumentum stellate. Leaves alternate, petiolate, peltate or epeltate, stipulate, simple, entire or repand-dentate, 3- or palminerved from the base, glanduligerous above at the base, white-tomentose beneath; stipules minute, soon falling. Inflorescences axillary, racemose or spiciform, shortly pedunculate, flowers 1 per bract, male above, with 1–more females below; bracts elongate, linear-subulate. Male flowers subsessile; calyx globose and closed in bud, valvately 5-lobate; petals 5, short; disc 0; stamens c. 30–100, borne on a convex receptacle, filaments free, erect, anthers dorsifixed, erect, oblong, thecae parallel, longitudinally dehiscent; pollen oblate-spheroidal, 3-colporate, colpus transversalis large, costae colpi 0, intectate, coarsely reticulate; pistillode 0. Female flowers shortly pedicellate; calyx deeply 6-partite, the lobes narrowly triangular, slightly imbricate; petals minute or 0; disc vestigial; ovary 3-locular, ovule 1 per locule; styles 3, basally connate, simple, recurved. Fruit trigonous,

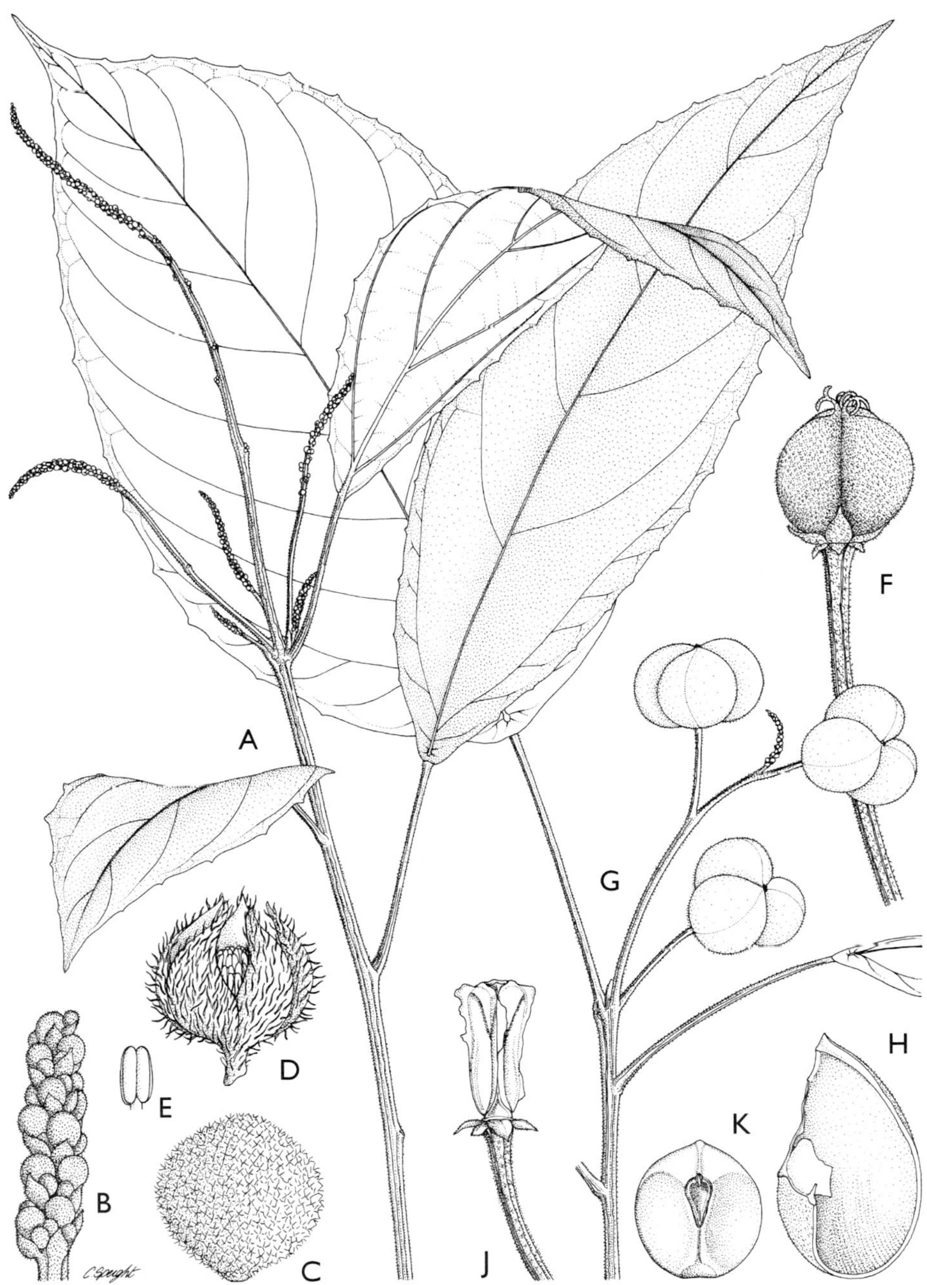

FIG. 13. *Sumbaviopsis albicans* (Blume) J. J. Sm. (B.: *Adisca albicans* Blume; S.: *Cephalocroton albicans* (Blume) Müll. Arg.). **A** Habit, male flowering specimen × 1; **B** Small portion of male inflorescence × 6; **C** Male flower-bud × 30; **D** Male flower × 30; **E** Stamen × 45; **F** Young fruit × 3; **G** Habit, fruiting specimen × 1; **H** Fruit-valve × 3; **J** Columella × 3; **K** Seed × 3. **A–E** from *Kostermans* 21432; **F** from *Kerr* 5704; **G** from *Geesink et al.* 6702; **H–K** from *Larsen et al.* 3294. Drawn by Camilla Speight.

septicidally dehiscent into 3 2-valved 1-seeded cocci, or loculicidally into 6 valves; epicarp pubescent; endocarp thinly woody; columella short, 3-alate. Seeds irregularly spheroidal, ecarunculate, arillate; aril fleshy, enveloping the seed; testa irregularly foveolate-rugulose; cotyledons broad. Fig.: See Pflanzenr. 57: 13, t. 2 A–D (1912) (as *Sumbavia rottleroides*).

A monotypic genus with a bimodal distribution: Indonesia (Lesser Sunda Is.: Bali, Sumbawa, Bima, Flores) and the Philippines (Luzon only).

116. **Sumbaviopsis** *J. J. Sm.*, Meded. Dept. Landb. Ned.-Indië 10: 13, 356 (1910); Pax & K. Hoffm., Pflanzenr. 57: 13 (1912); Gagnep., Fl. Indo-Chine 5: 418 (1926); Pax & K. Hoffm., Nat. Pflanzenfam. ed. 2, 19C: 95 (1931); Backer & Bakh. f., Fl. Java 1: 477 (1963); Airy Shaw, Kew Bull. 14: 357 (1960), and 26: 341 (1972); Whitmore, Tree Fl. Mal. 2: 132 (1973); Airy Shaw, Kew Bull. Addit. Ser. 4: 197 (1975), Kew Bull. 36: 346 (1981), and Euph. Philipp.: 45 (1983); G. L. Webster, Ann. Missouri Bot. Gard. 81: 73 (1994). T.: *S. albicans* (Blume) J. J. Sm. (B.: *Adisca* ? *albicans* Blume).

Monoecious shrub or small tree. Branches not spiny. Indumentum stellate. Leaves alternate, large, long-petiolate, shortly peltate, stipulate, simple, subentire to coarsely repand-dentate, palminerved, white-tomentose beneath; stipules minute, soon falling. Inflorescences terminal or subterminal, racemose, male above, usually with 3 flowers per bract, female or mixed below, when female then flowers solitary. Male flowers shortly pedicellate; calyx deeply 5-partite, segments valvate; petals (5)10, broad, imbricate; disc subobsolete or inconspicuous, 10-toothed; stamens 35–45, filaments free, erect, anthers dorsifixed, introrse, oblong, thecae parallel, longitudinally dehiscent; pollen suboblate, 3-colporate, colpus transversalis large, costae colpi +, intectate, coarsely reticulate; pistillode 0. Female flowers long-pedicellate; calyx deeply 5-partite, the lobes triangular-lanceolate, imbricate; petals 0; disc annular, obsolete or 0; ovary 3-locular, ovule 1 per locule; styles 3, basally slightly connate or ± free, 2-fid. Fruit large, strongly 3-lobate, or 2- or even 1-lobate by abortion, septicidally dehiscent into 2-valved 1-seeded cocci, then loculicidally into valves; epicarp tomentellous; endocarp thinly woody; columella irregularly 3-alate and/or 3-pronged. Seeds large, subglobose, ecarunculate, arillate; aril thinly fleshy, enveloping the seed; testa ± smooth. Fig.: See Pflanzenr. 57: 13, t. 2E (1912) (as *Sumbavia macrophylla*); Fl. Indo-Chine 5: 419, t. 49. 13–18 and 423, t. 50.1–5 (1926). See also Fig. 13, p. 147.

A monotypic genus ranging from Assam south through S.E. Asia to Palawan, Borneo and Java.

117. **Thyrsanthera** *Pierre ex Gagnep.*, Bull. Soc. Bot. France 71: 878 (1924); Gagnep., Fl. Indo-Chine 5: 299 (1925); Pax & K. Hoffm., Nat. Pflanzenfam. ed. 2, 19C: 90 (1931); Backer & Bakh. f., Fl. Java 1: 477 (1963); Airy Shaw, Kew Bull. 19: 308 (1965), and 26: 343 (1972); G. L. Webster, Ann. Missouri Bot. Gard. 81: 73 (1994). T.: *Th. suborbicularis* Pierre ex Gagnep.

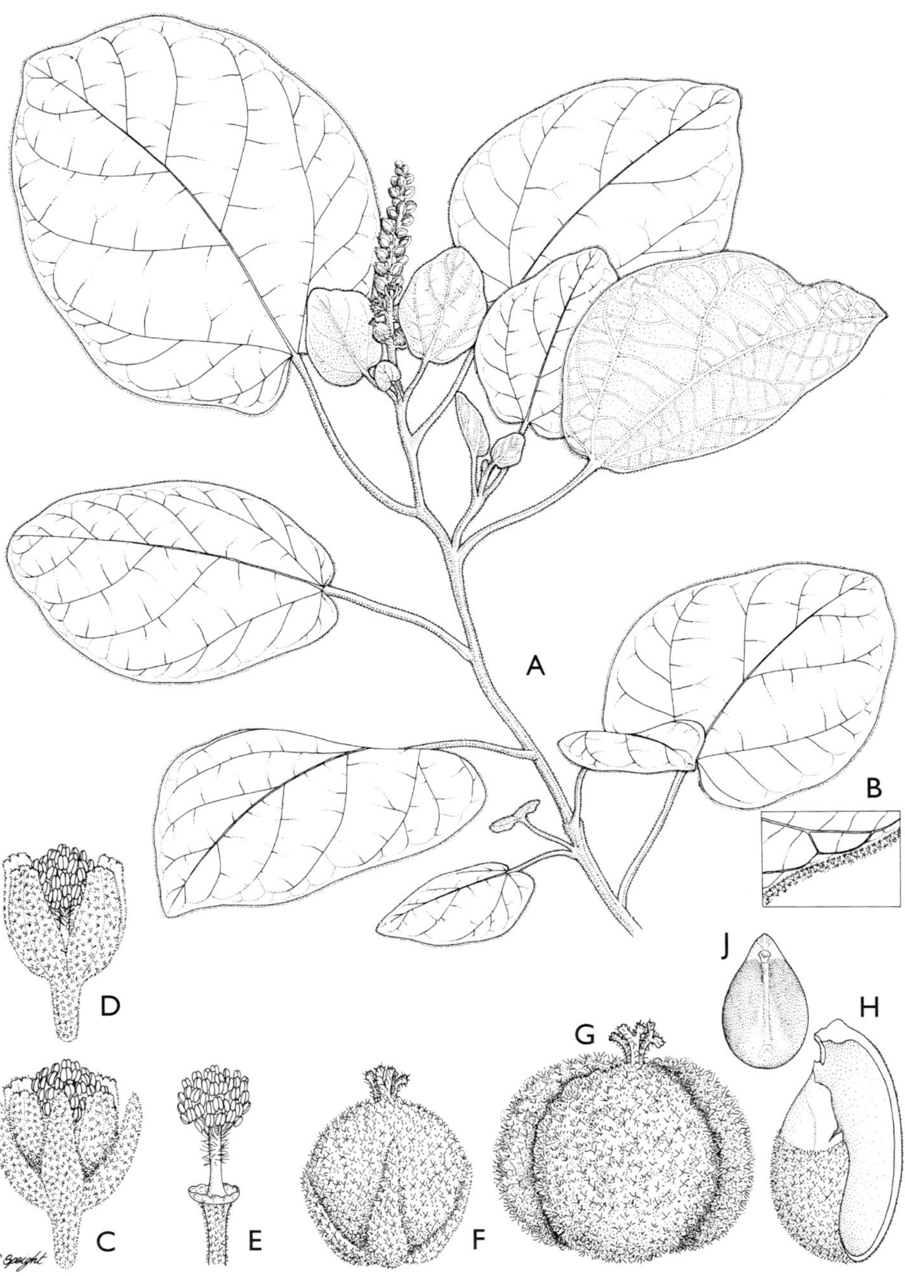

FIG. 14. *Thyrsanthera suborbicularis* Pierre ex Gagnep. **A** Habit × 1; **B** Leaf-margin × 4.5; **C** Male flower × 7.5; **D** Male flower, sepals removed × 7.5; **E** Male disc & androecium × 7.5; **F** Female flower × 7.5; **G** Fruit × 7.5; **H** Fruit-valve × 7.5; **J** Seed × 7.5. **A–F** from *Put* 3112; **G–J** from *Sørensen et al.* 3612. Drawn by Camilla Speight.

Monoecious dwarf shrub. Indumentum densely stellate, the hairs small, matted. Leaves alternate, long-petiolate, stipulate, simple, ovate-suborbicular, entire, margin ferrugineous, cordate, palminerved; petioles 2-glandular at the apex; stipules small, linear-setaceous, soon falling. Inflorescences terminal, spicate-racemose, male above, female below, flowers 1 per bract. Male flowers subsessile; calyx closed in bud, globose, valvately splitting into 5 segments; petals 5, ± imbricate; disc-glands 10, minute, at the base of the androphore; stamens 40–60, filaments united into a dendriform column, each with a short free terminal portion, anthers not whorled, crowded, introrse, ellipsoid, thecae parallel, longitudinally dehiscent; pollen suboblate, 3-colporate, colpus transversalis large, costae colpi 0, intectate, coarsely reticulate; pistillode 0. Female flowers sessile; sepals 5, elongate, linear-subulate, valvate; petals 0; disc shallowly cupular, small; ovary sessile, 3-locular, ovule 1 per locule; styles 3, ± free, 2-lobed, spreading. Fruit 3-lobate-subglobose, septicidally dehiscent into 3 2-valved 1-seeded cocci; epicarp roughly and unevenly tomentose; endocarp crustaceous; columella slender, 3-pronged, partially splitting longitudinally. Seeds ovoid, ecarunculate, arillate; aril thinly fleshy, enveloping the seed; testa ± smooth; cotyledons flat. Fig.: See Fl. Indo-Chine 5: 296–301, tt. 31.10, 11; 32.2–6 and 33.1–2' (1925). See also Fig. 14, p. 149.

A monotypic genus from Thailand and Cambodia.

118. **Melanolepis** *Rchb. f. & Zoll.*, Acta Soc. Regiae Sci. Indo-Neerl. 1: 22 (1856); Baill., Étude Euphorb.: 398 (1858); Pax & K. Hoffm., Pflanzenr. 63: 142 (1914); Gagnep., Fl. Indo-Chine 5: 347 (1925); Pax & K. Hoffm., Nat. Pflanzenfam. ed. 2, 19C: 113 (1931); Kaneh., Fl. Micron.: 180 (1933); Backer & Bakh. f., Fl. Java 1: 481 (1963); Airy Shaw, Kew Bull. 26: 309 (1972); Whitmore, Tree Fl. Mal. 2: 118 (1973); Airy Shaw, Kew Bull. Addit. Ser. 4: 175 (1975), 8: 174 (1980), Kew Bull. 36: 332 (1981), 37: 30 (1982), and Euph. Philipp.: 38 (1983); G. L. Webster, Ann. Missouri Bot. Gard. 81: 73 (1994). T.: *M. multiglandulosa* (Reinw. ex Blume) Rchb. f. & Zoll. (B.: *Croton multiglandulosus* Reinw. ex Blume).

Mallotus subgen. *Melanolepis* (Rchb. f. & Zoll.) Hurus., J. Fac. Sci. Univ. Tokyo, Sect. 3, Bot. 6: 308 (1954).

Monoecious or dioecious trees. Indumentum loosely and thinly floccose-stellate. Leaves alternate, large, long-petiolate, obsoletely 2-stipulate, simple, ovate-suborbicular, repand-dentate, sometimes 3(5)-lobate, palminerved, not granular-glandular beneath; stipules minute, vestigial. Inflorescences terminal, subterminal or pseudolateral, paniculate or racemose, elongate, unisexual, male bracts 3(5)-flowered, female flowers 1 per bract. Male flowers large, pedicellate; calyx closed in bud, globose, valvately splitting into 3–5 segments; petals 0; disc 0; stamens 200+, borne on a hemispherical pilose receptacle, filaments slender, free, anthers dorsifixed, introrse, oblong, emarginate, thecae parallel, elongate, free below, longitudinally dehiscent, connective appendiculate, subglobose, purple, not exceeding the thecae; pollen suboblate, 3-colporate, colpus transversalis large, costae transversales +, intectate, coarsely reticulate; pistillode 0. Female flowers pedicellate; sepals 5, triangular-lanceolate, valvate; petals 0; disc annular, crenulate; ovary sessile, 2(3)-locular, ovule 1 per locule; styles 2(3), free, entire, divaricate, shortly papillose. Fruit 2(3)-dymous, unadorned, septicidally dehiscent into 2(3) 2-valved 1-seeded cocci, or loculicidally dehiscent; epicarp ± smooth; endocarp thin; columella slender, expanded apically, subpersistent. Seeds subglobose, ecarunculate, arillate; aril fleshy, purple, enveloping the seed; testa foveolate; cotyledons broad, flat. Fig.: See Pflanzenr. 63: 143, t. 20 (1914); Fl. Indo-Chine 5: 348, t. 40.3–10 (1925); Nat. Pflanzenfam. ed. 2, 19C: 114, t. 58 (1931); Formos. Trees: 355, t. 310 (1936); Kew Bull. Addit. Ser. 8: 227, t. 5, f. 3 (1980).

An Eastern Asiatic genus of 2 species, 1 in Formosa, Ryu-Kyu Is. and the Marianas and ranging from S. Thailand throughout Malesia to the Bismarcks and Solomon Is., the other endemic to Cambodia.

Subtribe 26d. **Chrozophorinae** *G. L. Webster*, Taxon 24: 596 (1975), and Ann. Missouri Bot. Gard. 81(1): 73 (1994). T.: *Chrozophora* Neck. ex A. Juss.

Monoecious herbs or subshrubs; indumentum stellate; leaves alternate, stipulate, entire to lobate, 2-glandular at the base. Inflorescences mostly axillary, racemose, bisexual; male sepals and petals 5; disc inconspicuous, adnate to the staminal column; stamens 3–15, connate; anthers introrse; pollen oblate, stephanocolporate, colpi short, broad, distinctly heterobrochate; pistillode 0; female sepals 5; petals 5, small or 0; disc lobed or dissected; ovary 3-locular, stellate or lepidote, sometimes muricate; styles 2-fid. Fruit dehiscent. Seeds ecarunculate; exotesta somewhat fleshy; endotesta smooth or tuberculate.

This monogeneric subtribe is a herbaceous relative of the preceding but with stephanocolporate pollen and dry seeds, and linked to it via *Thyrsanthera* with its fused stamens.

119. **Chrozophora** *Neck. ex A. Juss.*, Euphorb. Gen.: 27 (1824) (as "*Crozophora*"), nom. and orth. cons.; Baill., Étude Euphorb.: 321 (1858) (as *Crozophora*); Müll. Arg. in DC., Prodr. 15(2): 746 (1866); Benth., Gen. Pl. 3: 305 (1880); Hook. f., Fl. Brit. India 5: 408 (1887); Pax & K. Hoffm., Pflanzenr. 57: 17 (1912); Prain in Fl. Trop. Afr. 6(1): 834 (1912), and Bull. Misc. Inform., Kew 1918: 49 (1918); Pax & K. Hoffm., Nat. Pflanzenfam. ed. 2, 19C: 90 (1931); Pojark., Fl. SSSR 14: 288 (1949); Vindt, Monogr. Euph. Maroc: 10 (1953); Keay (ed.), Fl. W. Trop. Afr., ed. 2, 1: 398 (1958); Backer & Bakh. f., Fl. Java 1: 477 (1963); Rech. f. & Schiman-Czeika, Fl. Iran. 6: 5 (1964); Tutin, Fl. Eur. 2: 211 (1968); Airy Shaw, Kew Bull. 26: 232 (1972); Berhaut, Fl. Ill. Sénég. 3: 397 (1975); Radcl.-Sm., Fl. Iraq 4 (1): 318 (1980), Fl. Turk. 7: 567 (1982), Fl. Cyp. 2: 1449 (1985), Fl. Pak. 172: 44 (1986), and Fl. Trop. E. Afr. Euph. 1: 160 (1987); G. L. Webster, Ann. Missouri Bot. Gard. 81: 73 (1994); Radcl.-Sm., Fl. Zamb. 9(4): 141 (1996); Philcox in Dassan., ed., Fl. Ceyl. XI: 121 (1997). T.: *Chr. tinctoria* (L.) A. Juss. (B.: *Croton tinctorius* L.)

Tournesol Adans., Fam. Pl. 2: 356 (1763). T.: (none designated).
Tournesol Scop., Intr. Hist. Nat.: 243 (1777). T.: *T. tinctoria* Scop. [=*Chr. tinctoria* (L.) A. Juss. (B.: *Croton tinctorius* L.)]
Lepidocroton C. Presl, Epimel. Bot.: 213 (1850). T.: *L. serratus* C. Presl [=*Chr. senegalensis* (Lam.) A. Juss. (B.: *Croton senegalensis* Lam.)]

Monoecious annual or perennial herbs or subshrubs. Milky latex 0. Indumentum stellate. Leaves alternate, long-petiolate, stipulate, simple, subentire, repand-dentate or sublobate, penni-, subtripli- or palminerved, often plicate or bullate at least when young, with 2 basal laminar glands; stipules filiform, chaffy. Inflorescences pseudoaxillary, supraaxillary, lateral or ± leaf-opposed, spicate, racemose or subpaniculate, bisexual, male above, female below; bracts 1-flowered; female flowers on 1 or more 1–4-flowered peduncles. Male flowers: pedicels short; calyx closed in bud, ovoid or globose, at length valvately 5-partite; petals 5, equalling or shorter than the sepals, distally imbricate, lightly coherent; disc 5-lobed, the lobes alternipetalous; stamens 3–15, filaments connate into a column, anthers 1–3-seriate, dorsifixed, introrse, oblong, erect, thecae parallel, longitudinally dehiscent; pollen suboblate-oblate, stephanocolporate, colpi short, colpus transversalis broad, isodiametric, costae +, intectate, very coarsely reticulate; pistillode 0. Female flowers long-pedicellate, pedicels often further elongating and becoming reflexed in fruit; sepals 5, narrow, open in bud; petals 5, smaller than in the male, or 0; disc 5-lobed, the lobes alternipetalous; staminodes 5 or 0; ovary 3-locular, ovule 1 per locule; styles 3, connate at the base, 2-fid, ± erect. Fruit 3-lobate, septicidally and loculicidally dehiscent into 3 2-valved 1-seeded cocci; epicarp stellate-tomentose or lepidote, often tuberculate; endocarp thinly woody; columella persistent. Seeds ovoid, somewhat angular, ecarunculate; exotesta arilloid, thin, pale, shiny; endotesta smooth or tuberculate; endosperm thick, fleshy; cotyledons broad, flat. Fig.: Euphorb. Gen.: t. 7, f. 25.1–11 (1824); Étude Euphorb.: t. 15, f. 12–22 (1858); Pflanzenr. 57: 22, t. 4 and 24, t. 5 (1912); Nat. Pflanzenfam. ed. 2, 19C: 89, t. 44F, G and 92, t. 46 (1931); Fl. SSSR 14: t. XVII.1–4 (1949); Monogr. Euph. Maroc: 11, t. 6 (1953); Fl. Ill. Sénég. 3: 396–402 (1975); Fl. Iraq 4 (1): 321, t. 59 (1980); Fl. Pak. 172: 48, t. 9 and 52, t. 10 (1986); Fl. Zamb. 9(4): 142, t. 26 (1996).

An Old World genus of c. 10 species from the Mediterranean Region to Tropical Africa and Central and southern Asia.

The fruits of several species are the source of purplish, bluish or greenish dyes.

Tribe 27. **CARYODENDREAE** *G. L. Webster*, Taxon 24: 596 (1975), and Ann. Missouri Bot. Gard. 81(1): 73 (1994). T.: *Caryodendron* H. Karst.

Dioecious trees; indumentum simple; leaves alternate, penninerved or 3-plinerved, 2-glandular at the base; stipules soon deciduous, or 0. Inflorescences terminal or axillary, spicate to paniculate; male sepals 3–5, valvate; petals 0; stamens 4–15, filaments free, anthers introrse, connective produced, usually conical; pollen mostly oblate, 3-colporate, colpi non-marginate, coarsely reticulate; disc usually intrastaminal, dissected or massive, pulviniform, pubescent; pistillode 0 or +; female sepals 4–6, imbricate or valvate; disc annular or dissected, pubescent; ovary 3-locular; fruit thin- or thick-walled, dehiscent; seed-coat dry or fleshy.

This tribe of 3 genera, 2 New World and 1 African, has pollen similar to that of the *Agrostistachydeae*, but differs in having apetalous flowers, fewer stamens, pubescent discs and simple styles.

KEY TO THE GENERA

1. Inflorescences axillary; pistillode +; fruit thin-walled; seed-coat fleshy; leaves 3-plinerved, exstipulate · 122. *Alchorneopsis*
1. Inflorescences terminal or subterminal, rarely axillary; pistillode 0; leaves stipulate · 2
2. Leaves penninerved; male disc massive, pulviniform; female sepals imbricate; fruit thick-walled; seed-coat dry · · · · · · · · 120. *Caryodendron*
2. Leaves 3-plinerved; male disc dissected; female sepals valvate; fruit thin-walled; seed-coat fleshy · 121. *Discoglypremna*

120. **Caryodendron** *H. Karst.*, Fl. Columb. 1: 91, t. 45 (1860); Müll. Arg. in DC., Prodr. 15(2): 765 (1866), and Mart., Fl. Bras. 11(2): 706 (1874); Benth., Gen. Pl. 3: 314 (1880); Pax & K. Hoffm., Pflanzenr. 63: 263 (1914), and Nat. Pflanzenfam. ed. 2, 19C: 122 (1931); Ducke, Trop. Woods 76: 18 (1943); G. L. Webster, Ann. Missouri Bot. Gard. 54: 287 (1968); Huft, *op. cit.* 76: 1077 (1989); A. H. Gentry, Woody Pl. NW. S. Amer.: 423 (1993); G. L. Webster, Ann. Missouri Bot. Gard. 81: 74 (1994); Murillo-Aldana & Franco-Rosselli, Euf. Reg. Ararac.: 49 (1995). T.: *C. orinocensis* H. Karst.

Centrodiscus Müll. Arg. in Mart., Fl. Bras. 11(2): 325 (1874), nom. inval. T.: *C. grandifolius* Müll. Arg.

Dioecious trees. Indumentum, present only on the inflorescences, simple. Leaves large, alternate, petiolate, stipulate, simple, entire, penninerved, 2-glandular at the base above; stipules soon deciduous. Inflorescences terminal, subterminal or occasionally axillary, spicate to subpaniculate, unisexual, the males sometimes branched, the females unbranched; bracts multiflorous. Male flowers sessile; calyx closed in bud, subglobose, at length valvately 3-lobate, lobes broad, ovate; petals 0; stamens 4–6(7), inserted around the disc, occasionally 1 arising from the centre of the disc, filaments free, anthers dorsifixed, introrse, ovoid, thecae subpendulous, parallel, contiguous, unequal, longitudinally dehiscent, connective produced; pollen oblate-spheroidal to suboblate, 3-colporate, convex-triangular in polar view, colpi narrow, colpus transversalis broad, isodiametric, costae 0, tectate, minutely perforate, intrareticulum coarse, lumina small; disc massive, pulviniform, central; pistillode 0. Female flowers sessile; sepals 5–6, imbricate, accrescent and persistent in fruit; petals 0; hypogynous disc

FIG. 15. *Caryodendron amazonicum* Ducke. **A** Habit × 1; **B** Small portion of male inflorescence × 9; **C** Male flower × 15; **D** Female flower × 6; **E** Young fruit × 4.5. **A–C** from *Ducke* 1070; **D, E** from *Ducke* 1071. Drawn by Camilla Speight.

annular, slightly pentagonal; ovary conic-ellipsoid, tapering into the style-base, 3-locular, ovule 1 per locule; styles 3, connate at the base, simple, recurved. Fruit large, ovoid-subglobose, (2)3(4)-locular, thick-walled, loculicidally dehiscent; endocarp hardened. Seeds ecarunculate; testa dry; endosperm fleshy; cotyledons broad, flat. Fig.: See Fl. Bras. 11(2): t. 102 (1874) (as *Centrodiscus*); Nat. Pflanzenfam. ed. 2, 19C: 126, t. 65G (1931); Woody Pl. NW. S. Amer.: 418, t. 120.6 (1993); Euf. Reg. Ararac.: 51, t. 5 (1995). See also Fig. 15, p. 154.

A neotropical genus of 3 species from Panama, Colombia, Venezuela, Brazil, Ecuador and Peru.

Because Müller (1874, p. 706) indicated that his genus *Centrodiscus*, which appeared on p. 325 of the same work, was the same as *Caryodendron* H. Karst., he thereby invalidated it.

121. **Discoglypremna** *Prain*, Bull. Misc. Inform., Kew 1911: 317 (1911), in Fl. Trop. Afr. 6(1): 931 (1912) and in Hooker's Icon. Pl. 30: t. 2988 (1913); Pax & K. Hoffm., Pflanzenr. 63: 18 (1914), and Nat. Pflanzenfam. ed. 2, 19C: 105 (1931); Keay (ed.), Fl. W. Trop. Afr. ed. 2, 1: 403 (1958); J.-G. Adam, Mém. Mus. Natl. Hist. Nat., B, Bot. 20: 469 (1971); Radcl.-Sm., Fl. Trop. E. Afr., Euph. 1: 222 (1987); G. L. Webster, Ann. Missouri Bot. Gard. 81: 74 (1994); J. Léonard, Bull. Jard. Bot. Belg. 64: 201 (1995), and Fl. Afr. Cent. Euph. 3: 11 (1996). T.: *D. caloneura* (Pax) Prain (B.: *Alchornea caloneura* Pax).

Dioecious trees. Indumentum simple. Leaves alternate, often subopposite or pseudoverticillate beneath the inflorescence, petiolate, stipulate, simple, subentire or shallowly toothed, subtriplinerved and penninerved, 2-glandular at the base; stipules soon deciduous. Inflorescences terminal, paniculate, the panicle branches pseudospicate or racemose; bracts small to minute, the male multiflorous, the female 1-flowered. Male flowers shortly pedicellate, the pedicels articulate; calyx closed in bud, ovoid-ellipsoid, apiculate, later splitting into 3–4(5) valvate lobes, becoming reflexed; petals 0; disc of c. 15 densely pubescent receptacular extra- and intrastaminal glands; stamens (5)8–12(15), filaments free, fairly long, flexuous, anthers subapicifixed, introrse, thecae free at base, pendulous, unequally 2-locellate, the larger locellus outside, longitudinally dehiscent, connective produced, conical; pollen prolate-spheroidal, 3-colporate, convex-triangular in polar view, colpi narrow, colpus transversalis large, costae sometimes indistinct, tectate, intrareticulum +; pistillode 0. Female flowers shortly pedicellate, the pedicels articulate; calyx (3)5-lobed, the lobes valvate, later strongly reflexed; petals 0; disc of 6–8(10) thick scale-like glands, truncate and pubescent at the apex; ovary subglobose, 3(4)-locular, ovule 1 per locule; styles 3(4), free, simple, recurved-spreading, fimbriate, persistent. Fruit 3(4)-lobate, loculicidally dehiscent into 3(4) valves, or septicidally dehiscent into 3(4) 2-valved cocci; exocarp slightly fleshy; endocarp crustaceous; columella narrow, triquetrous, persistent. Seeds ovoid, ecarunculate; exotesta pseudarillate, fleshy, orange-red; mesotesta thick, indurated, foveolate, shiny, black; endotesta distinct, grey; endosperm fleshy;

cotyledons broad, flat. Fig.: See Hooker's Icon. Pl. 30: t. 2988 (1913); Pflanzenr. 63: 18, t. 1 (1914); Fl. Trop. E. Afr., Euph. 1: 224, t. 44 (1987); Bull. Jard. Bot. Belg. 64: 205, t. 1 (1995); Fl. Afr. Cent. Euph. 3: 15, t. 1 (1996).

A monotypic W. and C. African genus ranging from Guinée to W. Uganda and Cabinda.

122. **Alchorneopsis** *Müll. Arg.*, Linnaea 34: 156 (1865) in DC., Prodr. 15(2): 764 (1866), and in Mart., Fl. Bras. 11(2): 327 (1874); Benth., Gen. Pl. 3: 315 (1880); Pax & K. Hoffm., Pflanzenr. 63: 267 (1914), and Nat. Pflanzenfam. ed. 2, 19C: 123 (1931); G. L. Webster, Ann. Missouri Bot. Gard. 54: 284 (1968); Philcox, Fl. Trin. Tob. II(X): 662 (1979); Alain, Fl. Español. 4: 76 (1986), and Descr. Fl. Puerto Rico 2: 352 (1988); A. H. Gentry, Woody Pl. NW. S. Amer.: 413 (1993); G. L. Webster, Ann. Missouri Bot. Gard. 81: 74 (1994); Murillo-Aldana & Franco-Rosselli, Euf. Reg. Ararac.: 43 (1995). T.: *A. floribunda* (Benth.) Müll. Arg. (B.: *Alchornea glandulosa* var. *floribunda* Benth.)

Dioecious trees. Indumentum, where present, simple. Leaves alternate, petiolate, exstipulate, simple, entire or remotely crenulate, triplinerved, 2-glandular at the base beneath. Inflorescences axillary, spiciform, the males often fasciculate, the females solitary; bracts inconspicuous, eglandular, the male many-flowered, the flowers glomerulate, the female 1-flowered. Male flowers shortly pedicellate to subsessile; calyx closed in bud, subglobose, later splitting into 3–4 valvate lobes; petals 0; disc of several large densely pubescent conoidal receptacular extra- and intrastaminal glands; stamens (5)6(8), filaments free, fairly long, flexuous, anthers dorsifixed, introrse, thecae free at base, pendulous, unequally 2-locellate, the larger locellus outside, longitudinally dehiscent, connective produced, conical; pollen oblate-spheroidal, 3-colporate, convex-triangular in polar view, colpi narrow, colpus transversalis elongated, costae 0, tectate, psilate; pistillode tripartite, glabrous. Female flowers shortly pedicellate; calyx (4)5-lobed, the lobes slightly imbricate; petals 0; disc annular, pulviniform, hirsutulous; ovary 3-locular, ovule 1 per locule; styles 3, free, short, simple, recurved, papillate. Fruit small, subglobose, septicidally dehiscent into 3 2-valved cocci; epicarp finely rugulose when dried; endocarp thinly woody; columella short, squat, trigonous, persistent. Seeds biconvex, dorsiventrally compressed, ecarunculate; exotesta pseudarillate, fleshy; mesotesta thick, indurated, striate-furrowed on the dorsal surface, rugulose-verruculose ventrally, shiny, black; endosperm fleshy; cotyledons broad, flat, orbicular. Fig.: See Fl. Bras. 11(2): t. 48 (1874); Pflanzenr. 63: 63, t. 8C–E (1914); Nat. Pflanzenfam. ed. 2, 19C: 108, t. 55C–E (1931); Fl. Español. 4: 79, t. 116-4 (1986); Descr. Fl. Puerto Rico 2: 354, t. 59-4 (1988); Woody Pl. NW. S. Amer.: 414, t. 118.4 (1993). See also Fig. 16, p. 157.

A neotropical genus of 3 species ranging from Costa Rica and Hispaniola to Ecuador, N. Brazil and Guyana.

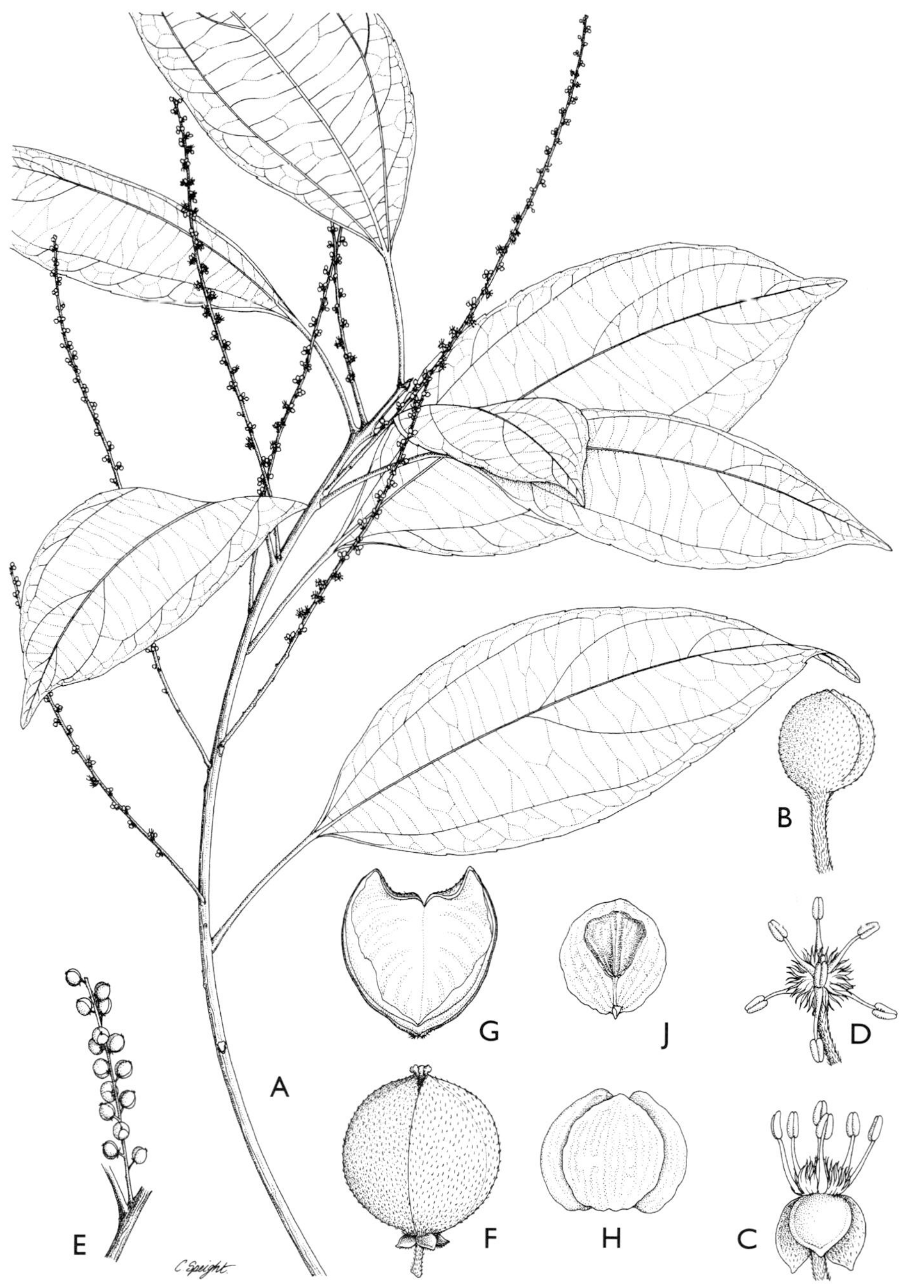

FIG. 16. *Alchorneopsis floribunda* (Benth.) Müll. Arg. (B.: *Alchornea glandulosa* var. *floribunda* Benth.). **A** Habit, male × 1; **B** Male flower-bud × 9; **C** Male flower × 9; **D** Male flower, sepals removed, viewed from above showing stamens, interstaminal glands & pistillode × 9; **E** Infructescence × 1; **F** Fruit × 9; **G** Fruit-coccus × 9; **H** Seed-triad × 9; **J** Seed × 9. **A–D** from *Harmon* 186; **E, F** from *Cerón* & *Palacios* 3039;**G–J** from *McPherson* 8750. Drawn by Camilla Speight.

Tribe 28. **BERNARDIEAE** *G. L. Webster*, Taxon 24: 596 (1975), and Ann. Missouri Bot. Gard. 81(1): 74 (1994). T.: *Bernardia* Houst. ex Mill.

Acalypheae subtribe *Mercurialinae* series *Bernardiiformes* Pax & K. Hoffm., Pflanzenr. 63: 13 (1914), p.p.

Monoecious or dioecious shrubs or rarely annual herbs. Indumentum simple, fasciculate or stellate. Leaves alternate, stipulate, simple, penninerved or palminerved, often with laminar glands. Inflorescences terminal, subterminal or axillary, racemose or spicate; male calyx valvately 3–4-partite; petals 0; disc various, not massive, sometimes 0; stamens (2)20–60(90), filaments free, anthers emarginate, muticous or apiculate, sometimes quasi-4-thecate; pollen 3-lobed, 3-colporate, colpi inoperculate, marginate, perforate-tectate or intectate; pistillode usually 0; female sepals 4–6, imbricate; disc annular or dissected; ovary 3-locular, ovule 1 per locule; styles usually 2-fid, lobes sometimes lacerate. Fruit dehiscent. Seeds mostly ecarunculate, subglobose, smooth, sometimes carinate.

This tribe of 6 genera includes only c. $^1/_2$ of the genera of Pax & Hoffmann's (1914) *Bernardiiformes*, together with *Adenophaedra* excised from their *Alchorneiformes*. For the rest, *Afrotrewia* cannot be satisfactorily evaluated, *Crotonogynopsis* has been moved to the *Adelieae* (q.v. infra) (Pax & Hoffmann's *Adeliiformes*), *Chondrostylis* has been moved to the *Agrostistachydeae* (q.v. supra), and *Clarorivinia* (= *Ptychopyxis*) and *Podadenia* have been moved to the *Pycnocomeae* (q.v. infra) (Pax & Hoffmann's *Wetriariiformes*). *Amyrea* has been brought back into this group from the *Claoxylinae* (Webster 1975, 1994), where it manifestly does not belong.

Key to the genera

1. Stamens 2 or 3, anther-connective greatly dilated; styles simple, dilated, forming a cap over the top of the ovary · · · · · · · · · · 128. *Adenophaedra*
1. Stamens 4–50, anther-connective not greatly enlarged; styles 2-fid · · · · · 2
2. Leaf-blades stipellate · 3
2. Leaf-blades not stipellate · 4
3. Leaves shortly petiolate, penninerved; anthers apiculate · 126. *Paranecepsia*
3. Leaves long-petiolate, palminerved; anthers muticous · · 127. *Discocleidion*
4. Stamens > 30, anthers apiculate; seeds not carinate · · · · · · · 124. *Necepsia*
4. Stamens < 30, anthers muticous or emarginate · · · · · · · · · · · · · · · · · · 5
5. Anthers emarginate; seeds carinate · · · · · · · · · · · · · · · · · · 123. *Bernardia*
5. Anthers muticous; seeds not carinate · · · · · · · · · · · · · · · · · · 125. *Amyrea*

123. **Bernardia** *Houst. ex Mill.*, Gard. Dict. Abr., ed. (1754), ex P. Browne, Civ. Nat. Hist. Jamaica: 361 (1756); Müll. Arg. in DC., Prodr. 15(2): 915 (1866), and in Fl. Bras., 11(2): 389 (1874); Benth., Gen. Pl. 3: 308 (1880); Pax & K. Hoffm., Pflanzenr. 63: 21 (1914); Fawc. & Rendle, Fl. Jamaica 4: 290 (1920); Pax & K. Hoffm., Nat. Pflanzenfam. ed. 2, 19C: 105 (1931); Standl. & Steyerm.,

Fieldiana, Bot. 24(6): 52 (1949); Alain, Fl. Cuba 3: 86 (1953); Buchheim, Willdenowia 2: 291 (1960), and 3: 217 (1962); G. L. Webster, Ann. Missouri Bot. Gard. 54: 276 (1968); Allem, Revista Brasil. Biol. 39: 529 (1979); Correll, Fl. Bah. Arch.: 781 (1982); Proctor, Kew Bull. Addit. Ser. 11: 541 (1984); Alain, Fl. Español. 4: 85 (1986), and Descr. Fl. Puerto Rico 2: 359 (1988); R. A. Howard, Fl. Lesser Antilles 5: 17 (1989); G. L. Webster, Ann. Missouri Bot. Gard. 81: 75 (1994). T.: *B. carpinifolia* Griseb.

Bivonia Spreng., Neue Entd. 2: 116 (1820). T.: *Bivonia axillaris* Spreng. [= *B. axillaris* (Spreng.) Müll. Arg.]

Traganthus Klotzsch, Arch. Naturgesch. 7(1): 188 (1841); Baill., Étude Euphorb.: 503 (1858). T.: *T. sidoides* Klotzsch [=*B. sidoides* (Klotzsch) Müll. Arg.]

Phaedra Klotzsch ex Endl., Gen. Pl. Suppl. 4(3): 88 (1850); Baill., Étude Euphorb.: 506 (1858). LT.: *Ph. caracasana* Klotzsch [=*B. jacquiniana* Müll. Arg.] (chosen by Webster, 1994).

Polyboea Klotzsch ex Endl., Gen. Pl. Suppl. 4(3): 88 (1850); Baill., Étude Euphorb.: 504 (1858). T.: *P. corensis* (Jacq.) Klotzsch ex Endl.(B.: *Acalypha corensis* Jacq.) [=*B. corensis* (Jacq.) Müll. Arg.]

Tyria Klotzsch ex Endl., Gen. Pl. Suppl. 4(3): 88 (1850); Baill., Étude Euphorb.: 506 (1858). LT.: *T. ovata* Klotzsch [=*B. mexicana* (Hook. & Arn.) Müll. Arg. (B.: *Hermesia ? mexicana* Hook. & Arn.)] (chosen by Webster, 1994).

Alevia Baill., Étude Euphorb.: 508 (1858). T.: *A. leptotaschia* Baill., sphalm. (altered to *leptostachya* in I.K.) [=*B. interrupta* (Schltdl.) Müll. Arg. (B.: *Acalypha interrupta* Schltdl.)]

Passaea Baill., Étude Euphorb.: 507 (1858). T.: *P. spartioides* Baill. [=*B. spartioides* (Baill.) Müll. Arg.]

Monoecious or dioecious shrubs, subshrubs or rarely annual herbs. Indumentum simple, fasciculate or stellate. Leaves alternate, petiolate or subsessile, stipulate, exstipellate, simple, dentate, penninerved or 3-nerved from the base, usually firmly chartaceous, 2-glandular at the base, rarely reduced and squamiform (§ *Passaea* (Baill.) Müll. Arg.); stipules small. Inflorescences usually unisexual, the males axillary, spicate, elongate or condensed and capituliform, bracts pluriflorous; the females few-flowered, terminal or subterminal, or sometimes many-flowered, racemose-spicate, bracts firm, coriaceous, concave. Male flowers shortly pedicellate; calyx closed in bud, globose, later valvately 3–4-partite; petals 0; disc-glands juxtastaminal, small, rarely 0; stamens 4–22, filaments free, short, somewhat thickened towards the base, anthers apicifixed, emarginate, thecae distinct, locelli also often distinct and anthers quasi-4-thecate, anther-connective not greatly enlarged; pollen prolate-spheroidal to subprolate, 3-lobed in polar view, 3-colporate, polar area index 0.15–0.2, colpi narrow, colpus transversalis small, costae +, intectate, finely reticulate; pistillode 0. Female flowers subsessile to shortly pedicellate; sepals 4–6, imbricate, sometimes augmented by 1–3 small sepaloid prophylls; disc hypogynous, annular or else composed of separate glands; ovary 3-locular, ovule 1 per locule; styles very short, discrete, erect or recurved, often continuous with the carpels and forming an open crown at their vertex, 2-fid, lobes smooth or lacerate. Fruit 3-dymous, septicidally and loculicidally dehiscent into 2-valved cocci; endocarp crustaceous to subligneous; columella short, 3-alate, persistent. Seeds prismatic-3-gonous, often flat at the base, dorsally ± acutely carinate, ecarunculate; testa crustaceous; endosperm fleshy; cotyledons broad, flat. Fig.: See Euphorb. Gen.: t. 9, f. 28.19 (1824) (as *Adelia*); Étude Euphorb.: t. 9, f. 1–2 (as *Polyboea*), and 18, f. 28–35 (1858) (as *Passaea*); Fl. Bras. 11(2): t. 57 (1874); Pflanzenr. 63: 26–40, tt. 3–5 (1914); Fl. Jamaica 4: 290, t. 94 (1920); Nat. Pflanzenfam. ed. 2, 19C: 106, t. 54 and 126, t. 65D, E (1931); Fl. Cuba 3: 87, t. 22 (1953); Fl. Bah. Arch.: 783, t. 327 (1982); Fl. Espanol. 4: 87, t. 116-7 (1986); Descr. Fl. Puerto Rico 2: 360, t. 59-7 (1988); Fl. Lesser Antilles 5: 16, t. 7 (1989).

A diverse mostly neotropical genus of 68 species, with the greatest number of species in S. Brazil. 1 species in SW USA (*B. myricifolia* (Scheele) Watson). Pax & Hoffmann (1914, 1931) recognize 7 sections, some of which, esp. *Passaea* (with broom-like habit) and *Traganthus* (an annual with simple hairs, penninerved lvs., no male disc and smooth styles), may be deserving of generic status.

124. **Necepsia** *Prain*, Bull. Misc. Inform., Kew 1910: 343 (1910), in Fl. Trop. Afr. 6(1): 923 (1912) and Hooker's Icon. Pl. 30: t. 2987 (1913); Pax & K. Hoffm., Pflanzenr. 63: 16 (1914), and Nat. Pflanzenfam. ed. 2, 19C: 105 (1931); Keay (ed.), Fl. W. Trop. Afr. ed. 2, 1: 405 (1958); Bouchat & J. Léonard, Bull. Jard. Bot. Belg. 56: 180 (1986); Radcl.-Sm., Fl. Trop. E. Afr., Euph. 1: 218 (1987); G. L. Webster, Ann. Missouri Bot. Gard. 81: 75 (1994); J. Léonard, Fl. Afr. Cent. Euph. 3: 32 (Oct. 1996); Radcl.-Sm., Fl. Zamb. 9(4): 143 (Nov. 1996). T.: *N. afzelii* Prain.

Palissya Baill., Étude Euphorb.: 502 (1858), non Endl. (1847). T.: *P. castaneifolia* Baill. [= *Necepsia castaneifolia* (Baill.) Bouchat & J. Léonard]
Neopalissya Pax in Pax & K. Hoffm., Pflanzenr. 63: 16 (1914); Leandri, Notul. Syst. (Paris) 9: 159 (1941). T.: *N. castaneifolia* (Baill.) Pax (B.: *Palissya castaneifolia* Baill.) [= *Necepsia castaneifolia* (Baill.) Bouchat & J. Léonard]

Dioecious or monoecious shrubs or small trees. Indumentum simple. Buds often perulate. Leaves alternate, petiolate, 2-stipulate, exstipellate, simple, subentire to crenate-serrate, penninerved, often subcoriaceous, sparingly gland-dotted beneath; petioles often pulvinate, geniculate. Inflorescences axillary, solitary or congested, pedunculate or sessile, spicate, racemose or subpaniculate, unisexual or bisexual; bracts rigid, ± scarious, males many-flowered, sometimes with a central female, females 1-flowered. Male flowers shortly pedicellate; calyx closed in bud, ovoid, apiculate, later valvately (3)4(5)-partite; petals 0; disc-glands free, numerous, elongate, pilose at apex, interstaminal; stamens 30–50(90), on a glabrous hemispherical receptacle, filaments free, erect in bud, anthers apicifixed, introrse, thecae oblique, pendulous, longitudinally dehiscent, connective produced; pollen prolate-spheroidal, convex-triangular in polar view, 3-

colporate, colpi narrow, colpus transversalis elongated, costae 0, tectate, psilate; pistillode 0. Female flowers subsessile or very shortly pedicellate; sepals (4)5(6), imbricate, persistent; petals 0; disc hypogynous, annular or cupuliform, entire or ± crenulate, thick, pubescent or setose; staminodes sometimes +; ovary 3-locular, verruculose, variously indumented, ovule 1 per locule; styles 3, ± free or connate at the base, once or twice 2-fid or -partite, subreflexed, pilose, stigmas papillose. Fruit depressed-3-lobate or 2-lobate by abortion, septicidally and loculicidally dehiscent into 3 2-valved cocci; endocarp thinly woody; columella persistent. Seeds subglobose, not carinate, ecarunculate; testa smooth, reddish-brown, white-spotted; hilum cordiform, white. Fig.: See Hooker's Icon. Pl. 30: t. 2987 (1913); Notul. Syst. (Paris) 9: 160, t. 1.1–5 (1941) (as *Neopalissya*); Bull. Jard. Bot. Belg. 56: 189, t. 2 (1986); Fl. Trop. E. Afr., Euph. 1: 219, t. 42 (1987); Fl. Afr. Cent. Euph. 3: 35, t. 5 (Oct. 1996); Fl. Zamb. 9(4): 144, t. 27 (Nov. 1996).

A tritypic genus with 1 species in W. Africa, 1 in Congo (Kinshasa) and the other in Tanzania, Zimbabwe and Madagascar, *fide* Bouchat & Léonard (1986).

125. **Amyrea** *Leandri*, Notul. Syst. (Paris) 9: 168 (1941); G. L. Webster, Taxon 24: 598 (1975), and Ann. Missouri Bot. Gard. 81(1): 88 (1994); Radcl.-Sm., Kew Bull. 53: 438 (1998). LT.: *A. humberti* Leandri (chosen by Radcliffe-Smith, 1998).

Dioecious shrubs or small trees. Wood hard. Indumentum simple, often confined to the petioles and inflorescence-bracts, sometimes 0. Leaves alternate, shortly petiolate, stipulate, simple, entire or dentate, elliptic, penninerved, tertiary nerves commonly transverse, parallel, sometimes reticulate, chartaceous to thinly coriaceous, eglandular; stipules very fugacious. Inflorescences terminal, subterminal or axillary in the uppermost axils, spicate or racemose, solitary, unisexual, the males several-flowered, the females occasionally 1-flowered; axis usually flexuous, agrostiform; bracts elliptic-ovate to triangular, strongly convex, scarious, glumiform, 1-flowered; bracteoles 4, unequal. Male flowers shortly pedicellate to subsessile; calyx closed in bud, ovoid, apiculate, later valvately 3–5-partite, becoming reflexed; petals 0; disc-glands ∞, interstaminal, separate or contiguous, densely pubescent; stamens 20–30(40), filaments free, anthers apicifixed, latrorse, muticous, thecae somewhat oblique, connate above, free and slightly divaricate below, subpendulous, longitudinally dehiscent, connective not produced; pollen prolate-spheroidal, 3-colporate, convex-triangular in polar view, colpus transversalis elongate, colpi narrow, costae 0, tectate, psilate; pistillode 0; receptacle hemispherical. Female flowers pedicellate; sepals 5, imbricate, persistent; petals 0; disc thick, fleshy, usually 8–10-lobed, lobes often in 2 series: 4–5 broader oppositisepalous, 4–5 smaller alternisepalous, or irregularly urceolate; ovary 3-locular, 3-lobate, smooth, ovule 1 per locule; styles ± free, short, accumbent, 2-fid, stigmas papillose. Fruit 3-lobate or 3-coccous, dehiscing septicidally into 3 2-valved cocci; endocarp crustaceous to thinly woody; columella triquetrous, persistent. Seeds ovoid, ecarunculate;

testa crustaceous, smooth, dry, slightly shiny, blackish; endosperm fleshy; cotyledons broad, flat. Fig.: See Notul. Syst. (Paris) 9: 160, t. 1.16–19 (1941); Kew Bull. 53: 441–450, tt. 1–6 (1998).

A genus of 11 spp. endemic to Madagascar (Radcliffe-Smith, 1998).

126. **Paranecepsia** *Radcl.-Sm.*, Kew Bull. 30: 684 (1976), and Fl. Trop. E. Afr., Euph. 1: 220 (1987); G. L. Webster, Ann. Missouri Bot. Gard. 81: 75 (1994); Radcl.-Sm., Fl. Zamb. 9(4): 145 (1996). T.: *P. alchorneifolia* Radcl.-Sm.

Small dioecious tree. Indumentum simple. Leaves alternate, clustered at the ends of the branches, shortly petiolate, 2-stipulate, stipellate, simple, shallowly serrate to crenate-serrate, penninerved. Inflorescences axillary in axils of leaves or scales, solitary, ± lax, pedunculate, racemose or subpaniculate; male flowers 1–5 per bract, females 1 per bract. Male flowers pedicellate, the pedicels articulate; calyx closed in bud, ovoid-subglobose, apiculate, later valvately 3–5-partite, the sepals reflexed; petals 0; disc-glands free, numerous, interstaminal; stamens 25–40, on a glabrous hemispherical receptacle, filaments free, anthers dorsifixed, introrse, shortly apiculate, thecae pendulous, longitudinally dehiscent, connective not enlarged; pollen not known; pistillode 0. Female flowers long-pedicellate; sepals 5–7, imbricate, very unequal, accrescent, persistent; petals 0; staminodes 8–10; disc annular, thin; ovary 3-locular, ovule 1 per locule; styles 3, connate at the base, 2-fid, divaricate, pilose, stigmas papillose. Fruit 3-lobate, septicidally and loculicidally dehiscent into 3 2-valved cocci; endocarp thinly woody; columella persistent. Seeds globose, ecarunculate; testa smooth, purplish-brown, lightly mottled. Fig.: See Kew Bull. 30: 685, t. 1 (1976); Fl. Trop. E. Afr., Euph. 1: 221, t. 43 (1987); Fl. Zamb. 9(4): 146, t. 28 (1996).

A monotypic genus confined to Tanzania and Mozambique.

127. **Discocleidion** *(Müll. Arg.) Pax & K. Hoffm.*, Pflanzenr. 63: 45 (1914), and Nat. Pflanzenfam. ed. 2, 19C: 107 (1931); G. L. Webster, Ann. Missouri Bot. Gard. 81: 75 (1994). T.: *D. ulmifolium* (Müll. Arg.) Pax & K. Hoffm. (B.: *Cleidion ulmifolium* Müll. Arg.)

Cleidion sect. *Discocleidion* Müll. Arg., Flora 47: 481 (1864), and in DC., Prodr. 15(2): 984 (1866).

Alchornea subgen. *Discocleidion* (Müll. Arg.) Hurus., J. Fac. Sci. Univ. Tokyo, Sect. 3, Bot. 6: 303 (1954).

Dioecious trees or shrubs. Indumentum simple. Leaves alternate, long-petiolate, 2-stipulate, stipellate, simple, dentate, palminerved, 2(4)-glandular near but not at the base beneath. Inflorescences terminal or subterminal, male paniculate, fairly dense, female racemose or subpaniculate, lax; male flowers 1–∞ per bract, females 1 per bract. Male flowers pedicellate; calyx closed in bud, globose, apiculate, later valvately 3–5-partite, the sepals

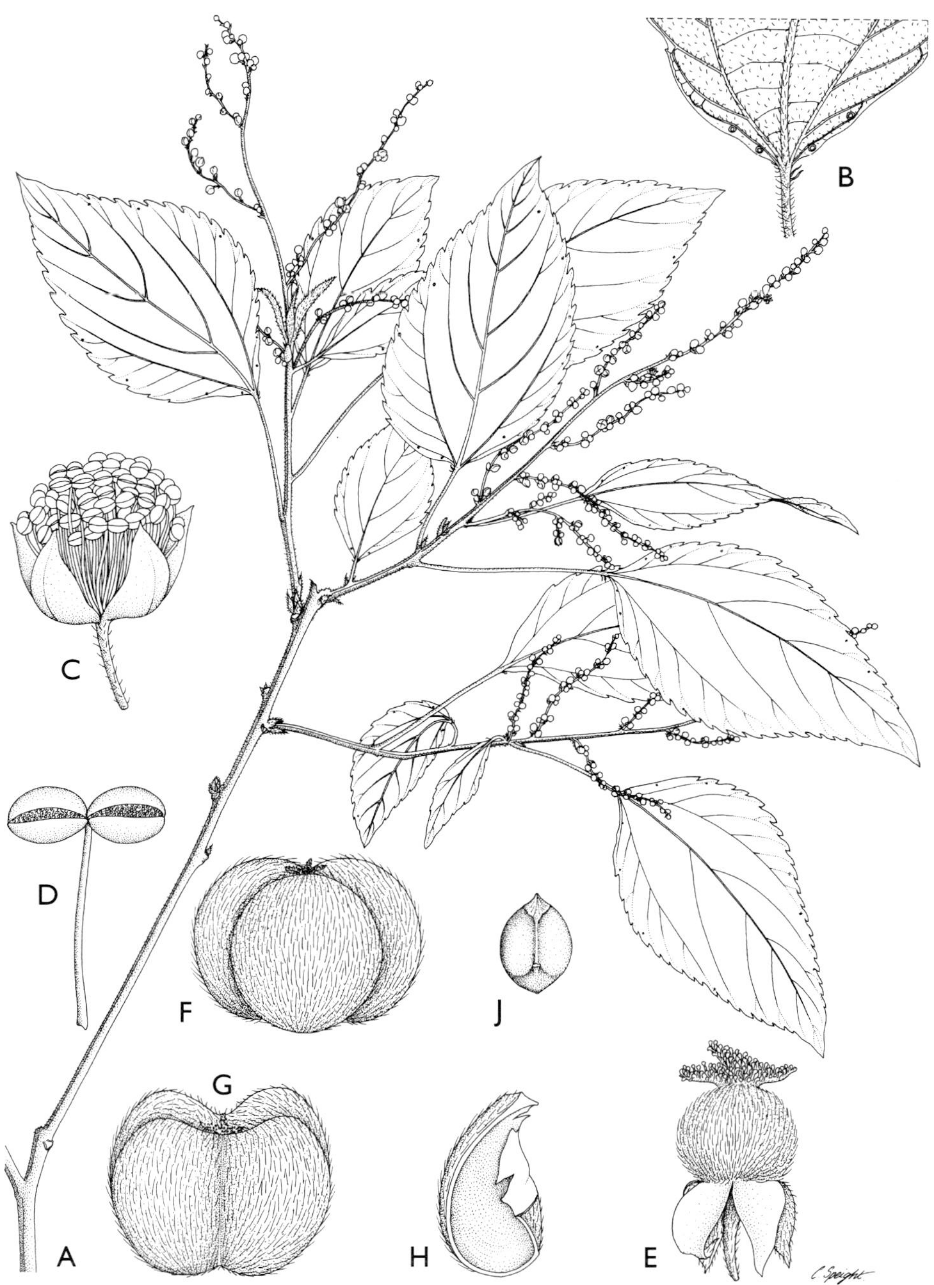

FIG. 17. *Discocleidion rufescens* (Franch.) Pax & K. Hoffm. (B.: *Alchornea rufescens* Franch.). **A** Habit, male × 1; **B** Leaf & glands × 3; **C** Male flower × 10.5; **D** Stamen × 16.5; **E** Female flower × 10.5; **F, G** Fruits × 6; **H** Fruit-valve × 6; **J** Seed × 6. **A–D** from *Wilson* 1991; **F, G** from *Henry* 20; **E, H, J** from *Ford* 160. Drawn by Camilla Speight.

becoming reflexed; petals 0; disc-glands free, numerous, interstaminal, clavate-cylindric, glabrous or pilose; stamens 35–60, on a convex receptacle, filaments free, anthers apicifixed, supertrorse to latrorse, muticous, thecae pendulous, 2-locellate, longitudinally dehiscent, appearing as if 4-thecous when dehisced, connective somewhat thickened, purplish, not produced; pollen suboblate-spheroidal, 3-colporate, polar area index < 0.2, colpi narrow, margo broad, colpus transversalis with broad costae, tectate, psilate to coarsely verrucate — the ornamentation can vary considerably between the grains of 1 specimen; pistillode 0. Female flowers pedicellate; sepals 5, imbricate, ± equal, not accrescent, subpersistent; petals 0; staminodes 8–10; disc annular, crenulate; ovary 3(4)-locular, ovule 1 per locule; styles 3(4), ± free, 2-fid or -partite, strongly recurved over the top of the ovary, stigmas densely papillose. Fruit 3(4)-lobate, septicidally and loculicidally dehiscent into 3(4) 2-valved cocci or 6(8) valves; endocarp thinly woody; columella thin, 3(4)-furcate at the apex, persistent. Seeds ± ovoid, carunculate, caruncle quickly becoming detached; testa shallowly tuberculate-rugulose, purplish-tinged. Fig.: See Pflanzenr. 63: 46, t. 6 (1914). See also Fig. 17, p. 163.

A small genus of 2 species, 1 from C. China, the other from the Ryukyus.

128. **Adenophaedra** *(Müll. Arg.) Müll. Arg.* in Mart., Fl. Bras., 11(2): 385, t. 101 (1874); Benth., Gen. Pl. 3: 314 (1880); Pax & K. Hoffm., Pflanzenr. 63: 261 (1914), and Nat. Pflanzenfam. ed. 2, 19C: 122 (1931); Croizat, Trop. Woods 88: 30 (1946); Jabl., Mem. New York Bot. Gard. 17: 140 (1967); Huft, Ann. Missouri Bot. Gard. 75: 1099 (1988); G. L. Webster, Ann. Missouri Bot. Gard. 81: 75 (1994). LT.: *A. grandifolia* (Klotzsch) Müll. Arg. (B.: *Tragia grandifolia* Klotzsch) (designated by Jablonski, 1967).

Bernardia sect. *Adenophaedra* Müll. Arg., Linnaea 34: 172 (1865), and in DC., Prodr. 15(2): 918 (1866).

Dioecious shrubs; indumentum simple, soon disappearing. Leaves alternate, often clustered at the ends of the branches, shortly petiolate, 2-stipulate, simple, shallowly repand-dentate, penninerved, chartaceous. Inflorescences terminal, subterminal or axillary in the uppermost leaf-axils, male spicate-racemose or longiramous-paniculate, female spicate, few-flowered, lax; male flowers several per bract, females 1 per bract. Male flowers very shortly pedicellate, pedicels articulate above; calyx closed in bud, globose, apiculate, later valvately (2)3-partite; petals 0; disc 0; stamens (2)3, alternisepalous, filaments connate at the base, short, thick, anthers subbasifixed, introrse, ovoid, thecae distinct, pendulous, ellipsoid, semiadnate, longitudinally dehiscent; connective somewhat thickened apically, subglobose, glandular, produced; pollen prolate-spheroidal to subprolate, 3-lobed in polar view, 3-colporate, polar area index 0.1–0.15, colpi and colpus transversalis narrow, costae +, intectate, finely reticulate; pistillode 0 or minute. Female flowers shortly pedicellate; sepals 6, biseriate, imbricate, the 3 inner smaller; disc annular, 3-lobate, lobes opposite outer sepals; ovary

3-locular, ovule 1 per locule; styles simple, dilated, connate into an obtusely 3-lobate stigmatoid cap closely adpressed to the top of the ovary. Fruit depressed-tridymous, septicidally and loculicidally dehiscent into 3 2-valved cocci. Seeds ecarunculate, globose; testa smooth, striate-mottled. Fig.: See Fl. Bras., 11(2): t. 101 (1874); Pflanzenr. 63: 289, t. 46C–E (1914); Nat. Pflanzenfam. ed. 2, 19C: 127C–E (1931).

A neotropical genus of 4 species, one ranging from Costa Rica to Peru and N. Brazil, 2 from Amazonian Brazil, and the other from Bahia.

On account of its stamens and styles, *Adenophaedra* is rather aberrant in this tribe, even though Müller (1865) treated it as a section of *Bernardia*. Palynologically, however, this does seem to be its affinity.

Tribe 29. **PYCNOCOMEAE** *Hutch.*, Amer. J. Bot. 56: 753 (1969); G. L. Webster, Taxon 24: 596 (1975), and Ann. Missouri Bot. Gard. 81(1): 75 (1994). T.: *Pycnocoma* Benth.

Monoecious or dioecious trees or shrubs; indumentum simple, glandular or 0; leaves alternate, subopposite or pseudoverticillate, simple, penninerved, rarely palminerved, eglandular, sometimes resinous; stipules small, fugacious or 0. Inflorescences axillary, racemose, paniculate or congested; male calyx valvately 3–4-partite; petals 0; disc-glands numerous, interstaminal; stamens ∞; thecae 2 or 4, pendulous or not; connective broad, sometimes large; pollen 3-colporate, colpi short, narrow, inoperculate, tectate-perforate or coarsely reticulate, scabrate or gemmate; pistillode 0; female flower terminal or not; sepals (3)5–6(9), imbricate; female disc annular, minute or 0, rarely massive; ovary (2)3(4)-locular, smooth or 6-horned, glabrous or pubescent; styles simple, elongate or stigmatiform. Fruit dehiscent, often tardily so, or indehiscent, smooth, horned or spiny, ridged or folded, flanged or glandular. Seeds ecarunculate; testa dry or fleshy.

As defined here, following Webster (1975, 1994), only the type-genus *Pycnocoma*, of the genera included by Hutchinson (1969) is retained in this tribe, the others being removed to the *Epiprineae* (*Epiprinus*) and the *Plukenetieae* (all the rest). The 2 subtribes recognized here may in fact on further investigation be found to be best regarded as constituting separate tribes.

Key to the subtribes

1. Monoecious; leaves shortly petiolate to subsessile; inflorescences racemose; pollen sexine scabrate-gemmate; seed-coat dry · · · · · 29a. *Pycnocominae*
1. Dioecious; leaves often long-petiolate; inflorescences ± paniculate; pollen sexine perforate-tectate or coarsely reticulate; seed-coat fleshy · 29b. *Blumeodendrinae*

Subtribe 29a. **Pycnocominae** *G. L. Webster*, Ann. Missouri Bot. Gard. 81(1): 76 (1994). T.: *Pycnocoma* Benth.

Acalypheae subtribe *Mercurialinae* series *Wetriariiformes* Pax & K. Hoffm., Pflanzenr. 63: 47(1914), and Nat. Pflanzenfam. ed. 2, 19C: 107(1931), p.p. T.: *Wetriaria* (Müll. Arg.) Kuntze [=*Argomuellera* Pax]

Monoecious or subdioecious shrubs or small trees; leaves shortly petiolate to subsessile; tertiary nerves reticulate. Inflorescences racemose; pollen scabrate-gemmate. Seed-coat dry.

An Afro-malagasy subtribe of 3 genera.

Key to the genera

1. Female flower terminating the inflorescence; female disc 0 · · · · · · · · · · 2
1. Female flower not terminating the inflorescence; female disc annular · 131. *Argomuellera*
2. Ovary 6-horned, pubescent; male flowers distinctly pedicellate; filaments inflexed in bud · 129. *Pycnocoma*
2. Ovary smooth, glabrous; male flowers subsessile; filaments erect in bud · 130. *Drocelonciа*

129. **Pycnocoma** *Benth.* in Hook., Niger Fl.: 508 (1849); Baill., Étude Euphorb.: 410 (1858); Müll. Arg. in DC., Prodr. 15(2): 951 (1866), p.p., excl. sect. *Wetriaria* Müll. Arg. (q.v. infra); Benth., Gen. Pl. 3: 326 (1880); Prain in Fl. Trop. Afr. 6(1): 955 (1912); Pax & K. Hoffm., Pflanzenr. 63: 52 (1914), and Nat. Pflanzenfam. ed. 2, 19C: 108 (1931); Keay (ed.), Fl. W. Trop. Afr. ed. 2, 1: 405 (1958); J. Léonard, Bull. Soc. Roy. Bot. Belg. 91: 273 (1959); Radcl.-Sm., Fl. Trop. E. Afr., Euph. 1: 228 (1987); G. L. Webster, Ann. Missouri Bot. Gard. 81: 76 (1994); J. Léonard, Bull. Jard. Bot. Belg. 65: 38 (May, 1996), and Fl. Afr. Cent. Euph. 3: 36 (Oct. 1996). T.: *P. macrophylla* Benth.

Monoecious little-branched subshrubs or small trees. Indumentum simple. Leaves alternate, sometimes pseudoverticillate, often clustered at the apex of the stem, elongated, large, subsessile or shortly petiolate, stipulate or not, simple, elobate or occasionally 3-lobate, shallowly toothed to subentire, penninerved, minutely glandular-punctate beneath. Inflorescences axillary in the upper leaf-axils, bisexual, racemose or spicate, solitary; bracts erect or reflexed, males concave, 1–∞-flowered, female flower terminal, usually solitary. Male flowers usually long-pedicellate; calyx closed in bud, ovoid, later valvately (2)3–4(5)-partite, the sepals reflexed; petals 0; disc-glands numerous, contiguous, disposed amongst the stamens; stamens 50–∞, filaments free, inflexed in bud, elongate, flexuous, anthers small, subdorsifixed, introrse, subglobose, 2-thecate, longitudinally dehiscent, connective rather broad; pollen oblate-spheroidal to suboblate, 3-colporate, polar area index 0.3–0.5, colpi narrow, elongate or not, colpus transversalis small, costae +, tectate, scabrate; pistillode 0; receptacle hemispherical. Female flowers usually sessile; sepals (4)5(8), biseriate, imbricate; petals 0; disc 0; ovary 3-locular, 6-winged or -horned, ovule 1 per locule; styles 3, usually connate into a column, free above, simple, erecto-patent, stigmas terminal, capitate or subcapitate. Fruits 3-lobate, 6-winged or -horned, dehiscing septicidally into 3 2-valved cocci, each valve 1-winged or -horned; exocarp separating; endocarp thinly woody; columella persistent. Seeds ovoid-trigonous or subglobose, ecarunculate; testa dry, shallowly ridged on 1 side, puberulent; endosperm fleshy; cotyledons broad, flat. Fig. : See Pflanzenr. 63: 54, t. 7 (1914); Notul. Syst. (Paris) 9: 160, t. 1.6–15, 163, t. 2.1–7 (1941); Fl. Trop. E. Afr., Euph. 1: 230, t. 46 (1987); Bull. Jard. Bot. Belg. 65: 39–61, tt. 1–4 (May 1996); Fl. Afr. Cent. Euph. 3: 41–57, tt. 6–9 (Oct. 1996).

A tropical African genus of 18 species.

130. **Droceloncia** *J. Léonard,* Bull. Soc. Roy. Bot. Belg. 91: 279 (1959); G. L. Webster, Ann. Missouri Bot. Gard. 81: 76 (1994). T.: *D. rigidifolia* (Baill.) J. Léonard (B.: *Pycnocoma rigidifolia* Baill.)

Very like 129. *Pycnocoma,* differing chiefly in having the terminal bud covered by a long acuminate perule, subsessile male flowers with the filaments erect in bud, and a smooth, glabrous ovary. Fig.: See Fig. 18, p. 168.

A monotypic genus of Madagascar and the Comoro Is.

131. **Argomuellera** *Pax,* Bot. Jahrb. Syst. 19: 90 (1894); Prain in Fl. Trop. Afr. 6(1): 925 (1912); Pax & K. Hoffm., Nat. Pflanzenfam. ed. 2, 19C: 107 (1931); Keay (ed.), Fl. W. Trop. Afr. ed. 2, 1: 405 (1958); J. Léonard, Bull. Soc. Roy. Bot. Belg. 91: 274 (1959); Radcl.-Sm., Fl. Trop. E. Afr., Euph. 1: 225 (1987); G. L. Webster, Ann. Missouri Bot. Gard. 81: 76 (1994); J. Léonard, Bull. Jard. Bot. Belg. 65: 24 (May 1996), and Fl. Afr. Cent. Euph. 3: 56 (Oct. 1996); Radcl.-Sm., Fl. Zamb. 9(4): 147 (Nov. 1996). T.: *A. macrophylla* Pax.

Pycnocoma Benth. sect. *Wetriaria* Müll. Arg. in DC., Prodr. 15(2): 952 (1866). LT.: *P. trewioides* Baill. (lectotype, chosen here) [= *A. trewioides* (Baill.) Pax & K. Hoffm.]

Neopycnocoma Pax, Bot. Jahrb. Syst. 43: 222 (1909); Prain in Fl. Trop. Afr. 6(1): 962 (1913). T.: *N. lancifolia* Pax [= *A. lancifolia* (Pax) Prain]

Wetriaria (Müll. Arg.) Kuntze sect. *Argomuellera* (Pax) Pax, Pflanzenr. 63: 50 (1914). T.: *W. macrophylla* (Pax) Pax [= *A. macrophylla* Pax]

Wetriaria (Müll. Arg.) Kuntze sect. *Neopycnocoma* (Pax) Pax, Pflanzenr. 63: 51 (1914).

Pycnocoma sensu Leandri, Notul. Syst. (Paris) 9: 161 (1941), non Benth.

Monoecious or subdioecious shrubs or small trees. Indumentum simple. Leaves alternate, crowded, elongated, large, very shortly petiolate or subsessile, stipulate or exstipulate, simple, elobate, dentate to subentire, penninerved, commonly eglandular. Inflorescences axillary, uni- or

FIG. 18. *Droceloncia rigidifolia* (Baill.) J. Léonard (B.: *Pycnocoma rigidifolia* Baill.). **A** Habit × 1; **B** Small portion of male inflorescence × 4.5; **C** Male flower × 7.5; **D** Female flower × 7.5. All from *Bernardi* 11793. Drawn by Camilla Speight.

bisexual, racemose, spicate or rarely subpaniculate, solitary or subfasciculate, ± lax; male flowers usually in fasciculate clusters along the axis, often with 1 female flower per cluster, or else the clusters either all male or all female, or rarely 1 female flower terminal. Male flowers pedicellate; calyx closed in bud, ovoid or globose, later valvately (2)3–4(5)-partite, the lobes reflexed; petals 0; disc-glands numerous, free, oblong, disposed amongst the stamens; stamens 15–120, filaments free, ± erect in bud, anthers small, basifixed, introrse, 2-thecate, thecae pendent, longitudinally dehiscent, connective broad, stout; pollen spheroidal, 3-colporate, convex triangular in polar view, polar area index 0.4, colpi narrow, costae +, colpus transversalis large, costae 0, tectate, gemmate; pistillode usually 0; receptacle hemispherical. Female flowers shortly pedicellate; sepals (3)5–6(9), indistinctly biseriate, imbricate; petals 0; disc annular; ovary 3(4)-locular, smooth, ovule 1 per locule; styles 3(4), slightly connate at the base, simple, rather stout, later spreading or reflexed, stigmas papillose. Fruits 3-lobate, smooth, dehiscing septicidally into 3 2-valved cocci; endocarp thinly woody; columella persistent. Seeds globose, ecarunculate; testa dry, smooth, glabrous; endosperm fleshy; cotyledons broad, flat. Fig.: See Pflanzenr. 63: 210, t. 30E–G (1914); Nat. Pflanzenfam. ed. 2, 19C: 119, t. 61 E–G (1931); Fl. Trop. E. Afr., Euph. 1: 226, t. 45 (1987); Bull. Jard. Bot. Belg. 65: 29, t. 1 and 33, t. 2 (May 1996); Fl. Afr. Cent. Euph. 3: 61, t. 10 and 63, t. 11 (Oct. 1996); Fl. Zamb. 9(4): 148, t. 29 (Nov. 1996).

An Afro-Malagasy genus of 12 species, of which 6 (including 1 as yet undescribed) occur in Tropical Africa and 6 in Madagascar and the Comoro Is.

Subtribe 29b. **Blumeodendrinae** *G. L. Webster*, Taxon 24: 596 (1975), and Ann. Missouri Bot. Gard. 81(1): 76 (1994). T.: *Blumeodendron* (Müll. Arg.) Kurz.

Dioecious trees; leaves usually long-petiolate; tertiary nerves sometimes scalariform. Inflorescences ± paniculate; pollen perforate-tectate or coarsely reticulate. Seed-coat fleshy.

An Asiatic subtribe of 4 genera, differing from the *Pycnocominae* most obviously in the long-petiolate leaves and ± paniculate inflorescences.

Key to the genera

1. Anther-connective umbraculiform; thecae 4, pendent; female disc pulviniform; styles stigmatiform · · · · · · · · · · · · · · · · 135. *Botryophora*
1. Anther-connective not umbraculiform; female disc minute or 0; styles elongate · 2
2. Thecae 4, unequal; fruit often spiny, ridged or folded · · · 134. *Ptychopyxis*
2. Thecae 2; fruit smooth, flanged or glandular · · · · · · · · · · · · · · · · · · · 3

3. Flowers without glandular hairs; thecae adnate to the connective; female disc annular ··························· 132. *Blumeodendron*
3. Flowers stipitate-glandular; thecae pendulous; female disc 0 ································· 133. *Podadenia*

132. **Blumeodendron** *(Müll. Arg.) Kurz,* J. Asiat. Soc. Bengal, Pt. 2, Nat. Hist. 42: 245 (1873); J. J. Sm., Meded. Dept. Landb. Ned.-Indië 10: 458 (1910); Pax & K. Hoffm., Pflanzenr. 63: 47 (1914), and Nat. Pflanzenfam. ed. 2, 19C: 107 (1931); Backer & Bakh. f., Fl. Java 1: 479 (1963); Airy Shaw, Kew Bull. 16: 348 (1963), and 26: 224 (1972); Whitmore, Tree Fl. Mal. 2: 68 (1973); Airy Shaw, Kew Bull. Addit. Ser. 4: 57 (1975), 8: 37 (1980), Kew Bull. 36: 267 (1981), 37: 9 (1982), and Euph. Philipp.: 10 (1983); G. L. Webster, Ann. Missouri Bot. Gard. 81: 77 (1994). T.: *B. tokbrai* (Blume) Kurz (B.: *Elateriospermum tokbrai* Blume).

Mallotus § *Blumeodendron* Müll. Arg. in DC., Prodr. 15(2): 956 (1866); Benth., Gen. Pl. 3: 319 (1880); Hook. f. Fl. Brit. India 5: 427 (1887); Pax in Pflanzenfam. 3(5): 53 (1890). LT.: *M. tokbrai* (Blume) Müll. Arg. (B.: *Elateriospermum tokbrai* Blume) (chosen by Wheeler, 1975).

Dioecious almost completely glabrous trees. Indumentum minutely stellate, on young growth only. Leaves alternate, subopposite or pseudoverticillate, fairly long-petiolate, exstipulate, simple, entire, penninerved or 3–5-nerved from the base, coriaceous, eglandular. Inflorescences axillary or subterminal, solitary or fasciculate, condensed-cymose, pseudoracemose or subpaniculate, female shorter than the male, often gummy when young. Male flowers pedicellate; calyx closed in bud, ellipsoid, obtuse, later valvately 3–4-fid; petals 0; disc-glands numerous, free, ovoid-subglobose, disposed amongst the stamens; stamens 15–40, filaments free, elongated, crisped in bud, anthers subdorsifixed, latrorse, subglobose, 2-thecate, thecae adnate to the broad connective, longitudinally dehiscent; pollen oblate-spheroidal, 3-colporate, colpi short, narrow, costae +, colpus transversalis small, narrow, costae +, coarsely reticulate, muri thick, consisting of many small columellae; pistillode 0; receptacle elevated, pulviniform. Female flowers pedicellate; sepals 3–5, small; petals 0; disc annular; ovary 2–3-locular, ovule 1 per locule; styles 2–3, subconnate at the base, divergent or recurved, linear-subulate, stigmas papillose. Fruits 2–3-locular, sometimes flanged, tardily dehiscing septicidally into 2-valved cocci; epicarp thin; mesocarp granular; endocarp woody, uneven within. Seeds large, laterally compressed, ecarunculate; exotesta somewhat fleshy; endosperm fleshy; cotyledons broad, flat. Fig.: See Tree Fl. Mal. 2: 69, t. 2 and 71, t. 3 (1973); Kew Bull. 36: 268, t. 3 A1–5; B1, 2 (1981).

An Asiatic genus of 5 or 6 species ranging from Myanmar [Burma] and the Andamans through Indonesia to the Bismarck Archipelago, with the greatest number of species in Borneo.

133. **Podadenia** *Thwaites,* Enum. Pl. Zeyl. 4: 273 (1861); Müll. Arg. in DC., Prodr. 15(2): 791 (1866); Benth., Gen. Pl. 3: 318 (1880); Hook. f., Fl. Brit. India 5: 422 (1887); Trimen, Handb. Fl. Ceyl. 4: 62 (1898); Pax & K. Hoffm., Pflanzenr. 63: 19

FIG. 19. *Podadenia thwaitesii* (Baill.) Müll. Arg. (B.: *Rottlera thwaitesii* Baill.; S.: *P. sapida* Thwaites, nom. superfl.; *Ptychopyxis thwaitesii* (Baill.) Croizat). **A** Habit, with female inflorescence × 1; **B** Small portion of male inflorescence × 3; **C** Male flower × 6; **D** Small portion of female inflorescence × 3; **E** Female flower × 6; **F** Young fruit × $1^1/_2$; **G** Fruit × $1^1/_2$. **A, D, E** from *Waas* 1526; **B, C** from *Kostermans* 28367; **F** from *Waas* 1690; **G** from *Jayasura* 2365. Drawn by Camilla Speight.

(1914), and Nat. Pflanzenfam. ed. 2, 19C: 105 (1931); G. L. Webster, Ann. Missouri Bot. Gard. 81: 77 (1994); Philcox in Dassan., ed., Fl. Ceyl. XI: 144 (1997) (as *Ptychopyxis*). T.: *P. thwaitesii* (Baill.) Müll. Arg. (B.: *Rottlera thwaitesii* Baill.) (S.: *P. sapida* Thwaites, nom. superfl.; *Ptychopyxis thwaitesii* (Baill.) Croizat).

Dioecious tree. Indumentum simple, pubescent, villous and glandular, ferrugineous. Leaves alternate, large, long-petiolate, exstipulate, simple, entire, penninerved, eglandular. Inflorescences large, terminal and axillary in the uppermost leaf-axils, paniculate; male bracts many-flowered, female flowers 1 per bract, bracts larger than the male. Male flowers: pedicels stipitate-glandular, persistent; calyx closed in bud, ovoid-subglobose, acuminate, later valvately 3–4-partite; petals 0; disc-glands numerous, free, minute, disposed amongst the stamens; stamens c. 25, filaments free, anthers apicifixed, latrorse, 2-thecate, thecae distinct, pendulous, longitudinally dehiscent, connective acuminate; pollen not known, but probably similar to that of 134. *Ptychopyxis*; pistillode 0; receptacle convex. Female flowers pedicellate; sepals 4–7, imbricate; petals 0; disc 0; ovary 3-locular, ovule 1 per locule; styles 3, ± free to shortly connate at the base, elongated, recurved-spreading, simple, stigmas often tightly coiled at the apex, densely papillose. Fruit subglobose, fleshy, indehiscent, densely covered with large sticky long-stipitate glands. Seeds 1–2 by abortion, large, oblong, ecarunculate; exotesta somewhat fleshy, arillate; endotesta bony; endosperm copious; cotyledons broad, flat. Fig.: See Pflanzenr. 63: 20, t. 2 (1914). See also Fig. 19, p. 171.

A monotypic genus endemic to Sri Lanka (Ceylon).

Croizat (1942a) combined 133. *Podadenia* with 134. *Ptychopyxis*, but, as Airy Shaw (1960) points out, the radically different fruit alone serves to set it apart.

134. **Ptychopyxis** *Miq.*, Fl. Ned. Ind. Suppl.: 402 (1861); Hook. f., Hooker's Icon. Pl. 18: t. 1703 (1887), and Fl. Brit. India 5: 454 (1887); Pax & K. Hoffm., Nat. Pflanzenfam. ed. 2, 19C: 230 (1931); Croizat, J. Arnold Arbor. 23: 47 (1942); Airy Shaw, Kew Bull. 14: 363 (1960) and 16: 347 (1963); Backer & Bakh. f., Fl. Java 1: 479 (1963); Airy Shaw, Kew Bull. 26: 327 (1972); Whitmore, Tree Fl. Mal. 2: 126 (1973); Airy Shaw, Kew Bull. Addit. Ser. 4: 188 (1975), 8: 197 (1980), Kew Bull. 36: 340 (1981), and Euph. Philipp.: 43 (1983); G. L. Webster, Ann. Missouri Bot. Gard. 81: 77 (1994). T.: *Pt. costata* Miq.

Clarorivinia Pax & K. Hoffm., Pflanzenr. 63: 17 (1914), and Nat. Pflanzenfam. ed. 2, 19C: 105 (1931). T.: *Cl. chrysantha* (K. Schum.) Pax & K. Hoffm. (B.: *Mallotus chrysanthus* K. Schum.) [=*Pt. chrysantha* (K. Schum.) Airy Shaw]

Dioecious small to medium trees. Indumentum simple, villous, ferrugineous. Leaves alternate, often clustered at the apex of the stem, large, often elongate, petiolate, exstipulate or minutely stipulate, simple, entire, penninerved, tertiary nerves often conspicuously scalariform, sometimes glandular; petioles bipulvinate. Inflorescences terminal, subterminal and axillary, pseudo-racemose to thyrsiform, robust or slender, sometimes short and congested; bracts small, male many-flowered, female 1-flowered. Male flowers: pedicels persistent; calyx

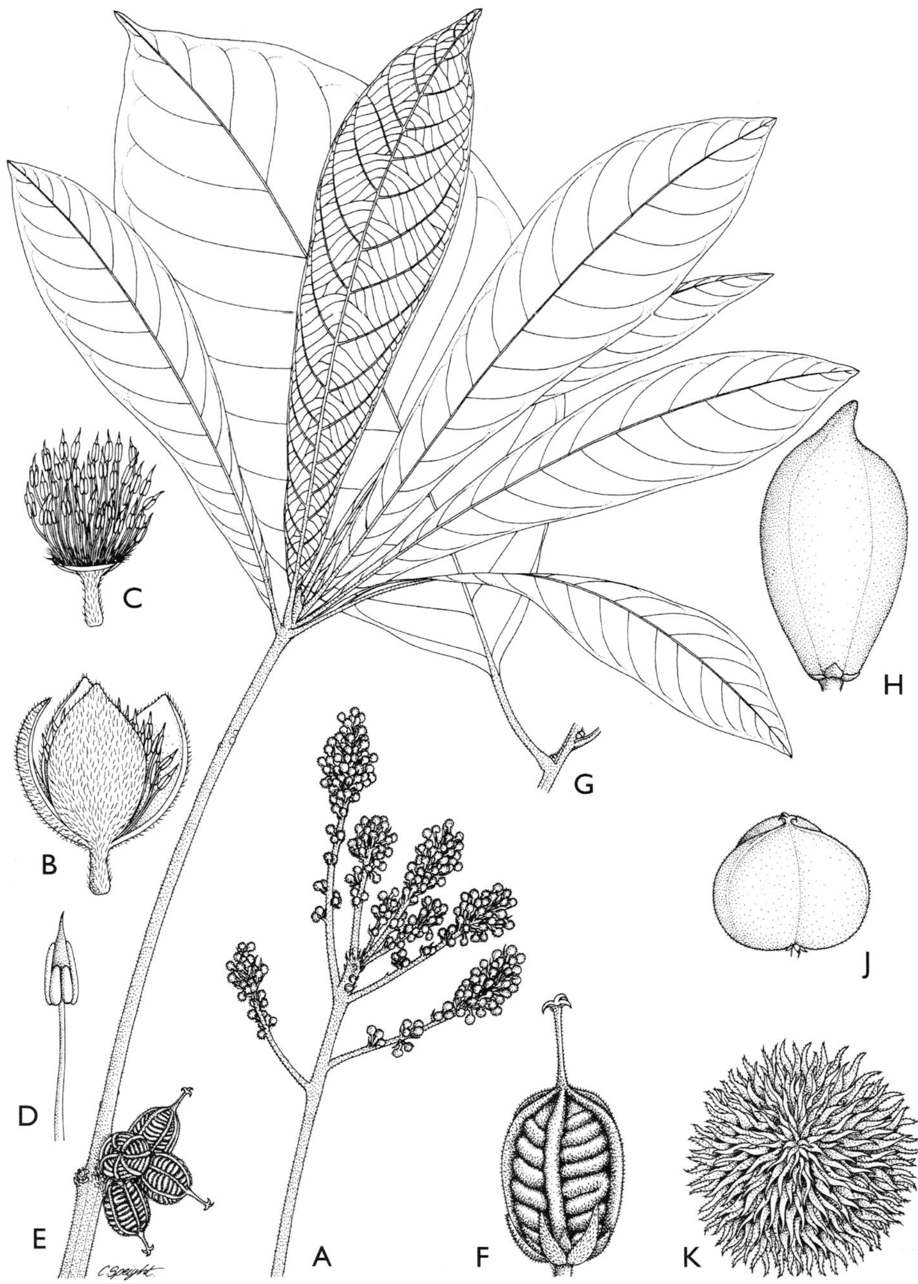

FIG. 20. *Ptychopyxis costata* Miq. **A** Male inflorescence × 1/2; **B** Male flower × 7.5; **C** Male flower, calyx removed to show androecium × 7.5; **D** Stamen × 21; **E** Habit, female × 1/2; **F** Fruit × 1.5; *Pt. kingii* Ridl. **G** leaf × 1/2; **H** Fruit × 1.5; *Pt. triradiata* Airy Shaw **J** Fruit × 1.5; *Pt. caput- medusae* Ridl. **K** Fruit × 1.5. **A–F** from *Scortechini* 1460; **G, H** from *Kepong Field No.* 63139; **J** from *Shah* 293; **K** from *Dr King's Collector* 7738. Drawn by Camilla Speight.

closed in bud, globose, apiculate, later splitting into 3–5 valvate segments; petals 0; disc-glands numerous, free, hispid, disposed amongst the stamens; stamens 35–65, filaments free, anthers dorsifixed, introrse, 4-thecate, thecae unequal, superposed or not, the outer larger than the inner, adnate to the broad, produced, triangular connective, longitudinally dehiscent; pollen suboblate to oblate-spheroidal, 3-colporate, colpi short, narrow, costae +, colpus transversalis small, narrow, costae +, coarsely reticulate, muri thick, consisting of many small columellae; pistillode 0; receptacle convex. Female flowers very shortly pedicellate to subsessile; sepals 4–5, lanceolate, persistent; petals 0; disc small, annular, pubescent; ovary 2–3-locular, pubescent, ovule 1 per locule; styles 2–3, very shortly connate at the base, recurved, simple, stigmas ± smooth to very shortly papillose. Fruit subglobose, shortly rostrate, indehiscent or subindehiscent, 2–3-locular, smooth, densely covered with numerous tomentose processes or longitudinally 6-ribbed with irregular transverse folds between the ridges; endocarp woody. Seeds 2–3, large, oblong, ecarunculate; exotesta ?arillate; endotesta bony; endosperm copious, indurated. Fig.: See Hooker's Icon. Pl. 18: t. 1703 (1887); and Fig. 20, p. 173.

A tropical Asiatic genus of 12 species ranging from Thailand through Indonesia to the Philippines and New Guinea, with the greatest number of species in the Malay Peninsula and Borneo.

135. **Botryophora** *Hook. f.*, Fl. Brit. India 5: 476 (1888), nom. cons.; Pax & K. Hoffm., Nat. Pflanzenfam. ed. 1, 3(5): 116 (1890), and ed. 2, 19C: 228 (1931); Airy Shaw, Kew Bull. 3: 484 (1948), 14: 374 (1960), Hooker's Icon. Pl. 36: t. 3576 (1962), and Kew Bull. 26: 224 (1972); Whitmore, Tree Fl. Mal. 2: 72 (1973); Airy Shaw, Kew Bull. Addit. Ser. 4: 61 (1975), and Kew Bull. 36: 269 (1981); G. L. Webster, Ann. Missouri Bot. Gard. 81: 77 (1994). T.: *B. kingii* Hook. f. [= *B. geniculata* (Miq.) Beaumée ex Airy Shaw (B.: *Sterculia geniculata* Miq.)]

Dioecious small tree; branchlets terete, slightly thickened where groups of leaves arise. Indumentum simple, hairs minute. Leaves alternate, crowded at the apex of the stem or at nodes along it, and with long bare internodes in between, large, somewhat elongate, acuminate, long-petiolate, exstipulate or possibly minutely stipulate, simple, entire, penninerved, eglandular; petioles bipulvinate, geniculate. Inflorescences terminal, subterminal, axillary in the upper leaf-axils or arising from the old wood, the males paniculate, pyramidal, pendulous, pluriflorous, branches horizontal, parallel, the females pseudospicate-racemose, pauciflorous, lax, axis angular, more robust than in the male; bracts small, male 1(2)-flowered, female 1-flowered. Male flowers shortly pedicellate to subsessile; calyx closed in bud, ovoid-subglobose, apiculate, later splitting irregularly into 2 cymbiform segments; petals 0; disc-glands numerous, minute, nodular, several surrounding each stamen; stamens c. 50–60, densely aggregated into a globose mass, filaments free, short, stout, anthers 4-thecate, thecae subequal, adnate in 2 pairs to 2 pendent lobules of the connective, longitudinally dehiscent, connectives pileiform, peltate, flat-topped, irregularly lobulate, subimbricately or tessellately disposed, concealing the thecae; pollen ±

as in 132. *Blumeodendron*, but with somewhat longer colpi; pistillode 0; receptacle subglobose. Female flowers sessile or subsessile; calyx 3–4-lobed, lobes occluded beneath the disc and closely adpressed to it; petals 0; disc massive, pulviniform, 6–8-sulcate; ovary 3–4-locular, ovule 1 per locule; styles 3–4, very shortly connate at the base, thick, recurved, simple, stigmas densely papillose. Fruit (immature) obpyramidal-3–4-gonous, (mature) subglobose, tardily loculicidally dehiscent into 3–4 valves; exocarp thinly crustaceous. Seeds 3–4, roundly subquadrate, subplano-convex, ecarunculate; testa brownish. Fig.: See Hooker's Icon. Pl. 36: t. 3576 (1962); Kew Bull. 36: 270, t. 4 (1981).

A monotypic tropical Asiatic genus ranging from Lower Burma (Myanmar) to Sumatra, Java and Borneo.

Tribe 30. **EPIPRINEAE** *(Müll. Arg.) Hurus.*, J. Fac. Sci. Univ. Tokyo, Sect. 3, Bot. 6: 309 (1954); G. L. Webster, Ann. Missouri Bot. Gard. 81: 77 (1994). T.: *Epiprinus* Griff.

Acalypheae subtribe *Epiprineae* Müll. Arg., Linnaea 34: 144 (1865), and in DC., Prodr. 15(2): 1024 (1866).

Monoecious trees or shrubs; indumentum stellate; leaves alternate, mostly stipulate, simple, penninerved or palminerved, glandular or not. Inflorescences terminal or axillary, paniculate, racemose, or spicate, mostly bisexual, male flowers often in pedunculate capitula; male calyx closed in bud, usually splitting into 2–6 valvate segments; petals 0; stamens (3)6–8(15), filaments free or connate, usually inflexed in bud; pollen 3-colpate or 3-colporate, angulaperturate, colpi inoperculate, scarcely marginate, coarsely reticulate or perforate-tectate; pistillode entire or 2–4-lobed; female sepals 4–10, mostly imbricate, entire or pinnatifid to dissected, often persistent, sometimes accrescent, foliose, sometimes involucrate with deciduous biglandular epicalyx-segments; disc mostly 0; ovary (1)3(4)-locular, styles free or connate, 2-fid or multifid. Fruit smooth or densely muricate, dehiscent or indehiscent. Seeds subglobose, ecarunculate; testa smooth, dry.

This tribe is ± equivalent to Pax & Hoffmann's series *Cladogynifomes* as modified by them for the second edition of Engler's 'Naturlichen Pflanzenfamilien' (1931), but with the removal of *Alchorneopsis* to the *Caryodendreae*, and the inclusion of *Epiprinus* and *Cleidiocarpon*.

KEY TO THE SUBTRIBES

1. Stipules +; inflorescences bisexual; male calyx splitting into distinct segments; pollen 3-colporate, not spinulose; female sepals often persistent, accrescent; fruit smooth or ± so · · · · · · · · · · 30a. *Epiprininae*
1. Stipules 0; inflorescences unisexual; male calyx turbinate, 2–5-lobed, lobes verrucose; pollen 3-colpate, spinulose; female sepals caducous; fruit densely muricate · 30b. *Cephalomappinae*

Subtribe 30a. **Epiprininae** *Müll. Arg.*, Linnaea 34: 144 (1865), and in DC., Prodr. 15(2): 1024 (1866); G. L. Webster, Ann. Missouri Bot. Gard. 81: 78 (1994). T.: *Epiprinus* Griff.

Acalypheae subtribe *Cephalocrotoneae* Müll. Arg., Linnaea 34: 143 (1865). T.: *Cephalocroton* Hochst.

Acalypheae subtribe *Mercurialinae* Series *Cladogynifomes* Pax & K. Hoffm., Pflanzenr. 63: 264 (1914). T.: *Cladogynos* Zipp. ex Span.

Epiprineae subtribe *Cleidiocarpinae* Thin, Tâp Chi Sinh Vât Hoc 10(2): 30 (1988). T.: *Cleidiocarpon* Airy Shaw.

Leaves entire or toothed, penninerved or palminerved; stipules +; inflorescences bisexual; male calyx splitting into (2)3–4(6) distinct segments; stamens (3)6–8(15), filaments mostly free; pollen 3-colporate, not spinulose; female sepals often persistent, accrescent; fruit smooth or ± so.

A palaeotropical subtribe of 8 genera.

Key to the genera

1. Male flowers in racemes or spikes; stipules (if present) glandular · · · · · 2
1. Male flowers in pedunculate capitula; stipules not glandular · · · · · · · · · 5
2. Female calyx accrescent, foliose, involucrate with deciduous biglandular epicalyx-segments; styles connate into a column, distally 2-fid or multifid; stamens (5)8–15; leaves with petiolar glands · · · · · · · · · 136. *Epiprinus*
2. Female calyx sometimes accrescent but never involucrate; styles ± free, distally multifid; stamens 3–6(8) · 3
3. Filaments erect in bud · 139. *Koilodepas*
3. Filaments inflexed in bud · 4
4. Leaves eglandular; fruits dehiscent · · · · · · · · · · · · · · · · · 137. *Symphyllia*
4. Leaves 2-glandular at the base; fruits indehiscent · · · · · 138. *Cleidiocarpon*
5. Inflorescences axillary · 140. *Cladogynos*
5. Inflorescences terminal · 6
6. Female sepals entire, not accrescent; styles free; leaves penninerved, coriaceous · 141. *Cephalocrotonopsis*
6. Female sepals pinnatifid to dissected, accrescent; styles connate; leaves palminerved, chartaceous · 7
7. Stamens (4)6–8; pollen intrareticulate; pistillode entire or 2–3-fid · 142. *Cephalocroton*
7. Stamens 4; pollen not as above; pistillode shallowly 4-lobed · 143. *Adenochlaena*

136. **Epiprinus** *Griff.*, Notul. Pl. Asiat. 4: 487 (1854); Müll. Arg. in DC., Prodr. 15(2): 1024 (1866); Benth., Gen. Pl. 3: 325 (1880); Hook. f., Fl. Brit. India 5: 463 (1887); Pax & K. Hoffm., Pflanzenr. 68: 109 (1919); Gagnep., Bull. Soc. Bot. France 72: 465 (1925), and Fl. Indo-Chine 5: 474 (1926); Pax & K. Hoffm., Nat. Pflanzenfam. ed. 2, 19C: 148 (1931); Croizat, J. Arnold Arbor. 23:

52 (1942); Airy Shaw, Kew Bull. 16: 356 (1963), and 26: 259 (1972); Whitmore, Tree Fl. Mal. 2: 95 (1973); Thin, Tâp Chi Sinh Vât Hoc 10(2): 30 (1988); G. L. Webster, Ann. Missouri Bot. Gard. 81: 78 (1994). T.: *E. malayanus* Griff.

Monoecious shrub or small tree. Indumentum densely minutely stellate. Leaves alternate, ± crowded at the ends of the branches, petiolate or subsessile, stipulate, variable in size, simple, entire, penninerved, biglandular at the base beneath; petioles bipulvinate; stipules reflexed, glandular beneath at the base. Inflorescences terminal or axillary in the upper leaf-axils, spicate or racemose, bisexual, male flowers densely clustered in the upper part, bracts many-flowered, female laxer in the lower part. Male flowers subsessile; calyx closed in bud, globose, later splitting irregularly into 2–4 membranaceous valvate segments; petals 0; disc 0; stamens (5)8–15, filaments free, folded at apex and base in bud, later erect, anthers large, oblong, subdorsifixed, pendulous, introrse, emarginate at the base, thecae parallel, contiguous, longitudinally dehiscent, connective minutely glandular-apiculate; pollen oblate-spheroidal, 3-colporate, colpi narrow, costae +, colpus transversalis long, costae +, tectate, columellae distinct, exine thick; pistillode thick, obconic-cylindric, excavate and sub-3-lobate at the apex, puberulous; receptacle 0. Female flowers shortly pedicellate; sepals 5–6, lanceolate, reduplicate-valvate, soon accrescent and becoming large and foliaceous, alternating with 5–6 much smaller involucral epicalyx-segments, biglandular at the base, which fall as a cup; petals 0; disc 0; ovary 3-locular, ovule 1 per locule; styles 3, connate into a tall column, free at the apex, bifid, segments lacerate, stigmas papillose. Infructescences lax, the male portion having fallen. Fruit large, obscurely 3-dymous-subglobose, dehiscing into 3 2-valved cocci; endocarp thick, subligneous. Seeds subglobose; testa thinly crustaceous. Fig.: See Pflanzenr. 68: 4, t. 1A, B and 110, t. 25 (1919); Fl. Indo-Chine 5: 476, t. 60.1–9 (1926); Nat. Pflanzenfam. ed. 2, 19C: 142, t. 73A,B (1931).

A rather variable genus of 4 tropical Asiatic species ranging from Assam to Malaya.

Müller (1866), followed by Pax & Hoffmann (1919), placed it in the subtribe by itself, on account of the presence of the epicalyx-segments, which they regarded as involucral.

137. **Symphyllia** *Baill.*, Étude Euphorb.: 473 (1858); Müll. Arg. in DC., Prodr. 15(2): 763 (1866); Pax, Pflanzenr. 44: 15 (1910); Gagnep., Fl. Indo-Chine 5: 477 (1926); Pax & K. Hoffm., Nat. Pflanzenfam. ed. 2, 19C: 123 (1931); G. L. Webster, Ann. Missouri Bot. Gard. 81: 78 (1994). T.: *S. siletiana* Baill.

Adenochlaena sect. *Symphyllia* (Baill.) Hook. f., Fl. Brit. India 5: 417 (1887).

Monoecious trees or shrubs. Indumentum stellate. Leaves alternate, sometimes crowded at the ends of the branches, petiolate or subsessile, stipulate, simple, entire, penninerved, eglandular; stipules each with a large gland at the base. Inflorescences terminal, subterminal or axillary in the upper leaf-axils, spicate, amentiform, bisexual, male flowers densely

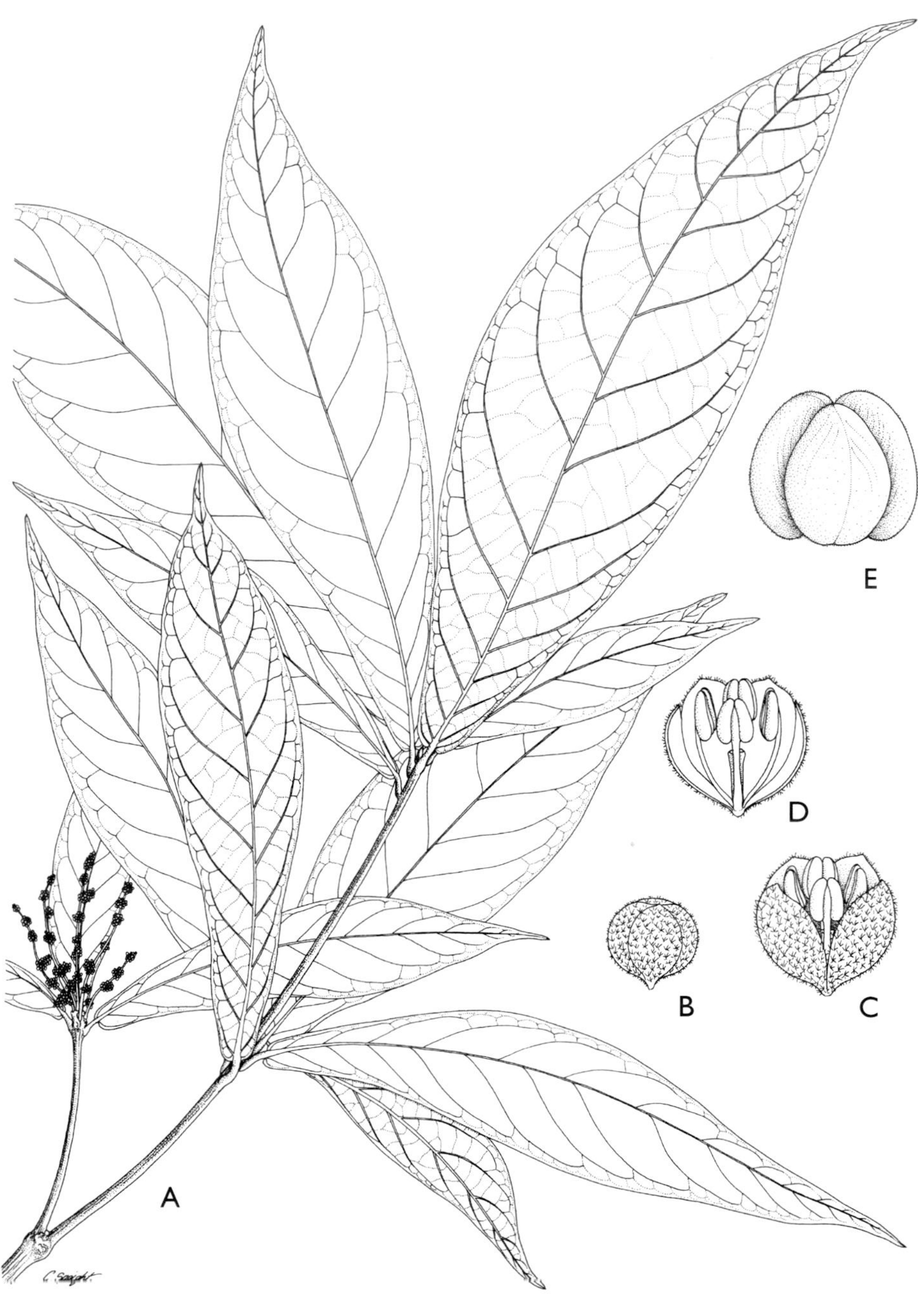

FIG. 21. *Symphyllia siletiana* Baill. **A** Habit, male flowering specimen × 3/4; **B** Male flower-bud × 20; **C** Male flower × 20; **D** Male flower, sepals partly cut away to show androecium × 20; **E** Fruit × 2. **A–D** from *Thakur Rup Chand* 2972; **E** from *Henry* 13161. Drawn by Camilla Speight.

clustered all along the axis, bracts many-flowered, females few, lax, usually in the lower part. Male flowers sessile or subsessile; calyx closed in bud, globose, later splitting valvately into 3–6 segments; petals 0; disc 0; stamens (3)4(6), filaments free, inflexed in bud, later erect, anthers dorsifixed, pendulous, introrse, thecae parallel, free below, longitudinally dehiscent; pollen prolate-spheroidal, otherwise ± as in 136. *Epiprinus*; pistillode thick, shortly columnar; receptacle 0. Female flowers shortly pedicellate; sepals 5–6, entire, not accrescent nor involucrate; petals 0; disc 0; ovary 3-locular, ovule 1 per locule; styles 3, shortly connate, long, spreading, twice bifid, stigmas densely fimbriate-papillose. Fruit trilobate, septicidally and loculicidally dehiscent into 3 2-valved cocci or 6 valves; exocarp thin, tomentose; endocarp woody. Seeds subglobose; testa crustaceous; endosperm fleshy; cotyledons flat. Fig.: See Étude Euphorb.: t. 11, f. 6–7 (1858); Fl. Indo-Chine 5: 476, t. 60.10–12 and 480, t. 61.1, 2 (1926). See also Fig. 21, p. 178.

A genus of 2 tropical Asiatic species ranging from India to S. China and peninsular Thailand.

Croizat (1942a) combined 137. *Symphyllia* with 136. *Epiprinus* as *E. balansae* (Pax & K. Hoffm.) Gagnep. appeared to him to be transitional between the two. However, the differences in the female flowers would seem to justify maintaining them as distinct.

138. **Cleidiocarpon** *Airy Shaw*, Kew Bull. 19: 313 (1965), and 32: 410 (1978); Thin, Tâp Chi Sinh Vât Hoc 10(2): 32 (1988); G. L. Webster, Ann. Missouri Bot. Gard. 81: 79 (1994). T.: *Cl. laurinum* Airy Shaw.

Sinopimelodendron Tsiang, Acta Bot. Sin. 15: 132 (1973). T.: *S. kwangsiense* Tsiang [= *Cl. cavaleriei* (H. Lév.) Airy Shaw (B.: *Baccaurea cavaleriei* H. Lév.)]

Monoecious trees. Indumentum stellate. Leaves alternate, petiolate, stipulate, simple, entire, penninerved, thinly coriaceous, 2-glandular at the base; stipules very fugacious. Inflorescences terminal or subterminal, shortly thyrsoid or pseudo-racemose or -spicate, male above, female below; male bracts many-flowered; female bracts 1-flowered. Male flowers sessile or subsessile; sepals 3–5, valvate; petals 0; disc 0; stamens 4–5, filaments incurved or inflexed in bud, anthers dorsifixed, 4-thecate, thecae subequal, latero-introrse, longitudinally dehiscent, connective not produced; pollen not known; pistillode angular-cylindric, glabrous, ± equalling the stamens; receptacle 0. Female flowers pedicellate; sepals 5–6, narrowly lanceolate, irregularly imbricate, neither accrescent nor involucrate; petals 0; disc 0; ovary ellipsoid, 1–2-locular, ovule 1 per locule, large; styles 3, slightly connate, persistent, stigmas deeply 3-fid, the branches flabellately 2–3-lobulate. Fruits 1–2-locular, fairly large, indehiscent; epicarp thinly subcrustaceous, 3-ribbed; mesocarp thinly fleshy. Seeds large, subglobose, ecarunculate; exotesta thinly crustaceous; endotesta thick, woody, formed of a single row of palisade-cells; endosperm thick. Fig.: See Acta Bot. Sin. 15: 135, t. I (1973).

A genus of 2 closely-related species from E. Burma (Myanmar) and W. China.

Thin (1988) placed it in a separate subtribe, the *Cleidiocarpinae*.

139. **Koilodepas** *Hassk.*, Verslagen Meded. Afd. Natuurk. Kon. Akad. Wetensch. 4: 139 (1856), and in Flora 40: 531 (1857) (as *Coelodepas*); Müll. Arg. in DC., Prodr. 15(2): 759 (1866); Benth., Hooker's Icon. Pl. 13: t. 1288 (1879), and Gen. Pl. 3: 313 (1880); Hook. f., Fl. Brit. India 5: 419 (1887) (as *Coelodepas*); Pax & K. Hoffm., Pflanzenr. 63: 268 (1914), and Nat. Pflanzenfam. ed. 2, 19C: 124 (1931); Croizat, J. Arnold Arbor. 23: 50 (1942); Airy Shaw, Kew Bull. 14: 382 (1960), and 16: 354 (1963); Backer & Bakh. f., Fl. Java 1: 486 (1963); Airy Shaw, Kew Bull. 26: 284 (1972); Whitmore, Tree Fl. Mal. 2: 103 (1973); Airy Shaw, Kew Bull. Addit. Ser. 4: 137 (1975), 8: 122 (1980) and Kew Bull. 36: 310, 609 (1981); G. L. Webster, Ann. Missouri Bot. Gard. 81: 79 (1994). T.: *Koilodepas bantamense* Hassk.

Calpigyne Blume, Ann. Mus. Bot. Lugduno-Batavum 2: 193 (1857). T.: *C. frutescens* Blume [=*K. frutescens* (Blume) Airy Shaw]

Nephrostylus Gagnep., Bull. Soc. Bot. France 72: 467 (1925), and Fl. Indo-Chine 5: 327 (1925). T.: *N. poilanei* Gagnep. [=*K. longifolium* Hook. f.]

Dioecious or monoecious shrubs or small trees. Indumentum stellate, soon evanescent. Leaves alternate, shortly petiolate, stipulate, oblong, entire or crenate-serrulate, penninerved, chartaceous or subcoriaceous; stipules often persistent, subulate, entire, dentate or pectinate. Inflorescences axillary, spicate, solitary or fasciculate, ± pendulous, male flowers clustered on the rhachis, several per bract, all male or with 1–2 female flowers at the base of the male spikes, the male part ultimately caducous, or females in few-flowered solitary spikes, 1 per bract. Male flowers sessile; calyx closed in bud, globose or obconic, later valvately 3–4-partite; petals 0; disc 0; stamens 4–5(8), 1-seriate, filaments thick, connate at the base, dilated beyond the middle, narrowing above, or subulate, erect in bud, sometimes later recurved, anthers apicifixed, introrse, small, thecae pendulous or laterally divergent from a small connective, longitudinally dehiscent; pollen suboblate-spheroidal to subprolate, 3-colporate, colpi very narrow, costae 0, colpus transversalis narrow, costae 0, tectate, psilate, columellae very short; pistillode shortly protruding out of the staminal column. Female flowers: calyx very variable: either cupular, adpressed to the ovary, broadly (4)8–10-lobed, lobes imbricate, persistent, sometimes glandular without, or occasionally large, urceolate, membranous, foliaceous, the lobes involute, or fruiting sepals narrowly linear, or spreading, crenate; petals 0; disc 0; ovary tomentose, (2)3-locular, ovule 1 per locule; styles (2)3, connate at the base, spreading, flattened, 2–more-fid, lobes adaxially and marginally fimbriate-multilobulate-plumose. Fruit (2)3-dymous, lobes rounded, surrounded by the ± enlarged calyx, dehiscing into 3 2-valved cocci; exocarp thinly crustaceous; endocarp thick, woody; columella squat, 3-alate, persistent. Seeds subglobose; testa crustaceous, smooth, mottled; endosperm fleshy; cotyledons broad, flat. Fig.: See Hooker's Icon. Pl.

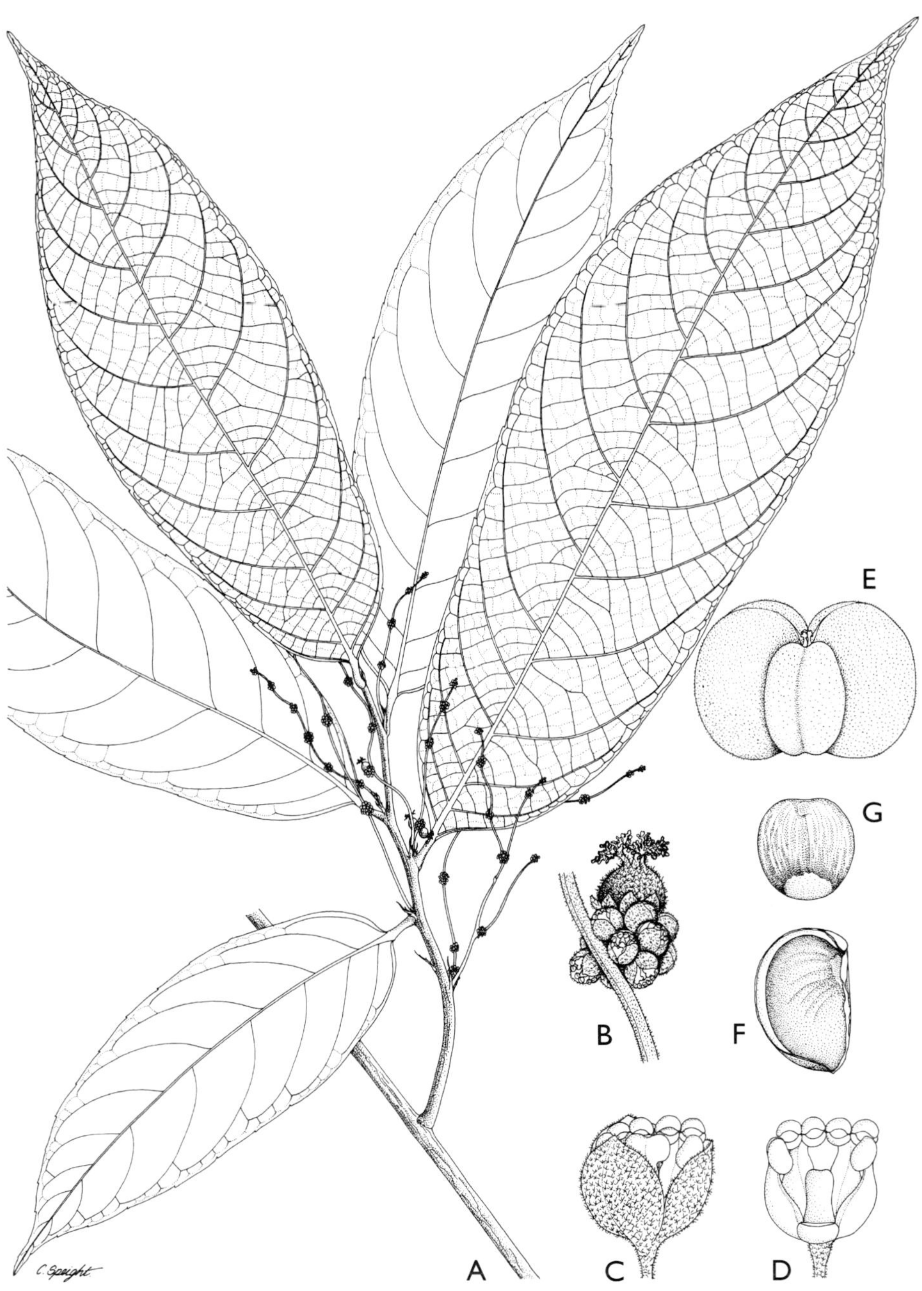

FIG. 22. *Koilodepas longifolium* Hook. f. **A** Habit × $^{3}/_{4}$; **B** Small portion of inflorescence showing male & female flowers × 6; **C** Male flower × 30; **D** Male flower, sepals removed to show androecium & pistillode × 30; **E** Fruit × 3; **F** Fruit-valve × 3; **G** Seed × 3. **A–D** from *Abdulkadir* 18304; **E–G** from *Hardial* 352. Drawn by Camilla Speight.

13: t. 1288 (1879); Pflanzenr. 63: 269, t. 42 (1914); Fl. Indo-Chine 5: 328, t. 37 (1925) (as *Nephrostylus*); Nat. Pflanzenfam. ed. 2, 19C: 125, t. 64 (1931). See also Fig. 22, p. 181.

A genus of c. 10–12 species centred on S.E. Asia and W. Malesia, with outliers in S. India, Hainan and Papua New Guinea.

140. **Cladogynos** *Zipp. ex Span.*, Linnaea 15: 349 (1841); Baill., Étude Euphorb.: 468 (1858); Müll. Arg. in DC., Prodr. 15(2): 895 (1866); Benth., Gen. Pl. 3: 323 (1880); J. J. Sm., Meded. Dept. Landb. Ned. Indië 10: 383 (1910); Pax & K. Hoffm., Pflanzenr. 63: 264 (1914); Gagnep., Fl. Indo-Chine 5: 478 (1926); Pax & K. Hoffm., Nat. Pflanzenfam. ed. 2, 19C: 123 (1931); Backer & Bakh. f., Fl. Java 1: 485 (1963); Airy Shaw, Kew Bull. 26: 232 (1972); Whitmore, Tree Fl. Mal. 2: 78 (1973); Airy Shaw, Kew Bull. 37: 11 (1982); G. L. Webster, Ann. Missouri Bot. Gard. 81: 79 (1994). T.: *Cl. orientalis* Zipp. ex Span.

Adenogynum Rchb. f. & Zoll., Acta Soc. Regiae Sci. Indo-Neerl. 1: 23 (1856). T.: *A. discolor* Rchb. f. & Zoll. [=*Cl. orientalis* Zipp. ex Span.]

Chloradenia Baill., Étude Euphorb.: 471 (1858). T.: *Chl. discolor* (Rchb. f. & Zoll.) Baill. (B.: *Adenogynum discolor* Rchb. f. & Zoll.) [=*Cl. orientalis* Zipp. ex Span.]

Monoecious shrub. Indumentum stellate, dense, white. Leaves alternate, long-petiolate, very shortly or scarcely peltate at the base, minutely stipulate, simple, dentate or bidentate, palminerved and penninerved, glabrous above, densely felted-tomentose beneath; stipules fugacious. Inflorescences axillary, spicate-racemose, solitary or geminate, shorter than the petioles, the male portion terminal, densely capitate, flowers small, the female of 1–2 much larger flowers at the base; male bracts small, narrow, female elongate, somewhat foliaceous. Male flowers subsessile; calyx closed in bud, globose, later valvately 3–4-partite; petals 0; disc 0; stamens (3)4(5), filaments free, inflexed above in bud, anthers dorsifixed, versatile, thecae contiguous, parallel, longitudinally dehiscent; pollen prolate-spheroidal, 3-colporate, colpi narrow, costae +, colpus transversalis narrow, costae +, tectum perforatum, columellae distinct, exine thick, intrareticulate; pistillode narrowly columnar; receptacle 0. Female flowers pedicellate; sepals 5–7, narrowed at the base, subfoliaceous, somewhat accrescent; petals 0; disc-glands alternisepalous; ovary tomentose, 3(4)-locular, ovule 1 per locule; styles 3(4), connate at the base, 2-fid, stigmas elongate, papillose. Fruit 3-dymous, dehiscing into 3 2-valved cocci; exocarp thin; endocarp crustaceous. Seeds subglobose; testa crustaceous, smooth, marbled; endosperm fleshy; cotyledons broad, flat. Fig.: See Étude Euphorb.: t. 19, f. 24–25 (1858) (as *Chloradenia*); Pflanzenr. 63: 265, t. 41 (1914); Fl. Indo-Chine 5: 480, t. 61.3–8 (1926); Nat. Pflanzenfam. ed. 2, 19C: 126, t. 65F (1931).

A monotypic genus from S. China, Indochina, Thailand, N. Malaya, Java, Celebes, Wetar, Timor, Tanimbar and the Philippines.

141. **Cephalocrotonopsis** *Pax*, Pflanzenr. 44: 15 (1910); Pax & K. Hoffm., Nat. Pflanzenfam. ed. 2, 19C: 123 (1931); G. L. Webster, Ann. Missouri Bot. Gard. 81: 79 (1994). T.: *C. socotrana* (Balf. f.) Pax (B.: *Cephalocroton socotranus* Balf. f.)

Cephalocroton sect. *Cephalocrotonopsis* (Pax) Radcl.-Sm., Kew Bull. 28: 131 (1973).

Monoecious shrub with a *Cotoneaster*-like habit. Indumentum stellate, dense, creamy-white. Leaves alternate, subsessile or shortly petiolate, minutely stipulate, simple, obovate, entire, penninerved, ± glabrous above, densely felted-tomentose beneath; stipules exceedingly fugacious. Inflorescences terminal or subterminal, spicate-racemose, the male portion terminal, pedunculate, densely capitate, flowers small, the female of (0)1–3(5) larger flowers at the base. Male flowers subsessile; calyx closed in bud, globose, later valvately 3-partite; petals 0; disc 0; stamens 7–10, filaments free, inflexed above in bud, later erect, anthers dorsifixed, oblong, thecae free at the base, parallel, longitudinally dehiscent, connective broad, produced, rounded-truncate; pollen ± as in 140. *Cladogynos*; pistillode ovoid, entire or 2-lobed; receptacle 0. Female flowers shortly pedicellate; sepals 6, subequal, connate at the base, entire, not or scarcely accrescent; petals 0; disc 0; ovary tomentose, 3-locular, ovule 1 per locule; styles 3, free, multifid, stigmas papillose. Fruit 3-dymous, dehiscing into 3 2-valved cocci; exocarp thinly crustaceous; endocarp woody. Seeds subglobose, ecarunculate; testa crustaceous, smooth, sparingly simply pubescent, dull, blackish. Fig.: See Trans. Roy. Soc. Edinburgh 31: t. XCIV (1887).

A monotypic genus of Socotra.

Radcliffe-Smith (1973) rather ill-advisedly reduced this distinct genus to sectional rank under 142. *Cephalocroton*.

142. **Cephalocroton** *Hochst.*, Flora 24: 370 (1841); Baill., Étude Euphorb.: 474 (1858); Müll. Arg. in DC., Prodr. 15(2): 760 (1866); Benth., Gen. Pl. 3: 307 (1880); Pax, Pflanzenr. 44: 7 (1910); Prain in Fl. Trop. Afr. 6(1): 843 (1912); Pax & K. Hoffm., Nat. Pflanzenfam. ed. 2, 19C: 123 (1931); Radcl.-Sm., Kew Bull. 28: 123 (1973), and in Fl. Trop. E. Afr., Euph. 1: 282 (1987); M. G. Gilbert, Kew Bull. 42: 365 (1987); G. L. Webster, Ann. Missouri Bot. Gard. 81: 79 (1994); Radcl.-Sm., Fl. Zamb. 9(4): 149 (1996). T.: *C. cordofanus* Hochst.

Monoecious shrubs or subshrubs. Indumentum stellate. Leaves alternate, petiolate or subsessile, stipulate, simple, entire or toothed, palminerved; stipules small, lacinulate. Inflorescences terminal, racemose, the male portion terminal, pedunculate, densely capitate, flowers small, the female of (0)1–7 larger flowers at the base; bracts small or minute. Male flowers pedicellate; calyx closed in bud, globose, later valvately 3–4-partite; petals 0; disc 0; stamens (4)6–8, 2-verticillate, filaments free, inflexed in bud, later erect, anthers dorsifixed, oblong, thecae free at the base, slightly divergent, longitudinally dehiscent, connective narrow, not produced; pollen subprolate, with broad colpus transversalis, otherwise ± as in 140. *Cladogynos*; pistillode columnar, short, entire or 2–3-fid; receptacle 0. Female flowers pedicellate, the pedicels elongating in fruit; sepals 5–6, free, unequal, elongate, 2-pinnatipartite, the segments gland-tipped or not, accrescent, persistent; petals 0; disc annular, subentire; ovary 3-locular, ovule 1 per locule; styles 3, shortly connate at the base, multifid, stigmas ± smooth. Fruit 3-lobate, dehiscing septicidally into 3 2-valved cocci; exocarp thinly crustaceous; endocarp thinly woody. Seeds ovoid-subglobose, ecarunculate; testa crustaceous, smooth, dull, brownish; endosperm fleshy; embryo straight, cotyledons broad, flat. Fig.: See Étude Euphorb.: t. 18, f. 24–27 (1858); Pflanzenr. 44: 9, t. 3 and 11, t. 4 (1910); Nat. Pflanzenfam. ed. 2, 19C: 124, t. 63 (1931); Fl. Trop. E. Afr., Euph. 1: 284, t. 55 (1987); Kew Bull. 42: 366, t. 6 (1987); Fl. Zamb. 9(4): 150, t. 30 (1996).

A tropical African genus of 3 or 9 species ranging from Nigeria to Somalia and south to northern South Africa, depending on whether one adopts Pax & Hoffmann's (1910) narrow view of the species, or Radcliffe-Smith's (1973, 1987, 1996b) subsumption of all their new taxa.

143. **Adenochlaena** *Boivin ex Baill.*, Étude Euphorb.: 472 (1858); Benth., Gen. Pl. 3: 307 (1880), p.p.; Pax, Nat. Pflanzenfam. 3(5): 78 (1890), p.p., and Pflanzenr. 44: 12 (1910); Pax & K. Hoffm., Nat. Pflanzenfam., ed. 2, 19C: 124 (1931), p.p.; G. L. Webster, Ann. Missouri Bot. Gard. 81: 78 (1994); Philcox in Dassan., ed., Fl. Ceyl. XI: 137 (1997) (as *Cephalocroton*). T.: *A. leucocephala* Baill.

Cephalocroton sect. *Adenochlaena* (Boivin ex Baill.) Müll. Arg. in DC., Prodr. 15(2): 762 (1866).

Centrostylis Baill., Étude Euphorb.: 469 (1858). T.: *C. zeylanica* Baill. [=*A. zeylanica* (Baill.) Thwaites]

Niedenzua Pax, Bot. Jahrb. Syst. 19: 106 (1894). T.: *N. cordata* Pax [=*A. leucocephala* Baill.]

Very like 142. *Cephalocroton,* differing in having only 4 stamens with the anthers sometimes apiculate, non-reticulate pollen, the pistillode shallowly 4-lobed and the female calyx-lobes sometimes pinnatifid. Fig.: See Étude Euphorb.: t. 2, f. 28–29 (1858) (as *Centrostylis*); Pflanzenr. 44: 13, t. 5 (1910); Notul. Syst. (Paris) 9: 184, t. 5.14–20 (1941).

A genus of 2 disjunct species in Sri Lanka (Ceylon) and Madagascar and the Comoro Is.

Although Webster (1994) keeps it separate, the treatment followed here, I still feel that Müller was probably nearer the mark in according it sectional status under *Cephalocroton* (see Radcliffe-Smith, 1973).

Subtribe 30b. **Cephalomappinae** *G. L. Webster,* Taxon 24: 597 (1975), and Ann. Missouri Bot. Gard. 81(1): 79 (1994). T.: *Cephalomappa* Baill.

Leaves entire, penninerved; stipules 0; inflorescences unisexual; male calyx turbinate, 2–5-lobed, lobes verrucose; stamens 2–4, filaments connate; pollen 3-colpate, spinulose; female sepals caducous; fruit densely muricate.

This subtribe includes only the type-genus.

FIG. 23. *Cephalomappa malloticarpa* J. J. Sm. **A** Habit × 1; **B** Male flower-bud × 15; **C** Male flower-bud beginning to open × 15; **D** Male flower × 15; **E** Androecium × 15; **F** Female flower × 8; **G** Fruit × 2. **A–F** from *de Wilde et al.* 14826; **G** from *de Wilde et al.* 12681. Drawn by Camilla Speight.

144. **Cephalomappa** *Baill.*, Adansonia 11: 130 (1874); Benth., Gen. Pl. 3: 323 (1880); Pax, Pflanzenr. 44:16 (1910); Pax & K. Hoffm., Nat. Pflanzenfam. ed. 2, 19C: 123 (1931); Airy Shaw, Kew Bull. 14: 378 (1960); Kosterm., Reinwardtia 5: 413 (1961); Backer & Bakh. f., Fl. Java 1: 486 (1963); Whitmore, Tree Fl. Mal. 2: 75 (1973); Airy Shaw, Kew Bull. Addit. Ser. 4: 66 (1975), and Kew Bull. 36: 274 (1981); G. L. Webster, Ann. Missouri Bot. Gard. 81: 79 (1994). T.: *C. beccariana* Baill.

Muricococcum Chun & F. C. How, Acta Phytotax. Sin. 5: 14 (1956). T.: *M. sinense* Chun & F. C. How [=*C. sinensis* (Chun & F. C. How) Kosterm.]

Monoecious shrubs or small trees. Indumentum stellate, ferrugineous. Leaves alternate, petiolate, exstipulate, simple, entire or crenulate-denticulate, penninerved, coriaceous or chartaceous, indistinctly glandular at the base. Inflorescences axillary, paniculate, unisexual, the branches of the males terminating in densely capitate pedunculate heads, those of the females in solitary flowers; bracts minute. Male flowers small, sessile; calyx turbinate, truncate, shortly 2–5-lobed, the lobes valvate, revolute-reflexed, verrucose; petals 0; disc 0; stamens 2–4, filaments connate at the base into a thick androphore, free above, exserted, inflexed at the apex, anthers dorsifixed, latrorse, thecae free at the base, longitudinally dehiscent, connective shortly produced, rounded, glandular at the apex; pollen oblate-spheroidal, 3-colpate, colpi very short, costae +, tectate, intrareticulum coarse, spinulose; pistillode columnar, short; receptacle 0. Female flowers pedicellate, the pedicels elongating in fruit; sepals 5–6, free, imbricate, soon caducous; petals 0; disc 0; ovary 3-locular, ovule 1 per locule; styles 3, short, thick, connate at the base, stigmas palmate or bifid. Fruit 3-coccous, densely muricate or almost echinate, dehiscing septicidally into 3 2-valved cocci; epicarp thinly crustaceous; mesocarp thin, granular; endocarp thick, woody; columella stout, 3-alate, 3-cornute, persistent. Seeds ovoid-subglobose, ecarunculate; testa hard, smooth, somewhat shiny, brownish. Fig.: See Tree Fl. Mal. 2: 75, t. 4 (1973). See also Fig. 23, p. 185.

A genus of 6 species in S. China, Malay Peninsula, Sumatra and Borneo.

Tribe 31. **ADELIEAE** *G. L. Webster*, Taxon 24: 597 (1975), and Ann. Missouri Bot. Gard. 81(1): 80 (1994). T.: *Adelia* L.

Acalypheae subtribe *Mercurialinae* series *Adeliiformes* Pax & K. Hoffm., Pflanzenr. 63: 59 (1914).

Dioecious or rarely monoecious trees or shrubs; indumentum simple, stellate or stellate-lepidote; leaves alternate, simple, entire or dentate, usually eglandular; stipules small or 0. Inflorescences axillary or sometimes cauliflorous, fasciculate or racemose, female 1–several-flowered; male calyx splitting into (2)4–5 valvate segments; petals 0; disk extrastaminal, interstaminal, annular, dissected or 0; stamens 8–30, filaments free or connate at the base, not inflexed in bud, anthers sometimes versatile; pollen mostly 3–4-colpate, operculate, perforate-tectate, polygonally nanospinulose;

pistillode + or 0; female sepals 5–7, usually valvate or open, often reflexed; disc usually +, annular, pulvinate, entire or lobate; ovary 2–3(4)-locular; styles usually laciniate. Fruit dehiscent. Seeds usually subglobose, mostly ecarunculate; testa dry, smooth.

A predominantly Neotropical (and especially Caribbean) (not *Crotonogynopsis*) tribe of 5 genera with operculate pollen.

Key to the genera

1. Indumentum simple; stipules deciduous; pollen tectum finely perforate · · 2
1. Indumentum stellate or stellate-lepidote; stipules 0; pollen tectum coarsely perforate · 4
2. Ovary 2-locular; male disk 0; stamens 20–30; seeds carunculate · 147. *Enriquebeltrania*
2. Ovary 3-locular; male disk +; stamens 8–20; seeds ecarunculate · · · · · · · 3
3. Flowers fasciculate; male disc annular or segmented; pistillode + · 145. *Adelia*
3. Flowers racemose; male disc of interstaminal glands; pistillode 0 · 146. *Crotonogynopsis*
4. Indumentum stellate; leaves 3-plinerved; filaments free; pollen 3-colpate; female inflorescences several-flowered; female disc entire · 148. *Lasiocroton*
4. Indumentum stellate-lepidote; leaves mostly penninerved; filaments ± connate; pollen 4-colpate; female inflorescences 1–2-flowered; female disc lobate · 149. *Leucocroton*

145. **Adelia** *L.*, Syst. Nat. ed. 10, 2: 1298 (1759) (nom. cons.); Baill., Étude Euphorb.: 417 (1858), excl. syn.; Benth., Gen. Pl. 3: 312 (1880); Pax & K. Hoffm., Pflanzenr. 63: 64 (1914); Fawc. & Rendle, Fl. Jamaica 4: 291 (1920); Pax & K. Hoffm., Nat. Pflanzenfam. ed. 2, 19C: 109 (1931); Alain, Fl. Cuba 3: 93 (1953); G. L. Webster, Ann. Missouri Bot. Gard. 54: 272 (1968); Philcox, Fl. Trin. Tob. II (X): 664 (1979); Proctor, Kew Bull. Addit. Ser. 11: 539 (1984); Sneep & de Roon, Fl. Neth. Ant.: 3: 254 (1984); Alain, Fl. Español. 4: 73 (1986), and Descr. Fl. Puerto Rico 2: 349 (1988); R. A. Howard, Fl. Lesser Antilles 5: 13 (1989); A. H. Gentry, Woody Pl. NW. S. Amer.: 417 (1993); G. L. Webster, Ann. Missouri Bot. Gard. 81: 80 (1994); Gillespie in Acev.-Rodr., ed., Fl. St. John: 203 (1996). T.: *A. ricinella* L. (typ. cons.)

Ricinella Müll. Arg., Linnaea 34: 153 (1865), in DC., Prodr. 15(2): 729 (1866), and in Mart., Fl. Bras. 11(2): 306 (1874). T.: *R. pedunculosa* (A. Rich.) Müll. Arg. (B.: *A pedunculosa* A. Rich.) [=*A. ricinella* L.]

Dioecious shrubs or trees, usually soon glabrescent; dried specimens of some species smelling of curry-powder. Indumentum simple. Branchlets often spinescent. Leaves alternate or clustered at the nodes on brachyblasts, shortly petiolate, stipulate, simple, entire, penninerved, rarely 3-plinerved (*A. oaxacana* (Müll. Arg.) Hemsl.), membranaceous or chartaceous, with domatial hairs in the angles of the midrib beneath, pellucid-puncticulate. Flowers axillary, males fasciculate on brachyblasts, females solitary or geminate. Male flowers shortly pedicellate; calyx closed in bud, ovoid, later valvately 4–5-partite; petals 0; disc extrastaminal, mostly annular, rarely composed of 5 segments, adnate to the base of the calyx; stamens 8–17, filaments free at first, soon becoming connate at the base or into a column, anthers versatile, subdorsifixed, latrorse, thecae parallel, longitudinally dehiscent, connective narrow; pollen spheroidal, (3)4-colpate, colpi wide, operculum narrow, costae 0, tectate, tectum finely perforate, psilate; pistillode 3-partite. Female flowers long-pedicellate, the pedicels further elongating in fruit, filiform; sepals 5–7, narrow, later reflexed; disc annular, pubescent, adnate to the calyx; ovary 3(4)-locular, ovule 1 per locule; styles 3(4), free or ± so, long-lacinulate, lying on the top of the ovary, or erecto-patent. Fruit 3-dymous, dehiscing into 3 2-valved cocci; pericarp thin, pubescent; endocarp woody; columella narrow, 3-cornute, usually persistent. Seeds globose to subglobose, ecarunculate; testa hard, smooth, greyish; endosperm fleshy, deep yellow or red; cotyledons broad, flat. Fig.: See Pflanzenr. 63: 68, t. 9 (1914); Fl. Jamaica 4: 292, t. 95 (1920); Fl. Cuba 3: 93, t. 25 (1953); Ann. Missouri Bot. Gard. 54: 274, t. 11 (1968); Kew Bull. Addit. Ser. 11: 540, t. 164 (1984); Fl. Español. 4: 75, t. 116-2 (1986); Descr. Fl. Puerto Rico 2: 351, t. 59-2 (1988); Fl. Lesser Antilles 5: 11, t. 3 (1989); Woody Pl. NW. S. Amer.: 418, t. 120.3 (1993); Fl. St. John: 204, t. 93P–U (1996).

A neotropical genus of 13 species ranging from Texas and Cuba to Paraguay.

146. **Crotonogynopsis** *Pax*, Bot. Jahrb. Syst. 26: 328 (1899); Prain, Fl. Trop. Afr. 6(1): 924 (1912); Pax & K. Hoffm., Pflanzenr. 63: 14 (1914), and Nat. Pflanzenfam. ed. 2, 19C: 104 (1931); Keay (ed.), Fl. W. Trop. Afr. ed. 2, 1: 404 (1958); Radcl.-Sm., Fl. Trop. E. Afr., Euph. 1: 213 (1987); G. L. Webster, Ann. Missouri Bot. Gard. 81: 80 (1994); J. Léonard, Bull. Jard. Bot. Belg. 65: 136 (May 1996), 341 (Nov. 1996), and Fl. Afr. Cent. Euph. 3: 28 (Oct. 1996). T.: *C. usambarica* Pax.

Dioecious or monoecious unbranched or little-branched shrub or small tree. Indumentum simple. Leaves alternate, crowded at the apices of the shoots, large, very shortly petiolate to subsessile, stipulate, simple, elongate, spathulate-lanceolate, shallowly repand-denticulate to subentire, narrowly cordulate, penninerved, minutely sparingly glandular beneath; stipules very fugacious. Inflorescences axillary or cauliflorous, racemose, pendulous, unisexual, the males fascicled, multiflorous, the females solitary, uni- to pauciflorous, lax; bracts subconcave, persistent, the males 1–2-flowered, the females 1-flowered. Male flowers pedicellate, pedicels articulated, 2-bracteolate; calyx closed in bud, ovoid, acute, later valvately 2–3(5)-partite, hyaline, becoming reflexed; petals 0; disc-glands numerous, free, large, glabrous, disposed amongst the stamens; stamens (10)12–15, filaments free, anthers dorsifixed, introrse and extrorse, elliptic-oblong, apiculate, 2-thecate, thecae partially free at the base, longitudinally dehiscent, connective narrow; pollen spheroidal, 3-colporate, convex-triangular in polar view, colpi narrow, costae 0, colpus transversalis elongated, costae 0, tectate, tectum finely perforate, psilate; pistillode 0. Female flowers pedicellate; pedicels jointed, 2-bracteolate, extending in fruit; sepals 4(5), subimbricate, hyaline; petals 0; disc hypocrateriform, subentire to 4(9)-lobate, sometimes irregularly so, lobes commonly alternisepalous; ovary 3(4)-locular, ovule 1 per locule; styles 3(4), free or slightly connate at the base, 2-partite, the segments recurved, laciniate, stigmas papillose. Fruit 3-lobate, nervose, dehiscing septicidally into 3 2-valved cocci; exocarp not separating; endocarp thinly woody; columella 3-quetrous, persistent. Seeds ovoid, ecarunculate. Fig.: See Fl. Trop. E. Afr., Euph. 1: 214, t. 40 (1987); Bull. Jard. Bot. Belg. 65: 139, t. 1 (May 1996), 343, t. 1 (Nov. 1996); Fl. Afr. Cent. Euph. 3: 31, t. 4 (Oct. 1996).

A ditypic African genus, occurring in Côte d'Ivoire, Ghana, SW Cameroun, Congo (Kinshasa), W. Uganda and E. Tanzania.

147. **Enriquebeltrania** *Rzed.*, Bol. Soc. Bot. México 38: 75 (1979); G. L. Webster, Ann. Missouri Bot. Gard. 81: 80 (1994). T.: *E. crenatifolia* (Miranda) Rzed. (B.: *Beltrania crenatifolia* Miranda).

Beltrania Miranda, Bol. Soc. Bot. México 21: 4 (1957), *non* Penzig (1882). T.: *B. crenatifolia* Miranda [=*E. crenatifolia* (Miranda) Rzed.]

Very like 145. *Adelia* in habit, but dried specimens not odoriferous, leaves crenate, male sepals 3(4), disc 0, stamens up to 30, multiseriate, filaments free, anthers basifixed, extrorse, pistillode 0, receptacle convex-columnar, female and fruiting pedicels much shorter, disc 0, ovary and fruit 2-locular, 2-dimpled, and seeds carunculate. Fig.: See Bol. Soc. Bot. México 21: 12, t. 2 (1957).

A monotypic genus endemic to S. Mexico.

148. **Lasiocroton** *Griseb.*, Fl. Brit. W. I.: 46 (1859), and Abh. Königl. Ges. Wiss. Göttingen 9: 20 (1861); Benth., Gen. Pl. 3: 317 (1880); Pax & K. Hoffm., Pflanzenr. 63: 60 (1914); Fawc. & Rendle, Fl. Jamaica 4: 293 (1920); Pax & K. Hoffm., Nat. Pflanzenfam. ed. 2, 19C: 108 (1931); Alain, Fl. Cuba 3: 87 (1953); Correll, Fl. Bah. Arch.: 825 (1982); Alain, Fl. Español. 4: 170 (1986); G. L. Webster, Ann. Missouri Bot. Gard. 81: 80 (1994). T.: *L. macrophyllus* (Sw.) Griseb. (B.: *Croton macrophyllus* Sw.)

Dioecious shrubs or small trees. Indumentum densely stellate-tomentose. Leaves alternate, petiolate, exstipulate, simple, entire or ± so, palminerved, tertiary nerves scalariform. Inflorescences axillary in the upper leaf-axils, solitary, the males short, spicate, rarely subpaniculate, dense-flowered, the females elongate, racemose, lax, fewer-flowered; male bracts 1-several-flowered, female 1–2- flowered. Male flowers subsessile; calyx closed in bud, globose, later valvately 5-partite; petals 0; disc annular; stamens (6)10–18, filaments free, anthers subdorsifixed, erect in bud, introrse, thecae parallel, separate, contiguous at apex, longitudinally dehiscent, connective broadly ovate, not produced; pollen oblate-spheroidal, 3-colpate, colpi wide, operculate, costae 0, tectate, tectum coarsely perforate, psilate, not reticulate; pistillode filiform, minute or 0. Female flowers pedicellate; sepals 5, open in bud, spreading below the ovary; petals 0; disc thick, pulvinate, entire; ovary 3(4)-locular, ovule 1 per locule; styles 3(4), very short, thick, shortly connate at the base, spreading, subentire, retuse or 2-fid, the lobes laciniate, stigmas papillose. Fruit 3-lobate, densely tomentose, dehiscing septicidally into 3 2-valved cocci; columella persistent. Seeds globose, ecarunculate, smooth, dorsally slightly keeled, shiny; endosperm fleshy; cotyledons broad, flat. Fig.: See Fl. Jamaica 4: 294, t. 96 (1920); Fl. Cuba 3: 88, t. 23 (1953); Fl. Bah. Arch.: 826, t. 342 (1982); Fl. Español. 4: 171, t. 116–23 (1986).

A West Indian genus of c. 6 species, confined to the Bahamas and the Greater Antilles.

149. **Leucocroton** *Griseb.*, Abh. Königl. Ges. Wiss. Göttingen 9: 20 (1861); Müll. Arg. in DC., Prodr. 15(2): 757 (1866); Benth., Gen. Pl. 3: 312 (1880); Pax & K. Hoffm., Pflanzenr. 63: 62 (1914); Urb., Ber. Deutsch. Bot. Ges. 36: 504 (1918); Pax & K. Hoffm., Nat. Pflanzenfam. ed. 2, 19C: 109 (1931); Alain, Fl. Cuba 3: 89 (1953); Borhidi, Acta Bot. Acad. Sci. Hung. 21: 222 (1975); Alain, Fl. Español. 4: 170 (1986); Borhidi, Acta Bot. Hung. 36: 13 (1990/1); G. L. Webster, Ann. Missouri Bot. Gard. 81: 80 (1994). T.: *L. wrightii* Griseb.

Very like 148. *Lasiocroton,* differing chiefly in having the indumentum usually stellate-lepidote, the leaves often narrow and sclerophyllous with revolute margins and mostly penninerved, the filaments often ± connate at the base, the pollen 4-colpate, a large convex receptacle, the female inflorescences 1–2-flowered and the female disc often lobate. Fig.: See Pflanzenr. 63: 63, t. 8A, B (1914); Nat. Pflanzenfam. ed. 2, 19C: 108, t. 55A, B (1931); Fl. Cuba 3: 92, t. 24 (1953); Acta Bot. Hung. 36: 25–40, tt. 1–16 (1990/1).

A West Indian genus of c. 30 species, confined to Cuba and Hispaniola.

Tribe 32. **ALCHORNIEAE** *(Hurus.) Hutch.*, Amer. J. Bot. 56: 752 (1969); G. L. Webster, Ann. Missouri Bot. Gard. 81: 80 (1994). T.: *Alchornea* Sw.

Acalypheae subtribe *Alchorneinae* Hurus., J. Fac. Sci. Univ. Tokyo, Sect. 3, Bot. 6: 302 (1954).

Dioecious or monoecious trees or shrubs; indumentum simple or stellate; leaves alternate, entire or dentate, rarely spiny, penninerved or palminerved, sometimes stipellate or with laminar glands; stipules mostly deciduous. Inflorescences terminal or axillary, spicate or paniculate, bracts glandular or eglandular; male calyx splitting into 2–5 valvate segments; petals 0; stamens 2–60, free or basally connate; anthers mostly introrse, apiculate or muticous;

pollen 3-colporate, operculate, rugulose to vermiculate, irregularly nanospinulose; male disk, if +, intrastaminal; pistillode 0 or +; female sepals 3–8(12), free or connate, mostly imbricate, sometimes glandular at the base; disc sometimes +; ovary (1)2–3(4)-locular, smooth or cristate; styles elongate, slender and entire or stigmatiform and dilated or 2-fid; stigmas smooth or papillose, obtuse or acute. Fruit dehiscent. Seeds usually ecarunculate, mostly smooth, testa dry.

According to Webster (1975, 1994) Hutchinson's (1969) tribe should be augmented by the incorporation of *Conceveiba* and its allies, excised from his tribe *Malloteae*, on account of their operculate pollen. This treatment is followed here:-

Key to the subtribes

1. Styles entire, or, if distinctly 2-fid, then indumentum simple; male inflorescences axillary; stamens 4–9 · · · · · · · · · · · · · 32a. *Alchorneinae*
1. Styles 2-fid; indumentum stellate; male inflorescences terminal or axillary; stamens 15–60 · 32b. *Conceveibinae*

Subtribe 32a. **Alchorneinae** *Hurus.*, J. Fac. Sci. Univ. Tokyo, Sect. 3, Bot. 6: 302 (1954); G. L. Webster, Ann. Missouri Bot. Gard. 81: 81 (1994). T.: *Alchornea* Sw.

Acalypheae subtribe *Mercurialinae* series *Alchorneiformes* Pax & K. Hoffm., Pflanzenr. 63: 220 (1914).

Leaves often shortly petiolate, sometimes stipellate; indumentum simple or stellate; male inflorescences axillary; stamens 2–9; female sepals mostly eglandular; ovary (1)2–3(4)-locular; styles mostly entire, often stigmatiform or dilated.

Key to the genera

1. Styles bifid, indumentum simple · 2
1. Styles unlobed, or bifid only at the tip · 3
2. Dioecious; pistillode 0; ovary smooth; seeds ecarunculate · · · · 150. *Orfilea*
2. Monoecious; pistillode +; ovary cristate; seeds carunculate · · · 151. *Bossera*
3. Indumentum simple or stellate; ovary 2–3-locular; styles elongate, slender · 152. *Alchornea*
3. Indumentum simple; ovary 3-locular; styles stigmatiform or dilated · · · · 4
4. Leaf-lobes often spine-tipped; stamens usually 8, free; stigmas smooth; female sepals glandular at the base · · · · · · · · · · · · · · · 153. *Caelebogyne*
4. Leaf-lobes not spine-tipped; stamens 2–4, basally connate · · · · · · · · · · · 5
5. Stamens 4; styles elongate, dilated; seeds ecarunculate · 154. *Aparisthmium*
5. Stamens 2 or 3; styles stigmatiform; seeds minutely carunculate · 155. *Bocquillonia*

150. **Orfilea** *Baill.*, Étude Euphorb.: 452 (1858); G. L. Webster, Ann. Missouri Bot. Gard. 81: 81 (1994). T.: *O. coriacea* Baill.

Laurembergia Baill., Étude Euphorb.: 451 (1858) (sphalm. *Lautembergia*); Pax & K. Hoffm., Pflanzenr. 63: 253 (1914), and Nat. Pflanzenfam. ed. 2, 19C: 121 (1931) (as *Lautembergia*); Coode, Fl. Masc. 160. Euph.: 55 (1982) (as *Lautembergia*), non *Laurembergia* Bergius (1767). T.: *L. multispicata* Baill. [=*O. multispicata* (Baill.) G. L. Webster]

Diderotia Baill., Adansonia I. 1: 274 (1861). T.: *D. multispicata* (Baill.) Baill. (B.: *Laurembergia multispicata* Baill.) [=*O. multispicata* (Baill.) G. L. Webster]

Laurembergia Baill. sect. *Orfilea* (Baill.) Pax in Engl., Pflanzenfam., ed. 1, 3(5): 56 (1890); Pax & K. Hoffm., Pflanzenr. 63: 254 (1914), and Nat. Pflanzenfam. ed. 2, 19C: 121 (1931) (as *Lautembergia*).

Dioecious or rarely submonoecious shrubs or small trees with the habit of *Alchornea.* Indumentum simple. Leaves alternate, shortly petiolate, stipulate, simple, entire or dentate, penninerved. Inflorescences terminal or axillary, solitary, the males interruptedly branched-spicate to paniculate, dense-flowered, the females racemose, lax, fewer-flowered; male bracts several-flowered, female 1-flowered. Male flowers subsessile; calyx closed in bud, subglobose, later valvately 2–4-partite; petals 0; disc 0; stamens 5–9, filaments unequally connate at the base, free above, anthers subdorsifixed, introrse, thecae deeply sulcate and 2-lobate, free at the base, ellipsoid-subglobose, longitudinally dehiscent, connective not produced; pollen suboblate to spheroidal, 3–4-colporate, colpi narrow, operculate, costae +, colpus transversalis costate, tectate, psilate; pistillode 0. Female flowers very shortly pedicellate; sepals (3)5(6), valvate, persistent in fruit; disc 0 (§ *Laurembergia* (Baill.) Pax) or of 5 hypogynous alternisepalous subpetaloid glands (§ *Orfilea* (Baill.) Pax); staminodes sometimes present; ovary 3-locular, ovule 1 per locule; styles 3, shortly connate at the base, recurved, 2-fid, stigmas papillose. Fruit 3-lobate, ± smooth, tardily dehiscing septicidally into 3 2-valved cocci; columella persistent. Seeds ellipsoid, ecarunculate, smooth. Fig.: See Notul. Syst. (Paris) 9: 184, t. 5.1–13 (1941) (as *Lautembergia*); Fl. Masc. 160. Euph.: 57, t. 11 (1982).

A genus of 4 species from Madagascar and Mauritius (Radcliffe-Smith & Govaerts, 1997).

151. **Bossera** *Leandri,* Adansonia II. 2: 216 (1962); G. L. Webster, Ann. Missouri Bot. Gard. 81: 81 (1994). T.: *B. cristatocarpa* Leandri.

Similar to 150. *Orfilea,* but monoecious, with 10 stamens, a pistillode in the male flowers, a cristate ovary and carunculate seeds. Fig.: See Adansonia II. 2: 217, t. 1 (1962).

A monotypic endemic genus from Madagascar.

152. **Alchornea** *Sw.*, Prodr.: 98 (1788), and Fl. Ind. Occid. 2 (2): 1153 (1800); A. Juss., Euphorb. Gen.: 42 (1824); Baill., Étude Euphorb.: 445 (1858); Müll.

Arg. in DC., Prodr. 15(2): 899 (1866), and in Mart., Fl. Bras. 11(2): 374 (1874); Benth., Gen. Pl. 3: 314 (1880); Hook. f. Fl. Brit. India 5: 420 (1887); Prain in Fl. Trop. Afr. 6(1): 914 (1912); Pax & K. Hoffm., Pflanzenr. 63: 220 (1914); Fawc. & Rendle, Fl. Jamaica 4: 295 (1920); Gagnep., Fl. Indo-Chine 5: 378 (1926); Pax & K. Hoffm., Nat. Pflanzenfam. ed. 2, 19C: 120 (1931); Robyns, Fl. Parc Nat. Alb. 1: 459 (1948); Alain, Fl. Cuba 3: 94 (1953); Hurus., J. Fac. Sci. Univ. Tokyo, Sect. 3, Bot. 6: 302 (1954); Keay (ed.), Fl. W. Trop. Afr. ed. 2, 1: 402 (1958); Backer & Bakh. f., Fl. Java 1: 485 (1963); Jabl., Mem. New York Bot. Gard. 17: 137 (1967); G. L. Webster, Ann. Missouri Bot. Gard. 54: 279 (1968), and 75: 1100 (1989); J.-G. Adam, Mém. Mus. Natl. Hist. Nat., B, Bot. 20: 450 (1971); Whitmore, Tree Fl. Mal. 2: 53 (1973); Dyer, Gen. S. Afr. Fl. Pl., ed. 3, 317 (1975); Airy Shaw, Kew Bull. Addit. Ser. 4: 28 (1975); Berhaut, Fl. Ill. Sénég. 3: 367 (1975); Philcox, Fl. Trin. Tob. II (X): 661 (1979); Airy Shaw, Kew Bull. Addit. Ser. 8: 25 (1980) and Kew Bull. 36: 249 (1981); Thin, Tâp Chi Sinh Vât Hoc 6(3): 26 (1984); Alain, Fl. Español. 4: 74 (1986); Grierson & D. G. Long, Fl. Bhut. 1: 798 (1987); Radcl.-Sm., Fl. Trop. E. Afr., Euph. 1: 251 (1987); Alain, Descr. Fl. Puerto Rico 2: 350 (1988); G. L. Webster, Ann. Missouri Bot. Gard. 81: 81 (1994); Murillo-Aldana & Franco-Rosselli, Euf. Reg. Ararac.: 38 (1995); Radcl.-Sm., Fl. Zamb. 9(4): 151 (1996). T.: *A. latifolia* Sw.

Cladodes Lour., Fl. Cochinch.: 574 (1790); A. Juss., Euphorb. Gen.: 61 (1824). T.: *Cl. rugosa* Lour. [=*A. rugosa* (Lour.) Müll. Arg.]

Hermesia Humb. & Bonpl. ex Willd., Sp. Pl. 4: 809 (1805). T.: *H. castanifolia* Humb. & Bonpl. ex Willd. [=*A. castanifolia* (Humb. & Bonpl. ex Willd.) A. Juss.]

Schousboea Schumach., Beskr. Guin. Pl.: 449 (1827), *non* Willd. (1799). T.: *Sch. cordifolia* Schumach. [=*A. cordifolia* (Schumach.) Müll. Arg.]

Stipellaria Benth., Hook. J. Bot. Kew Gard. Misc. 6: 2 (1854); Baill., Étude Euphorb.: 449 (1858). LT.: *St. trewioides* Benth. [=*A. trewioides* (Benth.) Müll. Arg.] (designated by Thin, 1984).

Lepidoturus Bojer ex Baill., Étude Euphorb.: 448 (1858); Benth., Hooker's Icon. Pl. 13: t. 1297 (1879); Prain in Fl. Trop. Afr. 6(1): 913 (1912). T.: *L. alnifolius* Bojer ex Baill. [=*A. alnifolia* (Bojer ex Baill.) Pax & K. Hoffm.]

Bleekeria Miq., Fl. Ned. Ind. 1(2): 407 (1859), *non* Hassk. (1855). T.: *B. zollingeri* (Hassk.) Miq. (B.: *A. zollingeri* Hassk.) [=*A. villosa* (Benth.) Müll. Arg. (B.: *Stipellaria villosa* Benth.)]

Dioecious or more rarely monoecious shrubs or trees. Indumentum simple or stellate. Leaves alternate, petiolate, stipulate, simple, crenate or serrate, often remotely so, palmi- or penninerved, sometimes glandular beneath at the base, sometimes stipellate. Inflorescences terminal, axillary or cauliflorous, spicate to paniculate, solitary or fasciculate, unisexual, the males interruptedly many-flowered, the females lax, fewer-flowered; male bracts several-flowered, female 1-flowered. Male flowers subsessile or shortly pedicellate; calyx closed in bud, ± globose, later valvately 2–5-partite; petals 0; disc 0; stamens (7)8, filaments free or ± connate into a small disc at the base, anthers dorsifixed, introrse, thecae ± parallel or slightly divergent and partially free at the base, oblong, longitudinally dehiscent, connective not produced; pollen oblate-spheroidal, 3-colporate, colpi narrow, operculate, costae +, colpus transversalis costate, tectate, psilate; pistillode minute, columnar, 2–3-lobed, or 0. Female flowers sessile to shortly pedicellate; sepals (3)4(6), imbricate; petals 0; disc 0; ovary (1)2–3(4)-locular, ovule 1 per locule; styles (1)2–3(4), free or slightly connate at the base, usually simple, linear, very rarely 2-fid at the apex. Fruit 2–4-lobate or subglobose, smooth or verrucose, dehiscing septicidally into 2–4 2-valved cocci; endocarp crustaceous, columella persistent. Seeds subglobose, ecarunculate or subecarunculate; exotesta sometimes somewhat fleshy; endotesta crustaceous; endosperm fleshy; cotyledons broad, flat. Fig.: See Euphorb. Gen.: t. 13, f. 41.1–6 (1824) (as *Siphonia*); Étude Euphorb.: t. 20, f. 8–12 (1858); Fl. Bras. 11(2): t. 56 (1874); Hooker's Icon. Pl. 13: t. 1297 (1879) (as *Lepidoturus*); Pflanzenr. 63: 224–246, tt. 33–37 (1914); Fl. Jamaica 4: 296, t. 97 (1920); Fl. Indo-Chine 5: 377, t. 43.4–10 and 383, t. 44.1 (1926); Notul. Syst. (Paris) 9: 178, t. 4.14–20 (1941); Fl. Cuba 3: 94, t. 26 (1953); Ann. Missouri Bot. Gard. 54: 280, t. 12 (1968); Mém. Mus. Natl. Hist. Nat., B, Bot. 20: 452–4, tt. 155–7 (1971); Fl. Ill. Sénég. 3: 366, 370 (1975); Kew Bull. 36: 250, t. 1B1-4 (1981); Fl. Español. 4: 77, t. 116-3 (1986); Fl. Bhut. 1: 801, t. 49f–h (1987); Fl. Trop. E. Afr., Euph. 1: 254, t. 50 and 258, t. 51 (1987); Descr. Fl. Puerto Rico 2: 353, t. 59-3 (1988); Árvores Brasileiras: 97, 98 (1992); Euf. Reg. Ararac.: 42, t. 2 (1995); Fl. Zamb. 9(4): 155, t. 31 (1996).

A pantropical genus of c. 60 species disposed in 3 well-marked sections depending on the type of indumentum and the presence or absence of stipels (Pax & Hoffmann, 1914, 1931).

153. **Caelebogyne** *J. Sm.*, Proc. Linn. Soc. London 1: 41 (1839) and Trans. Linn. Soc. London 18: 512 (1841); Baill., Étude Euphorb.: 416 (1858) (as *Coelebogyne*); Pax & K. Hoffm., Pflanzenr. 63: 255 (1914), and Nat. Pflanzenfam., ed. 2, 19C: 121 (1931); G. L. Webster, Ann. Missouri Bot. Gard. 81: 82 (1994) (as *Coelebogyne*). T.: *C. ilicifolia* J. Sm.

Alchornea sect. *Caelebogyne* (J. Sm.) Müll. Arg., Linnaea 34: 170 (1865), and in DC., Prodr. 15(2): 906 (1866); Benth., Gen. Pl. 3: 315 (1880); Pax, Pflanzenfam. 3(5): 56 (1890).

Very like 152. *Alchornea* Sw., but indumentum always simple, leaves penninerved, the lobes often spine-tipped, male calyx 4-partite, stamens usually 8, free, female sepals glandular at the base, ovary 3-locular, styles usually sessile, accumbent, cuneate-obovate, slightly emarginate, stigmatiform or dilated and stigmas smooth. Fig.: See Étude Euphorb.: t. 8, f. 32–36 (1858); Pflanzenr. 63: 255, t. 38 (1914); Nat. Pflanzenfam., ed. 2, 19C: 121, t. 62 (1931).

A ditypic genus endemic to Australia (Queensland and NSW).

154. **Aparisthmium** *Endl.*, Gen. Pl.: 1112 (1840), nom. cons.; Baill., Étude Euphorb.: 467 (1858); Pax & K. Hoffm., Pflanzenr. 63: 257 (1914), and Nat. Pflanzenfam., ed. 2, 19C: 122 (1931); Jabl., Mem. New York Bot. Gard. 17: 135

(1967); A. H. Gentry, Woody Pl. NW. S. Amer.: 413 (1993); G. L. Webster, Ann. Missouri Bot. Gard. 81: 82 (1994); Murillo-Aldana & Franco-Rosselli, Euf. Reg. Ararac.: 47 (1995). T.: *A. cordatum* (A. Juss.) Baill.

Conceveibum A. Rich ex A. Juss., Euphorb. Gen.: 42 (1824). T.: *C. cordatum* A. Juss. [=*A. cordatum* (A. Juss.) Baill.]
Alchornea sect. *Aparisthmium* (Endl.) Müll. Arg., Linnaea 34: 168 (1865), in DC., Prodr. 15(2): 900 (1866), and in Mart., Fl. Bras. 11(2): 375 (1874); Benth., Gen. Pl. 3: 315 (1880).

Very like 152. *Alchornea* Sw., but indumentum always simple, stamens 4, anthers mostly extrorse and styles dilated-2-lobate. Fig.: See Euphorb. Gen.: t. 13, f. 42A.1–6 (1824) (as *Conceveibum*); Étude Euphorb.: t. 21, f. 11 (1858); Pflanzenr. 63: 259, t. 40 (1914); Woody Pl. NW. S. Amer.: 414, t. 118.6 (1993); Euf. Reg. Ararac.: 48, t. 4 (1995).

A monotypic S. American genus ranging from Colombia to Bolivia and S. Brazil.

155. **Bocquillonia** *Baill.*, Adansonia I. 2: 225 (1861); Müll. Arg. in DC., Prodr. 15(2): 894 (1866); Benth., Gen. Pl. 3: 313 (1880); Pax & K. Hoffm., Pflanzenr. 63: 260 (1914), and Nat. Pflanzenfam. ed. 2, 19C: 122 (1931); Airy Shaw, Kew Bull. 29: 321 (1974); McPherson & Tirel, Fl. Nouv.-Caléd. 14(1): 114 (1987); G. L. Webster, Ann. Missouri Bot. Gard. 81: 82 (1994). T.: *B. sessiliflora* Baill. (designated by McPherson & Tirel, 1987).

Ramelia Baill., Adansonia I. 11: 132 (1874). T.: *R. codonostylis* Baill. [=*B. codonostylis* (Baill.) Airy Shaw]

Dioecious or rarely monoecious trees or shrubs. Leaves alternate, often shortly petiolate, stipulate, simple, entire, sinuate or dentate, penninerved, coriaceous, often gland-dotted beneath. Inflorescences generally cauliflorous, the males interrupted-spicate, -racemose or -paniculate, the numerous flowers often pulvinately congested into globose or oblong bracteate heads, the females subspicate or more often subcapitately cymose, fewer-flowered, very shortly pedunculate. Male flowers minute; calyx thin, closed in bud, ovoid, later valvately 2–3-partite; petals 0; disc 0; stamens 2–3(4), filaments shortly connate at the base, free above, anthers subextrorse, ovoid, subapicifixed, thecae distinct, parallel, longitudinally dehiscent; connective small; pollen ± as in 152. *Alchornea*; pistillode 0 or (*fide* Baillon & Müller) very minute; receptacle eglandular. Female flowers: calyx short, 4–5-partite, segments imbricate; petals 0; disc 0; ovary 3-locular, ovule 1 per locule; styles 3, sessile, flattened, fleshy, broad, subpetaloid-stigmatiform, plicate, papillose, closely adpressed to the top of the ovary. Fruit 3-lobate to subglobose, smooth, dehiscing septicidally into 3 2-valved cocci; endocarp woody; columella persistent. Seeds oblong, minutely carunculate; testa thinly crustaceous; endosperm fleshy; cotyledons broad, flat. Fig.: See Fl. Nouv.-Caléd. 14(1): 119–141, tt. 23–29 (1987).

A genus of 14 species endemic to New Caledonia (McPherson & Tirel, 1987).

Subtribe 32b. **Conceveibinae** *G. L. Webster*, Taxon 24: 597 (1975), and Ann. Missouri Bot. Gard. 81(1): 82 (1994). T.: *Conceveiba* Aubl.

Leaves long-petiolate, exstipellate; indumentum stellate, at least partly; male inflorescences terminal or axillary; stamens 15 or more; female sepals often glandular; ovary 2–3-locular; styles 2-fid.

This predominantly Neotropical subtribe was included by Pax & Hoffmann (1914) in their series *Trewiiformes*, an otherwise Palaeotropical group.

Key to the genera

1. Inflorescences axillary; anthers muticous · · · · · · · · · · · · · 158. *Polyandra*
1. Inflorescences terminal. · 2
2. Female sepals separate, often glandular; anthers muticous; ovary 3-locular; style-arms blunt · 156. *Conceveiba*
2. Female sepals connate, eglandular, lobes scarcely evident; anthers apiculate; ovary 2-locular; style-arms acute · · · · · · · · · · 157. *Gavarretia*

156. **Conceveiba** *Aubl.*, Hist. Pl. Guiane: 923 (1775); Baill., Étude Euphorb.: 414 (1858); Müll. Arg., Linnaea 34: 166 (1865), in DC., Prodr. 15(2): 895 (1866), and in Mart., Fl. Bras. 11(2): 370 (1874); Benth., Gen. Pl. 3: 316 (1880); Pax & K. Hoffm., Pflanzenr. 63: 214 (1914), and Nat. Pflanzenfam., ed. 2, 19C: 118 (1931); Jabl., Mem. New York Bot. Gard. 17: 131 (1967); D. W. Thomas, Ann. Missouri Bot. Gard. 77: 856 (1990); A. H. Gentry, Woody Pl. NW. S. Amer.: 413 (1993); G. L. Webster, Ann. Missouri Bot. Gard. 81: 82 (1994); Breteler & Mennega, Bull. Jard. Bot. Belg. 63: 209 (Nov. 1994); Murillo-Aldana & Franco-Rosselli, Euf. Reg. Ararac.: 57 (1995); Murillo-Aldana, Caldasia 18: 239 (1996). T.: *C. guianensis* Aubl.

Conceveiba § *Veconcibea* Müll. Arg., Linnaea 34: 167 (1865), in DC., Prodr. 15(2): 897 (1866), and in Mart., Fl. Bras. 11(2): 373 (1874).

Alchornea § *Conceveibastrum* Müll. Arg. in Mart., Fl. Bras. 11(2): 375 (1874).

Conceveibastrum (Müll. Arg.) Pax & K. Hoffm., Pflanzenr. 63: 217 (1914), and Nat. Pflanzenfam., ed. 2, 19C: 119 (1931); A. H. Gentry, Woody Pl. NW. S. Amer.: 415 (1993). T.: *C. martianum* (Baill.) Pax & K. Hoffm. [=*Conceveiba martiana* Baill.]

Veconcibea (Müll. Arg.) Pax & K. Hoffm., Pflanzenr. 63: 218 (1914), and Nat. Pflanzenfam., ed. 2, 19C: 119 (1931). T.: *V. latifolia* (Benth.) Pax & K. Hoffm. [=*C. latifolia* Benth.]

Dioecious trees, soon glabrescent. Indumentum, at least in part, stellate. Leaves alternate, petiolate, stipulate, simple, entire, subentire or remotely dentate, sometimes cordate, penninerved, reticulate-venose, coriaceous or chartaceous, glandular at the base and on the margins; stipules sometimes large (§ *Conceveibastrum*). Inflorescences terminal, tomentellous, the males usually paniculate, ample, the females smaller, simpler, fewer-flowered or rarely only 1-flowered; rhachides thick, sometimes very robust (§ *Conceveibastrum*); bracts with 2 large glands at the base, the males subtending glomerules of small flowers, the females 1-flowered. Male flowers shortly pedicellate to subsessile; calyx closed in bud, ovoid, later valvately 3–4-partite; petals 0; disc 0; stamens c. 16, free, the outer 6–8 fertile, short, filaments broad, the inner 6–8 sterile, filaments elongate, flexuous, or stamens numerous and all fertile (§ *Veconcibea*), anthers erect, latrointrorse, thecae distinct, usually separated by a wide connective, muticous, longitudinally dehiscent; pollen ± as in 152. *Alchornea*; pistillode 0; receptacle convex. Female flowers shortly pedicellate; pedicels thick, fleshy; sepals (4)5–8(12), rigid, imbricate, with as many large external glands alternating with and adnate to them; ovary 3-locular, ovule 1 per locule; styles 3, very shortly connate at the base, 2-lobate, the lobes obtuse, thick, stigmas papillose. Fruit large, subglobose, 3-gonous, 3-costate or 3-lobate, smooth to echinate-rugulose, dehiscing septicidally into 3 2-valved cocci; endocarp woody. Seeds large, ellipsoid, usually carunculate; testa smooth, crustaceous; endosperm fleshy; cotyledons broad, flat. Fig.: See Euphorb. Gen.: t. 13, f. 42B.1–5 (1824); Étude Euphorb.: t. 21, f. 12–13 (1858); Fl. Bras. 11(2): t. 55 (1874); Pflanzenr. 63: 215, t. 32 (1914); Ann. Missouri Bot. Gard. 77: 857, t. 1 (1990); Woody Pl. NW. S. Amer.: 414, t. 118.3, 5 (1993); Bull. Jard. Bot. Belg. 63: 211, t. 1 (Nov. 1994); Euf. Reg. Ararac.: 59–60, tt. 8, 9 (1995); Caldasia 18: 244, t. 1 (1996).

A genus of c. 12 species, of which c. 10 Neotropical, from Costa Rica to Amazonian Peru and Brazil, and 2 W. African, recently described from Cameroun and Gabon (Thomas, 1990; Breteler & Mennega, 1994).

157. **Gavarretia** *Baill.*, Adansonia I. 1: 185, t. 7 (1861); Benth., Gen. Pl. 3: 316 (1880); Pax & K. Hoffm., Pflanzenr. 63: 213 (1914), and Nat. Pflanzenfam., ed. 2, 19C: 118 (1931); Jabl., Mem. New York Bot. Gard. 17: 130 (1967); A. H. Gentry, Woody Pl. NW. S. Amer.: 411 (1993); G. L. Webster, Ann. Missouri Bot. Gard. 81: 83 (1994); Murillo-Aldana & Franco-Rosselli, Euf. Reg. Ararac.: 75 (1995). T.: *G. terminalis* Baill.

Conceveiba § *Gavarretia* Müll. Arg. in Mart., Fl. Bras. 11(2): 372 (1874).

Like 156. *Conceveiba* Aubl., but sometimes with axillary inflorescences, with the anthers apiculate, the female sepals connate, eglandular, and the lobes scarcely evident, the ovary 2-locular and the style-arms acute. Fig.: See Adansonia I. 1: t. VII.3–4' (1861); Woody Pl. NW. S. Amer.: 412, t. 117.4 (1993); Euf. Reg. Ararac.: 77, t. 16 (1995). See also Fig. 24, p. 198.

A monotypic S. American genus from Amazonian Colombia, Peru, Venezuela, Guiana and Brazil.

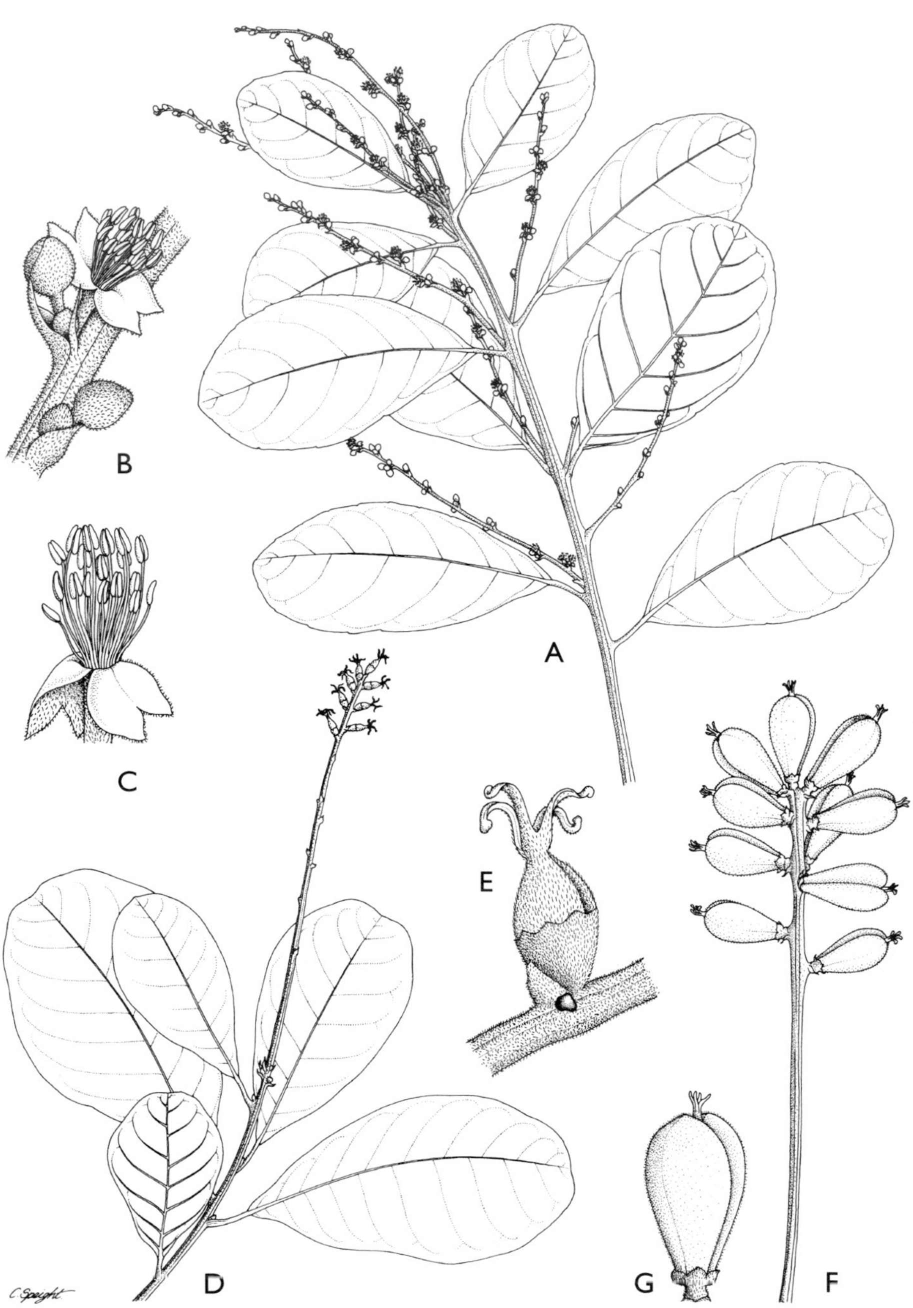

FIG. 24. *Gavarretia terminalis* Baill. **A** Habit, male × $^2/_3$; **B** Portion of male inflorescence × 6; **C** Male flower × 8; **D** Habit, female × $^2/_3$; **E** Female flower × 8; **F** Young infructescence × 1; **G** Young fruit × 2. **A–C** from *Ducke* 53; **D, E** from *Spruce* 3087; **F, G** from *Morawetz & Wallnöfer* 27-18988. Drawn by Camilla Speight.

158. **Polyandra** *Leal,* Arch. Jard. Bot. Rio de Janeiro, 11: 63 (1961); G. L. Webster, Ann. Missouri Bot. Gard. 81: 83 (1994). T.: *P. bracteosa* Leal.

Like 156. *Conceveiba* Aubl., but with axillary pseudo-racemose male inflorescences. The female inflorescences and fruits are still unknown. Fig.: See Arch. Jard. Bot. Rio de Janeiro, 11: 64, tt. 1 and 2.1 (1961).

A monotypic S. American genus from Amazonian Brazil.

When more material is forthcoming, it may well show that this genus is referable to *Conceveiba* § *Veconcibea,* although Webster & Huft (1988) suggest that it may be synonymous with *Cleidion prealtum* Croizat.

Tribe 33. **ACALYPHEAE** *Dumort.,* Anal. Fam. Pl.: 45(1829); G. L. Webster, Ann. Missouri Bot. Gard. 81: 83 (1994). T.: *Acalypha* L.

Monoecious or dioecious herbs, shrubs or trees; indumentum simple, stellate, lepidote or 0; leaves alternate or opposite, stipulate, stipellate or not, simple, entire or dentate, penninerved or palminerved, eglandular, with embedded laminar glands and/or granulose-glandular; stipules free or connate. Inflorescences terminal, oppositifolious or axillary, unisexual or bisexual; bracts sometimes glandular, female sometimes large and foliaceous; male calyx closed in bud, splitting into 2–5 valvate segments, sepals free or connate; petals 0; disc 0 or +, interstaminal or receptacular, entire or dissected; stamens 2–1000+; filaments free or connate into fascicles, not inflexed in bud; anthers muticous or apiculate, 2–4-thecate, thecae separated to the connective or not, sometimes vermiform, erect or pendulous, connective enlarged or not; pollen globose, inaperturate, brevicolporate or porate, colpi usually inoperculate, finely perforate-tectate to striate or rugulose; pistillode usually 0; female sepals 3–6, free, rarely accrescent or 0; disc +, minute or 0; ovary (1)2–4(9)-locular; styles free or basally connate, simple, 2-fid or lacerate. Fruit smooth, tuberculate, muricate-setose, echinate, horned or winged, mostly dehiscent. Seeds carunculate or ecarunculate, testa dry or fleshy; cotyledons mostly broader than the radicle.

The largest tribe of the *Acalyphoideae,* according to Webster (1975, 1994), with 11 subtribes, 30 genera and over 1,000 species.

Key to the subtribes

1. Indumentum stellate or 0; inflorescences terminal; styles 2-fid; seeds carunculate · 2
1. Indumentum simple or stellate; seeds ecarunculate, or, if carunculate, then inflorescences axillary, or styles unlobed · · · · · · · · · · · · · · · · 3
2. Monoecious; stipules connate; filaments connate into fascicles; female sepals valvate, caducous · 33a. *Ricininae*
2. Dioecious; stipules free; filaments free; female sepals imbricate, 2-seriate, persistent · 33b. *Adrianinae*

3. Stamens many, connate into fascicles; shrubs or trees with alternate eglandular leaves; indumentum simple or lepidote · · 33l. *Lasiococcinae*
3. Stamens not connate into fascicles · 4
4. Anther-thecae vermiform, pendulous; male disc 0; pollen porate; female bracts usually large and foliaceous; styles lacerate; seeds subecarunculate · 33k. *Acalyphinae*
4. Anther-thecae not so; pollen colporate; seeds usually ecarunculate · · · 5
5. Indumentum usually stellate; leaves often opposite, with embedded laminar glands and granulose-glandular; usually dioecious; inflorescences often terminal; male disc 0 or +; pollen ± spinulose-rugulose · 33j. *Rottlerinae*
5. Indumentum usually simple; leaves alternate or opposite, eglandular or with embedded laminar glands, not granulose-glandular · · · · · · · · · 6
6. Herbs, often with opposite leaves; indumentum simple; styles undivided · 7
6. Trees or shrubs, or, if herbs, then styles 2-fid · · · · · · · · · · · · · · · · · · 8
7. Male sepals free; stamens free; anthers not apiculate; pollen 3-colporate; ovary 2-locular; cotyledons broader than the radicle · · 33c. *Mercurialinae*
7. Male sepals connate; filaments connate; anthers apiculate; pollen inaperturate; ovary 3-locular; cotyledons scarcely broader than the radicle · 33d. *Dysopsidinae*
8. Anther-thecae not separated to the connective; male disc 0; leaves mostly with laminar glands · 9
8. Anther-thecae separated to the connective, erect or pendulous; interstaminal disc +; styles unlobed · 10
9. Anthers 2-thecate; pollen vermiculate-rugulose; styles 2-fid; testa dry or fleshy · 33e. *Cleidiinae*
9. Anthers 3–4-thecate; pollen spinulose-rugulose; styles simple; testa usually fleshy · 33f. *Macaranginae*
10. Seed-exotesta not fleshy · 33i. *Mareyinae*
10. Seed-exotesta fleshy · 11
11. Indumentum simple · 33g. *Claoxylinae*
11. Indumentum stellate · 33h. *Lobaniliinae*

Subtribe 33a. **Ricininae** *Griseb.*, Fl. Brit. W. I.: 37 (1859), and Abh. Königl. Ges. Wiss. Göttingen 9: 15 (1861); Müll. Arg., Linnaea 34: 143 (1865), in DC., Prodr. 15(2): 1016 (1866), and in Mart., Fl. Bras. 11(2): 419–420 (1874); Pax & K. Hoffm., Pflanzenr. 68: 112 (1919), and Nat. Pflanzenfam. ed. 2, 19C: 149 (1931); G. L. Webster, Taxon 24: 597 (1975), and Ann. Missouri Bot. Gard. 81(1): 84 (1994). T.: *Ricinus* L.

Monoecious annual or perennial large herb or shrub; indumentum 0; leaves alternate, long-petiolate, palmatilobate; stipules connate into a sheath, deciduous. Inflorescences leaf-opposed or subterminal, paniculate, proximal cymules male, distal female; bracts glandular; stamens up to c. 1000, filaments united into branching fascicles; pollen 3-colporate, colpi very narrow, inoperculate; pistillode 0; ovary usually muricate; styles 2-partite; Fruit mostly echinate. Seeds carunculate.

As defined by Webster (1975, 1994), this subtribe is monogeneric. Pax & Hoffmann (1919, 1931) also included *Homonoia* and *Lasiococca* on account of androecial similarities. However, *Ricinus* differs from these in its palmately-lobed leaves with glandular petioles, terminal inflorescences, 2-fid styles and carunculate seeds. The androecial similarities are therefore deemed to be of no significance. *Adriana* appears to be its closest relative, as indicated here.

159. **Ricinus** *L.*, Sp. Pl.: 1007 (1753), and Gen. Pl. ed. 5: 437 (1754); A. Juss., Euphorb. Gen.: 36 (1824); Baill., Étude Euphorb.: 289 (1858); Müll. Arg. in DC., Prodr. 15(2): 1016 (1866), and in Mart., Fl. Bras. 11(2): 419–20 (1874); Benth., Gen. Pl. 3: 321 (1880); Hook. f., Fl. Brit. India 5: 457 (1887); Pax & K. Hoffm., Pflanzenr. 68: 119 (1919); Fawc. & Rendle, Fl. Jamaica 4: 306 (1920); Gagnep., Fl. Indo-Chine 5: 327 (1925); Nat. Pflanzenfam., ed. 2, 19C: 149 (1931); Schischk. in Fl. URSS XIV: 300 (1949); Alain, Fl. Cuba 3: 107 (1953); Vindt, Monogr. Euph. Maroc: 21 (1953); de Carvalho, Anais 11(4,1): 11 (1956); Backer & Bakh. f., Fl. Java 1: 492 (1963); Rech. f. & Schiman-Czeika, Fl. Ir. 6: 8 (1964); G. L. Webster, J. Arnold Arbor. 48: 379 (1967), and Ann. Missouri Bot. Gard. 54: 298 (1968); Tutin, Fl. Eur. 2: 212 (1968); Berhaut, Fl. Ill. Séneg. 3: 587 (1975); Radcl.-Sm. in Fl. Iraq 4 (1): 325 (1980), and Fl. Turk. 7: 570 (1982); Scott, Fl. Masc. 160. Euph.: 80 (1982); Correll, Fl. Bah. Arch.: 840 (1982); Radcl.-Sm. in Fl. Cyp. 2: 1452 (1985); Proctor, Kew Bull. Addit. Ser. 11: 541 (1984); Alain, Fl. Español. 4: 208 (1986); Radcl.-Sm. in Fl. Pak. 172: 69 (1986), and Fl. Trop. E. Afr., Euph. 1: 322 (1987); Grierson & D. G. Long, Fl. Bhut. 1: 808 (1987); Alain, Descr. Fl. Puerto Rico 2: 423 (1988); R. A. Howard, Fl. Lesser Antilles 5: 83 (1989); G. L. Webster, Ann. Missouri Bot. Gard. 81: 84 (1994); Radcl.-Sm., Fl. Zamb. 9(4): 156 (1996); Gillespie in Acev.-Rodr., ed., Fl. St. John: 220 (1996); Philcox in Dassanayake, ed., Fl. Ceyl. XI: 174 (1997); van Welzen, Blumea 43: 151 (1998). T.: *R. communis* L.

Cataputia Ludw., Gen.: 81 (1760). T.: *C. minor* Ludw.

Monoecious, glabrous, often glaucous annual or perennial herb or shrub, often tree-like in its dimensions. Leaves alternate, long-petiolate, stipulate, peltate, palmatilobate, the lobes 7–more, glandular-serrate; petioles glanduliferous at apex and base; stipules connate into a sheath, deciduous, leaving a scar. Inflorescences leaf-opposed or subterminal, paniculate, lower cymules male, upper female or else all female; bracts glandular, soon caducous. Male flowers: pedicels jointed, 2-bracteolate; buds globose; calyx membranous, closed at first, later splitting into 3–5 valvate lobes; petals 0; disc 0; stamens up to c. 1000, filaments variously united into branching fascicles, anthers basifixed, thecae subglobose, distinct, divaricate, longitudinally dehiscent; pollen spheroidal, 3-colporate, colpi very narrow, inoperculate, costae 0, colpus transversalis narrow, costae 0, tectate, psilate;

pistillode 0. Female flowers: pedicels jointed, elongating in fruit, persistent; buds conical; sepals 5, valvate, soon caducous; petals 0; disc 0; ovary 3-locular, echinate, muricate or smooth, 1 ovule per locule; styles 3, ± free or slightly connate at the base, 2-partite, stigmas papillose-plumose, usually dark red. Fruit 3-lobate, echinate or smooth, the spines accrescent, dehiscing into 3 2-valved cocci; endocarp crustaceous to thinly woody; columella prominent, 3-alate, persistent. Seeds dorsiventrally compressed-ovoid, carunculate; testa crustaceous, smooth, marmorate; endosperm fleshy; cotyledons broad, flat. Fig.: See Étude Euphorb.: t. 10 and 11, f. 1–5 (1858); Fl. Bras. 11(2): t. 60 (1874); Fl. Jamaica 4: 307, t. 101 (1920); Fl. Cuba 3: 107, t. 32 (1953); Monogr. Euph. Maroc: 20, t. 8 and 22, t. 9 (1953); Anais 11(4,1): ff. 1–60 and tt. 1–5 (1956); J. Arnold Arbor. 48: 380, t. 4 (1967); Amer. J. Bot. 56: 756, t. 18A–M (1969); Fl. Ill. Sénég. 3: 586 (1975); Fl. Masc. 160. Euph.: 81, t. 16 (1982); Fl. Bah. Arch.: 841, t. 349 (1982); Kew Bull. Addit. Ser. 11: 542, t. 165 (1984); Fl. Español. 4: 209, t. 116-32 (1986); Fl. Pak. 172: 70, t. 14 (1986); Fl. Bhut. 1: 807, t. 50g–j (1987); Fl. Lesser Antilles 5: 84, t. 29 (1989); Fl. Zamb. 9(4): 158, t. 32 (1996); Fl. St. John: 221, t. 100A–G (1996).

A monotypic genus native to N.E. Africa, *fide* Radcliffe-Smith (1984), now widely cultivated and ruderal throughout the tropics, subtropics and warm temperate regions (See Frontispiece).

The seeds yield Castor Oil, used medicinally and as a lubricant.

Subtribe 33b. **Adrianinae** *Benth.*, Gen. Pl. 3: 250 (1880); G. L. Webster, Taxon 24: 597 (1975), and Ann. Missouri Bot. Gard. 81(1): 84 (1994). T.: *Adriana* Gaudich.

Adrianeae (Benth.) Pax & K. Hoffm., Pflanzenr. 44: 1 (1910).

Dioecious; indumentum stellate or 0; leaves alternate or opposite, simple or 3(5)-lobate; stipules glandular. Inflorescences terminal or leaf-opposed, spicate; bracts glandular; stamens numerous, free, anthers apiculate, connective liguliform; ovary 3-locular, smooth or ± muricate; styles 2-partite. Fruit dehiscent. Seeds carunculate.

Webster (1975, 1994) upholds Bentham's subtribe, but as a monogeneric taxon, having some features in common with the preceding. His treatment is followed here. Pax (1910) had followed Bentham (1880) in associating with it such other genera as *Manihot* and *Cephalocroton,* with the addition of yet others such as *Symphyllia, Cephalomappa* and *Pachystroma,* but in 1931 he had abandoned it and included *Adriana* in his subtribe Mercurialinae.

160. **Adriana** *Gaudich.*, Ann. Sci. Nat. (Paris) 5: 223 (1825); Baill., Étude Euphorb.: 405 (1858); Müll. Arg. in DC., Prodr. 15(2): 889 (1866); Benth., Fl. Austral. 6: 133 (1873), and Gen. Pl. 3: 306 (1880); Pax & K. Hoffm., Pflanzenr. 44: 17 (1910), and Nat. Pflanzenfam., ed. 2, 19C: 118 (1931); Airy Shaw, Kew

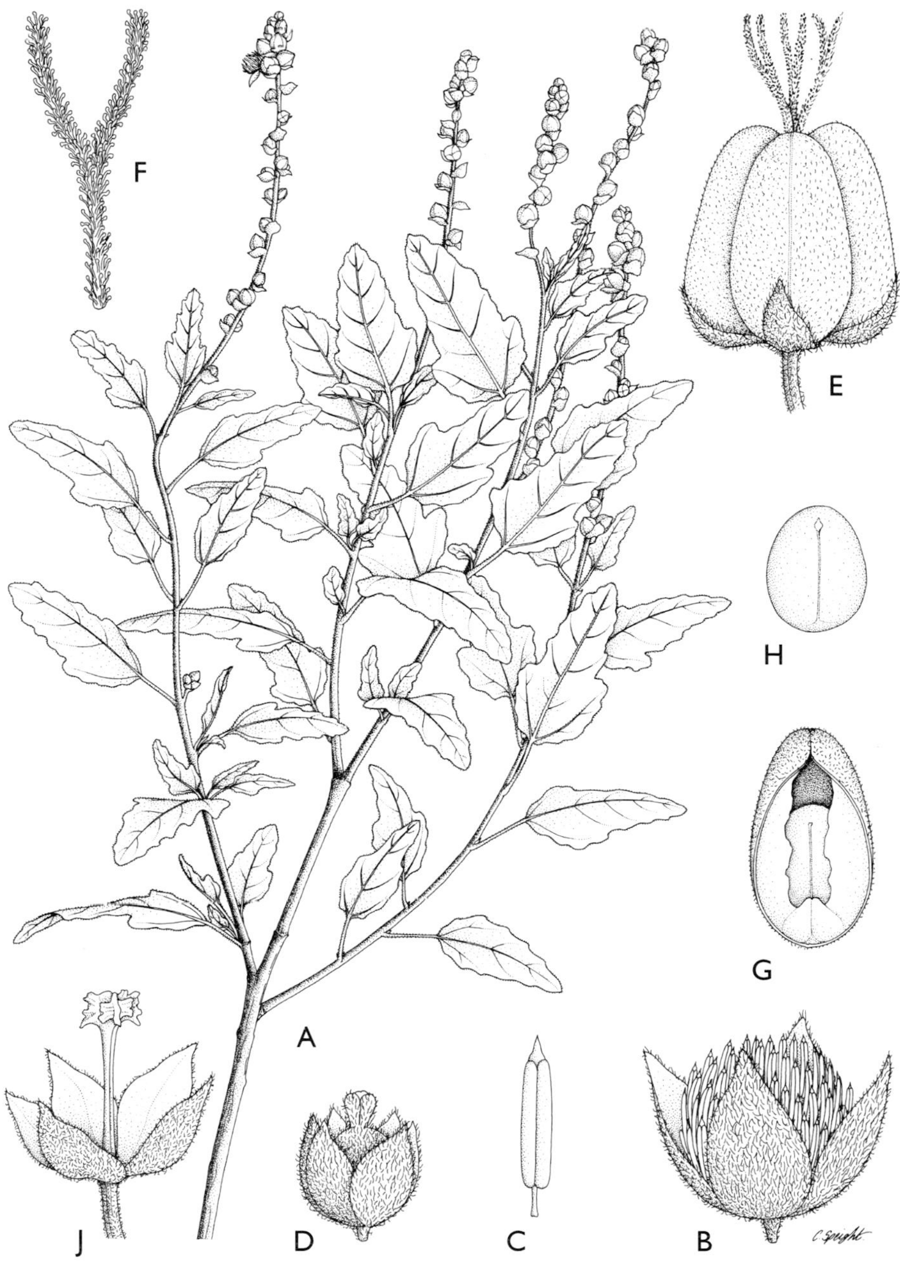

FIG. 25. *Adriana hookeri* (F. Muell.) Müll. Arg. (B.: *Trachycaryon hookeri* F. Muell.). **A** Habit, male × 1; **B** Male flower × 7.5; **C** Stamen × 18; **D** Female flower × 7.5; **E** Fruit × 4.5; **F** Style- arm × 10.5; **G** Coccus × 4.5; **H** Seed × 4.5; **J** Columella × 4.5; **A–C** from *Lazarides* 6105; **D–F** from *Lazarides* & *Palmer* 236; **G–J** from *Hicks* 189. Drawn by Camilla Speight.

Bull. 35: 589 (1980); G. L. Webster, Ann. Missouri Bot. Gard. 81: 84 (1994). T.: *A. tomentosa* Gaudich. [lectotype, chosen by Webster, 1994]

Meialisa Raf., Sylva Tellur. : 63 (1838). T.: *M. australis* Raf. [= *A. quadripartita* (Labill.) Gaudich. (B.: *Croton quadripartitus* Labill.)]
Trachycaryon Klotzsch in Lehm., Pl. Preiss. 1: 175 (1844). T.: *T. labillardieri* Klotzsch [= *A. quadripartita* (Labill.) Gaudich. (B.: *Croton quadripartitus* Labill.)]

Dioecious erect shrubs. Indumentum stellate or 0. Leaves alternate or opposite, petiolate or subsessile, stipulate, simple or 3(5)-lobate, mostly coarsely toothed, palminerved; stipules small, free, variable, glandular, discoid or shortly cylindric and umbilicate at the apex as though hydathodal. Inflorescences terminal or leaf-opposed, pseudospicate-narrowly thyrsiform, unisexual, male elongate, interrupted, bracts glandular, subtending a 3–5-flowered centrifugal cymule, female short, dense, few-flowered, sessile or shortly pedunculate. Male flowers subsessile or very shortly pedicellate; calyx closed in bud, globose, later splitting into 4–5 valvate lobes; petals 0; disc 0; stamens numerous, densely congested, filaments free, very short, anthers basifixed, erect, thecae parallel, curvilinear, longitudinally dehiscent, connective produced into a short, liguliform, cellular-papillose, amber-coloured, glistening appendage; pollen oblate-spheroidal, 3–4-colporate, colpi narrow, costae +, colpus transversalis small, costae +, tectate, psilate, intra-reticulate; pistillode 0; receptacle convex or subglobose. Female flowers shortly pedicellate; sepals (5)6–8, 2-seriate, imbricate, persistent; petals 0; disc 0; ovary 3-locular, ± muricate, 1 ovule per locule; styles 3, ± free or slightly connate at the base, 2-partite, stigmas coarsely papillose-plumose, crimson. Fruit 3-coccous, dehiscing into 3 2-valved cocci; epicarp scurfily pubescent; endocarp thinly woody; columella narrow, 3-gonous at the apex, persistent. Seeds ovoid, shortly or minutely carunculate, caruncle readily caducous; testa smooth, somewhat shiny, marmorate-mottled; endosperm fleshy; cotyledons broad, flat. Fig.: See Étude Euphorb.: t. 2, f. 19–22 (as *Trachycarion*), and t. 18, f. 12 (1858); Pflanzenr. 44: 19, t. 6 and 21, t. 7 (1910). See also Fig. 25, p. 203.

An endemic Australian genus of 5 species, *fide* Airy-Shaw (1980c).

Subtribe 33c. **Mercurialinae** *Pax*, Nat. Pflanzenfam. ed. 1, 3(5): 46 (1890); G. L. Webster, Ann. Missouri Bot. Gard. 81(1): 84 (1994). T.: *Mercurialis* L.

Acalypheae subtribe *Mercurialinae* series *Mercurialiiformes* Pax & K. Hoffm., Pflanzenr. 63: 270 (1914), and Nat. Pflanzenfam. ed. 2, 19C: 125 (1931).

Dioecious or monoecious herbs; indumentum simple; leaves opposite or rarely alternate, penninerved, eglandular. Inflorescences axillary, spicate or glomerulate; stamens 2–12(20), free, anthers extrorse, muticous; pollen 3-colporate, colpi operculate; pistillode 0; ovary 2-locular, smooth; styles 2, simple. Fruit 2-coccous, sometimes setose, dehiscent. Seeds carunculate; cotyledons much broader than the radicle.

With the removal of all the other of Pax's "Series" from his subtribe, together with *Dysopsis*, only three genera are left — one Eurasian and two S. African.

KEY TO THE GENERA

1. Dioecious; stamens 8–20; female flower with 2 subulate disc-segments (? staminodia); seeds caruncular · · · · · · · · · · · · · · · · · · 161. *Mercurialis*
1. Monoecious; stamens 2–7; female disc-segments (? staminodia) 0 or minute; seeds ecarunculate · 2
2. Leaves entire or denticulate; female sepals 3; fruit smooth · · 162. *Seidelia*
2. Leaves crenulate; female sepals nearly 0; fruit setose-muricate · 163. *Leidesia*

161. **Mercurialis** *L.*, Sp. Pl.: 1035 (1753), and Gen. Pl. ed. 5: 457 (1754); A. Juss., Euphorb. Gen.: 46 (1824); Baill., Étude Euphorb.: 488 (1858); Müll. Arg. in DC., Prodr. 15(2): 794 (1866); Benth., Gen. Pl. 3: 309 (1880); Pax & K. Hoffm., Pflanzenr. 63: 271 (1914); W. Zimm. *et al.* in Hegi, Ill. Fl. Mitteleur. 5(1): 126 (1925); Gagnep., Fl. Indo-Chine 5: 349 (1925); Pax & K. Hoffm., Nat. Pflanzenfam., ed. 2, 19C: 125 (1931); Pojark. in Fl. URSS. XIV: 295 (1949); Vindt, Monogr. Euph. Maroc: 13 (1953); Hurus., J. Fac. Sci. Univ. Tokyo, Sect. 3, Bot. 6: 302 (1954); Rech. f. & Schiman-Czeika, Fl. Ir. 6: 8 (1964); G. L. Webster, J. Arnold Arbor. 48: 366 (1967); Tutin, Fl. Eur. 2: 212 (1968); Radcl.-Sm. in Fl. Iraq 4 (1): 322 (1980), and Fl. Turk. 7: 569 (1982); Correll, Fl. Bah. Arch.: 830 (1982); Radcl.-Sm. in Fl. Cyp. 2: 1450 (1985); Alain, Fl. Español. 4: 176 (1986); Grierson & D. G. Long, Fl. Bhut. 1: 796 (1987); G. L. Webster, Ann. Missouri Bot. Gard. 81: 85 (1994). LT.: *M. perennis* L. (designated by Small, 1913).

Dioecious or more rarely monoecious annual or perennial herbs. Indumentum simple. Roots imparting a purple tinge to water. Leaves opposite, petiolate or sessile, stipulate, simple, mostly dentate or crenate, penninerved. Inflorescences axillary, spicate or racemose, mostly unisexual, male elongate, interrupted-glomerulate, pedunculate or sessile, female fasciculate or shortly spiciform, few-flowered, sometimes with male flowers intermixed. Male flowers subsessile or very shortly pedicellate; calyx thinly membranaceous, closed in bud, subglobose, obtuse or acute, later splitting into 3 valvate lobes; petals 0; disc 0; stamens 8–12(20), filaments free, anthers subdorsifixed, extrorse, globose or ovoid, muticous, thecae distinct at the base, divaricate or subpendulous, later ascending, longitudinally dehiscent; pollen subprolate to prolate, 3-colporate, colpi narrow, operculate, operculum very narrow, consisting of only 1 row of columellae, costae +, colpus transversalis small, costae +, tectate, psilate, columellae distinct; pistillode 0; receptacle 0. Female flowers shortly pedicellate; sepals 3, imbricate; petals 0; disc-glands 2, filiform, alternating with the carpidia; ovary 2-locular, 1 ovule per locule; styles 2, free or scarcely connate at the base, short, erect or divaricate, simple, stigmas prominently papillose. Fruit 2-coccous, dehiscing into 2 2-valved cocci; endocarp crustaceous. Seeds ovoid or globose, carunculate; testa smooth or foveolate; endosperm fleshy; cotyledons broad, flat, much broader than the radicle. Fig.: See Euphorb. Gen.: t. 14, f. 47.1–7

(1824); Étude Euphorb.: t. 9, f. 12–29 (1858); Fl. URSS XIV: t. XVII.5 (1949); Monogr. Euph. Maroc: 16, t. 7 (1953); Amer. J. Bot. 56: 750, t. 10A–H (1969); Fl. Iraq 4 (1): 323, t. 60 (1980); Fl. Bah. Arch.: 831, t. 345 (1982); Fl. Español. 4: 179, t. 116-26 (1986); Fl. Bhut. 1: 801, t. 49a–c (1987).

A Eurasian genus of 8 species, mostly Mediterranean, a few C. European, and 1 in E. Temperate Asia; *M. annua* L. naturalized elsewhere.

M. leiocarpa Sieb. & Zucc., a perennial Japanese species, yields a blue dye once used there for printing on cloth.

162. **Seidelia** *Baill.*, Étude Euphorb.: 465 (1858); Benth., Gen. Pl. 3: 310 (1880); Prain, Ann. Bot. (London) 27: 398 (1913); Pax & K. Hoffm., Pflanzenr. 63: 282 (1914); Prain, Fl. Cap. 5(2): 464 (1920); Pax & K. Hoffm., Nat. Pflanzenfam., ed. 2, 19C: 126 (1931); Dyer, Gen. S. Afr. Fl. Pl. ed. 3, 316 (1975); G. L. Webster, Ann. Missouri Bot. Gard. 81: 85 (1994). LT.: *S. mercurialis* Baill. (nom. illeg.) [= *S. triandra* (E. Mey.) Pax (B.: *Mercurialis triandra* E. Mey.) (designated by Pfeiffer, 1874).

Monoecious or more rarely dioecious low annual herbs. Indumentum, where present, simple. Leaves alternate, the lowest opposite, petiolate, stipulate, simple, dentate to subentire, narrow, penninerved. Flowers axillary in upper leaf-axils, fasciculate, male uppermost, lower female, or with male and female intermixed. Male flowers subsessile or very shortly pedicellate; calyx thinly membranaceous, closed in bud, depressed-globose, later splitting into 3 valvate lobes; petals 0; disc 0; stamens 2–5, filaments very shortly connate at the base, anthers apicifixed, latrorse, thecae distinct, deeply grooved before dehiscence, almost 2-globose, apically dehiscent, cruciately 4-valved after dehiscence; pollen prolate-spheroidal, distinctly 3-lobate in polar view, 3-colporate, colpi narrow, exoperculate, costae 0, colpus transversalis small, costae +, tectate, psilate, columellae short; pistillode 0; receptacle 0. Female flowers shortly pedicellate; calyx 3(4)-fid; petals 0; disc-glands 2, minute, alternating with the carpidia; ovary 2-locular, 1 ovule per locule; styles 2, short, simple. Fruit 2-coccous, smooth, thin-walled, dehiscing into 2 2-valved cocci. Seeds ovoid, ecarunculate; testa crustaceous; endosperm fleshy; cotyledons narrow, ovate, flat, scarcely twice as wide as the radicle. Fig.: See Étude Euphorb.: t. 9, f. 7 (1858); Pflanzenr. 63: 283, t. 43 (1914); Nat. Pflanzenfam., ed. 2, 19C: 126, t. 65C (1931).

A S. African genus of 2 species.

163. **Leidesia** *Müll. Arg.* in DC., Prodr. 15(2): 792 (1866); Benth., Hooker's Icon. Pl. 13: 66, t. 1284 (1879), and Gen. Pl. 3: 310 (1880); Prain, Ann. Bot. (London) 27: 399 (1913); Pax & K. Hoffm., Pflanzenr. 63: 284 (1914); Prain, Fl. Cap. 5(2): 462 (1920); Pax & K. Hoffm., Nat. Pflanzenfam., ed. 2, 19C: 126 (1931); Dyer, Gen. S. Afr. Fl. Pl. ed. 3, 316 (1975); G. L. Webster, Ann. Missouri Bot. Gard. 81: 85 (1994); Radcl.-Sm., Fl. Zamb. 9(4): 159, t. 33 (1996). T.: *L. procumbens* (L.) Prain (B.: *Mercurialis procumbens* L.)

Like 162. *Seidelia* Baill., but leaves ovate, crenate, palminerved, bisexual inflorescences racemose, female bracts foliaceous, stamens 4–7, pollen oblate-spheroidal, colpi narrowly operculate, colpus transversalis long, costae 0, female calyx 0 or of 1 bracteiform lobe, disc 0 and fruit setose. Fig.: See Hooker's Icon. Pl. 13: 66, t. 1284 (1879); Pflanzenr. 63: 285, t. 44 (1914); Nat. Pflanzenfam., ed. 2, 19C: 126, t. 65A, B (1931); Fl. Zamb. 9(4): 160, t. 33 (1996).

A monotypic southern African genus in SE Congo (Kinshasa), Zimbabwe, Mozambique, Swaziland and S. Africa.

Subtribe 33d. **Dysopsidinae** *Hurus.*, J. Fac. Sci. Univ. Tokyo, Sect. 3, Bot. 6: 302 (1954); G. L. Webster, Ann. Missouri Bot. Gard. 81: 85 (1994). T.: *Dysopsis* Baill.

Monoecious herb; indumentum simple; leaves alternate, simple, crenate, eglandular; stipules deciduous. Flowers axillary, mostly solitary; male flowers long-pedicellate; calyx gamosepalous; stamens 3 or 6, inner connate at the base, anthers apiculate; pollen globose, porate; pistillode 0; female calyx 3-partite; ovary 3-locular; styles 3, simple, lacerate. Fruit dehiscent. Seeds subecarunculate.

A monogeneric S. American subtribe, chiefly differing from the *Mercurialinae* in the fusion of male floral whorls, the produced anther-connective, the quasi-3-colpate pollen, the 3-locular ovary and the narrow cotyledons.

164. **Dysopsis** *Baill.*, Étude Euphorb.: 435 (1858); Müll. Arg. in DC., Prodr. 15(2): 949 (1866); Benth., Gen. Pl. 3: 264 (1880); Grüning, Pflanzenr. 58: 10 (1913); Pax & K. Hoffm., Pflanzenr. 63: 286 (1914), and Nat. Pflanzenfam., ed. 2, 19C: 127 (1931); G. L. Webster, Ann. Missouri Bot. Gard. 81: 85 (1994). T.: *D. glechomoides* (A. Rich.) Müll. Arg. (B.: *Hydrocotyle "glocomoides"* (sic) A. Rich.)

Molina Gay, Fl. Chil. 5: 345 (1841), *non* Cav. (1790). T.: *M. chilensis* Gay (*D. gayana* Baill., nom. illeg. superfl.) [= *D. glechomoides* (A. Rich.) Müll. Arg. (B.: *Hydrocotyle "glocomoides"* (*sic*) A. Rich.)]

A weak decumbent or prostrate monoecious herb rooting at the nodes, in habit recalling *Glechoma hederacea.* Indumentum simple. Leaves alternate, petiolate, stipulate, simple, orbicular-subreniform, coarsely crenate to crenate-dentate, palminerved, membranaceous, eglandular; stipules subulate, deciduous. Flowers co-axillary in upper leaf-axils with reduced branchlets, solitary or in few-flowered fascicles, male and female irregularly intermixed. Male flowers long-pedicellate; calyx minute, campanulate or urceolate, 3-fid, pellucid; petals 0; disc 0; stamens 3 or 6 and then 2-verticillate, inner filaments longer than the outer, connate at the base, anthers basifixed, introrse, ellipsoid, thecae 2-rimous, slightly divergent at the base, longitudinally

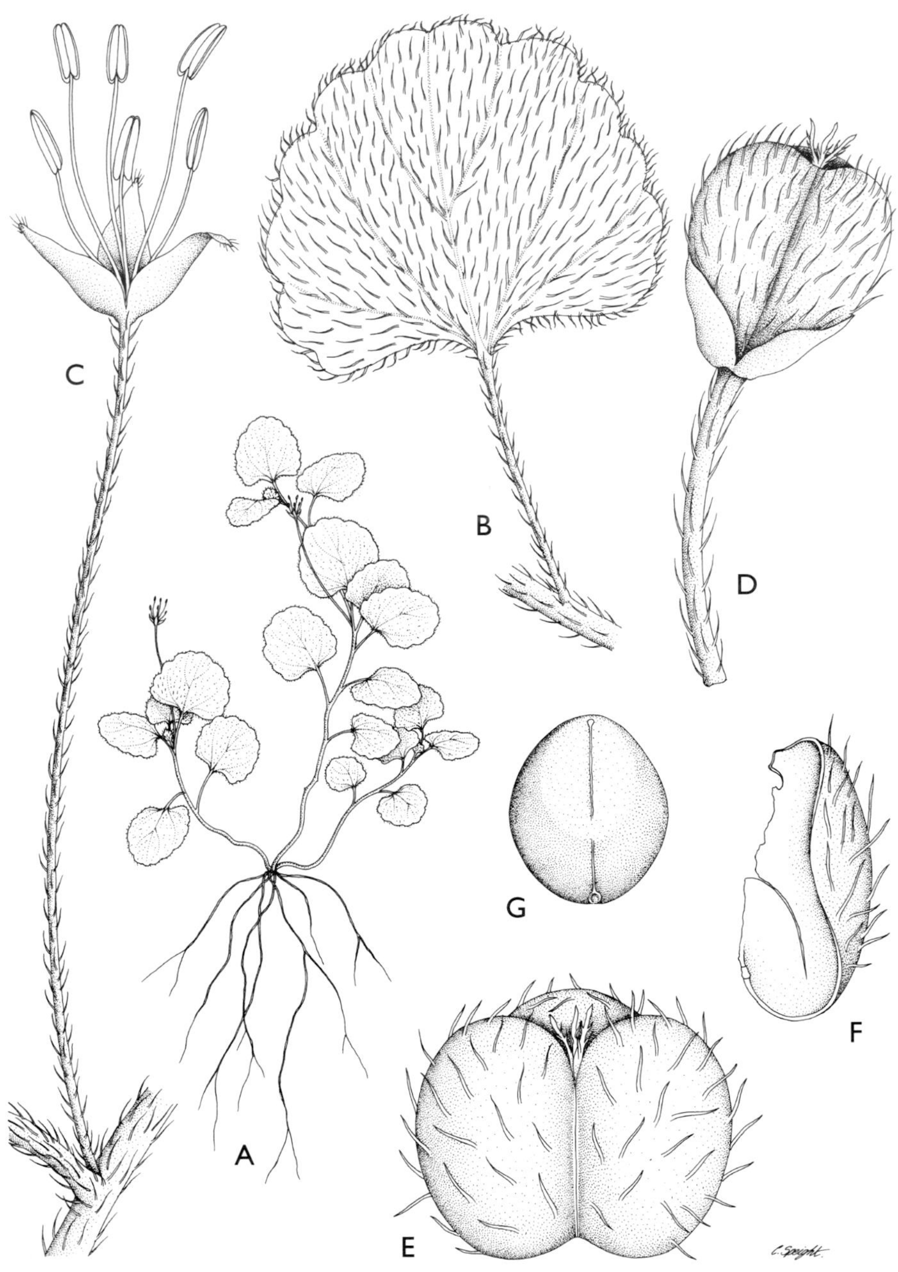

FIG. 26. *Dysopsis glechomoides* (A. Rich.) Müll. Arg. (B.: *Hydrocotyle "glocomoides"* (sic) A. Rich.). **A** Habit × 1; **B** Leaf × 4.5; **C** Male Flower × 9; **D** Young fruit × 30; **E** Fruit × 30; **F** Fruit-valve × 30; **G** Seed × 30. **A–C** from *Donat* 333; **D** from *Werdermann* 906; **E–G** from *Sparre* 2098. Drawn by Camilla Speight.

dehiscent, connective thin, slightly produced; pollen oblate-spheroidal, 3-lobate, quasi-3-colpate, inaperturate, tectate, psilate, columellae fairly coarse; pistillode 0. Female flowers long-pedicellate, the pedicels extending in fruit; calyx 3-partite, the segments subvalvate; petals 0; disc 0; ovary 3-locular, locules oppositisepalous, 1 ovule per locule; styles 3, suberect, lanceolate, slightly fimbriate or incised-asperous. Fruit 3-coccous, hirsute, thin-walled, dehiscing into 3 2-valved cocci; endocarp chartaceous. Seeds subglobose, ± ecarunculate; testa ± smooth, somewhat shiny, dark grey, crustaceous; endosperm fleshy; embryo linear or slightly clavate, straight, compressed, situated in mid-endosperm. Fig.: See Pflanzenr. 63: 287, t. 45 (1914). See also Fig. 26, p. 208.

A monotypic disjunct C. and S. American genus from Costa Rica, Ecuador to Chile and Argentina, and Juan Fernandez Is.

Subtribe 33e. **Cleidiinae** *G. L. Webster*, Taxon 24: 598(1975), and Ann. Missouri Bot. Gard. 81(1): 85 (1994). T.: *Cleidion* Blume.

Monoecious or dioecious trees or shrubs; indumentum simple; leaves alternate, stipulate, simple, penninerved, glandular. Inflorescences axillary, spicate or racemose; bracts eglandular; male sepals 3–4; petals 0; disc 0; stamens 15–110, free or basally connate, anthers muticous or apiculate; pollen 3–4-colporate, colpi exoperculate, rugulose-vermiculate, vestigially spinulose; pistillode 0; receptacle convex; female sepals 3–5, imbricate; ovary (2)3-locular; styles 2-fid, arms elongate. Fruit dehiscent. Seeds ecarunculate; testa dry or fleshy.

A small group of woody mostly palaeotropical *Acalypheae* with characteristic globose pollen having vermiculate sexine ornamentation. Pax (1914) had placed *Wetria* with *Conceveiba* and its allies, but Webster (1994) showed that it differs from these both palynologically and in having a simple indumentum.

Key to the genera

1. Male flowers sessile, 1 per bract; stamens 15–25, anthers apiculate, connective not enlarged; fruit-lobes shouldered · · · · · · 167. *Sampantaea*
1. Male flowers pedicellate, usually 2 or more per bract; stamens 25–110 · · 2
2. Stamens 25–30, anthers muticous, connective scarcely enlarged; female pedicel < 1cm long · 165. *Wetria*
2. Stamens 30–110, anthers apiculate, connective enlarged; female pedicel >1cm long · 166. *Cleidion*

165. **Wetria** *Baill.*, Étude Euphorb.: 409 (1858); J. J. Sm., Meded. Dept. Landb. Ned.-Indië 10: 470 (1910); Pax & K. Hoffm., Pflanzenr. 63: 219 (1914) and Nat. Pflanzenfam., ed. 2, 19C: 119 (1931); Backer & Bakh. f., Fl. Java 1: 485 (1963); Airy Shaw, Kew Bull. 26: 350 (1972); Whitmore, Tree Fl. Mal. 2:

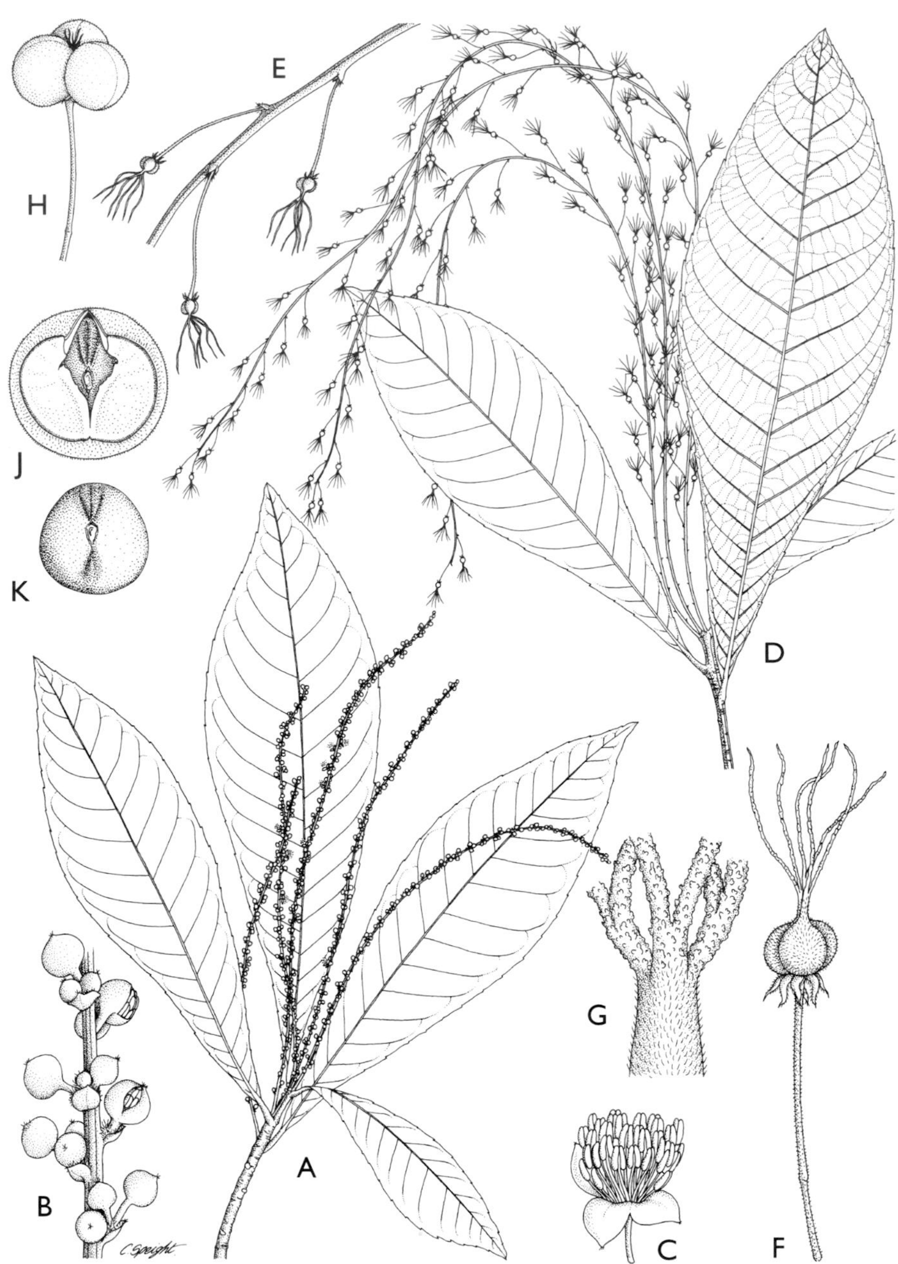

FIG. 27. *Wetria insignis* (Steud.) Airy Shaw (B.: *Trewia insignis* Steud.). **A** Habit, male flowering specimen × $^1/_2$; **B** Small portion of male inflorescence × 4.5; **C** Male flower × 4.5; **D** Habit, female flowering specimen × $^1/_2$; **E** Small portion of female inflorescence × $2^1/_4$; **F** Female flower × 4.5; **G** Lower portion of style, showing branching-pattern of stigmas × 27; **H** Fruit × $1^1/_2$; **J** Fruit-coccus × 3; **K** Seed × 3. **A–C** from *Alston* 14559; **D–G** from SAN 93138; **H–K** from SAN 35274. Drawn by Camilla Speight.

136 (1973); Airy Shaw, Kew Bull. Addit. Ser. 4: 206 (1975), 8: 206 (1980), Kew Bull. 36: 358 (1981), and Euph. Philipp.: 48 (1983); P. I. Forst., Austrobaileya 4(2): 141 (1994); G. L. Webster, Ann. Missouri Bot. Gard. 81: 86 (1994); van Welzen, Blumea 43: 156 (1998). T.: *W. trewioides* Baill., nom. illeg. [=*W. insignis* (Steud.) Airy Shaw (B.: *Trewia insignis* Steud.)]

Pseudotrewia Miq., Fl. Ned. Ind. I. 2: 414 (1859). T.: *P. macrophylla* Miq. [=*W. insignis* (Steud.) Airy Shaw]

Dioecious trees. Indumentum simple, soon lost. Leaves alternate, large, shortly petiolate, stipulate, simple, denticulate, attenuate to the base, penninerved, subcoriaceous, gland-dotted towards the base. Inflorescences axillary, spiciform-racemose, elongate, many-flowered, unisexual, male flowers fasciculate, females 1 per bract. Male flowers subsessile or shortly pedicellate; calyx closed in bud, globose, later valvately 3–4-partite; petals 0; disc 0; stamens 25–30, filaments almost free, anthers subbasifixed, introrse, ellipsoid, muticous, thecae contiguous at the apex, free at the base, curved, divaricate or subpendulous, longitudinally dehiscent, connective fairly broad; pollen spheroidal, (3)4-colporate, colpi narrow, costae +, colpus transversalis small, costae +, tectate, psilate; pistillode 0; receptacle convex. Female flowers shortly pedicellate; sepals 5, imbricate, persistent; petals 0; disc 0; ovary 3-locular, 1 ovule per locule; styles 3, shortly connate at the base, elongate, spreading, 2-fid, style-arms linear, stigmas coarsely papillose. Fruit 3-dymous, dehiscing into 3 2-valved cocci; epicarp tomentellous; endocarp woody; columella 3-quetrous, 3-cornute at the apex, persistent. Seeds globose, ecarunculate; testa smooth, dull; endosperm fleshy; cotyledons broad and flat. Fig.: See Austrobaileya 4(2): 142, t. 1 (1994); Blumea 43: 158, t. 5 (1998). See also Fig. 27, p. 210.

A tropical Asiatic genus of 2 species, one ranging from Lower Burma (Myanmar) to Java, Borneo and the Philippines, with an outlier in Papua, and the other, recently-described, from N. Queensland.

166. **Cleidion** *Blume*, Bijdr. Fl. Ned. Ind.: 612 (1826); Baill., Étude Euphorb.: 404 (1858); Müll. Arg. in DC., Prodr. 15(2): 983 (1866) and in Mart., Fl. Bras. 11(2): 418 (1874); Benth., Gen. Pl. 3: 320 (1880); Hook. f., Fl. Brit. India 5: 444 (1887); Prain in Fl. Trop. Afr. 6(1): 930 (1912); Pax & K. Hoffm., Pflanzenr. 63: 288 (1914); Gagnep., Fl. Indo-Chine 5: 449 (1926); Pax & K. Hoffm., Nat. Pflanzenfam., ed. 2, 19C: 127 (1931); Croizat, J. Arnold Arbor. 24: 166 (1943); Hurus., J. Fac. Sci. Univ. Tokyo, Sect. 3, Bot. 6: 304 (1954); Keay (ed.), Fl. W. Trop. Afr. ed. 2, 1: 406 (1958); Backer & Bakh. f., Fl. Java 1: 487 (1963); G. L. Webster, Ann. Missouri Bot. Gard. 54: 278 (1968); J. Léonard, Bull. Jard. Bot. Belg. 42: 297 (1972); Leandri, Adansonia II. 12: 193 (1972); Airy Shaw, Kew Bull. 26: 234 (1972); Whitmore, Tree Fl. Mal. 2: 79 (1973); Airy Shaw, Kew Bull. Addit. Ser. 4: 74 (1975), 8: 57 (1980), Kew Bull. 35: 607 (1980), and 36: 279 (1981); A. C. Sm., Fl. Vit. Nov. 2: 514 (1981); Airy Shaw, Kew Bull. 37: 12 (1982), and Euph. Philipp.: 13 (1983); Grierson & D. G. Long, Fl. Bhut. 1: 803 (1987); McPherson & Tirel, Fl. Nouv.-Caléd. 14(1): 143 (1987); G. L. Webster & Huft,

Ann. Missouri Bot. Gard. 75: 1103 (1989); A. H. Gentry, Woody Pl. NW. S. Amer.: 420 (1993); G. L. Webster, Ann. Missouri Bot. Gard. 81: 86 (1994); Murillo-Aldana & Franco-Rosselli, Euf. Reg. Ararac.: 53 (1995); Philcox in Dassan., ed., Fl. Ceyl. XI: 166 (1997). T.: *Cl. javanicum* Blume.

Reidia Casar., Nov. Stirp. Bras.: 51 (1843); Baill., Étude Euphorb.: 407 (1858) ("*Redia*"). T.: *R. tricocca* Casar. [=*Cl. tricoccum* (Casar.) Baill.]

Psilostachys Turcz., Bull. Soc. Imp. Naturalistes Moscou 16: 58 (1843). T.: *P. axillaris* Turcz. [=*Cl. tricoccum* (Casar.) Baill.]

Lasiostyles C. Presl, Abh. Königl. Böhm. Ges. Wiss. V. 3: 579 (1845); Baill., Étude Euphorb.: 653 (1858). T.: *L. salicifolia* C. Presl [=*Cl. javanicum* Blume.]

Tetraglossa Bedd., Madras J. Lit. Sci. II. 22: 70 (1861). T.: *T. indica* Bedd. [=*Cl. javanicum* Blume.]

Monoecious or dioecious mostly glabrous trees or shrubs. Indumentum, where present, simple. Leaves alternate, large, petiolate, stipulate, simple, obovate-oblanceolate, often dentate, penninerved, 2–more gland-dotted near the base beneath; stipules inconspicuous, soon deciduous. Inflorescences axillary, males interruptedly glomerate- or fasciculate-spicate or -racemose, sometimes subpaniculate, mostly elongate, rarely abbreviated and subcapituliform, females in simple racemes or else solitary in the leaf-axils. Male flowers small, pedicellate; calyx closed in bud, globose or ovoid, subacute, later valvately 3–4-partite; petals 0; disc 0; stamens 35–80(110), crowded on the receptacle, filaments free, in whorls alternating with each other, anthers disposed in vertical series, dorsifixed, introrse, obliquely peltate, 4-locellate, 2 locelli on each side, superposed, subcruciately dehiscent and then locelli confluent, connective produced, usually more darkly coloured than the thecae; pollen oblate-spheroidal, 3-colporate, colpi narrow, costae +, colpus transversalis small, costae +, tectate, psilate; pistillode 0; receptacle convex or conical. Female flowers sometimes long-pedicellate, pedicels often rigid, flattened and expanded at the apex; sepals 3–4(5), imbricate; petals 0; disc 0; ovary (2)3-locular, 1 ovule per locule; styles 3, usually shortly connate at the base, elongate, slender, spreading, 2-crurate, segments filiform, stigmas densely and minutely papillose. Fruit (2)3-dymous, smooth, dehiscing into (2)3 2-valved cocci, or by abortion 1-coccous. Seeds subglobose, ecarunculate; testa smooth, dull, often mottled; endosperm fleshy; cotyledons broad and flat. Fig.: See Étude Euphorb.: t. 9, f. 3–5; t. 21, f. 1–2 (1858) (as *Redia*); Fl. Bras. 11(2): t. 59 (1874); Pflanzenr. 63: 289, t. 46F–H and 291, t. 47 (1914); Fl. Indo-Chine 5: 445, t. 54.6–11 (1926); Nat. Pflanzenfam., ed. 2, 19C: 127, t. 66F–H (1931); Fl. Nouv.-Caléd. 14(1): 147–165, tt. 30–34 (1987); Woody Pl. NW. S. Amer.: 418, t. 120.1 (1993); Euf. Reg. Ararac.: 56, t. 7 (1995).

A pantropical genus of c. 25 species, of which 5 are Neotropical, 1 W. African, 7 tropical Asian and 12 endemic to New Caledonia.

167. **Sampantaea** *Airy Shaw*, Kew Bull. 26: 350 (1972), and Hooker's Icon. Pl. 38: t. 3717 (1974); G. L. Webster, Ann. Missouri Bot. Gard. 81: 86 (1994). T.: *S. amentiflora* (Airy Shaw) Airy Shaw (B.: *Alchornea ? amentiflora* Airy Shaw).

Very like 165. *Wetria,* but with fewer pairs of not conspicuously parallel lateral nerves, shorter somewhat catkin-like inflorescences, sessile male flowers only 1 per bract, fewer stamens (15–25), shorter filaments, apiculate anthers and high-shouldered almost conoidal fruit-lobes. Fig.: See Hooker's Icon. Pl. 38: t. 3717 (1974).

A monotypic genus restricted to Thailand and Cambodia (Kampuchea).

Subtribe 33f. **Macaranginae** *(Hutch.) G. L. Webster,* Taxon 24: 598(1975), and Ann. Missouri Bot. Gard. 81(1): 86 (1994). T.: *Macaranga* Thouars.

Macarangeae Hutch., Amer. J. Bot. 56: 755 (1969)

Dioecious or rarely monoecious trees or shrubs; indumentum simple; leaves alternate, simple or lobate, penninerved or palminerved, granulate-glandular beneath; stipules minute to large and conspicuous. Inflorescences axillary, racemose or paniculate; bracts glandular; male calyx valvately 2–5-partite; petals 0; disc 0; stamens (1)2–50, free, anthers muticous, (2)3–4-valved; pollen 3–4-colporate, finely to coarsely rugulose-spinulose; female sepals 3–5, free or connate; ovary (1)2- or 3(6)-locular; styles simple. Fruit dehiscent. Seeds ecarunculate; exotesta fleshy.

Webster (1975, 1994) has a much narrower view of this group than had Hutchinson (1969), since he removes all the other genera that the latter included with *Macaranga* (i.e. *Hasskarlia, Tetrorchidium, Cleidiocarpon, Cleidion, Endospermum, Ptychopyxis* and *Bernardia*) and places them either in another acalyphean subtribe (*Cleidion*), in other acalyphoid tribes (*Cleidiocarpon, Ptychopyxis* and *Bernardia*) or else outside the *Acalyphoideae* altogether (*Endospermum, Hasskarlia* and *Tetrorchidium*).

168. **Macaranga** *Thouars,* Gen. Nov. Madagasc: 26 (1806); A. Juss., Euphorb. Gen.: 43 (1824); Baill., Étude Euphorb.: 431 (1858); Müll. Arg. in DC., Prodr. 15(2): 987 (1866); Benth., Gen. Pl. 3: 320 (1880); Hook. f., Fl. Brit. India 5: 445 (1887); Prain in Fl. Trop. Afr. 6(1): 932 (1912); Pax & K. Hoffm., Pflanzenr. 63: 298 (1914); Gagnep., Fl. Indo-Chine 5: 434 (1926); Pax & K. Hoffm., Nat. Pflanzenfam., ed. 2, 19C: 128 (1931); Leandri, Notul. Syst. (Paris) 10: 138 (1942); Robyns, Fl. Parc Nat. Alb. 1: 460 (1948); L. M. Perry, J. Arnold Arbor. 34: 191 (1953); Hurus., J. Fac. Sci. Univ. Tokyo, Sect. 3, Bot. 6: 309 (1954); Keay (ed.), Fl. W. Trop. Afr. ed. 2, 1: 406 (1958); Backer & Bakh. f., Fl. Java 1: 487 (1963); J.-G. Adam, Mém. Mus. Natl. Hist. Nat., B, Bot. 20: 483 (1971); Airy Shaw, Kew Bull. 26: 286 (1972); Whitmore, Tree Fl. Mal. 2: 105 (1973); Airy Shaw, Hooker's Icon. Pl. 38: t. 3718 (1974); Berhaut, Fl. Ill. Séneg. 3: 517 (1975); Whitmore, Kew Bull. Addit. Ser. 4: 140 (1975); Coode, Taxon 25: 184 (1976); Whitmore, Kew Bull. Addit. Ser. 8: 123 (1980); Airy Shaw, Kew Bull. 35: 646 (1980); A. C. Sm., Fl. Vit. Nov. 2: 500 (1981); Whitmore, Kew Bull. 36: 312 (1981); Coode, Fl. Masc. 160. Euph.: 53 (1982); Airy Shaw, Kew Bull. 37: 26 (1982), and Euph. Philipp.: 33 (1983); Grierson &

2. Racemes interrupted; fruits crustaceous; leaves stipellate · 173. *Micrococca*
2. Racemes usually uniformly floriferous; fruits coriaceous; leaves exstipellate · 3
3. Male disc urceolate; stamens 6–12; styles recurved, lacerate · 172. *Discoclaoxylon*
3. Male disc of interstaminal segments · 4
4. Stamens mostly 20 or more; styles recurved, papillose but hardly lacerate · 170. *Claoxylon*
4. Stamens up to 15; styles erect, lacerate · · · · · · · · · · · · · · 171. *Claoxylopsis*

169. **Erythrococca** *Benth.* in Hook., Niger Fl.: 506 (1849); Baill., Étude Euphorb.: 437 (1858); Müll. Arg. in DC., Prodr. 15(2): 790 (1866); Benth., Gen. Pl. 3: 308 (1880); Prain, Ann. Bot. (London) 25: 606 (1911), and Fl. Trop. Afr. 6(1): 847 (1912); Pax & K. Hoffm., Pflanzenr. 63: 86 (1914), and Nat. Pflanzenfam., ed. 2, 19C: 110 (1931); Robyns, Fl. Parc Nat. Alb. 1: 456 (1948); Keay (ed.), Fl. W. Trop. Afr. ed. 2, 1: 400 (1958); J.-G. Adam, Mém. Mus. Natl. Hist. Nat., B, Bot. 20: 478 (1971); Berhaut, Fl. Ill. Sénég. 3: 437 (1975); Radcl.-Sm., Kew Bull. 33: 235 (1978); M. G. Gilbert, Kew Bull. 42: 363 (1987); Radcl.-Sm., Fl. Trop. E. Afr., Euph. 1: 265 (1987); G. L. Webster, Ann. Missouri Bot. Gard. 81: 87 (1994); Radcl.-Sm., Fl. Zamb. 9(4): 165 (1996). T.: *E. aculeata* Benth., nom. illeg. [=*E. anomala* (Juss. ex Poir.) Prain (B.: *Adelia anomala* Juss. ex Poir.)]

Claoxylon A. Juss. sect. *Athroandra* Hook. f., J. Proc. Linn. Soc., Bot. 6: 21 (1862); Müll. Arg. in DC., Prodr. 15(2): 776 (1866). T.: *Cl. mannii* Hook. f. [=*E. mannii* (Hook. f.) Prain]

Claoxylon A. Juss. sect. *Adenoclaoxylon* Müll. Arg., Flora 47: 436 (1864), and in DC., Prodr. 15(2): 775 (1866). T.: *Cl. kirkii* Müll. Arg. [=*E. kirkii* (Müll. Arg.) Prain]

Deflersia Schweinf. ex Penz., Atti Congr. Bot. Genova 1892: 359 (1893). T.: *D. erythrococca* Schweinf. ex Penz. [=*E. abyssinica* Pax]

Poggeophyton Pax, Bot. Jahrb. Syst. 19: 88 (1894). T.: *P. aculeatum* Pax [=*E. poggeophyton* Prain]

Chloropatane Engl., Bot. Jahrb. Syst. 26: 383 (1899). T.: *Chl. africana* (Baill.) Engl. (B.: *Trewia africana* Baill.) [=*E. africana* (Baill.) Prain]

Athroandra (Hook. f.) Pax & K. Hoffm., Pflanzenr. 63: 76 (1914), and Nat. Pflanzenfam., ed. 2, 19C: 109 (1931); Robyns, Fl. Parc Nat. Alb. 1: 455 (1948). T.: *A. mannii* (Hook. f.) Pax & K. Hoffm. (B.: *Claoxylon mannii* Hook. f.) [=*E. mannii* (Hook. f.) Prain]

Dioecious shrubs. Indumentum simple. Buds perulate, the perulae crustaceous, persistent. Leaves alternate, shortly petiolate, stipulate, simple, glandular-crenate or -dentate, penninerved, sometimes purplish-tinged; stipules often accrescent, indurated, sometimes spinescent, usually persistent. Inflorescences axillary, solitary or fasciculate, sessile or pedunculate, glomerulate or racemose, rarely subpaniculate, usually bracteate and bracteolate; male bracteoles usually many-flowered, females 1(3)-flowered. Male flowers pedicellate; pedicels capillary, articulate near the base; calyx closed in bud, subglobose, apiculate, later valvately 3–4(5)-partite; petals 0; disc-glands extrastaminal, free or connate and annular, interstaminal and free, or both, generally hirsute; stamens (2)10–40(60), filaments free, anthers erect, basifixed, extrorse, 2-thecate, thecae obovoid-subglobose, free except at the base, apically dehiscent; pollen suboblate to prolate-spheroidal, stephanocolporate, colpi narrow, colpus transversalis small, costae indistinct, tectate, psilate, columellar capita distinct; pistillode 0. Female flowers pedicellate; calyx 2–4(5)-partite, the lobes imbricate, smaller than in the male; disc-glands 2–3(5), free or contiguous, with or without minute intercalated supernumerary glands, or disc shallowly urceolate, bilobate to subentire; ovary 2–3(5)-locular, 1 ovule per locule; styles 2–3(5), free or slightly connate at the base, spreading, simple, stigmas smooth, papillose, lobulate, fimbriate, laciniate or plumulose. Fruit 1–3(5)-coccous, cocci often subglobose, each loculicidally dehiscent into 2 valves; endocarp coriaceous; columella persistent. Seeds subglobose, ecarunculate, thinly arillate; aril fleshy, viscid, scarlet; testa crustaceous, foveolate-reticulate or ± smooth; endosperm fleshy; embryo axile, radicle conical, cotyledons broad, flat. Fig.: See Étude Euphorb.: t. 21, f. 10 (1858); Pflanzenr. 63: 80–97, tt. 11–15 (1914); Nat. Pflanzenfam., ed. 2, 19C: 110, t. 56 and 111, t. 57 (1931); Fl. Parc Nat. Alb. 1: 457, t. 45 (1948); Fl. Ill. Sénég. 3: 436 (1975); Kew Bull. 42: 364, t. 5 (1987); Fl. Trop. E. Afr., Euph. 1: 268, t. 53 (1987); Fl. Zamb. 9(4): 169, t. 35, and 172, t. 36 (1996).

A genus of c. 50 species, mostly confined to tropical Africa, with some extending into S. Africa and 1 into S. Arabia.

170. **Claoxylon** *A. Juss.*, Euphorb. Gen.: 43 (1824); Baill., Étude Euphorb.: 491 (1858); Müll. Arg. in DC., Prodr. 15(2): 775 (1866); Benth., Gen. Pl. 3: 309 (1880); Hook. f., Fl. Brit. India 5: 410 (1887); Pax & K. Hoffm., Pflanzenr. 63: 100 (1914); Merr., Enum. Philipp. Fl. Pl. 2: 429 (1923); Gagnep., Fl. Indo-Chine 5: 420 (1926); Pax & K. Hoffm., Nat. Pflanzenfam., ed. 2, 19C: 111 (1931); Hurus., J. Fac. Sci. Univ. Tokyo, Sect. 3, Bot. 6: 301 (1954); Backer & Bakh. f., Fl. Java 1: 480 (1963); Airy Shaw, Kew Bull. 26: 233 (1972); Whitmore, Tree Fl. Mal. 2: 78 (1973); Airy Shaw, Hooker's Icon. Pl. 38: t. 3714 (1974), Kew Bull. Addit. Ser. 4: 69 (1975), 8: 46 (1980), Kew Bull. 35: 605 (1980), and 36: 277 (1981); A. C. Sm., Fl. Vit. Nov. 2: 516 (1981); Coode, Fl. Masc. 160. Euph.: 58 (1982); Airy Shaw, Kew Bull. 37: 11 (1982), and Euph. Philipp.: 12 (1983); Grierson & D. G. Long, Fl. Bhut. 1: 795 (1987); McPherson & Tirel, Fl. Nouv.-Caléd. 14 (1): 169 (1987); G. L. Webster, Ann. Missouri Bot. Gard. 81: 88 (1994); Susila Rani & N. P. Balakr., Rheedea 5(2): 113 (1995). T.: *Cl. parviflorum* A. Juss.

Erythrochilus Reinw. ex Blume, Bijdr. Fl. Ned. Ind.: 614 (1826). T.: *E. indicus* Reinw. ex Blume [=*Cl. indicum* (Reinw. ex Blume) Hassk.]

Quadrasia Elmer, Leafl. Philipp. Bot. 7: 2656 (1915). T.: *Q. euphorbioides* Elmer [=*Cl. euphorbioides* (Elmer) Merr.]

Dioecious or rarely monoecious trees or shrubs. Indumentum simple, hairs usually adpressed. Buds not perulate. Leaves alternate, petiolate, stipulate, simple, glandular-dentate to subentire, penninerved, usually coriaceous, opaque, often scabrid, usually purplish when young; stipules minute, herbaceous, soon deciduous. Inflorescences axillary or lateral, racemose, usually elongate, fasciculate or racemulose, sessile or pedunculate, glomerulate or racemose, bisexual; male bracts 1–∞-flowered, females mostly 1-flowered. Male flowers shortly pedicellate; calyx closed in bud, subglobose or obovoid, apiculate, later valvately 2–4-partite; petals 0; disc-glands interstaminal, (0)∞, free, obovoid, glabrous or pilose; stamens (10)20–30(200{∞}), filaments free, anthers erect, basifixed, extrorse, 2-thecate, thecae obovoid-cylindric, free except at the base, apically dehiscent; pollen spheroidal, 3(4)-colporate, sometimes scabrate, otherwise ± as in 169. *Erythrococca*; pistillode 0. Female flowers shortly pedicellate; sepals (2)3(4), valvate; petals 0; disc-glands 2–3(4), free, or urceolate and deeply 2–3(4)-lobate; ovary 2–3(4)-locular, 1 ovule per locule; styles 2–3(4), free or slightly connate at the base, recurved-spreading, simple, stigmas ± papillose, rarely ± smooth. Fruit 2–3(4)-coccous, loculicidally dehiscent into 2–3(4) valves; endocarp coriaceous; columella persistent. Seeds globose, ecarunculate, thinly arillate; aril fleshy; testa crustaceous, foveolate or reticulate; endosperm fleshy; cotyledons broad, flat. Fig.: See Euphorb. Gen.: t. 14, f. 43.1–10 (1824); Étude Euphorb.: t. 20, f. 20–24 (1858); Pflanzenr. 63: 109, t. 16 and 123, t. 17 (1914); Fl. Indo-Chine 5: 423, t. 50.6–13 (1926); Formos. Trees: 335, t. 289 (1936); Notul. Syst. (Paris) 9: 172, t. 3.12–33 (1941); Hooker's Icon. Pl. 38: t. 3714 (1974); Kew Bull. Addit. Ser. 8: 227, t. 5.4 (1980); Fl. Masc. 160. Euph.: 64, t.12 and 66, t. 13 (1982); Fl. Nouv.-Caléd. 14 (1): 171, t. 35.1–5 (1987); Rheedea 5(2): 117–126, tt. 1–4 and 139, t. 8 (1995).

A palaeotropical genus of c. 80 species in Madagascar, the Mascarenes and the Indo-Pacific, extending as far as Hawai'i.

171. **Claoxylopsis** *Leandri*, Bull. Soc. Bot. France 85: 526 (1938); Radcl.-Sm., Kew Bull. 43: 642 (1988); G. L. Webster, Ann. Missouri Bot. Gard. 81: 88 (1994). T.: *C. perrieri* Leandri.

Very like 170. *Claoxylon*, but monoecious, often climbing shrubs with smaller leaves having only 4–7 pairs of lateral nerves, stamens up to 15, styles erect, lacerate-plumose, and fruits commonly 2-coccous. Fig.: See Bull. Soc. Bot. France 85: 525, t. 1.20, 21 (1938); Notul. Syst. (Paris) 9: 163, t. 2.8–11 (1941); Kew Bull. 43: 641, t. 5 and 644, t. 6 (1988).

A genus of 3 species endemic to E. Madagascar, *fide* Radcliffe-Smith (1988).

172. **Discoclaoxylon** *(Müll. Arg.) Pax & K. Hoffm.*, Wiss Ergebn. Deutsch. Zentr.-Afr. Exp. 2: 452 (1912), in Pflanzenr. 63: 137 (1914) and in Nat. Pflanzenfam. ed. 2, 19C: 112 (1931); Radcl.-Sm., Fl. Trop. E. Afr., Euph. 1: 279 (1987); G. L. Webster, Ann. Missouri Bot. Gard. 81: 88 (1994). LT.: *D. hexandrum* (Müll. Arg.) Pax & K. Hoffm. (B.: *Cl. hexandrum* Müll. Arg.) (chosen by Webster, 1994).

Claoxylon sect. *Discoclaoxylon* Müll. Arg., Flora 47: 437 (1864), and in DC., Prodr. 15(2): 779 (1866); Pax in Engl. & Prantl, Nat. Pflanzenfam. ed. 1,

3(5): 48 (1890); Prain, Ann. Bot. (London) 25: 604 (1911), and in Fl. Trop. Afr. 6(1): 874 (1912); Keay (ed.), Fl. W. Trop. Afr. ed. 2, 1: 401 (1958); J.-G. Adam, Mém. Mus. Natl. Hist. Nat., B, Bot. 20: 462 (1971).

Dioecious shrubs or small trees. Indumentum simple, evanescent. Buds not perulate. Twigs stout. Leaves alternate, long-petiolate, stipulate, large, simple, glandular-crenate, -serrate or -dentate, membranous, penninerved, often purplish-tinged; stipules very small, soon deciduous. Inflorescences axillary, solitary, pedunculate, interruptedly racemose, elongate; male bracts 3–11-flowered, female 1-flowered. Male flowers shortly pedicellate; calyx closed in bud, subglobose or obovoid, apiculate, later valvately 3–4-partite; petals 0; disc urceolate, extrastaminal, entire or lobulate; stamens 6–12, filaments free, very short, anthers erect, basifixed, extrorse, 2-thecate, thecae obovoid, free except at the base, longitudinally dehiscent; pollen 3-colporate, otherwise ± as in 170. *Claoxylon*; pistillode 0; receptacle slightly raised. Female flowers shortly pedicellate; sepals 4, imbricate, smaller than in the male; petals 0; disc-glands 2, broadly 3-angular, petaloid; ovary 2-locular, 1 ovule per locule; styles 2, free, broad at the base, recurved and accumbent on the ovary, lacerate to $^1/_2$-way, plumose. Fruit 2-coccous, or 1-coccous by abortion, cocci subglobose, loculicidally dehiscent from the base into 2 valves; endocarp coriaceous; columella short, clavate, persistent. Seeds globose, ecarunculate, thinly arillate; aril fleshy; testa crustaceous, foveolate-reticulate or ± smooth; endosperm fleshy; cotyledons broad, flat. Fig.: See Pflanzenr. 63: 138, t. 19 (1914); Mém. Mus. Natl. Hist. Nat., B, Bot. 20: 464, t. 163 (1971) (as *Claoxylon*); Fl. Trop. E. Afr., Euph. 1: 281, t. 54 (1987).

A W. African genus of 4 species, 3 from the islands in the Gulf of Guinea, the other extending from Sierra Leone to W. Uganda.

173. **Micrococca** *Benth.* in Hook., Niger Fl.: 503 (1849); Baill., Étude Euphorb.: 436 (1858); Benth., Gen. Pl. 3: 309 (1880); Pax in Engl. & Prantl, Nat. Pflanzenfam. ed. 1, 3(5): 48 (1890); Prain, Ann. Bot. (London) 25: 628 (1911), and Fl. Trop. Afr. 6(1): 876 (1912); Pax & K. Hoffm., Pflanzenr. 63: 131 (1914), and in Nat. Pflanzenfam. ed. 2, 19C: 112 (1931); Robyns, Fl. Parc Nat. Alb. 1: 458 (1948); Keay (ed.), Fl. W. Trop. Afr. ed. 2, 1: 402 (1958); Airy Shaw, Kew Bull. 25: 524 (1971); Whitmore, Tree Fl. Mal. 2: 118 (1973); Berhaut, Fl. Ill. Sénég. 3: 535 (1975); Dyer, Gen. S. Afr. Fl. Pl., ed. 3. 1: 315 (1975); Radcl.-Sm., Fl. Trop. E. Afr., Euph. 1: 260 (1987); G. L. Webster, Ann. Missouri Bot. Gard. 81: 88 (1994); Radcl.-Sm., Fl. Zamb. 9(4): 176 (1996); Philcox in Dassanayake, ed., Fl. Ceyl. XI: 146 (1997). T.: *M. mercurialis* (L.) Benth. (B.: *Tragia mercurialis* L.)

Claoxylon sect. *Micrococca* (Benth.) Müll. Arg., Linnaea 34: 166 (1865), and in DC., Prodr. 15(2): 789 (1866); Hook. f., Fl. Brit. India 5: 412 (1887); Susila Rani & N. P. Balakr., Rheedea 5(2): 113 (1995).

Dioecious or monoecious annual or perennial herbs or shrubs, rarely small trees. Indumentum simple. Buds not perulate. Leaves alternate or sometimes also the lowest opposite, petiolate, stipulate, stipellate, simple, crenate to serrate, membranaceous, penninerved, often purplish when young; stipules minute, not hardened. Inflorescences axillary, solitary or fasciculate, pedunculate or not, interruptedly spicate or racemose, mostly unisexual, rarely bisexual; unisexual: male bracts many-flowered, female 1–2- flowered; bisexual: all bracts male with 1 terminal female bract, or bracts androgynous, each subtending a few male flowers with 1 central female. Male flowers shortly pedicellate; calyx closed in bud, ± globose, apiculate, later valvately 3–4-partite; petals 0; disc-glands (0)3–30, free, interstaminal; stamens 3–30, usually 2-seriate, filaments free, short, anthers erect, basifixed, extrorse, 2-thecate, thecae obovoid, free except at the base, longitudinally dehiscent; pollen prolate-spheroidal, otherwise ± as in 169. *Erythrococca*; pistillode 0; receptacle slightly convex. Female flowers pedicellate; calyx-lobes 3(5), imbricate, larger than in the male; petals 0; disc-glands (2)3(6), staminodial, alternating with the carpidia; ovary (2)3(4)-locular, 1 ovule per locule; styles (2)3(4), free, spreading, linear, stigmas fimbriate- or plumose-laciniate. Fruit (2)3(4)-coccous, sometimes 1-coccous by abortion, cocci subglobose, septicidally, loculicidally and septifragally dehiscent; exocarp thin; endocarp thinly crustaceous; columella crustaceous or woody, persistent. Seeds subglobose, ecarunculate, thinly arillate; testa crustaceous, foveolate-reticulate, muricate or smooth; endosperm fleshy; cotyledons suborbicular, sometimes scarcely larger than the radicle. Fig.: See Pflanzenr. 63: 134, t. 18 (1914); Fl. Ill. Sénég. 3: 534 (1975); Fl. Trop. E. Afr., Euph. 1: 262, t. 52 (1987); Rheedea 5(2): 129–137, tt. 5–7 (1995) (as *Claoxylon* spp.); Fl. Zamb. 9(4): 178, t. 37 (1996).

A palaeotropical genus of 12 species: 1 widespread, 4 in Africa, 1 in the Comoro Is., 1 in Madagascar, 3 in S. India and Sri Lanka and 2 in the Malay Peninsula.

Opinions have long been divided on the question of the distinctness of *Micrococca* from *Claoxylon*, with Bentham in Hooker (1849), Pax (1890), Prain (1911, 1912), Pax & Hoffmann (1914, 1931), Airy Shaw (1971), Whitmore (1973) and Webster (1994) favouring recognition at the generic level, and Müller (1865, 1866), Hooker (1887) and Susila Rani & Balakrishnan (1995) favouring subsumption. The former and majority view is being adopted here.

Subtribe 33h. **Lobaniliinae** *Radcl.-Sm.*, Kew Bull. 44: 339 (1989); G. L. Webster, Ann. Missouri Bot. Gard. 81(1): 88 (1994). T.: *Lobanilia* Radcl.-Sm.

Acalypheae subtribe *Mercurialinae* series *Claoxyliformes* Pax & K. Hoffm., Pflanzenr. 63: 75 (1914), p.p.
Claoxyleae Hutch., Amer. J. Bot. 56: 752 (1969), p.p.

Dioecious trees; indumentum stellate and sometimes also simple; dried specimens often yellowish-brown in colour; leaves sometimes palminerved, otherwise ± as in *Claoxylinae*; stipules minute, soon deciduous. Inflorescences axillary, racemiform, the male usually elongate, dense, the female much shorter or flowers solitary; male sepals 3, becoming reflexed; petals 0; disc-glands interstaminal; stamens 17–30, free, anther-thecae free to the base; pollen ?; pistillode 0; female sepals 3; petals 0; disc-glands 3, free, petaloid, or disc annular; ovary 3–4-locular; styles simple, papillose or plumose. Fruit dehiscent. Seeds ecarunculate.

A monogeneric subtribe close to the *Claoxylinae*, but differing primarily in the stellate indumentum.

174. **Lobanilia** *Radcl.-Sm.*, Kew Bull. 44: 334 (1989); G. L. Webster, Ann. Missouri Bot. Gard. 81(1): 88 (1994). T.: *L. luteobrunnea* (Baker) Radcl.-Sm. (B.: *Croton luteo-brunneus* Baker).

Claoxylon sect. *Luteobrunnea* Pax & K. Hoffm., Pflanzenr. 63: 126 (1914).

Dioecious small to medium trees. Indumentum stellate. Dried specimens often yellowish-brown in colour. Leaves alternate, petiolate, stipulate, simple, subentire or finely glandular-serrulate, -denticulate or -crenate, penni- or sometimes palminerved; stipules minute, soon deciduous. Inflorescences axillary, racemiform, solitary, pedunculate, unisexual, the male usually elongate, dense, the female much shorter or even represented only by a single flower; male bracts 1–3-flowered, females 1-flowered. Male flowers shortly pedicellate; calyx closed in bud, subglobose, later valvately 3-partite, the segments becoming reflexed; petals 0; disc-glands interstaminal, ∞, free, obovoid, pilose at the apex; stamens 17–30, filaments free, elongate, anthers erect, basifixed, extrorse, 2-thecate, thecae obovoid, free except at the base, longitudinally dehiscent; pollen not known; pistillode 0. Female flowers shortly pedicellate; calyx open in bud, 3-partite; petals 0; disc-glands 3, free, petaloid, or disc annular; ovary 3–4-locular, 1 ovule per locule; styles 3–4, shortly connate at the base, reflexed, simple, stigmas papillose-plumose. Fruit 3–4-coccous, loculicidally dehiscent into 3–4 valves. Seeds subglobose, ecarunculate, thinly arillate; aril fleshy. Fig.: See Notul. Syst. (Paris) 9: 163, t. 2.12–16 and 172, t. 3.1–11 (1941) (as *Claoxylon*); Kew Bull. 44: 336, t. 1 (1989).

A genus of 8 species confined to Madagascar.

Subtribe 33i. **Mareyinae** *(Hutch.) Radcl.-Sm.*, **stat. nov.**

Mareyeae Hutch., Amer. J. Bot. 56: 751 (1969), p.p. . T.: *Mareya* Baill.

Dioecious or monoecious trees or shrubs; leaves alternate, stipulate, simple, denticulate, penninerved, often glandular; stipules deciduous. Inflorescences axillary or terminal, spicate or racemose; flowers fragrant; male sepals (2)3–4(5), valvate; disc-glands 8–∞; stamens (7)8–40, anther-thecae pendulous in bud, connective branched or not; pistillode + or 0; female sepals 3–5, imbricate or open; disc 3–5-lobed; ovary (2)3(4)-locular; styles (2)3(4), simple, papillose to laciniate-plumulose. Fruits dehiscent or indehiscent. Seed-exotesta not fleshy.

This subtribe, subsumed in my view unwarrantably by Webster (1994) into the *Claoxylinae* on account of superficial similarities in the androecium, is here considered only to consist of 2 genera, and thus is much reduced in concept from Hutchinson's tribe *Mareyeae,* which also included *Afrotrewia, Amyrea, Benoistia, Chondrostylis, Crotonogynopsis, Discoglypremna, Necepsia* (incl. *Neopalissya*), and *Podadenia.* Of these, the first is unplaceable owing to the type having been destroyed and the genus never having been re-collected, the next two are here accorded suprageneric categories of their own, the fourth is placed in the *Agrostistachydeae,* the fifth in the *Adelieae,* the sixth in the *Caryodendreae,* the seventh and eighth (which are in fact congeneric) in the *Bernardieae,* and the last in the *Pycnocomeae-Blumeodendrinae*!

KEY TO THE GENERA

1. Monoecious; leaves stipellate; male disc-glands ∞; stamens 9–40, not whorled; connective branched; pistillode usually 0; stigmas laciniate-plumulose; fruits dehiscent ························ 175. *Mareya*
1. Dioecious; leaves exstipellate; male disc-glands 8; stamens (7)8, in 2 whorls; connective not branched; pistillode +; stigmas lightly papillose; fruits indehiscent ································ 176. *Mareyopsis*

175. **Mareya** *Baill.,* Adansonia I. 1: 73 (1860); Müll. Arg. in DC., Prodr. 15(2): 792 (1866); Benth., Hooker's Icon. Pl. 13: 63, t. 1281 (1879), and Gen. Pl. 3: 312 (1880); Prain, Fl. Trop. Afr. 6(1): 910 (1912); Pax & K. Hoffm., Pflanzenr. 68.XIV(Addit. VI): 11 (1919), and in Nat. Pflanzenfam. ed. 2, 19C: 104 (1931); Keay (ed.), Fl. W. Trop. Afr. ed. 2, 1: 404 (1958); J.-G.Adam, Mém. Mus. Natl. Hist. Nat., B, Bot. 20: 495 (1971); Radcl.-Sm., Fl. Trop. E. Afr., Euph. 1: 216 (1987); G. L. Webster, Ann. Missouri Bot. Gard. 81: 88 (1994); J. Léonard, Bull. Jard. Bot. Belg. 65: 4 (May 1996), and Fl. Afr. Cent. Euph. 3: 18 (Oct. 1996). T.: *M. spicata* Baill. [=*M. micrantha* (Benth.) Müll. Arg. (B.: *Acalypha micrantha* Benth.)]

Monoecious trees or shrubs. Indumentum simple. Leaves alternate, shortly petiolate, stipulate, often stipellate, simple, entire to remotely glandular-denticulate or -serrulate, penninerved, often basally 2-glandular; stipules and stipels soon deciduous. Inflorescences axillary or terminal, spicate or racemose, solitary or paucifasciculate, usually elongate, bisexual, lax; male bracts many-flowered, females mostly 1-flowered. Flowers small or minute, fragrant. Male flowers subsessile or pedicellate; calyx closed in bud, subglobose, apiculate, later valvately (2)3–4-partite; petals 0; disc-glands interstaminal, ∞, free, minute; stamens 9–40, not whorled, inserted on the receptacle, filaments free, anthers apicifixed, 2-thecate, thecae distinct, fusiform, stipitate, pendulous in bud, later flexuous-divaricate, distally dehiscent, connective 2-ramous; pollen small (c. 20μ diam.), oblate-spheroidal, convex-triangular in polar view, 3-colporate, colpi narrow, colpus-membrane granulate, costae 0, colpus transversalis elongated, costae 0, tectate, psilate; pistillode usually 0; receptacle convex. Female flowers subsessile or long-pedicellate; sepals 3–5, imbricate or open in bud; petals 0; disc flattened, deeply somewhat irregularly 3–5-lobate, the lobes oppositisepalous; ovary (2)3(4)-locular, 1 ovule per locule; styles (2)3(4), free or slightly connate at the base, recurved and accumbent on the ovary, simple, stigmas fimbriate-plumose, persistent. Fruit deeply (2)3(4)-lobate, depressed, septicidally dehiscent into 3 2-valved cocci; endocarp woody; columella persistent, 3-cornute at the apex. Seeds subglobose, ecarunculate; testa thinly crustaceous, smooth, shiny, brown; endosperm fleshy; cotyledons broad, flat. Fig.: See Hooker's Icon. Pl. 13: 63, t. 1281 (1879); Mém. Mus. Natl. Hist. Nat., B, Bot. 20: 498, t. 184 (1971); Fl. Trop. E. Afr., Euph. 1: 217, t. 41 (1987); Bull. Jard. Bot. Belg. 65: 9, t. 1 and 13, t. 2 (May 1996); Fl. Afr. Cent. Euph. 3: 22, fig. 1 and 27, t. 3 (Oct. 1996).

A tritypic tropical African genus ranging from Guinea Bissau to Uganda and Angola.

176. **Mareyopsis** *Pax & K. Hoffm.*, Pflanzenr. 68. XIV(Addit. VI): 13 (1919), and in Nat. Pflanzenfam. ed. 2, 19C: 105 (1931); Keay (ed.), Fl. W. Trop. Afr. ed. 2, 1: 403 (1958); J. Léonard, Bull. Jard. Bot. Belg. 65: 15 (May 1996), and Fl. Afr. Cent. Euph. 3: 16 (Oct. 1996); Breteler *et al.* Bull. Jard. Bot. Belg. 66: 131 (Jul. 1997). T.: *M. longifolia* (Pax) Pax & K. Hoffm. (B.: *Mareya longifolia* Pax).

Very like 175. *Mareya*, but dioecious; leaves exstipellate; male sepals (3)4(5); disc-glands 4 pairs, situated among the outermost stamens, oppositisepalous; stamens (7)8, in 2 whorls of 4, filaments slightly inflected in bud and dilated at the base, connective slightly enlarged but not 2-branched, not or very slightly produced; pistillode (2)3-ampulliform; female sepals open in bud; stigmas lightly papillose; fruits indehiscent. Fig.: See Bull. Jard. Bot. Belg. 65: 17, t. 3 (1996); Fl. Afr. Cent. Euph. 3: 19, t. 2 (Oct. 1996); Bull. Jard. Bot. Belg. 66: 133, t. 1 and 137, t. 3 (1997).

A ditypic African genus, 1 from Nigeria, Cameroun, Gabon and Congo (Kinshasa), the other only in Gabon.

Subtribe 33j. **Rottlerinae** *Meisn.*, Pl. Vasc. Gen. 1: 339 (1841); G. L. Webster, Taxon 24: 598 (1975), and Ann. Missouri Bot. Gard. 81(1): 89 (1994). T.: *Rottlera* Roxb. [=*Mallotus* L.]

Trewiaceae Lindl., Intr. Nat. Syst. ed. 2 : 174 (1835). T.: *Trewia* L.
Acalypheae subtribe *Mercurialinae* series *Trewiiformes* Pax & K. Hoffm., Pflanzenr. 63: 139 (1914), and Nat. Pflanzenfam. ed. 2, 19C: 113(1931), p.p. T.: *Trewia* L.
Acalypheae subtribe *Coelodisceae* Müll. Arg., Linnaea 34: 143 (1865), and in DC., Prodr. 15(2): 757 (1866). T.: *Coelodiscus* Baill. [=*Mallotus* Lour.]
Malloteae Hutch., Amer. J. Bot. 56: 752 (1969), p.p. T.: *Mallotus* Lour.

Dioecious or more rarely monoecious trees or shrubs; indumentum stellate, rarely simple; leaves alternate or opposite, simple or lobate, penninerved or palminerved, glandular or not, often granulose-glandular beneath; stipules + or 0. Inflorescences terminal, oppositifolious or axillary, racemose or paniculate; male sepals (2)3–4(5); petals 0; disc +, inter- or intrastaminal or 0; stamens 15–300, free, anthers usually muticous, thecae not pendulous; pollen 3(4)-colporate, tectate-perforate to rugulose, spinulose; pistillode usually 0; female sepals 3–6(10), free or calyx gamosepalous; disc 0; ovary (1)2–4(9)-locular, smooth, echinate, horned or winged; styles usually simple, stigmas papillose or plumose-laciniate. Fruit dehiscent or indehiscent. Seeds usually ecarunculate; exotesta often fleshy.

Pax's revised *Trewiiformes* (1931) and Hutchinson's *Malloteae* were substantially the same, except that the latter had pulled in *Discocleidion* from the former's *Bernardiiformes* and *Blumeodendron* from his *Wetriariiformes.* However, in the current treatment these are back in the affinities which Pax had indicated for them. Furthermore, other excisions were made by Webster (1994): *Melanolepis* (removed to the *Doryxylinae*), *Gavarretia* and *Conceveiba* (to the *Conceveibinae*), *Adriana* (to the *Adrianinae*) and *Wetria* (to the *Cleidiinae*). *Avellanita* has been kept in, despite Webster's (1994) having included it amongst his *'Incertae sedis'*, for although imperfect, this seems to be where its affinities lie, and not with *Argythamnia,* which it vegetatively resembles.

Key to the genera

1. Ovary (7)8(9)-locular; fruit baccate; leaves alternate, glandular; stamens 15–20 · 185. *Octospermum*
1. Ovary (1)2–4(5)-locular; fruit regmataceous or drupaceous · · · · · · · · · 2
2. Styles 2-fid or simple; leaves eglandular · · · · · · · · · · · · 184. *Rockinghamia*
2. Styles simple; leaves usually glandular · 3
3. Fruit indehiscent; female calyx gamosepalous · · · · · · · · · · · · · · · · · · · 4
3. Fruit dehiscent; female sepals free or connate · 5
4. Leaves palminerved; male disc 0; ovary 2–4-locular · · · · · · · · 182. *Trewia*
4. Leaves penninerved; male disc +, partly pistillodioid; ovary 1(2)-locular · 183. *Neotrewia*
5. Ovary horned or winged; inflorescences axillary · · · · · · · · · · · · · · · · · 6
5. Ovary smooth to echinate; inflorescences terminal, oppositifolious or axillary; bracts small; stamens 20–200; pollen tectate-perforate to rugulose · 7
6. Stamens over 200; seeds ecarunculate · · · · · · · · · · · · · · · 179. *Cordemoya*
6. Stamens 15–20; seeds carunculate · · · · · · · · · · · · · · · · · · 180. *Coccoceras*
7. Leaves usually granulose-glandular, alternate or opposite; styles somewhat elongate; fruits smooth to echinate · · · · · · · · · · · · · · · · 177. *Mallotus*
7. Leaves eglandular · 8
8. Leaves opposite; dioecious or monoecious; styles contracted; fruit echinate or ± smooth · 178. *Deuteromallotus*
8. Leaves alternate; monoecious; styles elongate; fruit smooth · · 181. *Avellanita*

177. **Mallotus** *Lour.*, Fl. Cochinch.: 635 (1790); Müll. Arg. in DC., Prodr. 15(2): 956 (1866), p.p.; Benth., Fl. Austral. 6: 138 (1873), and Gen. Pl. 3: 319 (1880); Hook. f., Fl. Brit. India 5: 427 (1887), p. max. p., excl. Sect. I; F. B. Forbes & Hemsl., J. Linn. Soc., Bot. 26: 439 (1894); K. Schum. & Lauterb., Fl. Schutzgeb. Südsee Nachtr.: 394 (1901); J. J. Sm., Meded. Dept. Landb. Ned.-Indië 10: 394 (1910); Prain, Fl. Trop. Afr. 6(1): 927 (1912); Pax & K. Hoffm., Pflanzenr. 63: 145 (1914); Gagnep., Fl. Indo-Chine 5: 350, 366 (1925); Pax & K. Hoffm., Nat. Pflanzenfam. ed. 2, 19C: 113 (1931); Robyns, Fl. Parc Nat. Alb. 1: 459 (1948); Merr. & Steenis, Webbia 8: 405 (1951); Hurus., J. Fac. Sci. Univ. Tokyo, Sect. 3, Bot. 6: 304 (1954); Keay (ed.), Fl. W. Trop. Afr. ed. 2, 1: 402 (1958); Backer & Bakh. f., Fl. Java 1: 482 (1963); Airy Shaw, Kew Bull. 20: 38 (1966), 21: 379 (1968), 25: 526 (1971), and 26: 292 (1972); Whitmore, Tree Fl. Mal. 2: 113 (1973); Berhaut, Fl. Ill. Sénég. 3: 523 (1975); Airy Shaw, Hooker's Icon. Pl. 38: t. 3715 (1974), Kew Bull. Addit. Ser. 4: 160 (1975), 8: 162 (1980), Kew Bull. 35: 649 (1980), 36: 323 (1981), 37: 28 (1982), and Euph. Philipp.: 35 (1983); Grierson & D. G. Long, Fl. Bhut. 1: 799 (1987); McPherson & Tirel, Fl. Nouv.-Caléd. 14 (1): 104 (1987); Radcl.-Sm., Fl. Trop. E. Afr., Euph. 1: 235 (1987); G. L. Webster, Ann. Missouri Bot. Gard. 81: 89 (1994); Radcl.-Sm., Fl. Zamb. 9(4): 179 (1996); Philcox in Dassan., ed., Fl. Ceyl. XI: 148 (1997). T.: *M. cochinchinensis* Lour.

Echinus Lour., Fl. Cochinch.: 633 (1790); A. Juss., Euphorb. Gen.: 61 (1824). T.: *E. trisulcus* Lour. [=*M. cochinchinensis* Lour.]

Rottlera Roxb., Pl. Coromandel. 2: 36 (1798), *non* Willd. (1797) (= *Trewia* L.); A. Juss., Euphorb. Gen.: 32 (1824), p.p.; Baill., Étude Euphorb.: 421 (1858). T.: *R. tinctoria* Roxb. [=*M. philippensis* (Lam.) Müll. Arg. (B.: *Croton philippense* Lam.) (sic)]

Adisca Blume, Bijdr. Fl. Ned. Ind.: 609 (1826). LT.: *A. floribunda* Blume [=*M. floribundus* (Blume) Müll. Arg.] (chosen by Webster, 1994)

Lasipana Raf., Sylva Tellur.: 21 (1838). T.: *L. tricuspis* Lour. [=*M. cochinchinensis* Lour.]

Plagianthera Rchb. f. & Zoll., Acta Soc. Regiae Sci. Indo-Neerl. 1(4): 19 (1856). T.: *P. oppositifolia* (Blume) Rchb. f. & Zoll. (B.: *Rottlera oppositifolia* Blume) [=*M. blumeanus* Müll. Arg.]

Hancea Seem., Bot. Voy. Herald: 409 (1857). T.: *H. hookeriana* Seem. [=*M. hookerianus* (Seem.) Müll. Arg.]

Axenfeldia Baill., Étude Euphorb.: 419 (1858). T.: *A. intermedia* Baill. [=*M. muricatus* (Wight) Müll. Arg. (B.: *Claoxylon muricatus* Wight)]

Coelodiscus Baill., Étude Euphorb.: 293 (1858); Hook. f., Fl. Brit. India 5: 425 (1887). T.: *C. dioicus* (Wall. ex Baill.) Baill. (B.: *Ricinus dioicus* Wall. ex Baill. [=*M. eriocarpus* (Thwaites) Müll. Arg. (B.: *Rottlera eriocarpa* Thwaites)]

Aconceveibum Miq., Fl. Ned. Ind. 1(2): 389 (1859). T.: *A. trinerve* Miq. [=*M. philippensis* (Lam.) Müll. Arg. (B.: *Croton philippense* Lam.) (sic)]

Echinocroton F. Muell., Fragm. 1: 31 (1859). T.: *E. claoxyloides* F. Muell. [=*M. claoxyloides* (F. Muell.) Müll. Arg.]

Diplochlamys Müll. Arg., Flora 47: 539 (1864). T.: *D. griffithianus* Müll. Arg. [=*M. griffithianus* (Müll. Arg.) Hook. f.]

Dioecious or more rarely monoecious trees or shrubs, rarely scramblers. Indumentum simple, stellate or mixed. Leaves alternate or opposite, often unequal when the latter, petiolate, sometimes peltate, stipulate, simple or lobate, entire or toothed, penninerved or palminerved, tertiary nerves commonly transverse, parallel, often with 2 or more basal macular glands, usually pellucid- or granulose-glandular beneath; stipules mostly subulate, often small. Inflorescences terminal, subterminal, lateral, oppositifolious or axillary, spiciform, racemiform, paniculate or thyrsiform, unisexual, few- or more usually many-flowered; bracts small, males (1)–many-flowered, females 1(2)-flowered. Male flowers usually pedicellate; calyx closed in bud, ± globose or ellipsoid, later valvately (2)3–4(5)-partite; petals 0; disc-glands 0–∞, free, interstaminal; stamens 20–200(250), inserted on the receptacle, filaments free, anthers dorsifixed near the base, thecae distinct, parallel, longitudinally dehiscent, connective narrow or broad, truncate, subpeltate or produced, fimbriate, 2-fid or branched, often varying in the same flower; pollen oblate-spheroidal, 3-colporate, colpi narrow, costae +, colpus transversalis small, costae +, tectate-perforate to rugulose; pistillode usually 0, rarely +, minute and disciform; receptacle slightly elevated. Female flowers usually pedicellate; calyx shallowly or deeply 3–5(10)-lobed, the lobes imbricate or valvate, persistent, or spathaceous, caducous; disc 0; ovary (2)3(4)-locular, smooth to echinate, ovule 1 per locule; styles (2)3(4), ± elongate, free or slightly connate at the base, simple, recurved, stigmas papillose or plumose. Fruit globose or (2)3(4)-dymous, smooth, muricate or echinate, tomentose or glandular-lepidote, dehiscing septicidally into 2-valved cocci; endocarp crustaceous; columella 3-alate, persistent. Seeds globose or ovoid, ecarunculate; exotesta thinly fleshy or softly membranaceous; endotesta indurated, crustaceous, smooth or rugulose; endosperm fleshy; cotyledons broad, flat. Fig.: See Euphorb. Gen.: t. 9, f. 21A.1–9 and B.1–9 (1824) (as *Rottlera*); Étude Euphorb.: t. 19, f. 29–31 (1858) (as *Rottlera*); Pflanzenr. 63: 150–200, tt. 22–29 (1914); Fl. Indo-Chine 5: 360–377, tt. 41–43.1 (1925/6); Nat. Pflanzenfam. ed. 2, 19C: 116, t. 59 and 117, t. 60 (1931); Formos. Trees: 352–354, tt. 307–309 (1936); Notul. Syst. (Paris) 9: 178, t. 4.1–6 (1941); J. Fac. Sci. Univ. Tokyo, Sect. 3, Bot. 6: 306–7, tt. 42, 43 (1954); Hooker's Icon. Pl. 38: t. 3715 (1974); Fl. Ill. Séneg. 3: 522 (1975); Kew Bull. Addit. Ser. 8: 225, t. 3.4 (1980); Kew Bull. 35: 660, t. 5B1–4 (1980); 36: 314, t. 7B1–5 and 331, t. 8 (1981); Fl. Pak. 172: 58, t. 11C–F (1986); Fl. Bhut. 1: 801, t. 49i–k (1987); Fl. Nouv.-Caléd. 14 (1): 171, t. 35.6–9 (1987); Fl. Trop. E. Afr., Euph. 1: 238, t. 48 (1987); Fl. Zamb. 9(4): 181, t. 38 (1996).

A palaeotropical genus of c. 150 species, 2 in Tropical Africa, the rest Indo-Pacific. The species are grouped into 8 well-characterized sections. For a key to these, see Airy Shaw (1972).

The light wood of *M. cochinchinensis* Lour. is used for matches and packing-cases; a bright yellow dye is obtained from *M. discolor* F. Muell., and a reddish dye from *M. philippensis* (Lam.) Müll. Arg., the Kamala Tree.

178. **Deuteromallotus** *Pax & K. Hoffm.*, Pflanzenr. 63: 212 (1914), and in Nat. Pflanzenfam. ed. 2, 19C: 118 (1931); Leandri in H. Perrier, Cat. Pl. Madag.,

FIG. 28. *Deuteromallotus acuminatus* (Baill.) Pax & K. Hoffm. (B.: *Boutonia acuminata* Baill.). **A** Habit × $^{3}/_{4}$; **B** Leaf-undersurface × 9; **C** Male flower-bud × 7.5; **D** Male flower × 7.5; **E** Female flower × 7.5; **F** Fruit × 1. **A–D** from *McNicoll et al.* 550; **E** from *McNicoll et al.* 586; **F** from *Scott-Elliot* 1773. Drawn by Camilla Speight.

Euph.: 38 (1935), and in Bull. Soc. Bot. France 103: 605 (1956); G. L. Webster, Ann. Missouri Bot. Gard. 81: 90 (1994). T.: *D. acuminatus* (Baill.) Pax & K. Hoffm. (B.: *Boutonia acuminata* Baill.)

Dioecious or rarely monoecious trees. Indumentum simple and stellate. Leaves opposite or subopposite, unequal at each node, petiolate, stipulate, simple, entire, penninerved, eglandular. Inflorescences axillary, usually unisexual, the males racemiform, many-flowered, females 1(2)-flowered; bracts 1-flowered. Male flowers pedicellate; calyx closed in bud, globose, apiculate or not, later valvately 3(4)-partite; petals 0; disc 0; stamens 75–150, filaments free, anthers basifixed or subapicifixed, latrorse, muticous, thecae distinct, free at base, divergent or not, longitudinally dehiscent, connective narrow; pollen ± as in 177. *Mallotus*, but with the colpi very short; pistillode 0; receptacle hemispherical. Female flowers shortly pedicellate to subsessile; sepals 4–6(7), free, imbricate, linear-subulate; petals 0; disc 0; ovary 2–3-locular, smooth or echinate, ovule 1 per locule; styles 2–3, elongate, or stigmas sessile, forming a 4–6-lobed corona atop the ovary, ± smooth. Fruit 2–3-coccous, ± smooth or echinate, dehiscing into 2–3 2-valved cocci; endocarp very thin. Seeds subglobose, ecarunculate; exotesta soft, whitish; endotesta crustaceous, brown; endosperm copious; embryo small, radicle centrifugal, cotyledons orbicular or elliptic, flat. Fig.: See Notul. Syst. (Paris) 9: 178, t. 4.7–13 (1941); Bull. Soc. Bot. France 103: 606, t. 1 (1956). See also Fig. 28, p. 227.

A Malagasy genus of 2 species.

179. **Cordemoya** *Baill.*, Adansonia I. 1: 255 (1861); Pax & K. Hoffm., Pflanzenr. 63: 208 (1914), and in Nat. Pflanzenfam. ed. 2, 19C: 117 (1931); Coode, Fl. Masc. 160. Euph.: 50 (1982); G. L. Webster, Ann. Missouri Bot. Gard. 81: 90 (1994). LT.: *C. integrifolia* (Willd.) Baill. (B.: *Ricinus integrifolius* Willd.) (chosen by Webster, 1994).

Boutonia Bojer ex Bouton, Trav. Soc. Hist. Nat. Maurice 1: 151 (1846); Baill., Étude Euphorb.: 400 (1858), non DC. (1838). T.: as above.

Mallotus sect. *Cordemoya* (Baill.) Müll. Arg., Linnaea 34: 186 (1865), and in DC., Prodr. 15(2): 960 (1866), p.p.

Mallotus sect. *Boutonia* (Bojer ex Bouton) Benth., Gen. Pl. 3: 319 (1880); Pax in Engl. & Prantl, Nat. Pflanzenfam. 3(5): 54 (1890).

Monoecious trees. Indumentum simple and stellate, on young growth only. Leaves alternate, petiolate, stipulate, simple, entire, penninerved, subcoriaceous, subeglandular. Inflorescences axillary, males racemose or paniculate, females racemose, fewer-flowered than the males or flowers solitary; bracts 1-flowered. Male flowers shortly pedicellate; calyx closed in bud, globose, later valvately 3(5)-partite; petals 0; disc 0; stamens 200–300, filaments free, anthers basifixed, latrorse, subglobose, muticous, thecae parallel, ± free at base, longitudinally dehiscent, connective scarcely visible; pollen ± as in 177. *Mallotus*; pistillode 0; receptacle convex. Female flowers shortly pedicellate; sepals (3)5–6, free, linear-subulate; petals 0; disc 0; ovary (2)3-locular, 6-horned or winged, ovule 1 per locule; styles (2)3, shortly connate at the base, elongate, stigmas minutely papillulose. Fruit 3-coccous, cocci 2-horned, dehiscing into 3 2-valved cocci; exocarp thinly crustaceous; endocarp thinly woody; columella obtriangular, 3-alate, persistent. Seeds subglobose, ecarunculate; testa shiny, dark brown. Fig.: See Fl. Masc. 160. Euph.: 52, t. 9 (1982).

A monotypic genus endemic to Mauritius and Réunion.

180. **Coccoceras** *Miq.*, Fl. Ned. Ind. Suppl.: 455 (1860); Müll. Arg. in DC., Prodr. 15(2): 949 (1866); Benth., Gen. Pl. 3: 308 (1880); Hook. f., Fl. Brit. India 5: 424 (1887); Pax in Engl. & Prantl, Nat. Pflanzenfam. 3(5): 55 (1890); Pax & K. Hoffm. in Pflanzenr. 63: 209 (1914); Gagnep., Fl. Indo-Chine 5: 376 (1926); Pax & K. Hoffm. in Nat. Pflanzenfam. ed. 2, 19C: 118 (1931); G. L. Webster, Ann. Missouri Bot. Gard. 81: 90 (1994). T.: *C. sumatranum* Miq.

Mallotus sect. *Coccoceras* (Miq.) Airy Shaw, Kew Bull. 16: 351 (1963).

Very like 177. *Mallotus* Lour., but with leaves sometimes glandular-punctate above as well as beneath, stamens 15–20, pollen 3–4-colporate, the colpi very short, fruits depressed, the fruit-cocci characteristically produced into long horn-like processes, winged or angled and carinate, rarely rounded, loculicidally dehiscent into 3 valves or else indehiscent, and seeds carunculate. Fig.: See Pflanzenr. 63: 210, t. 30A–D (1914); Fl. Indo-Chine 5: 377, t. 43.2, 3 (1926); Nat. Pflanzenfam. ed. 2, 19C: 119, t. 61 A–D (1931).

A Tropical Asiatic genus of 5 species, ranging from Lower Burma (Myanmar), Thailand and Indochina to Sumatra and Borneo, with an outlier in the Solomon Is.

Treated as a section of *Mallotus* by Airy Shaw (1963) and in all his subsequent publications.

181. **Avellanita** *Phil.*, Linnaea 33: 237 (1864/65); Benth., Gen. Pl. 3: 289 (1880); Pax in Engl. & Prantl, Nat. Pflanzenfam. ed. 1, 3(5): 76 (1890), and in Pflanzenr. 42: 15 (1910); Pax & K. Hoffm. in Nat. Pflanzenfam. ed. 2, 19C: 113 (1931); G. L. Webster, Ann. Missouri Bot. Gard. 81(1): 130 (1994). T.: *A. bustillosii* Phil.

A monoecious shrub. Indumentum simple. Leaves alternate, the upper sometimes verticillate, shortly petiolate, stipulate, simple, entire, penninerved, eglandular, somewhat purplish-tinged; stipules triangular, persistent. Inflorescences terminal, cymose, few-flowered, each cyme with 1 median female flower and 2–3 peripheral male flowers; bracts linear. Male flowers shortly pedicellate; sepals 5, valvate, connate at the base; petals 0; disc 0; stamens numerous, forming a globose mass, filaments very short, free, anthers almost sessile, small, subglobose; pollen resembling that of 177. *Mallotus*, but

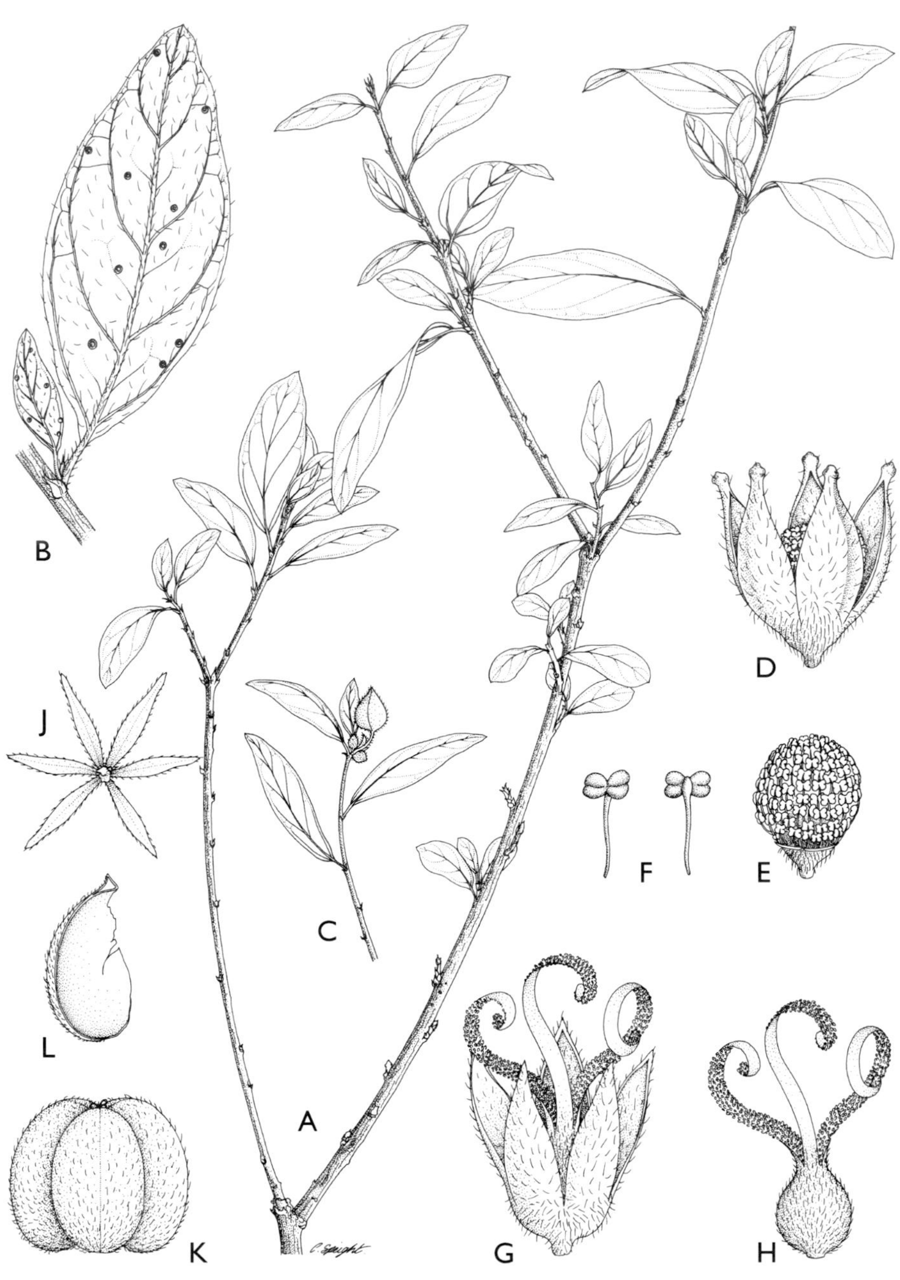

FIG. 29. *Avellanita bustillosii* Phil. **A** Habit, substerile × 1; **B** Leaf × 3; **C** Male inflorescence × 1; **D** Male flower × 9; **E** Androecium × 9; **F** Stamens × 45; **G** Female flower × 9; **H** Ovary × 9; **J** Fruiting-calyx × 3; **K** Fruit × 3; **L** Fruit-valve × 3. All from *Philippi* s.n. Drawn by Camilla Speight.

with indistinct costae; pistillode 0. Female flowers usually pedicellate; sepals almost free to the base, narrower than in the male, persistent, but otherwise resembling them; disc 0; ovary globose, 3-locular, smooth, ovule 1 per locule; styles 3, erect, elongate, free, simple, stigmas densely papillose. Fruit 3-coccous, smooth, pubescent, dehiscing septicidally into 3 2-valved cocci; exocarp thinly crustaceous; endocarp crustaceous; columella short, persistent. Seeds globose, ecarunculate; testa smooth. Fig.: See Pflanzenr. 42: 15, t. 4 (1910). See also Fig. 29, p. 230.

A monotypic genus from Chile.

Webster (1994) had this genus among his 'Incertae sedis', but with the observation that "evidence from pollen and floral characters clearly points to.... *Acalyphoideae*". In habit, the genus strongly recalls the subtribe *Ditaxinae*, but the author considers that from other standpoints it belongs among the *Rottlerinae*, conceding however that further investigation may warrant the establishment of a distinct subtribe to accommodate it.

182. **Trewia** *L.*, Sp. Pl.: 1193 (1753) ("*Trevia*"), and Gen. Pl. ed. 5: 500 (1754); Klotzsch, Arch. Naturgesch. 7: 255 (1841); Baill., Étude Euphorb.: 408 (1858); Müll. Arg. in DC., Prodr. 15(2): 953 (1866), p.p.; Benth., Gen. Pl. 3: 318 (1880); Hook. f., Fl. Brit. India 5: 423 (1887); J. J. Sm., Meded. Dept. Landb. Ned.-Indië 10: 389 (1910); Pax in Engl. & Prantl, Nat. Pflanzenfam. 3(5): 53 (1890); Pax & K. Hoffm. in Pflanzenr. 63: 140 (1914); Gagnep., Fl. Indo-Chine 5: 343 (1925); Pax & K. Hoffm. in Nat. Pflanzenfam. ed. 2, 19C: 113 (1931); Backer & Bakh. f., Fl. Java 1: 481 (1963) (as *Trevia*); Airy Shaw, Kew Bull. 20: 405 (1967), and 26: 343 (1972); Whitmore, Tree Fl. Mal. 2: 134 (1973); Airy Shaw, Kew Bull. Addit. Ser. 4: 200 (1975), Kew Bull. 36: 350 (1981), and Euph. Philipp.: 46 (1983); Radcl.-Sm. in Fl. Pak. 172: 56 (1986); Grierson & D. G. Long, Fl. Bhut. 1: 798 (1987); G. L. Webster, Ann. Missouri Bot. Gard. 81: 90 (1994); Philcox in Dassan., ed., Fl. Ceyl. XI: 142 (1997). T.: *T. nudiflora L.*

Dioecious deciduous trees. Wood soft. Indumentum stellate. Leaves opposite, petiolate, stipulate, simple, entire, palminerved, often with 1–2 basal macular glands. Stipules soon falling. Inflorescences axillary, racemose, males longer than the females, which may sometimes consist of only a solitary flower; male bracts 3-flowered, females 1-flowered. Flowers appearing with or before the leaves. Male flowers large, pedicellate; calyx closed in bud, globose, later valvately 3(5)-partite, the sepals broad, becoming reflexed; petals 0; disc 0; stamens 70–100, filaments free, anthers dorsifixed near the base, latrorse, oblong, emarginate, thecae parallel, contiguous, longitudinally dehiscent, connective scarcely visible; pollen ± as in 178. *Deuteromallotus*; pistillode 0; receptacle thick, convex. Female flowers pedicellate; calyx-lobes 3–5, slightly imbricate, separating somewhat irregularly and often unilaterally, soon caducous; petals 0; disc 0; ovary 2–4(5)-locular, smooth, ovule 1 per locule; styles connate at the base, elongate, simple, stigmas grossly papillose to plumose-lacinulate. Fruit globose, 2–5-locular, drupaceous, indehiscent or tardily loculicidally dehiscent; exocarp ± fleshy; endocarp thinly crustaceous. Seeds ovoid, smooth, ecarunculate; exotesta fleshy; endotesta hard;

endosperm fleshy; cotyledons broad, flat. Fig.: See Euphorb. Gen.: t. 9, f. 21C1–4 (1824) (as *Rottlera*); Étude Euphorb.: t. 18, f. 18–23 (1858); Pflanzenr. 63: 211, t. 31F,G (1914); Fl. Indo-Chine 5: 342, t. 39.3–5 (1925); Fl. Pak. 172: 58, t. 11A,B (1986).

A tropical Asiatic genus of 1 (Pax) or 2 (Bentham & Hooker) species, one wide-ranging from India and Sri Lanka to S. China, SE. Asia and W. Malesia, the other confined to peninsular India.

183. **Neotrewia** *Pax & K. Hoffm.*, Pflanzenr. 63: 211 (1914); Merr., Enum. Philipp. Fl. Pl. 2: 437 (1923); Pax & K. Hoffm., Nat. Pflanzenfam. ed. 2, 19C: 118 (1931); Airy Shaw, Kew Bull. 37: 31(1982), and Euph. Philipp.: 39 (1983); G. L. Webster, Ann. Missouri Bot. Gard. 81: 90 (1994). T.: *N. cumingii* (Müll. Arg.) Pax & K. Hoffm. (B.: *Mallotus cumingii* Müll. Arg.)

Like 182. *Trewia* L., but evergreen, leaves penninerved, male bracts many-flowered, male flowers smaller, disc +, of several glands amongst the inner stamens towards the apex of the receptacle, with the 3 at the apex pistillodioid, anthers dorsifixed above the middle, thecae pendulous, connective broad, produced, pollen operculate (opercula consisting of a single row of pila), ovary 1(2)-locular, style(s) subcircinate at the apex, and fruit 1-locular. Fig.: See Pflanzenr. 63: 211, t. 31 A–E (1914). See also Fig. 30, p. 233.

A monotypic genus confined to the N. and C. Celebes (Sulawesi) and the Philippines.

Although Webster (1994) says of this genus, 'dubiously separable from *Trewia*', the differences enumerated above would appear to justify Pax & Hoffmann's treatment.

184. **Rockinghamia** *Airy Shaw*, Kew Bull. 20: 29 (1966), and 35: 667 (1980); G. L. Webster, Ann. Missouri Bot. Gard. 81: 90 (1994). T.: *R. angustifolia* (Benth.) Airy Shaw (B.: *Mallotus angustifolius* Benth.)

Like 177. *Mallotus* Lour., but monoecious, leaves pseudoverticillate, the verticels separated by long, bare internodes, coriaceous, usually prominently nervose above and beneath, eglandular, terminal inflorescences strongly divaricate-fasciculate-thyrsoid, with male and female flowers intermingled, interstaminal glands +, pilose, anthers 4-locellate, pistillode sometimes +, ovary closely tuberculate, each tubercle tipped with a terminal bristle, styles commonly 2-fid and fruits closely muriculate. Fig.: See Kew Bull. 35: 668, t. 6C1–4 (1980).

A ditypic Australian genus, confined to Queensland.

185. **Octospermum** *Airy Shaw*, Kew Bull. 19: 311 (1965), Hooker's Icon. Pl. 38: t. 3716 (1974), and Kew Bull. Addit. Ser. 8: 176 (1980); G. L. Webster, Ann. Missouri Bot. Gard. 81: 90 (1994). T.: *O. pleiogynum* (Pax & K. Hoffm.) Airy Shaw (B.: *Mallotus pleiogynus* Pax & K. Hoffm.)

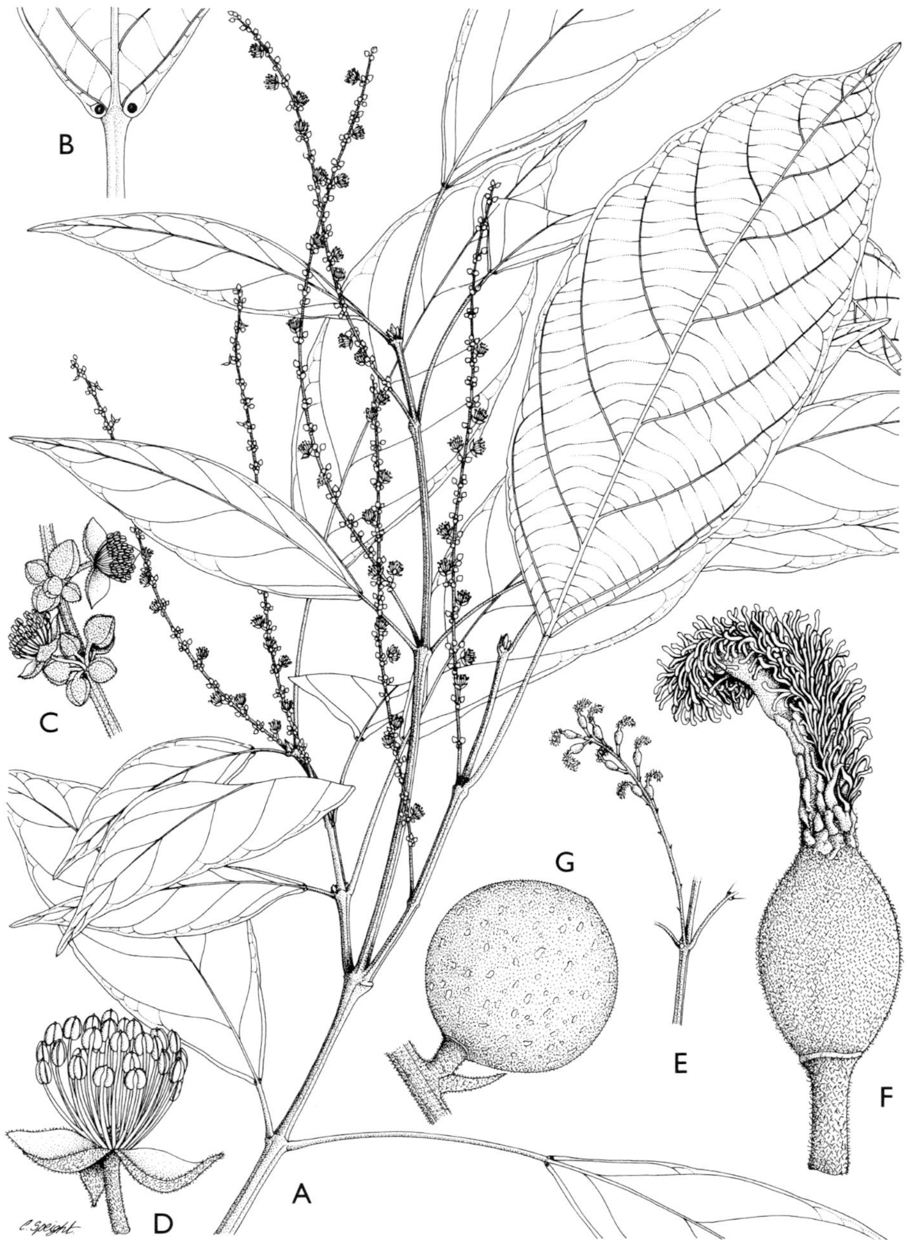

FIG. 30. *Neotrewia cumingii* (Müll. Arg.) Pax & K. Hoffm. (B.: *Mallotus cumingii* Müll. Arg.). **A** Habit × 1; **B** Leaf-base × 4.5; **C** Portion of male inflorescence × 4.5; **D** Male flower × 9; **E** Female inflorescence × 1; **F** Female flower × 13.5; **G** Fruit × 4.5. **A–D** from *Kostermans* 11061; **E, F** from *Vidal* 3890; **G** from *de Vogel &Vermeulen* 6492. Drawn by Camilla Speight.

Like 177. *Mallotus* Lour. (sect. *Rottlera*), but a very tall primary forest tree up to 36 m tall with 15–20 stamens, a conspicuously peltate anther-connective apex, a (7)8(9)-locular ovary, baccate indehiscent fruit and flattened lenticular sharp-edged radially-disposed seeds. Fig.: See Hooker's Icon. Pl. 38: t. 3716 (1974); Kew Bull. Addit. Ser. 8: 230, t. 8 (1980).

A monotypic genus endemic to New Guinea.

Subtribe 33k. **Acalyphinae** *Griseb.*, Fl. Brit. W. I.: 45 (1859); Pax & K. Hoffm., Pflanzenr. 85: 1 (1924), and in Nat. Pflanzenfam. ed. 2, 19C: 134 (1931); G. L. Webster, Taxon 24: 598 (1975), and Ann. Missouri Bot. Gard. 81(1): 90 (1994). T.: *Acalypha* L.

Monoecious or rarely dioecious annual or perennial herbs, shrubs or rarely trees; indumentum simple, rarely stellate, often glandular; leaves alternate, stipulate, sometimes stipellate, simple, crenate or serrate, palminerved or penninerved. Inflorescences terminal, axillary or both, unisexual or bisexual, spicate, rarely racemose or paniculate; female bracts usually accrescent; male sepals 4; petals 0; disc 0; stamens 8(12), filaments free, anther-thecae pendulous, flexuous-vermiform; pollen quasiporate, rugulose; pistillode 0; female sepals 3–5, imbricate; petals 0; disc 0; ovary 2–3-locular, sometimes muricate; styles 2–3, mostly free, laciniate, reddish and conspicuous; allomorphic flowers sometimes +. Fruits dehiscent (allomorphic sometimes indehiscent). Seeds minutely carunculate or ecarunculate.

A monogeneric subtribe. The genus *Acalypha* L. is distinctive in the tribe *Acalypheae* on account of its vermiform, pendulous anther-thecae, porate pollen, usually large, foliaceous female bracts and deeply lacerate often scarlet or crimson styles.

186. **Acalypha** *L.* Sp. Pl.: 1003 (1753), and Gen. Pl. ed. 5: 436 (1754); A. Juss., Euphorb. Gen.: 45 (1824); Baill., Étude Euphorb.: 440 (1858); Müll. Arg. in DC., Prodr. 15(2): 799 (1866); Benth., Hooker's Icon. Pl. 13: t. 1291 (1879), and Gen. Pl. 3: 311 (1880); Hook. f., Fl. Brit. India 5: 414 (1887); Hutch. in Fl. Trop. Afr. 6(1): 880 (1912); Pax in Engl. & Prantl, Nat. Pflanzenfam. 3(5): 60 (1890); Fawc. & Rendle, Fl. Jamaica 4: 296 (1920); Pax & K. Hoffm. in Pflanzenr. 85: 12 (1924); Gagnep., Fl. Indo-Chine 5: 333 (1925); Pax & K. Hoffm. in Nat. Pflanzenfam. ed. 2, 19C: 134 (1931); Staner, Bull. Jard. Bot. État. 15: 132 (1938); Leandri, Notul. Syst. (Paris) 10: 252 (1942); Robyns, Fl. Parc Nat. Alb. 1: 462 (1948); Pojark. in Fl. URSS XIV: 298 (1949); Alain, Fl. Cuba 3: 95 (1953); Hurus., J. Fac. Sci. Univ. Tokyo, Sect. 3, Bot. 6: 295 (1954); Keay (ed.), Fl. W. Trop. Afr. ed. 2, 1: 408 (1958); Paul G. Wilson, Hooker's Icon. Pl. 36: t. 3588 (1962); Backer & Bakh. f., Fl. Java 1: 489 (1963); Jabl., Mem. New York Bot. Gard. 17: 140 (1967); G. L. Webster, J. Arnold Arbor. 48: 370 (1967); Tutin, Fl. Eur. 2: 212 (1968); G. L. Webster, Ann. Missouri Bot. Gard. 54: 299 (1968); J.-G. Adam, Mém. Mus. Natl. Hist. Nat., B, Bot. 20: 450 (1971); Airy Shaw, Kew Bull.

26: 205 (1972); Whitmore, Tree Fl. Mal. 2: 51 (1973); Airy Shaw, Hooker's Icon. Pl. 38: t. 3719 (1974), and Kew Bull. Addit. Ser. 4: 23 (1975); Berhaut, Fl. Ill. Sénég. 3: 355 (1975); Philcox, Fl. Trin. Tob. II (X): 665 (1979); Airy Shaw, Kew Bull. Addit. Ser. 8: 13 (1980), Kew Bull. 35: 583 (1980), and 36: 246 (1981); A. C. Sm., Fl. Vit. Nov. 2: 522 (1981); Coode, Fl. Masc. 160. Euph.: 68 (1982); Correll, Fl. Bah. Arch.: 772 (1982); Airy Shaw, Kew Bull. 37: 2 (1982), and Euph. Philipp.: 2 (1983); Proctor, Kew Bull. Addit. Ser. 11: 543 (1984); Seberg, Nordic J. Bot. 4: 159 (1984); Alain, Fl. Español. 4: 62 (1986); Radcl.-Sm. in Fl. Pak. 172: 61 (1986), and in Fl. Trop. E. Afr., Euph. 1: 185 (1987); M. G. Gilbert, Kew Bull. 42: 360 (1987); Grierson & D. G. Long, Fl. Bhut. 1: 796 (1987); McPherson & Tirel, Fl. Nouv.-Caléd. 14(1): 108 (1987); G. L. Webster & Huft, Ann. Missouri Bot. Gard. 75: 1104 (1988); Alain, Descr. Fl. Puerto Rico 2: 344 (1988); R. A. Howard, Fl. Lesser Antilles 5: 5 (1989); Radcl.-Sm., Kew Bull. 45: 677 (1990); Lebrun & Stork, Enum. Pl. Afr. Trop. 1: 204 (1991); P. I. Forst., Austrobaileya 4: 209 (1994); G. L. Webster, Ann. Missouri Bot. Gard. 81: 90 (1994); Radcl.-Sm. in Fl. Zamb. 9(4): 182 (1996); Gillespie in Acev.-Rodr., ed., Fl. St. John: 202 (1996); Philcox in Dassan., ed., Fl. Ceyl. XI: 130 (1997). LT.: *A. virginica L.* (designated by Small, 1913).

Mercuriastrum Heist. ex Fabr., Enum.: 202 (1759). T.: (none designated).
Cupameni Adans., Fam. Pl. 2: 356 (1763). T.: (none designated).
Caturus L., Syst. Nat. ed. 12, 2: 650 (1767), and Mant. Pl.: 19, 127 (1767); A. Juss., Euphorb. Gen.: 45 (1824). T.: *C. spiciflora* L. [=*A. hispida* Burm. f.]
Usteria Dennst., Schlüssel Hortus Malab.: 31 (1818). T.: *U. racemosa* Dennst. [=*A. racemosa* Wall. ex Baill.]
Galurus Spreng., Syst. Veg. 1: 362 (1825). T.: *G. spiciflorus* Spreng. [=*A. spiciflora* Burm. f.]
Cupamenis Raf., Sylva Tellur.: 67 (1838). T.: *C. chamaedrifolia* Raf. [=*A. chamaedrifolia* (Raf.) Müll. Arg.]
Linostachys Klotzsch ex Schltdl., Linnaea 19: 235 (1847). T.: *L. padifolia* Schltdl. [=*A. schlechtendaliana* Müll. Arg.]
Odonteilema Turcz., Bull. Soc. Imp. Naturalistes Moscou 21: 587 (1848); Baill., Étude Euphorb.: 500 (1858). T.: *O. claussenii* Turcz. [=*A. claussenii* (Turcz.) Müll. Arg.]
Calyptrospatha Klotzsch ex Baill., Étude Euphorb.: 440 (1858), pro syn. T.: *C. pubiflora* (Baill.) Klotzsch [=*A. pubiflora* Baill.]
Gymnalypha Griseb., Bonplandia 6: 2 (1858). T.: *G. jacquinii* Griseb., nom. illeg. [=*A. villosa* Jacq.]
Corythea S. Watson, Proc. Amer. Acad. Sci. 22: 451 (1887). T.: *C. filipes* S. Watson [=*A. filipes* (S. Watson) McVaugh]
Schizogyne Ehrenb. ex Pax in Pflanzenr. 85: 98 (1924). T.: *S. ciliata* (Forssk.) Ehrenb. ex Pax, pro syn. [=*A. ciliata* Forssk.]
Acalyphopsis Pax & K. Hoffm. in Pflanzenr. 85: 178 (1924). T.: *A. celebica* Pax & K. Hoffm. [=*A. hoffmanniana* Hurus.]

Monoecious or rarely dioecious annual or perennial herbs, shrubs or rarely trees. Indumentum simple, rarely stellate, often glandular, the glands sometimes sessile, refringent. Leaves alternate, petiolate or subsessile, stipulate, sometimes stipellate, simple, crenate or serrate, palminerved or penninerved. Inflorescences terminal, axillary or both, solitary or paired, unisexual or bisexual, spicate, rarely racemose or paniculate; male inflorescences always axillary; bisexual inflorescences commonly male above, female at the base, rarely the reverse; male bracts small, usually subtending glomerules of small flowers; female bracts usually accrescent, toothed or lobed, 1–5-flowered; allomorphic female flowers terminal, axillary or basal, ebracteate. Male flowers shortly pedicellate; calyx closed in bud, later valvately 4-partite; petals 0; disc 0; stamens 8(12), filaments free, filiform or flattened, anthers apicifixed, 2-thecate, thecae distinct, divaricate or pendulous, oblong or linear, later flexuous-vermiform, extrorse, longitudinally dehiscent, connective scarcely perceptible; pollen oblate-spheroidal to suboblate, 3–5-colporate, quasiporate, colpi and colpi transversales small, short, isometric, costae +, tectate, psilate, rugulose; pistillode 0; receptacle slightly elevated. Female flowers sessile or rarely pedicellate; calyx-lobes 3–4(5), imbricate, small; petals 0; disc 0; ovary 2-3-locular, often muricate, ovule 1 per locule; styles 2–3, free or variously connate, usually laciniate, reddish and conspicuous, rarely 2-lobate or entire, yellowish-green and inconspicuous. Allomorphic female flowers sessile to long-pedicellate; calyx ± as in the normal female flowers; ovary 1(2)-locular; styles subbasal. Fruits 3-lobate, small, dehiscing septicidally into 3 2-valved cocci; endocarp thinly crustaceous. Fruits of the allomorphic female flowers 1(2)-coccous, dehiscent or not. Seeds ellipsoid or subglobose, small, carunculate or not; testa crustaceous; endosperm fleshy; cotyledons broad, flat. Fig.: See Euphorb. Gen.: t. 14, f. 45.1–5 (as *Caturus*), f. 46.1–8 (1824); Étude Euphorb.: t. 20, f. 13–19 (1858); Hooker's Icon. Pl. 13: t. 1291(1879); Fl. Jamaica 4: 298, t. 98 (1920); Pflanzenr. 85: 2, t. 1A–E, 3, t. 2 and 15, t. 3 (1924); Fl. Indo-Chine 5: 331, t. 38.9–16 and 342, t. 39.1, 2 (1925); Nat. Pflanzenfam. ed. 2, 19C: 135, t. 70, 137, t. 71A–D and 139, t. 72 (1931); Notul. Syst. (Paris) 10: 254–285, tt. 1–9 (1942); Fl. Parc Nat. Alb. 1: 465, t. 46 (1948); Fl. URSS XIV: t. XVII.6 (1949); Fl. Cuba 3: 97, t. 27 (1953); J. Fac. Sci. Univ. Tokyo, Sect. 3, Bot. 6: 299, t. 41 (1954); Hooker's Icon. Pl. 36: t. 3588 (1962); Ann. Missouri Bot. Gard. 54: 301, t. 17 (1968); Amer. J. Bot. 56: 750, t. 11A–E (1969); Hooker's Icon. Pl. 38: t. 3719 (1974); Fl. Ill. Sénég. 3: 354–364 (1975); Kew Bull. Addit. Ser. 8: 225, t. 3.1 (1980); Fl. Masc. 160. Euph.: 75, t. 14 and 77, t. 15 (1982); Fl. Bah. Arch.: 775, t. 323 (1982); Nordic J. Bot. 4: 170–183, tt. 5–13 (1984); Fl. Español. 4: 66, t. 116-1 (1986); Fl. Pak. 172: 64, t. 12 and 67, t. 13 (1986); Kew Bull. 42: 362, t. 4 (1987); Fl. Bhut. 1: 801, t. 49d,e (1987); Fl. Nouv.-Caléd. 14(1): 111, t. 22 (1987); Fl. Trop. E. Afr., Euph. 1: 191, t. 37, 201, t. 38 and 208, t. 39 (1987); Descr. Fl. Puerto Rico 2: 347, t. 59-1 (1988); Fl. Lesser Antilles 5: 11, t. 4 (1989); Kew Bull. 45: 678, t. 1 (1990); Fl. Zamb. 9(4): 188–201, tt. 39–43 (1996); Fl. St. John: 204, t. 93H–O (1996).

A largely pantropical genus of 450 species, with a few extratropical species in the Americas and E. Asia; absent from Hawai'i and a few other Pacific archipelagi (See Frontispiece).

The 4th largest genus of the family after *Euphorbia, Croton* and *Phyllanthus.*

Subtribe 33l. **Lasiococcinae** *G. L. Webster*, Taxon 24: 597 (1975), and Ann. Missouri Bot. Gard. 81(1): 91 (1994). T.: *Lasiococca* Hook. f.

Dioecious or monoecious shrubs or trees; indumentum simple or lepidote; leaves alternate, stipulate, simple, penninerved, eglandular. Inflorescences axillary or cauliflorous, unisexual, racemose, spicate or female flowers solitary; bracts eglandular; male sepals 3, petals 0; disc 0; stamens many, connate into branching fascicles; anther-connective obscure or distinct; pollen 3-colporate, exoperculate, striate or not; female sepals 5–8, imbricate; disc 0; ovary 3-locular; styles simple, papillose to plumose. Fruit dehiscent. Seeds ecarunculate; exotesta often thinly fleshy.

This subtribe comprises 3 genera included by Pax & Hoffmann (1890, 1919, 1931) in the subtribe Ricininae on account of their polyadelphous dendroid androecia. However, they differ from *Ricinus* in many ways, especially in their simple penninerved leaves, axillary spikes and ecarunculate seeds. Airy Shaw (1974) has suggested an affinity to *Mallotus*.

Key to the genera

1. Monoecious; indumentum simple; female sepals persistent, accrescent; ovary muricate-setose · 187. *Lasiococca*
1. Dioecious or monoecious; indumentum simple or lepidote; female sepals caducous · 2
2. Peltate scales 0; anther-connective arching over the thecae; pollen not striate; ovary muricate, tuberculate or smooth · · · · · 188. *Spathiostemon*
2. Peltate scales +; anther-connective obscure; pollen striate; ovary smooth · 189. *Homonoia*

187. **Lasiococca** *Hook. f.*, Hooker's Icon. Pl. 16: t. 1587 (1887), and Fl. Brit. India 5: 456 (1887); Pax & K. Hoffm., Pflanzenr. 68: 118 (1919); Haines, Bull. Misc. Inform., Kew 1920: 70 (1920); Pax & K. Hoffm., Nat. Pflanzenfam. ed. 2, 19C: 149 (1931); Airy Shaw, Kew Bull. 16: 358 (1963), and 21: 406 (1968); Whitmore, Tree Fl. Mal. 2: 104 (1973); Thin, J. Biol. (Vietnam) 8(3): 36 (1986); Grierson & D. G. Long, Fl. Bhut. 1: 806 (1987); G. L. Webster, Ann. Missouri Bot. Gard. 81: 91 (1994); van Welzen, Blumea 43: 142 (1998). T.: *L. symphylliifolia* (Kurz) Hook. f. (B.: *Homonoia symphylliifolia* Kurz).

Small monoecious trees. Indumentum simple. Leaves alternate or subternately verticillate, shortly petiolate, stipulate, simple, entire, narrow, penninerved, eglandular. Inflorescences axillary, the males racemose, the females each consisting of a solitary female flower. Male flowers shortly pedicellate; calyx closed in bud, globose, later valvately 3-partite, the segments strongly concave; petals 0; disc 0; stamens very numerous, arising from the centre of the flower, anthers arranged in numerous much-branched phalanges forming a dense globose head, apicifixed, thecae globose, subdivaricate, longitudinally dehiscent, connective arching over their apices and margins; pollen spheroidal, 3-colporate, colpi and colpi transversales very narrow, ecostate, tectate, psilate, columellae very short; pistillode 0; receptacle 0. Female flowers pedicellate; sepals 5–7, imbricate, unequal, linear to elliptic-oblong, glandular-pubescent, persistent, accrescent; petals 0; disc 0; ovary 3-locular, globose, muricate-setose, ovule 1 per locule; styles 3, shortly connate at the base, filiform, erect, simple, stigmatic surfaces papillose. Fruit 3-coccous, densely beset with rigid hispidly setose compressed-conoidal tubercles or ± muricate-setose, dehiscing septicidally into 3 2-valved cocci; endocarp thinly crustaceous; columella narrowly triquetrous, 3-cornute at apex, subpersistent. Seeds ovoid-subglobose, ecarunculate; testa papyraceous, shiny, light brown. Fig.: See Hooker's Icon. Pl. 16: t. 1587 (1887); Pflanzenr. 68: 118, t. 28 (1919); Nat. Pflanzenfam., ed. 2, 19C: 150, t. 76 (1931); Blumea 43: 144, t. 3 (1998).

A pentatypic genus ranging discontinuously from N. India to Hainan, Vietnam and Malaya.

188. **Spathiostemon** *Blume*, Bijdr. Fl. Ned. Ind.: 621 (1826); Baill., Étude Euphorb.: 292 (1858); Benth., Gen. Pl. 3: 322 (1880), in obs.; Airy Shaw, Kew Bull. 16: 357 (1963), 20: 408 (1966), and 26: 341 (1972); Whitmore, Tree Fl. Mal. 2: 132 (1973); Airy Shaw, Hooker's Icon. Pl. 38: t. 3720 (1974), Kew Bull. Addit. Ser. 4: 196 (1975), Kew Bull. 32: 407 (1978), Kew Bull. Addit. Ser. 8: 202 (1980), Kew Bull. 36: 345 (1981), 37: 35 (1982), and Euph. Philipp. 45 (1983); G. L. Webster, Ann. Missouri Bot. Gard. 81: 91 (1994); van Welzen, Blumea 43: 145 (1998). T.: *S. javensis* Blume.

Spathiostemon sect. *Euspathiostemon* Baill., Étude Euphorb.: 293 (1858).
Homonoia sect. *Euspathiostemon* (Baill.) Müll. Arg. in DC., Prodr. 15(2): 1022 (1866).
Polydragma Hook. f., Hooker's Icon. Pl. 18: t. 1701(1887), and Fl. Brit. India 5: 456 (1887). T.: *P. mallotiforme* Hook. f. [=*S. javensis* Blume]
Homonoia subgen. *Spathiostemon* (Blume) Pax & K. Hoffm., Pflanzenr. 68: 117 (1919), and Nat. Pflanzenfam., ed. 2, 19C: 149 (1931).
Clonostylis S. Moore, J. Bot. 63 (Suppl.): 101 (1925); van Welzen, Blumea 43: 150 (1998). T.: *C. forbesii* S. Moore [=*S. forbesii* (S. Moore) Airy Shaw]

Like 187. *Lasiococca*, but differing primarily in being sometimes dioecious, in having the leaves always alternate and longer-petiolate, the lamina broader, staminal phalanges 6(7), pollen prolate-spheroidal, the female sepals deciduous and the ovary and fruit sometimes smooth. Fig.: See Hooker's Icon. Pl. 18: t. 1701(1887) (as *Polydragma*), and 38: t. 3720 (1974); Kew Bull. 36: 347, t. 9 (1981); Blumea 43: 148, t. 4 (1998).

A tritypic Tropical Asiatic genus ranging from Thailand to the Philippines and New Guinea.

Airy Shaw (1978), describing how he suddenly came to realize that *Clonostylis* was congeneric with *Spathiostemon*, having grappled with the problem of the identity of the former for many years, demonstrates perfectly the important part played by intuition in taxonomic research.

189. **Homonoia** *Lour.*, Fl. Cochinch.: 636 (1790); A. Juss., Euphorb. Gen.: 60 (1824); Müll. Arg. in DC., Prodr. 15(2): 1022 (1866); Benth., Gen. Pl. 3: 322 (1880); Hook. f., Fl. Brit. India 5: 455 (1887); Pax in Engl. & Prantl, Nat. Pflanzenfam. 3(5): 71 (1890); J. J. Sm., Meded. Dept. Landb. Ned.-Indië 10: 542 (1910); Pax & K. Hoffm., Pflanzenr. 68: 114 (1919); Gagnep., Fl. Indo-Chine 5: 330 (1925); Pax & K. Hoffm., Nat. Pflanzenfam. ed. 2, 19C: 149 (1931); Backer & Bakh. f., Fl. Java 1: 491 (1963); Airy Shaw, Kew Bull. 26: 282 (1972); Whitmore, Tree Fl. Mal. 2: 102 (1973); Airy Shaw, Kew Bull. Addit. Ser. 4: 136 (1975), 8: 121 (1980), Kew Bull. 36: 310 (1981), 37: 25 (1982), and Euph. Philipp. 45 (1983); Grierson & D. G. Long, Fl. Bhut. 1: 806 (1987); G. L. Webster, Ann. Missouri Bot. Gard. 81: 91 (1994); Philcox in Dassan., ed., Fl. Ceyl. XI: 173 (1997); van Welzen, Blumea 43: 137 (1998). T.: *H. riparia* Lour.

Lumanaja Blanco, Fl. Filip.: 821 (1837). T.: *L. fluviatilis* Blanco [=*H. riparia* Lour.]

Spathiostemon sect. *Haematospermum* Baill., Étude Euphorb.: 293 (1858). T.: *S. salicinum* (Hassk.) Hassk. (B.: *Ricinus salicinus* Hassk.) [=*H. riparia* Lour.]

Homonoia sect. *Haematospermum* (Baill.) Müll. Arg. in DC., Prodr. 15(2): 1022 (1866).

Dioecious or rarely submonoecious rheophytic shrubs. Indumentum simple and lepidote. Leaves alternate, shortly petiolate to subsessile, stipulate, simple, entire or dentate, often narrow, penninerved, pellucid- or granulose-glandular beneath. Inflorescences axillary or cauliflorous, spicate, solitary or fascicled, mostly unisexual. Male flowers very shortly pedicellate to subsessile; calyx closed in bud, ± globose, later valvately 3-partite; petals 0; disc 0; stamens very numerous, filaments in several stipitate fascicles, united into a common column, anthers arranged in numerous much-branched phalanges forming a dense globose head, apicifixed, thecae small, globose, subdivaricate, longitudinally dehiscent, connective obscure; pollen oblate-spheroidal to suboblate, striate, otherwise ± as in 187. *Lasiococca*; pistillode 0; receptacle 0. Female flowers sessile; sepals 5–8, imbricate, unequal, narrow, caducous; petals 0; disc 0; ovary 3-locular, subglobose, smooth, appressed-pubescent, ovule 1 per locule; styles 3, shortly connate at the base, filiform, spreading, simple, stigmatic surfaces papillose-plumose. Fruit 3-dymous, ± smooth, dehiscing loculicidally into 3 valves; endocarp coriaceous; columella fusiform, persistent. Seeds ovoid, ecarunculate; exotesta thinly fleshy or membranous; endotesta hard, crustaceous; endosperm fleshy; cotyledons broad, flat. Fig.: See Pflanzenr. 68: 115, t. 27A–E (1919); Fl. Indo-Chine 5: 331, t. 38.5–8 (1925); Formos. Trees: 349, t. 304 (1936); Ill. Fl. Tam. Carn., t. 641 (1982); Fl. Bhut. 1: 807, t. 50d–f (1987); Blumea 43: 139, t. 2 (1998).

A ditypic Tropical Asiatic genus ranging from peninsular India to Taiwan, the Philippines and New Guinea.

Tribe 34. **PLUKENETIEAE** *(Benth.) Hutch.*, Amer. J. Bot. 56: 753 (1969). T.: *Plukenetia* L.

Crotoneae subtribe *Plukenetieae* Benth., Gen. Pl. 3: 252 (1880).

Monoecious or rarely dioecious trees, shrubs or herbs, often scandent; indumentum simple, sometimes or rarely glandular (*Tragia*), sometimes urticating; leaves alternate, stipulate, simple or palmatipartite, penninerved or palminerved, sometimes glandular or stipellate. Inflorescences terminal, axillary or leaf-opposed, capitate, racemose, spicate or thyrsiform, rarely paniculate, rarely flowers fasciculate in leaf-axils (*Angostylis*), usually bisexual; bracts eglandular or very rarely 2-glandular (some *Plukenetia*); male calyx splitting into 3–5(6) valvate segments; petals 0; disc interstaminal, extrastaminal or 0; stamens (2)3–100, usually free, anthers introrse or extrorse, muticous or apiculate; pollen 3-colpate, porate or inaperturate, tectate or intectate; pistillode 0 or rarely +; female sepals 3–6, imbricate; petals 0; disc 0; ovary 3–4-locular; styles simple, slender or dilated, sometimes papillose or plumose, usually connate into a distinct column. Fruit dehiscent or rarely indehiscent; columella persistent. Seeds ecarunculate; testa usually dry; endosperm copious; cotyledons much longer and broader than the radicle.

This distinctive pantropical tribe of 17 genera and over 250 species has 9 neotropical, 3 pantropical and 5 palaeotropical genera, and c. 150 neotropical and c. 100 palaeotropical species, but with only a few pantropical species.

Key to the subtribes

1. Flowers in 2-bracteate capitate pseudanthia; pollen prolate; urticating hairs + ········ 34c. *Dalechampiinae*
1. Flowers often in racemose or spicate thyrses; pollen oblate to spheroidal; urticating hairs + or 0 ········ 2
2. Urticating hairs 0; pollen 3-colpate; fruit variously ornamented or else indehiscent ········ 34a. *Plukenetiinae*
2. Urticating hairs often +; pollen various; fruit usually unadorned and dehiscent ········ 34b. *Tragiinae*

Subtribe 34a. **Plukenetiinae** *Benth.*, Gen. Pl. 3: 253 (1880); Pax in Engl. & Prantl, Nat. Pflanzenfam. 3(5): 62 (1890); Pax & K. Hoffm., Pflanzenr. 68: 1 (1919), and in Nat. Pflanzenfam. ed. 2, 19C: 141(1931); G. L. Webster, Taxon 24: 598 (1975), and Ann. Missouri Bot. Gard. 81(1): 92 (1994). T.: *Plukenetia* L.

Trees, shrubs, lianas or vines; indumentum simple; leaf-blades simple, entire or toothed, often glandular, sometimes stipellate. Inflorescences terminal, axillary or leaf-opposed, usually in racemose or spicate thyrses; stamens 4–50, filaments free or anthers subsessile; pollen suboblate-spheroidal, tricolpate, tectate-perforate to semitectate-reticulate; ovary 3–4-

locular, often winged; styles massive, partly to entirely connate. Fruit dehiscent or rarely indehiscent, often ornamented.

Webster (1975, 1994) has a narrower view of this group than Bentham (1880) or Pax & Hoffmann (1919, 1931). Apart from a few African species of *Plukenetia* (*Pterococcus, Tetracarpidium*), 1 Asiatic species of *Plukenetia* (*Pterococcus*) and 3 Madagascan species of *Plukenetia* s. str., the rest of this subtribe is neotropical.

KEY TO THE GENERA

1. Ovary 3-locular; female sepals 4–6; twining or non-twining · · · · · · · · · · 2
1. Ovary 4-locular; female sepals 4; twining vines or lianes · · · · · · · · · · · · 5
2. Stamens 4; male disc cupular or 0, or pseudodisc 4-lobed or 0; leaves penninerved; non-twining trees or shrubs · 3
2. Stamens 10 or more; male disc-segments 0 or pseudodisc 5-lobed · · · · · 4
3. Disc 0; pseudodisc sometimes +; style ± infundibuliform · 190. *Haematostemon*
3. Disc extrastaminal; style obovoid · 191. *Astrococcus*
4. Stamens 15–25; male disc 0; style ± infundibuliform; flowers fasciculate in leaf-axils; leaves penninerved; trees or shrubs · · · · · · · · 192. *Angostylis*
4. Stamens 10; male pseudodisc 5-lobed; style cylindrical, clavate, 3-lobate at the apex; flowers in axillary racemes; leaves palminerved; twining vine · 193. *Romanoa*
5. Styles connate for $^1/_2$ or < $^1/_2$ their length, 2-fid; male calyx 4–5(6)-partite; stamens (24)30–50; fruit subglobose · · · · · · · · · · · 194. *Eleutherostigma*
5. Styles connate for $^1/_2$ or >$^1/_2$ their length or completely connate, sometimes 2-fid; male calyx 4-partite; stamens 8–50 · 6
6. Fruit mostly strongly 4-lobate, dehiscent, or, if indehiscent, then male receptacle not globose; stylar column mostly elongate · · 195. *Plukenetia*
6. Fruit globose, indehiscent; male receptacle globose with anthers widely-scattered on the surface; stylar column obovoid · · · · · · · · · · 196. *Vigia*

190. **Haematostemon** *(Müll. Arg.) Pax & K. Hoffm.*, Pflanzenr. 68: 31, t. 11, f. C, D (1919), and Nat. Pflanzenfam. ed. 2, 19C: 144 (1931); Sandwith, Kew Bull. 1950: 133 (1951); Jabl., Mem. New York Bot. Gard. 17: 143 (1967); G. L. Webster, Ann. Missouri Bot. Gard. 81: 92 (1994); Gillespie, Smithsonian Contr. Bot. 86: 28 (1997). T.: *H. coriaceus* (Baill.) Pax & K. Hoffm. (B.: *Astrococcus coriaceus* Baill.)

Astrococcus sect. *Haematostemon* Müll. Arg., Linnaea 34: 157 (1865).

Monoecious trees or shrubs, non-twining. Indumentum simple; urticating hairs 0. Leaves alternate, shortly petiolate, minutely stipulate, simple, entire, serrate or dentate, penninerved, coriaceous, eglandular. Inflorescences terminal, subterminal or axillary, pseudoracemose, solitary, pedunculate, bisexual, female at base, male above, but often terminating in a female flower; male bracts fasciculiflorous, the flowers in condensed cymes, female 1-flowered, or sometimes cymes bisexual with a central female flower. Male flowers shortly

A monotypic genus from Amazonian Venezuela and Brazil.

192. **Angostylis** *Benth.*, Hooker's J. Bot. Kew Gard. Misc. 6: 328 (1854) (as *Angostyles*); Baill., Étude Euphorb.: 498 (1858) (as *Angostyles*); Müll. Arg. in DC., Prodr. 15(2): 767 (1866), and in Mart., Fl. Bras. 11(2): 331, t. 50 (1874); Benth., Gen. Pl. 3: 327 (1880); Pax & K. Hoffm., Pflanzenr. 68: 29 (1919), and Nat. Pflanzenfam. ed. 2, 19C: 144 (1931); G. L. Webster, Ann. Missouri Bot. Gard. 81: 93 (1994). T.: *A. longifolia* Benth.

Monoecious slender trees or tall shrubs, non-twining. Indumentum simple; urticating hairs 0. Leaves alternate, subsessile, stipulate, large, elongate, crowded at the apices of the shoots, simple, serrate-dentate, penninerved, up to 12-glandular at the base; stipules subulate-filiform. Flowers larger than in 190. *Haematostemon* or in 191. *Astrococcus*. Male flowers in 2–3-flowered fascicles in the axils of fallen leaves, females solitary in upper leaf-axils. Male flowers pedicellate; calyx closed in bud, ovoid, apiculate; sepals 3(4), valvate, thick, papillose within; petals 0; disc 0; stamens 15–25, filaments short, anthers dorsifixed, extrorse, thecae distinct, divergent towards the base, free at the base, longitudinally dehiscent, connective dorsally pubescent; pollen very like that of 191. *Astrococcus*; pistillode 0; receptacle conical to columnar, pubescent. Female flowers pedicellate; sepals 5, unequal, imbricate; petals 0; hypogynous disc small, annular, c. 30-lobulate; ovary 3-locular, muricate, exalate, ovule 1 per locule; stylar column ± infundibuliform, hollow, persistent, styles free at apex, 2-fid, lobes erecto-patulous, stigmas multiseriately cristate-papillose. Fruit 3-dymous, dehiscent into 2-valved cocci; exocarp softly muricate, endocarp crustaceous; columella short, squat. Seeds subglobose, ecarunculate; endosperm fleshy; cotyledons broad, flat. Fig.: See Étude Euphorb.: t. 9, f. 8–11 (1858) (as *Angostyles*); Fl. Bras. 11(2): t.50 (1874); Pflanzenr. 68: 5, t. 2H and 29, t. 10 (1919); Nat. Pflanzenfam. ed. 2, 19C: 143, t. 74H (1931).

A monotypic genus from Amazonian Brazil. *A. tabulamontana* Croizat, described from Surinam, is based on a poorly-preserved specimen in immature fruit; it is not possible to assign it to a genus, but it is definitely not an *Angostylis* (Gillespie, pers. comm.).

193. **Romanoa** *Trevis.*, Sagg. Algh. Coccot.: 99 (1848); Radcl.-Sm., Kew Bull. 34: 589 (1980); G. L. Webster, Ann. Missouri Bot. Gard. 81: 93 (1994). T.: *R. tamnoides* (A. Juss.) Radcl.-Sm. (B.: *Anabaena tamnoides* A. Juss.)

Anabaena A. Juss., Euphorb. Gen.: 46 (1824), non *Anabaina* Bory (1821). T.: *A. tamnoides* A. Juss. [= *R. tamnoides* (A. Juss.) Radcl.-Sm.]

Anabaenella Pax & K. Hoffm., Pflanzenr. 68: 27 (1919), & Nat. Pflanzenfam. ed. 2, 19C: 144 (1931). T.: *A. tamnoides* (A. Juss.) Pax & K. Hoffm. [= *R. tamnoides* (A. Juss.) Radcl.-Sm.]

Monoecious twining shrubs, with the habit of some *Dioscoreaceae*. Indumentum simple; urticating hairs 0. Leaves alternate, long-petiolate, stipulate, simple,

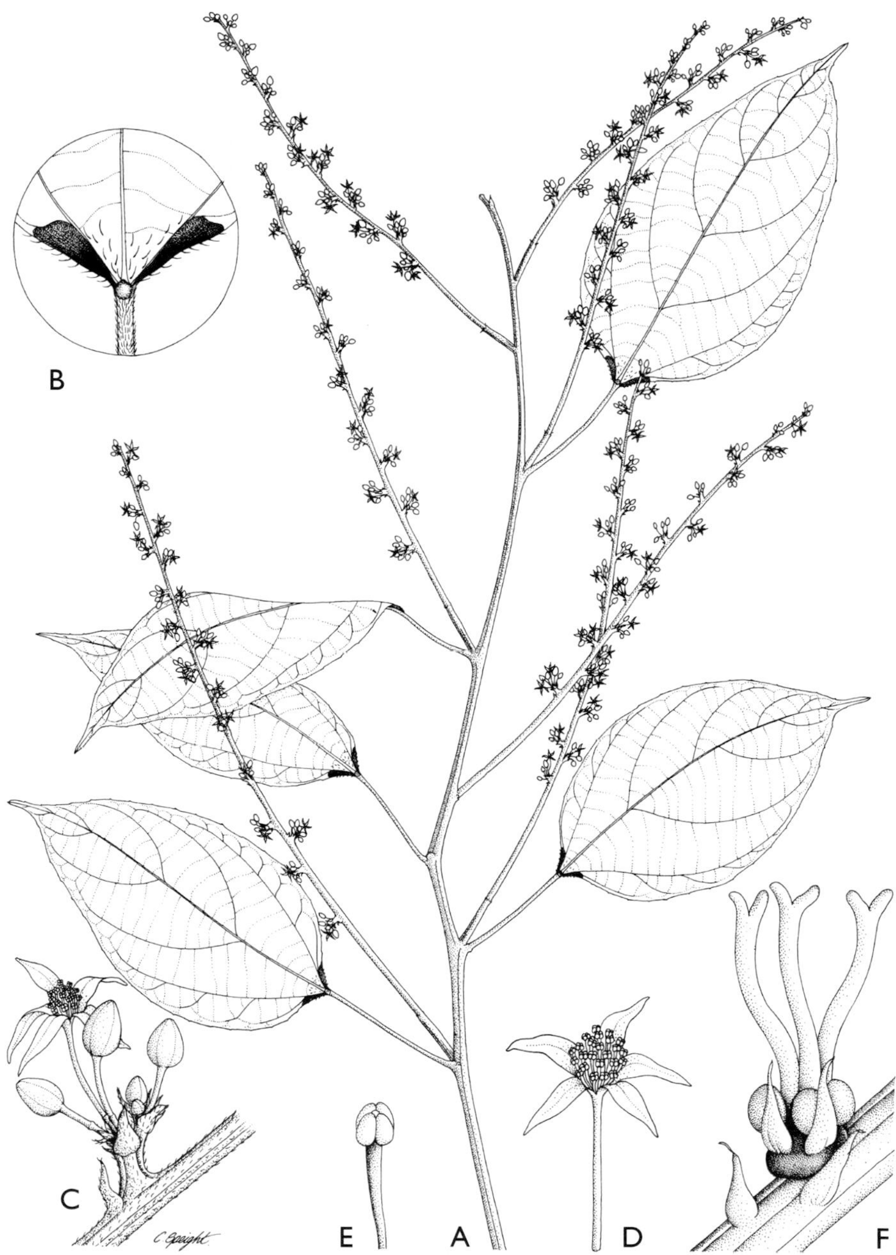

FIG. 32. *Eleutherostigma lehmannianum* Pax & K. Hoffm. **A** Habit, male × 1; **B** Basal foliar glands × 4.5; **C** Small portion of male inflorescence × 4.5; **D** Male flower × 6; **E** Stamen × 30; **F** Female flower × 6. **A–E** from *Cazalet* & *Pennington* 5089; **F** from *Lehmann* 5158, after Pax. Drawn by Camilla Speight.

triangular-cordate, subentire, dentate or sinuate, palminerved, 2-glandular at the base and maculate-glandular in 2 rows beneath; stipules small, fugacious. Inflorescences axillary, pseudoracemose, solitary, bisexual, 1–2 female flowers at the base, the rest male; male bracts 2–3(or more)-flowered. Male flowers very shortly pedicellate; calyx closed in bud, ovoid-cylindric, acute, later valvately 5-partite; petals 0; pseudodisc 5-lobed, lobes interstaminal, torulose, episepalous, arising from the base of the receptacle; stamens 10, outer 5 alternisepalous, filaments free, anthers medifixed, extrorse or introrse, 4-thecate, thecae at first terminal, later inflexed, basally dehiscent, connective papillulose; pollen like that of 195. *Plukenetia*, but the tectum is fossulate-foveolate; pistillode 0 or +, filiform, apically thickened and hispid; receptacle columnar. Female flowers pedicellate, pedicels elongating in fruit; sepals 5–6, narrow; petals 0; disc 0; ovary 3-locular, smooth, ovule 1 per locule; stylar column elongate, cylindrical, clavate, 3-lobate at the apex, lobes erect; stigmatic surface smooth. Fruit obtusely 3-gonous, smooth, irregularly dehiscent; epicarp rugose when dry; mesocarp somewhat fleshy, easily separating when dry. Seeds ecarunculate, somewhat compressed; endosperm fleshy; cotyledons broad, flat. Fig.: See Euphorb. Gen.: t. 15, f. 48.1–9 (1824) (as *Anabaena tamnoides*); Fl. Bras. 11(2): t. 51 (1874) (as *Plukenetia tamnoides*); Pflanzenr. 68: 28, t. 9 (1919) (as *Anabaenella tamnoides*).

A monotypic genus from E. and S. Brazil and Paraguay.

194. **Eleutherostigma** *Pax & K. Hoffm.*, Pflanzenr. 68: 11, t.3 (1919), and Nat. Pflanzenfam. ed. 2, 19C: 141 (1931); G. L. Webster, Ann. Missouri Bot. Gard. 81: 93 (1994). T.: *E. lehmannianum* Pax & K. Hoffm.

Monoecious twining vine or liane. Indumentum simple; urticating hairs 0. Leaves alternate, petiolate, stipulate, simple, triangular-ovate, subentire to remotely denticulate, 3-nerved at the base, 1-glandular on the midrib and 2-glandular on the lamina at the base; stipules glandular. Inflorescences axillary, pseudoracemose to narrowly paniculate, solitary, bisexual with 1–2 female flowers at the base, the rest male; male bracts several-flowered, females 1-flowered. Male flowers pedicellate; calyx closed in bud, ovoid, acute, somewhat torulose, later valvately 4–5(6)-partite; petals 0; disc-glands juxtastaminal, filiform, thickened at the apex; stamens (24)30–50, filaments free, stout, anthers medifixed, terminal, 4-thecate, thecae subglobose-ellipsoid, connective convex; pollen suboblate, very like that of 195. *Plukenetia*; pistillode 0; receptacle cylindric to conical. Female flowers shortly pedicellate; sepals 4; petals 0; disc 0; ovary 4-locular, 4-alate, ovule 1 per locule; styles connate for $^1/_2$ or $<^1/_2$ their length into a short column, 2-fid at the apex. Fruit subglobose, large, fleshy, carinate. Seeds not known. Fig.: See Pflanzenr. 68: 11, t. 3 (1919); Syst. Bot. 18(4): 577, t. 1A (1993) (as *Plukenetia*). See also Fig. 32, p. 245.

A monotypic genus from Colombia and Ecuador.

Huft & Gillespie (in Gillespie, 1993) combined this under the following genus without even according it infrageneric status, as they considered that it could not be adequately separated from *Plukenetia*, but Webster (1994) keeps it up.

195. **Plukenetia** *L.* Sp. Pl.: 1192 (1753), and Gen. Pl. ed. 5: 438 (1754); A. Juss., Euphorb. Gen.: 47 (1824) (as *Pluknetia*); Müll. Arg. in DC., Prodr. 15(2): 768 (1866), and in Mart., Fl. Bras. 11(2): 332 (1874); Benth., Gen. Pl. 3: 327 (1880); Hook. f., Fl. Brit. India 5: 464 (1888); Prain in Fl. Trop. Afr. 6(1): 949 (1912); Pax in Engl. & Prantl, Nat. Pflanzenfam. 3(5): 66 (1890), p.p.; Pax & K. Hoffm. in Pflanzenr. 68: 12 (1919), and Nat. Pflanzenfam. ed. 2, 19C: 141 (1931); Jabl., Mem. New York Bot. Gard. 17: 142 (1967); G. L. Webster, Ann. Missouri Bot. Gard. 54: 293 (1968); Philcox, Fl. Trin. Tob. II (X): 669 (1979); Huft, Ann. Missouri Bot. Gard. 75: 1105 (1988); R. A. Howard, Fl. Lesser Antilles 5: 82 (1989); Gillespie, Syst. Bot. 18(4): 575 (1993); G. L. Webster, Ann. Missouri Bot. Gard. 81: 93 (1994); Murillo-Aldana & Franco-Rosselli, Euf. Reg. Ararac.: 137 (1995); Gillespie, Smithsonian Contr. Bot. 86: 30 (1997). T.: *Pl. volubilis* L.

Pterococcus Hassk., Flora 25 (2), Beibl. 3: 41 (1842), nom. cons., non Pall. (1773); Pax in Pflanzenr. 68: 21 (1919), and in Nat. Pflanzenfam. ed. 2, 19C: 143 (1931); Croizat, J. Arnold Arbor. 22: 423 (1941); Backer & Bakh. f., Fl. Java 1: 490 (1963); Airy Shaw, Kew Bull. 26: 327 (1972); Whitmore, Tree Fl. Mal. 2: 126 (1973); Airy Shaw, Kew Bull. Addit. Ser. 4: 187 (1975), Kew Bull. 36: 340 (1981), 37: 34 (1982), and Fl. Philipp.: 43 (1983); Grierson & D. G. Long, Fl. Bhut. 1: 811 (1987); Radcl.-Sm., Fl. Zamb. 9(4): 210 (1996). T.: *Pt. glaberrimus* Hassk., nom. illeg. [=*Pl. corniculata* Sm.]
Ceratococcus Meisn., Pl. Vasc. Gen. 2: 369 (1843). T.: (none designated).
Sajorium Endl., Gen. Pl. Suppl. 3: 98 (1843); Baill., Étude Euphorb.: 480 (1858). T.: *S. corniculatum* (Sm.) D. Dietr. [=*Pl. corniculata* Sm.]
Hedraiostylus Hassk., Tijdschr. Natuurl. Gesch. Physiol. 10: 141 (1843), and Cat. Hort. Bot. Bogor: 234 (1844). T.: *H. glaberrimus* (Hassk.) Hassk. (B.: *Pterococcus glaberrimus* Hassk.) [=*Pl. corniculata* Sm.]
Tetracarpidium Pax in Engl., Bot. Jahrb. Syst. 26: 329 (1899), and in Nat. Pflanzenfam. ed. 2, 19C: 142 (1931); Keay (ed.), Fl. W. Trop. Afr. ed. 2, 1: 410 (1958). T.: *T. staudtii* Pax [=*Pl. conophora* Müll. Arg.]
Pseudotragia Pax, Bull. Herb. Boissier, Sér. 2, 8: 635 (1908). T.: *Ps. scandens* Pax [=*Pl. corniculata* Sm.]
Angostylidium (Müll. Arg.) Pax & K. Hoffm., Pflanzenr. 68: 17 (1919). T.: *A. conophorum* (Müll. Arg.) Pax & K. Hoffm. [=*Pl. conophora* Müll. Arg.]
Apodandra Pax & K. Hoffm., Pflanzenr. 68: 20 (1919), and in Nat. Pflanzenfam. ed. 2, 19C: 142 (1931). T.: *A. loretensis* (Ule) Pax & K. Hoffm. [=*Pl. loretensis* Ule]
Elaeophora Ducke, Arch. Jard. Bot. Rio de Janeiro 4: 112 (1925). T.: *E. abutifolia* Ducke (S.: *Pl. abutifolia* (Ducke) Pax & K. Hoffm. [=*Pl. polyadenia* Müll. Arg.]

Monoecious twining vines or lianes. Indumentum simple; urticating hairs 0. Leaves alternate, petiolate, stipulate, sometimes stipellate, simple, ovate or sometimes elliptic, subentire to shallowly crenate or serrate, penninerved, palminerved or 3-plinerved at the base, 2–more-glandular on the lamina and sometimes 1-glandular on the midrib at the base, maculate-glandular on one or both surfaces; stipules minute, soon falling. Inflorescences terminal, axillary, leaf-opposed, pseudoracemose, rarely paniculate, solitary, bisexual with 1–2 female flowers at the base, the rest male, or rarely unisexual; bracts small, males several-flowered, females 1-flowered. Male flowers pedicellate; calyx closed in bud, subglobose, ovoid or ellipsoid, later valvately 4-partite; petals 0; disc-glands juxtastaminal, few, minute, indistinct or 0; stamens 8–50, filaments free, long or short or both, or very short or 0, slender or basally swollen, anthers basifixed, terminal, cruciately 4-thecate, thecae subglobose-ellipsoid, longitudinally dehiscent, connective wide or narrow; pollen suboblate to oblate, 3-colpate, colpi broad, margins uneven, tectate-foveolate or semitectate-reticulate; pistillode 0; receptacle conical, cylindrical, ellipsoid, globose or hemispherical. Female flowers pedicellate to long-pedicellate, pedicels extending in fruit; sepals 4, small, imbricate; petals 0; disc 0; ovary 4-locular, 4-alate, ovule 1 per locule; styles 4, connate for >$^1/_2$ or $^1/_2$ their length or else completely connate into a tall or short and globose or obovate column and then the stigmas obcordate-obovate, cruciately-disposed at the top, sometimes 2-fid. Fruit large or moderate-sized, strongly 4-lobate or sometimes globose, the lobes carinate, alate, cornute or 2-verrucose, dehiscent into 4 2-valved cocci, rarely indehiscent; exocarp coriaceous; endocarp woody or sometimes fleshy; columella short. Seeds laterally compressed, lenticular, perialate, or subglobose, rugulose, ecarunculate; testa crustaceous; endosperm fleshy; cotyledons ovate-suborbicular, flat, cordate, 3-nerved from the base, slightly longer than the radicle. Fig.: See Étude Euphorb.: t. 21, f. 3–4 (1858) (as *Hedraiostylus*); Pflanzenr. 68: 14, t. 4, 18, t. 5 and 23, t. 7 (1919); Mem. New York Bot. Gard. 17: 144, t. 23 (1967); Ann. Missouri Bot. Gard. 54: 295, t. 15 (1968); Fl. Lesser Antilles 5: 88, t. 33 (1989); Syst. Bot. 18(4): 580, 582, tt. 9, 10 (1993); Euf. Reg. Ararac.: 138–140, tt. 39, 40 (1995); Fl. Zamb. 9(4): 211, t. 44 (1996) (as *Pterococcus*); Smithsonian Contr. Bot. 86: 32–36, tt. 10–12 (1997).

A pantropical genus of 15 species, of which 10 are Neotropical. Of the rest, 3 are from Africa, 1 from Madagascar and 1 from SE Asia.

Several recent workers, up to and including Grierson & Long (1987) and Radcliffe-Smith (1996b), have continued to accept *Pterococcus* as a distinct genus, but perhaps unjustifiably so.

196. **Vigia** *Vell.*, Fl. Flumin. 9: t. 128 (1832), ex Benth., Gen. Pl. 3: 328 (1880), pro syn.; G. L. Webster, Ann. Missouri Bot. Gard. 81: 93 (1994). T.: *V. serrata* Vell.

Fragariopsis A. St.-Hil., Leçons Bot.: 426 (1840); Baill., Étude Euphorb.: 497 (1858); Müll. Arg. in DC., Prodr. 15(2): 773 (1866), and in Mart., Fl. Bras. 11(2): 336, t. 52 (1874); Pax in Pflanzenr. 68: 19 (1919), and Nat. Pflanzenfam. ed. 2, 19C: 142 (1931). T.: *Fr. scandens* A. St.-Hil. [=*V. serrata* Vell.]

Accia A. St.-Hil., Leçons Bot.: 426 (1840). T.: *A. scandens* A. St.-Hil. [=*V. serrata* Vell.]

Botryanthe Klotzsch, Arch. Naturgesch. 7(1): 190 (1841). T.: *B. discolor* Klotzsch [=*V. serrata* Vell.]

Monoecious scandent or subscandent shrubs or lianas. Indumentum simple; urticating hairs 0. Leaves alternate, shortly petiolate, stipulate, stipellate, simple, ovate, subentire to shallowly and remotely denticulate or serrate, penninerved, 2-glandular on the lamina at the base above; stipules minute, soon falling. Inflorescences terminal, lateral or leaf-opposed, pseudoracemose, solitary or paired, bisexual, female in the lower third, the upper two-thirds male; bracts small, subulate, males 1–2-flowered, females 1-flowered. Male flowers pedicellate; calyx closed in bud, shortly conical, acute, later induplicate-valvately 4-partite; petals 0; disc 0; stamens 9–15(30), filaments very short or 0, anthers inserted directly on the receptacle, widely-scattered on its surface, 2-thecate, thecae subglobose, apically dehiscent, connective 0; pollen suboblate, 3-colpate, colpi broad, margins uneven, semitectate-reticulate, muri often crenate, fragmented and gemmate at the colpus-margin; pistillode 0; receptacle spherical, prominent, smooth or plicate-rugose. Female flowers pedicellate, pedicels elongating somewhat in fruit; sepals 4, oblong; petals 0; disc 0; ovary 4-locular, 4-angular or not, ovule 1 per locule; stylar column obovoid-clavate, papillose-verruculose or ± smooth, stigmas 4, immersed and cruciately-disposed at the top. Fruit moderately large, globose, indehiscent, with 4 small warts at the base; pericarp fleshy to corky. Seeds globose. Fig.: See Étude Euphorb.: t. 12, f. 45; t. 13, f. 29–36 (1858) (as *Accia*); Fl. Bras. 11(2): t. 52 (1874); Pflanzenr. 68: 5, t. 2 F,G and 20, t. 6 (1919); Nat. Pflanzenfam. ed. 2, 19C: 143, t. 74F,G (1931); Amer. J. Bot. 56: 750, t. 12A–G (1969).

A monotypic genus restricted to SE Brazil (Bahia, Espirito Santo, Minas Gerais, Rio de Janeiro and São Paulo).

Gillespie (1993) combined this under the previous genus without even according it infrageneric status, as she considered that it could not be adequately separated from *Plukenetia*, but Webster (1994) keeps it up.

Subtribe 34b. **Tragiinae** *G. L. Webster*, Taxon 24: 598 (1975), and Ann. Missouri Bot. Gard. 81(1): 93 (1994). T.: *Tragia* L.

Shrubs or herbs, often twining; indumentum simple, urticating hairs often +; leaf-blades simple, lobed or not, entire or toothed, without laminar glands, exstipellate. Inflorescences terminal, axillary or leaf-opposed, usually racemose or spicate; stamens 2(3)–50, filaments free or connate; pollen suboblate to spheroidal, 3-colpate, 3-porate or inaperturate, intectate-baculate, semitectate-reticulate or tectate-perforate; ovary 3-locular; styles ± free or connate. Fruit usually unadorned, dehiscent.

Apart from *Platygyne, Acidoton, Gitara* and the neotropical species of the pantropical genus *Tragia*, the rest of this subtribe is palaeotropical.

Key to the genera

1. Male sepals usually inflexed to form a pseudodisc; stamens 3–4, anthers introrse; styles massive · 2
1. Male sepals not inflexed; stamens 2–55, anthers introrse or extrorse; styles usually slender · 5

2. Female sepals pinnatifid; pistillode +; leaves with urticating hairs · · · · · · · 200. *Tragiella*
2. Female sepals entire or toothed; pistillode 0 · · · · · · · 3
3. Female sepals toothed; anther-connective not enlarged or caudate; stamens connate at the base; stylar column globose; inflorescence terminal and leaf-opposed; leaves subglabrous; stipules suborbicular · · · · · 199. *Sphaerostylis*
3. Female sepals entire; anther-connective enlarged, triangular, sometimes caudate; stamens free; styles free or fused into a column; stipules mostly triangular-lanceolate · · · · · · · 4
4. Styles free or ± so; stigmas plumose; inflorescence terminal and leaf-opposed; leaves with urticating hairs; anther-connective caudate · · 197. *Cnesmone*
4. Stylar column globose or clavate, epapillose; inflorescence axillary; leaves subglabrous; anther-connective usually ecaudate · · · 198. *Megistostigma*
5. Anthers 2, subsessile; male calyx flat, lobes reflexed; stylar column thick · · · · · · · 205. *Pachystylidium*
5. Anthers 3 or more, or if 2 then not subsessile; male calyx concave; styles free to partially connate · · · · · · · 6
6. Anther-connective without a tuft of urticating hairs; stamens (2)3–50; monoecious or rarely to sometimes dioecious subshrubs, herbs or twining vines · · · · · · · 7
6. Anther-connective ending in a tuft of urticating hairs; stamens 30–55; dioecious erect shrubs; leaves glabrescent · · · · · · · 8
7. Staminal receptacle usually flat; stamens 2–8(50); styles commonly slender, epapillose · · · · · · · 201. *Tragia*
7. Staminal receptacle convex; stamens 3–14; styles thick, papillose · · · · · · · 202. *Platygyne*
8. Leaves mostly entire; pollen inaperturate · · · · · · · 203. *Acidoton*
8. Leaf-blades toothed in the upper half; pollen 3-colpate · · · · · 204. *Gitara*

197. **Cnesmone** *Blume*, Bijdr. Fl. Ned. Ind.: 630 (1826) (as *Cnesmosa*), and Fl. Javae: vi (1828); Baill., Étude Euphorb.: 458 (1858); Müll. Arg. in DC., Prodr. 15(2): 926 (1866); Benth., Gen. Pl. 3: 330 (1880); Hook. f., Fl. Brit. India 5: 466 (1888); J. J. Sm., Meded. Dept. Landb. Ned.-Indië 10: 513 (1910); Pax & K. Hoffm., Pflanzenr. 68: 102, t. 23, A–E (1919); Gagnep., Fl. Indo-Chine 5: 385 (1926); Pax & K. Hoffm., Nat. Pflanzenfam. ed. 2, 19C: 147 (1931); Croizat, J. Arnold Arbor. 22: 427 (1941); Backer & Bakh. f., Fl. Java 1: 490 (1963) (as *Cnesmosa*); Airy Shaw, Kew Bull. 23: 117 (1969), and 26: 240 (1972); Whitmore, Tree Fl. Mal. 2: 83 (1973); Airy Shaw, Kew Bull. Addit. Ser. 4: 88 (1975), Kew Bull. 36: 282 (1981), 37: 13 (1982) and Euph. Philipp.: 16 (1983); N. P. Balakr. & N. G. Nair, Gard. Bull. Singapore 31: 49 (1978); G. L. Webster, Ann. Missouri Bot. Gard. 81: 94 (1994). T.: *Cn. javanica* Blume.

Cenesmon Gagnep., Bull. Soc. Bot. France 71: 866 (1924), and Fl. Indo-Chine 5: 386 (1926). LT.: *C. tonkinense* Gagnep. [=*Cn. tonkinense* (Gagnep.) Croizat] (designated by Wheeler, 1975)

Monoecious villous twining shrubs. Indumentum of simple and urticating hairs. Leaves alternate, petiolate, stipulate, simple, oblong, denticulate, cordate, penninerved and 3-plinerved from the base; stipules broad, persistent. Inflorescences terminal or leaf-opposed, racemose, solitary, bisexual, 1 or more females at the base, the rest male; bracts 1-flowered, males ebracteolate, females 2-bracteolate. Male flowers shortly pedicellate; calyx closed in bud, globose, at length valvately 3-lobed, tube depressed, contracted at the mouth, lobes broad; petals 0; disc 0; stamens 3, alternisepalous, scarcely exserted, filaments free, thick, short, anthers subdorsifixed, thecae separate, subparallel, introrse to latrorse, laterally longitudinally dehiscent, connective thick, produced beyond the thecae into a linear, deflexed appendage; pollen ellipsoid-spheroidal, sometimes irregular, weakly 3-colpate, colpi large, elliptic, margins uneven, tectum punctate, microverrucate; pistillode 0; receptacle flat. Female flowers pedicellate; sepals 3, entire, accrescent; petals 0; disc 0; ovary 3-locular, depressed, strigose, ovule 1 per locule; styles fleshy, connate at the base into a short, broad mass, incurved, stigmas fimbriate-plumose, margins almost pinnatifid. Fruit 3-dymous, dehiscing into 3 2-valved cocci; exocarp thin, pubescent and strigose; endocarp woody; columella slender, expanded and 3-cornute at the apex. Seeds globose, ecarunculate; exotesta thinly fleshy, minutely papillulose, lineate-mottled; endotesta crustaceous; endosperm fleshy; cotyledons broad, flat. Fig.: See Étude Euphorb.: t. 4, f. 14–17 (1858); Pflanzenr. 68: 4, t. 1 G, H and 103, t. 23A–E (1919); Fl. Indo-Chine 5: 383, t. 44.2–8 and 388, t. 45.3–9 (1926); Nat. Pflanzenfam. ed. 2, 19C: 142, t. 73G, H (1931).

A SE Asiatic genus of 12 species ranging from Assam and Bangladesh to S. China, the Philippines, Borneo and Java, *fide* Croizat (1941).

198. **Megistostigma** *Hook. f.*, Hooker's Icon. Pl. 16: t. 1592 (1887), and Fl. Brit. India 5: 466 (1888); Merr., Philipp. J. Sci., 16C: 563 (1920); Croizat, J. Arnold Arbor. 22: 425 (1941); Backer & Bakh. f., Fl. Java 1: 491 (1963); Airy Shaw, Kew Bull. 23: 119 (1969), and 26: 309 (1972); Whitmore, Tree Fl. Mal. 2: 117 (1973); Airy Shaw, Kew Bull. Addit. Ser. 4: 175 (1975), Kew Bull. 36: 330 (1981), and Euph. Philipp.: 38 (1983); G. L. Webster, Ann. Missouri Bot. Gard. 81: 94 (1994). T.: *M. malaccense* Hook. f.

Clavistylus J. J. Sm., Meded. Dept. Landb. Ned.-Indië 10: 517 (1910); Pax & K. Hoffm., Pflanzenr. 68: 104, t. 23F (1919), and Nat. Pflanzenfam. ed. 2, 19C: 148 (1931). T.: *Cl. peltatus* J. J. Sm. [= *M. peltatum* (J. J. Sm.) Croizat]

Like 197. *Cnesmone*, but leaves subglabrous, stipules triangular-lanceolate, inflorescences axillary, anther-connective usually ecaudate and the stylar column globose or clavate and epapillose. Fig.: See Hooker's Icon. Pl. 16: t. 1592 (1887); Pflanzenr. 68: 4, t. 1J, K (as *Sphaerostylis malaccensis*) and 103, t. 23F (as *Clavistylus peltatus*) (1919); Fl. Indo-Chine 5: 383, t. 44.9–12 and 388, t. 45.1–2 (1926) (as *Cenesmon peltatum*); Nat. Pflanzenfam. ed. 2, 19C: 142, t. 73 J, K (1931) (as *Sphaerostylis malaccensis*).

A pentatypic tropical SE Asiatic genus ranging from Burma [Myanmar] to SW. China, the Philippines, Borneo and Java.

Pax & Hoffmann (1919) merged this with 199. *Sphaerostylis*, but Croizat (1941) maintains it as distinct.

199. **Sphaerostylis** *Baill.*, Étude Euphorb.: 466 (1858); Müll. Arg. in DC., Prodr. 15(2): 768 (1866); Benth., Gen. Pl. 3: 327 (1880); Baill. in Grandid., Hist. Phys. Madagascar 4 (29): t. 196 (1891); Pax & K. Hoffm., Pflanzenr. 68: 106 (1919), and Nat. Pflanzenfam. ed. 2, 19C: 148 (1931); Croizat, J. Arnold Arbor. 22: 430 (1941); G. L. Webster, Ann. Missouri Bot. Gard. 81: 94 (1994). T.: *S. tulasneana* Baill.

Like 198. *Megistostigma*, but stipules suborbicular, inflorescences terminal and leaf-opposed, stamens connate at the base, anther-connective not enlarged and female sepals toothed; like 197. *Cnesmone*, but leaves subglabrous, stamens connate at the base, anther-connective not enlarged or caudate, female sepals toothed and stylar column globose. Fig.: See Étude Euphorb.: t. 21, f. 19–21 (1858); Pflanzenr. 68: 5, t. 2E (1919); Nat. Pflanzenfam. ed. 2, 19C: 143, t. 74E (1931).

A ditypic genus endemic to Madagascar.

200. **Tragiella** *Pax & K. Hoffm.*, Pflanzenr. 68: 104 (1919), and Nat. Pflanzenfam. ed. 2, 19C: 148 (1931); Radcl.-Sm., Kew Bull. 35: 777, t. 60 (1981), and Fl. Trop. E. Afr. Euph. 1: 318, t. 60 (1987); G. L. Webster, Ann. Missouri Bot. Gard. 81: 94 (1994); Radcl.-Sm., Fl. Zamb. 9(4): 212, t. 45 (1996). LT.: *Tr. natalensis* (Sond.) Pax & K. Hoffm. (B.: *Tragia natalensis* Sond.) (chosen by Webster, 1994)

Monoecious erect or twining herbs. Indumentum of simple and urticating hairs. Leaves alternate, petiolate to subsessile, stipulate, simple, denticulate, cordate, palminerved. Inflorescences terminal or leaf-opposed, racemose, solitary, bisexual, 1–2 females at the base, the rest male; bracts conspicuous, entire or toothed, 1-flowered, males ebracteolate, females bracteolate. Male flowers pedicellate; calyx closed in bud, tube campanulate, usually inwardly-folded at the throat, later valvately 3-lobed, lobes patent; petals 0; disc 0; stamens 3(4), alternisepalous, filaments short, thickened and sometimes connate at the base, anthers dorsifixed, thecae separate, parallel, introrse, longitudinally dehiscent, connective thickened, rounded at the apex, not or scarcely produced; pollen oblate-spheroidal, 3-colpate, colpi narrow, margins uneven, semitectate-reticulate, muri microverrucate, warts transversely oblong; pistillode +, small, circular or 3-lobate, adnate to the filament-bases; receptacle flat. Female flowers pedicellate; calyx-lobes 6, 1-seriate, imbricate, pinnatifid, usually with a narrow rhachis, 1 terminal and several lateral lobules often longer than the width of the rhachis, accrescent, the rhachis indurescent; petals 0; disc 0; ovary 3-locular, ovule 1 per locule; styles 3, almost completely connate into an excavate conical or infundibuliform column or subglobose, hemispherical or obpyriform mass. Fruit 3-lobate, dehiscing into 3 2-valved cocci; exocarp thinly crustaceous; endocarp thick, woody; columella 3-fid, non-persistent. Seeds globose, ecarunculate; exotesta minutely papillulose to subtomentellous, mottled; endotesta crustaceous; endosperm fleshy; cotyledons broad, flat. Fig.: See Pflanzenr. 68: 105, t. 24 (1919); Fl. Trop. E. Afr., Euph. 1: 319, t. 60 (1987); Fl. Zamb. 9(4): 213, t. 45 (1996).

A tetratypic African genus.

Webster (1994) considers this genus to be doubtfully separable from 199. *Sphaerostylis*, but Radcliffe-Smith (1981) had already demonstrated the distinctness of the two genera.

201. **Tragia** *Plum. ex L.*, Sp. Pl.: 980 (1753), and Gen. Pl. ed. 5: 421 (1754); A. Juss., Euphorb. Gen.: 47 (1824); Baill., Étude Euphorb.: 459 (1858); Müll. Arg. in DC., Prodr. 15(2): 927 (1866), and in Mart., Fl. Bras. 11(2): 403 (1874); Benth., Gen. Pl. 3: 329 (1880); Hook. f., Fl. Brit. India 5: 464 (1888); Pax in Engl. & Prantl, Nat. Pflanzenfam. 3(5): 64 (1890); Prain in Fl. Trop. Afr. 6(1): 964 (1913); Pax & K. Hoffm., Pflanzenr. 68: 32 (1919); Prain in Fl. Cap. 5(2): 502 (1920); Fawc. & Rendle, Fl. Jamaica 4: 304 (1920); Pax & K. Hoffm., Nat. Pflanzenfam. ed. 2, 19C: 144 (1931); Lourteig & O'Donnell, Lilloa 6: 347 (1941); Robyns, Fl. Parc Nat. Alb. 1: 469 (1948); Alain, Fl. Cuba 3: 101 (1953); Keay (ed.), Fl. W. Trop. Afr. ed. 2, 1: 410 (1958); M. C. Johnst., Rhodora 64: 137 (1962); K. I. Mill. & G. L. Webster, Rhodora 69: 241 (1967); G. L. Webster, J. Arnold Arbor. 48: 376 (1967), and Ann. Missouri Bot. Gard. 54: 291 (1968); J.-G. Adam, Mém. Mus. Natl. Hist. Nat., B, Bot. 20: 513 (1971); Berhaut, Fl. Ill. Sénég. 3: 599 (1975); Philcox, Fl. Trin. Tob. II (X): 673 (1979); Radcl.-Sm., Kew Bull. 37: 684 (1983), and 40: 231 (1985); Alain, Fl. Español. 4: 218 (1986); Radcl.-Sm. in Fl. Trop. E. Afr., Euph. 1: 291, tt. 57–59 (1987), and Kew Bull. 42: 395 (1987); G. L. Webster & Huft, Ann. Missouri Bot. Gard. 75: 1106 (1988); Alain, Descr. Fl. Puerto Rico 2: 430 (1988); R. A. Howard, Fl. Lesser Antilles 5: 89 (1989); J.-P. Lebrun & Stork, Enum. Pl. Afr. Trop. 1: 238 (1991); G. L. Webster, Ann. Missouri Bot. Gard. 81: 95 (1994); Radcl.-Sm. in Fl. Zamb. 9(4): 216, tt. 46–48 (1996); Gillespie in Acev.-Rodr., ed., Fl. St. John: 222 (1996), and Smithsonian Contr. Bot. 86: 37 (1997); Philcox in Dassan., ed., Fl. Ceyl. XI: 138 (1997). LT.: *Tr. volubilis* L. (designated by Small, 1913)

Schorigeram Adans., Fam. Pl. 2: 355 (1763). T.: (none designated).

Bia Klotzsch, Arch. Naturgesch. 7(1): 189 (1841); Baill., Étude Euphorb.: 501 (1858). NT.: *B. alienata* Didr. [=*Tr. alienata* (Didr.) Múlgura & M. M. Gut.] (S.: *B. sellowiana* Klotzsch [=*Tr. sellowiana* (Klotzsch) Müll. Arg.]) (designated by Wheeler, 1975).

Leptorrhachis Klotzsch, Arch. Naturgesch. 7(1): 189 (1841); Baill., Étude Euphorb.: 495 (1858). T.: *L. hastata* Klotzsch [=*Tr. leptorrhachis* Radcl.-Sm. & R. Govaerts; S.: *Tr. hastata* (Klotzsch) Müll. Arg. (1874), non Reinw. ex Hassk. (1868)]

Leucandra Klotzsch, Arch. Naturgesch. 7(1): 188 (1841); Baill., Étude Euphorb.: 477 (1858). T.: *L. betonicifolia* Klotzsch [=*Tr. leucandra* Pax & K. Hoffm.; S.: *Tr. betonicifolia* (Klotzsch) Müll. Arg. (1865), non Nutt. (1837)]

Ctenomeria Harv., London J. Bot. 1: 29 (1842); Endl., Gen. Pl. Suppl. 3: 98 (1843); Baill., Étude Euphorb.: 494 (1858); Prain in Fl. Cap. 5(2): 500 (1920). T.: *Ct. cordata* Harv. [=*Tr. capensis* Thunb.]

Lassia Baill., Étude Euphorb.: 464 (1858). T.: *L. scandens* Baill. [=*Tr. lassia* Radcl.-Sm. & R. Govaerts; S.: *Tr. scandens* (Baill.) Müll. Arg. (1865), non L. (1754), nec Aubl. (1775)]

Leptobotrys Baill., Étude Euphorb.: 478 (1858). T.: *L. discolor* Baill. [=*Tr. urens* L.]

Zuckertia Baill., Étude Euphorb.: 495 (1858). T.: *Z. cordata* Baill. [=*Tr. bailloniana* Müll. Arg.]

Leptorrhachis sect. *Ctenomeria* (Harv.) Müll. Arg. in DC., Prodr. 15(2): 925 (1866).

Monoecious or rarely dioecious erect, suberect, scrambling or twining perennial or rarely annual herbs or subshrubs. Indumentum simple, usually mixed with urticating hairs. Leaves alternate, petiolate or sessile, stipulate, simple to palmatisect, dentate, serrate or entire, often triangular, cordate and palminerved, rarely elliptic, cuneate and penninerved. Inflorescences terminal, leaf-opposed or lateral, rarely axillary, racemose or biracemose [§ *Bia* (Klotzsch) Müll. Arg.], solitary, pedunculate, mostly androgynous, 1–3 females at the base, or with 5–30 females on a basal branch [§ *Bia* (Klotzsch) Müll. Arg.], the rest male; bracts conspicuous, persistent, 2-bracteolate, males 1-flowered or subtending small cymules, females 1-flowered. Male flowers pedicellate; calyx closed in bud, globose or ovoid, later valvately 3(5)-lobed; petals 0; disc-glands among the outer stamens, free or adnate to the filaments, but more often 0; stamens (1)3(50), when few alternisepalous, filaments short or very short, free or sometimes connate at the base (very rarely completely connate), anthers dorsifixed or subbasifixed, ovate or oblong, thecae contiguous, parallel, introrse or rarely extrorse, longitudinally dehiscent, connective narrow; pollen variable: ellipsoid-spheroidal or oblate-spheroidal, inaperturate [§ *Bia* (Klotzsch) Müll. Arg.], 3-colpate or 3-porate (very rarely 5-colpate), intectate or tectate, reticulate, fossulate, foveolate, punctate or baculate, apertural insulae + or 0; pistillode minute, often 3-radiate, or obsolete; receptacle usually flat. Female flowers pedicellate, pedicels usually shorter and stouter than in the male, but sometimes elongate, slender; calyx-lobes 3 or 6(8), 1- or 2-seriate, imbricate, entire and not or scarcely accrescent, but persistent, or else pinnatifid or subpalmatifid, rhachis narrow or broad, with 1 terminal and several lateral lobules longer or shorter than the width of the rhachis, or with numerous peripheral lobules, accrescent, the rhachis indurescent and often incrassate, the margins reflexed, persistent; petals 0; disc 0; ovary 3-locular, ovule 1 per locule; styles 3, connate at the base into a short, slender column, free above, simple, entire, rarely stigmas thick, sessile. Fruit 3-lobate to 3-coccous or 3-dymous, depressed, dehiscing into 3 2-valved cocci (rarely also monocarpous allomorphic cornute fruits developed, as in the type-species), cocci subglobose, occasionally slightly angled dorsally; endocarp crustaceous; columella 3-fid. Seeds globose, ecarunculate; testa crustaceous, minutely papillulose to subtomentellous-squamulose, mottled, marbled or lineate; endosperm fleshy; cotyledons broad, flat. Fig.: See Euphorb. Gen.: t. 15, f. 49A.1–6, B.1–3 (1824); Étude Euphorb.: t. 2, f. 17–18 (as *Leptobotrys*); t. 4, f. 6–9 (as *Leucandra*), f. 10–13 (as *Zuckertia*), f. 23–28 (as *Lassia*) (1858); Pflanzenr. 68: 4, t. 1 E,F and 35–97, tt. 12–22 (1919); Fl. Jamaica 4: 305, t. 100

(1920); Nat. Pflanzenfam. ed. 2, 19C: 142, t. 73E,F and 145, t. 75 (1931); Fl. Cuba 3: 102, t. 29 (1953); Ann. Missouri Bot. Gard. 54: 292, t. 14 (1968); Fl. Ill. Sénég. 3: 598, 600 (1975); Fl. Español. 4: 221, t. 116-37 (1986); Fl. Trop. E. Afr., Euph. 1: 299–312, tt. 57–59 (1987); Descr. Fl. Puerto Rico 2: 433, t. 59-30 (1988); Fl. Lesser Antilles 5: 88, t. 35 (1989); Novon 4: 334–337, tt. 7–10 (1994); Fl. Zamb. 9(4): 219–231, tt. 46–48 (1996); Fl. St. John: 223, t. 101E–I (1996); Smithsonian Contr. Bot. 86: 38 and 40, tt. 13 and 14 (1997).

A large and diverse pantropical genus of c. 150 species, mainly in Tropical America and Africa, with only 3 species in Tropical Asia (India, Myanmar (Burma) and China).

The morphological and palynological heterogeneity indicated above may prove on further study to indicate recognition at the generic level for some of the more disparate elements. For further palynological details, see Gillespie (1994).

202. **Platygyne** *P. Mercier* in Seringe, Bull. Bot. 1: 168 (1830); Baill., Étude Euphorb.: 453 (1858); Müll. Arg. in DC., Prodr. 15(2): 913 (1866); Benth., Gen. Pl. 3: 328 (1880); Pax in Engl. & Prantl. Pflanzenfam. ed. 1, 3(5): 64 (1890); Pax & K. Hoffm., Pflanzenr. 68: 26, t. 8, A–C (1919), and Nat. Pflanzenfam. ed. 2, 19C: 144 (1931); Alain, Fl. Cuba 3: 100 (1953), and Mem. New York Bot. Gard. 21: 132 (1971); Borhidi, Ann. Hist.-Nat. Mus. Natl., Hung. 64: 89 (1972); G. L. Webster, Ann. Missouri Bot. Gard. 81: 94 (1994) (as *Platygyna*). T.: *P. urens* P. Mercier [=*P. hexandra* (Jacq.) Müll. Arg. (B.: *Tragia hexandra* Jacq.)]

Acanthocaulon Klotzsch ex Endl., Gen. Pl. Suppl. 4(3): 88 (1850). T.: *A. pruriens* Klotzsch ex Endl. (sphalm. *pruricus*) (S.: *Tragia pruriens* Klotzsch ex Endl.) [=*P. hexandra* (Jacq.) Müll. Arg. (B.: *Tragia hexandra* Jacq.)]

Monoecious twining shrubs. Indumentum of large, urticating hairs. Leaves alternate, shortly petiolate to subsessile, stipulate, simple, dentate, linear, oblong, oblanceolate or obovate, penninerved. Inflorescences terminal, leaf-opposed or lateral, sessile, males fasciculate, depauperate-cymose, females shortly spicate-racemose, 1–4-flowered. Male flowers long-pedicellate; calyx closed in bud, ovoid, apiculate, later valvately (3)4–5(6)-lobed, lobes 1–3-nerved, becoming reflexed; petals 0; disc 0; stamens (3)5–9(14), the outer alternisepalous, filaments short, subrecurved at the apex, anthers dorsifixed, ovate-suborbicular, thecae parallel, contiguous, extrorse or sublatrorse, longitudinally dehiscent, connective dorsally broad, slightly produced; pollen ellipsoid-spheroidal, inaperturate, tectate-perforate, intra-reticulate; pistillode 0; receptacle subglobose, densely ferrugineous-villous, -pilose or -setulose. Female flowers shortly pedicellate to subsessile; sepals (5)6–7(9), imbricate, linear, lanceolate or ovate; petals 0; disc 0; ovary 3(4)-locular, ovule 1 per locule; styles 3(4), shortly connate at the base, contiguous in a thick, infundibuliform column, crenate at the apex, pubescent and setulose without, papillose within. Fruit 3-dymous, dehiscing into 3 2-valved cocci; exocarp thinly crustaceous, pubescent and setose; endocarp woody, hard; columella 3-

FIG. 33. *Platygyne hexandra* (Jacq.) Müll. Arg. (B.: *Tragia hexandra* Jacq.). **A** Habit × 1; **B** Detail of leaf-apex × 2.25; **C** Male flower × 7.5; **D** Female flower × 7.5; **E** Fruit × 7.5; **F** Mericarp × 7.5. **A–C** from *Arias et al.* 70911; **D–F** from *Jack* 7146. Drawn by Camilla Speight.

fid, persistent. Seeds globose, ecarunculate; testa smooth, crustaceous; endosperm fleshy; cotyledons broad, flat. Fig.: See Étude Euphorb.: t. 4, f. 18–22 (1858); Pflanzenr. 68: 4, t. 1C,D and 25, t. 8A–C (1919); Nat. Pflanzenfam. ed. 2, 19C: 142, t. 73C, D (1931); Fl. Cuba 3: 101, t. 28 (1953). See also Fig. 33, p. 256.

A heptatypic genus endemic to Cuba, *fide* Borhidi (1972).

Some Caribbean *Tragia* species (e.g. *Tr. biflora* Urb. and Ekman) have convex or globose male receptacles, and are therefore somewhat intermediate between *Tragia* and *Platygyne*.

203. **Acidoton** *Sw.*, Prodr. 6: 83 (1788) (nom. cons.), and Fl. Ind. Occid. 2: 952, t. 18 (1800); A. Juss., Euphorb. Gen.: 32 (1824); Baill., Étude Euphorb.: 401 (1858); Müll. Arg. in DC., Prodr. 15(2): 914 (1866); Benth., Gen. Pl. 3: 328 (1880); Urb., Symb. Antill. 7: 513 (1913); Pax in Engl. & Prantl., Pflanzenfam. ed. 1, 3(5): 64 (1890); Pax & K. Hoffm., Pflanzenr. 68: 24, t. 8D (1919); Fawc. & Rendle, Fl. Jamaica 4: 303 (1920); Pax & K. Hoffm., Nat. Pflanzenfam. ed. 2, 19C: 143 (1931); Alain, Fl. Español. 4: 71 (1986); G. L. Webster, Ann. Missouri Bot. Gard. 81: 95 (1994), p.p. excl. syn. T.: *A. urens* Sw.

Dioecious erect shrubs or small trees; branches sometimes spinescent. Indumentum of short urticating hairs. Leaves alternate, shortly petiolate, stipulate, simple, usually entire, sometimes toothed, elliptic, penninerved, mostly coriaceous, glabrescent; stipules narrow, chaffy, persistent. Male inflorescences axillary, racemose or fasciculate, short, sessile, few-flowered, females axillary, terminal or lateral, longer than the male, lax-flowered, at length pendulous, or flowers of both sexes on spinescent short shoots; bracts 1-flowered. Male flowers long-pedicellate; calyx closed in bud, ovoid, apiculate, later valvately (3)4–5(6)-lobed, lobes 3–5-nerved, becoming reflexed, somewhat *Clematis*-like; petals 0; disc 0; stamens (20)30–55, pluriverticillate, filaments short, free, anthers subdorsifixed, ovate, thecae parallel, distinct, extrorse, longitudinally dehiscent, connective broad, produced, the apex with a tuft of small urticating hairs; pollen ellipsoid-spheroidal, inaperturate, tectate-rugulate, not reticulate; pistillode 0; receptacle thick and elevated, or sublevel, glabrous. Female flowers sessile or subsessile, becoming shortly pedicellate in fruit; sepals 5–6, narrow, imbricate or open in bud; petals 0; disc 0; ovary 3-locular, ovule 1 per locule; styles 3, connate into a thick column at the base, shortly spreading above, entire, pubescent without, stigmas plumose-papillose. Fruit 3-dymous, dehiscing into 3 2-valved cocci; exocarp thin, setose; endocarp woody, hard; columella subpersistent. Seeds globose, ecarunculate; testa smooth, marmorate; endosperm fleshy; cotyledons broad, flat. Fig.: See Étude Euphorb.: t. 18, f. 10–11 (1858); Pflanzenr. 68: 25, t. 8D (1919); Fl. Jamaica 4: 304, t. 99 (1920). See also Fig. 34, p. 258.

An octotypic genus confined to Jamaica and Hispaniola.

FIG. 34. *Acidoton urens* Sw. **A** Habit, male × 1; **B** Male flower × 4.5; **C** Stamens × 15; **D** Female flower × 4.5; **E** Fruit × 4.5; **F** Mericarp × 4.5. **A–C** from *Robertson* 5486; **D** from *Harris* 12643; **E–F** from *Alexander* s.n., Jamaica 1850. Drawn by Camilla Speight.

FIG. 35. *Gitara panamensis* Croizat. **A** Habit, female × 1; **B** Female flower × 7.5; **C** Male flower × 7.5; **D** Fruit × 6. **A–C** from *Pittier* 6543; **D** from *Fonnegra et al.* 2016. Drawn by Camilla Speight.

204. **Gitara** *Pax & K. Hoffm.*, Pflanzenr. 85, Addit. VII: 187 (1924), and Nat. Pflanzenfam. ed. 2, 19C: 147 (1931); Croizat, J. Arnold Arbor. 26: 192 (1945). T.: *G. venezolana* Pax & K. Hoffm. (S.: *Acidoton venezolanus* (Pax & K. Hoffm.) G. L. Webster)

Acidoton sensu G. L. Webster, Ann. Missouri Bot. Gard. 54: 191 (1967), non Sw.

Very like 203. *Acidoton,* but sometimes monoecious, leaves large, oblanceolate, dentate in the upper half, crowded at the stem-apices, pollen oblate-spheroidal, narrowly 3-colpate with scattered 'islands' on the apertural membrana, tectate-perforate. Fig.: See Ann. Missouri Bot. Gard. 54: 290, t. 13 (1967) (as *Acidoton nicaraguensis*). See also Fig. 35, p. 259.

A ditypic neotropical genus with 1 species in Nicaragua, Panama and Colombia (Antioquia), the other in Venezuela.

Webster (1967b, 1994) and Webster & Burch (1968) placed this in synonymy with 203. *Acidoton,* but the different chorology, *gestalt* and palynology seem to me to warrant its upkeep.

205. **Pachystylidium** *Pax & K. Hoffm.*, Pflanzenr. 68: 108 (1919), and Nat. Pflanzenfam. ed. 2, 19C: 148 (1931); Backer & Bakh. f., Fl. Java 1: 491 (1963); Airy Shaw, Kew Bull. 23: 115 (1969), 26: 311 (1972) and Euph. Philipp.: 39 (1983); G. L. Webster, Ann. Missouri Bot. Gard. 81: 95 (1994). T.: *P. hirsutum* (Blume) Pax & K. Hoffm. (B.: *Tragia hirsuta* Blume) (S.: *Tragia delpyana* Gagnep.)

Monoecious scandent shrub. Indumentum simple, admixed with urticating hairs. Leaves alternate, petiolate, stipulate, simple, elliptic, cordate, coarsely toothed, 3-plinerved at the base; stipules triangular, persistent. Inflorescences terminal or oppositifolious, racemose, pedunculate, basal female portion 2-flowered, the remainder male; bracts 1-flowered. Male flowers pedicellate; calyx closed in bud, globose, later valvately 4–5(6)-lobed, the lobes becoming patent-reflexed; petals 0; disc annular, with 2 central depressions; stamens 2, connivent, filaments extremely short, inserted in depressions on the disc, anthers basifixed, apicitrorse, thecae basally divergent, longitudinally dehiscent, connective narrow, not produced; pollen spheroidal, tectate, tectum punctate, psilate, weakly 3-porate, apertures circular, depressed, indistinct, densely insulate; pistillode 0; receptacle plane. Female flowers shortly pedicellate, pedicels shorter and stouter than in the male; sepals 6(7), narrowly triangular, open; petals 0; disc 0; ovary 3-locular, setose, ovule 1 per locule; styles 3, short, connate into a broadly conic-cylindric to cylindric base, 3-lobed at apex or apical free portions strongly recurved, simple, entire, stigmas smooth to minutely papillose. Fruit 3-dymous, dehiscing into 3 2-valved cocci; exocarp thin, sparingly setose; endocarp thinly woody; columella persistent, 3-radiate. Seeds globose, ecarunculate; testa smooth, marmorate; endosperm fleshy; cotyledons broad, flat. Fig.: See Fl. Indo-Chine 5: 388, t. 45.10–14 and 397, t. 46.1 (1926) (as *Tragia delpyana*). See also Fig. 36, p. 261.

A monotypic tropical Asiatic genus ranging from E. Peninsular India to Indochina, Java and the Philippines, but not reported from either Sumatra or Borneo.

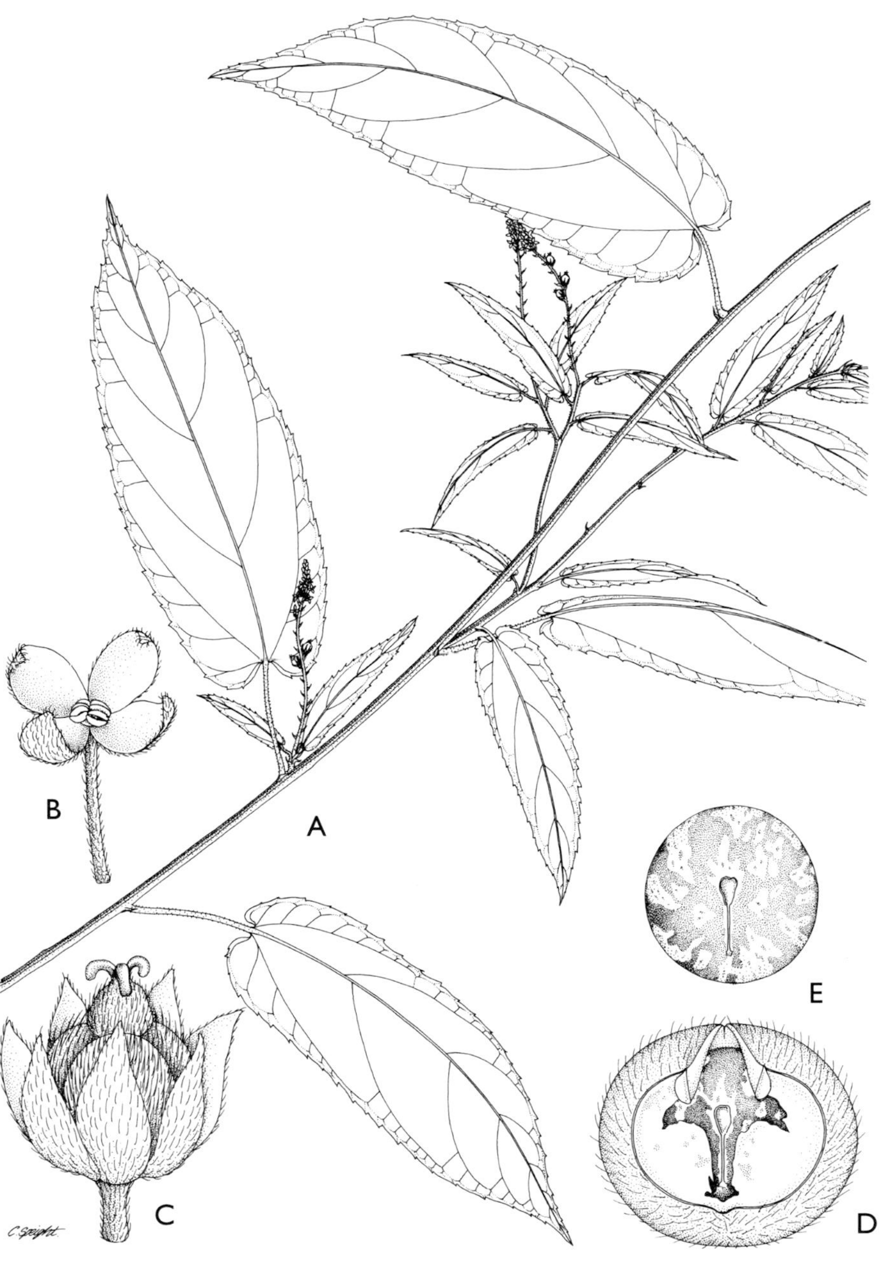

FIG. 36. *Pachystylidium hirsutum* (Blume) Pax & K. Hoffm. (B.: *Tragia hirsuta* Blume; S.: *Tragia delpyana* Gagnep.). **A** Habit × 1; **B** Male flower × 18; **C** Female flower × 18; **D** Mericarp × 7.5; **E** Seed × 7.5. **A–C** from *Kerr* 17893; **D–E** from *Kerr* 14097. Drawn by Camilla Speight.

Subtribe 34c. **Dalechampiinae** *(Müll. Arg.) G. L. Webster*, Ann. Missouri Bot. Gard. 81(1): 95 (1994). T.: *Dalechampia* Plum. ex L.

Dalechampieae Müll. Arg., Bot. Zeitung (Berlin) 22: 324 (1864), and in DC., Prodr. 15(2): 1232 (1866).

Monoecious shrubs or herbs, more often twining; indumentum simple; urticating hairs often +; leaves stipulate, usually stipellate; stipules persistent; leaf-blades simple or compound, palmatilobate to palmatisect, entire or toothed. Flowers in 2-bracteate terminal capitate pseudanthia; male pleiochasium terminal, usually 8–12-flowered; bracts mostly resiniferous; female cymule basal, (1)3-flowered; male flowers: stamens 8–100, filaments connate; pollen prolate, 3-colporate, reticulate; pistillode 0; female flowers: sepals entire to pinnatifid; ovary 3-locular; stylar column elongate, stigma extending down to upper $^3/_4$, sometimes confined to tip. Fruit unadorned, dehiscent.

A monogeneric subtribe. Although most workers since Müller (1866) have treated this as a tribe, *Dalechampia* has links to both the *Tragiinae* (urticating hairs) and the *Plukenetieae* (elongate stylar column), so the above treatment better reflects this linkage.

206. **Dalechampia** *Plum. ex L.*, Sp. Pl.: 1054 (1753), and Gen. Pl. ed. 5: 473 (1754); A. Juss., Euphorb. Gen.: 55 (1824); Baill., Étude Euphorb.: 485 (1858); Müll. Arg. in DC., Prodr. 15(2): 1232 (1866), and in Mart., Fl. Bras. 11(2): 633 (1874); Benth., Gen. Pl. 3: 330 (1880); Hook. f., Fl. Brit. India 5: 467 (1888); Pax in Engl. & Prantl, Nat. Pflanzenfam. 3(5): 67 (1890); Prain in Fl. Trop. Afr. 6(1): 952 (1912); Pax & K. Hoffm. in Pflanzenr. 68.XII: 3 (1919); Prain in Fl. Cap. 5(2): 497 (1920); Gagnep., Fl. Indo-Chine 5: 344 (1926); Pax & K. Hoffm. in Nat. Pflanzenfam. ed. 2, 19C: 151 (1931); Leandri, Notul. Syst. (Paris) 11: 35 (1943); Alain, Fl. Cuba 3: 103 (1953); Hurus., J. Fac. Sci. Univ. Tokyo, Sect. 3, Bot. 6: 293 (1954); Keay (ed.), Fl. W. Trop. Afr. ed. 2, 1: 412 (1958); J. Léonard, Fl. Cong. Rwa.-Bur. 8(1): 194, t. 13 (1962); Backer & Bakh. f., Fl. Java 1: 492 (1963); Jabl., Mem. New York Bot. Gard. 17: 145 (1967); G. L. Webster, Ann. Missouri Bot. Gard. 54: 193, 308 (1967, 1968); J.-G. Adam, Mém. Mus. Natl. Hist. Nat., B, Bot. 20: 467 (1971); Airy Shaw, Kew Bull. 26: 251 (1972); Berhaut, Fl. Ill. Sénég. 3: 425 (1975); Philcox, Fl. Trin. Tob. II (X): 682 (1979); G. L. Webster & Armbr., Brittonia 31: 352 (1979), and Syst. Bot. 7: 484 (1982); Correll, Fl. Bah. Arch.: 793 (1982); Armbr., Syst. Bot. 9: 272 (1984); Alain, Fl. Español. 4: 134 (1986); Radcl.-Sm. in Fl. Pak. 172: 72 (1986), and in Fl. Trop. E. Afr., Euph. 1: 285, t. 56 (1987); Alain, Descr. Fl. Puerto Rico 2: 382 (1988); Armbr., Syst. Bot. 13: 303 (1988); G. L. Webster & Huft, Ann. Missouri Bot. Gard. 75: 1108 (1988); G. L. Webster, Brittonia 41: 1 (1989); Armbr., Brittonia 41: 44 (1989); R. A. Howard, Fl. Lesser Antilles 5: 42 (1989); G. L. Webster & Armbr., Bot. J. Linn. Soc. 105: 137 (1991); G. L. Webster, Ann. Missouri Bot. Gard. 81: 95 (1994); Radcl.-Sm. in Fl. Zamb. 9(4): 238, t. 49 (1996); Gillespie in Acev.-Rodr., ed., Fl. St. John: 210 (1996); Philcox in Dassan., ed., Fl. Ceyl. XI: 176 (1997); Armbr., Smithsonian Contr. Bot. 86: 14 (1997). T.: *D. scandens* L.

Cremophyllum Scheidw., Bull. Acad. Roy. Sci. Bruxelles 9(1): 23 (1842). T.: *C. spathulatum* Scheidw. [=*D. spathulatum* (Scheidw.) Baill.]
Rhopalostylis Klotzsch ex Baill., Adansonia 5: 317 (1865). T.: *R. buettnerioides* Klotzsch ex Baill.[=*D. micrantha* Poepp. & Endl.]
Megalostylis S. Moore, J. Bot. 54: 208 (1916). T.: *M. poeppigii* S. Moore [=*D. micrantha* Poepp. & Endl.]

Monoecious shrubs, subshrubs or perennial herbs, mostly twining. Indumentum simple, sometimes urticating. Leaves alternate, petiolate, stipulate, simple, palmatilobate, palmatifid, palmatisect, palmatipartite, trifoliolate or palmate, dentate, serrate or entire, often triangular, cordate and palminerved, rarely elliptic, cuneate and penninerved, basal glands usually stipelliform; stipules persistent. Inflorescences axillary or terminal, capitate, pseudanthial, mostly solitary, bisexual, with 2 unisexual subsidiary coflorescences, pedunculate, 2-bracteate; bracts usually large, foliaceous, equal or subequal, sessile, stipulate, entire, toothed or 3-lobate, pale green, yellow, white or pink; male coflorescence terminal, pleiochasial, involucrate, (7)8–12(20)-flowered, comprising 5(7) sessile, (1)3-flowered bracteate cymules, with an adjacent mass of fused resiniferous or wax-encrusted bracts and/or sterile flowers; female a basal, subsessile, (1)3-flowered cymule. Male flowers: pedicels articulate; calyx closed in bud, globose, later valvately 4–6-lobed; petals 0; disc 0; stamens (8)10–30(100), filaments short, partially connate into a column, anthers erect, often 2-dymous, longitudinally dehiscent; pollen large, prolate, 3-colporate, colpi short, narrow, colpus transversalis elongate, tectate, intra-reticulate, reticulum very coarse; pistillode 0; receptacle convex or columnar. Female flowers shortly pedicellate, pedicels elongating in fruit; sepals 5–12, imbricate, linear to narrowly lanceolate, entire to pinnatifid, glabrous or hirsute, glandular or not, often later accrescent, indurated and enclosing the fruit; petals 0; disc 0; ovary 3(4)-locular, ovule 1 per locule; styles connate into an elongate column, obtuse, dilated or obliquely excavate and radiate-lobulate at the apex, stigmatic surface extending down to as far as the upper $^{3}/_{4}$ of the column, but sometimes confined to tip. Fruit 3(4)-lobate to 3(4)-coccous or 3(4)-dymous, dehiscing into 3(4) 2-valved cocci; endocarp crustaceous or woody; columella persistent or subpersistent. Seeds globose to ellipsoid, ecarunculate; testa ± smooth, lineate or mottled; endosperm fleshy; cotyledons broad, flat. Fig.: See Euphorb. Gen.: t. 17, f. 59.1–10 (1824); Étude Euphorb.: t. 3, f. 16–33 (partly as *Cremophyllum*), t. 4, f. 1–5 (1858); Fl. Bras. 11(2): tt. 88–91 (1874); Pflanzenr. 68.XII: 6–47, tt. 1–9 (1919); Fl. Indo-Chine 5: 342, t. 39.6–12 and 348, t. 40.1, 2 (1926); Nat. Pflanzenfam. ed. 2, 19C: 152, t. 77 (1931); Notul. Syst. (Paris) 11: 37–41, tt. 1, 2 (1943); Fl. Cuba 3: 104, t. 30 (1953); Ann. Missouri Bot. Gard. 54: 312, t. 18 (1968); Amer. J. Bot. 56: 754, t. 15A–H (1969); Fl. Ill. Séneg. 3: 424 (1975); Fl. Bah. Arch.: 794, t. 332 (1982); Fl. Pak. 172: 73, t. 15 A–C (1986); Fl. Español. 4: 136, t. 116-12 (1986); Fl. Trop. E. Afr., Euph. 1: 288, t. 56 (1987); Descr. Fl. Puerto Rico 2: 384, t. 59-12 (1988); Fl. Lesser Antilles 5: 43, t. 14 (1989); Fl. Zamb. 9(4): 239, t. 49 (1996); Fl. St. John: 211, t. 95A–E (1996); Smithsonian Contr. Bot. 86: 18–26, tt. 4–8 (1997).

A pantropical but predominantly neotropical genus of 110 species, of which c. 90 are American, with a few in Africa, Madagascar and India, and 1 extending to S. China and Java (See Frontispiece).

The genus exhibits considerable diversity. Pax & Hoffmann (1919) recognized 13 sections in it (one of which, however, subsequently proved to be a *Cissus*!), but Webster & Armbruster (1991) have whittled these down to six.

Tribe 35. **OMPHALEAE** *(Pax & K. Hoffm.) G. L. Webster*, Taxon 24: 598 (1975), and Ann. Missouri Bot. Gard. 81(1): 96 (1994). T.: *Omphalea* L.

Hippomaneae subtribe *Omphaleinae* Pax & K. Hoffm. in Pflanzenr. 52: 14 (1912).

Monoecious trees, shrubs or lianas; stems with a reddish latex; indumentum simple; leaves alternate, simple or lobate, penninerved or palminerved, usually with 2 sessile basal glands; stipules deciduous. Inflorescences terminal, spicate-, racemose- or paniculate-thyrsiform, bracts ± foliaceous and 2-glandular; flowers apetalous; male sepals 4–5, decussate or imbricate; disc annular or lobed, often massive, rarely 0; stamens 2–3, filaments connate, anther connectives free or fused; pollen oblate, 3-colporate, finely punctate-spinulose; pistillode 0; female sepals 4–5, imbricate; disc usually 0, tenuous when +; ovary 3-locular; styles connate into a stout column, stigmas undivided. Fruit thick-walled, dehiscent or indehiscent. Seeds large, globose, ecarunculate; testa usually dry, sometimes fleshy; endosperm copious.

A monogeneric tribe. The affinities of *Omphalea* have long been controversial: A. Jussieu (1824) and Baillon (1858) associated it with the *Hippomaneae*, Müller (1866) had it in the subtribe *Gelonieae* of the tribe *Hippomaneae*, whilst Pax & K. Hoffmann (1912a) created a special subtribe for it under the *Hippomaneae*, but later (1931) re-assigned this to the tribe *Gelonieae;* Croizat (1942a), on the other hand, pointed out affinities with the tribe *Plukenetieae*, and Punt (1962) noted that pollen morphology supports this. Gillespie (1997) considers that there may be a link with the *Crotonoideae.*

207. **Omphalea** *L.*, Syst. Nat. ed. 10: 1264 (1759), and Gen. Pl. ed. 6: 479 (1764), nom. cons.; A. Juss., Euphorb. Gen.: 54 (1824); Baill., Étude Euphorb.: 527 (1858); Müll. Arg. in DC., Prodr. 15(2): 1134 (1866), and in Mart., Fl. Bras. 11(2): 513, t. 72 (1874); Benth., Gen. Pl. 3: 332 (1880); Pax in Engl. & Prantl, Nat. Pflanzenfam. 3(5): 92 (1890); Hemsl., Hooker's Icon. Pl. 26: t. 2537 (1897); Pax & K. Hoffm., Pflanzenr. 52.V: 14 (1912); Fawc. & Rendle, Fl. Jamaica 4: 319 (1920); Pax & K. Hoffm., Nat. Pflanzenfam. ed. 2, 19C: 186 (1931); Croizat, Bull. Jard. Bot. Buitenzorg 3. 17: 204 (1941); Alain, Fl. Cuba 3: 109 (1953); Jabl., Mem. New York Bot. Gard. 17: 162 (1967); G. L. Webster, Ann. Missouri Bot. Gard. 54: 295 (1968); Airy Shaw, Kew Bull. 23: 130 (1969); Alain, Mem. New York Bot. Gard. 21(2): 124 (1971); Airy Shaw, Kew Bull. 26: 310 (1972); Whitmore, Tree Fl. Mal. 2: 120 (1973); Airy Shaw,

Kew Bull. Addit. Ser. 4: 179 (1975); Philcox, Fl. Trin. Tob. II (X): 670 (1979); Airy Shaw, Kew Bull. Addit. Ser. 8: 177 (1980), Kew Bull. 35: 661 (1980), 37: 31 (1982), and Euph. Philipp.: 39 (1983); Alain, Fl. Español. 4: 178 (1986); Radcl.-Sm. in Fl. Trop. E. Afr., Euph. 1: 371, t. 69 (1987); Gillespie, Rev. Phyl. Anal. *Omphalea* (diss.): 144 (1988); R. A. Howard, Fl. Lesser Antilles 5: 67 (1989); A. H. Gentry, Woody Pl. NW. S. Amer.: 407 (1993); G. L. Webster, Ann. Missouri Bot. Gard. 81: 96 (1994); P. I. Forst., Austrobaileya 4(3): 381 (1995); Murillo-Aldana & Franco-Rosselli, Euf. Reg. Ararac.: 122 (1995); Gillespie, Novon 7: 127 (1997), and Smithsonian Contr. Bot. 86: 6 (1997). T.: *O. triandra* L. (typ. cons.).

Omphalandria P. Browne, Civ. Nat. Hist. Jamaica: 335 (1756), nom. rej. T.: (none designated).
Duchola Adans., Fam. Pl. 2: 357 (1763), nom. superfl. T.: (none designated).
Ronnowia Buc'hoz, Pl. Nouv. Découv. 6: t.4 (1779). T.: *R. domingensis* Buc'hoz, nom. illeg. [=*O. triandra* L.]
Hecatea Thouars, Hist. Vég. Îles France 27: t.5 (1804). LT.: *H. oppositifolia* Willd. [=*O. oppositifolia* (Willd.) Gillespie] (designated by Gillespie, 1988).
Hebecocca Beurl., Kongl. Vetensk. Acad. Handl. 1854: 146 (1856). T.: *H. panamensis* Beurl. [=*O. diandra* L.]
Neomphalea Pax & K. Hoffm., Pflanzenr. 68 (xiv: Addit. VI): 54 (1919). T.: *N. papuana* (Pax & K. Hoffm.) Pax & K. Hoffm. [=*O. papuana* Pax & K. Hoffm.]

Monoecious lianes, shrubs or trees; stems with a reddish latex. Indumentum simple. Leaves alternate, often long-petiolate, stipulate, large, simple or lobate, entire, oblong and penninerved or cordate and 3–5-palminerved, often coriaceous, with 2 sessile glands at the apex of the petiole or beneath the base of the lamina; stipules small, triangular, deciduous. Inflorescences terminal, coaëtaneous or precocious, spicate-, racemose- or paniculate-thyrsiform, bisexual, with bracteate cymules of 1(3) central female flower(s) and several peripheral males, or else all male; bracts stipitate to subsessile, usually elongate or linear-spathulate, attenuate, foliaceous or subpetaloid, membranaceous, pallid, (1)2-glandular at the apex of the stipe; intracymular bractlets small. Male flowers shortly pedicellate; sepals 4–5, free, broad, decussate or strongly imbricate; petals 0; disc glands 5, fleshy, inserted at the base of the sepals, or connate into a cone or ring surrounding the staminal column, rarely 0; stamens 2–3, filaments connate into a short, slender column, anther connectives thick, broad, free or connate into a peltate, pileiform, 2–3-lobate mass, anthers peripheral, on the margins of the lobes, small, reniform, vertically 2-dymous, thecae extrorse, longitudinally dehiscent; pollen oblate, 3-colpate, colpi broad, tectum finely punctate-spinulose; pistillode 0; receptacle 0. Female flowers: pedicels and sepals ± as in the male; petals 0; hypogynous disc 0; ovary 2–3-locular, ovule 1 per locule; stylar column thick, continuous with the top of the ovary, obtuse or very shortly 2–3-lobate and slightly excavate, stigmas undivided. Fruit 2–3-lobate to subglobose, often large, thin- or thick-walled, dehiscing into 3 2-valved cocci, or subdrupaceous and indehiscent or tardily dehiscent; exocarp fleshy; endocarp hard, woody. Seeds large, globose or 3-gono-subglobose, ecarunculate; testa dry or occasionally thinly fleshy; endosperm copious; cotyledons broad, flat. Fig.: See Euphorb. Gen.: t. 17, f. 58.1–11 (1824); Étude Euphorb.: t. 7, f. 1–9 (1858); Fl. Bras. 11(2): t. 72 (1874); Hooker's Icon. Pl. 26: t. 2537 (1897); Pflanzenr. 52.V: 18, t. 1 and 21, t. 2 (1912); Fl. Jamaica 4: 321, t. 107 (1920); Bull. Soc. Bot. France 85: 525, t. 1.22–30 (1938); Fl. Cuba 3: 110, t. 34 (1953); Ann. Missouri Bot. Gard. 54: 297, t. 16 (1968); Fl. Español. 4: 182, t. 116-27 (1986); Fl. Trop. E. Afr., Euph. 1: 372, t.69 (1987); Fl. Lesser Antilles 5: 68, t. 26 (1989); Woody Pl. NW. S. Amer.: 408, t. 115.2 (1993); Austrobaileya 4(3): 383, t. 1 (1995); Euf. Reg. Ararac.: 123, t. 33 (1995); Novon 7: 129, t. 1 and 130, tt. 2–7 (1997); Smithsonian Contr. Bot. 86: 7, t. 1 (1997).

A pantropical genus of c. 20 species, with 1 widespread in the neotropics, 5 in the W. Indies, 1 in Central America, 1 in Brazil, 1 in E. Africa, 4 in Madagascar, 3 in SE Asia, ranging from Burma (Myanmar) to Indochina, the Philippines, Borneo and the Lesser Sunda Is., and 3 in New Guinea, Queensland and the Solomon Is.

Subfamily IV. **CROTONOIDEAE** Pax, Bot. Jahrb. Syst. 5: 413 (1884), and Nat. Pflanzenfam. ed. 1, 3(5): 14 (1890); G. L. Webster, Ann. Missouri Bot. Gard. 81: 97 (1994). T.: *Croton* L.

Trees, shrubs or herbs; latex coloured or 0; indumentum simple, stellate or lepidote; leaves alternate, opposite or rarely verticillate, stipulate or not, simple, palmatilobate or compound, often glandular. Inflorescences axillary or terminal, dichasial, spicate, racemose, thyrsiform or paniculate; male sepals imbricate or valvate; petals and disc + or 0; stamens 3–5(∞), filaments free or connate; pollen 2- or 3-nucleate, 3-colporate, porate or inaperturate, mostly with hexagonally-disposed clavae, baculae or echinae; pistillode + or 0; female sepals 2–3(6), imbricate or valvate, often connate; petals and disc + or 0; ovary mostly 2–4-locular, ovules 1 per locule, anatropous; styles 2-fid to multifid, rarely entire. Fruit mostly dehiscent. Seeds carunculate or not; testa sometimes fleshy; endosperm usually copious, often oily.

Following Webster (1975, 1994) the subfamily *Crotonoideae* here comprises 12 tribes, a narrower circumscription than that of Pax (1931), who included in it nearly all of the 1-ovulate *Euphorbiaceae*, but much wider than that of Hurusawa (1954), who included only the tribe *Crotoneae*. Müller (1866) had the genera of the subfamily dispersed among the tribes *Crotoneae*, *Acalypheae* and *Hippomaneae*. Bentham (1880) placed many crotonoids in the first two subtribes of his tribe *Crotoneae*, but had a mixture of acalyphoids and crotonoids in the rest. Hutchinson (1969) placed many crotonoids in his tribes 13–20, 38 and 39, but 5 of them also included either acalyphoids or euphorbioids.

The distinctive pollen morphology is useful for establishing membership of the subfamily, but except in a few cases does not help to clarify tribal demarcation within it.

KEY TO THE TRIBES

1. Pollen colporate or porate, reticulate to clavate; petals 0; indumentum mostly simple · 2
1. Pollen inaperturate, croton-pattern +; petals usually +; indumentum simple, malpighiaceous, stellate or lepidote; laticifers not articulated, except in some *Jatropheae* · 5
2. Laticifers articulated; pollen 3-nucleate; mostly monoecious; x=9 · · · · · 3
2. Laticifers not articulated; pollen 2-nucleate; mostly dioecious; x=11 · · · 4
3. Pollen colporate; styles 2-fid; seeds carunculate or not; endosperm oily · 36. *Micrandreae*
3. Pollen periporate; styles mostly multifid; seeds carunculate; endosperm starchy · 37. *Manihoteae*
4. Pollen colporate; leaves not pellucid-punctate; stipules free; inflorescences axillary, mostly spicate or paniculate · · · · · · · · · · · · · 38. *Adenoclineae*
4. Pollen periporate; leaves usually pellucid-punctate; stipules connate; inflorescences usually leaf-opposed, glomerulate · · · · · · 39. *Gelonieae*
5. Endosperm scanty; cotyledons massive; inflorescences dichasial; petals 0; styles simple, dilated; latex milky · · · · · · · · · · · · · 40. *Elateriospermeae*
5. Endosperm copious; cotyledons thin; inflorescences dichasial; petals mostly +, at least in male flowers; styles 2-fid to multifid, less often entire · · · · 6
6. Male sepals fused in bud, splitting valvately or irregularly into segments; fruit dehiscent or drupaceous; seeds mostly ecarunculate · · 47. *Aleuritideae*
6. Male sepals imbricate, free or if connate then ± open in bud; seeds carunculate or ecarunculate · 7
7. Indumentum mostly simple, occasionally malpighiaceous · · · · · · · · · · 8
7. Indumentum , at least in part, stellate or lepidote · · · · · · · · · · · · · · · 10
8. Stamens 3–5, filaments connate; pollen finely clavate; monoecious; inflorescences terminal or axillary, racemose or thyrsiform; seeds ecarunculate. · 43. *Trigonostemoneae*
8. Stamens mostly >5, free or connate; pollen coarsely clavate · · · · · · · · · 9
9. Inflorescences terminal, mostly dichasial; leaves mostly palmatilobate or compound; fruit dehiscent or drupaceous; seeds carunculate or ecarunculate · 41. *Jatropheae*
9. Inflorescences terminal or axillary, racemose or spicate to paniculate; leaves elobate, usually eglandular; fruit dehiscent · · · · · · 42. *Codiaeae*
10. Fruit drupaceous; seeds ecarunculate; petals ± coherent · 46. *Ricinodendreae*
10. Fruit dehiscent; seeds carunculate; petals free or 0 · · · · · · · · · · · · · · 11
11. Pollen with spinose or reduced processes; leaves often exstipulate; filaments often connate, at least in part; cotyledons broad or narrow · 44. *Ricinocarpeae*
11. Pollen processes not spinose or reduced; leaves usually stipulate; filaments free or connate; cotyledons broad · · · · · · · · · · · · · · · · · 45. *Crotoneae*

Shaw, Kew Bull. Addit. Ser. 4: 109 (1975), 8: 78 (1980), Kew Bull. 35: 628 (1980), 36: 293 (1981), 37: 18 (1982), and Euph. Phil.: 24 (1983); Grierson & D. G. Long, Fl. Bhut. 1: 808 (1987); G. L. Webster, Ann. Missouri Bot. Gard. 81: 101 (1994). T.: *E. chinense* Benth.

Capellenia Teijsm. & Binn., Natuurk. Tijdschr. Ned.-Indië 29: 238 (1867)). T.: *C. moluccana* Teijsm. & Binn. [=*E. moluccanum* (Teijsm. & Binn.) Kurz]

Dioecious, very rarely monoecious, often tall trees. Indumentum thin, stellate. Twigs cicatricose, sometimes hollow and myrmecophilous. Leaves crowded at the branch-apices, long-petiolate, stipulate, simple, ovate-orbicular, often peltate or cordate, entire, often coriaceous, 3–8-palminerved from the base, then penninerved, often with either a globose gland or a domatium at each nerve-junction; petiole usually 2-glandular at the apex; stipules small, soon deciduous. Inflorescences axillary or lateral, elongate, males paniculate, multiflorous, sometimes long-pedunculate, females simple or weakly paniculate, mostly pedunculate; bracts acute, males several-flowered, females 1-flowered; rarely (in 2 species) flowers hermaphrodite. Male flowers subsessile; calyx globose in bud, shortly cupular, shortly unequally 3–4-dentate, lobes slightly imbricate to subvalvate; petals 0; disc extrastaminal, 4–5-angled, lobulate; stamens (5)6–10, filaments short, ± connate, anthers dorsifixed, extrorse, thecae distinct, 2-dymous, 4-locellate; pollen as in 215. *Klaineanthus*, but colpus membrane 0; pistillode 0. Female flowers subsessile; calyx 5-toothed; petals 0; disc annular; ovary globose, (1)2–6(7)-locular, ovules 1 per locule; styles stigmatiform, connate into a sessile, (1)2–6(7)-lobate disc. Fruit subindehiscent, baccate, (1)2–6(7)-locular, breaking up into indehiscent cocci; pericarp ± fleshy; columella 0. Seeds ecarunculate, subglobose or compressed; testa smooth or reticulate-rugose. Fig.: See Malesia 2: 45, t. 2 (1884); Pflanzenr. 52.IV: 37, t. 10 and 38, t. 11 (1912); Treubia 10: 431, tt. 12, 13 (1919); Fl. Indo-Chine 5: 445, t. 54.12–16 and 455, t. 55.1–4 (1926); Nat. Pflanzenfam. ed. 2, 19C: 185, t. 97 (1931); Tree Fl. Mal. 2: 94, t. 7 (1973); Kew Bull. Addit. Ser. 8: 226, t. 4.2 (1980).

A palaeotropical genus of 12 species, ranging from Assam to China, Malaysia, Indonesia, Philippines, New Guinea, Solomon Is., tropical Australia (Queensland) and Fiji.

The wood of *E. chinense* Benth., *E. peltatum* Merr. and *E. moluccanum* Becc., being very light, has a variety of uses, e.g. for packing-cases, matches and floats.

Tribe 39. **GELONIEAE** *(Müll. Arg.) Pax*, Nat. Pflanzenfam. ed. 1, 3(5): 88 (1890); G. L. Webster, Ann. Missouri Bot. Gard. 81: 102 (1994). T.: *Gelonium* Roxb. ex Willd. [=*Suregada* Roxb. ex Rottler]

Hippomaneae subtribe *Gelonieae* Müll. Arg., Linnaea 34: 202 (1865), and in DC., Prodr. 15(2): 1124 (1866).

Dioecious or very rarely monoecious trees or shrubs; branches terete or flattened and cladode-like; latex not evident; indumentum simple, usually scanty or 0; leaves alternate, petiolate or subsessile, stipulate, simple, penninerved; lamina usually pellucid-punctate; stipules deciduous or persistent and foliaceous. Inflorescences leaf-opposed or terminal, glomerulate; male sepals usually 5–6, free, imbricate, sometimes abaxially glanduliferous; petals 0; disc extrastaminal or intrastaminal, annular or dissected; stamens (6)10–25(60), filaments free, anthers dehiscing longitudinally; pollen 2-nucleate, subglobose, 3–6-porate, crotonoid; pistillode 0 or +; female sepals (4)5(8), imbricate, sometimes abaxially glanduliferous; petals 0; disc annular, staminodes sometimes +; ovary (2)3(4)-locular; styles 2-fid, rarely multifid, or stigmas sessile, cordiform. Fruit dehiscent or sometimes indehiscent and drupaceous. Seeds ecarunculate; testa ± fleshy; endosperm +.

A digeneric palaeotropical tribe. Pax & K. Hoffmann (1931) included in it 18 genera in 7 subtribes; several of these are here included in the previous tribe.

KEY TO THE GENERA

1. Branches terete; leaves petiolate, lamina usually pellucid-punctate, teeth and marginal glands not as below; stipules deciduous; inflorescences leaf-opposed; pistillode 0; styles +, stigmas linear-subulate · · · · 220. *Suregada*
1. Branches flattened, cladode-like; leaves subsessile, lamina not pellucid-punctate, teeth 2-dentate, marginal glands excavate beneath; stipules foliaceous, persistent; inflorescences terminal; pistillode +; stigmas sessile, cordiform · 221. *Cladogelonium*

220. **Suregada** *Roxb. ex Rottler*, Ges. Naturf. Freunde Berlin Neue Schriften 4: 206 (1803); A. Juss., Euphorb. Gen.: 60 (1824); Baill., Étude Euphorb.: 395 (1858); Croizat, Bull. Jard. Bot. Buitenzorg III. 17: 212 (1942); Merr., J. Arnold Arbor. 37: 79 (1951); J. Léonard, Bull. Jard. Bot. État. 28: 443 (1958), and Fl. Cong. Rwa.-Bur. 8(1): 124 (1962); Backer & Bakh. f., Fl. Java 1: 497 (1963); Airy Shaw, Kew Bull. 23: 128 (1969), and 26: 342 (1972); Whitmore, Tree Fl. Mal. 2: 133 (1973); Airy Shaw, Kew Bull. Addit. Ser. 4: 198 (1975), 8: 203 (1980), Kew Bull. 35: 689 (1980), 36: 346 (1981), 37: 36 (1982), and Euph. Phil.: 46 (1983); Grierson & D. G. Long, Fl. Bhut. 1: 809 (1987); Radcl.-Sm., Fl. Trop. E. Afr. Euph. 1: 376 (1987), and Kew Bull.: 46: 711 (1991); G. L. Webster, Ann. Missouri Bot. Gard. 81: 102 (1994); Radcl.-Sm., Fl. Zamb. 9(4): 249 (1996); Philcox in Dassan., ed., Fl. Ceyl. XI: 118 (1997). T.: *S. roxburghii* Rottler.

Gelonium Roxb. ex Willd., Sp. Pl. 4: 831 (1806), non Gaertn. (1791); A. Juss., Euphorb. Gen.: 34 (1824); Müll. Arg. in DC., Prodr. 15(2): 1126 (1866); Benth., Gen. Pl. 3: 324 (1880); Hook. f., Fl. Brit. India 5: 458 (1887); Pax & K. Hoffm., Pflanzenr. 52.IV: 14 (1912); Prain in Fl. Trop. Afr. 6(1): 947 (1912); Gagnep., Fl. Indo-Chine 5: 424 (1926); Pax & K. Hoffm., Nat.

Pflanzenfam. ed. 2, 19C: 182 (1931); Hurus., J. Fac. Sci. Univ. Tokyo, Sect. 3, Bot. 6: 318 (1954); Keay (ed.), Fl. W. Trop. Afr., ed. 2, 1: 413 (1958). T.: *G. lanceolatum* Willd. [=*S. lanceolata* (Willd.) Croizat]

Erythrocarpus Blume, Bijdr. Fl. Ned. Ind.: 604 (1825). LT.: *E. glomerulatus* Blume [=*S. glomerulata* (Blume) Baill.]

Ceratophorus Sond., Linnaea 23: 120 (1850). T.: *C. africanus* Sond. [=*S. africana* (Sond.) Kuntze]

Dioecious, sometimes monoecious, often glabrous small trees or shrubs. Branches terete. Indumentum, when present, simple. Leaves alternate, shortly petiolate, stipulate, simple, entire, denticulate or crenate-serrate, vesicular-punctate or -zonate, penninerved, occasionally deciduous; stipules connate, deciduous, leaving a prominent annular scar. Inflorescences leaf-opposed, cymulose; cymules fasciculate or glomerulate, or flowers subsolitary, sessile or pedunculate, 1- or 2-sexual, often resinous when immature; bracts usually minute. Male flowers usually pedicellate; sepals (4)5–6(7), free, imbricate, broad, unequal, outer sometimes glandular, occasionally cornute and cucullate, inner petaloid; petals 0; disc-glands usually interstaminal, confluent, rarely extrastaminal, annular; stamens (6)10–25(60), filaments free, anthers subdorsifixed, introrse, oblong, 2-thecous, thecae parallel, contiguous, longitudinally dehiscent; pollen as in 212. *Manihot* and 213. *Cnidoscolus*; pistillode 0. Female flowers pedicellate; sepals (4)5–6(8), free, imbricate, subequal, glandular or not, narrower than in the male; petals 0; disc annular, sometimes lobate; staminodes 5–10, staminiform or minute, subulate; ovary (2)3(4)-locular, ovules 1 per locule; styles (2)3(4), connate at the base, short, divaricate, 2-fid or multifid, stigmas linear or subulate, not persistent. Fruit (1)3(4)-lobate or subglobose, septicidally dehiscent into as many 2-valved cocci, loculicidal or indehiscent, drupaceous; exocarp smooth or reticulate; endocarp subligneous, thinly crustaceous or firmly coriaceous; columella persistent. Seeds ovoid, ecarunculate; exotesta thinly pulpy; endotesta crustaceous, smooth or foveolate-reticulate; endosperm fleshy; cotyledons broad, flat. Fig.: See Euphorb. Gen.: t. 10, f. 31A.1–10, B.1–10(1824); Pflanzenr. 52. IV: 17, t. 4 and 22, t. 5 (1912) (as *Gelonium*); Fl. Indo-Chine 5: 423, t. 50.14–16 and 431, t. 51.1–5 (1926); Formos. Trees: 341, t. 295 (1936) (as *Gelonium*); Kew Bull. 35: 668, t. 6B1–3 (1980); Fl. Trop. E. Afr. Euph. 1: 379, t. 71 (1987); Fl. Bhut. 1: 807, t. 50k,l (1987); Kew Bull.: 46: 714–724, tt. 1–6 (1991); Fl. Zamb. 9(4): 252, t. 53 (1996).

A palaeotropical genus of 31 species, 8 in Africa, 14 in Madagascar and adjacent islands and 9 in tropical Asia.

221. **Cladogelonium** *Leandri*, Bull. Soc. Bot. France 85: 530 (1938); G. L. Webster, Ann. Missouri Bot. Gard. 81: 131 (1994). T.: *Cl. madagascariense* Leandri.

Like 220. *Suregada*, but branches flattened, cladode-like; leaves subsessile, lamina not pellucid-punctate, teeth 2-dentate, marginal glands excavate beneath; stipules foliaceous, persistent; inflorescences terminal; pistillode + in

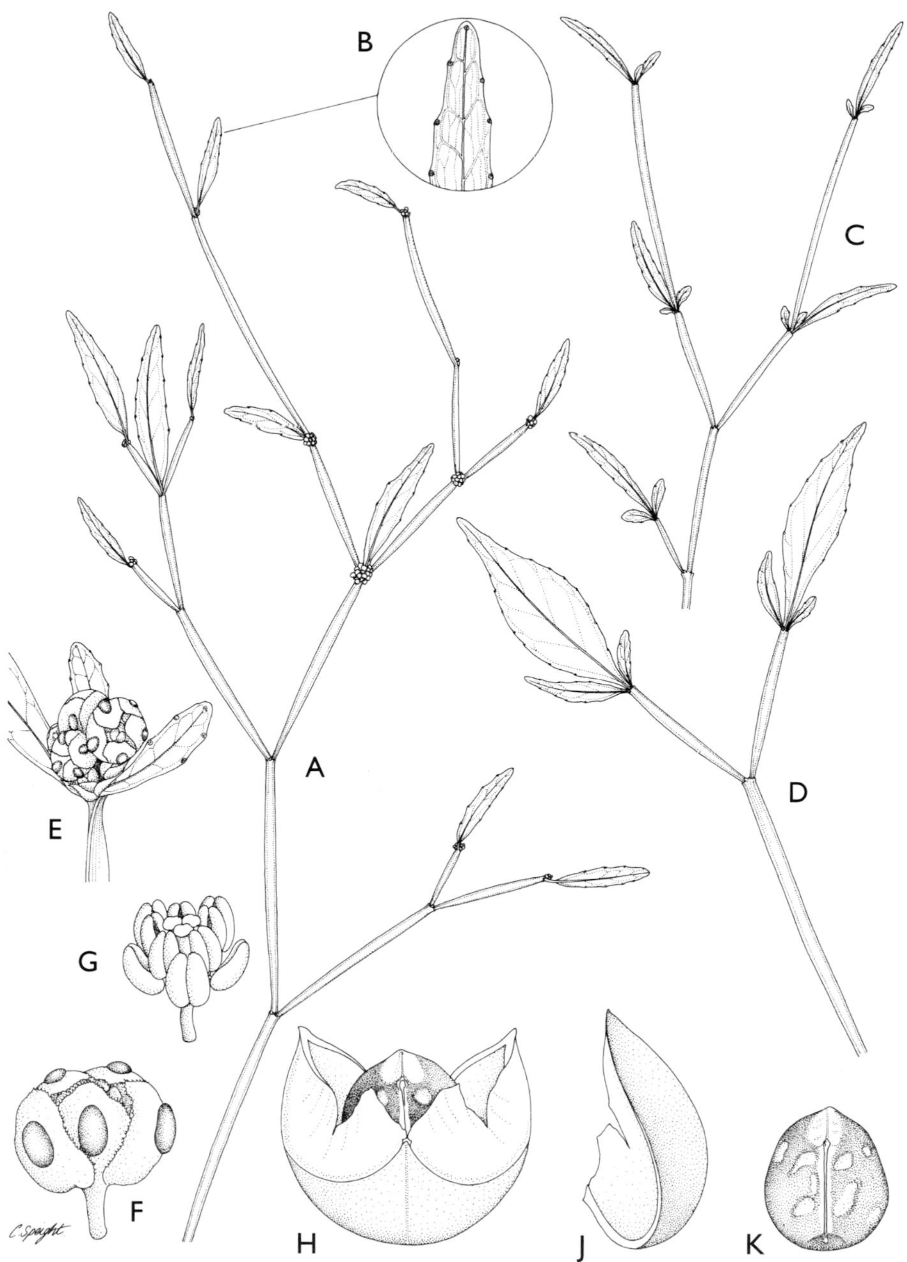

FIG. 38. *Cladogelonium madagascariense* Leandri. **A** Habit, male, leaves monomorphic × 1; **B** Detail of leaf-apex × 7.5; **C** Habit, leaves dimorphic × 1; **D** Habit, leaves trimorphic × 1; **E** Male inflorescence × 10.5; **F** Male flower × 30; **G** Stamens & pistillode × 30; **H** Mericarp × 10.5; **J** Fruit-valve × 10.5; **K** Seed × 10.5. **A, B** from *Humbert & Capuron* 23949; **C** from *Capuron* 22710bis-SF; **D** from *Cours* 3153; **E–K** from *Perrier de la Bâthie* 9696. Drawn by Camilla Speight.

the male flower; stigmas sessile, cordiform. Fig.: See Bull. Soc. Bot. France 85: 525, t.1.15–19bis (1938). See also Fig. 38, p. 285.

A monotypic genus from Madagascar.

Despite the diffidence of Webster (1994), *Cladogelonium* definitely belongs in the *Gelonieae*, but it should certainly be kept distinct from *Suregada*, although the differences are mainly vegetative, which fact led Hutchinson (ined.) to subsume it under the latter.

Tribe 40. **ELATERIOSPERMEAE** *G. L. Webster*, Taxon 24: 599 (1975), and Ann. Missouri Bot. Gard. 81: 102 (1994). T.: *Elateriospermum* Blume.

Monoecious trees; latex whitish; indumentum simple; leaves alternate, petiolate, stipulate, simple, entire, penninerved; petiole 2-glandular at apex; stipules deciduous. Inflorescences subterminal or axillary, cymose, 2-sexual; male sepals 5, free, imbricate; petals 0; disc lobulate, pubescent; stamens 10–20, filaments free, anthers apiculate; pollen globose, inaperturate, crotonoid; pistillode minute or 0; female sepals 5, imbricate, caducous; petals 0; disc cupular, pubescent; ovary 2–4-locular; styles elobate, stigmas dilated. Fruit dehiscent, but exocarp fleshy. Seeds ecarunculate; endosperm scanty; cotyledons massive.

A monogeneric tribe. *Elateriospermum* was placed between *Suregada* and *Endospermum* by Müller (1866), whilst Bentham (1880) and Pax (1910) placed it with *Micrandra* and *Cunuria*. However, the inaperturate pollen does not accord with either position. Airy Shaw (1975) included it in his subtribe *Jatrophinae* along with *Aleurites, Annesijoa, Tapoides* and *Loerzingia*. *Elateriospermum* appears to be without close relatives, but is possibly distantly related to *Micrandra* and *Manihot*.

222. **Elateriospermum** *Blume*, Bijdr. Fl. Ned. Ind.: 620 (1825); Baill., Étude Euphorb.: 397 (1858); Müll. Arg. in DC., Prodr. 15(2): 1130 (1866); Benth., Hooker's Icon. Pl. 13, t. 1294 (1879), and Gen. Pl. 3: 288 (1880); Hook. f., Fl. Brit. India 5: 381 (1887); Pax, Pflanzenr. 42: 17 (1910); J. J. Sm., Meded. Dept. Landb. Ned.-Indië 10: 571 (1910); Pax & K. Hoffm., Nat. Pflanzenfam. ed. 2, 19C: 181 (1931); Backer & Bakh. f., Fl. Java 1: 496 (1963); Airy Shaw, Kew Bull. 26: 258 (1972); Whitmore, Tree Fl. Mal. 2: 91 (1973); Airy Shaw, Kew Bull. Addit. Ser. 4: 108 (1975), and Kew Bull. 36: 292 (1981); G. L. Webster, Ann. Missouri Bot. Gard. 81: 102 (1994). LT.: *E. tapos* Blume (chosen by Webster, 1994).

Monoecious trees; crown dense; laticifers not articulated, latex whitish. Indumentum simple. Leaves alternate, long-petiolate, stipulate, simple, oblong, entire, penninerved, subcoriaceous, glossy; petiole 2-glandular at apex; stipules deciduous. Inflorescences subterminal or axillary, cymose, 2-chasial, 2-chotomous, lax, long-pedunculate, 2-sexual; central flower of the cyme female, the rest male, not numerous. Male flowers: sepals (4)5, free, strongly imbricate, broad; petals 0; disc thick, fleshy, lobulate, pubescent;

stamens 10–20, inserted in the excavations of the disc, filaments free, short, anthers introrse, narrowly oblong, erect, thecae parallel, connective apiculate, glanduliform; pollen globose, inaperturate, tectate, crotonoid, clavate on ridges; pistillode minute, pubescent, or 0. Female flowers: sepals 5(6), imbricate, larger than the male, caducous; petals 0; disc cupular, entire, pubescent; staminodes numerous, minute, subulate, immersed in excavations of the disc; ovary 2–4-locular, villous, ovules 1 per locule; styles 2–4, shortly subconnate and erect at first, later elongating, elobate, simple, stigmas short, dilated, thick, convex. Fruit large, 3-coccous, dehiscent into 3 2-valved cocci; exocarp somewhat fleshy; endocarp hardened. Seeds large, ecarunculate, oblong-ellipsoid, 3-gonous, shiny; endosperm very scanty; cotyledons massive, fleshy. Fig.: See Étude Euphorb.: t. 19, f. 26–28 (1858); Hooker's Icon. Pl. 13, t. 1294 (1879); Tree Fl. Mal. 2: 92, t. 6 (1973).

A monotypic Asiatic genus ranging from peninsular Thailand through Malaya to Sumatra, Java and Borneo.

The hard, durable wood is much-used in general carpentry. The seeds are eaten after roasting, but are poisonous when raw.

Tribe 41. **JATROPHEAE** *(Meisn.) Pax*, Nat. Pflanzenfam. ed. 1, 3(5): 72 (1890), and Pflanzenr. 42: 1(1910); G. L. Webster, Ann. Missouri Bot. Gard. 81: 102 (1994). T.: *Jatropha* L.

Crotoneae subtribe *Jatropheae* Meisn., Pl. Vasc. Gen. 1: 341 (1841).
Jatropheae subtribe *Jatrophinae* Pax, Pflanzenr. 42: 21(1910).
Joannesieae (Müll. Arg.) Pax, Bot. Jahrb. Syst. 59: 142 (1924).
Acalypheae subtribe *Joannesieae* ("*Johanneseae* ") Müll. Arg., Linnaea 34: 201 (1865). T.: *Joannesia* Vell.

Monoecious or rarely dioecious trees, shrubs or herbs; laticifers articulated or not; latex clear, white, greenish or reddish; indumentum simple, sometimes glandular; leaves elobate, palmatilobate, palmatipartite or compound; petioles usually glandular at the apex; stipules persistent or deciduous. Inflorescences terminal, sometimes axillary as well, dichasial-paniculate, sometimes reduced; male sepals usually 5, distinct, imbricate or valvate; petals 5, free or sometimes coherent; disc entire or dissected; stamens (6)8–30, filaments at least partly connate; pollen globose, 2-nucleate, inaperturate, crotonoid; pistillode + or 0; female sepals 5(6), free, imbricate, ± persistent in fruit; disc annular, lobate or dissected; staminodia sometimes +; ovary mostly 2–3-locular; styles elobate or 2-fid, stigmas sometimes dilated. Fruit dehiscent or drupaceous. Seeds carunculate or ecarunculate; endosperm +.

Following Webster (1994), only 2 of the 13 genera assigned by Pax (1910) to this tribe are retained, i.e. *Jatropha* L. and *Joannesia* Vell. None of those assigned to it by Hutchinson (1969), apart from the type-genus, appear to belong. Webster (1975) formerly included 10 genera in 3 subtribes in it under the later name *Joannesieae*, but later reduced *Loerzingia* Airy Shaw under

Deutzianthus Gagnep. and accorded tribal status to *Ricinodendron* Müll. Arg. & *Givotia* Griff., following Hutchinson. *Loerzingia* is resurrected in this treatment, however.

Key to the genera

1. Leaves simple, lobate or elobate; monoecious or dioecious; male sepals free or connate; latex usually reddish · 2
1. Leaves palmately compound; monoecious; male sepals connate; seeds ecarunculate · 6
2. Styles 2 × 2-fid or 6, 2-fid; male sepals connate at the base; dioecious · · · 3
2. Styles 3, 2-fid or elobate; sepals distinct; monoecious or rarely dioecious · · 4
3. Leaf-glands disciform, persistent; sepals valvate; pistillode 0 · 226. *Deutzianthus*
3. Leaf-glands subulate, deciduous; sepals imbricate; pistillode + · 227. *Loerzingia*
4. Fruit drupaceous; sepals cornute · · · · · · · · · · · · · · · · · · 225. *Oligoceras*
4. Fruit dehiscent; sepals ecornute · 5
5. Seeds carunculate; leaves often lobate or dentate; petioles not 2-glandular at the apex; stipules usually persistent, not infra-axillary · 223. *Jatropha*
5. Seeds ecarunculate; leaves elobate, entire; petioles 2-glandular at the apex; stipules infra-axillary, deciduous · · · · · · · · · · · · · · · · · · 224. *Vaupesia*
6. Styles slender, 2-fid; ovary 3-locular; leaves glabrous; anthers muticous; male calyx 5-lobate; stamens 15–25, inner filaments connate; male disc dissected; fruit dehiscent · 230. *Annesijoa*
6. Styles dilated, subentire to laciniate; ovary 2-locular; leaves indumented; anthers apiculate · 7
7. Male calyx cupular, open in bud; stamens 7–10, all filaments connate; male disc dissected; fruit drupaceous · · · · · · · · · · · · · · · · · · · 228. *Joannesia*
7. Male calyx fused in bud; stamens 18–33, inner filaments connate; male disc annular, intrastaminal; fruit not known · · · · · · · · · · 229. *Leeuwenbergia*

223. **Jatropha** *L.*, Sp. Pl.: 1006 (1753), and Gen. Pl., ed. 5: 437 (1754); A. Juss., Euphorb. Gen.: 37 (1824); Baill., Étude Euphorb.: 294 (1858); Müll. Arg. in DC., Prodr. 15(2): 1076 (1866), and in Mart., Fl. Bras. 11(2): 485 (1874); Benth., Gen. Pl. 3: 290 (1880); Hook. f., Fl. Brit. India 5: 382 (1887); Pax & K. Hoffm., Pflanzenr. 42: 21 (1910); Hutch. in Fl. Trop. Afr. 6(1): 775 (1912); Prain in Fl. Cap. 5(2): 418 (1920); Fawc. & Rendle, Fl. Jamaica 4: 310 (1920); Gagnep., Fl. Indo-Chine 5: 323 (1926); Pax & K. Hoffm., Nat. Pflanzenfam. ed. 2, 19C: 160 (1931); McVaugh, Bull. Torrey Bot. Club 72: 271 (1944); Keay (ed.), Fl. W. Trop. Afr., ed. 2, 1: 396 (1958); J. Léonard, Fl. Cong. Rwa.-Bur. 8(1): 88 (1962); Backer & Bakh. f., Fl. Java 1: 494 (1963); Jabl., Mem. New York Bot. Gard. 17: 155 (1967); G. L. Webster, J. Arnold Arbor. 48: 340 (1967), and Ann. Missouri Bot. Gard. 54: 234 (1968); J.-G. Adam, Mém. Mus. Natl. Hist. Nat., B, Bot. 20: 483 (1971); Berhaut, Fl. Ill.

Sénég. 3: 501 (1975); Dehgan & G. L. Webster, Univ. Calif. Publ. Bot. 74: 1 (1979); Philcox, Fl. Trin. Tob. II(X): 657 (1979); Correll, Fl. Bah. Arch.: 821 (1982); Proctor, Kew Bull. Addit. Ser. 11: 531 (1984); Alain, Fl. Español. 4: 164 (1986); Grierson & D. G. Long, Fl. Bhut. 1: 790 (1987); Hemming & Radcl.-Sm., Kew Bull. 42: 103 (1987); Radcl.-Sm., Fl. Trop. E. Afr. Euph. 1: 343 (1987); Alain, Dcscr. Fl. Puerto Rico 2: 401 (1988); R. A. Howard, Fl. Lesser Antilles 5: 60 (1989); J.-P. Lebrun & Stork, Enum. Pl. Afr. Trop. 1: 226 (1991); Radcl.-Sm., Kew Bull.: 46: 141, 327 (1991); G. L. Webster, Ann. Missouri Bot. Gard. 81: 103 (1994); Gillespie in Acev.-Rodr., ed., Fl. St. John: 216 (1996); Radcl.-Sm., Fl. Zamb. 9(4): 253 (1996), and Kew Bull.: 52: 177 (1997); Philcox in Dassan., ed., Fl. Ceyl. XI: 83 (1997). LT.: *J. gossypiifolia* L.

Curcas Adans., Fam. Pl. 2: 356 (1763); Baill., Étude Euphorb.: 313 (1858). T.: *C. adansonii* Endl. ex Heynh. [=*J. curcas* L.]

Castigliona Ruiz & Pav., Fl. Peruv. Prodr.: 139 (1794). T.: *C. lobata* Ruiz & Pav., nom. illeg. [=*J. curcas* L.]

Mozinna Ortega, Nov. Pl. Descr. Dec. 8: 104 (1798); A. Juss., Euphorb. Gen.: 35 (1824); Hook., Icon. Pl. 4: t. 357 (1841). T.: *M. spathulata* Ortega [=*J. dioica* Sessé]

Loureira Cav., Icon. 5: 17 (1799). T.: *L. glandulifera* Cav. [=*J. cordata* (Ort.) Müll. Arg. (B.: *Mozinna cordata* Ortega)]

Adenoropium Pohl, Pl. Bras. Icon. Descr. 1: 12 (1827). T.: *A. gossypiifolium* (L.) Pohl [=*J. gossypiifolia* L.]

Zimapania Engl. & Pax, Nat. Pflanzenfam. ed. 1, 3(5): 119 (1891)). T.: *Z. schiedeana* Engl. & Pax [=*J. dioica* Sessé]

Collenucia Chiov., Fl. Somala 1: 177 (1929). T.: *C. paradoxa* Chiov. [=*J. paradoxa* (Chiov.) Chiov.]

Monoecious or rarely dioecious trees, shrubs, suffrutices or herbs with the stems arising from a thick perennial root. Indumentum simple, sometimes glandular. Leaves alternate, fasciculate or lax, petiolate or sessile, stipulate, simple, entire, 3-more palmatilobate or -partite or occasionally pinnatilobate or -partite, segments entire or lobulate; petiole occasionally glanduliferous; stipules subulate, 2-fid or often multifid, segments commonly setaceous and glanduliferous, sometimes laciniate, sometimes rigid, spiny, branched or not. Inflorescences commonly terminal, occasionally subterminal, often corymbiform, dichotomously cymose, commonly androgynous, protogynous, with 1 female flower terminating each primary axis, lateral cymules male; bracts usually entire, sometimes glandular-fimbriate. Male flowers shortly pedicellate; sepals (4)5(6), often ± connate at the base, imbricate; petals 5, free or sometimes ± laterally coherent, imbricate or contorted; disc-glands 5, free, or disc entire; stamens (6)8–10, 1–2-verticillatc, commonly 5+3, outer oppositipetalous, filaments commonly partially connate into a column, anthers dorsifixed, introrse, latrorse and extrorse, ovoid or oblong, erect, thecae parallel, contiguous, longitudinally dehiscent; pollen globose, 2-nucleate, inaperturate, tectate, with massive hexagonally-arranged clavae on ridges; pistillode 0. Female flowers: pedicels, sepals and petals ± as in the male, but sepals sometimes foliaceous; disc annular, 5-lobate, or sometimes disc-glands 5; staminodes sometimes +, filiform; ovary (1)3(5)-locular, ovules 1 per locule; styles (1)3(5), connate at the base, erect or spreading, stigmas usually 2-fid and somewhat tumid. Fruit ovoid-subglobose to shallowly 3-lobate, septicidally dehiscent into 3 2-valved cocci, loculicidally into 3 valves, or rarely subdrupaceous and ± indehiscent; endocarp crustaceous or indurate; columella persistent. Seeds ovoid or oblong, caruncular; caruncle usually 2-fid and fimbriate or multifid; testa crustaceous, shiny; endosperm fleshy; cotyledons broad, flat. Fig.: See Euphorb. Gen.: t. 11, f. 34A.1–8 (1824); Icon. Pl. 4: t. 357 (1841) (as *Mozinna*); Étude Euphorb.: t. 13, 1–18 (as *Mozinna*), t. 14, f. 10–27; t. 19, f. 10–11(as *Curcas*) (1858); Fl. Bras. 11(2): tt. 68, 69 (1874); Pflanzenr. 42: 29–84, tt. 8–32 (1910); Fl. Jamaica 4: 311, t. 103 (1920); Nat. Pflanzenfam. ed. 2, 19C: 161, t. 83 and 163, t. 84 (1931); Ann. Missouri Bot. Gard. 54: 236, t. 5 (1968); Amer. J. Bot. 56: 748, t. 8A–J (1969); Mém. Mus. Natl. Hist. Nat., B, Bot. 20: 485–6, tt. 175–6 (1971); Fl. Ill. Séneg. 3: 500–514 (1975); Univ. Calif. Publ. Bot. 74: tt. 1–33 (1979); Kew Bull. 35: 254, t. 1 (1980); Correll, Fl. Bah. Arch.: 824, t. 341 (1982); Fl. Español. 4: 168, t. 116-22 (1986); Fl. Bhut. 1: 789, t. 48s–u (1987); Kew Bull. 42: 112, t. 1 and 117, t. 2 (1987); Fl. Trop. E. Afr. Euph. 1: 347–362, tt. 65–67 (1987); Descr. Fl. Puerto Rico 2: 404, t. 59-20 (1988); Fl. Lesser Antilles 5: 68, t. 25 (1989); Kew Bull. 46: 143–155, tt. 1–5 and 328, t. 1 (1991); Fl. Zamb. 9(4): 256, t. 54 and 263, t. 55 (1996); Fl. St. John: 217, t. 98 (1996); Kew Bull.: 52: 179, t. 1 (1997).

A diverse, widespread genus of c. 175 species, especially common in Tropical America and Africa, but with only a few native species in SW. Asia, 1 in Madagascar and none in SE. Asia, Australasia or Oceania.

Dehgan & Webster (1979) recognized 2 subgenera, 10 sections and 10 subsections in their treatment of the genus, and Hemming & Radcliffe-Smith (1987) added 1 section and 4 subsections to these in their account of the Somali species; Radcliffe-Smith (1991) subsequently added a further new subsection, also based on a Somali species, and evaluated his new Zambesian species on the basis of Pax's (1910) and Dehgan & Webster's (1979) treatments.

J. curcas L., the Physic Nut, yields a powerful purgative. The bark of *J. dioica* Sessé is used for tanning and dyeing. Several species are in cultivation, mainly in the tropics (See Frontispiece).

224. **Vaupesia** *R. E. Schult.*, Bot. Mus. Leafl. 17: 27 (1955); G. L. Webster, Ann. Missouri Bot. Gard. 81: 103 (1994); Murillo-Aldana & Franco-Rosselli, Euf. Reg. Ararac.: 157 (1995). T.: *V. cataractarum* R. E. Schult.

Tall monoecious glabrous tree; latex watery, reddish, or 0. Leaves alternate, petiolate, stipulate, simple, ovate, entire, penninerved, coriaceous, 2-glandular at the base of the lamina; stipules infra-axillary, soon deciduous. Inflorescences axillary or subterminal, paniculate, multiflorous, robust, 2-sexual. Male flowers shortly pedicellate; sepals 5, strongly imbricate, quincuncial, subcucullate; petals 5, 2× as long as the sepals, oblanceolate; disc-glands 5, commonly free, alternipetalous; stamens 8, 2-seriate, outer 5 oppositipetalous, filaments connivent, anthers basifixed, latrorse, 2-thecous, longitudinally dehiscent; pollen globose, 2-nucleate, inaperturate, tectate, crotonoid; pistillode 0. Female flowers: pedicels, sepals and petals ± as in the male; disc annular, 5-lobate; ovary 3-locular, ovules 1 per locule; styles 3, sessile, stigmas 2-fid, erect or spreading. Fruit large, subglobose, dehiscing into 3 2-valved cocci; exocarp fleshy; endocarp thickly woody; columella large, 3-quetrous, persistent, later splitting vertically into 3 segments. Seeds large, compressed-ovoid, ecarunculate; testa crustaceous, smooth, concolorous; endosperm fleshy; cotyledons broad, flat. Fig.: See Bot. Mus. Leafl. 17: 29b, t. 12 (1955); Euf. Reg. Ararac.: 159, t. 48 (1995).

A monotypic genus from Vaupés Province, SE Colombia and adjacent parts of Brazil (upper Rio Negro, Amazonas).

Vaupesia provides a link between the *Micrandreae* and the *Jatropheae*, having vegetative characteristics of the former, but the pollen of the latter. Its author prefers to include it in the former, but Webster (1975, 1994) weighs the pollen-character the more heavily.

225. **Oligoceras** *Gagnep.*, Bull. Soc. Bot. France 71: 872 (1924), and Fl. Indo-Chine 5: 467 (1926); Pax & K. Hoffm., Nat. Pflanzenfam. ed. 2, 19C: 90 (1931); Airy Shaw, Kew Bull. 14: 392 (1960); G. L. Webster, Ann. Missouri Bot. Gard. 81: 104 (1994). T.: *O. eberhardtii* Gagnep.

Monoecious glabrous tree. Twigs cicatricose, scars circular. Leaves alternate, long-petiolate, exstipulate, ovate-deltate, 3-nerved from the base, then penninerved; petiole 2-glandular at the apex, glands contiguous, discoid. Inflorescences terminal, paniculate, bisexual, with 1 female flower terminating each axis, male flowers lateral; bracts small, deltate, scale-like. Male flowers pedicellate; calyx campanulate, 5-lobed, the lobes subimbricate to imbricate, each with a dorsal cylindric-cornute truncate appendage; petals 5, free, unguiculate; disc 5-partite, the segments lanceolate, connate around the base of the staminal column, alternistemonous; stamens and staminodes connate in 2 whorls into a cylindrical column, outer whorl of 5 stamens, inner of 3 staminodes, anthers subbasifixed, introrse, suborbicular; pollen smaller than, but otherwise similar to, that of 224. *Vaupesia*; pistillode 0. Female flowers shortly pedicellate; calycine appendages obtuse, otherwise calyx and petals as in the males; disc annular; staminodes linear; ovary 3-locular, ovules 1 per locule; styles 3, connate at the base, divergent, 2-fid, stigmas coiled. Fruit drupaceous. Seeds not known. Fig.: See Fl. Indo-Chine 5: 468, t. 58.3–9 and 472, t. 59.1–2 (1926).

A monotypic genus from Vietnam.

Like 226. *Deutzianthus*, but monoecious and with appendiculate sepals.

226. **Deutzianthus** *Gagnep.*, Bull. Soc. Bot. France 71: 139 (1924), and Fl. Indo-Chine 5: 297 (1925); Pax & K. Hoffm., Nat. Pflanzenfam. ed. 2, 19C: 90 (1931); C. Y. Wu, Acta Phytotax. Sin. 6: 245 (1957); Airy Shaw, Kew Bull. 14: 362 (1960), and 16: 346 (1963); G. L. Webster, Ann. Missouri Bot. Gard. 81: 104 (1994). T.: *D. tonkinensis* Gagnep.

Dioecious trees or shrubs. Latex yellowish-orange, verniceous. Indumentum simple, largely confined to the inflorescences and fruits. Leaves alternate, long-petiolate, simple, entire, palminerved at the base, then penninerved; petiole 2-glandular at the apex; glands disciform, persistent. Inflorescences terminal, broadly corymbiform-paniculate, 1-sexual; bracts filiform, persistent. Male flowers pedicellate; calyx-lobes 5, narrowly triangular, valvate; petals 5, linguiform, induplicate-valvate, much exceeding the calyx; disc-glands 5, deltate, squamiform, alternipetalous; stamens 7, 2-seriate, 5 outer oppositipetalous, free, 2 inner united to $^1/_2$-way, filaments pilose $^1/_2$-way up, otherwise glabrous, anthers subdorsifixed, introrse, cordulate at the base, longitudinally dehiscent, connective broad, brownish; pollen typically crotonoid; pistillode 0. Female flowers pedicellate; calyx-lobes and petals ± as in the male; disc annular or cupular, 5-lobate, lobes deltate, persistent in fruit; ovary 3-locular, densely sericeous and setulose, ovules 1 per locule; styles 3, ± free, 2× 2-fid, puberuolous except for the stigmas. Fruit asymmetrically compressed-ovoid, sericeous and setulose, subindehiscent, 1–3-seeded. Mature seeds not known. Fig.: See Fl. Indo-Chine 5: 296, t. 31.3–9 and 298, t. 32.1 (1925).

A monotypic Asiatic genus from northern Vietnam (formerly Tonkin).

227. **Loerzingia** *Airy Shaw*, Kew Bull. 16: 365 (1963), 25: 543 (1971), and 36: 312 (1981). T.: *L. thyrsiflora* Airy Shaw (S.: *D. thyrsiflorus* (Airy Shaw) G. L. Webster).

Like 226. *Deutzianthus* in some ways, but leaves elliptic-lanceolate, penninerved, basal glands subulate, stipelliform, deciduous, petiole ± 2-pulvinate; inflorescences laxly pyramidal, narrowly paniculate-thyrsiform, female long-pedunculate; male calyx-lobes imbricate, connate to $^1/_2$-way, rounded, petals spathulate, disc extrastaminal, annular below, 8–9-lobate above, stamens 8, 5 outer and 3 inner, shortly connate at the base, filaments lanuginous, anthers shortly sagittate at the base, pistillode +, minute, subulate, glabrous; female pedicels bracteolate, bracteoles deciduous, calyx thickened at the base, petals fleshy, styles 6, 2-fid. Fig.: See Kew Bull. 36: 314, t. 7A1–3 (1980).

A monotypic Asiatic genus from Sumatra.

Webster (1994) placed this in synonymy with the Tonkinese *Deutzianthus*, unjustifiably so in my opinion, in view of the many differences as indicated, which seem to me to be more than merely of a specific nature, as well as the considerable geographical separation.

228. **Joannesia** *Vell.*, Alogr. Alk.: 199 (1798); Müll. Arg. in DC., Prodr. 15(2): 715 (1866), and in Mart., Fl. Bras. 11(2): 295 (1874); Benth., Gen. Pl. 3: 290 (1880); Pax, Pflanzenr. 42: 116 (1910); Ducke, Arch. Jard. Bot. Rio de Janeiro 3: 198 (1922); Pax & K. Hoffm., Nat. Pflanzenfam. ed. 2, 19C: 101 (1931); Pittier, Bol. Soc. Venez. Ci. Nat. 6: 8 (1940); R. E. Schult., Bot. Mus. Leafl. 17: 25 (1955); Radcl.-Sm., Fl. Trop. E. Afr. Euph. 1: 181 (1987); L. B. Smith *et al.*, Fl. Ilustr. Catar. (Euph.): 165 (1988); G. L. Webster, Ann. Missouri Bot. Gard. 81: 104 (1994). T.: *J. princeps* Vell.

Anda A. Juss., Euphorb. Gen.: 39 (1824); Baill., Étude Euphorb.: 316 (1858). T.: *A. gomesii* A. Juss., nom. illeg. [=*J. princeps* Vell.]
Andicus Vell., Fl. Flum.: 80 (1825), t. 86 (1827). T.: *A. pentaphyllus* Vell. [=*J. princeps* Vell.]

Tall monoecious trees. Indumentum simple. Leaves alternate, long-petiolate, stipulate, digitate; leaflets petiolulate, entire, penninerved; petiole 2-glandular at the apex, the glands stipitate. Inflorescences terminal, cymose, paniculiform, 2-sexual, male above, female below, rarely 1-sexual; bracts accrescent, foliascent, glandular or not at the base. Male flowers shortly pedicellate; calyx globose and open in bud, broadly campanulate, truncate, shortly 5-toothed, teeth valvate; petals 5, free, imbricate, much longer than the calyx; disc-glands 5, free, alternipetalous; stamens (7)8(10), unequally 2-seriate, outer oppositipetalous, united in the lower $^1/_2$, inner $^3/_4$ united, anthers subdorsifixed, introrse in bud, later extrorse when reflexed, cordulate at the base, longitudinally dehiscent, connective subapiculate; pollen crotonoid, with clavae on the ridges; pistillode minute, 2-fid, or 0. Female flowers shortly pedicellate; calyx 5-lobed, the lobes valvate, accrescent; petals 5, free, imbricate, narrower than in the male, soon caducous; disc-glands 5, free; ovary 2-locular, ovules 1 per locule; styles 2, short, stigmas thickened, reflexed, lobate, not persistent. Fruit large, ovoid, drupaceous or subsequently tardily subindehiscent into 4 valves from the base; exocarp fleshy; endocarp woody. Seeds compressed-ovoid; exotesta somewhat fleshy; endotesta crustaceous; cotyledons broad, flat. Fig.: See Euphorb. Gen.: t. 12, f. 37.1–5 (1824) (as *Anda*); Fl. Flum.: t. 86 (1827); Étude Euphorb.: t. 12, f. 28–34 (1858) (as *Anda*); Fl. Bras. 11(2): t. 43 (1874); Pflanzenr. 42: 117, t. 42 (1910); Nat. Pflanzenfam. ed. 2, 19C: 101, t. 52 (1931); Árvores Brasileiras: 105 (1992).

A tritypic neotropical genus; 1 from Venezuela, 1 from Amazonian Brazil and the other from coastal Brazil.

229. **Leeuwenbergia** *Letouzey & N. Hallé*, Adansonia II. 14: 380 (1974); G. L. Webster, Ann. Missouri Bot. Gard. 81: 104 (1994). T.: *L. letestui* Letouzey & N. Hallé.

Like 228. *Joannesia*, but male calyx fused in bud, splitting at anthesis as in the *Aleuritideae*; stamens 18–33, with only the inner filaments connate, and male disc annular, intrastaminal. Fig.: See Adansonia II. 14: 385, t. 2 and 387, t. 3 (1974).

A ditypic W. African genus from Cameroon, Congo and Gabon.

230. **Annesijoa** *Pax & K. Hoffm.*, Pflanzenr. 68: 9 (1919), and Nat. Pflanzenfam. ed. 2, 19C: 101 (1931); Airy Shaw, Kew Bull. 16: 345 (1963), Hooker's Icon. Pl. 38: t. 3713 (1974), and Kew Bull. Addit. Ser. 8: 27 (1980); G. L. Webster, Ann. Missouri Bot. Gard. 81: 104 (1994). T.: *A. novoguineensis* Pax & K. Hoffm.

Like 228. *Joannesia* and 229. *Leeuwenbergia* but having very readily deciduous cylindric stipelliform petiolar glands leaving scars, a 5-lobate male calyx with rounded, imbricate lobes, muticous anthers, a 3-locular ovary, and slender, 2-fid styles; differing furthermore from 228. *Joannesia* in having the inflorescence-bracts much less conspicuous, 15–25 stamens with only the inner filaments connate, and a readily dehiscent fruit, and from 229. *Leeuwenbergia* by the dissected male disc with 5 oppositisepalous glands. Fig.: See Hooker's Icon. Pl. 38: t. 3713 (1974); Kew Bull. Addit. Ser. 8: 229, t. 7 (1980).

A monotypic genus endemic to New Guinea.

Tribe 42. **CODIAEAE** *(Pax) Hutch.*, Amer. J. Bot. 56: 747 (1969); G. L. Webster, Ann. Missouri Bot. Gard. 81: 104 (1994). T.: *Codiaeum* Rumph. ex A. Juss.

Clutieae subtribe *Codiaeinae* Pax, Pflanzenr. 47: 10 (1911).

Monoecious or dioecious trees or shrubs; laticifers not articulated; latex clear, sometimes reddish, non-toxic; indumentum mostly simple, sometimes malpighiaceous; leaves elobate, penninerved or 3-plinerved, usually eglandular; stipules often deciduous, sometimes 0. Inflorescences terminal or axillary, rarely leaf-opposed, racemose or paniculate; male sepals 4–6, free or connate, imbricate or valvate; petals mostly 5, free, usually imbricate; disc dissected or lobate; stamens (5)10–100 or more, free or basally connate; pollen globose, 2-nucleate, inaperturate, crotonoid; pistillode 0; female sepals 4, usually imbricate, sometimes accrescent; petals 5 or 0; disc annular, rarely dissected or 0; ovary mostly 3-locular; styles elobate to 2-partite. Fruit dehiscent; columella usually persistent. Seeds carunculate or ecarunculate; testa sometimes fleshy; endosperm +.

This tribe of 16 genera and over 100 species is generically the most diverse in the subfamily. Relationships between the genera are not clear, however, and a thorough study of this group is needed before formal subtribes are recognized in it.

Key to the genera

1. Petals 0; styles 2-fid; seeds carunculate; leaves glandular · 246. *Baliospermum*
1. Petals +, at least in the male flowers · 2
2. Petals 0 in the female flowers · 3
2. Petals + in the female flowers · 8
3. Inflorescences axillary · 4
3. Inflorescences mostly terminal · 6
4. Flowers clustered; stipules persistent, ± spinose; anthers apiculate; styles 2-fid or dilated · 5
4. Flowers racemose; stipules deciduous or 0; anthers muticous · 240. *Codiaeum*
5. Stipules commonly strongly spinous; female calyx-lobes slightly accrescent; styles almost sessile; stigmas usually dilated · · · · · · · · 241. *Acidocroton*

5. Stipules scarcely spinous; female calyx-lobes strongly accrescent and venose; styles elongate, 2-fid · · · · · · · · · · · · · · · · · · 242. *Ophellantha*
6. Female sepals recurved, accrescent; disc 0 · · · · · · · · · · · · · 245. *Sagotia*
6. Female sepals accrescent or not, not recurved; disc + · · · · · · · · · · · · · 7
7. Female sepals eglandular; seeds mostly ecarunculate · · · · · · 243. *Blachia*
7. Female sepals glandular; seeds carunculate · · · · · · · 244. *Strophioblachia*
8. Male calyx almost truncate · 9
8. Male calyx lobate · 10
9. Female calyx accrescent; petals glabrous; fruit dehiscent · 238. *Dimorphocalyx*
9. Female calyx not accrescent; petals tomentose; fruit drupaceous · 239. *Fontainea*
10. Leaves with bead-like glands beneath; anthers and petals with apical glands; inflorescences leaf-opposed · 11
10. Leaves not as above; anthers and petals eglandular; inflorescences terminal or axillary · 12
11. Ovary 3-locular; fruit dehiscent, 3-seeded · · · · · · · · · · · 236. *Pantadenia*
11. Ovary 2-locular; fruit indehiscent, 1-seeded · · · · · · · 237. *Parapantadenia*
12. Inflorescences usually terminal; male disc and receptacle glabrous; seeds carunculate · 13
12. Inflorescences axillary; male disc or receptacle pilose · · · · · · · · · · · · 14
13. Ovary 3-locular; petals glabrous outside · · · · · · · · · · · · · · 231. *Baloghia*
13. Ovary 2-locular; petals golden-sericeous outside · · · · · · · · 232. *Hylandia*
14. Indumentum simple; styles 2-fid · 233. *Ostodes*
14. Indumentum malpighiaceous (inflorescence only); male disc cupular · · 15
15. Styles 2-fid; male disc glabrous; stamens mostly 5–7; petals pubescent adaxially; seeds carunculate · 234. *Pausandra*
15. Styles multifid; male disc pubescent; stamens (8)14–16; petals not pubescent adaxially; seeds ecarunculate · · · · · · · · · 235. *Dodecastigma*

231. **Baloghia** *Endl.*, Prodr. Fl. Norfolk: 84 (1833); Baill., Étude Euphorb.: 344 (1858); Benth., Gen. Pl. 3: 300 (1880); Maiden, Forest Fl. N.S.W. 1: 165, t. 28 (1904); Pax & K. Hoffm., Pflanzenr. 47: 12 (1911), and Nat. Pflanzenfam. ed. 2, 19C: 156 (1931); C. T. White, Proc. Roy. Soc. Queensland 53: 226 (1942); Airy Shaw, Kew Bull. 35: 598 (1980); P. S. Green, Kew Bull. 41: 1026 (1986); McPherson & Tirel, Fl. Nouv. Caléd. 14: 43 (1987); T. A. James & G. J. Harden, Fl. N.S.W. 1: 410 (1990); G. L. Webster, Ann. Missouri Bot. Gard. 81: 105 (1994). T.: *B. inophylla* (G. Forst.) P. S. Green (B.: *Croton inophyllus* G. Forst.) (S.: *B. lucida* Endl.)

Steigeria Müll. Arg., Linnaea 34: 215 (1865), and in DC., Prodr. 15(2): 1121 (1866). T.: *S. montana* Müll. Arg. [=*B. montana* (Müll. Arg.) Pax]
Codiaeum sect. *Baloghia* (Endl.) Müll. Arg. in DC., Prodr. 15(2): 1116 (1866).

Dioecious or monoecious glabrous or subglabrous pseudodichotomous shrubs or small trees. Latex yellowish, orange or red. Perulae pseudostipuliform, soon deciduous. Leaves alternate or opposite, often clustered towards the ends of the branches, shortly petiolate, exstipulate, simple, entire, penninerved, tertiary nerves reticulate, coriaceous, glandular. Inflorescences usually terminal, occasionally lateral or axillary, usually racemose, rarely narrowly paniculate, usually lax, elongate or abbreviated, 1- or 2-sexual; female flowers at the base, males above; flowers large; bracts 1-flowered. Male flowers pedicellate; calyx-lobes (4)5(6), shortly connate, imbricate, later spreading or reflexed; petals (4)5(6), free, obovate to spathulate or oblong, usually exceeding the calyx-lobes; disc annular, or glands small, alternipetalous, or 0; stamens (10)40–50(100), arising from the receptacle, filaments connate at the base or free, thick, anthers dorsifixed, extrorse, thecae oblong, parallel, longitudinally dehiscent; pollen typically crotonoid; pistillode 0; receptacle convex. Female flowers: pedicels, calyx-lobes and petals ± as in the male; disc 0; ovary 3(4)-locular, ovules 1 per locule; styles 3, short, free or very shortly connate, usually spreading, deeply 2-fid, the arms simple or multifid. Fruit 3-lobate-subglobose, dehiscing into 3 2-valved cocci; exocarp separating; endocarp hard, woody or bony; columella stout, 3-dentate at the apex, persistent. Seeds globose to oblong, caruncle minute or 0; endosperm fleshy; cotyledons broad, flat. Fig.: See Forest Fl. N.S.W. 1: 165, t. 28 (1904); Pflanzenr. 47: 14, t. 3 and 15, t. 4 (1911); Fl. Nouv. Caléd. 14: 35, t. 5.4,5 and 49–71, tt. 8–13 (1987).

An Australasian genus of 15 species, 12 of which are endemic to New Caledonia.

B. inophylla, the type-species, known as the "Brush Bloodwood" in Australia, is the source of a red pigment which leaves an indelible stain when the tree is wounded.

232. **Hylandia** *Airy Shaw*, Kew Bull. 29: 329 (1974), and 35: 643, t. 4 (1980); G. L. Webster, Ann. Missouri Bot. Gard. 81: 105 (1994). T.: *H. dockrillii* Airy Shaw.

Very like 231. *Baloghia*, but a tall tree with ample paniculate inflorescences, petals golden-sericeous outside and reddish-pilose within, and ovary 2-locular. Fig.: See Kew Bull. 35: 644, t. 4 (1980).

A monotypic Australian genus, confined to northern Queensland.

Although its author considered it as being most closely affined to *Loerzingia, Ostodes and Dimorphocalyx*, Webster (1994) considered it to be closest to *Baloghia*.

233. **Ostodes** *Blume*, Bijdr. Fl. Ned. Ind.: 619 (1825); Baill., Étude Euphorb.: 391 (1858); Müll. Arg. in DC., Prodr. 15(2): 1114 (1866); Benth., Gen. Pl. 3: 299 (1880); Hook. f., Fl. Brit. India 5: 400 (1887); Pax & K. Hoffm., Pflanzenr. 47: 17 (1911); Gagnep., Fl. Indo-Chine 5: 322 (1925); Pax & K. Hoffm., Nat. Pflanzenfam. ed. 2, 19C: 157 (1931); Backer & Bakh. f., Fl. Java 1: 493 (1963); Airy Shaw, Kew Bull. 20: 411 (1967), 26: 311 (1972), and 36: 334 (1981); Chakrab. & N. P. Balakr., Bull. Bot. Surv. India 27: 259 (1985, publ. 1987); Grierson & D. G. Long, Fl. Bhut. 1: 795 (1987); G. L. Webster, Ann. Missouri Bot. Gard. 81: 105 (1994). T.: *O. paniculata* Blume.

Dioecious often mostly glabrous trees or shrubs. Indumentum, where present, simple. Leaves alternate, crowded at the tips of the branches, long-petiolate, stipulate, simple, ovate, conspicuously glandular-crenate-dentate, 3-plinerved from the base, then penninerved, 2-glandular at the apex of the petiole; stipules quickly deciduous. Inflorescences axillary or extra-axillary, in axils of fallen leaves on older shoots, arising below a terminal leaf-tuft, thyrsoid, the males elongate, slender, lax, much-branched, multiflorous, pendulous, the females shorter, ± racemose, robust, few-flowered; bracts small to minute. Male flowers long-pedicellate; calyx (3)5-partite, the segments broad, unequal, imbricate; petals 5(6), free, exceeding the calyx; disc-glands 5(6), free, oppositisepalous; stamens (8)30–35(40), outer filaments shortly connate at the base or ± free, inner more markedly connate, villous at the base, anthers subdorsifixed, introrse, thecae oblong, parallel, subsagittate at the base, longitudinally dehiscent; pollen typically crotonoid; pistillode 0; receptacle pilose. Female flowers long-pedicellate; calyx ± as in the male, not accrescent in fruit; petals 8–10, otherwise ± as in the male; ovary 3-locular, sericeous, ovules 1 per locule; styles 3, stout, 2-fid at the apex. Fruit rounded-3-gonous to subglobose, apiculate, 6-ribbed, somewhat tardily loculicidally dehiscent into 3 valves, or septicidally into 3 2-valved cocci; endocarp bony; columella slender, splitting longitudinally into 3 strands, subpersistent. Seeds large, ovoid; endosperm fleshy; cotyledons broad, flat. Fig.: See Fl. Bhut. 1: 801, t. 49 p–r (1987).

A tritypic tropical Asiatic genus ranging from the E. Himalaya to S. China, SE. Asia and Java, according to Airy Shaw (1967). However, Chakrabarty & Balakrishnan (1987), consider 2 of the remaining species as deserving only of varietal status, thus returning *Ostodes* to being a monotype.

Airy Shaw (1967) removed several species formerly included in the genus to *Dimorphocalyx* and *Fahrenheitia* [=*Paracroton*, q.v. infra]

234. **Pausandra** *Radlk.*, Flora 53: 92, t. 2 (1870); Baill., Adansonia I. 11: 92 (1873); Müll. Arg. in Mart., Fl. Bras. 11(2): 503, t. 99 (1874); Benth., Gen. Pl. 3: 298 (1880); Pax & K. Hoffm., Pflanzenr. 47: 41 (1911), and in Nat. Pflanzenfam. ed. 2, 19C: 173 (1931); Lanj., Recueil Trav. Bot. Néerl. 33: 758 (1936); Jabl., Mem. New York Bot. Gard. 17: 153 (1967); Secco, Bol. Mus. Paraense Emilio Goeldi, N.S., Bot. 3: 59 (1987); Huft, Ann. Missouri Bot. Gard. 75: 1114 (1988); Secco, Rev. Gen. *Euph.-Crot.* Amer. Sul: 58 (1990); A. H. Gentry, Woody Pl. NW. S. Amer.: 417 (1993); G. L. Webster, Ann. Missouri Bot. Gard. 81: 105 (1994); Murillo-Aldana & Franco-Rosselli, Euf. Reg. Ararac.: 124 (1995). T.: *P. morisoniana* (Casar.) Radlk. (B.: *Thouinia morisoniana* Casar.)

Dioecious trees or shrubs. Latex reddish. Indumentum malpighiaceous, sometimes also simple. Leaves alternate, petiolate, stipulate, large, elongate-oblanceolate, simple, serrate, penninerved; petioles slightly tumid and 2-glandular at the apex; stipules quickly deciduous. Inflorescences axillary, male spiciform or subpaniculate, female racemose; bracts inconspicuous, 2-glandular, males few-flowered, females 1-flowered. Male flowers subsessile; calyx short, campanulate, 4–5-lobate, the lobes imbricate;

petals 4–5, connate at the base, imbricate-contorted, adaxially barbate ½-way up; disc extrastaminal, cupular, shallowly lobate, glabrous; stamens (3)5–7, filaments free, erect, anthers subdorsifixed, introrse, thecae oblong, parallel, subsagittate at the base, longitudinally dehiscent, connective shortly produced, apiculate; pollen globose, inaperturate, clavate, crotonoid; pistillode 0. Female flowers subsessile; sepals 5, imbricate; petals 5, free, adaxially villous; disc urceolate, sometimes lobate, glabrous; ovary 3-locular, ovules 1 per locule; styles 3, free, 2-fid. Fruit 3-dymous, dehiscent into 3 2-valved cocci; endocarp somewhat hardened. Seeds ovoid-subglobose, carunculate; testa smooth, blotched; endosperm copious; embryo straight; cotyledons palminerved, much longer than the radicle. Fig.: See Flora 53: 92, t. 2 (1870); Fl. Bras. 11(2): t. 99 (1874); Pflanzenr. 47: 42, t. 13A–B and 43, t. 14 (1911); Nat. Pflanzenfam. ed. 2, 19C: 173, t. 90A–B (1931); Woody Pl. NW. S. Amer.: 418, t. 120.4 (1993); Euf. Reg. Ararac.: 127, t. 34 (1995).

A neotropical genus of 8 (Secco, 1990) or 14 (Ind. Kew.) species ranging from Honduras south to Brazil.

235. **Dodecastigma** *Ducke*, Notizbl. Bot. Gart. Berlin-Dahlem 11: 343 (1932), and Arch. Jard. Bot. Rio de Janeiro 6: 58, t. 5 (1933); Sandwith, Kew Bull. 5: 134 (1951); Jabl., Mem. New York Bot. Gard. 17: 154 (1967); Secco, Rev. Gen. *Euph.-Crot.* Amer. Sul: 42 (1990); G. L. Webster, Ann. Missouri Bot. Gard. 81: 106 (1994); Murillo-Aldana & Franco-Rosselli, Euf. Reg. Ararac.: 70 (1995). T.: *D. amazonicum* Ducke.

Very like 234. *Pausandra*, e.g. in having a malpighiaceous indumentum, but differing in having entire leaves, in not having adaxially pubescent petals, in having a pubescent male disc, (8)14–16 stamens, multifid styles and ecarunculate seeds. Fig.: See Arch. Bot. Rio de Janeiro 6: 58, t. 5 (1933); Euf. Reg. Ararac.: 73, t. 15 (1995).

A tritypic genus from Amazonian Brazil and the Guianas.

236. **Pantadenia** *Gagnep.*, Bull. Soc. Bot. France 71: 873 (1924), and Fl. Indo-Chine 5: 470 (1926); Pax & K. Hoffm., Nat. Pflanzenfam. ed. 2, 19C: 155 (1931); Airy Shaw, Kew Bull. 23: 122 (1969), and 26: 312 (1972); G. L. Webster, Ann. Missouri Bot. Gard. 81: 106 (1994). T.: *P. adenanthera* Gagnep.

Dioecious shrubs. Indumentum, where present, simple. Leaves alternate, petiolate, minutely stipulate, simple, entire, palminerved then penninerved, copiously granulate-glandular beneath, the glands orange-yellow. Inflorescences leaf-opposed, males thyrsiform, females laxly pseudoracemose, pedunculate; bracts and bracteoles linear. Male flowers pedicellate, pedicels articulate above; buds globose; sepals 5, quincuncially imbricate; petals 5, 1–3-glandular at the apex; disc extrastaminal, annular to cupular, undulate; stamens 13–15, filaments free, inserted on the receptacle, anthers

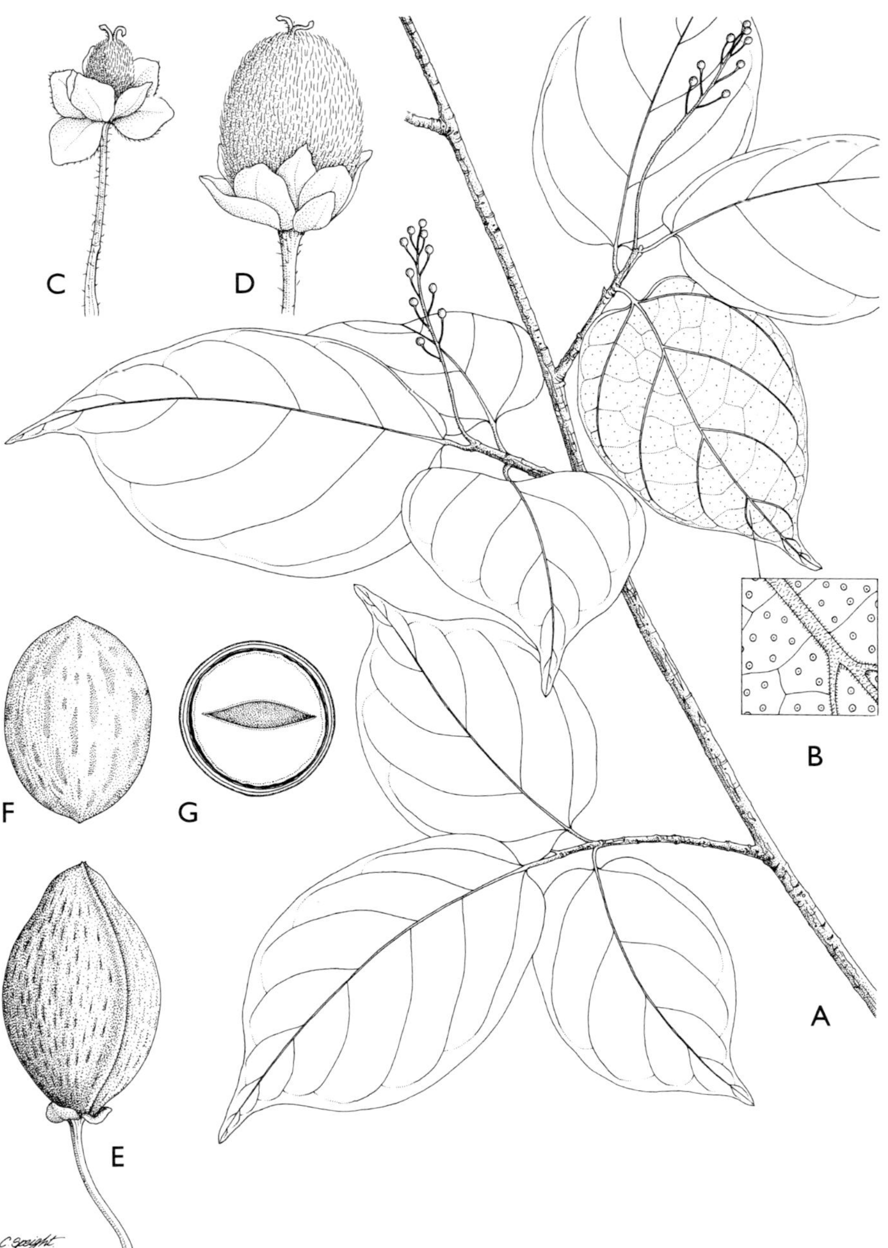

FIG. 39. *Parapantadenia chauvetiae* Leandri (S.: *Pantadenia chauvetiae* (Leandri) G. L. Webster). **A** Habit, male × 1; **B** Portion of leaf-undersurface × 10; **C** Female flower × 6; **D** Young fruit × 6; **E** Mature fruit × 3; **F** Seed × 3; **G** Transverse section of fruit & seed × 3. **A, B** from *Capuron* 22163-SF; **C, D** from *Capuron* 22164-SF; **E–G** from 15.544- SF. Drawn by Camilla Speight.

subbasifixed, introrse, ovoid-subglobose, 2-thecate, thecae parallel, contiguous, longitudinally dehiscent, connective with a subspherical apical gland; pollen typically crotonoid; pistillode small, conic, velutinous, or 0; receptacle conic. Female flowers pedicellate; sepals 5–6, imbricate, not accrescent; petals 2–3, 5-glandular at the apex; disc annular, slightly concave; ovary 3-locular, ovules 1 per locule; styles 3, free, divergent, 2-fid, segments slender, acute, stigmas smooth to shortly papillose. Fruit 3-gonous, septicidally and loculicidally dehiscent into 6 valves; exocarp not separating; endocarp thinly woody; columella capitate, later splitting, persistent. Seeds subglobose, ?carunculate; testa smooth; endosperm +; embryo straight; cotyledons orbicular, flat; radicle conic. Fig.: See Fl. Indo-Chine 5: 472, t. 59.3–11 (1926).

A monotypic genus from Thailand, Laos, Cambodia and Vietnam.

237. **Parapantadenia** *Capuron*, Adansonia II. 12(2): 206 (1972). T.: *P. chauvetiae* Leandri (S.: *Pantadenia chauvetiae* (Leandri) G. L. Webster).

Very like 236. *Pantadenia* in many respects, but differing markedly in having a 2-locular ovary and a 1-seeded indehiscent fruit. Fig.: See Adansonia II. 12(2): 207, t. 1 (1972). See also Fig. 39, p. 299.

A monotypic genus from Madagascar.

Webster (1994) considered that as 237. *Parapantadenia* otherwise had so many features in common with 236. *Pantadenia*, the two genera should be united, but I feel that the very different ovary and fruit, taken together with the vast geographical separation, militates against this view.

238. **Dimorphocalyx** *Thwaites*, Enum. Pl. Zeyl.: 278 (1861); Benth., Gen. Pl. 3: 301 (1880); Hook. f., Fl. Brit. India 5: 403 (1887); Trimen, Handb. Fl. Ceylon 4: 54, t. 84 (1898); Pax & K. Hoffm., Pflanzenr. 47: 31 (1911); Gagnep., Fl. Indo-Chine 5: 295 (1925); Pax & K. Hoffm., Nat. Pflanzenfam. ed. 2, 19C: 158 (1931); Airy Shaw, Kew Bull. 20: 412 (1967), 23: 123 (1969), and 26: 251 (1972); Whitmore, Tree Fl. Mal. 2: 86 (1973); Airy Shaw, Kew Bull. 29: 328 (1974), Kew Bull. Addit. Ser. 4: 95 (1975), 8: 73 (1980), Kew Bull. 35: 624 (1980), 36: 286 (1981), 37: 16 (1982), and Euph. Philipp.: 20 (1983); Chakrab. & N. P. Balakr., Proc. Indian Acad. Sci. 100: 286 (1990); G. L. Webster, Ann. Missouri Bot. Gard. 81: 106 (1994); Philcox in Dassan., ed., Fl. Ceyl. XI: 107 (1997). T.: *D. glabellus* Thwaites.

Trigonostemon sect. *Dimorphocalyx* (Thwaites) Müll. Arg., Linnaea 34: 212 (1865), and in DC., Prodr. 15(2): 1105 (1866).

Dioecious or rarely monoecious glabrescent shrubs or small trees, rarely scandent. Indumentum simple, confined to the innovations and flowers. Leaves alternate, petiolate, stipulate, simple, entire or glandular-crenulate or -denticulate, penninerved; stipules shortly deltate, deciduous. Inflorescences axillary or terminal, racemose, umbellate or cymose, few-flowered, pedunculate, mostly 1-sexual, female flowers sometimes subsolitary; bracts resembling the stipules. Male flowers pedicellate; calyx cupular, almost truncate, shortly 5-lobate, the lobes imbricate or open; petals 5, imbricate, glabrous, white, longer than the calyx; disc-glands 5, free, oppositisepalous; stamens (8)10–15(16), 2-seriate, outer 5 usually free, oppositipetalous, inner connate into a column with the anthers sometimes capitate, anthers basifixed and dorsifixed (in the same flower), extrorse, 2-thecate, thecae parallel or divergent, longitudinally dehiscent, connective broad, not produced, eglandular; pollen typically crotonoid; pistillode 0. Female flowers pedicellate; sepals 5, free or shortly connate, imbricate, persistent, often considerably accrescent in fruit; petals ± as in the male; disc annular to shallowly cupular; ovary 3-locular, ovules 1 per locule; styles 3, free or connate at the base, 2-fid, erect or spreading, segments slender, obtuse, stigmas ± smooth to minutely papillulose. Fruit 3-lobate-subglobose, closely enveloped by the accrescent calyx, septicidally dehiscent into 3 2-valved cocci; endocarp crustaceous; columella subcapitate, persistent. Seeds obovoid-3-gonous, ecarunculate; testa crustaceous, smooth, marmorate; endosperm copious, fleshy; cotyledons broad, flat. Fig.: See Pflanzenr. 47: 32, t. 8 (1911); Fl. Indo-Chine 5: 292, t. 30.11–14 and 296, t. 31.1,2 (1925); Proc. Indian Acad. Sci. 100: 289–297, tt. 1–4 (1990).

A predominantly tropical Asiatic genus of c. 20 species ranging from India and Sri Lanka to Hainan, the Philippines, Moluccas and Lesser Sunda Is., with 1 outlying species in Australasia.

239. **Fontainea** *Heckel, Fontainea pancheri* (1870), ex Baill., Adansonia I. 11: 80 (1873), and Hist. Pl. 5: 194 (1874); Benth., Gen. Pl. 3: 301 (1880); Pax in Engl. & Prantl, Nat. Pflanzenfam. ed. 1, 3(5): 301 (1890); Pax & K. Hoffm., Pflanzenr. 47: 30 (1911), and Nat. Pflanzenfam. ed. 2, 19C: 158 (1931); Airy Shaw, Kew Bull. 29: 326 (1974), Kew Bull. Addit. Ser. 8: 91 (1980), Kew Bull. 35: 632 (1980); Jessup & Guymer, Austrobaileya 2: 112 (1985); McPherson & Tirel, Fl. Nouv. Caléd. 14: 74 (1987); G. L. Webster, Ann. Missouri Bot. Gard. 81: 106 (1994). T.: *F. pancheri* (Baill.) Heckel ex Baill. (B.: *Baloghia pancheri* Baill.) (S.: *Codiaeum pancheri* (Baill.) Müll. Arg.)

Similar to 238. *Dimorphocalyx*, but petals densely white-pubescent to lanuginous, stamens more numerous (18–32), female calyx not accrescent, fruit drupaceous, and endocarp 3–5(6)-angled. Fig.: See Kew Bull. Addit. Ser. 8: 225, t. 3.3 (1980); Fl. Nouv. Caléd. 14: 35, t. 5.9,10 and 75, t. 14. 6–12 (1987).

A hexatypic Australasian genus, 1 species of which occurs in E. New Guinea, Queensland, New Caledonia and Vanuatu, whilst the other 5 are endemic to E. Australia.

240. **Codiaeum** *Rumph. ex A. Juss.*, Euphorb. Gen.: 33 (1824); Baill., Étude Euphorb.: 384 (1858); Müll. Arg. in DC., Prodr. 15(2): 1116 (1866); Benth., Gen. Pl. 3: 299 (1880); Hook. f., Fl. Brit. India 5: 399 (1887); Pax & K. Hoffm., Pflanzenr. 47: 23 (1911); Gagnep., Fl. Indo-Chine 5: 408 (1926); Pax & K.

Hoffm., Nat. Pflanzenfam. ed. 2, 19C: 157 (1931); Backer & Bakh. f., Fl. Java 1: 493 (1963); Berhaut, Fl. Ill. Sénég. 3:407 (1975); Airy Shaw, Kew Bull. Addit. Ser. 4: 88 (1975), Kew Bull. 33: 74 (1978), Kew Bull. Addit. Ser. 8: 62 (1980), and Kew Bull. 35: 613 (1980); A. C. Sm., Fl. Vit. Nov. 2: 549 (1981); Airy Shaw, Kew Bull. 37: 14 (1982), and Euph. Philipp.: 16 (1983); Proctor, Kew Bull. Addit. Ser. 11: 538 (1984); Grierson & D. G. Long, Fl. Bhut. 1: 794 (1987); McPherson & Tirel, Fl. Nouv. Caléd. 14: 95 (1987); R. A. Howard, Fl. Lesser Antilles 5: 32 (1989); G. L. Webster, Ann. Missouri Bot. Gard. 81: 106 (1994). T.: *C. variegatum* (L.) A. Juss. (B.: *Croton variegatus* L.)

Phyllaurea Lour., Fl. Cochinch.: 575 (1790). T.: *Ph. codiaeum* Lour., nom. illeg. [=*C. variegatum* (L.) A. Juss. (B.: *Croton variegatus* L.)]

Synaspisma Endl., Gen. Pl.: 1110 (1840); Baill., Étude Euphorb.: 387 (1858). T.: *S. peltatum* (Labill.) Endl. (B.: *Chrozophora peltata* Labill.) [=*C. peltatum* (Labill.) P. S. Green]

Junghuhnia Miq., Fl. Ned. Ind. 1(2): 412 (1859), non Corda (1842). T.: *J. glabra* Miq. [=*C. variegatum* (L.) A. Juss. (B.: *Croton variegatus* L.)]

Monoecious or occasionally dioecious shrubs or small trees. Indumentum, when present, simple. Leaves alternate, petiolate, minutely stipulate or exstipulate, simple, entire or occasionally 3-lobate, sometimes the lamina elongated, recurved, twisted or interrupted, penninerved, eglandular or subeglandular at the base, sometimes variegated; stipules deciduous. Inflorescences axillary in the upper axils or subterminal, sometimes subtended by a small orbicular leaf, laxly elongate-racemose, solitary or 2-nate, usually 1-sexual, pedunculate; male bracts 1–6-flowered, female 1-flowered. Male flowers often repeatedly produced over a long period, pedicellate, pedicels slender, articulate; sepals (3)5(6), free, imbricate; petals (3)5(6), small to minute, rarely 0; disc-glands 5–15, small, free, alternipetalous; stamens 15–100, filaments free, inserted on the receptacle, anthers erect, dorsifixed, extrorse, muticous, 2-thecate, thecae apically confluent, longitudinally dehiscent, connective broad, not produced, eglandular; pollen typically crotonoid; pistillode 0; receptacle slightly raised. Female flowers pedicellate, pedicels not articulate; sepals ± as in the male; petals 0; disc shallowly cupular, ± entire; ovary 3(4)-locular, ovules 1 per locule; styles 3(4), free or shortly connate at the base, simple, rarely 2-partite, spreading or recurved, segments usually slender. Fruit globose or 3(4)-dymous, septicidally dehiscent into 3(4) 2-valved cocci; endocarp crustaceous; columella narrow, subcapitate, usually persistent. Seeds obovoid, caruncluate; testa crustaceous, smooth, shiny, marmorate; endosperm fleshy; cotyledons broad, flat. Fig.: See Euphorb. Gen.: t. 9, f. 30.1–5 (1824); Étude Euphorb.: t. 16, f. 26–35 (1858); Pflanzenr. 47: 27, t. 6 and 29, t. 7 (1911); Nat. Pflanzenfam. ed. 2, 19C: 157, t. 81 (1931); Fl. Ill. Sénég. 3: 406–412 (1975); Fl. Pak. 172: 20, t. 4.A–B (1986); Fl. Nouv. Caléd. 14: 35, t. 5.11,12 & 97, t. 19 (1987); Fl. Lesser Antilles 5: 33, t. 13 (1989).

A genus of 16 species ranging from Borneo and Java to the Philippines, New Guinea, tropical Australia, New Caledonia and the SW. Pacific islands.

Müller (1866) adopted a rather broad view of the genus, including *Baloghia* and *Blachia* in it as sections. However, recent workers all concur with the narrower definition.

The type species is widely cultivated in the tropics and subtropics generally, where it often becomes naturalized, and as a houseplant in temperate regions; it exists in a number of forms or cultivars, mostly based on the leaf-shape (See Frontispiece).

241. **Acidocroton** *Griseb.*, Fl. Brit. W. I.: 42 (1859) (nom. cons.); Müll. Arg. in DC., Prodr. 15(2): 1042 (1866); Benth., Gen. Pl. 3: 291 (1880); Pax, Pflanzenr. 42: 13 (1910); Urb., Symb. Antill. 7: 513 (1913); Fawc. & Rendle, Fl. Jamaica 4: 315 (1920); Urb., Symb. Antill. 9: 208 (1924), and Ark. Bot. 20A(15): 62 (1926); Pax & K. Hoffm., Nat. Pflanzenfam. ed. 2, 19C: 172 (1931); Alain, Fl. Cuba 3: 73 (1953) and Fl. Español. 4: 69 (1986); G. L. Webster, Ann. Missouri Bot. Gard. 81: 107 (1994). T.: *A. adelioides* Griseb.

Monoecious much-branched usually spiny shrubs. Brachyblasts pulviniform, minutely spinous. Indumentum simple. Leaves small, alternate on lead-shoots, fasciculate on brachyblasts, shortly petiolate, stipulate, simple, entire, sometimes revolute, obovate, thick, penninerved, eglandular; stipules mostly ± spinose, persistent. Inflorescences axillary, few-flowered, flowers small, clustered. Male flowers subsessile to pedicellate; calyx 5–6-lobate, lobes short, subimbricate; petals 5–7, larger than the calyx-lobes, free, imbricate-contorted, later spreading; disc-glands interstaminal, numerous, obconical, fleshy, pilose at the apex; stamens 20–30, inserted on the receptacle, filaments free, glabrous, anthers subbasifixed, introrse, short, erect, thecae distinct, longitudinally dehiscent, connective broad, caudate-apiculate; pollen typically crotonoid; pistillode 0; receptacle disciform, pubescent. Female flowers shortly pedicellate, pedicels extending in fruit; calyx ± as in the male, but lobes slightly accrescent, persistent; petals commonly 0 or, if present, very minute; disc annular, pubescent; ovary 3(5)-locular, ovules 1 per locule; styles very short or stigmas sessile or almost so, flat, spreading, thick, dilated, broadly obovate, retuse, emarginate or 2-fid. Fruit 3-lobate-subglobose, septicidally dehiscent into 3 2-valved cocci, or loculicidally into 6 valves; endocarp crustaceous; columella subpersistent. Seeds ovoid, carunculate; testa smooth, marmorate; endosperm fleshy; cotyledons broad, flat. Fig.: See Pflanzenr. 42: 14, t. 2 (1910); Fl. Jamaica 4: 316, t. 104 (1920); Fl. Cuba 3: 74, t. 14 (1953).

A Greater Antillean genus of 11 species, confined to Cuba (7 species), Hispaniola (3 species) and Jamaica (1 species).

Webster (1994) includes *Ophellantha* Standl. (q.v. infra) in *Acidocroton*, which would give it 3 more species and make it a mainland genus as well. For the nomenclatural complexities involved in the correct application of the name *Acidocroton*, see Webster (1994). Furthermore, there has been considerable divergence of opinion concerning the affinities of the genus.

242. **Ophellantha** *Standl.*, J. Wash. Acad. Sci. 14: 97 (1924); Pax & K. Hoffm., Nat. Pflanzenfam. ed. 2, 19C: 160 (1931); A. H. Gentry, Woody Pl. NW. S. Amer.: 407 (1993). T.: *O. spinosa* Standl. (S.: *Acidocroton spinosus* (Standl.) G. L. Webster)

Acidocroton sect. *Ophellantha* (Standl.) G. L. Webster, Ann. Missouri Bot. Gard. 81: 107 (1994).

Differs from 241. *Acidocroton* in its open few-branched habit, larger membranaceous remotely denticulate leaves, smaller often scarcely spinous stipules, long-pedicellate flowers, stamens 50 or more, venose strongly accrescent female calyx-lobes, 2-locular ovary and fruit, 2 elongate 2-fid styles, flattened columella and large ecarunculate seeds. Fig.: See Woody Pl. NW. S. Amer.: 408, t. 115.3 (1993).

A tritypic genus from C. America and Colombia.

The differences between *Ophellantha* and 241. *Acidocroton* are greater and more numerous than stated by Webster (1994), and so I feel that this, taken together with the geographical separation, militates against the union of the two genera.

243. **Blachia** *Baill.*, Étude Euphorb.: 385 (1858), *nom. cons.*; Benth., Gen. Pl. 3: 301 (1880); Hook. f., Fl. Brit. India 5: 402 (1887); Pax & K. Hoffm., Pflanzenr. 47: 36 (1911); Gagnep., Fl. Indo-Chine 5: 410 (1926); Pax & K. Hoffm., Nat. Pflanzenfam. ed. 2, 19C: 159 (1931); Airy Shaw, Kew Bull. 23: 121 (1969), and 26: 223 (1972); Whitmore, Tree Fl. Mal. 2: 68 (1973); Airy Shaw, Kew Bull. Addit. Ser. 4: 57 (1975), Kew Bull. 37: 9 (1982), and Euph. Philipp.: 10 (1983); N. P. Balakr. & Chakrab., Proc. Indian Acad. Sci. 99: 568 (1989); Thin, J. Biol. (Vietnam) 11(3): 16 (1989); G. L. Webster, Ann. Missouri Bot. Gard. 81: 107 (1994); Philcox in Dassan., ed., Fl. Ceyl. XI: 105 (1997). T.: *B. umbellata* (Willd.) Baill. (B.: *Croton umbellatus* Willd.)

Bruxanel(l)ia Dennst. ex Kostel., Allg. Med.-Pharm. Fl. 5: 2002 (1830), *nom. rej.*
Codiaeum sect. *Blachia* (Baill.) Müll. Arg. in DC., Prodr. 15(2): 1118 (1866).

Monoecious or sometimes subdioecious glabrous or glabrescent shrubs or small trees. Brachyblasts sometimes present. Indumentum, when present, simple. Leaves alternate, rarely the upper subopposite, shortly petiolate, stipulate or exstipulate, simple, rarely pandurate-lobate, entire, mostly penninerved; stipules minute, inconspicuous or 0. Inflorescences terminal, sometimes also axillary, racemose, umbellate or verticillate, solitary, usually few-flowered, pedunculate, peduncles filiform, usually 2-sexual, male above, female below; bracts 0. Male flowers pedicellate, pedicels slender, filiform; calyx globose in bud, 4–5-partite, lobes concave, imbricate; petals 4–5, shorter than the calyx, rounded, hyaline; disc-glands 4–5, squamiform, equalling the petals, oppositisepalous; stamens 10–40, inserted on the receptacle, filaments free, anthers basifixed, latrorse, ovoid or orbicular, 2-thecate, thecae encircling the margins of the connective, apices confluent, longitudinally dehiscent; pollen typically crotonoid; pistillode 0; receptacle convex. Female flowers pedicellate, pedicels thickened above; sepals 4–6, lanceolate, often accrescent, sometimes reflexed, usually persistent, sometimes caducous, eglandular; petals 0; disc annular or cupular, often scarcely prominent; ovary 3(5)-locular, ovules 1 per locule, anatropous; styles 3(5), free, filiform, usually

2-fid or -partite, reflexed, recurved or revolute. Fruit 3(5)-dymous-subglobose, often enveloped by the accrescent calyx, septicidally dehiscent into 3 2-valved cocci; endocarp crustaceous; columella persistent. Seeds subglobose-oblong to 3-gonous, mostly ecarunculate; testa crustaceous, smooth, shiny, marmorate; endosperm thick, fleshy; cotyledons broad, flat. Fig.: See Étude Euphorb.: t. 19, f. 18–20 (1858); Pflanzenr. 47: 10, t. 1.G,H and 38, t. 11 (1911); Fl. Indo-Chine 5: 414, t. 48.5–13 and 419, t. 49.1 (1926); Nat. Pflanzenfam. ed. 2, 19C: 155, t. 79 G,H (1931); Proc. Indian Acad. Sci. 99: 570–577, tt. 1–4 (1989).

A tropical Asiatic genus of 15–16 species ranging from peninsular India and Sri Lanka to China, Indo-China, the Andaman Is., Thailand, Malaya, Celebes and the Philippines; doubtfully present in Borneo.

244. **Strophioblachia** *Boerl.*, Handl. Fl. Ned. Ind. 3(1): 235 (1900); Pax & K. Hoffm., Pflanzenr. 47: 35 (1911); Merr., Univ. Calif. Publ. Bot. 10: 425 (1925); Gagnep., Fl. Indo-Chine 5: 408 (1926); Pax & K. Hoffm., Nat. Pflanzenfam. ed. 2, 19C: 159 (1931); Airy Shaw, Kew Bull. 25: 544 (1971), 26: 341 (1972), 37: 36 (1982), and Euph. Philipp.: 45 (1983); Thin, J. Biol. (Vietnam) 11(3): 15 (1989); G. L. Webster, Ann. Missouri Bot. Gard. 81: 107 (1994); Thin *et al.*, Blumea 43: 484 (1998). T.: *S. fimbricalyx* Boerl.

Very like 243. *Blachia*, but male calyx ciliate with gland-tipped hairs, petals equalling the calyx, denticulate, filaments elongated, anther-connective narrow, female sepals long-stipitate-glandular or fimbriate and more strongly accrescent, and seeds ovoid, carunculate. Fig.: See Pflanzenr. 47: 35, t. 10 (1911); Fl. Indo-Chine 5: 405, t. 47.8–14 and 414, t. 48.1–4 (1926); Blumea 43: 480–483, tt. 1–4 (1998).

A ditypic tropical Asiatic genus ranging from S. China through Indo-China to Celebes and the Philippines.

245. **Sagotia** *Baill.*, Adansonia I. 1: 53 (1860) (nom. cons.); Müll. Arg. in DC., Prodr. 15(2): 1113 (1866), and in Mart., Fl. Bras. 11(2): 504, t. 70 (1874); Benth., Gen. Pl. 3: 302 (1880); Pax & K. Hoffm., Pflanzenr. 47: 39 (1911), and in Nat. Pflanzenfam. ed. 2, 19C: 160 (1931); Jabl., Mem. New York Bot. Gard. 17: 151 (1967); Secco, Acta Amazon. 15(1–2, Suppl.): 81 (1985), and Rev. Gen. *Euph.-Crot.* Amer. Sul: 99 (1990); G. L. Webster, Ann. Missouri Bot. Gard. 81: 107 (1994). T.: *S. racemosa* Baill.

Monoecious glabrous or subglabrous small trees. Indumentum, when present, usually only on the female flowers, simple. Leaves alternate, petiolate, stipulate, simple, sometimes large, entire, penninerved, tertiary nerves reticulate. Inflorescences terminal, racemose, solitary, short, usually few-flowered, 1- or 2-sexual, the latter usually male above, female below. Male flowers long-pedicellate, pedicels filiform, 2-bracteolate; sepals 5, broad, strongly quincuncially imbricate; petals 5, longer than the calyx, imbricate; disc 0; stamens 20+, clustered on the receptacle, filaments free, very short, continuous with the connective, anthers erect, basifixed, ± introrse, thecae distinct, parallel, longitudinally dehiscent; pollen echinate or clavate, clavae sharply pointed, otherwise typically crotonoid; pistillode 0; receptacle convex. Female flowers pedicellate; sepals 5, narrow, short or elongate, recurved-spreading, accrescent, usually persistent; petals 0; disc 0; ovary 3-locular, ovules 1 per locule; styles 3, spreading, thickened, 2-partite. Fruit depressed-3-lobate-subglobose, surrounded by the accrescent calyx, septicidally dehiscent into 3 2-valved cocci, or loculicidally into 6 valves; endocarp woody; columella thick, 3-quetrous, persistent. Seeds ovoid, carunculate; testa crustaceous, smooth, shiny, dark brown; endosperm fleshy; cotyledons broad, flat. Fig.: See Fl. Bras. 11(2): t. 70 (1874); Pflanzenr. 47: 10, t. 1 E, F and 40, t. 12 (1911); Nat. Pflanzenfam. ed. 2, 19C: 155, t. 79.E,F (1931); Bol. Mus. Paraense Emilio Goeldi, N.S. Bot. 3: 165–170, tt. 5D–I, 6C–F, 9 and 10 (1987).

A polymorphic monotypic or ditypic (Secco, 1985) genus of Costa Rica, Panama, Amazonian Colombia, Venezuela and Brazil, and the Guianas.

246. **Baliospermum** *Blume*, Bijdr. Fl. Ned. Ind.: 603 (1826); Decne. in Jacquem., Voy. Inde Atlas 2: t. 155 (1844); Baill., Étude Euphorb.: 394 (1858); Müll. Arg. in DC., Prodr. 15(2): 1125 (1866); Benth., Gen. Pl. 3: 324 (1880); Hook. f., Fl. Brit. India 5: 461 (1887); J. J. Sm., Meded. Dept. Landb. Ned.-Indië 10: 599 (1910); Pax & K. Hoffm., Pflanzenr. 52.IV: 24 (1912); Gagnep., Fl. Indo-Chine 5: 429 (1926); Pax & K. Hoffm., Nat. Pflanzenfam. ed. 2, 19C: 182 (1931); Backer & Bakh. f., Fl. Java 1: 497 (1963); Airy Shaw, Kew Bull. 26: 222 (1972); Whitmore, Tree Fl. Mal. 2: 68 (1973); Airy Shaw, Kew Bull. 36: 267 (1981), and 37: 9 (1982); Radcl.-Sm. in Fl. Pak. 172: 83 (1986); Grierson & D. G. Long, Fl. Bhut. 1: 809 (1987); Thin, J. Biol. (Vietnam) 11(3): 16 (1989); G. L. Webster, Ann. Missouri Bot. Gard. 81: 108 (1994). T.: *B. montanum* (Willd.) Müll. Arg. (B.: *Jatropha montana* Willd.) (S.: *B. axillare* Blume).

Monoecious or dioecious erect few-branched shrubs, subshrubs or woody herbs. Indumentum simple. Leaves alternate, petiolate, stipulate, simple, subentire, sinuate-dentate or pinnately lobate, palminerved at the base, then penninerved, 2-glandular at the base; stipules very small. Inflorescences axillary, sometimes terminal, racemose to paniculate, dense or lax, pedunculate to ± sessile, 1- or rarely 2-sexual, males many-flowered, females few-flowered. Male flowers pedicellate, pedicels capillary; sepals (4)5(6), ± orbicular, imbricate; petals 0; disc extrastaminal, annular, (4)5(6)-lobate or of (4)5(6) free glands; stamens 10–30, inserted on the receptacle, outer oppositisepalous, filaments slender, free, anthers erect, basifixed, introrse, thecae contiguous to distinct, sometimes confluent at the apex, longitudinally dehiscent, connective broad; pollen typically crotonoid; pistillode minute or 0; receptacle convex. Female flowers shortly pedicellate; sepals 5–6, imbricate, sometimes shallowly denticulate, sometimes accrescent; petals 0; disc annular, entire; ovary 3(4)-locular, ovules 1 per locule; styles 3(4), short, stout, recurved-spreading, shortly 2-fid. Fruit 3-lobate to 3-dymous, septicidally dehiscent into 3 2-valved cocci, or loculicidally into 6 valves; endocarp woody; columella persistent. Seeds ovoid, carunculate; testa crustaceous, smooth, shiny, marmorate; endosperm fleshy; cotyledons broad, flat. Fig.: See Voy. Inde Atlas 2: t. 155 (1844); Pflanzenr. 52.IV: 25, t. 6 and 27, t. 7 (1912); Fl. Indo-Chine 5: 431, t. 51.6–19 and 436, t. 52.1–6 (1926); Nat. Pflanzenfam. ed. 2, 19C: 183, t. 96 (1931); Fl. Pak. 172: 7, t. 1.E–I (1986); Fl. Bhut. 1: 807, t. 50m–o (1987).

A tropical Asiatic genus of 7–12 species ranging from the W. Himalaya and India through Assam to SW. China, SE. Asia, Malaya, Sumatra, Java and Sumbawa.

Associated by Müller (1866), Pax & K. Hoffmann (1912a) and others with *Suregada*, but Webster (1994), following a suggestion of Airy Shaw (1972, 1975), and on account of the inaperturate pollen, places it here despite the absence of petals from both male and female flowers.

A decoction of the leaves of *B. montanum* is used as a purgative; the seeds are a drastic purgative.

Tribe 43. **TRIGONOSTEMONEAE** *G. L. Webster*, Taxon 24: 599 (1975), and Ann. Missouri Bot. Gard. 81: 108 (1994). T.: *Trigonostemon* Blume.

Monoecious trees or shrubs; indumentum simple; leaves alternate, elobate, penninerved, sometimes stipellate at the base, but not 2-glandular at the apex of the petiole; stipules persistent or deciduous, often small. Inflorescences terminal or axillary, racemose or thyrsiform; male sepals 5, free, imbricate; petals 5, exceeding the calyx; disc cupular or dissected; stamens 3(5), filaments connate, anther-connective enlarged, often elongate; pollen globose, inaperturate, crotonoid, processes small; pistillode 0; female sepals 5, free, imbricate; disc usually cupular; ovary 3-locular; styles free or ± so, elobate or 2-partite. Fruit dehiscent. Seeds ecarunculate; testa dry; endosperm +.

A monogeneric tropical Asiatic tribe.

247. **Trigonostemon** *Blume*, Bijdr. Fl. Ned. Ind.: 600 (1826) (as *Trigostemon*), and Fl. Javae: viii (1828), nom. cons.; Baill., Étude Euphorb.: 340 (1858); Müll. Arg. in DC., Prodr. 15(2): 1105 (1866); Benth., Gen. Pl. 3: 298 (1880); Hook. f., Fl. Brit. India 5: 395 (1887); Pax & K. Hoffm., Pflanzenr. 47: 85 (1911); Gagnep., Fl. Indo-Chine 5: 309 (1925); Quisumb., Philipp. J. Sci. 41: 329 (1930); Pax & K. Hoffm., Nat. Pflanzenfam. ed. 2, 19C: 169 (1931); Jabl., Brittonia 15: 151 (1963); Backer & Bakh. f., Fl. Java 1: 495 (1963); Airy Shaw, Kew Bull. 20: 47 (1966), 21: 407 (1968), 23: 126 (1969), 25: 545 (1971); 26: 344 (1972); Whitmore, Tree Fl. Mal. 2: 134 (1973); Airy Shaw, Hooker's Icon. Pl. 38: t. 3721 (1974), Kew Bull. Addit. Ser. 4: 201 (1975), Kew Bull. 32: 415 (1978), 33: 534 (1979), Kew Bull. Addit. Ser. 8: 205 (1980), Kew Bull. 35: 690 (1980), 36: 352 (1981), 37: 36, 121 (1982), and Euph. Philipp.: 46 (1983); G. L. Webster, Ann. Missouri Bot. Gard. 81: 108 (1994); R. Milne, Kew Bull. 49: 445 (1994), and 50: 25, 51 (1995); Philcox in Dassan., ed., Fl. Ceyl. XI: 110 (1997). T.: *Tr. serratus* Blume.

Enchidium Jack, Malayan Misc. 2(7): 89 (1822), nom. rej.; Baill., Étude Euphorb.: 652 (1858). T.: *E. verticillatum* Jack [=*Tr. verticillatus* (Jack) Pax]

Silvaea Hook. & Arn., Bot. Beechey Voy.: 211 (1837); Baill., Étude Euphorb.: 341 (1858). T.: *S. semperflorens* (Roxb.) Hook. & Arn. (B.: *Cluytia semperflorens* Roxb.) [=*Tr. semperflorens* (Roxb.) Müll. Arg.]

Athroisma Griff., Not. Pl. Asiat. 4: 477 (1854), non DC. (1833). T.: (none designated).

Telogyne Baill., Étude Euphorb.: 327 (1858). T.: *T. indica* Baill. [=*Tr. verticillatus* (Jack) Pax]

Tritaxis Baill., Étude Euphorb.: 342 (1858); Hook. f., Fl. Brit. India 5: 384 (1887); Gagnep., Fl. Indo-Chine 5: 302 (1925). T.: *Trit. gaudichaudii* Baill. [=*Trig. gaudichaudii* (Baill.) Müll. Arg.]

Tylosepalum Kurz ex Teijsm. & Binn., Natuurk. Tijdschr. Ned.-Indië 27: 50 (1864). T.: *T. aurantiacum* Kurz ex Teijsm. & Binn. [=*Tr. aurantiacus* (Kurz ex Teijsm. & Binn.) Boerl.]

Nepenthandra S. Moore, J. Bot. 43: 149 (1905). T.: *N. lanceolata* S. Moore [=*Tr. lanceolatus* (S. Moore) Pax]

Actephilopsis Ridl., Bull. Misc. Inform., Kew 1923: 360 (1923). T.: *A. malayana* Ridl. (S.: *Tr. malayanus* (Ridl.) Airy Shaw) [=*Tr. aurantiacus* (Kurz ex Teijsm. & Binn.) Boerl. (B.: *Tylosepalum aurantiacum* Kurz ex Teijsm. & Binn.)]

Prosartema Gagnep., Bull. Soc. Bot. France 71: 875 (1924), and Fl. Indo-Chine 5: 304 (1925). T.: *P. stellaris* Gagnep. [=*Tr. stellaris* (Gagnep.) Airy Shaw]

Poilaniella Gagnep., Bull. Soc. Bot. France 72: 467 (1925), and Fl. Indo-Chine 5: 307 (1925). T.: *P. fragilis* Gagnep. [=*Tr. fragilis* (Gagnep.) Airy Shaw]

Neotrigonostemon Pax & K. Hoffm., Notizbl. Bot. Gart. Berlin-Dahlem 10: 385 (1928), and Nat. Pflanzenfam. ed. 2, 19C: 169 (1931). T.: *N. diversifolius* Pax & K. Hoffm. [=*Tr. viridissimus* (Kurz) Airy Shaw (B.: *Sabia viridissima* Kurz)]

Kurziodendron N. P. Balakr., Bull. Bot. Surv. India 8: 68 (1966). T.: *K. viridissimum* (Kurz) N. P. Balakr. (B.: *Sabia viridissima* Kurz) (S.: *Blachia viridissima* (Kurz) King) [=*Tr. viridissimus* (Kurz) Airy Shaw]

Monoecious shrubs or small trees. Indumentum simple. Leaves alternate or subopposite, often crowded at the tops of the branches, petiolate or subsessile, stipulate or exstipulate, sometimes stipellate, sometimes large, simple, elobate, entire or denticulate, often 3-plinerved at the base, then penninerved, mostly eglandular; stipules mostly minute, subulate, or 0. Inflorescences axillary or terminal, occasionally cauliflorous, cymose, racemose, thyrsiform or paniculate, mostly elongate, less often abbreviate, protogynous and often apparently 1-sexual, occasionally male with a terminal female; bracts often conspicuous, males many-flowered, females few- or 1-flowered. Male flowers pedicellate; sepals 5, free or very shortly connate at the base, imbricate; petals 5, free, mostly exceeding the sepals, occasionally 2-lobate, often brightly-coloured (yellow, orange, red, deep purple or blackish); disc annular, cupular or of 5 free glands; stamens 3–5(13), filaments connate or free and spreading above, anthers dorsifixed, extrorse, erect or horizontally-held, 2-thecate, thecae distinct, parallel, divergent or divaricate towards the apex and ± 2-fid, connective sometimes much-produced and cornute; pollen spheroidal, inaperturate, tectate or not, gemmate or reticulate; pistillode 0. Female flowers shortly pedicellate, pedicels extending somewhat in fruit; sepals 5, free, imbricate, occasionally fringed with capitate glands, rarely accrescent in fruit; petals similar to those in the male flowers, but sometimes differently-coloured, or 0; disc glands free or variously united, or disc cupular; ovary 3-locular, ovules 1 per locule; styles free or or very shortly connate at the base, elobate, 1× or 2× 2-lobate, -fid or -partite, spreading. Fruit 3-coccous or -dymous, smooth or verruculose, dehiscent into 3 2-valved cocci; endocarp crustaceous; columella 3-quetrous, subpersistent. Seeds subglobose to 3-gonous-ovoid, ecarunculate; testa dry, crustaceous; endosperm fleshy; cotyledons broad, flat. Fig.: See Étude Euphorb.: t. 11, f. 8–13 (1858) (partly as *Tritaxis*, partly as *Telogyne*); Pflanzenr. 47: 10, t. 1.A, 89, t. 28 and 93, t. 29 (1911); Fl. Indo-Chine 5: 301–308, tt. 33.10–12 (as *Tritaxis*), 34 (as *Tritaxis* and *Prosartema*), 35 (& also as *Prosartema*), 36 (as *Poilaniella*), and 331, t. 38.1–4 (1925); Nat. Pflanzenfam. ed. 2, 19C: 155, t. 79.A (1931); Hooker's Icon. Pl. 38: t. 3721 (1974); Kew Bull. 36: 354, t. 11 (1981); Bull. Bot. Surv. India 8: 70, t. 1–7 (1966); Kew Bull. 49: 447 t. 1 and 449, t. 2 (1994); 50: 34–43, tt. 1–3 and 52, t. 1 (1995).

A diverse tropical Asiatic genus of c. 50–60 species ranging from India and Sri Lanka to China, New Guinea and Queensland; a species described from New Caledonia is in fact referable to *Baloghia*.

Variously treated by Müller (1866) in his *Jatropheae*, by Pax & K. Hoffmann (1911) in their *Cluytieae*, and by Webster (1975, 1994) as a distinct tribe, it is possible that *Trigonostemon* really belongs to the *Codiaeae*.

Tribe 44. **RICINOCARPEAE** *Müll. Arg.*, Bot. Zeitung (Berlin) 22: 324 (1864), and in DC., Prodr. 15(2): 199 (1866); G. L. Webster, Ann. Missouri Bot. Gard. 81: 109 (1994). T.: *Ricinocarpos* Desf.

Subfam. *Ricinocarpoideae* (Müll. Arg.) Pax, Nat. Pflanzenfam. ed. 1, 3(5): 113 (1890).

Fam. *Ricinocarpaceae* (Müll. Arg.) Hurus., J. Fac. Sci. Univ. Tokyo, Sect. 3, Bot. 6: 224 (1954).

Monoecious or rarely dioecious trees, shrubs or herbs; latex scanty or 0; indumentum stellate and often glandular; leaves alternate, exstipulate, entire, often ericoid, eglandular. Inflorescences terminal or axillary, racemose, fasciculate or flowers solitary; male sepals 4–6, free, imbricate, often petaloid; petals usually 5, or 0; disc dissected or 0; stamens mostly 15 or more, filaments connate into a column or short and apparently free, anthers extrorse; pollen 3-nucleate, globose, inaperturate, modified-crotonoid, processes clavate, spinose or reduced; pistillode 0; female sepals (4)5, imbricate; petals 5, sometimes reduced or 0; disc dissected or 0; ovary (1)3-locular; styles elobate, 2-fid or multifid. Fruit dehiscent, 1–3-locular. Seeds carunculate; endosperm +; embryo cylindrical; cotyledons mostly narrow.

A heptageneric largely Australasian tribe. On palynological grounds, Webster (1994) has shown that it belongs in the *Crotonoideae*, and not with other stenocotyledonous groups as associated by Müller (1866), Bentham (1880), Grüning (1913) and Pax & Hoffmann (1931).

KEY TO THE SUBTRIBES

1. Leaves alternate; petals and disc usually +; pollen processes large, clavate; ovary 3-locular · 44a. *Ricinocarpinae*
1. Leaves alternate, opposite or verticillate; petals and disc 0; pollen processes reduced; ovary 1–3-locular · 44b. *Bertyinae*

Subtribe 44a. **Ricinocarpinae** *G. L. Webster*, Taxon 24: 599 (1975), and Ann. Missouri Bot. Gard. 81: 109 (1994). T.: *Ricinocarpos* Desf.

Trees, shrubs or undershrubs; leaves alternate, sometimes resinous. Inflorescences terminal or axillary, racemose, fasciculate or flowers solitary; sepals free or connate; petals usually +, longer or shorter than the sepals; stamens 20 or more, filaments free, connate or inner connate at the base; pollen with large clavate processes; ovary 3-locular; styles 2-fid, multifid or stigma elobate. Fruit dehiscent, 3-spermous.

A trigeneric Australasian subtribe, with 2 genera endemic to Australia.

KEY TO THE GENERA

1. Stamens free; stigma mostly elobate, dilated, capping the ovary · 250. *Beyeria*
1. Stamens at least partially connate; styles 2-fid to multifid · · · · · · · · · · · · · 2
2. Petioles short; male calyx lobate to $^1/_2$-way or more; filaments connate; fruit smooth, tuberculate, muricate, echinate or rugose · · · 248. *Ricinocarpos*
2. Petioles long; male calyx lobate to < $^1/_2$-way; outer filaments free; fruit smooth · 249. *Alphandia*

248. **Ricinocarpos** *Desf.*, Mém. Mus. Hist. Nat. 3: 459, t. 22 (1817); A. Juss., Euphorb. Gen.: 36 (1824); Baill., Étude Euphorb.: 343 (1858); Hook. f., Fl. Tasm. 1: 338 (1860); Müll. Arg. in DC., Prodr. 15(2): 203 (1866) (as *Ricinocarpus*); Benth. & F. Muell., Fl. Austral. 6: 68 (1873); Benth., Gen. Pl. 3: 263 (1880); Maiden, Ill. N.S.W. Pl., t. 16 (1908); Grüning, Pflanzenr. 58: 37 (1913); Pax & K. Hoffm., Nat. Pflanzenfam. ed. 2, 19C: 226 (1931); Stanley, Fl. SE Queensl. 1: 432 (1983); G. L. Webster, Ann. Missouri Bot. Gard. 81: 109 (1994). T.: *R. pinifolius* Desf.

Roeperia Spreng., Syst. Veg. 3: 13, 147 (1826). T.: *Roep. pinifolia* (Desf.) Spreng. [=*R. pinifolius* Desf.]

Echinosphaera Sieber ex Steud., Nomencl. Bot., ed. 2, 1: 538 (1840). T.: *E. rosmarinoides* Sieber ex Steud. [=*R. pinifolius* Desf.]

Monoecious often glabrous shrubs. Indumentum, when present, stellate. Leaves alternate, shortly petiolate, exstipulate, simple, entire, mostly linear or lanceolate, margins recurved or revolute, pallid or white-tomentose beneath. Inflorescences terminal or pseudoaxillary on short lateral shoots, males fasciculate or flowers solitary, rarely racemose, females mostly solitary, occasionally associated with the males; bracts usually 2 per flower, small, squamiform. Male flowers often long-pedicellate, pedicels minutely bracteolate; calyx-lobes (4)5(6), subequal, imbricate; petals commonly longer than the calyx-lobes, contorted, rarely 0; disc-glands (4)5(6), alternipetalous; stamens numerous, filaments mostly connate into a central column, the outer mostly free, anthers dorsifixed, extrorse, reflexed, thecae distinct, parallel, longitudinally dehiscent; pollen crotonoid, processes clavate; pistillode 0. Female flowers: pedicels shorter and stouter than those in the male; sepals, petals and disc ± as in the male, but petals more often caducous; ovary 3-locular, sessile, ovules 1 per locule; styles 3, connate at the base, 2-fid, 2× 2-fid or 3–4-fid or -partite. Fruit 3-lobate, smooth, tuberculate, muricate, echinate or rugose, dehiscent into 3 2-valved cocci; endocarp thinly woody; columella 3-quetrous, fimbriate, persistent. Seeds oblong, carunculate; testa smooth, shiny, often marmorate or blotched; endosperm +; embryo linear, in the middle of the endosperm; cotyledons longer than the radicle. Fig.: See Mém. Mus. Hist. Nat. 3: 459, t. 22 (1817); Étude Euphorb.: t. 12, f. 39–44 (1858); Ill. N.S.W. Pl., t. 16 (1908); Pflanzenr. 58: 7, t. 2.A, B, 45, t. 8 and 49, t. 9 (1913); Nat. Pflanzenfam. ed. 2, 19C: 227, t. 122.A, B (1931); Amer. J. Bot. 56: 750, t. 9A–K (1969).

An Australian genus of 16 species, grouped in 4 sections by Grüning (1913).

249. **Alphandia** *Baill.*, Adansonia I. 11: 85 (1873); Benth., Gen. Pl. 3: 300 (1880); Pax & K. Hoffm., Pflanzenr. 47: 22 (1911), and Nat. Pflanzenfam. ed. 2, 19C: 100 (1931); Guillaumin, J. Arnold Arbor. 13: 92 (1932), and Ann. Inst. Bot.-Géol. Colon. Marseille 55/6: 31 (1948); Airy Shaw, Kew Bull. 20: 395 (1967), and Kew Bull. Addit. Ser. 8: 27 (1980); McPherson & Tirel, Fl. Nouv. Caléd. 14: 86 (1987); G. L. Webster, Ann. Missouri Bot. Gard. 81: 109 (1994). T.: *A. furfuracea* Baill.

Monoecious shrubs or small trees. Latex often yellowish or red. Indumentum, confined to innovations, of rusty-furfuraceous fascicled hairs, or of glands or minute scales exuding a resinous or verniciform sap. Leaves alternate, long-petiolate, exstipulate, simple, entire, broadly obovate, coriaceous, glossy above, glaucous beneath, penninerved, 2-glandular at the base of the lamina. Inflorescences terminal or axillary in the uppermost axils, pseudoracemose or very narrowly thyrsiform with pedunculate few-flowered cymules, 1- or 2-sexual, if the latter then females below the males; bracts small, subulate. Male flowers shortly pedicellate; calyx cupular, shortly 5-lobate, lobes open in bud; petals 5, free, oblong-elliptic, imbricate, undulate-twisted, somewhat fleshy, yellow, spreading or reflexed; disc-glands small, free or connate; stamens 20–35, ± fasciculate, inserted on the receptacle, filaments fairly rigid, outer free, inner longer and shortly connate at the base, recurved-geniculate at the apex, anthers small, subbasifixed, extrorse, thecae distinct, ovoid, longitudinally dehiscent, connective convex on the back; pollen crotonoid, processes clavate; pistillode 0; receptacle conical. Female flowers: pedicels stouter than those in the male; sepals and petals ± as in the male, but petals smaller, more readily caducous or sometimes 0; disc annular; ovary 3-locular, 3-lobate-subglobose, finely puberulous, later apparently glabrous because becoming covered with varnish, ovules 1 per locule; styles 3, very short, connate at the base, stigmas short, thick, 2-fid, recurved. Fruit ellipsoid-subglobose, smooth, dehiscent into 3 2-valved cocci or 6 valves; endocarp thick, woody; columella large, narrowly 3-alate, fimbriate at the apex, persistent. Seeds ellipsoid-cylindric, apiculate, carunculate; testa smooth, somewhat shiny, ochraceous or castaneous, often marmorate or maculate; endosperm +; cotyledons broad, flat. Fig.: See Fl. Nouv. Caléd. 14: 37, t. 6.1,2 and 89, t. 17 (1987).

A tritypic genus from Irian Jaya, Nouvelle-Calédonie and Vanuatu.

250. **Beyeria** *Miq.*, Ann. Sci. Nat. III. 1: 350, t. 15 (1844); Baill., Étude Euphorb.: 402 (1858); Müll. Arg. in DC., Prodr. 15(2): 201 (1866); Benth. & F. Muell., Fl. Austral. 6: 63 (1873); Benth., Gen. Pl. 3: 263 (1880); Grüning, Pflanzenr. 58: 63 (1913); Pax & K. Hoffm., Nat. Pflanzenfam. ed. 2, 19C: 227 (1931); T. A. James & G. J. Harden, Fl. N.S.W. 1: 412 (1990); G. L. Webster, Ann. Missouri Bot. Gard. 81: 109 (1994). T.: *B. viscosa* (Labill.) Miq. (B.: *Croton viscosus* Labill.)

Calyptrostigma Klotzsch in Lehm., Pl. Preiss. 1: 175 (1845). T.: *C. viscosum* (Labill.) Klotzsch [=*B. viscosa* (Labill.) Miq. (B.: *Croton viscosus* Labill.)]
Beyeriopsis Müll. Arg., Linnaea 34: 56 (1865). LT.: *B. brevifolia* Müll. Arg. [=*Beyeria brevifolia* (Müll. Arg.) Benth.] (designated by Wheeler, 1975).

Monoecious or rarely dioecious often resinous or viscous shrubs. Indumentum, when present, stellate or fasciculate. Leaves alternate, shortly petiolate to subsessile, exstipulate, simple, entire, mostly narrowly elliptic to ericoid, margins recurved or revolute, mostly pallid or whitish-tomentose beneath. Flowers axillary, males fasciculate in few-flowered fascicles or solitary, rarely racemose, females usually solitary. Male flowers pedicellate, pedicels minutely bracteolate; calyx-lobes (4)5, broad, concave, often rigid, ± petaloid, imbricate; petals (4)5, sometimes fewer or 0, mostly minute, rarely longer than the calyx-lobes; disc-glands (4)5, mostly small, alternipetalous, occasionally 0; stamens numerous, filaments very short, free, crowded on the receptacle, anthers subbasifixed, extrorse, thecae distinct, parallel, longitudinally dehiscent, connective short, narrow; pollen crotonoid, processes clavate, clavae sharply acute; pistillode 0; receptacle hemispherical. Female flowers: pedicels apically dilated; sepals (4)5, mostly narrower and more rigid than in the male, sometimes accrescent; petals as in the male; disc-glands (2)5, or disc annular, entire or crenate, or 0; ovary 3-locular, sessile, ovules 1 per locule; common stigma sessile, elobate or shallowly 3-lobate, dilated, peltate or conic-calyptriform, capping the ovary, very rarely deeply 3-fid (*B. tristigma*). Fruit 3-lobate, smooth, tardily dehiscent into 3 2-valved cocci or 6 valves, 3-seeded, rarely 1-seeded by abortion; endocarp thick, hard, woody; columella short, 3-alate, persistent. Seeds oblong, carunculate; testa smooth, shiny, often maculate or marmorate; endosperm +; embryo linear, in the middle of the endosperm; cotyledons longer than the radicle. Fig.: See Étude Euphorb.: t. 18, f. 13–17 (1858); Pflanzenr. 58: 7, t. 2.D and 74, t. 12 (1913); Nat. Pflanzenfam. ed. 2, 19C: 227, t. 122.D (1931).

An Australian endemic genus of 19 species.

Subtribe 44b. **Bertyinae** *Müll. Arg.*, Linnaea 34: 56 (1865), and in DC., Prodr. 15(2): 208 (1866); G. L. Webster, Ann. Missouri Bot. Gard. 81: 110 (1994). T.: *Bertya* Planch.

Trees or shrubs; leaves alternate, opposite or verticillate. Inflorescences axillary or terminal, racemose, fasciculate or flowers solitary; sepals mostly +; petals 0; disc 0, or very rarely +; stamens numerous, filaments connate into a column, anthers often pubescent; pollen with reduced processes; ovary 2–3-locular; styles 2-fid to multifid. Fruit dehiscent, 1–3-spermous.

A quadrigeneric Australasian subtribe, with 1 genus endemic to Australia, 2 genera in New Caledonia and 1 in Borneo.

Key to the genera

1. Male sepals 0; ovary 2-locular; male flowers in terminal bracteate catkins; female solitary, axillary; anthers pubescent; styles 2-fid · 254. *Borneodendron*
1. Male sepals +; ovary 1- or 3-locular · 2
2. Leaves mostly alternate; anthers glabrous; fruit 1-seeded · · · · · 251. *Bertya*
2. Leaves opposite or whorled; anthers pubescent; fruit 3-seeded · · · · · · · 3
3. Leaves opposite; inflorescences terminal, racemose; styles multifid · 252. *Myricanthe*
3. Leaves whorled; inflorescences axillary, 1-flowered; styles 2–4-fid · 253. *Cocconerion*

251. **Bertya** *Planch.*, London J. Bot. 4: 472, t. 16A (1845); Baill., Étude Euphorb.: 347 (1858); Hook. f., Fl. Tasman. 1: 339 (1860); Müll. Arg. in DC., Prodr. 15(2): 208 (1866); Benth. & F. Muell., Fl. Austral. 6: 74 (1873); Benth., Gen. Pl. 3: 264 (1880); Grüning, Pflanzenr. 58: 49 (1913); Pax & K. Hoffm., Nat. Pflanzenfam. ed. 2, 19C: 226 (1931); Guymer, Austrobaileya 2: 427 (1988); T. A. James & G. J. Harden, Fl. N.S.W. 1: 414 (1990); G. L. Webster, Ann. Missouri Bot. Gard. 81: 110 (1994). LT.: *B. rosmarinifolia* Planch. (chosen by Webster, 1994).

Monoecious often glutinous shrubs or rarely trees. Indumentum, when present, stellate. Leaves alternate or sometimes opposite, shortly petiolate to subsessile, exstipulate, simple, entire, linear with margins revolute to oblong with margins plane or recurved, mostly pallid or whitish-tomentose beneath, penninerved. Flowers axillary, fasciculate in few-flowered sessile or pedunculate fascicles or solitary, rarely terminal and racemose, almost always involucrate; bracts 2–8 per peduncle, small, imbricate, rarely decussate, calyciform, deciduous or persistent. Male flowers shortly pedicellate or subsessile; calyx-lobes (4)5, subequal, ± petaloid, imbricate; petals 0; disc 0; stamens 18–70, filaments very short or anthers sessile or subsessile, glabrous, adnate to, or connate into, a central column, subbasifixed, extrorse, thecae distinct, parallel, longitudinally dehiscent, connective 2-fid at the apex; pollen indistinctly crotonoid; pistillode 0; receptacle columnar. Female flowers: pedicels not or rarely apically dilated; calyx-lobes (4)5, mostly smaller, narrower and more acute than in the male, sometimes accrescent; petals 0; disc 0 or very rarely +, of 5 glands; ovary 3-locular, sessile or shortly stipitate, ovules 1 per locule; styles free or shortly connate, entire or deeply (2)3(7)-fid. Fruit oblong-ovoid, smooth, mostly 1-locular by abortion, dehiscent from the apex into 6 valves, 1-seeded; endocarp thinly woody; columella excentric, narrowly claviform, persistent. Seeds oblong or subglobose, caruncuIate; testa smooth, slightly shiny, brownish; endosperm +; embryo linear, straight; cotyledons longer and scarcely wider than the radicle. Fig.: See Étude Euphorb.: t. 18, f. 8–9 (1858); Pflanzenr. 58: 7, t. 2. G–J, 55, t. 10 and 58, t. 11 (1913); Nat. Pflanzenfam. ed. 2, 19C: 227, t. 122.G–J (1931); Austrobaileya 2: 428, t. 1 (1988).

An Australian endemic genus of 25–30 species.

B. cunninghamii is the source of a gum-resin.

252. **Myricanthe** *Airy Shaw*, Kew Bull. 35: 390 (1980); McPherson & Tirel, Fl. Nouv. Caléd. 14: 72 (1987); G. L. Webster, Ann. Missouri Bot. Gard. 81: 110 (1994). T.: *M. discolor* Airy Shaw.

Monoecious shrub; branching subverticillate. Latex 0. Indumentum, when present, stellate. Leaves opposite, decussate, petiolate, exstipulate, simple, entire, elliptic-oblong, margins revolute, mostly pallid beneath, penninerved. Inflorescences terminal, racemose, usually lax, few-flowered, 1- or 2-sexual, male above, female below in the latter; bracts ovate, deciduous. Male flowers pedicellate; calyx-lobes 3, unequal, 2-fid, 3-fid or 3-partite, open in bud; petals 0; disc 0; stamens 60–80, filaments very short, connate into a central column, anthers basifixed, extrorse, ovoid, 2-thecate, thecae distinct, pilosulous, longitudinally dehiscent, connective not produced; pollen spinulose; pistillode 0; receptacle columnar. Female flowers pedicellate; calyx-lobes 6, unequal, imbricate; petals 0; disc 0; ovary 3-locular, ovules 1 per locule; styles 3, sessile, spreading, deeply 10-palmatifid. Fruit 3-gonous-subglobose, smooth, 3-locular, dehiscent into 3 2-valved cocci or 6 valves, 3-seeded; exocarp separating; endocarp woody; columella narrowly obtriangular, 3-alate, persistent. Seeds ellipsoid, carunculate; testa smooth, dull, dark brown, streaked or blotched with darker brown. Fig.: See Fl. Nouv. Caléd. 14: 37, t. 6.3,4 and 75, t. 14.1–5 (1987).

A monotypic genus endemic to Nouvelle-Calédonie, where it is restricted to the extreme NW.

253. **Cocconerion** *Baill.*, Adansonia I. 11: 87 (1873); Benth., Gen. Pl. 3: 323 (1880); Pax & K. Hoffm., Nat. Pflanzenfam. ed. 2, 19C: 229 (1931); Guillaumin, Fl. Anal. Synopt. Nouv. Caléd.: 189 (1948); Airy Shaw, Kew Bull. 25: 503 (1971), and 32: 382 (1978); McPherson & Tirel, Fl. Nouv. Caléd. 14: 38 (1987); G. L. Webster, Ann. Missouri Bot. Gard. 81: 110 (1994). T.: *C. balansae* Baill.

Monoecious trees or shrubs; branching subverticillate, branches thick, nodose, cicatricose. Latex translucent, dark red. Indumentum, when present, stellate. Leaves verticillate, 6–10 per node, often crowded at the ends of the branches, petiolate or sessile, stipulate, simple, entire, elliptic-oblanceolate, coriaceous, margins sometimes revolute, penninerved, the lateral nerves 20–25, perpendicular to the midrib; stipules intrapetiolar, elongate, fused, mitriform, quickly deciduous. Flowers axillary, solitary, males 3(4) per node, females 3–6 per node. Male flowers long-pedicellate; calyx-lobes (4)6(7), rarely shortly 2-fid at apex, imbricate, at length reflexed; petals 0; disc 0; stamens 30–100, filaments very short, connate into a central column, anthers ventrally medifixed, extrorse, oblong, curved, 2-thecate, thecae contiguous, pilosulous, longitudinally dehiscent, connective 2-fid at the apex; pollen microechinulate; pistillode 0; receptacle columnar. Female flowers long-pedicellate; calyx-lobes (5)6(7), imbricate, persistent; petals 0; disc 0; staminodes sometimes +; ovary (2)3-locular, ovules 1 per locule; styles (2)3, deeply 2-fid or 2× 2-fid, segments linear-subulate, acute, deflexed-patent. Fruit 6-lobate-subglobose, smooth, 3-locular, dehiscent into 3 2-valved cocci or 6 valves, 3-seeded; exocarp not separating; endocarp woody; columella narrow, 3-alate, persistent. Seeds ellipsoid-cylindric, caruncuIate; testa smooth, shiny, dark brown. Fig.: See Fl. Nouv. Caléd. 14: 37, t. 6.5,6 and 41, t. 7 (1987).

A ditypic genus endemic to New Caledonia.

254. **Borneodendron** *Airy Shaw*, Kew Bull. 16: 359 (1963), Hooker's Icon. Pl. 7(2): t. 3633 (1967), and Kew Bull. Addit. Ser. 4: 60 (1975); G. L. Webster, Ann. Missouri Bot. Gard. 81: 110 (1994). T.: *B. aenigmaticum* Airy Shaw

Monoecious tree; trunk buttressed; branching subverticillate, branches fairly thick, cicatricose. Latex watery, reddish. Indumentum, when present, stellate. Leaves verticillate, 3 per node, often crowded at the ends of the branches, petiolate, stipulate, simple, entire, obovate-oblanceolate, somewhat coriaceous, margins narrowly revolute, penninerved, the lateral nerves 20–30, perpendicular to the midrib; stipules intrapetiolar, elongate, fused, mitriform, quickly deciduous. Male inflorescences terminal, subcapitate-amentiform, cernuous, pedunculate, flowers in 2–4 3-flowered verticels; bracts large, convex, 1- flowered; female flowers axillary, solitary. Male flowers very shortly pedicellate to subsessile; calyx 0; petals 0; disc 0; stamens 25–30, forming a ± spherical mass, filaments very short, connate into a central column, anthers basifixed, latrorse, oblong, 2-thecate, thecae contiguous, pilosulous, somewhat sinuously longitudinally dehiscent, connective scarcely visible; pollen not known; pistillode 0; receptacle columnar. Female flowers long-pedicellate, pedicels straight, flattened, elongating and becoming rigid in fruit; calyx-lobes 3, small, very shortly connate, imbricate, caducous; petals 0; disc 0; ovary 2-locular, ovules 1 per locule; styles 2, ± free, 2-fid, segments filiform-subulate, acute, flexuous. Fruit 2-coccous, 2-locular, dehiscent into 2 2-valved cocci or 4 valves, 2-seeded; exocarp thin, exfoliating; endocarp thick, woody; columella narrower than the pedicel, somewhat flattened, persistent. Seeds ovoid, carunculate; testa ± smooth, dull, dark brown, ochraceous-lineate or longitudinally marmorate; cotyledons oblong. Fig.: See Hooker's Icon. Pl. 7(2): t. 3633 (1967); Kew Bull. Addit. Ser. 4: cover illustration (1975).

A monotypic genus endemic to NE Borneo.

In addition to an affinity with *Cocconerion* etc., *Borneodendron* possesses some features also found in genera of the Oldfieldioid subtribe *Mischodontinae*; thus in its leaf-shape and nervation, and in having the columella narrower than the fruiting pedicel, it resembles *Aristogeitonia*; in its whorled leaves, it recalls *Mischodon*; and in the soon-deciduous fused stipules capping the terminal bud, as well as the nature of the androecium and pubescent anthers, it parallels *Androstachys*.

Tribe 45. **CROTONEAE** *Dumort.*, Anal. Fam. Pl.: 45 (1829) (as "Crotonieae"); G. L. Webster, Ann. Missouri Bot. Gard. 81: 110 (1994). T.: *Croton* L.

Monoecious or dioecious trees, shrubs or herbs; laticifers non-articulated; latex clear or reddish, often scanty or apparently 0; indumentum stellate or lepidote; leaves alternate, simple to palmatilobate, sometimes pellucid-punctate; petiole often 2-glandular at the apex; stipules + or 0. Inflorescences terminal or axillary, racemose or spicate; male sepals mostly 5, free, imbricate to valvate; petals 5, imbricate, sometimes 0; disc annular or dissected; stamens 3–400, filaments free or connate, often inflexed in bud, anthers muticous; pollen globose, inaperturate, crotonoid; pistillode 0; receptacle often villous; female sepals 4–7(10), ± free, mostly homogeneous, sometimes accrescent, imbricate to reduplicate-valvate, rarely 0; petals 5, mostly reduced or 0; ovary mostly 3-locular; styles free or ± so, once to several times 2-fid. Fruit mostly dehiscent. Seeds carunculate; testa mostly dry; endosperm copious.

A heptageneric tribe, consisting of the huge genus *Croton* and 6 small satellites, 4 of which are very closely related to it and have in the recent past been merged with it, but are being recognized here. The other 2, *Mildbraedia* and *Paracroton*, with a very different androecium from the rest, are doubtfully included here on the basis of the indumentum character *inter alia*.

Key to the genera

1. Filaments not distinctly inflexed in bud; female flowers mostly petaliferous 2
1. Filaments ± distinctly inflexed in bud; female petals mostly reduced or 0 . . 4

2. Stamens 3–6; male petals similar to the sepals; filaments free; female petals 0; styles simple or emarginate · · · · · · · · · · · · · · · · · · · 257. *Moacroton*
2. Stamens 10–30; male petals thinner than the sepals; inner filaments connate; female petals +; styles 2-fid; · 3
3. Inflorescences axillary; male petals free; testa dry · · · · · · 255. *Mildbraedia*
3. Inflorescences terminal; male petals coherent; testa fleshy · · 256. *Paracroton*
4. Female calyx-lobes heterogeneous · · · · · · · · · · · · · · · · · · 261. *Julocroton*
4. Female calyx-lobes, where present, homogeneous · · · · · · · · · · · · · · · 5
5. Ovary (2)3(4)-locular · 258. *Croton*
5. Ovary 1-locular · 6
6. Male petals and female sepals +; fruits indehiscent · · · · · · 259. *Crotonopsis*
6. Male petals and female sepals 0; fruits dehiscent · · · · · · 260. *Eremocarpus*

255. **Mildbraedia** *Pax,* Bot. Jahrb. Syst. 43: 319 (1909); Pax & K. Hoffm., Pflanzenr. 47: 11 (1911); Hutch., Fl. Trop. Afr. 6(1): 798 (1912); Pax & K. Hoffm., Nat. Pflanzenfam. ed. 2, 19C: 167 (1931); Keay (ed.), Fl. W. Trop. Afr., ed. 2, 1: 397 (1958); J. Léonard, Fl. Cong. Rwa.-Bur. 8(1): 85 (1962); Radcl.-Sm., Kew Bull.: 27: 507 (1972), and Fl. Trop. E. Afr. Euph. 1: 340 (1987); G. L. Webster, Ann. Missouri Bot. Gard. 81: 111 (1994); Radcl.-Sm., Fl. Zamb. 9(4): 273 (1996). T.: *M. paniculata* Pax.

Neojatropha Pax in Pflanzenr. 42: 114 (1910). T.: *N. carpinifolia* (Pax) Pax (B.: *Jatropha carpinifolia* Pax) [=*M. carpinifolia* (Pax) Hutch.]

Dioecious shrubs or small trees. Indumentum stellate, sometimes also simple. Leaves alternate, long-petiolate, stipulate, simple or occasionally deeply 2–3-lobate, entire to repand-dentate, palminerved at the base, then penninerved; stipules subulate, deciduous. Inflorescences axillary or supra-axillary, racemose or cymose, solitary, few-flowered, long-pedunculate to scapose; bracts subulate. Male flowers pedicellate; sepals 5(6), free or shortly connate at the base, imbricate; petals 5(6), free, imbricate, equalling or exceeding the calyx; disc-glands 5, free, fleshy, oppositisepalous; stamens 10–25, filaments erect in bud, outer free, short, inner connate into a column, anthers small, dorsifixed, introrse, longitudinally dehiscent; pollen typically crotonoid; pistillode 0; receptacle fleshy, pilose. Female flowers pedicellate; sepals and petals ± as in the male, but slightly larger; disc annular, shallowly 5-lobate; ovary 3-locular, ovules 1 per locule; styles 3, shortly connate at the base, deeply 2-fid or -partite, spreading. Fruit 3-lobate, septicidally dehiscent into 3 2-valved cocci or loculicidally into 6 valves; exocarp thin, crustaceous; endocarp thinly woody or crustaceous; columella narrowly obtriangular, persistent. Seeds ovoid-subglobose, carunculate; testa crustaceous, smooth, shiny, marmorate; caruncle foliaceous, adpressed. Fig.: See Pflanzenr. 47: 11, t. 2 (1911); Fl. Trop. E. Afr. Euph. 1: 342, t. 64 (1987); Fl. Zamb. 9(4): 274, t. 56 (1996).

A tritypic tropical African endemic genus.

The connate inner stamens of this and the next genus set them apart from the rest of the tribe.

256. **Paracroton** *Miq.*, Fl. Ned. Ind. I. 2: 382 (1859); Müll. Arg. in DC., Prodr. 15(2): 1112 (1866); Benth., Gen. Pl. 3: 299 (1880); Pax in Engl. & Prantl, Pflanzenfam. ed. 1, 3(5): 85 (1890); J. J. Sm., Meded. Dept. Landb. Ned.-Indië 10: 513 (1910); Pax & K. Hoffm., Pflanzenr. 47: 12 (1911), and Nat. Pflanzenfam. ed. 2, 19C: 155 (1931); N. P. Balakr. & Chakrab., Kew Bull. 48: 716 (1993). T.: *P. pendulus* (Hassk.) Miq. (B.: *Croton pendulus* Hassk.)

Fahrenheitia Rchb. f. & Zoll., Linnaea 28: 599 (1857), nom. prov. inval.; Baill., Étude Euphorb.: 652 (1858); Müll. Arg. in DC., Prodr. 15(2): 1256 (1866); Benth., Gen. Pl. 3: 299 (1880), in obs.; Pax & K. Hoffm., Pflanzenr. 47: 17, 21 (1911), in obs.; Airy Shaw, Kew Bull. 20: 409 (1967), and 26: 270 (1972); Whitmore, Tree Fl. Mal. 2: 97 (1973); Airy Shaw, Kew Bull. 29: 325 (1974), and Kew Bull. Addit. Ser. 4: 114 (1975); Ramamoorthy in C. J. Saldanha & Nicholson, Fl. Hassan Distr.: 340 (1976); Airy Shaw, Kew Bull. Addit. Ser. 8: 90 (1980), Kew Bull. 36: 298 (1981), and Euph. Philipp.: 27 (1983); G. L. Webster, Ann. Missouri Bot. Gard. 81: 111 (1994); Philcox in Dassan., ed., Fl. Ceyl. XI: 112 (1997). T.: *F. collina* Rchb. f. & Zoll. (S.: *F. pendula* (Hassk.) Airy Shaw) (B.: *Croton pendulus* Hassk.) [= *P. pendulus* (Hassk.) Miq.]

Desmostemon Thwaites, Enum. Pl. Zeyl. 4: 278 (1861). T.: *D. zeylanicus* Thwaites [=*P. pendulus* subsp. *zeylanicus* (Thwaites) N. P. Balakr. & Chakrab.]

Monoecious or dioecious trees. Sap red. Indumentum stellate or lepidote. Leaves alternate, long-petiolate, stipulate, simple, elliptic, glandular-dentate or -serrate, rarely entire, penninerved, 2-glandular at the base; petioles slightly pulvinate at apex and base; stipules minute, triangular to subulate, or 0. Inflorescences terminal and axillary, usually elongate and pendulous, unisexual, males pseudoracemose to narrowly thyrsiform, flowers small, fragrant, borne in distant crowded cymules along the angular rhachis, females ± racemose, flowers larger, 1(3) per node; bracts resembling the stipules. Male flowers shortly pedicellate; calyx cupular, shortly 5-lobate, lobes imbricate, dorsally appendiculate, appendage gongyloid or cornute; petals 5, free, but sometimes apparently connate owing to coherent margins, imbricate, white; disc-glands 5–10, rectangular or trapeziform, alternipetalous when equal, or disc annular or 0; stamens 12–32, outer filaments free, inner connate into a column or 2 concentric columns, anthers 2–4-verticillate, dorsifixed, extrorse, thecae longitudinally dehiscent, connective broad; pollen typically crotonoid; pistillode 0. Female flowers shortly pedicellate; calyx and petals ± as in the male, but petals caducous; disc annular or shallowly 5-lobate, pilose above; ovary 3-locular, ovules 1 per locule; styles 3, ± free, short, erect or recurved, 2-fid. Fruit depressed-globose or 3-lobate, 3-locular, dehiscing loculicidally into 3 valves or septicidally into 3 2-valved cocci; exocarp sometimes separating; endocarp thickly or thinly woody; columella slender, capitate, persistent. Seeds large, subglobose, mostly broader than long, carunculate; exotesta thinly fleshy; endotesta crustaceous, smooth, dull when dry, marmorate; caruncle foliaceous, adpressed; endosperm copious; cotyledons broad, flat. Fig.: See Pflanzenr. 47: 19, t. 5 (1911) (as *Ostodes*); Nat. Pflanzenfam. ed. 2, 19C: 156, t. 80 (1931) (as *Ostodes*); Kew Bull. 48: 720–724, tt. 1–3 (1993).

A tetratypic tropical Asiatic genus ranging from S. India and Sri Lanka to Lower Myanmar (Burma), Lower Thailand, W. Malesia and the Philippines, with an outlier in Irian Jaya, *fide* Balakrishnan & Chakrabarty (1993). The Irian Jaya species may represent a distinct genus, however.

257. **Moacroton** *Croizat,* J. Arnold Arbor. 26: 189 (1945); Alain, Fl. Cuba 3: 85 (1953); Borhidi, Acta Bot. Acad. Sci. Hung. 36: 7 (1990/1); G. L. Webster, Ann. Missouri Bot. Gard. 81: 111 (1994). T.: *M. leonis* Croizat.

Cubacroton Alain, Candollea 17: 116 (1960); G. L. Webster, Ann. Missouri Bot. Gard. 81: 131 (1994), **synon. nov.** T.: *C. maestrensis* Alain, Candollea 17: 116 (1960) [=*M. maestrensis* (Alain) Radcl.-Sm., **comb. nov.**]

Very like 258. *Croton,* but differing in often having narrow coriaceous revolute leaves, male petals similar to the sepals, fewer (3–6) stamens, filaments very short, clavate, glandular, or almost 0, anthers subsessile and almost horizontal, no female petals, simple or emarginate styles, and fruit-cocci dorsally angled. Fig.: See J. Arnold Arbor. 26: 189, t. 1.5,6 (1945); Fl. Cuba 3: 86, t. 21 (1953); Acta Bot. Acad. Sci. Hung. 36: 11, t. 1 and 12, t. 2 (1990/1).

An octotypic genus endemic to Cuba.

258. **Croton** *L.* Sp. Pl.: 1004 (1753), and Gen. Pl., ed. 5: 436 (1754); A. Juss., Euphorb. Gen.: 28 (1824); Klotzsch, London J. Bot. 2: 48 (1843); Baill., Étude Euphorb.: 349 (1858); Müll. Arg. in DC., Prodr. 15(2): 512 (1866), and in Mart., Fl. Bras. 11(2): 81 (1873); Benth., Gen. Pl. 3: 293 (1880); Hook. f., Fl. Brit. India 5: 385 (1887); A. M. Ferguson, Annual Rep. Missouri Bot. Gard. 12: 33 (1901); Hutch. in Fl. Trop. Afr. 6(1): 746 (1912); Prain in Fl. Cap. 5(2): 410 (1920); Fawc. & Rendle, Fl. Jamaica 4: 275 (1920); Gagnep., Fl. Indo-Chine 5: 256 (1925); Pax & K. Hoffm., Nat. Pflanzenfam. ed. 2, 19C: 83 (1931); Leandri, Ann. Inst. Bot.-Géol. Colon. Marseille V. 7(1): 5 (1939); Carabia, Caribbean Forest. 3: 114 (1942); Croizat, Darwiniana 5: 417 (1941), and 6: 442 (1944); Alain, Fl. Cuba 3: 63 (1953); Keay (ed.), Fl. W. Trop. Afr., ed. 2, 1: 393 (1958); M. C. Johnst. & Warnock, Southw. Naturalist 7: 1 (1962); J. Léonard, Fl. Cong. Rwa.-Bur. 8(1): 50 (1962); Backer & Bakh. f., Fl. Java 1: 475 (1963); Jabl., Mem. New York Bot. Gard. 12: 150 (1965); G. L. Webster, J. Arnold Arbor. 48: 353 (1967), and Ann. Missouri Bot. Gard. 54: 247 (1968); Airy Shaw, Kew Bull. 21: 374 (1968), 23: 69 (1969), 25: 514 (1971), 26: 241 (1972), 27: 78 (1972), 29: 310 (1974), Hooker's Icon. Pl. 38: t. 3712 (1974), Kew Bull. Addit. Ser. 4: 89 (1975), Kew Bull. 32: 387 (1978), 33: 55 (1978), Kew Bull. Addit. Ser. 8: 65 (1980), Kew Bull. 35: 392, 614 (1980), 36: 283, 604 (1981), 37: 14, 379 (1982), and Euph. Philipp.: 17 (1983); Leandri, Adansonia II. 9: 497 (1969), 10: 183, 309 (1970), 12: 65, 403 (1972), 13: 173, 295, 423 (1973), and 15: 331 (1976); J.-G. Adam, Mém. Mus. Natl. Hist. Nat., B, Bot. 20: 465 (1971); Radcl.-Sm., Hooker's Icon. Pl. 37: t. 3699 (1971), and Kew Bull.: 27: 505 (1972); Whitmore, Tree Fl. Mal. 2: 84 (1973); Berhaut, Fl. Ill. Sénég. 3: 415 (1975); Philcox, Fl.

Trin. Tob. II(X): 644 (1979); Correll, Fl. Bah. Arch.: 787 (1982); Proctor, Kew Bull. Addit. Ser. 11: 534 (1984); Alain, Fl. Español. 4: 108 (1986); Thin, J. Biol. (Vietnam) 8(2): 28 (1986); Grierson & D. G. Long, Fl. Bhut. 1: 791 (1987); McPherson & Tirel, Fl. Nouv. Caléd. 14: 78 (1987); Radcl.-Sm., Fl. Trop. E. Afr. Euph. 1: 135 (1987); Alain, Descr. Fl. Puerto Rico 2: 375 (1988); G. L. Webster, Ann. Missouri Bot. Gard. 75: 1116 (1988); R. A. Howard, Fl. Lesser Antilles 5: 34 (1989); Radcl.-Sm., Kew Bull.: 45: 555 (1990); J.-P. Lebrun & Stork, Enum. Pl. Afr. Trop. 1: 209 (1991); G. L. Webster, Novon 2: 270 (1992), Taxon 42: 800 (1993), and Ann. Missouri Bot. Gard. 81: 111 (1994); Murillo-Aldana & Franco-Rosselli, Euf. Reg. Ararac.: 61 (1995); Radcl.-Sm., Fl. Zamb. 9(4): 275 (1996); Gillespie in Acev.-Rodr., ed., Fl. St. John: 208 (1996); Philcox in Dassan., ed., Fl. Ceyl. XI: 87 (1997). LT.: *Cr. tiglium* L. (chosen by Small, 1913).

Cascarilla Adans., Fam. Pl. 2: 355 (1763). LT.: *C. officinalis* Raf. (chosen here) [=*Cr. linearis* Jacq. (S.: *Cr. cascarilla* (L.)L.; B.: *Clutia cascarilla* L.)]

Aroton Neck., Elem. Bot. 2: 336 (1790). T.: (none designated).

Cinogasum Neck., Elem. Bot. 2: 336 (1790). T.: (none designated).

Brunsvia Neck., Elem. Bot. 2: 337 (1790). T.: (none designated).

Tridesmis Lour., Fl. Cochinch.: 576 (1790). LT.: *Tr. tomentosa* Lour. (chosen here) [=*Cr. crassifolius* Geiseler]

Schradera Willd., Gött. J. Naturwiss. 1: 1 (1797). T.: *Schr. scandens* Willd. [=*Cr. lobatus* L.]

Lascadium Raf., Fl. Ludov.: 114 (1817); A. Juss., Euphorb. Gen.: 62 (1824). T.: *L. lanatum* Raf.

Decarinium Raf., Neogenyton: 1 (1825). T.: *D. glandulosum* (L.) Raf. [=*Cr. glandulosus* L.]

Heptallon Raf., *l.c.* (1825). LT.: *H. graveolens* Raf. [=*Cr. capitatus* Michx.] (chosen by Webster, 1994)

Drepadenium Raf., *op. cit.*: 2 (1825). T.: *D. maritimum* (Walter) Raf. [=*Cr. maritimus* Walter]

Hendecandra Eschsch., Mém. Acad. Imp. Sci. St. Pétérsbourg Hist. Acad. 10: 287 (1826); Baill., Étude Euphorb.: 371 (1858). T.: *H. procumbens* Eschsch. [=*Cr. californicus* Müll. Arg.]

Halecus [Rumph. ex] Raf., Sylva Tellur.: 62 (1838). T.: *H. mauritianus* (Lam.) Raf. [=*Cr. mauritianus* Lam.]

Kurkas Raf., *l.c.* (1838). LT.: *K. tiglium* (L.) Raf. (chosen here) [=*Cr. tiglium* L.]

Semilta Raf., *op. cit.*: 63 (1838). T.: *S. althaeifolia* (Mill.) Raf. [=*Cr. althaeifolius* Mill.]

Luntia Neck. ex Raf., *l.c.* (1838). T.: *L. sericea* (Lam.) Raf., nom. illeg. (B.: *Cr. sericeus* Lam., nom. illeg.) [=*Cr. matourensis* Aubl.]

Penteca Raf., *l.c.* (1838). T.: *P. tomentosa* Raf. [=*Cr. dioicus* Cav.]

Triplandra Raf., *l.c.* (1838). T.: *Tr. lanata* (Lam.) Raf. (B.: *Cr. lanatus* Lam.) [=*Cr. lasianthus* Pers.]

Astrogyne Benth., Pl. Hartw.: 14 (1839). T.: *A. crotonoides* Benth. [=*Cr. dioicus* Cav.]

Merleta Raf., Autik. Bot.: 49 (1840). T.: *M. microphylla* Raf. [?=*Cr. nummulariifolius* A. Rich.]

Pleopadium Raf., Autik. Bot.: 50 (1840). T.: *Pl. ciliatum* Raf. [?=*Cr. ciliatoglandulifer* Ortega]

Medea Klotzsch, Arch. Naturgesch. 7: 193 (1841). T.: *M. hirta* Klotzsch [=*Cr. timandroides* (Didr.) Müll. Arg. (B.: *M. timandroides* Didr.)]

Podostachys Klotzsch, *l.c.* (1841). NT.: *P. subfloccosa* Didr. [=*Cr. lundianus* (Didr.) Müll. Arg. var. *subfloccosus* (Didr.) Müll. Arg.] (designated by Wheeler, 1975).

Astraea Klotzsch, *op. cit.*: 194 (1841). T.: *A. lobata* (L.) Klotzsch [=*Cr. lobatus* L.]

Ocalia Klotzsch, *op. cit.*: 195 (1841). NT.: *O. sellowiana* Klotzsch ex Baill. [=*Cr. perdicipes* A. St.-Hil.; = *Cr. antisyphiliticus* Mart.] (designated by Wheeler, 1975).

Cleodora Klotzsch, *op. cit.*: 196 (1841). T.: *C. sellowiana* Klotzsch [=*Cr. sphaerogynus* Baill.]

Eutropia Klotzsch, *l.c.* (1841). T.: *E. brasiliensis* (Spreng.) Klotzsch, nom. illeg. (B.: *Rottlera brasiliensis* Spreng., nom. illeg.) [=*Cr. polyandrus* Spreng.] (designated by Wheeler, 1975).

Timandra Klotzsch, *op. cit.*: 197 (1841). LT.: *T. serrata* Klotzsch [=*Cr. serratus* (Klotzsch) Müll. Arg.] (designated by Wheeler, 1975).

Engelmannia Klotzsch, in *op. cit.*: 253 (1841), non A. Gray ex Nutt. (1840). T.: *E. nuttalliana* Klotzsch, nom. illeg. [=*Cr. lindheimerianus* Scheele]

Geiseleria Klotzsch, *op. cit.*: 254 (1841). T.: *G. chamaedryfolia* Klotzsch [=*Cr. trinitatis* Millsp.]

Pilinophytum Klotzsch, *op. cit.*: 255 (1841); Baill., Étude Euphorb.: 377 (1858). T.: *P. capitatum* (Michx.) Klotzsch [=*Cr. capitatus* Michx.]

Lasiogyne Klotzsch, Nov. Actorum Acad. Caes. Leop.-Carol. Nat. Cur. 19, Suppl. 1: 418 (1843). T.: *L. brasiliensis* Klotzsch [=*Cr. compressus* Lam.]

Tiglium Klotzsch, *l.c.* (1843). T.: *T. officinale* Klotzsch, nom. illeg. [=*Cr. tiglium* L.]

Brachystachys (L'Hér.) Klotzsch, London J. Bot. 2: 47 (1843); Baill., Étude Euphorb.: 373 (1858). T.: *Br. hirta* (L'Hér.) Klotzsch [=*Cr. hirtus* L'Hér.]

Macrocroton Klotzsch in M. R. Schomb., Faun. & Fl. Br. Guiana: 1186 (1848). LT.: *M. cuneatus* (Klotzsch) Klotzsch [=*Cr. cuneatus* Klotzsch] (chosen here).

Angelandra Endl., Gen. Pl. Suppl. 5: 91 (1850). T.: *A. elliptica* (Nutt.) Baill. (B.: *Cr. ellipticus* Nutt., non Geiseler) [=*Cr. lindheimerianus* Scheele]

Barhamia Klotzsch in Seem., Bot. Voy. Herald: 104 (1853). LT.: *B. panamensis* Klotzsch [=*Cr. hircinus* Vent.] (designated by Wheeler, 1975).

Cyclostigma Klotzsch, *l.c.* (1853), non Hochst. ex Endl. (1842). LT.: *C. panamense* Klotzsch (designated by Wheeler, 1975) [=*Cr. draco* Schltdl. subsp. *panamensis* (Klotzsch) G. L. Webster]

Gynamblosis Torr. in Marcy, Explor. Red River Louisiana: 295 (1853); Baill., Étude Euphorb.: 368 (1858). T.: *G. monanthogyna* (Michx.) Torr. [=*Cr. monanthogynus* Michx.]

Crotonanthus Klotzsch ex Schltdl., Linnaea 26: 634 (1853/5). T.: *Cr. padifolius* Klotzsch [=*Croton heliotropiifolius* Kunth]

Calypteriopetalon Hassk., Flora 40: 531 (1857). T.: *C. brasiliense* Hassk. [=*Cr. urticifolius* Lam.]

Myriogomphos Didr., Vidensk. Meddel. Dansk Naturhist. Foren. Kjøbenhavn 1857: 142 (1857). T.: *M. fuscus* Didr. [=*Cr. fuscus* (Didr.) Müll. Arg.]

Croton sect. *Eutropia* (Klotzsch) Baill., Étude Euphorb.: 357(1858).

Croton sect. *Geiseleria* (Klotzsch) Baill., Étude Euphorb.: 359 (1858.)

Croton sect. *Astraea* (Klotzsch) Baill., Étude Euphorb.: 363 (1858).

Croton sect. *Podostachys* (Klotzsch) Baill., Étude Euphorb.: 365 (1858).

Croton sect. *Ocalia* (Klotzsch) Baill., Étude Euphorb.: 366 (1858).
Croton sect. *Barhamia* (Klotzsch) Baill., Étude Euphorb.: 367 (1858).
Croton sect. *Medea* (Klotzsch) Baill., Étude Euphorb.: 368 (1858).
Croton sect. *Timandra* (Klotzsch) Baill., Étude Euphorb.: 368 (1858).
Croton sect. *Cleodora* (Klotzsch) Baill., Étude Euphorb.: 369 (1858).
Croton sect. *Codonocalyx* (Klotzsch ex Baill.) Baill., Étude Euphorb.: 369 (1858).
Croton sect. *Lasiogyne* (Klotzsch) Baill., Étude Euphorb.: 370 (1858).
Croton sect. *Tiglium* (Klotzsch) Baill., Étude Euphorb.: 361 (1858).
Klotzschiphytum Baill., Étude Euphorb.: 382 (1858), nom. illeg. T.: *K. mauritianum* (Lam.) Baill. [=*Cr. mauritianus* Lam.]
Comatocroton H. Karst., Wochenschr. Gärtnerei Pflanzenk. 2: 6 (1859). T.: *C. ovalifolius* (Vahl) H. Karst. [=*Cr. ovalifolius* Vahl]
Monguia Chapel. ex Baill., Adansonia I. 1: 147 (1860), pro syn. LT.: *M. lanceolata* Chapel. ex Baill. [=*Cr. chrysodaphne* Baill.] (chosen here).
Anisophyllum Boivin ex Baill., Adansonia I. 1: 153 (1860), pro syn., non Haw. (1812). LT.: *A. acutifolium* Boivin ex Baill. [=*Cr. muricatus* Vahl] (chosen here).
Argyra Noronha ex Baill., Adansonia I. 1: 162 (1860), pro syn. T.: *A. fasciculata* Noronha ex Baill. [=*Cr. noronhae* Baill.]
Aubertia Chapel. ex Baill., Adansonia I. 1: 162, 166 (1860), pro syn. LT.: *A. glandulosa* Chapel. ex Baill. (chosen here) [=*Cr. chapelieri* Baill.]
Argyrodendron Klotzsch in Peters, Naturw. Reise Mossambique 6, 1: 100 (1861), non F. Muell. (1858). T.: *A. bicolor* Klotzsch [=*Cr. menyharthii* Pax]
Leucadenia Klotzsch ex Baill., Adansonia I. 4: 338 (1864). T.: *L. pilosa* Klotzsch ex Baill. [=*Cr. fuscus* (Didr.) Müll. Arg.]
Oxydectes [L. ex] Kuntze, Revis. Gen. Pl.: 610 (1891). T.: (none designated).
Heterocroton S. Moore, Trans. Linn. Soc. London, Bot. 4: 461 (1895). T.: *H. mentiens* S. Moore [=*Cr. mentiens* (S. Moore) Pax]

Monoecious or rarely dioecious trees, shrubs or occasionally herbs or lianes. Indumentum stellate and/or lepidote, scales often shiny. Leaves alternate or occasionally opposite or subverticillate, petiolate, mostly stipulate, simple or lobate, entire or dentate, penninerved, 3-plinerved or palminerved from the base, usually 2-glandular at the top of the petiole or the base of the lamina; stipules sometimes foliaceous. Inflorescences terminal, rarely axillary, spicate or racemose, rarely subpaniculate, 1- or 2-sexual, male flowers often fasciculate on the axis, females usually in the lower $\frac{1}{2}$ or at the base; bracts usually small, persistent or deciduous. Male flowers shortly pedicellate; calyx-lobes (4)5(6), ± equal, valvate or narrowly imbricate; petals (4)5(6), rarely more, usually equalling or shorter than the sepals, or 0; disc-glands or scales small, oppositisepalous, free or connate, fleshy, rarely 0; stamens 5–∞, filaments free, inserted on a receptacle, tips inflexed in bud, soon becoming erect, anthers basifixed, pendulous in bud, muticous, thecae parallel, longitudinally dehiscent; pollen 3-nucleate, crotonoid, clavate, clavae usually on ridges, sometimes acuminate; pistillode 0; receptacle usually pilose. Female flowers usually shortly pedicellate, pedicels sometimes elongating in fruit; calyx-lobes often narrower than in the male, usually ± homogeneous, generally persistent, sometimes accrescent in fruit; petals often smaller than in the male, sometimes reduced or

0; staminodes sometimes +; disc annular, or scales distinct, sometimes minute; ovary (2)3(4)-locular, ovules 1 per locule; styles (2)3(4), recurved, once to several × 2-lobate, -fid or -partite or occasionally multifid, rarely laciniate. Fruit usually 3-lobate-subglobose, septicidally dehiscent into 3 2-valved cocci, loculicidally into 3 valves, or sometimes subindehiscent and drupaceous or bacciform; endocarp woody or crustaceous; columella usually persistent, often slender. Seeds mostly ellipsoid, carunculate; caruncle small; testa smooth or very shallowly rugulose, woody or crustaceous; endosperm copious, fleshy; embryo straight; cotyledons broad, flat. Fig.: See Euphorb. Gen.: t. 8, f. 26A.1–11 (1824); Étude Euphorb.: t. 16, f. 36–38; t. 17; t. 18, f. 1–7 and 19, f. 1–2 (1858); Fl. Bras. 11(2): tt. 14.I–39 (1873); Fl. Jamaica 4: 277, t. 91 (1920); Fl. Indo-Chine 5: 261, t. 28 and 267, t. 29 (1925); Nat. Pflanzenfam. ed. 2, 19C: 84, t. 41 and 85, t. 42 (1931); Ann. Inst. Bot.-Géol. Colon. Marseille V. 7(1): 9–70, tt. 1–11 (1939); Fl. Cuba 3: 67, t. 12 and 68, t. 13 (1953); Fl. W. Trop. Afr., ed. 2, 1: 395, t. 138 (1958); Fl. Cong. Rwa.-Bur. 8(1): 58–79, tt. 5–6 and figs. 3–5 (1962); J. Arnold Arbor. 48: 356, t. 2 (1967); Ann. Missouri Bot. Gard. 54: 251, t. 9 (1968); Amer. J. Bot. 56: 756, t. 19A–J (1969); Adansonia II. 9: 503, t. 1 (1969), 10: 185, t. 1, 188, t. 2 and 311, t. 1 (1970), 12: 67, t. 1 and 70, t. 2 (1972), 13: 174, t. 1, 296, t. 1 and 424, t. 1 (1973), and 15: 331, t. 1 (1976); Hooker's Icon. Pl. 37: t. 3699 (1971), and 38: t. 3712 (1974); Fl. Ill. Séneg. 3: tt. 414–422 (1975); Kew Bull. Addit. Ser. 8: 223, t. 1.4 (1980); Kew Bull. 35: 619, t. 2A (1980); Fl. Bah. Arch.: 789, t. 330 and 791, t. 331 (1982); Kew Bull. Addit. Ser. 11: 535, t. 162 and 536, t. 163 (1984); Fl. Español. 4: 125, t. 116-11 (1986); Fl. Bhut. 1: 789, t. 48.n–r (1987); Fl. Nouv. Caléd. 14: 81, t. 15 (1987); Fl. Trop. E. Afr. Euph. 1: 140–158, tt. 25–27 (1987); Descr. Fl. Puerto Rico 2: 378, t. 59-11 (1988); Ann. Missouri Bot. Gard. 75: 1122, t. 3 (1988); Fl. Lesser Antilles 5: 22, t. 11 (1989); Bol. Mus. Paraense Emilio Goeldi, N.S., Bot. 7: 40–42, tt. 12–14 (1991); Árvores Brasileiras: 99–101 (1992); Euf. Reg. Ararac.: 63–69, tt. 10–13 (1995); Fl. Zamb. 9(4): 291, t. 57 and 293, t. 58 (1996); Fl. St. John: 207, t. 94K–R (1996); Árvores Brasileiras 2: 94 (1998).

A large and diverse but nevertheless very natural and distinctive pantropical genus of over 800 species, with the strongest representation in the Neotropical realm and Madagascar.

Webster (1993) has recognized 40 sections in the genus, 3 of which have been reassigned generic status here.

Croton colobocarpus Airy Shaw from NE Thailand is not a *Croton* at all, but appears to constitute a new genus of the tribe *Codiaeae* (Esser, pers. comm. Oct. 2000).

Cr. cascarilla Benn. and *Cr. eluteria* Benn. yield Cascarilla bark, used as a tonic; the seeds of *Cr. tiglium* L. yield Croton Oil, a powerful purgative; *Cr. lacciferus* L. hosts a lac insect, and produces a substance used in varnish-making.

259. **Crotonopsis** *Michx.*, Fl. Bor.-Amer. 2: 185, t. 46 (1803);); A. Juss., Euphorb. Gen.: 31 (1824); Baill., Étude Euphorb.: 380 (1858); Müll. Arg. in DC., Prodr. 15(2): 707 (1866); Benth., Gen. Pl. 3: 296 (1880); Pennell, Bull. Torrey Bot. Club 45: 477 (1918); Pax & K. Hoffm., Nat. Pflanzenfam. ed. 2, 19C: 88 (1931); G. L. Webster, J. Arnold Arbor. 48: 360 (1967). T.: *C. linearis* Michx. (S.: *Croton michauxii* G. L. Webster).

Leptemon Raf., Med. Repos. Ser. 2, 5: 352 (1808), nom. illeg. superfl. T.: *L. lineare* (Michx.) Raf. [=*C. linearis* Michx. (S.: *Croton michauxii* G. L. Webster)]
Friesia Spreng., Anleit. Kenntn. Gew., ed. 2, 2(2): 885 (1818). T.: *F. argentea* Spreng. [=*C. linearis* Michx. (S.: *Croton michauxii* G. L. Webster)]
Croton sect. *Crotonopsis* (Michx.) G. L. Webster, Novon 2: 270 (1992), and Taxon 42: 814 (1993).
Croton subgen. *Crotonopsis* (Michx.) Radcl.-Sm. & Govaerts, Kew Bull.: 52: 183 (1997).

Like 258. *Croton* in many respects, but differing in having a 1-locular ovary, an indehiscent achene-like 1-locular fruit with 1 ecarunculate seed. Fig.: See Fl. Bor.-Amer. 2: 185, t. 46 (1803); Euphorb. Gen.: t. 8, f. 27.1–5 (1824); Étude Euphorb.: t. 12, f. 23–27 (1858).

A ditypic genus restricted to the SE USA.

Webster (1993) considered this taxon to have a close affinity to *Croton* sect. *Gynamblosis* (Torr.) A. Gray, but in addition to the fruit, there are also differences in the indumentum- and style-characters.

260. **Eremocarpus** *Benth.*, Bot. Voy. Sulphur: 53 (1844); Baill., Étude Euphorb.: 381 (1858); Müll. Arg. in DC., Prodr. 15(2): 708 (1866); Benth., Gen. Pl. 3: 297 (1880); Pax & K. Hoffm., Nat. Pflanzenfam. ed. 2, 19C: 88 (1931). T.: *E. setigerus* (Hook.) Benth. (B.: *Croton setigerus* Hook.)

Piscaria Piper, Contr. U.S. Natl. Herb. 11: 382 (1906), nom. illeg. superfl. T.: *P. setigera* (Hook.) Piper [=*E. setigerus* (Hook.) Benth. (B.: *Croton setigerus* Hook.)]
Croton sect. *Eremocarpus* (Benth.) G. L. Webster, Novon 2: 270 (1992), and Taxon 42: 813 (1993).
Croton subgen. *Eremocarpus* (Benth.) Radcl.-Sm. & Govaerts, Kew Bull.: 52: 184 (1997).

Like 259. *Crotonopsis* in many respects, but with the leaves mostly clustered at the forks of dichotomous stems, male flowers apetalous, female flowers achlamydeous, style elobate, and with the unilocular fruit dehiscent into 2 valves. Fig.: See Fig. 40, p. 325.

A monotypic genus from Pacific N. America.

Webster (1993) considered this taxon to have an affinity to *Croton* sect. *Pilinophytum* (Klotzsch) A. Gray, but the flowers are much simpler than those in any section of that genus.

261. **Julocroton** *Mart.*, Flora 20 (2), Beibl.: 119 (1837), nom. cons.; Baill., Étude Euphorb.: 374 (1858); Müll. Arg. in DC., Prodr. 15(2): 700 (1866), and in Mart., Fl. Bras. 11(2): 274 (1873); Benth., Gen. Pl. 3: 296 (1880); Pax & K. Hoffm., Nat. Pflanzenfam. ed. 2, 19C: 87 (1931); Croizat, Revista Argent. Agron. 10: 121 (1943); Jabl., Mem. New York Bot. Gard. 12: 169 (1965); Cordeiro, Acta Bot. Brasil. 4: 83 (1990). T.: *J. phagedaenicus* Mart. [=*J. triqueter* (Lam.) Didr. (B.: *Croton triqueter* Lam.)]

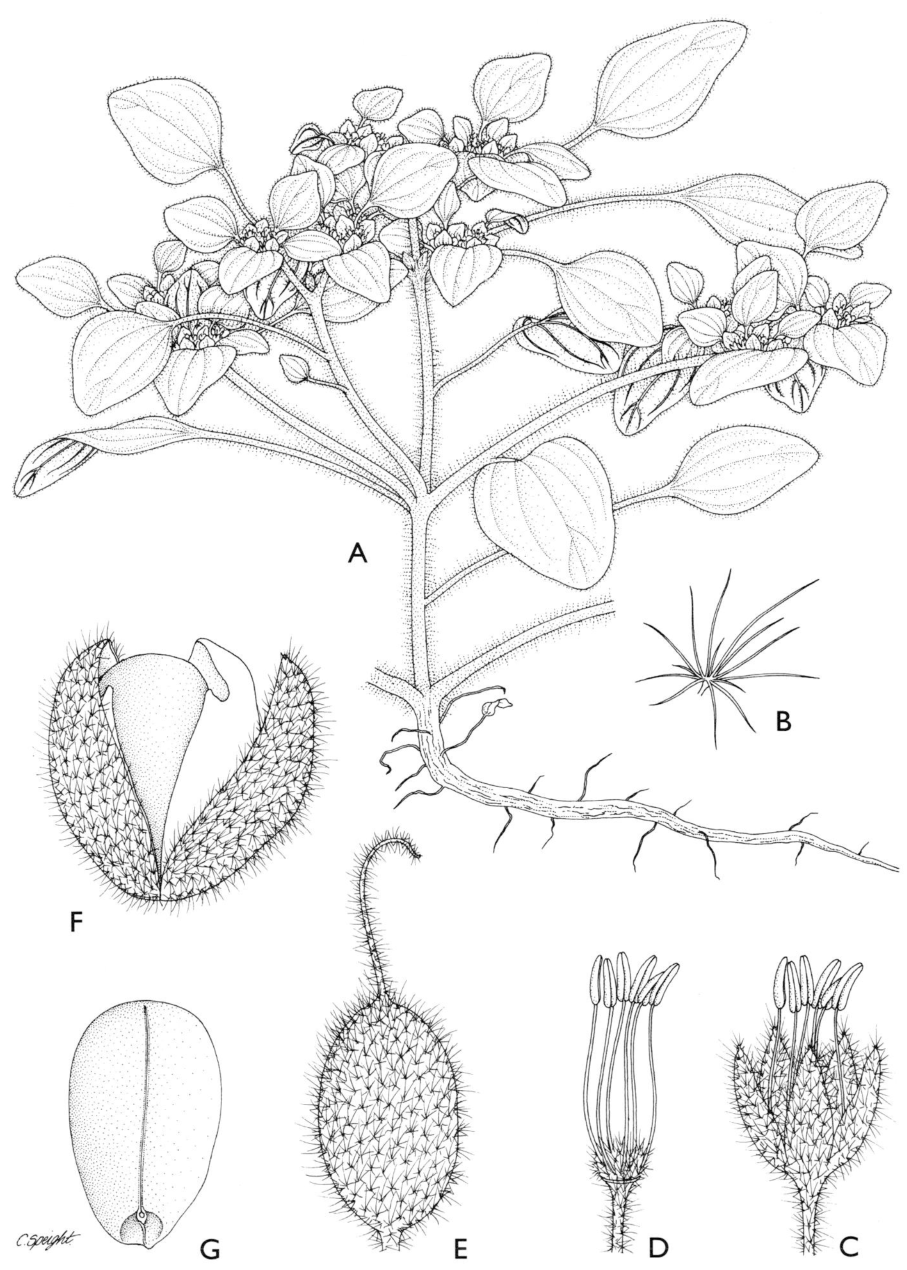

FIG. 40. *Eremocarpus setigerus* (Hook.) Benth. (B.: *Croton setigerus* Hook.). **A** Habit × 1; **B** Stellate hair × 80; **C** Male flower × 20; **D** Stamens × 20; **E** Fruit × 13.5; **F** Fruit, dehisced × 13.5; **G** Seed × 13.5. All from *Rose* 41457. Drawn by Camilla Speight.

Cieca Adans., Fam. Pl. 2: 355 (1763), nom. rej. T.: *Croton argenteus* L. [=*J. argenteus* (L.) Didr.]
Heterochlamys Turcz., Bull. Soc. Imp. Naturalistes Moscou 16: 61 (1843). T.: *H. quinquenervia* Turcz. [=*J. argenteus* (L.) Didr. (B.: *Croton argenteus* L.)]
Centrandra H. Karst., Linnaea 28: 440 (1856). T.: *C. hondensis* H. Karst. [=*J. hondensis* (H. Karst.) Müll. Arg. (S.: *Croton hondensis* (H. Karst.) G. L. Webster)]
Croton sect. *Julocroton* (Mart.) G. L. Webster, J. Arnold Arbor. 48: 354 (1967), Novon 2: 270 (1992), and Taxon 42: 815 (1993).
Croton subgen. *Julocroton* (Mart.) Radcl.-Sm. & Govaerts, Kew Bull.: 52: 184 (1997).

Like 258. *Croton* in many respects, but differing in having the male flowers sometimes slightly zygomorphic with the larger sepals appendiculate, the female flowers strongly zygomorphic with the 5 calyx-lobes very markedly heterogeneous, namely 3 large, multifid to laciniate-fimbriate, and 2 small, few-lobulate to entire or else almost obsolete, and in having an asymmetric female disc. Fig.: See Étude Euphorb.: t. 2, f. 23–24 (1858); Fl. Bras. 11(2): tt. 40, 41 (1873).

A neotropical genus of 25–50 poorly-characterized species occurring in most C. and S. American countries, but with the greatest concentration in Brazil.

There is no general agreement on species-boundaries, since the compact and often densely-indumented inflorescence serves to make the proper evaluation of taxonomically significant characters extremely difficult.

Tribe 46. **RICINODENDREAE** *(Pax) Hutch.*, Amer. J. Bot. 56: 749 (1969); G. L. Webster, Ann. Missouri Bot. Gard. 81: 112 (1994). T.: *Ricinodendron* Müll. Arg.

Cluytieae subtribe *Ricinodendrinae* Pax, Nat. Pflanzenfam. ed. 1, 3(5): 87 (1890).

Dioecious trees or shrubs; latex clear, viscous, indurating and obfuscating when desiccated; indumentum stellate; leaves alternate, elobate to palmatilobate, palmatisect or palmate; stipules entire, lobate or 0. Inflorescences axillary, or female terminal, paniculate, fasciculate or flowers solitary; male sepals 4–5, imbricate; petals 5, ± coherent into a tube; disc dissected or lobate; stamens (3)5–20, filaments connate below; pistillode 0; female sepals 4–5, imbricate; petals 5, coherent; disc lobate; ovary 1–3-locular; styles 2-fid. Fruit drupaceous, 1–3-seeded. Seeds ecarunculate; endosperm copious, oily.

A trigeneric palaeotropical tribe in Africa, Madagascar and India. An isolated group which has been considered related to the *Jatropheae*, but the similarities are only superficial.

Key to the genera

1. Leaves elobate to 5-lobate; stipules flabelliform or 0 · · · · · · · 262. *Givotia*
1. Leaves palmatisect or palmate · 2.
2. Stipules flabelliform, persistent; ovary mostly 2–3-locular; endocarp smooth, thin-walled · 263. *Ricinodendron*
2. Stipules small, elobate, deciduous; ovary mostly 1-locular; endocarp pitted, thick-walled · 264. *Schinziophyton*

262. **Givotia** *Griff.*, Calcutta J. Nat. Hist. 4: 88 (1844); Wight, Icon. Pl. Ind. Orient. 5: 24, t. 1889 (1852); Baill., Étude Euphorb.: 389 (1858); Müll. Arg. in DC., Prodr. 15(2): 1112 (1866); Benth., Gen. Pl. 3: 297 (1880); Hook. f., Fl. Brit. India 5: 395 (1887); Pax in Engl. & Prantl, Pflanzenfam. ed. 1, 3(5): 87 (1890); Trimen, Fl. Ceyl. 4: 50 (1898); Pax & K. Hoffm., Pflanzenr. 47: 44 (1911); Gamble, Fl. Madras: 1341 (1925); Pax & K. Hoffm., Nat. Pflanzenfam. ed. 2, 19C: 173 (1931); Radcl.-Sm., Kew Bull. 22: 497 (1968), and Fl. Trop. E. Afr. Euph. 1: 329 (1987); G. L. Webster, Ann. Missouri Bot. Gard. 81: 112 (1994); Philcox in Dassan., ed., Fl. Ceyl. XI: 103 (1997). T.: *G. rottleriformis* Griff. [=*G. moluccana* (L.) Sreem. (B.: *Croton moluccanus* L., p.p.)]

Dioecious trees or shrubs; wood soft. Indumentum stellate, dense. Leaves alternate, petiolate, stipulate or exstipulate, simple or lobate, entire or sinuate- to repand-glandular-dentate, broadly ovate, cordate, palminerved; petiole 2-glandular; stipules foliaceous, flabelliform or 0. Inflorescences pseudopaniculate, ultimate ramifications cymulose, male axillary in the upper leaf-axils, multiflorous, female terminal, pauciflorous, or flowers solitary; bracts filiform-subulate. Male flowers shortly pedicellate to subsessile; buds globose; calyx-lobes 5, subequal, broad, imbricate; petals (4)5, free at first, later becoming partially adnate, exceeding the calyx; disc-glands 5, extrastaminal, free or connate, large, lobulate, or disc annular, thick, 5-lobate; stamens (3)5–15(25), 1- or 2-seriate, outer filaments, where present, ± free, inner connate at the base or to $^{1}/_{2}$-way, pilose at the junction, anthers dorsifixed, introrse, muticous, 2-thecate, thecae longitudinally dehiscent, connective broad; pollen typically crotonoid; pistillode 0; receptacle villous. Female flowers pedicellate; calyx-lobes and petals ± as in the male, sometimes slightly larger, and petals soon caducous; disc sinuate-5-lobate; ovary 1–3-locular, ovules 1 per locule; styles 1–3, short, spreading, 2-fid, flattened, stigmas thickened, dark. Fruit drupaceous, subglobose, 1-locular and 1-seeded by abortion; exocarp thick, fleshy; endocarp hardened. Seed globose, ecarunculate; testa osseous; endosperm fleshy; cotyledons broad, flat. Fig.: See Icon. Pl. Ind. Orient 5: 24, t. 1889 (1852); Pflanzenr. 47: 45, t. 15 (1911); Kew Bull. 22: 498, t. 2 and 500, t. 3 (1968); Fl. Trop. E. Afr. Euph. 1: 330, t. 62 (1987).

A tetratypic palaeotropical genus with 1 species in NE. Africa, 2 in W. Madagascar and 1 in S. India and Sri Lanka, *fide* Radcliffe-Smith (1968).

263. **Ricinodendron** *Müll. Arg.*, Flora 47: 533 (1864), and in DC., Prodr. 15(2): 1111 (1866); Benth., Hooker's Icon. Pl. 13: t. 1300 (1879), and Gen. Pl. 3: 297 (1880); Pax in Engl. & Prantl, Pflanzenfam. ed. 1, 3(5): 88 (1890); Pax & K. Hoffm., Pflanzenr. 47: 45 (1911); Hutch. in Fl. Trop. Afr. 6(1): 744 (1912); Pax & K. Hoffm., Nat. Pflanzenfam. ed. 2, 19C: 173 (1931); Mildbr., Notizbl. Bot. Gart. Berlin-Dahlem 12: 516 (1935); Robyns, Fl. Parc Nat. Alb. 1: 473 (1948); Keay (ed.), Fl. W. Trop. Afr., ed. 2, 1: 393 (1958); J. Léonard, Fl. Cong. Rwa.-Bur. 8(1): 116 (1962); J.-G. Adam, Mém. Mus. Natl. Hist. Nat., B, Bot. 20: 507 (1971); Berhaut, Fl. Ill. Sénég. 3: 583 (1975); Radcl.-Sm., Fl. Trop. E. Afr. Euph. 1: 325 (1987); G. L. Webster, Ann. Missouri Bot. Gard. 81: 113 (1994); Radcl.-Sm., Fl. Zamb. 9(4): 294 (1996). T.: *R. africanum* Müll. Arg. [=*R. heudelotii* (Baill.) Pierre ex Heckel subsp. *africanum* (Müll. Arg.) J. Léonard]

Dioecious pachycaul trees. Indumentum stellate, detersible. Leaves alternate, long-petiolate, stipulate, large, palmatipartite to palmatisect; segments sometimes pseudopetiolulate, glandular-denticulate, penninerved; petiole sometimes 2-glandular at the apex; stipules sessile, large, foliaceous, palmatifid, glandular-dentate, persistent. Inflorescences terminal or subterminal, pseudopaniculate, ultimate ramifications cymulose, male larger than female; bracts linear-setaceous. Male flowers shortly pedicellate; buds globose; calyx-lobes 4–5, unequal, imbricate; petals 5, free at base and apex, laterally coherent to form a tube, imbricate; disc-glands 4–6, extrastaminal, free, entire, fleshy; stamens 6–14, 1-seriate, filaments connate at the base, pilose at the junction, anthers dorsifixed, introrse, versatile, muticous, 2-thecate, thecae longitudinally dehiscent, connective narrow; pollen typically crotonoid; pistillode 0. Female flowers pedicellate; calyx-lobes and petals ± as in the male, sometimes slightly larger; disc crenellate, fleshy; ovary 2–3-locular, ovules 1 per locule; styles 2–3, short, 2-fid to -partite, stigmas adaxial. Fruit indehiscent, 2–3-lobate, usually broader than long; pericarp coriaceous; mesocarp fleshy; endocarps 2–3, rarely 1 by abortion, distinct, each subglobose, 1-locular, 1-seeded, thinly woody, smooth. Seed subglobose, ecarunculate; exotesta fleshy; endotesta ligneous, brownish-black; endosperm copious, fleshy or oily; cotyledons broad, flat, palminerved. Fig.: See Hooker's Icon. Pl. 13: t. 1300 (1879); Pflanzenr. 47: 47, t. 16 and 48, t. 17E (1911); Nat. Pflanzenfam. ed. 2, 19C: 173, t. 90C–D (1931); Fl. W. Trop. Afr., ed. 2, 1: 392, t. 137 (1958); Mém. Mus. Natl. Hist. Nat., B, Bot. 20: 510, t. 191 (1971); Fl. Ill. Sénég. 3: 582 (1975); Fl. Trop. E. Afr. Euph. 1: 327, t. 61 (1987); Fl. Zamb. 9(4): 295, t. 59 (1996).

A monotypic tropical African genus ranging from Guinea-Bissau to Tanzania, Mozambique and Angola.

264. **Schinziophyton** *Hutch. ex Radcl.-Sm.*, Kew Bull. 45: 157 (1990); G. L. Webster, Ann. Missouri Bot. Gard. 81: 113 (1994). T.: *Sch. rautanenii* (Schinz) Radcl.-Sm. (B.: *Ricinodendron rautanenii* Schinz).

Like 263. *Ricinodendron*, but indumentum persistent; stipules small, stipitate, elobate, deciduous; leaves digitately 3–7-foliolate, leaflets distinctly petiolulate; male disc-glands 3-lobate or 3-fid; stamens 13–21; female calyx larger; ovary 1(2)-locular; fruit longer than wide, ellipsoid-ovoid; endocarp single, 1(2)-locular, thick-walled, deeply pitted. Fig.: See Pflanzenr. 47: 48, t. 17A–D (1911) (as *Ricinodendron*); Fl. Cong. Rwa.-Bur. 8(1): opp. 120, photos. 1 and 2 (1962) (as *Ricinodendron*); Fl. Zamb. 9(4): 298, t. 60 (1996).

A monotypic tropical African genus found in S. Congo (Kinshasa), S. Tanzania, Mozambique, Malawi, Zambia, Zimbabwe, Botswana, Angola, Namibia and the NW. Transvaal.

The "Manketti" nut; the seed-kernel has a high oil-content and is edible.

Tribe 47. **ALEURITIDEAE** *Hurus.*, J. Fac. Sci. Univ. Tokyo, Sect. 3, Bot. 6: 309 (1954); G. L. Webster, Ann. Missouri Bot. Gard. 81: 113 (1994). T.: *Aleurites* J. R. & G. Forst.

Monoecious, dioecious or polygamous trees or shrubs; laticifers non-articulated, latex scanty or 0; indumentum simple or stellate; leaves alternate, elobate to palmatilobate, penninerved or palminerved, glandular or not; stipules + or 0. Inflorescences terminal or axillary, dichasial-paniculate or cymose-fasciculate; male calyx closed in bud, later splitting into valvate segments; petals mostly 5(6–13), free, imbricate; disc dissected or of intrastaminal segments, or 0; stamens 8–20(100), filaments free; pollen globose, inaperturate, crotonoid; pistillode 0; female sepals and petals as male; disc lobate, dissected or 0; ovary 2–5-locular; styles 2-fid. Fruit drupaceous or tardily dehiscent. Seeds mostly ecarunculate; testa dry; endosperm copious, oily.

A largely palaeotropical tribe of 6 subtribes and 17 genera having links with the *Jatropheae* via *Joannesia*.

KEY TO THE SUBTRIBES

1. Monoecious · 2
1. Dioecious · 3
2. Petals 5, glabrous; stamens 8–20, filaments connate; inflorescence paniculate; leaves palminerved or palmatilobate · · · · · · 47a. *Aleuritinae*
2. Petals 6–13, sericeous; stamens 30–100, filaments free; inflorescence cymose-fasciculate; leaves penninerved · · · · · · · · · · · · 47b. *Garciinae*
3. Petals +, at least in the male flower; leaves not gland-dotted · · · · · · · · · 4
3. Petals 0; indumentum simple, stellate or lepidote; leaves gland-dotted · · 5
4. Indumentum simple or 0; petals free; inflorescence mostly terminal · 47c. *Grosserinae*
4. Indumentum stellate or lepidote; petals often coherent or connate; inflorescence terminal or axillary · · · · · · · · · · · · 47d. *Crotonogyninae*
5. Leaves palminerved; indumentum simple and stellate; male disc +; female calyx not or slightly accrescent; seeds carunculate · · 47e. *Neoboutoniinae*
5. Leaves penninerved; indumentum simple and lepidote; male disc + or 0; female calyx strongly accrescent; seeds ecarunculate · · 47f. *Benoistiinae*

Subtribe 47a. **Aleuritinae** *(Hurus.) G. L. Webster*, Taxon 24: 599 (1975), and Ann. Missouri Bot. Gard. 81: 113 (1994). T.: *Aleurites* J. R. & G. Forst.

Monoecious trees or shrubs; indumentum simple or stellate; leaf-blades palminerved or palmatilobate, glandular at the apex of the petiole. Inflorescences ± paniculate; male calyx-segments 2 or 3; petals 5, free, glabrous; disc-glands 5; stamens 7–20, inner filaments connate; ovary 2- or 3(5)-locular. Fruit drupaceous or dehiscent. Seeds ecarunculate.

A trigeneric subtribe, following Airy Shaw's (1967) treatment of *Aleurites*.

KEY TO THE GENERA

1. Indumentum stellate; stamens 15–20, anthers introrse; ovary 2-locular; fruit drupaceous · 265. *Aleurites*
1. Indumentum stellate, malpighiaceous or simple; stamens 7–12; ovary 3–5-locular; fruit dehiscent · 2
2. Indumentum stellate; inflorescence densely pubescent; bracts conspicuous; anthers extrorse · 266. *Reutealis*
2. Indumentum malpighiaceous or simple; inflorescence not densely pubescent; bracts inconspicuous; anthers introrse · · · · · · 267. *Vernicia*

265. **Aleurites** *J. R. & G. Forst.*, Char. Gen. Pl.: 111, t. 56 (1776); A. Juss., Euphorb. Gen.: 38 (1824); Baill., Étude Euphorb.: 345 (1858); Müll. Arg. in DC., Prodr. 15(2): 722 (1866), and in Mart., Fl. Bras. 11(2): 303 (1874); Benth., Gen. Pl. 3: 292 (1880); Hook. f., Fl. Brit. India 5: 384 (1887); Pax, Pflanzenr. 42: 128 (1910), p.p.; Hutch. in Fl. Trop. Afr. 6(1): 813 (1912), p.p.; Fawc. & Rendle, Fl. Jamaica 4: 314 (1920); Gagnep., Fl. Indo-Chine 5: 290 (1925), p.p.; Pax & K. Hoffm., Nat. Pflanzenfam. ed. 2, 19C: 99 (1931), p.p.; Pojark. in Fl. URSS. XIV: 303 (1949); Alain, Fl. Cuba 3: 80 (1953), p.p.; Hurus., J. Fac. Sci. Univ. Tokyo, Sect. 3, Bot. 6: 309 (1954); Backer & Bakh. f., Fl. Java 1: 477 (1963); Airy Shaw, Kew Bull. 20: 26, 393 (1967); G. L. Webster, J. Arnold Arbor. 48: 342 (1967), p.p.; E. Walker, Fl. Okin.: 644 (1976); Airy Shaw, Kew Bull. 26: 213 (1972), Kew Bull Addit. Ser. 4: 29 (1975), 8: 25 (1980), Kew Bull. 35: 595 (1980), 36: 252 (1981); A. C. Sm., Fl. Vitiensis. Nov. 2: 547 (1981); Correll, Fl. Bah. Arch.: 777 (1982); Airy Shaw, Kew Bull. 37: 5 (1982), and Euph. Philipp.: 4 (1983); Proctor, Kew Bull. Addit. Ser. 11: 532 (1984); Alain, Fl. Español. 4: 78 (1986), p.p.; McPherson & Tirel, Fl. Nouv. Caléd. 14: 83 (1987); Radcl.-Sm., Fl. Trop. E. Afr. Euph. 1: 176 (1987); Alain, Descr. Fl. Puerto Rico 2: 354 (1988); G. L. Webster, Ann. Missouri Bot. Gard. 81: 114 (1994); Radcl.-Sm., Fl. Zamb. 9(4): 299 (1996); Philcox in Dassan., ed., Fl. Ceyl. XI: 116 (1997). T.: *A. triloba* J. R. & G. Forst. [=*A. moluccana* (L.) Willd. (B.: *Jatropha moluccana* L.)]

Camirium [Rumph. ex] Gaertn., Fruct. Sem. Pl. 2: 194, t. 125.ii (1791). T.: *C. cordifolium* Gaertn. [=*A. moluccana* (L.) Willd. (B.: *Jatropha moluccana* L.)]

Telopea Sol. ex Baill., Étude Euphorb.: 345 (1858), pro syn., non R. Br. (1810). T.: *T. perspicua* Sol. ex Seem., pro syn. [=*A. moluccana* (L.) Willd. (B.: *Jatropha moluccana* L.)]

Monoecious or subdioecious evergreen trees. Indumentum stellate. Leaves alternate, long-petiolate, stipulate, simple or palmatilobate, entire or irregularly dentate, palminerved, cuneate, 2-glandular at the base of the lamina; stipules minute, fugacious. Inflorescences terminal, paniculate-thyrsiform, androgynous, protogynous, some cymules with terminal female flowers, others all male; bracts subulate, readily deciduous. Male flowers pedicellate, pedicels slender; calyx closed in bud, globose, later valvately rupturing into 2–3 lobes; petals 5, free, imbricate, longer than the calyx; disc-glands 5, free, alternipetalous; stamens 15–20, 3–4-verticillate, the outer filaments free, oppositipetalous, the inner ± connate into a short column, anthers basifixed, introrse, erect, ovoid, 2-thecate, thecae distinct, parallel, longitudinally dehiscent, connective wide, slightly produced and emarginate at the apex; pollen typically crotonoid; pistillode 0; receptacle conical. Female flowers pedicellate, pedicels short, stout; calyx calyptriform, 2-lobate, splitting laterally; petals as in the male; disc shallowly 5-lobate, inconspicuous; ovary 2-locular, ovules 1 per locule; styles 2, thick, erect, 2-lobate, -fid or -partite. Fruit usually 2-lobate, drupaceous, indehiscent; exocarp fleshy; endocarp thinly woody, 2-locular or 1-locular by abortion. Seeds broadly ovoid, ecarunculate; testa thick, woody; endosperm thick, indurated; embryo straight; cotyledons broad, flat. Fig.: See Euphorb. Gen.: t. 12, f. 36.1–16 (1824); Étude Euphorb.: t. 11, f. 19–20 and t. 12, f. 1–15 (1858); Fl. Bras. 11(2): t. 45 (1874); Pflanzenr. 42: 130, t. 45 (1910); Fl. Indo-Chine 5: 292, t. 30.6–10 (1925); Nat. Pflanzenfam. ed. 2, 19C: 99, t. 50 (1931); Fl. Cuba 3: 81, t. 18 (1953); Fl. Bah. Arch.: 776, t. 324 (1982); Fl. Español. 4: 81, t. 116-5 (1986); Fl. Nouv. Caléd. 14: 85, t. 16 (1987); Fl. Trop. E. Afr. Euph. 1: 177, t. 34 (1987); Descr. Fl. Puerto Rico 2: 356, t. 59-5 (1988); Fl. Zamb. 9(4): 300, t. 61 (1996).

A ditypic palaeotropical genus with 1 widespread species extending from India east to the S. Pacific Islands, the other endemic to Hawai'i, in the strict sense of Airy Shaw (1967).

A. moluccana is the "Candlenut Tree".

266. **Reutealis** *Airy Shaw*, Kew Bull. 20: 394 (1967), and Euph. Philipp.: 43 (1983); G. L. Webster, Ann. Missouri Bot. Gard. 81: 114 (1994). T.: *R. trisperma* (Blanco) Airy Shaw (B.: *A. trisperma* Blanco).

Very like 265. *Aleurites*, but leaves elobate, cordate at the base, inflorescence densely tomentellous, bracts conspicuous, stamens fewer (7–12), 2-verticillate, anthers extrorse, ovary 3–5-locular and fruit dehiscent. Fig.: See Fig. 41, p. 332.

A monotypic genus endemic to the Philippines.

267. **Vernicia** *Lour.*, Fl. Cochinch.: 586 (1790); Hemsl., Hooker's Icon. Pl. 29: t. 2801, 2802 (1906) (as *Aleurites*); Airy Shaw, Kew Bull. 20: 394 (1967), and 26: 349 (1972); Berhaut, Fl. Ill. Séneg. 3: 375 (1975) (as *Aleurites*); Radcl.-Sm., Fl. Trop. E. Afr. Euph. 1: 178 (1987); Grierson & D. G. Long, Fl. Bhut. 1: 791 (1987); G.

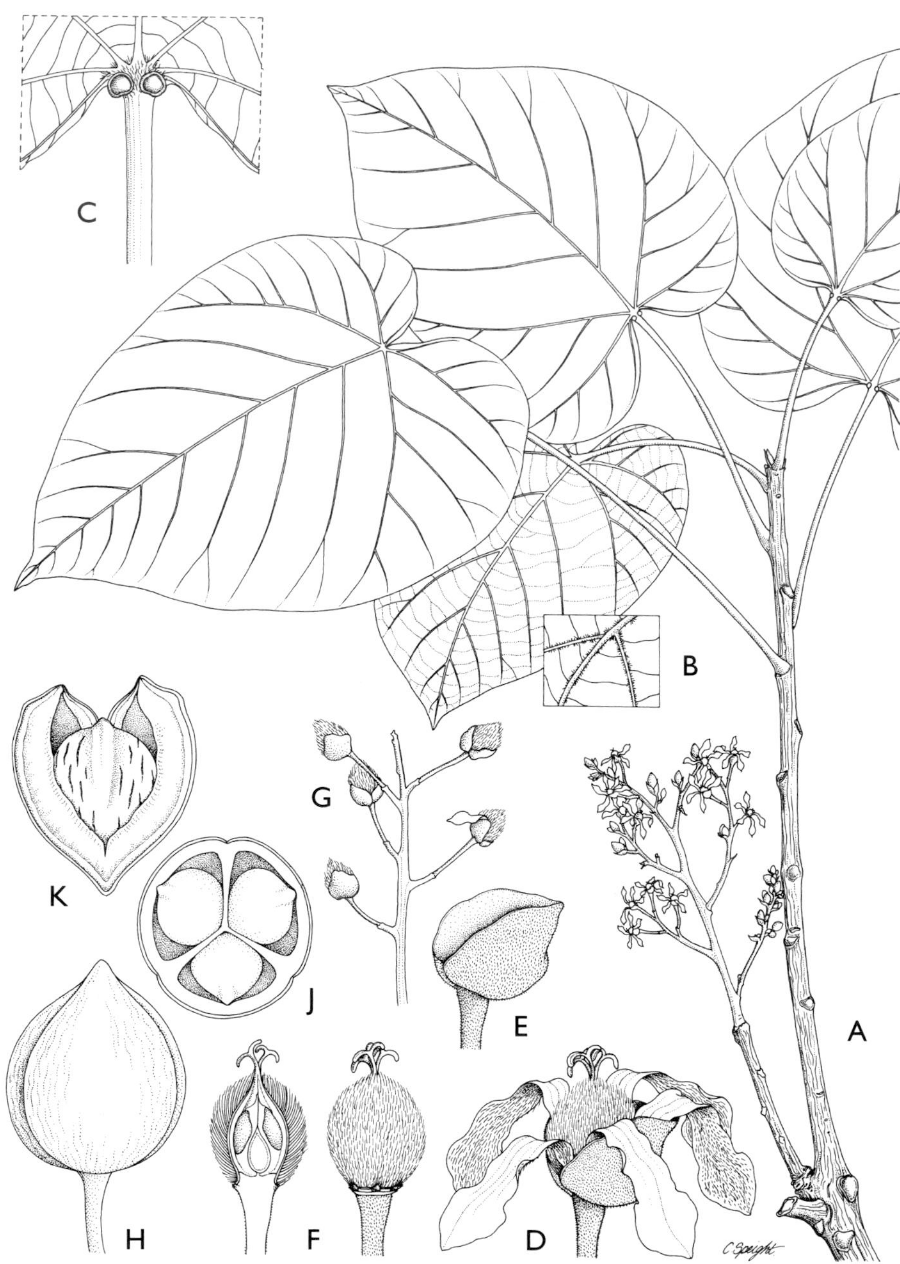

FIG. 41. *Reutealis trisperma* (Blanco) Airy Shaw (B.: *Aleurites trisperma* Blanco). **A** Habit × $^{2}/_{3}$; **B** Portion of leaf-undersurface × 6; **C** Basal leaf-glands × 6; **D** Female flower × 4.5; **E** Persistent calyx-lobe × 4.5; **F** Ovary × 4.5; **G** Portion of young infructescence × $^{2}/_{3}$; **H** Fruit × 1; **J** TS Fruit × 1; **K** Fruit-valve & seed × 1. **A–G** from FB 2440 (*Ahern's Collector* 25668); **H–K** from *Vidal* 3732. Drawn by Camilla Speight.

L. Webster, Ann. Missouri Bot. Gard. 81: 114 (1994); Radcl.-Sm., Fl. Zamb. 9(4): 301 (1996). T.: *V. montana* Lour. (S.: *Aleurites montana* (Lour.) E. H. Wilson)

Dryandra Thunb., Fl. Jap.: 13, t. 27 (1784), non R. Br. (1810) (nom. cons.). T.: *D. cordata* Thunb. [=*V. cordata* (Thunb.) Airy Shaw]
Elaeococca A. Juss., Euphorb. Gen.: 38 (1824); Baill., Étude Euphorb.: 318 (1858). T.: *E. verrucosa* A. Juss. [=*V. cordata* (Thunb.) Airy Shaw + *V. fordii* (Hemsl.) Airy Shaw (B.: *A. fordii* Hemsl.)]

Like 265. *Aleurites* and 266. *Reutealis,* but indumentum malpighiaceous or simple; differs from the former also in its ovoid flower-buds, subulate disc-glands and tardily dehiscent fruit, and from the latter also in having the inflorescences fewer-flowered and not densely tomentellous, the bracts inconspicuous, the flowers much larger and more conspicuous, the petals nervose and the anthers introrse. Fig.: See Fl. Jap.: 13, t. 27 (1784); Euphorb. Gen.: t. 11, f. 35.1–12 (1824); Étude Euphorb.: t. 12, f. 35–38 (1858) (as *Elaeococca*); Hooker's Icon. Pl. 29: t. 2801, 2802 (1906) (as *Aleurites*); Fl. Indo-Chine 5: 292, t. 30.1–5 (1925) (as *Aleurites*); Fl. Ill. Sénég. 3: 374 (1975) (as *Aleurites*); Fl. Trop. E. Afr. Euph. 1: 180, t. 35 (1987); Fl. Zamb. 9(4): 302, t. 62 (1996).

A tritypic genus from E. Asia, ranging from Myanmar and N. Thailand through Indochina and China to Japan.

Vernicia is found in more temperate latitudes or elevations than 265. *Aleurites* or 266. *Reutealis.*

The "Tung Oil Tree" (*V. fordii* (Hemsl.) Airy Shaw) and the "Wood Oil Trees" (*V. montana* Lour. and *V. cordata* (Thunb.) Airy Shaw) all yield a seed-oil with a variety of applications.

Subtribe 47b. **Garciinae** *Müll. Arg.*, Linnaea 34: 143 (1865), and in DC., Prodr. 15(2): 719 (1866); Pax & K. Hoffm., Pflanzenr. 68. XIV (Add. VI): 9 (1919); G. L. Webster, Taxon 24: 600 (1975), and Ann. Missouri Bot. Gard. 81: 114 (1994). T.: *Garcia* Vahl ex Rohr.

Monoecious trees; indumentum simple; leaves exstipulate, entire, penninerved, eglandular. Inflorescences terminal, cymose-fasciculate; petals 6–13, sericeous; male disc dissected, intrastaminal; stamens 30–100, filaments free; ovary 3-locular, styles 2-fid. Fruit dehiscent. Seeds ecarunculate.

Although Müller (1866) included 3 other genera in his subtribe *Garciinae* (*Aleurites, Crotonogyne* and *Manniophyton*), Pax & K. Hoffmann (1919) removed them, leaving only *Garcia,* a course also followed by Webster (1975, 1994).

268. **Garcia** *Vahl ex Rohr,* Skr. Naturhist.-Selsk. 2: 217, t. 9 (1792); A. Juss., Euphorb. Gen.: 41 (1824); Baill., Étude Euphorb.: 392 (1858); Müll. Arg. in DC., Prodr. 15(2): 721 (1866); Benth., Gen. Pl. 3: 292 (1880); Pax, Pflanzenr. 42: 14 (1910); Pax & K. Hoffm., Nat. Pflanzenfam. ed. 2, 19C: 100 (1931); Lundell, Wrightia 1: 1 (1945); G. L. Webster, Ann. Missouri Bot. Gard. 54: 238

(1968), and 81: 114 (1994); R. A. Howard, Fl. Lesser Antilles 5: 50 (1989); A. H. Gentry, Woody Pl. NW. S. Amer.: 424 (1993). T.: *G. nutans* Vahl ex Rohr.

Monoecious trees. Latex scanty. Indumentum simple. Leaves alternate, long-petiolate, exstipulate, entire, penninerved, eglandular. Inflorescences terminal, cymose-fasciculate, of several male and 1–2 female flowers. Flowers large. Male flowers pedicellate; calyx closed in bud, globose, later rupturing into 2–3 persistent valvate segments; petals 6–13, narrow, densely long-sericeous, exceeding the calyx; disc-glands free or connate at the base, glabrous; stamens 30–100, filaments free, anthers dorsifixed, introrse, erect, thecae distinct, longitudinally dehiscent; pollen globose, inaperturate, crotonoid, with massive hexagonal processes; pistillode 0; receptacle convex, pilose. Female flowers pedicellate, pedicels thickening considerably in fruit; calyx caducous, but otherwise ± as in male; petals often fewer than in male, quickly caducous; disc deeply lobate; ovary 3-locular, ovules 1 per locule; styles short, stigmas thick, broad, spreading-reflexed, 2-fid or emarginate. Fruit large, dehiscing into 3 2-valved cocci; endocarp woody; columella stout, capitate and lobulate at the apex. Seeds ovoid-subglobose, ecarunculate; endosperm copious. Fig.: See Skr. Naturhist-Selsk. 2: 217, t. 9 (1792); Euphorb. Gen.: t. 13, f. 40.1–6 (1824); Étude Euphorb.: t. 14, f. 28–38 (1858); Pflanzenr. 42: 15, t. 3 (1910); Nat. Pflanzenfam. ed. 2, 19C: 100, t. 51 (1931); Ann. Missouri Bot. Gard. 54: 239, t. 6 (1968); Fl. Lesser Antilles 5: 51, t. 15 (1989); Woody Pl. NW. S. Amer.: 418, t. 120.2 (1993).

A ditypic neotropical genus, with one species in E. Mexico, the other ranging from Mexico south to Colombia.

Subtribe 47c. **Grosserinae** *G. L. Webster*, Taxon 24: 600 (1975), and Ann. Missouri Bot. Gard. 81: 114 (1994). T.: *Grossera* Pax.

Dioecious or rarely monoecious trees or shrubs; indumentum simple or 0; leaf-blades penninerved or 3-plinerved, glandular or eglandular; stipules deciduous or 0. Inflorescences terminal or axillary, racemose or paniculate; male petals 4–5, free; disc-glands 5; stamens 6–40, filaments free or connate; female petals 4–5, free; ovary 3–5-locular; styles 2-fid. Fruit dehiscent. Seeds usually ecarunculate.

A octogeneric mostly Afro-Malagasy subtribe, but including 2 neotropical genera, and 1 genus from Borneo.

Key to the genera

1. Leaves mostly 3-plinerved; male disc dissected; pollen echinate · · · · · · · 2
1. Leaves penninerved; pollen processes rounded to obtuse · · · · · · · · · · · 4
2. Female sepals 2–3; male receptacle pilose · · · · · · · · · · · · 276. *Domohinea*
2. Female sepals 4–5, imbricate; male receptacle glabrous · · · · · · · · · · · · 3
3. Female petals exceeding the calyx · · · · · · · · · · · · · · · · · · 274. *Tannodia*

3. Female petals exceeded by the calyx · · · · · · · · · · · · · · · · 275. *Neoholstia*
4. Male petals glabrous; leaves pellucid-punctate · · · · · · · · · · · · · · · · · · · 5
4. Male petals pubescent; leaves not or only sparingly pellucid-punctate · · 6
5. Bracts large, imbricate, pre-anthetically strobiliform, deciduous; leaves entire; stipular scars conspicuous, subannular; female sepals unthickened · 269. *Cavacoa*
5. Bracts small, not pre-anthetically strobiliform, persistent; leaves denticulate; stipular scars very small; female sepals medially thickened · · 270. *Grossera*
6. Inflorescences terminal or subterminal; male disc dissected; stamens >20; fruit tomentose; seeds carunculate · · · · · · · · · · · · · · · 273. *Sandwithia*
6. Inflorescences axillary; male disc annular; seeds ecarunculate · · · · · · · ·7
7. Stamens 6–8; fruit tomentose · 271. *Tapoïdes*
7. Stamens >20; fruit glabrous · 272. *Anomalocalyx*

269. **Cavacoa** *J. Léonard*, Bull. Jard. Bot. État. 25: 320 (1955), and Fl. Cong. Rwa.-Bur. 8(1): 191 (1962); Elffers & P. Taylor, Hooker's Icon. Pl. 36: t. 3561 (1956); Radcl.-Sm., Fl. Trop. E. Afr. Euph. 1: 174 (1987); G. L. Webster, Ann. Missouri Bot. Gard. 81: 115 (1994); Radcl.-Sm., Fl. Zamb. 9(4): 304 (1996). T.: *C. quintassii* (Pax & K. Hoffm.) J. Léonard (B.: *Grossera quintassii* Pax & K. Hoffm.)

Grossera sect. *Racemiformes* Pax & K. Hoffm., Pflanzenr. 57: 108 (1912).
Grossera subgen. *Eugrossera* Cavaco, Bull. Mus. Natl. Hist. Nat., Sér. 2, 21: 274 (1949), p.p.
Grossera subgen. *Quadriloculastrum* Cavaco, *l.c.* (1949).

Dioecious trees or shrubs. Indumentum simple. Leaves alternate, petiolate, stipulate, simple, entire, penninerved, minutely pellucid-punctate; petiole eglandular at the apex; stipules quickly deciduous, leaving conspicuous subannular scars. Inflorescences terminal, racemose, strobiliform at first; bracts broad, quickly deciduous. Male flowers long-pedicellate, pedicels capillary, articulate near the base; calyx closed in bud, subglobose, later splitting into 2–3(4) unequal valvate lobes; petals 4–5, free, imbricate; disc-glands 4–5(6), free, alternipetalous, fleshy; stamens 15–30, erect, filaments free, inserted on the receptacle, anthers dorsifixed, extrorse, subcordate, 2-thecate, thecae soleiform, adnate to the connective, longitudinally dehiscent, connective broad, thick; pollen typically crotonoid; pistillode 0; receptacle convex or columnar. Female flowers long-pedicellate, pedicels stout, articulate near the base; sepals and petals 4–5, free, imbricate, not thickened; disc cupular, fleshy; ovary 3–5-locular, ovules 1 per locule; styles 3–5, connate at the base, 2–3-fid. Fruit 3–5-lobate, dehiscing into 3–5 2-valved cocci; exocarp thin, crustaceous; endocarp thick, woody; columella large, persistent. Seeds subglobose, ecarunculate; testa smooth, somewhat shiny, dark brown, marmorate, crustaceous. Fig.: See Hooker's Icon. Pl. 36: t. 3561 (1956); Bull. Jard. Bot. État. 25: 321, 322, tt. 53, 54 (1955); Fl. Cong. Rwa.-Bur. 8(1): 192, t. 4 and 193, fig. 16 (1962); Fl. Trop. E. Afr. Euph. 1: 175, t. 33 (1987); Fl. Zamb. 9(4): 305, t. 63 (1996).

A tritypic genus from tropical and S. subtropical Africa.

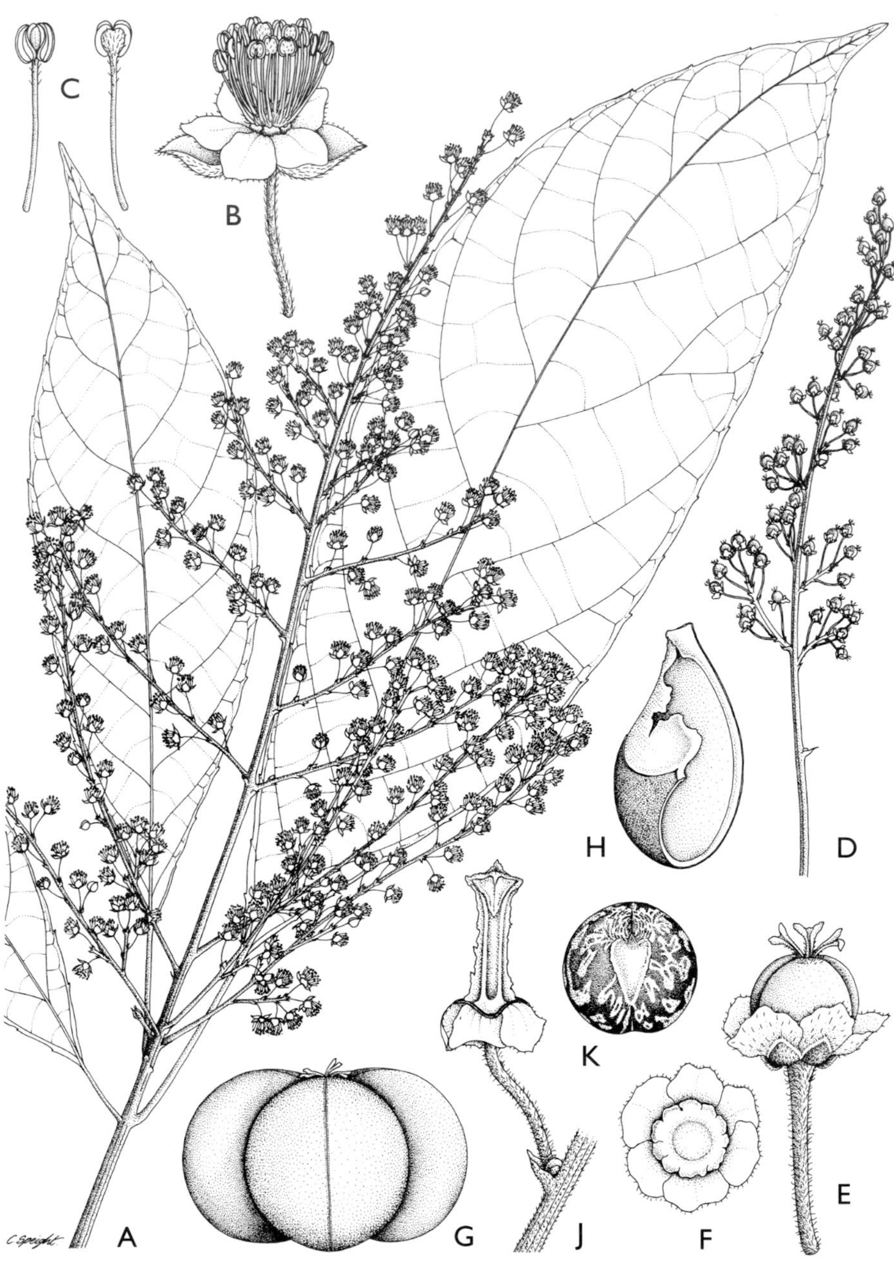

FIG. 42. *Grossera vignei* Hoyle. **A** Habit, male × 1; **B** Male flower × 6; **C** Stamens × 15; **D** Female inflorescence × 1; **E** Female flower × 6; **F** Female petals & disc × 6; **G** Fruit × 4.5; **H** Fruit- valve × 4.5; **J** Columella × 4.5; **K** Seed × 4.5. **A–C** from *Vigne* 4094; **D–F** from *Vigne* 4343; **G** from *Lock & Hall* in *GC* 44150; **H–K** from *Bamps* 2363. Drawn by Camilla Speight.

270. **Grossera** *Pax*, Bot. Jahrb. Syst. 33: 281 (1903); Pax & K. Hoffm., Pflanzenr. 57: 105 (1912), and Nat. Pflanzenfam. ed. 2, 19C: 97 (1931); Léandri, Bull. Soc. Bot. France 85: 524 (1938); Cavaco, Bull. Mus. Natl. Hist. Nat., Sér. 2, 21: 272 (1949); J. Léonard, Bull. Jard. Bot. État. 25: 316 (1955), and 28: 118 (1958); Keay (ed.), Fl. W. Trop. Afr., ed. 2, 1: 398 (1958); J. Léonard, Fl. Cong. Rwa.-Bur. 8(1): 188 (1962); G. L. Webster, Ann. Missouri Bot. Gard. 81: 115 (1994). LT.: *G. paniculata* Pax (designated by J. Léonard, 1955).

Dioecious or monoecious trees or shrubs. Indumentum simple. Leaves alternate, petiolate, stipulate, simple, denticulate or glandular-denticulate, penninerved, minutely pellucid-punctate; petiole very variable in length, sometimes glandular at the apex; stipules small, deciduous, leaving small scars. Inflorescences terminal or sometimes axillary, males usually broadly paniculate, rarely spicate-glomerulate, females paniculate, racemose, or spicate-fasciculate; bracts small, glandular, not strobiliform, persistent; bracteoles 2, minute. Male flowers pedicellate; calyx closed in bud, ovoid, apiculate, later splitting into 2–3(4) valvate lobes, membranaceous; petals 5, free, imbricate, exceeding the calyx; disc-glands 5, free, alternipetalous; stamens 13–40, erect, filaments irregularly connate at the base, inserted on the receptacle, anthers dorsifixed, extrorse, subcordate, 2-thecate, thecae soleiform, adnate to the connective, longitudinally dehiscent, connective broad, thick; pollen typically crotonoid; pistillode 0; receptacle convex or columnar. Female flowers pedicellate; sepals (4)5, free, imbricate, basally and medially thickened; petals (4)5(6), free, imbricate; disc cupular; ovary 3-locular, ovules 1 per locule; styles 3, short, 2-partite. Fruit depressed-3-lobate, often humerate, shiny, dehiscing into 3 2-valved cocci; exocarp thin, not separating; endocarp thinly woody; columella short, 3-alate, persistent. Seeds subglobose, ecarunculate; testa smooth, somewhat shiny, dark brown, marmorate, crustaceous. Fig.: See Pflanzenr. 57: 107, t. 21 (1912). See also Fig. 42, p. 336.

An octotypic tropical W. and C. African and Malagasy genus.

271. **Tapoïdes** *Airy Shaw*, Kew Bull. 14: 473 (1960), Hooker's Icon. Pl. 37: t. 3632 (1967), and Kew Bull. Addit. Ser. 4: 200 (1975); G. L. Webster, Ann. Missouri Bot. Gard. 81: 115 (1994). T.: *T. villamilii* (Merr.) Airy Shaw (B.: *Ostodes villamilii* Merr.)

Dioecious tree. Indumentum, when present, simple. Leaves alternate, crowded towards the apices of the branches, 2-morphic, inflorescence-leaves smaller than those of the vegetative shoots, petiolate, stipulate, simple, entire, penninerved, not pellucid-punctate; petioles weakly 2-pulvinate, eglandular; stipules of inflorescence-leaves small, subulate, patent or reflexed, deciduous, of the vegetative leaves 0. Inflorescences axillary, crowded towards the apices of the branches, irregularly thyrsiform, males laxly many-flowered, females 1–3-flowered; bracts subulate, linear or narrowly spatulate. Male flowers pedicellate; calyx closed in bud, subglobose, later splitting into 3 unequal imbricate 1–2-lobulate nervose lobes; petals 5, free, patent, venose, subentire, densely pubescent within at the base, white; disc annular, fleshy, glabrous, irregularly lobate; stamens 6–8, 2-seriate (5 + 1–3), filaments shortly connate at the base, short, thick, pubescent, anthers basifixed, latrorse, subcordate, 2-thecate, thecae ovoid-subrotund, laterally dehiscent, connective thick, pigmented; pollen not known; pistillode 0. Female flowers pedicellate; sepals larger than the male, patent-reflexed and persistent, otherwise similar; petals, disc, ovary and styles not known. Fruit 3-lobate, rugulose, densely tomentellous, dehiscing septicidally into 3 2-valved cocci or loculicidally into 6 valves; exocarp thin, not separating; endocarp thick, woody. Seeds ovoid, ecarunculate; testa smooth, dull, dark brown, crustaceous. Fig.: See Hooker's Icon. Pl. 37: t. 3632 (1967).

A monotypic genus from Borneo (NE Sarawak, Sabah).

272. **Anomalocalyx** *Ducke*, Notizbl. Bot. Gart. Berlin-Dahlem 11: 344 (1932), and Arch. Jard. Bot. Rio de Janeiro 6: 60 (1933); Secco, Rev. Gen. *Euph.-Crot.* Amer. Sul: 39 (1990); G. L. Webster, Ann. Missouri Bot. Gard. 81: 115 (1994). T.: *A. uleanus* (Pax) Ducke (B.: *Cunuria uleana* Pax).

Very like 271. *Tapoïdes*, but leaves very sparingly pellucid-punctate beneath, flowers rather larger, male calyx fused for 7/8 of its length at first, but later splitting as in 271. *Tapoïdes*, stamens more than 20, female calyx more shallowly-lobate and fruit glabrous. Fig.: See Bol. Mus. Paraense Emilio Goeldi, N.S. Bot. 3: 165–170, tt. 5A–C, 6B and 8 (1987). See also Fig. 43, p. 339.

A monotypic genus of Amazonian Brazil, apparently only in the vicinity of Manáus [Manáos].

According to Webster (1994), this genus links the *Aleuritideae* with the *Codiaeae* via *Dodecastigma* and *Sagotia.*

273. **Sandwithia** *Lanj.*, Bull. Misc. Inform., Kew 1932: 184 (1933); Jabl., Mem. New York Bot. Gard. 17: 152 (1967); Secco, Bol. Mus. Paraense Emilio Goeldi, N.S. Bot. 3: 157 (1987), and 4: 177 (1988); G. L. Webster, Ann. Missouri Bot. Gard. 81: 115 (1994); Murillo-Aldana & Franco-Rosselli, Euf. Reg. Ararac.: 145 (1995). T.: *S. guianensis* Lanj.

Differing from 271. *Tapoïdes* and 272. *Anomalocalyx* in being either dioecious or monoecious, in having terminal or subterminal inflorescences, dissected male disc and caruncalate seeds, but resembling the former in its eglandular leaves and tomentose fruit, and the latter in having more than 20 stamens. Fig.: See Bol. Mus. Paraense Emilio Goeldi, N.S. Bot. 3: 163–164, tt. 3–4 and 171–172, tt. 11–14 (1987); 4: 182–184, tt. 3–5 (1988); Euf. Reg. Ararac.: 147, t. 43 (1995).

A ditypic S. American genus from Venezuela, Guyana and N. Brazil.

Secco (1988) regards this genus as being most closely related to *Sagotia.*

274. **Tannodia** *Baill.*, Adansonia I. 1: 251 (1861); Müll. Arg. in DC., Prodr. 15(2): 728 (1866); Benth., Gen. Pl. 3: 304 (1880); Prain, J. Bot. 50: 125

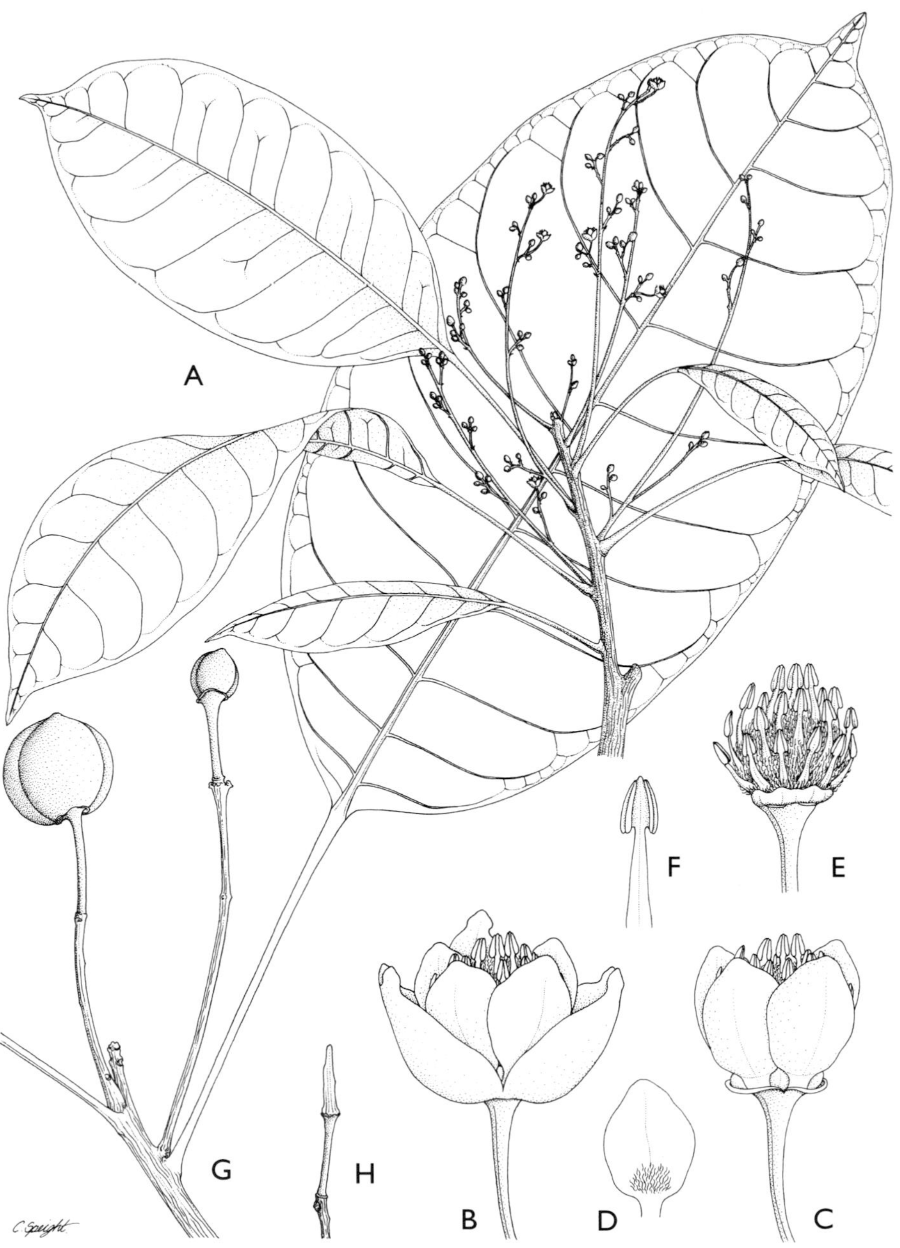

FIG. 43. *Anomalocalyx uleanus* (Pax) Ducke (B.: *Cunuria uleana* Pax). **A** Habit, male × 1/2; **B** Male flower × 6; **C** Male flower with calyx removed × 6; **D** Male petal × 6; **E** Male flower with calyx and petals removed × 6; **F** Stamen × 14; **G** Habit, young fruiting branch × 1; **H** Columella × 1. **A–F** from *Ducke* 23518; **G** from *Ducke* 23517; **H** from *Ule* 8905. Drawn by Camilla Speight.

(1912), and in Fl. Trop. Afr. 6(1): 826 (1912), p.p.; Pax & K. Hoffm., Pflanzenr. 57: 110 (1912), and Nat. Pflanzenfam. ed. 2, 19C: 97 (1931); J. Léonard, Bull. Jard. Bot. État. 25: 300 (1955), and Fl. Cong. Rwa.-Bur. 8(1): 186 (1962); Radcl.-Sm., Fl. Trop. E. Afr. Euph. 1: 172 (1987); G. L. Webster, Ann. Missouri Bot. Gard. 81: 115 (1994) p.p., excl. syn.; Radcl.-Sm., Fl. Zamb. 9(4): 306 (1996), and Kew Bull. 53(1): 173 (1998). T.: *T. cordifolia* (Baill.) Baill. (B.: *Tandonia cordifolia* Baill.)

Tandonia Baill., Adansonia I. 1: 184 (1860), non Moq. (1849). T.: *T. cordifolia* Baill. [=*Tannodia cordifolia* (Baill.) Baill.]

Dioecious or polygamo-dioecious evergreen trees or shrubs. Indumentum simple. Leaves alternate, petiolate, stipulate, simple, entire, mostly palminerved at the base, then penninerved, eglandular; stipules small, soon deciduous. Inflorescences terminal or occasionally also axillary, solitary, interruptedly spicate or racemose; bracts small, glandular, 1–several-flowered; bracteoles minute. Male flowers pedicellate, pedicels articulate near the base; calyx closed in bud, ovoid-subglobose, later valvately 2–5-partite; petals (4)5, free, imbricate, exceeding the calyx, pubescent in the middle within; disc-glands 4–5, free, alternipetalous; stamens 6–14, filaments erect, 2-seriate, the outer short, oppositipetalous, ± free, the inner longer, oppositisepalous, connate into a short column, inserted on the receptacle, anthers subdorsifixed, cordulate, the outer introrse, the inner extrorse, 2-thecate, thecae oblong, longitudinally dehiscent, connective broad; pollen spherical, inaperturate, crotonoid, echinate; pistillode scarcely-developed or 0; receptacle glabrous. Female flowers pedicellate, pedicels articulate near the base; calyx-lobes 4–5, subequal, imbricate; petals 4–5, free, imbricate, much exceeding the calyx; disc annular or shallowly cupuliform, ovary 3-locular, pubescent, ovules 1 per locule; styles 3, shortly connate at the base, 2-fid. Fruit 3-lobate-subglobose, smooth to verrucose or muricate, then puberulous between the warts, pubescent on them, with the wart-hairs in bunches and quasi-stellate, dehiscing septicidally into 3 2-valved cocci; endocarp thinly woody; columella 3-fid, 3-alate. Seeds ovoid, ecarunculate; testa smooth, somewhat shiny, dark brown, crustaceous; endosperm fleshy; cotyledons broad, flat. Fig.: See Adansonia I. 1: t. VII.1, 2 (1861); Fl. Trop. E. Afr. Euph. 1: 173, t. 32 (1987); Fl. Zamb. 9(4): 307, t. 64 (1996).

A tritypic Afro-Malagasy genus, with 2 species in Tropical Africa and 1 in the Comoro Is. and Madagascar.

275. **Neoholstia** *Rauschert,* Taxon 31: 559 (1982); Radcl.-Sm., Fl. Trop. E. Afr. Euph. 1: 169 (1987), and Fl. Zamb. 9(4): 308 (1996). LT.: *N. tenuifolia* (Pax) Rauschert (B.: *Holstia tenuifolia* Pax) (S.: *Tannodia tenuifolia* (Pax) Prain) (chosen by Webster, 1994).

Holstia Pax, Bot. Jahrb. Syst. 43: 220 (1909); Pax & K. Hoffm., Pflanzenr. 57: 108 (1912), and Nat. Pflanzenfam. ed. 2, 19C: 97 (1931). T.: *H. tenuifolia* Pax [=*N. tenuifolia* (Pax) Rauschert]

Tannodia sect. *Holstia* (Pax) Prain, J. Bot. 50: 127 (1912).
Tannodia sect. *Neoholstia* (Rauschert) Radcl.-Sm., Kew Bull. 53: 176 (1998).

Very like 274. *Tannodia*, but buds perulate, leaves membranaceous, sometimes dentate to lobate, deciduous, and with the petals exceeded by the calyx, especially in the female flowers. Fig.: See Pflanzenr. 57: 109, t. 22 (1912); Fl. Trop. E. Afr. Euph. 1: 170, t. 31 (1987); Fl. Zamb. 9(4): 309, t. 65 (1996).

A monotypic genus from SE Tropical Africa (SE Kenya, Tanzania, Mozambique, Malawi, Zambia and Zimbabwe).

276. **Domohinea** Leandri, Bull. Soc. Bot. France 87: 285 (1940). T.: *D. perrieri* Leandri (S.: *Tannodia perrieri* (Leandri) Radcl.-Sm.)

Tannodia subgen. *Domohinea* (Leandri) Radcl.-Sm., Kew Bull. 53: 176 (1998).

Very like 274. *Tannodia*, but male receptacle pilose, and female sepals 2–3. Fig.: See Bull. Soc. Bot. France 87: 280, t. 1 (1940); Kew Bull. 53: 179–185, tt. 1–4 (1998) (as *Tannodia*).

Previously considered a monotypic genus from Madagascar, but 4 new species have recently been recognized (Radcliffe-Smith, 1998).

Subtribe 47d. **Crotonogyninae** *G. L. Webster*, Taxon 24: 600 (1975), and Ann. Missouri Bot. Gard. 81: 115 (1994). T.: *Crotonogyne* Müll. Arg.

Dioecious trees or shrubs, sometimes scandent; indumentum stellate or lepidote; leaves petiolate, stipulate, penninerved or palminerved, 2-glandular at the base of the lamina. Inflorescences mostly axillary, spicate or racemose to paniculate; male petals mostly coherent or connate; disc dissected; stamens 7–40, filaments free or connate; anthers often apiculate; female sepals 4 or 5, imbricate or valvate; disc cupular; ovary 3-locular; styles 2-fid or multifid. Fruit dehiscent. Seeds ecarunculate.

A trigeneric W. African subtribe, formerly placed with *Garcia* (Müller, 1866), but differing considerably from that genus in indumentum, oecy and corollar characters.

Key to the genera

1. Lianas; leaves palminerved; petioles with inflated trichomes; styles 2-fid; male petals connate · 279. *Manniophyton*
1. Trees or shrubs; leaves penninerved; petioles lacking inflated trichomes · · 2
2. Styles 2-fid; inflorescences terminal, paniculate; male petals free; disc receptacular and extrastaminal, of >10 segments · · · · · 277. *Cyrtogonone*
2. Styles multifid; inflorescences axillary, racemose or spicate; male petals mostly coherent or connate; disc extrastaminal, of 5–8 segments · 278. *Crotonogyne*

277. **Cyrtogonone** *Prain*, Bull. Misc. Inform., Kew 1911: 231 (1911); Pax & K. Hoffm., Pflanzenr. 57: 111 (1912); Prain, Hooker's Icon. Pl. 31: t. 3008 (1915); Pax & K. Hoffm., Nat. Pflanzenfam. ed. 2, 19C: 97 (1931); Keay (ed.), Fl. W. Trop. Afr., ed. 2, 1: 399 (1958); G. L Webster, Ann. Missouri Bot. Gard. 81: 116 (1994). T.: *C. argentea* (Pax) Prain (B.: *Crotonogyne argentea* Pax).

Dioecious tree. Latex yellowish. Indumentum silvery-lepidote. Leaves alternate, long-petiolate, minutely stipulate, large, simple, entire, lobulate-dentate or lobate, penninerved; petiole 2-glandular at the apex, lacking inflated trichomes. Inflorescences terminal, paniculate; bracts minute. Male flowers pedicellate; calyx closed in bud, globose, apiculate, later valvately irregularly (2)3(4)-partite; petals 5(6), free, contorted; extrastaminal disc-glands c. 10, receptacular glands ∞, resembling sterile sessile anthers, carinate, stellate-pubescent; stamens c. 30, 2-seriate, erect in bud, filaments free, inserted on the receptacle, outer oppositisepalous, anthers dorsifixed, introrse, erect, 2-thecate, thecae longitudinally dehiscent, connective broad; pollen typically crotonoid; pistillode 0; receptacle shallowly convex. Female flowers pedicellate; calyx-lobes 4 or 5, imbricate; petals ± as in the male; disc shallowly cupular; ovary 3-locular, ovules 1 per locule; styles 2-fid. Fruit 3-lobate, large, brown-tomentellous, dehiscing into 3 2-valved cocci. Seeds subglobose, ecarunculate. Fig.: See Pflanzenr. 57: 112, t. 23 (1912); Hooker's Icon. Pl. 31: t. 3008 (1915).

A monotypic W. African genus in Nigeria, Cameroun and Equatorial Guinea.

278. **Crotonogyne** *Müll. Arg.*, Flora 47: 535 (1864), and in DC., Prodr. 15(2): 720 (1866); Benth., Gen. Pl. 3: 305 (1880); Prain in Fl. Trop. Afr. 6(1): 819 (1912); Pax & K. Hoffm., Pflanzenr. 57: 111 (1912); Prain, Hooker's Icon. Pl. 31: t. 3019 (1915); Pax & K. Hoffm., Nat. Pflanzenfam. ed. 2, 19C: 97 (1931); Keay (ed.), Fl. W. Trop. Afr., ed. 2, 1: 399 (1958); J. Léonard, Fl. Cong. Rwa.-Bur. 8(1): 174 (1962); J.-G. Adam, Mém. Mus. Natl. Hist. Nat., B, Bot. 20: 467 (1971); G. L. Webster, Ann. Missouri Bot. Gard. 81: 116 (1994). T.: *Cr. manniana* Müll. Arg.

Neomanniophyton Pax & K. Hoffm., Pflanzenr. 57: 115 (1912). LT.: *N. impeditum* (Prain) Pax (B.: *Crotonogyne impeditum* Prain) (chosen by Webster, 1994).

Dioecious or rarely monoecious shrubs or small trees. Indumentum stellate or lepidote, sometimes also simple. Leaves alternate, very shortly petiolate, stipulate, large, simple, narrowly oblanceolate to obovate-spathulate, entire, penninerved, 2-glandular at the base; petiole lacking inflated trichomes; stipules subpersistent or deciduous. Inflorescences axillary, males interruptedly spicate, bracts many-flowered, females racemose to subpaniculate, few-flowered, bracts 1–few-flowered; bracts small, often 2-glandular; bracteoles minute. Male flowers subsessile; calyx closed in bud, globose, later splitting irregularly into (2)3(4) valvate lobes; petals (4)5(7), imbricate, sometimes free, mostly connate into a wide corolla-tube, contorted; disc-glands 5(8), free or connate, extrastaminal, oppositisepalous, glabrous; stamens 7–19(28), 1–2-seriate, erect in bud, filaments free, elongate, inserted on the receptacle, anthers subbasifixed, subintrorse, erect, cordate, apiculate, 2-thecate, thecae parallel, longitudinally dehiscent, connective very broad; pollen typically crotonoid; pistillode 0; receptacle thick, convex to columnar, glabrous. Female flowers larger than the male, stoutly pedicellate; calyx-lobes (4)5, imbricate, soon open, slightly accrescent, sometimes glandular in the sinuses; petals (4)5(6), free, imbricate or contorted, broad, equalling the calyx; disc cupular, sometimes lobulate, glabrous; ovary 3-locular, stellate, lepidote or setose, ovules 1 per locule; styles 3, free or connate at the base, 2-partite, arms 2-fid, or styles (3)4(8)-fid, segments elongate, filiform. Fruit 3-lobate-subglobose, dehiscing into 3 2-valved cocci; endocarp thinly crustaceous; columella persistent. Seeds oblong, usually ecarunculate. Fig.: See Pflanzenr. 57: 114, t. 24 (1912); Hooker's Icon. Pl. 31: t. 3019 (1915); Nat. Pflanzenfam. ed. 2, 19C: 98, t. 49 (1931); Fl. Cong. Rwa.-Bur. 8(1): 177, t. 14 (1962); Mém. Mus. Natl. Hist. Nat., B, Bot. 20: 468, t. 165 (1971).

A W. African genus of 15 species, ranging from Sierra Leone to CAR, Congo (Kinshasa) and Angola.

279. **Manniophyton** *Müll. Arg.*, Flora 47: 530 (1864), and in DC., Prodr. 15(2): 719 (1866); Benth., Hooker's Icon. Pl. 13: tt. 1267, 1268 (1878), and Gen. Pl. 3: 297 (1880); Prain in Fl. Trop. Afr. 6(1): 818 (1912); Pax & K. Hoffm., Pflanzenr. 57: 120 (1912), and Nat. Pflanzenfam. ed. 2, 19C: 99 (1931); Keay (ed.), Fl. W. Trop. Afr., ed. 2, 1: 400 (1958); J. Léonard, Bull. Jard. Bot. État. 25: 290 (1955), and Fl. Cong. Rwa.-Bur. 8(1): 171 (1962); J.-G. Adam, Mém. Mus. Natl. Hist. Nat., B, Bot. 20: 495 (1971); G. L. Webster, Ann. Missouri Bot. Gard. 81: 116 (1994). T.: *M. africanum* Müll. Arg.

Dioecious woody lianas or climbing shrubs. Indumentum ferrugineous to fulvous, stellate, fasciculate or simple. Leaves alternate, long-petiolate, stipulate, stipellate, simple to deeply 3–5-lobate, entire, palminerved, tertiary nerves araneiform, 3–4-glandular at the base, glands lageniform; petioles with bulbous-based trichomes; stipules soon deciduous; stipels 2–3, subulate. Inflorescences sometimes terminal, mostly axillary in the upper leaf-axils, usually geminate, males paniculate, elongate, females racemiform-subpaniculate; male bracts several-flowered, female fewer. Male flowers sessile to shortly pedicellate; calyx closed in bud, ovoid-subglobose, later splitting into 2–3 valvate entire or 2-fid lobes; petals 5(6), connate into a broad, cupular, sinuate, scarcely-lobate corolla-tube; disc-glands 5(6), oppositisepalous, pilose; stamens 10–20, 2-seriate, outer epipetalous, filaments free, erect in bud, unequal, inserted on the receptacle, anthers dorsifixed, introrse, erect, cordulate, 2-apiculate, 2-thecate, thecae parallel, longitudinally dehiscent, connective broad; pollen typically crotonoid; pistillode 0; receptacle convex, pilose. Female flowers pedicellate; calyx-lobes 5, almost free, short, valvate or imbricate, soon open; petals 5, free, imbricate, much longer than the sepals; disc annular, hirsute or pilose; ovary 3-locular, densely hispid, ovules 1 per locule; styles free or ± so, 2-fid or -partite, arms linear, spreading. Fruit large, deeply 3-lobate or -dymous, dehiscing into 3 2-valved cocci; exocarp thinly crustaceous, not separating; endocarp thick, indurated;

columella 3-quetrous, large, persistent. Seeds compressed-ovoid, ecarunculate; testa crustaceous, shiny, brown, marmorate or not; endosperm fleshy. Fig.: See Hooker's Icon. Pl. 13: tt. 1267, 1268 (1878); Pflanzenr. 57: 122, t. 25 (1912); Fl. Cong. Rwa.-Bur. 8(1): 173, t. 13 (1962); Mém. Mus. Natl. Hist. Nat., B, Bot. 20: 497, t. 183 (1971).

A widespread monotypic W. African genus, ranging from Sierra Leone to CAR, Congo (Kinshasa), Sudan and Angola; also on Principe.

Bark a source of fibre used in Congo for ropes and nets.

Subtribe 47e. **Neoboutoniinae** *(Hutch.) G. L. Webster*, Taxon 24: 600 (1975), and Ann. Missouri Bot. Gard. 81: 116 (1994). T.: *Neoboutonia* Müll. Arg.

Acalypheae subtribe *Mercurialinae* ser. *Neoboutoniiformes* Pax & K. Hoffm., Pflanzenr. 63: 71 (1914).
Tribe *Neoboutonieae* Hutch., Amer. J. Bot. 56: 752 (1969).

Dioecious trees or shrubs; indumentum stellate and simple; leaves elobate, petiolate, stipulate, palminerved; petiole 2-glandular at the apex; stipules persistent. Inflorescences terminal, sometimes also axillary, paniculate; male calyx 2–3-partite; petals 0; disc dissected; stamens 15–40, filaments free, not longer than the anthers, connective glandular; ovary 3-locular; styles 2-fid or -partite. Fruit dehiscent. Seeds carunculate.

A widespread monogeneric African subtribe.

280. **Neoboutonia** *Müll. Arg.*, J. Bot. 2: 336 (1864), and in DC., Prodr. 15(2): 892 (1866); Benth., Hooker's Icon. Pl. 13: tt. 1298, 1299 (1879), and Gen. Pl. 3: 317 (1880); Prain in Fl. Trop. Afr. 6(1): 918 (1912); Pax & K. Hoffm., Pflanzenr. 63: 71 (1914), and Nat. Pflanzenfam. ed. 2, 19C: 109 (1931); Robyns, Fl. Parc Nat. Alb. 1: 451 (1948); Keay (ed.), Fl. W. Trop. Afr., ed. 2, 1: 404 (1958); Radcl.-Sm., Fl. Trop. E. Afr. Euph. 1: 231 (1987); G. L. Webster, Ann. Missouri Bot. Gard. 81: 116 (1994); Radcl.-Sm., Fl. Zamb. 9(4): 310 (1996). T.: *N. africana* Müll. Arg. [=*N. melleri* (Müll. Arg.) Prain (B.: *Mallotus melleri* Müll. Arg.)]

Conceveiba sect. *Convecibea* Müll. Arg., Flora 47: 530 (1864), and in DC., Prodr. 15(2): 897 (1866). T.: *C. africana* Müll. Arg. (S.: *N. africana* (Müll. Arg.) Pax) [=*N. mannii* Benth.]

Dioecious or rarely monoecious shrubs or small trees. Indumentum stellate, sometimes also simple. Leaves alternate, long-petiolate, stipulate, large, simple, elobate, orbicular, cordate, entire, subentire or denticulate, palminerved, tertiary nerves araneiform, minutely pellucid-punctate; petiole 2-glandular at the apex, lacking inflated trichomes; stipules fairly large, somewhat foliaceous, persistent. Inflorescences terminal, paniculate, sometimes also axillary, racemose, males larger and more floribund than the females; bracts minute, males many-flowered, females 1-flowered. Male flowers subsessile or shortly pedicellate; calyx membranaceous, closed in bud, globose, later splitting irregularly into 2–3 valvate lobes; petals 0; disc-glands 8–10, minute, ± connate at the base, extrastaminal; stamens 15–30(40), filaments free, short, inserted on the receptacle, anthers basifixed, introrse, erect, oblong, 2-thecate, thecae parallel, longitudinally dehiscent, connective 1–3-glandular at the apex, glands waxy, soon falling; pollen typically crotonoid; pistillode 0; receptacle shallowly convex. Female flowers shortly pedicellate; calyx-lobes 5(6), lobes imbricate; petals 0; disc annular; ovary 3-locular, densely stellate-pubescent, ovules 1 per locule; styles 3, shortly connate at the base, rigid, 2-partite, recurved-patent. Fruit 3-lobate, dehiscing into 3 2-valved cocci; endocarp woody; columella persistent. Seeds ovoid-ellipsoid to -subglobose, carunculate; testa crustaceous, brown, marmorate or not; caruncle small, adpressed, 2-lobate; endosperm fleshy; cotyledons broad, flat. Fig.: See Hooker's Icon. Pl. 13: tt. 1298, 1299 (1879); Pflanzenr. 63: 73, t. 10 (1914); Fl. Parc Nat. Alb. 1: 453, t. 44 (1948); Fl. Trop. E. Afr. Euph. 1: 233, t. 47 (1987); Fl. Zamb. 9(4): 312, t. 66 (1996).

A tritypic widespread tropical African genus.

Subtribe 47f. **Benoistiinae** *(Radcl.-Sm.) Radcl.-Sm.*, **stat. nov.** T.: *Benoistia* H. Perrier & Leandri.

Tribe *Benoistieae* Radcl.-Sm., Kew Bull. 43: 632 (1988).

Dioecious or polygamo-dioecious trees; indumentum simple and glandular-lepidote; leaves elobate, petiolate, stipulate, penninerved, eglandular; stipules inconspicuous, deciduous or 0. Inflorescences axillary, racemose or paniculate; male calyx 2–3-lobate; petals 0; disc-glands ∞, interstaminal, or 0; stamens 28–30, filaments free, shorter than the anthers, anthers linear, connective glandular; female calyx strongly accrescent; ovary 3-locular; styles shortly 2-lobate. Fruit dehiscent. Seeds ecarunculate.

A monogeneric Malagasy subtribe.

281. **Benoistia** *H. Perrier & Leandri*, Bull. Soc. Bot. France 85: 528 (1938); Radcl.-Sm., Kew Bull. 43: 632 (1988); G. L. Webster, Ann. Missouri Bot. Gard. 81: 117 (1994). LT.: *B. perrieri* Leandri (chosen by Webster, 1994).

Dioecious or polygamo-dioecious trees. Indumentum simple and glandular-lepidote. Leaves alternate, shortly petiolate, stipulate, elobate, entire, elliptic-lanceolate, penninerved, pellucid-punctate, petioles eglandular at the apex; stipules inconspicuous, deciduous or 0. Inflorescences axillary, shortly racemose or paniculate. Male flowers pedicellate; calyx membranaceous, closed in bud, globose or ellipsoid, apiculate, later splitting imperfectly into 2–3 valvate lobes; petals 0; disc-glands ∞, interstaminal, or 0; stamens 28–30, filaments free, much shorter than the anthers, anthers basifixed, introrse, linear, 2-thecate, thecae parallel, adnate to the connective, longitudinally dehiscent, connective 1-glandular at the apex; pollen typically crotonoid; pistillode 0. Female flowers pedicellate, pedicels articulate; calyx-lobes 5–7, imbricate, subequal, strongly accrescent; petals 0; disc annular, thick, irregularly lobate, villous; ovary (2)3(4)-locular, velutinous, ovules 1 per locule; styles (2)3(4), ± free, recurved, shortly 2-lobate or 2-fid, stigmas canaliculate. Fruit ovoid-subglobose, (2)3(4)-gonous or (2)3(4)-lobate, smooth to tuberculate-rugulose, tomentose, dehiscing septicidally into (2)3(4) 2-valved cocci or (4)6(8) valves, or 1-coccous by abortion; exocarp thinly crustaceous, not separating; endocarp thinly woody; columella narrowly obtriangular or capitate, 3-quetrous, subpersistent. Seeds ovoid-ellipsoid, ecarunculate; testa soft, smooth, shiny, dark brown, maculate; endosperm oily; cotyledons thin, broad, cordate at the base; radicle superior. Fig.: See Bull. Soc. Bot. France 85: 525, t. 1.4–7 (1938); Kew Bull. 43: 638, t. 4 (1988).

A tritypic genus endemic to Madagascar.

Subfamily V. **EUPHORBIOIDEAE** *(Boiss.) G. L. Webster,* Taxon 24: 600 (1975), and Ann. Missouri Bot. Gard. 81: 117 (1994). Type: *Euphorbia* L.

Subordo *Euphorbieae* Boiss. in DC., Prodr. 15(2): 3 (1862). Type: *Euphorbia* L.

Subfamily *Sapioideae* Hurus., J. Fac. Sci. Univ. Tokyo, Sect. 3, Bot. 6: 310 (1954). Type: *Sapium* Jacq.

Monoecious or less commonly dioecious trees, shrubs or herbs, rarely scandent; laticifers non-articulate; latex usually whitish, sometimes scanty or apparently 0; indumentum simple or 0, dendritic in *Mabea* and *Dendrothrix*, malpighiaceous in *Rhodothyrsus* ined.; leaves alternate, opposite or verticillate, stipulate or exstipulate, simple, elobate, entire or dentate, usually penninerved, often glandular at the base of the lamina. Inflorescences terminal or axillary, pseudospicate, -racemose or -paniculate, thyrsiform, pseudanthial (cyathium) or pseudopleiochasial (aggregation of cyathia); bracts usually 2-glandular at the base; flowers erect or inclinate in bud; male sepals (1)3–6(8), imbricate, valvate or open in bud, commonly minute or 0; petals and disc 0; stamens 1–20(70), filaments free or connate; pollen oblate to prolate, 3-colporate, colpi usually marginate, perforate-tectate; pistillode + or 0; female sepals (1)3–6, free or connate imbricate or open in bud, sometimes minute or 0; petals and disc 0; ovary 2–3(20)-locular; ovules anatropous; styles mostly elobate, free or commonly connate into a column. Fruit mostly dehiscent. Seeds caruncuIate or not; testa dry or fleshy; endosperm copious.

The circumscription of this subfamily ± corresponds to the tribes *Hippomaneae* and *Euphorbieae* in the sense of Pax & Hoffmann (1931), except that the tribe *Stomatocalyceae*, which they included in the *Gelonieae*, is added, following Webster (1975, 1994).

Key to the tribes

1. Pollen reticulate; inflorescences racemose or paniculate; bracts eglandular; stamens 10–32, free; male calyx 3–8-lobate; styles simple; seeds ecarunculate; dioecious trees or lianes · · · · · · · · · · 48. *Stomatocalyceae*
1. Pollen perforate-tectate; inflorescences mostly spicate, racemose or pseudanthial; bracts mostly glandular or adnate to the rachis; stamens 1–∞, free or connate; monoecious (rarely dioecious) trees, shrubs or herbs, not scandent · 2
2. Inflorescences pseudanthial (cyathial), usually with 1 terminal female flower and 4 or 5 lateral male mono- or dichasia; perianth minute or 0; styles mostly 2-fid; seeds carunculate or ecarunculate · · 52. *Euphorbieae*
2. Inflorescences mostly racemose or spicate (if capitate, then not pseudanthial), styles mostly simple; male calyx usually developed · · · 3
3. Floral bracts eglandular, peltate or adnate to the rachis and covering the flowers; seeds ecarunculate · 51. *Hureae*
3. Floral bracts usually 2-glandular at the base, not peltate or adnate to the rachis; stamens free or connate · 4
4. Flowers inclinate in bud; male calyx open in bud; leaves mostly not spinose-dentate; stamens free or filaments connate; seeds carunculate or ecarunculate · 49. *Hippomaneae*
4. Flowers erect in bud; male calyx closed in bud, splitting valvately into 2 segments; leaves spinose-dentate; filaments and anthers connate; seeds ecarunculate · 50. *Pachystromateae*

Tribe 48. **STOMATOCALYCEAE** *(Müll. Arg.) G. L. Webster*, Taxon 24: 600 (1975), and Ann. Missouri Bot. Gard. 81: 117 (1994). T.: *Stomatocalyx* Müll. Arg. [=*Pimelodendron* Hassk.]

Hippomaneae subtribe *Stomatocalyceae* Müll. Arg., Linnaea 34: 202 (1865), and in DC., Prodr. 15(2): 1142 (1866)

Dioecious trees, shrubs or lianes; latex whitish, often scanty; indumentum usually 0; leaves alternate, elobate, penninerved, eglandular; stipules small and deciduous or 0. Inflorescences axillary or sometimes terminal, racemose or paniculate, bracts 1-flowered, eglandular; flowers erect in bud; male sepals 3–8, free and imbricate or connate and calyx 2-lipped; stamens 10–32, free, filaments short; pollen coarsely reticulate or finely reticulate-perforate; female sepals 5 or 6, free or connate and calyx 2-lipped; ovary (1)2–10-locular; styles elobate, sometimes dilated or stigmatiform. Fruit dehiscent or indehiscent, 1–3-seeded. Seeds ecarunculate; endosperm copious, oily.

This tribe comprises 3 palaeotropical genera and 1 neotropical genus, arranged in 2 subtribes. Some stomatocalyceous characteristics, such as the scandent habit, coloured latex and oily endosperm suggest a possible affinity to the acalyphoid genus *Omphalea* (q.v.).

KEY TO THE SUBTRIBES

1. Pollen finely reticulate-perforate; fruits indehiscent; styles short; ovary (1)2–10-locular · 48a. *Stomatocalycinae*
1. Pollen coarsely reticulate; fruits dehiscent; styles ± elongate; ovary 2–3-locular · 48b. *Hamilcoinae*

Subtribe 48a. **Stomatocalycinae** *Müll. Arg.*, Linnaea 34: 202 (1865), and in DC., Prodr. 15(2): 1142 (1866); G. L. Webster, Ann. Missouri Bot. Gard. 81: 118 (1994). T.: *Stomatocalyx* Müll. Arg. [=*Pimelodendron* Hassk.]

Trees or shrubs; latex scanty, whitish; leaves entire or dentate. Inflorescences simply racemose or slightly branched; male sepals free or connate; stamens 10–20(32); pollen finely reticulate; ovary 1–10-locular; styles stigmatiform. Fruit indehiscent, 1-seeded.

This palaeotropical subtribe consists of 2 genera, 1 W African and 1 Malesian and Australasian.

KEY TO THE GENERA

1. Sepals 6–8, free; ovary 1-locular · · · · · · · · · · · · · · · · · · · 282. *Plagiostyles*
1. Sepals connate, male calyx 2-lipped; ovary 2–10-locular · · 283. *Pimelodendron*

282. **Plagiostyles** *Pierre*, Bull. Mens. Soc. Linn. Paris 2: 1326 (1897); Prain, Bull. Misc. Inform., Kew 1912: 107 (1912), Fl. Trop. Afr. 6(1): 170 (1909), and 1001 (1913); Pax & K. Hoffm., Pflanzenr. 63, Addit. V: 420 (1914); Stapf, Hooker's Icon. Pl. 31: t. 3010 (1915); Pax & K. Hoffm., Nat. Pflanzenfam. ed. 2, 19C: 190 (1931); Keay (ed.), Fl. W. Trop. Afr., ed. 2, 1: 414 (1958); J. Léonard, Fl. Cong. Rwa.-Bur. 8(1): 131 (1962); G. L. Webster, Ann. Missouri Bot. Gard. 81: 118 (1994). T.: *Pl. klaineana* Pierre [=*Pl. africana* (Müll. Arg.) Prain (B.: *Daphniphyllum africanum* Müll. Arg.)]

Dioecious glabrous tree or arborescent shrub. Latex scanty, white. Wood white. Leaves alternate, petiolate, stipulate, simple, entire or subdentate, penninerved, coriaceous, eglandular; petioles pulvinate and geniculate at the apex; stipules fugacious. Inflorescences axillary, solitary; male racemose or subpaniculate, many-flowered; female racemose, few-flowered (often c. 6-flowered); bracts 1-flowered, eglandular; bracteoles minute. Male flowers pedicellate; buds depressed-subglobose; sepals (5)6–8, unequal, free, open in bud at the apex, laterally imbricate; petals 0; disc 0; stamens (10)15–20(32), agglomerated into a capituliform mass, filaments free, short, multiseriate, outer whorl alternisepalous, anthers subsessile, subdorsifixed, extrorse, rounded, 2-thecate, thecae parallel, contiguous, longitudinally dehiscent, connective narrow; pollen oblate-spheroidal, 3-colporate, circular to slightly convex-triangular in polar view, colpus transversalis narrow, colpi narrow, short, costae indistinct, finely intra-reticulate, columellae small; pistillode 0;

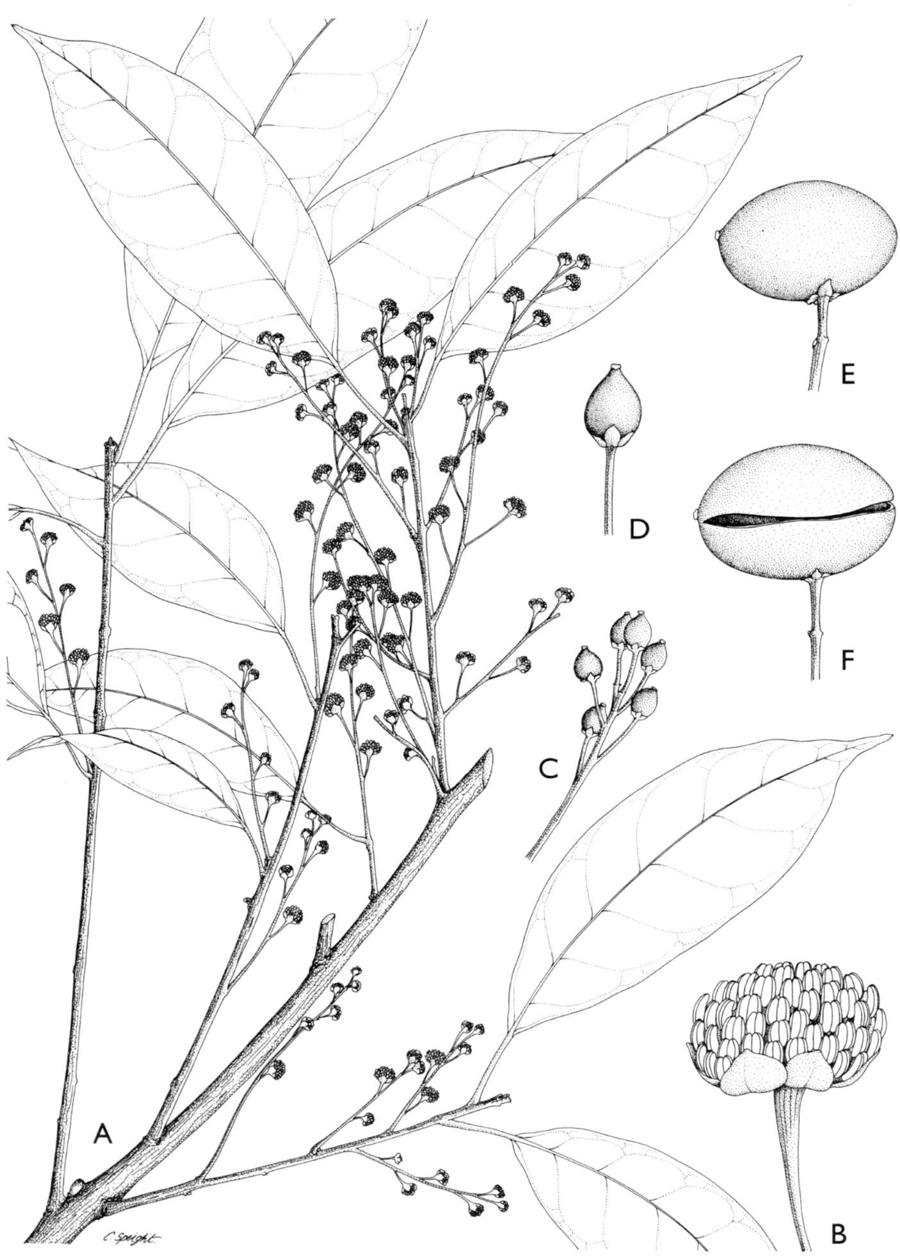

FIG. 44. *Plagiostyles africana* (Müll. Arg.) Prain (B.: *Daphniphyllum africanum* Müll. Arg.). **A** Habit, male × 1; **B** Male flower × 9; **C** Female inflorescence × 1; **D** Young fruit × 2.25; **E** Fruit × 2.25; **F** Fruit showing transverse dehiscence × 2.25. **A, B** from *de Wilde* 8453; **C, D** from *Thomas* 6681; **E** from *Brenan* 9330; **F** from *Etuge & Thomas* 22. Drawn by Camilla Speight.

receptacle ± flat. Female flowers pedicellate; calyx 5-lobate, lobes unequal, thickened at the base, glandular without, imbricate; petals 0; disc 0; ovary 1(2)-locular, locule slightly excentric, ovules 1(2) per locule; styles very short, slightly excentric at first, later lateral, stigma thick, discoid, papillulose, persistent. Fruit transversely oblong, subdrupaceous, indehiscent or tardily dehiscent, 1-seeded; exocarp fleshy-coriaceous; endocarp thinly membranous, adhering to the seed. Seed asymmetrically ellipsoid, large, ecarunculate; exotesta crustaceous, smooth, venose; endotesta spongiform, lacunar; endosperm fleshy, oily, yellowish; embryo transverse; cotyledons reniform, cordate, flat, venose; radicle short. Fig.: See Hooker's Icon. Pl. 31: t. 3010 (1915). See also Fig. 44, p. 349.

A monotypic W African genus ranging from S Nigeria to Gabon and S Congo (Kinshasa).

283. **Pimelodendron** *Hassk.*, Verslagen Meded. Afd. Natuurk. Kon. Akad. Wetensch. 4: 140 (1856); Benth., Gen. Pl. 3: 331 (1880); Hook. f., Fl. Brit. India 5: 468 (1888) (as *Pimeleodendron*); Pax, Pflanzenr. 52: 54 (1912); J. J. Sm., Bull. Jard. Bot. Buitenzorg III. 6: 100 (1924); Pax & K. Hoffm., Nat. Pflanzenfam. ed. 2, 19C: 190 (1931); Airy Shaw, Kew Bull. 25: 551 (1971); Whitmore, Tree Fl. Mal. 2: 124 (1973); Airy Shaw, Kew Bull. Addit. Ser. 4: 186 (1975), 8: 196 (1980), Kew Bull. 35: 666 (1980), 36: 339 (1981), and 37: 34 (1982); G. L. Webster, Ann. Missouri Bot. Gard. 81: 118 (1994). T.: *P. amboinicum* Hassk.

Stomatocalyx Müll. Arg. in DC., Prodr. 15(2): 1142 (1866). T.: *S. griffithianus* Müll. Arg. [=*P. griffithianum* (Müll. Arg.) Benth. ex Hook. f.]
Carumbium sect. *Pimelodendron* (Hassk.) Müll. Arg., DC., Prodr. 15(2): 1143 (1866).

Dioecious tall glabrous trees. Latex scanty, white, becoming yellowish. Leaves alternate, often crowded towards the tips of the branches, petiolate, mostly exstipulate, simple, entire or crenate, penninerved, coriaceous, eglandular; petioles unequal, bipulvinate, geniculate at the apex; stipules minute or 0. Inflorescences racemose, male axillary or extra-axillary, mostly fasciculate, female axillary, mostly solitary; bracts 1-flowered, minute, eglandular. Male flowers shortly pedicellate; buds laterally compressed; calyx 2-lobed, lobes short, broad, flat, subequal, reniform, slightly imbricate; petals 0; disc 0; stamens (10)12(16), filaments free or coherent, very short, anthers subsessile, dorsifixed, extrorse, rounded, 2-thecate, thecae parallel, contiguous, longitudinally dehiscent, connective narrow; pollen spheroidal, not reticulate, otherwise ± as in 282. *Plagiostyles*; pistillode 0; receptacle flat. Female flowers pedicellate; calyx shallowly cupular, 2–3-lobate; petals 0; disc 0; ovary 2–10-locular, shortly cylindric-oblong, ovules 1 per locule; stigmas sessile or subsessile, pulvinate or 2–10-lobulate, slightly broader than the ovary. Fruit ellipsoid, obtusely angulate, drupaceous, indehiscent, 1-seeded; exocarp thin; endocarp thinly crustaceous. Seed depressed-globose, large, half-enclosed by a broad, shallow multilobulate aril; testa radiate-striolate. Fig.: See Pflanzenr. 52: 55, t. 9 (1912); Tree Fl. Mal. 2: 125, t. 12 (1973); Kew Bull. 35: 668, t. 6A (1980).

A tropical Asiatic and Australasian genus of 6–8 species ranging from Malaya to New Guinea, the Bismarcks, Solomon Is. and Queensland.

Subtribe 48b. **Hamilcoinae** *Pax & K. Hoffm.*, Pflanzenr. 63, Addit. V: 419 (1914); G. L. Webster, Ann. Missouri Bot. Gard. 81: 118 (1994). T.: *Hamilcoa* Prain.

Trees or lianas; leaves long-petiolate, entire or crenate. Inflorescences simply racemose or branched; male sepals 4–5, free, imbricate; stamens 10–20(30), free, filaments very short; pollen coarsely reticulate; female sepals 4–6, free; ovary 2–3-locular; styles connate at the base, ± elongate, erect or spreading. Fruit dehiscent.

This subtribe consists of 2 genera, 1 W African and 1 Neotropical. Although included in 2 different subtribes of the *Gelonieae* by Pax & Hoffmann (1931), *Hamilcoa* and *Nealchornea* share similar distinctive male flowers and coarsely reticulate pollen which somewhat resembles that of *Dalechampia* in ornamentation.

KEY TO THE GENERA

1. Stamens 18–20(30); female sepals eglandular; ovary 3-locular; styles erect; plant ± scandent · 284. *Hamilcoa*
1. Stamens 10–15; female sepals with large basal glands; ovary 2-locular; styles spreading; plant not scandent · · · · · · · · · · · · · · · · · · 285. *Nealchornea*

284. **Hamilcoa** *Prain*, Bull. Misc. Inform., Kew 1912: 107 (1912), and Fl. Trop. Afr. 6(1): 1000 (1913); Pax & K. Hoffm., Pflanzenr. 63, Addit. V: 419 (1914); Stapf, Hooker's Icon. Pl. 31: t. 3009 (1915); Pax & K. Hoffm., Nat. Pflanzenfam. ed. 2, 19C: 184 (1931); Keay (ed.), Fl. W. Trop. Afr., ed. 2, 1: 413 (1958); G. L. Webster, Ann. Missouri Bot. Gard. 81: 118 (1994). T.: *H. zenkeri* (Pax) Prain (B.: *Plukenetia zenkeri* Pax).

Dioecious glabrous scandent shrub. Leaves alternate, petiolate, often crowded towards the tips of otherwise leafless branches, stipulate, simple or 2–3-lobate, entire or remotely and obscurely toothed, penninerved, thinly coriaceous, eglandular; petioles unequal, pulvinate and geniculate at the apex; stipules minute, soon deciduous. Inflorescences racemose, solitary, shorter than the laminae, few-flowered, male axillary, female terminal or subterminal; bracts small, 1-flowered, lower infertile; bracteoles minute. Male flowers long-pedicellate, pedicels thickened, asperulous; calyx globose in bud, sepals 5, broad, strongly imbricate, partially asperulous without; petals 0; disc 0; stamens 18–20(30), crowded on the receptacle, filaments very short, broad, free, anthers subsessile, subdorsifixed, outer extrorse, broadly elliptic, smooth, 2-thecate, thecae parallel, contiguous, longitudinally dehiscent; pollen suboblate to oblate-spheroidal and coarsely reticulate, otherwise ± as in 282. *Plagiostyles*; pistillode 0; receptacle hemispherical, fleshy. Female flowers long-pedicellate, pedicels flexuous or nutant, extending in fruit; sepals 6, smaller than in the male, 2-seriate, imbricate, eglandular; petals 0; disc 0; ovary 3-locular, ovules 1 per locule; styles 3, connate into a short, thick column, stigmas free, erect, simple, connivent, oblong, stout, margins recurved, confluent at the base. Fruit deeply 3-coccous, dehiscing loculicidally from the base upwards into 3 or 6 valves; pericarp coriaceous; columella short, 3-quetrous, 3-cornute at the apex, persistent. Seeds large, globose, ecarunculate; exotesta crustaceous, smooth, ochreous, marbled chestnut; mesotesta spongiose, lacunar; endotesta indurated; endosperm fleshy; embryo minute; cotyledons broad, flat. Fig.: See Hooker's Icon. Pl. 31: t. 3009 (1915).

A monotypic W African genus confined to Cameroon.

285. **Nealchornea** *Huber*, Bol. Mus. Paraense Hist. Nat. 7: 297 (1913); Pax & K. Hoffm. in Pflanzenr. 68.XIV (Add.VI): 51 (1919); Ducke, Arch. Jard. Bot. Rio de Janeiro 4: 107 (1925); Pax & K. Hoffm., Nat. Pflanzenfam. ed. 2, 19C: 181 (1931); J. F. Macbr., Publ. Field Mus. Nat. Hist., Bot. Ser. 13(IIIA,1): 180 (1951); B. Walln., Linzer Biol. Beitr. 23(2): 777 (1991); A. H. Gentry, Woody Pl. NW. S. Amer.: 422 (1993); G. L. Webster, Ann. Missouri Bot. Gard. 81: 118 (1994); Murillo-Aldana & Franco-Rosselli, Euf. Reg. Ararac.: 118 (1995). T.: *N. yapurensis* Huber.

Like 284. *Hamilcoa*, but plant not scandent; indumentum +, simple, papillose-puberulous; leaves mostly long-petiolate, elobate; male inflorescences racemose-paniculate; male and female calyx open in bud, lobes 4, slightly imbricate; stamens 10–15, 2-seriate, outer 4 alternisepalous; pollen-apertures large; female sepals with large, basal, sometimes stipitate disc-glands; staminodes +; ovary 2-locular, sometimes stipitate, styles 2, large, recurved-spreading and fruit more tardily dehiscent. Fig.: See Linzer Biol. Beitr. 23(2): 784, t. A1,2 and 785, t. B3–5 (1991); Woody Pl. NW. S. Amer.: 421, t. 121.2 (1993); Euf. Reg. Ararac.: 121, t. 32 (1995).

A ditypic S American genus from Amazonian Colombia, Peru and Brazil, *fide* Wallnöfer (1991).

Tribe 49. **HIPPOMANEAE*** *A. Juss. ex Bartl.*, Ord. Nat. Pl.: 372 (1830); Spach, Hist. Nat. Veg. 2: 488, 522 (1834); Müll. Arg., Linnaea 32: 82 (1863); Benth., J. Linn. Soc., Bot. 17: 239 (1878); in Benth. & Hook. f., Gen. Pl. 3(1): 254 (1880); Pax & K. Hoffm. in Engl., Pflanzenr. 52: 1 (1912); Pax, Bot. Jahrb. Syst. 59: 149 (1924); Pax & K. Hoffm. in Engl. & Harms, Nat. Pflanzenfam. ed. 2, 19c: 187 (1931); Hurus., J. Fac. Sci. Univ. Tokyo, Sect. 3, Bot. 6: 310 (1954); Hutch., Amer. J. Bot. 56: 755 (1969); G. L. Webster, Taxon 24: 600 (1975); Ann. Missouri Bot. Gard. 81: 118 (1994); Esser, Syst. Hipp.: 7 (1994). T.: *Hippomane* L.

---

* by H.-J. Esser, Harvard University Herbaria, 22 Divinity Avenue, Cambridge, MA 02138-2020, U.S.A.

Series *Stillingiidae* Baill., Étude Euphorb.: 509 (1858). T.: *Stillingia* Garden ex L.

Series *Excaecarieae* Baill., Hist. Pl. 5: 156, 227 (1874), pro parte.

Subtribe *Hippomaninae* Pax in Engl. & Prantl, Nat. Pflanzenfam. 3(5): 91 (1890).

Subfamily *Sapioideae* Hurus., J. Fac. Sci. Univ. Tokyo, Sect. 3, Bot. 6: 310 (1954), pro max. parte. T.: *Sapium* Jacq.

Monoecious, very rarely dioecious trees, shrubs, lianas, or herbs. Indumentum present or absent, not stellate. Latex white. Leaves petiolate, stipulate or very rarely exstipulate, solitary or rarely palmately lobed, penninerved and sometimes triplinerved, nearly always glandular above and/or below. Inflorescences terminal or axillary, elongate to glomerulate thyrsal, basal part female, apical part male, bracts glandular to rarely eglandular, basal ones with 1 female flower, apical ones with 1–c. 30 male flowers. Floral buds inclinate, initially usually covered by their large bracts. Male flowers sessile to pedicellate; calyx open in bud, (1)2–5-lobed, actinomorphic or bilateral or zygomorphic, free or fused; stamens 1–c. 100, free to connate. Female flowers sessile to pedicellate; calyx open in bud, (1)2–6(9)-lobed, actinomorphic or bilateral, free or fused; ovary 1–9-locular; stigmas undivided or rarely apically bifid, eglandular or glandular below. Fruits opening septicidally or loculicidally or indehiscent; exocarp dry or fleshy, endocarp woody. Seeds 1–9 per fruit, dry or with aril or with sarcotesta, carunculate or ecarunculate.

Key to the subtribes

1. Calyx of male flowers laterally compressed or zygomorphic-inclinate, with 1 or 2 distinct lobes; female flowers with glandular, apically often bifid stigmas · 49a. *Carumbiinae*
1. Calyx of male flowers radially symmetric or rarely zygomorphic-inclinate, with 2 or more lobes or completely fused; female flowers with eglandular, apically undivided stigmas · · · · · · · · · · · · · · · · · · · 49b. *Hippomaninae*

Subtribe 49a. **Carumbiinae** *Müll. Arg.*, Linnaea 34: 203 (1865) (as *Carumbieae*); in DC., Prodr. 15(2): 1034, 1142 (1866); G. L. Webster, Taxon 24: 600 (1975); Ann. Missouri Bot. Gard. 81: 120 (1994); Esser *et al.*, Syst. Bot. 22: 625 (1998). T.: *Carumbium* Reinw. [=*Homalanthus* A. Juss.]

Subtribe *Homalanthinae* Pax & K. Hoffm. in Engl., Pflanzenr. 52: 42 (1912); in Engl. & Harms, Nat. Pflanzenfam. ed. 2, 19c: 188 (1931); Hurus., J. Fac. Sci. Univ. Tokyo, Sect. 3, Bot. 6: 311 (1954). T.: *Homalanthus* A. Juss.

Monoecious trees or shrubs. Indumentum present or absent, uniseriate. Leaves with large stipules enclosing terminal buds. Inflorescences solitary; bracts of male part with 1–3(8) flowers. Male flowers pedicellate; calyx 2-lobed and bilaterally compressed or 1-lobed and zygomorphic. Female flowers pedicellate; stigmas undivided or apically bifid, glandular below. Seeds with an enlarged caruncle forming an arilloid.

286. **Homalanthus** *A. Juss.*, Euphorb. Gen.: 50 (1824), orth. cons.; Benth. in Benth. & Hook. f., Gen. Pl. 3(1): 254, 331 (1880); Hook. f., Fl. Brit. India 5: 468 (1887); Pax in Engl. & Prantl, Nat. Pflanzenfam. 3(5): 96 (1890); Pax & K. Hoffm. in Engl., Pflanzenr. 52: 42 (1912); in Engl. & Harms, Nat. Pflanzenfam. ed. 2, 19c: 188 (1931); Airy Shaw, Kew Bull. 21: 409 (1968); Radcl.-Sm. in Fl. Trop. East Africa, Euph. 1: 381 (1987); Esser, Taxon 45: 555 (1996), orth. cons. prop.; Philcox in Dassan., ed., Fl. Ceyl. XI: 185 (1997). T.: *H. leschenaultianus* A. Juss. [=*H. populneus* (Geiseler) Pax (B.: *Stillingia populnea* Geiseler)]

Omalanthus A. Juss., Euphorb. Gen.: 50 (1824), orth. rejic.; Baill., Étude Euphorb.: 537 (1858); Backer & Bakh. f., Fl. Java 1: 498 (1963); G. L. Webster, Ann. Missouri Bot. Gard. 81: 120 (1994); P. I. Forst., Telopea 6: 169 (1994); Esser, Blumea 42: 421 (1997). T.: *O. leschenaultianus* A. Juss. [=*H. populneus* (Geiseler) Pax (B.: *Stillingia populnea* Geiseler)]

Carumbium Reinw., Elench. Sem. Hort. Leiden ex Isis 1: 319 (1823) (nomen); Syll. Pl. Nov. 2(1): 6 (1826); Benth., Fl. Austr. 6: 149 (1873); Baill., Hist. Pl. 5: 229 (1874) pro parte excl. *Pimelodendron.* T.: *C. populifolium* Reinw. [=*O. populneus* (Geiseler) Pax (B.: *Stillingia populnea* Geiseler)]

Wartmannia Müll. Arg., Linnaea 34: 218 (1865); in DC., Prodr. 15(2): 1147 (1866). T.: *W. stillingiifolia* (F. Muell.) Müll. Arg. (B.: *O. stillingiifolious* F. Muell.).

Dibrachion Regel, Index Sem. Hort. Bot. Petropol. 1865: 51 (1866), Gartenflora 15: 100 (1866), nom. illeg. [non Tul., Ann. Sci. Nat. Bot. sér. 2, 10: 139 (1843)]. T.: *D. peltatum* Regel. [=*H. fastuosus* (Linden) Fern.-Vill. (B.: *Mappa fastuosa* Linden)]

Carumbium sect. *Eucarumbium* Müll. Arg. in DC., Prodr. 15(2): 1143 (1866), nom. inval.

Monoecious shrubs or trees. Indumentum absent, shortly papillate, or multicellular uniseriate and usually coloured. Latex white. Apical buds enclosed by the conspicuous large stipules. Leaves alternate, with moderately short to very long, eglandular or apically variously glandular petioles, stipulate, simple or shallowly and remotely crenate, peltate at base or not, penninerved, chartaceous to coriaceous, above eglandular or with one or two basilaminar glands, below with a submarginal to laminar row of glands or only basal glands, sometimes more on the petiole than on the lamina, lower surface smooth and shiny or more often whitish-farinose; stipules at least 5 mm long, eglandular, entire, with distinct parallel venation, enclosing late buds and later caducous. Inflorescences terminal, solitary, bracts sometimes elevated by a peduncle, with one to several pairs of disc- to cup-shaped or spheroidal to elliptical glands, rarely eglandular, (0)2–c. 30 basal ones with 1 female flower, numerous apical ones with 1–3(8) male flowers; bracteoles absent. Floral buds inclinate, sexual organs initially covered by the large flattened calyx. Male flowers pedicellate; calyx 1- or 2-lobed, free, bilaterally compressed; petals and disc 0; stamens 4–c. 30, filaments nearly as long as anthers, free, anthers basifixed, extrorse, 2-thecate, longitudinally dehiscent; pollen 3-colporate, hardly to distinctly 3-lobate in polar view, colpus transversalis+, costae+, colpi narrow, tectate-perforate, columellae distinct; pistillode 0; receptacle flat to elevated. Female flowers pedicellate; calyx 1–3-lobed, free, eglandular; petals and disc 0; ovary smooth, glabrous, with 2 or 3 locules, ovules 1 per locule; style short to distinct, stigmas 2 or 3, glandular below, apically often bifid. Fruits pedicellate, 2- or 3-locular, club-shaped or spheroidal, glabrous, dehiscing regularly to irregularly, septicidally or more often loculicidally; pericarp dry or with fleshy exocarp, thinly membranaceous to very thick and woody; columella 2- or 3-quetrous, alate, persistent. Seeds 2 or 3 per fruit, irregularly flattened-elliptic, irregularly foveolate, reddish-brownish, with fleshy arillus partly to completely covering the seed. Fig.: See Euphorb. Gen.: t. 16, f. 53.1–11 (1824); Bot. Mag. 54: t. 2780 (1827); Étude Euphorb. Atlas: t. 8 fig. 22–31 (1858); Gartenflora 15: t. 504 (1866) (as *Dibrachion*); Adansonia 6: t. 7 following p. 351 (1867) (as *Carumbium*); Nat. Pflanzenfam. 3(5): 96, t. 60 (1890); Icon. Horti Then. 5: t. 169 (1904); Pflanzenr. 52: 47, t. 7 and 52, t. 8 (1912); Nova Guinea 8: tt. 140 and 141 (1912); Nova Guinea 12: tt. 184, 229 (1917); Nat. Pflanzenfam. ed. 2, 19c: 189, t. 100 (1931); Wayside Tr. Malaya 1: 257, t. 81 (1940); Woody Fl. Taiwan: 424, t. 165 (1963); Mountain Fl. Java: color pl. 23, t. 19 fig. 5 (1972); Kew Bull. 35: 639, t. 3 fig. B (1981); Fl. Vitiens. Nov. 2: 561, t. 149 (1981); Alpine Fl. New Guinea 2: 2180 and 2183, tt. 662 and 663 (1982); Nordic J. Bot. 4: 54 and 55, tt. 1 and 2 (1984); Fl. Nouv.-Caléd. 14(1): 31, t. 4 fig. 1–3 (1987); Tropenbos ser. 7: 344, t. 96 (1994); Blumea 42: 440, t. 1 (1997).

A genus of 20–25 species, ranging from the S border of Thailand through Malesia to Australia, New Zealand and the Pacific Islands, also widely cultivated (naturalized in Sri Lanka).

Subtribe 49b. **Hippomaninae** *Griseb.*, Fl. Brit. W. I.: 49 (1859) (as *Hippomaneae*); G. L. Webster, Taxon 24: 600 (1975); Ann. Missouri Bot. Gard. 81: 120 (1994); Esser, Syst. Hipp.: 22 (1994). T.: *Hippomane* L.

Subtribe *Euhippomaneae* Müll. Arg., Linnaea 34: 203 (1865), nom. inval.; in DC., Prodr. 15(2): 1035, 1147 (1866); in Mart., Fl. Bras. 11(2): 515 (1874).

Subtribe *Adenopeltinae* Pax & K. Hoffm. in Engl., Pflanzenr. 52: 263 (1912); in Engl. & Prantl, Nat. Pflanzenfam. ed. 2, 19c: 204 (1931); Hurus., J. Fac. Sci. Univ. Tokyo, Sect. 3, Bot. 6: 311 (1954). T.: *Adenopeltis* Bertero ex A. Juss.

Subtribe *Excoecariinae* Pax & K. Hoffm. in Engl., Pflanzenr. 52: 153 (1912); in Engl. & Prantl, Nat. Pflanzenfam. ed. 2, 19c: 195 (1931); Hurus., J. Fac. Sci. Univ. Tokyo, Sect. 3, Bot. 6: 311 (1954). T.: *Excoecaria* L.

Subtribe *Gymnanthinae* Pax & K. Hoffm. in Engl., Pflanzenr. 52: 57 (1912); in Engl. & Prantl, Nat. Pflanzenfam. ed. 2, 19c: 190 (1931). T.: *Gymnanthes* Sw.

Subtribe *Mabeinae* Pax & K. Hoffm. in Engl., Pflanzenr. 52: 22 (1912); in Engl. & Prantl, Nat. Pflanzenfam. ed. 2, 19c: 187 (1931); Hurus., J. Fac. Sci. Univ. Tokyo, Sect. 3, Bot. 6: 311 (1954); G. L. Webster, Taxon 24: 600 (1975); Ann. Missouri Bot. Gard. 81: 119 (1994); Esser, Syst. Hipp.: 20 (1994). T.: *Mabea* Aubl.

Subtribe *Stillingiinae* Pax & K. Hoffm. in Engl., Pflanzenr. 52: 174 (1912); in Engl. & Prantl, Nat. Pflanzenfam. ed. 2, 19c: 197 (1931); Hurus., J. Fac. Sci. Univ. Tokyo, Sect. 3, Bot. 6: 313 (1954); Esser *et al.*, Syst. Bot. 22: 625 (1998). T.: *Stillingia* Garden ex L.

Monoecious or rarely dioecious trees, shrubs, lianas or herbs. Indumentum absent or uniseriate, dendritic or dibrachiate. Leaves with small and inconspicuous, rarely with large and enclosing stipules. Inflorescence solitary or compound; bracts of male part with 1–c. 30 male flowers. Male flowers sessile to pedicellate; calyx (1)2–5 lobed, actinomorphic or rarely zygomorphic. Female flowers sessile to pedicellate; stigmas undivided, eglandular. Seeds with or without an enlarged caruncle.

The circumscription of this subtribe includes the *Mabeinae* as accepted by Webster (1975, 1994). The latter is probably polyphyletic and should be united with the otherwise paraphyletic *Hippomaninae* (Esser, 1994). Hitherto no phylogenetically founded subdivision of this huge and diverse subtribe is apparent.

Key to the genera

1. Leaves present only on young branchlets, later leafless; succulent, thorny shrub, flowering when leafless; Patagonia · · · · 298. *Spegazziniophytum*
1. Leaves present on older branches; flowering with leaves, or if flowering when leafless then not a succulent, thorny shrub and not in Patagonia · · · 2
2. Leaves with distinct gland(s) on petiole apex or lamina base visible from above · 3
2. Leaves without distinct gland(s) visible from above, sometimes with indistinctly glandular auricles · 15
3. Leaves distinctly serrate · 4
3. Leaves entire (rarely indistinctly and irregularly serrate) · · · · · · · · · · 10
4. Leaves with one disc-shaped gland above; ovary and fruit 6–9-locular · 301. *Hippomane*
4. Leaves with two differently shaped glands above; ovary and fruit (2)3-locular · 5
5. Floral bracts eglandular; filaments of stamens mostly fused · 294. *Grimmeodendron*
5. Floral bracts glandular; filaments of stamens free (compare also note under 312. *Anomostachys*) · 6

6. Calyx of male flowers free to nearly free; seeds initially with a caruncle that separates and remains at the columella · · · · · · · · · · · 288. *Excoecaria*
6. Calyx of male flowers distinctly fused; seeds ecarunculate or with persistent caruncle · 7
7. Male and female flowers in separate, elongate inflorescences; fruits irregularly dehiscent; seeds pale arillate · · · · · · · · · · · 299. *Falconeria*
7. Male and female flowers in same inflorescences (sometimes one sex absent); fruits regularly dehiscent; seeds dry and brownish or reddish arillate · · 8
8. Bract glands often in several pairs; male flowers with 2–5 stamens; base of fruit not remaining as carpidiophore; seeds greyish-brownish, ecarunculate, exarillate · · · · · · · · · · · · · · · · · · · 304. *Pleradenophora*
8. Bract glands in only one pair; male flowers always with 2 stamens; base of fruit remaining as a carpidiophore or seeds reddish arillate · · · · · · 9
9. Glands of upper leaf surface on lamina base; seeds dry and brownish; basal part of fruit remaining on the plant as woody carpidiophore · 297. *Stillingia*
9. Glands of upper leaf surface on petiole-apex, rarely on lamina base; seeds reddish arillate; basal part of fruit inconspicuous, remaining columella regularly alate · 300. *Sapium*
10. Inflorescences compound; male flowers with 3–13 stamens · · · · · · · · 11
10. Inflorescences solitary; male flowers with 2 or 3 stamens · · · · · · · · · · 12
11. Leaves alternate; male flowers with 3–5 stamens; female flowers and young fruits pubescent; fruits leathery-fleshy outside · · · · 303. *Senefelderopsis*
11. Leaves pseudoverticillate, rarely subalternate; male flowers with 6–13 stamens; female flowers and young fruits glabrous; fruits dry · 316. *Senefeldera*
12. Flowers in dense catkins; female and male flowers in separate thyrses; caruncule separating from seed and remaining at columella · 288. *Excoecaria*
12. Female and male flowers together in elongate inflorescences; caruncle remaining on seed or absent · 13
13. Glands on lower leaf surface submarginal to laminar; male flowers with (2)3 stamens; female flowers and fruits long-pedicellate · · · · · · · · 14
13. Glands on lower leaf surface strictly marginal or absent; male flowers with 2 stamens; female flowers and fruits subsessile · · · · · · · · · · · · · · · · 8
14. Stipules more than 5 mm long, conspicuous; glands on leaf base stipitate; seeds brownish, dry · 313. *Neoshirakia*
14. Stipules less than 5 mm long, inconspicuous; glands on leaf base sessile; seeds with whitish sarcotesta · 307. *Triadica*
15. Herbs to small shrubs; leaves very densely glandular-serrate (distance less than 0.5 mm) or entire; inflorescences often leaf-opposite; female flowers and fruits often with rows of multiple appendages · 292. *Microstachys*
15. Shrubs to trees (rarely herbs in 297. *Stillingia*); leaves more distantly serrate (distance 0.5–50 mm) or entire; inflorescences not leaf-opposite (sometimes leaf-opposite in 293. *Conosapium*); female flowers and fruits smooth or with one pair of appendages per carpel · · · · · · · · · · · · 16

16. Leaves often palmately lobed; male flowers with 1 stamen · 318. *Dalembertia*
16. Leaves simple; male flowers with at least 2 stamens · · · · · · · · · · · · · · 17
17. Floral bracts very small and inconspicuous; male flowers sessile, with a disc represented by a fringe of hairs · · · · · · · · · · · · · · 287. *Dendrocousinsia*
17. Floral bracts distinct and conspicuous (if indistinct like in 315. *Actinostemon* then flowers pedicellate); male flowers sessile to pedicellate, without any disc · 18
18. Inflorescence buds covered by numerous sterile scales; floral bracts irregular and weak, often nearly absent · · · · · · · · · · 315. *Actinostemon*
18. Inflorescence buds covered by the stiff, scaly floral bracts, not by sterile scales · 19
19. Inflorescences and infructescences compound · · · · · · · · · · · · · · · · · 20
19. Inflorescences and infructescences solitary · · · · · · · · · · · · · · · · · · · 26
20. Hairs partly malpighiaceous-dibrachiate; leaf glands laminar, very close to midrib; seed surface sculptured · · · · · · · · · · · · · · · 317. *Rhodothyrsus*
20. Hairs multicellular, unbranched or dendritic, not malpighiaceous-dibrachiate; leaf glands closer to margin than to midrib; seed surface smooth · 21
21. Hairs usually dendritic, very rarely unbranched; female flowers and young fruits pubescent · 22
21. Hairs absent or unbranched; female flowers and fruits always glabrous · 23
22. Bract-glands cup- to disc-shaped; male flowers with 2 fused stamens; fruits with very short pedicel, glabrescent · · · · · · · · · · · · · 302. *Dendrothrix*
22. Bract-glands elliptic or 0; male flowers with 3–c. 100 free stamens; fruits with long pedicel, not glabrescent · · · · · · · · · · · · · · · · · · 309. *Mabea*
23. Male flowers with 2 stamens; fruits indehiscent berries with fleshy pericarp · 306. *Balakata*
23. Male flowers with at least 3 stamens; fruits dehiscent, with dry woody pericarp · 24
24. Leaves pseudoverticillate, rarely crowded-alternate; glands of lower leaf surface laminar; lateral branches of inflorescences initially completely covered by a single large bract; male flowers strongly zygomorphic when flowering; seeds with dorsal chalaza · · · · · · · · · · 314. *Pseudosenefeldera*
24. Leaves alternate; glands of lower leaf surface marginal to submarginal; lateral branches of inflorescences covered not by a single large bract but by the numerous floral bracts; male flowers not or slightly inclinate when flowering; seeds with basal chalaza · · · · · · · · · · · · · · · · · · 25
25. Leaves very shortly petiolate; floral bracts eglandular; male flowers sessile · 312. *Anomostachys*
25. Leaves shortly but distinctly petiolate; floral bracts glandular, very rarely eglandular; male flowers pedicellate, very rarely subsessile only if bracts pedunculate · 310. *Gymnanthes*
26. Plants pubescent, at least on some floral parts · · · · · · · · · · · · · · · · · 27
26. Plants glabrous · 29
27. Female flowers with (3)6(9) sepals; ovaries and fruits densely pubescent · 309. *Mabea*
27. Female flowers with 2 or 3 sepals; ovaries and fruits glabrous · · · · · · · 28

28. Male flowers (3)5–7 per bract; female flowers with 2 or 3 locules; fruits either 2-locular or, if 3-locular, with conspicuously thick pericarp (pericarp-thickness more than $^1/_{10}$ of fruit length) · · · 308. *Shirakiopsis*
28. Male flowers 1–3(5) per bract; female flowers with 3 locules; fruits 3-locular, with moderately thick pericarp (thickness c. $^1/_{10}$ of fruit length) · 310. *Gymnanthes*
29. Leaves long pedicellate (compared to lamina), entire; male flowers in a glomerulate head, female flowers long pedicellate, sometimes plants even dioecious; upper part of seeds covered by an enlarged caruncle · 319. *Maprounea*
29. Leaves shortly pedicellate (compared to lamina), entire or serrate; male flowers in an elongated inflorescence (rarely glomerulate in 288. *Excoecaria*), female flowers at base of male part but not long pedicellate (rarely distinctly pedicellate in 310. *Gymnanthes*, or rarely dioecious); caruncle absent or small on top of seeds · · · · · · · · · · · · · · · · · · · 30
30. Calyx of male flowers absent or minute (stamens nearly naked in bud) · 31
30. Calyx of male flowers well-developed (stamens largely enclosed in bud) · 33
31. Bark scabrate-papillate; leaf margin with enlarged glandular teeth; male flowers with 2 stamens; base of female flowers and fruits thickened, woody; opened fruits leaving a 3-cornute carpidiophore on the plant · 296. *Adenopeltis*
31. Bark nearly smooth; leaf margin serrate to entire but without enlarged glandular teeth; male flowers with (2)3–12 stamens; base of female flowers and fruits not thickened or woody; opened fruits leaving an alate columella without conspicious base · · · · · · · · · · · · · · · · · · · 32
32. Stipules usually divided into several ciliae; floral bracts eglandular or glandular-serrate but without distinct pair(s) of glands; male flowers subsessile · 291. *Colliguaja*
32. Stipules undivided; floral bracts with distinct glands (rarely eglandular); male flowers pedicellate · 310. *Gymnanthes*
33. Leaves below with submarginal to laminar glands distinctly separated from the margin · 34
33. Leaves below eglandular or with strictly marginal glands · · · · · · · · · · 36
34. Male flowers (3)5–8 per bract; ovaries and fruits with 3 pairs of appendages · 305. *Sclerocroton*
34. Male flowers 1–3(5) per bract; ovaries and fruits smooth · · · · · · · · · · 35
35. Glands of floral bracts stipitate, kidney-shaped; male flowers 1 per bract · 311. *Ditrysinia*
35. Glands of floral bracts sessile, not kidney-shaped; male flowers 3–5 per bract · 310. *Gymnanthes*
36. Male flowers (1)7–18 per bract; male flowers with 2 or 3–5 stamens · · · 8
36. Male flowers 1–3(10) per bract; male flowers with (2)3 stamens · · · · · 37
37. Male flowers distinctly pedicellate, rarely subsessile when bracts pedunculate · 310. *Gymnanthes*
37. Male flowers sessile to subsessile; bracts not pedunculate · · · · · · · · · · 38

38. Stipules conspicuous, more than 5 mm long; stigmas conspicuously compressed; endemic to Madagascar · · · · · · · · · · · · 293. *Conosapium*
38. Stipules inconspicuous, less than 5 mm long; stigmas not conspicuously compressed; in all tropical regions · 39
39. Floral bracts eglandular; calyx of male flowers partly fused · · · · · · · · 40
39. Floral bracts usually glandular, rarely eglandular; calyx of male flowers free to nearly free · 41
40. Inflorescences on lateral short shoots · · · · · · · · · · · · · · · · 295. *Bonania*
40. Inflorescences terminal · 294. *Grimmeodendron*
41. Calyx of male flowers 5-lobed (or 3-lobed but 2 bracteoles very similar?) · 289. *Spirostachys*
41. Calyx of male flowers 3-lobed · 42
42. Glands of bracts irregularly elliptical, sometimes fragmented · 290. *Sebastiania*
42. Glands of bracts disc- to cup-shaped, not fragmented · · · 288. *Excoecaria*

287. **Dendrocousinsia** Millsp., Publ. Field Columb. Mus., Bot. Ser. 2: 374 (1913); Fawc. & Rendle, Fl. Jamaica 4: 322 (1920). T.: *D. spicata* Millsp.

Sebastiania sect. *Dendrocousinsia* (Millsp.) Pax & K. Hoffm. in Engl., Pflanzenr. 63: 422 (1914); in Engl. & Harms, Nat. Pflanzenfam. ed. 2, 19c: 193 (1931); C. D. Adams, Flow. Pl. Jamaica: 418 (1972).

Dioecious (?), glabrous shrubs or trees. Latex not described. Leaves alternate, opposite or whorled, stipulate, shortly petiolate to sessile, simple, entire or serrate, penninerved, chartaceous to coriaceous, eglandular on both sides; stipules glandular. Inflorescences terminal or axillary, solitary or clustered, floral bracts very small, with a pair of disc- to cup-shaped, sometimes even stipitate glands, all 1-flowered. Male flowers sessile; calyx 3-lobed, free, open before flowering; petals 0; disc represented by a fringe of glandular hairs; stamens 3, filaments long, free, anthers basifixed, extrorse, 2-thecate, longitudinally dehiscent; pistillode 0; receptacle flat. Female flowers shortly pedicellate to sessile; calyx 3-lobed, free or partly fused, fleshy, often cupuliform; petals 0; disc represented by a fringe of scales or hairs; ovary smooth, with 3 locules, ovules 1 per locule; style short, stigmas short and thick, undivided, eglandular. Fruits smooth, septicidally dehiscent into 3 mericarps. Seeds 3 per fruit, elliptic, carunculate. Fig.: See Fl. Jamaica 4: 322, t. 108 (1920).

A genus of three species, endemic to Jamaica.

The position of this genus is highly doubtful. It certainly is not part of *Sebastiania*, and even its affiliation with the *Hippomaneae* is uncertain. Dioecy, the minute floral bracts, floral discs and the fleshy female calyx make the latter improbable.

288. **Excoecaria** *L.*, Syst. Nat. ed. 10: 1288 (1759), Sp. Pl. ed. 2: 1451 (1763); A. Juss., Euphorb. Gen.: 52 (1824), pro parte excl. *Gymnanthes*; Baill., Hist. Pl. 5: 227 (1874), pro parte; Benth. in Benth. & Hook. f., Gen. Pl. 3(1): 255, 337 (1880), pro parte; Hook. f., Fl. Brit. India 5: 472 (1887); Pax in Engl. & Prantl,

Nat. Pflanzenfam. 3(5): 95 (1890), pro parte; Hemsl., Hooker's Icon. Pl. 28: t. 2741 (1902); Prain, Fl. Trop. Afr. 6(1): 1018 (1913); Hurus., J. Fac. Sci. Univ. Tokyo, Sect. 3, Bot. 6: 311 (1954); J. Léonard, Bull. Jard. Bot. État 29: 133 (1959); Fl. Cong. Rwa.-Bur. 8(1): 148 (1962); Backer & Bakh. f., Fl. Java 1: 499 (1963); A. C. Sm., Fl. Vitiens. Nov. 2: 562 (1981); Radcl.-Sm. in Fl. Trop. East Africa, Euph. 1: 382 (1987); G. L. Webster, Ann. Missouri Bot. Gard. 81: 122 (1994), pro parte excl. *Anomostachys*; Chakrab. & M.G. Gangop., J. Econ. Taxon. Bot. 18: 193 (1994); Radcl.-Sm. in Fl. Zamb. 9(4): 315 (1996); Philcox in Dassan., ed., Fl. Ceyl. XI: 187 (1997); Esser *et al.*, Syst. Bot. 22: 625 (1998). T.: *Excoecaria agallocha* L.

Commia Lour., Fl. Cochinch.: 605 (1790); A. Juss., Euphorb. Gen.: 53 (1824). T.: *C. cochinchinensis* Lour. [=*E. agallocha* L.]
Stillingia sect. *Excoecaria* (L.) Baill., Étude Euphorb.: 517 (1858).
Excoecaria sect. *Euexcoecaria* Müll. Arg., Linnaea 32: 123 (1863), nom. inval.; in DC., Prodr. 15(2): 1218 (1866).
Taeniosapium Müll. Arg. in DC., Prodr. 15(2): 1200 (1866); Benth. in Benth. & Hook. f., Gen. Pl. 3(1): 337 (1880); Pax in Engl. & Prantl, Nat. Pflanzenfam. 3(5): 95 (1890). T.: *T. goudotianum* Müll. Arg.
Excoecaria sect. *Commia* (Lour.) Pax & K. Hoffm. in Engl., Pflanzenr. 52: 157 (1912), nom. inval.; in Engl. & Harms, Nat. Pflanzenfam. ed. 2, 19c: 196 (1931).
Glyphostylus Gagnep., Bull. Soc. Bot. France 71: 871 (1925) and in Lecomte, Fl. Indo-Chine 5: 402 (1926); Pax & K. Hoffm. in Engl. & Harms, Nat. Pflanzenfam. ed. 2, 19c: 195 (1931); G. L. Webster, Ann. Missouri Bot. Gard. 81: 123 (1994); Esser, Nordic J. Bot. 16: 579 (1997). T.: *G. laoticus* Gagnep. [=*E. laotica* (Gagnep.) Esser]

Monoecious or dioecious, glabrous trees or shrubs. Latex white. Leaves alternate, pseudoverticillate, or opposite, with very short to moderately long eglandular petioles, stipulate, serrate to entire, penninerved, chartaceous to coriaceous, above eglandular or with a pair of basal, disc-shaped glands, below eglandular or with strictly marginal glands, lower surface smooth or minutely papillate; stipules small, eglandular, irregularly scaly. Inflorescences axillary, solitary, bracts eglandular or with a pair of disc- to cup-shaped, sometimes stipitate glands, 0–3 basal ones with 1 female flower and numerous apical ones with 1 or 3 male flowers, or sexes separated into many-flowered female and male thyrses; bracteoles present or absent. Floral buds inclinate and initially covered by their large bracts, later on erect. Male flowers sessile or very shortly pedicellate, pedicel hardly elongating when flowering; calyx 3-lobed, free to nearly free; petals and disc 0; stamens 3, filaments present and often very long, ± free, anthers basifixed, extrorse, 2-thecate, longitudinally dehiscent; pollen 3-colporate, hardly to distinctly 3-lobate in polar view, colpus transversalis+, costae+, colpi narrow, tectate-perforate, columellae distinct; pistillode 0; receptacle flat. Female flowers shortly pedicellate; calyx 3-lobed, free, eglandular or with few basal, disc-shaped glands; petals and disc 0; ovary smooth, with 3 locules, ovules 1 per locule; style short to absent, stigmas 3, eglandular, undivided. Fruits shortly pedicellate, elliptic to globose, smooth, dehiscing septicidally into 3 mericarps; pericarp dry, woody, thin (thickness much less than 1/10 of fruit length); septa of mericarps with 3 or more vascular strands; columella 3-quetrous, in particular apically alate, with adhering caruncles of the seeds. Seeds 3 per fruit, elliptic, smooth or with patches of smooth and others of sculptured cells, greyish-brownish, with apical caruncle that separates from the seed and leaves a tiny scar. Fig.: See Icon. Pl. Ind. Orient. 5(2): t. 1865 (1853); Étude Euphorb.: t. 7, f. 31–34 (1858); Hooker's Icon. Pl. 28: t. 2741 (1902); Pflanzenr. 52: 163–169, tt. 29–31 (1912); Fl. Indo-Chine 5: 397, t. 46 fig. 9–17 (1926) (as *Glyphostylus*); Nat. Pflanzenfam. ed. 2, 19c: 197, t. 104 (1931); Fl. Congo Rwa.-Bur. 8(1): 149, t. 10 (1962); Woody Fl. Taiwan: 424, t. 162 (1963); Illustr. Guide Trop. Pl.: 352 and 353 (1969); Fl. Vitiens. Nov. 2: 563, t. 150 (1981); Fl. Trop. East Africa, Euph. 1: 384, t. 72 (1987); Fl. Nouv.-Caléd. 14(1): 31, t. 4 fig. 4–6 (1987); Fl. Zamb. 9(4): 317, t. 68 (1996).

A genus of c. 35 species, growing from W to S Africa, in Madagascar and the Comores, in Asia from Sri Lanka and India through the Himalaya to China, Malesia, and Polynesia, and also in N and E Australia and New Caledonia (See Frontispiece).

289. **Spirostachys** *Sond.*, Linnaea 23: 106 (1850); Benth. in Benth. & Hook. f., Gen. Pl. 3(1): 337 (1880) (under *Excoecaria*); Pax & K. Hoffm. in Engl., Pflanzenr. 52: 153 (1912); Prain, Fl. Trop. Afr. 6(1): 1005 (1913); Pax & K. Hoffm. in Engl. & Harms, Nat. Pflanzenfam. ed. 2, 19c: 195 (1931); Palgrave, Trees S. Afr.: 435 (1977); Radcl.-Sm. in Fl. Trop. East Africa, Euph. 1: 385 (1987); G. L. Webster, Ann. Missouri Bot. Gard. 81: 121 (1994); Radcl.-Sm. in Fl. Zamb. 9(4): 313 (1996). T.: *S. africana* Sond.

Excoecaria subsect. *Spirostachys* (Sond.) Müll. Arg. in DC., Prodr. 15(2): 1213 (1866).

Excoecariopsis Pax, Bot. Jahrb. Syst. 45: 239 (1910). T.: *E. dinteri* Pax [=*S. africana* Sond.]

Monoecious or dioecious shrubs or trees. Latex white. Deciduous, often flowering when leafless. Leaves alternate, shortly but distinctly petiolate with a pair of petiolar glands below, stipulate, simple, serrate, penninerved, membranaceous to chartaceous, eglandular above, below eglandular or with few marginal glands (compare petiolar glands), lower surface smooth; stipules small, eglandular, divided into several ciliae. Inflorescences terminal and axillary, solitary, bracts eglandular or with several pairs of cup-shaped glands, all bracts 1-flowered, if monoecious 1–3 basal ones female, numerous apical ones male; bracteoles present, undivided. Floral buds inclinate and initially covered by their large bracts, later on erect. Male flowers sessile; calyx 3-lobed, nearly free (but very similar to bracteoles, therefore seemingly 5-parted); petals and disc 0; stamens 3, filaments long and free to largely fused, anthers free, anthers basifixed, extrorse, 2-thecate, longitudinally dehiscent;

pollen 3-colporate, hardly 3-lobate in polar view, colpus transversalis+, costae+, colpi narrow, margo absent, tectate-perforate, columellae present; pistillode 0; receptacle flat. Female flowers very shortly pedicellate; calyx 3-lobed; petals and disc 0; ovary smooth, with (2)3 locules, ovules 1 per locule; style short but evident, stigmas 3, undivided, eglandular. Fruits short-pedicellate, spheroidal, smooth, dehiscing septicidally into (2)3 mericarps; pericarp dry, woody, thin (thickness much less than $^1/_{10}$ of fruit length); septa of mericarps with 3 or 4 vascular strands; columella (2)3-quetrous, apically alate with adherent caruncles. Seeds (2)3 per fruit, ovoid-subglobose, brownish mottled, with caruncle separating from seed. Fig.: See Trees South. Afr. 2: 1156 (1972); Fl. Trop. East Africa, Euph. 1: 388, t. 73 (1987); Fl. Zamb. 9(4): 314, t. 67 (1996).

A genus of 2 or 3 species, growing in Africa from Tanzania and Kenya to Namibia and South Africa.

Very similar to *Excoecaria* and only doubtfully distinct.

290. **Sebastiania** *Spreng.*, Neue Entd. 2: 118 (1821); A. Juss., Euphorb. Gen.: 51 (1824); Klotzsch, Arch. Naturgesch. 7(1): 183 (1841); Benth., J. Linn. Soc., Bot. 17: 242 (1878), pro parte; G. L. Webster, J. Arnold Arbor. 48: 385 (1967), pro parte; Ann. Missouri Bot. Gard. 81: 121 (1994), pro parte; Esser, Syst. Hipp.: 55 (1994); Esser *et al.*, Syst. Bot. 22: 625 (1998). T.: *S. brasiliensis* Spreng.

Sebastiania sect. *Eusebastiania* Müll. Arg. in DC., Prodr. 15(2): 1185 (1866), nom. inval.; in Mart., Fl. Bras. 11(2): 582 (1874); Benth. in Benth. & Hook. f., Gen. Pl. 3(1): 336 (1880); Pax in Engl. & Prantl, Nat. Pflanzenfam. 3(5): 94 (1890); Pax & K. Hoffm. in Engl., Pflanzenr. 52: 135 (1912); in Engl. & Harms, Nat. Pflanzenfam. ed. 2, 19c: 195 (1931).

Monoecious, glabrous shrubs or trees. Leaves alternate, with short eglandular petioles, stipulate, simple, serrate, penninerved, rarely triplinerved, chartaceous, eglandular above, below with strictly marginal glands or eglandular, lower surface smooth; stipules small, lineal to apically divided. Inflorescences terminal, solitary, bracts with a pair of irregularly elliptical, sometimes fragmented glands, few basal ones with 1 female flower, numerous apical ones with 1–3(7) male flowers. Floral buds inclinate and initially covered by their large bracts, later on erect. Male flowers sessile to subsessile; calyx 3-lobed, basally slightly fused; stamens 3, filaments long, free, anthers basifixed, extrorse, 2-thecate, longitudinally dehiscent; pistillode 0; receptacle flat. Female flowers sessile to shortly pedicellate; calyx 3-lobed, eglandular; ovary smooth (echinate in a single species of dubious affinity), with 3 locules, ovules 1 per locule; styles short, stigmas 3, undivided, eglandular. Fruits subsessile, spheroidal to elliptic, smooth (echinate in a single species of dubious affinity), dehiscing septicidally into 3 mericarps; pericarp dry, woody, very thin (thickness much less than $^1/_{10}$ of fruit length); septa of mericarps with 3 or more vascular strands; columella 3-quetrous, especially apically alate, often (always?) with adherent caruncle.

Seeds 3 per fruit, elliptic, smooth or with patches of sculptured cells, greyish-brownish, carunculate but caruncle separating from seed and leaving a tiny scar (known in only few species). Fig.: See Neue Entd. 2: t. 3 (1821); Fl. Bras. 11(2): t. 80 fig. 2 (1874); Nat. Pflanzenfam. 3(5): 94, t. 59 fig. a–c (1890); Pflanzenr. 52: 140, t. 26 (1912); Nat. Pflanzenfam. ed. 2, 19c: 194, t. 103 (1931); Sellowia 11: t. 29 fig. l–p (1959); Fl. Ilustr. Catar. EUFO: 306, t. 40 fig. l–p (1988); Árvores Brasileiras 2: 98 (1998).

A Neotropical genus of c. 25 or less species, ranging from Mexico to Paraguay and Uruguay.

291. **Colliguaja** *Molina*, Sag. Stor. Nat. Chili: 158 (1782); A. Juss., Euphorb. Gen.: 62 (1824); Hook., Bot. Misc. 1: 138 (1830); A. Juss., Ann. Sci. Nat. (Paris) 25: 22 (1832); Baill., Étude Euphorb.: 534 (1858); Müll. Arg. in DC., Prodr. 15(2): 1226 (1866); in Mart., Fl. Bras. 11(2): 630 (1874); Benth. in Benth. & Hook. f., Gen. Pl. 3(1): 255, 338 (1880); Pax in Engl. & Prantl, Nat. Pflanzenfam. 3(5): 99 (1890); Pax & K. Hoffm. in Engl., Pflanzenr. 52: 265 (1912); in Engl. & Harms, Nat. Pflanzenfam. ed. 2, 19c: 205 (1931); L. B. Sm. *et al.*, Fl. Ilustr. Catar. EUFO: 326 (1988); G. L. Webster, Ann. Missouri Bot. Gard. 81: 122 (1994). T.: *C. odorifera* Molina.

Monoecious, glabrous shrubs. Latex white. Leaves alternate or opposite, with short eglandular petioles, stipulate, simple, entire or serrate, penninerved, chartaceous to coriaceous, eglandular above and below but often with glandular, marginal teeth, lower surface smooth; stipules eglandular, divided into several ciliae. Inflorescences terminal, solitary, bracts sometimes elevated by a peduncle, often with numerous minute glandular teeth but eglandular otherwise, 0–2 basal ones with 1 female flower, numerous apical ones with 3 male flowers; bracteoles absent. Floral buds inclinate and initially covered by their large bracts, later on erect. Male flowers shortly pedicellate to sessile; calyx consisting of 1 or 2 minute, free lobes or completely absent; petals and disc 0; stamens 2–4, filaments distinct, free or partly fused, anthers basifixed, extrorse, 2-thecate, longitudinally dehiscent; pollen 3-colporate, hardly 3-lobate in polar view, colpus transversalis+, costae+, margo indistinct, colpi narrow, tectate-perforate, columellae distinct; pistillode 0; receptacle flat. Female flowers shortly pedicellate; calyx 3-lobed, lobes free?; petals and disc 0; ovary smooth, with 3 locules, ovules 1 per locule; style short but evident, stigmas 3, undivided, eglandular. Fruits short-pedicellate, subglobose with usually distinctly carinate lobes, dehiscing septicidally into 3 mericarps; pericarp dry, woody; septa of mericarps with at least 3 vascular strands; columella 3-quetrous, apically alate. Seeds 3 per fruit, globose-elliptic, with elongated preraphe, greyish to brown, smooth, ecarunculate. Fig: See Bot. Misc. 1: tt. 39 and 40 (1830); Étude Euphorb.: t. 7, f. 10–14 (1858); Fl. Bras. 11(2): t. 85, fig. 2 (1874); Pflanzenr. 52: 267, t. 53 (1912); Nat. Pflanzenfam. ed. 2, 19c: 191, t. 100 fig. c–e (1931).

A genus of 4 species, growing in temperate South America from S Brazil to Chile.

292. **Microstachys** *A. Juss.*, Euphorb. Gen.: 48 (Feb. 1824); Griseb., Fl. Brit. W. I.: 49 (1859); Müll. Arg., Linnaea 32: 90 (1863); G. L. Webster, Ann. Missouri Bot. Gard. 81: 122 (1994); Esser, Kew Bull. 53: 957 (1998), and Blumea 44: 173 (1999). LT.: *M. bicornis* A. Juss. (B.: *Tragia bicornis* Vahl) (designated by G. L. Webster, Ann. Missouri Bot. Gard. 81: 121 (1994)) [=*M. corniculata* (Vahl) Griseb. (S.: *Sebastiania corniculata* (Vahl) Pax; B.: *Tragia corniculata* Vahl)]

Cnemidostachys Mart. & Zucc., Flora 7(1) Beibl. 4: 136 (Apr.–June 1824); Nov. Gen. Sp. Pl. 1: 66 (Oct. 1824). LT.: C. *myrtilloides* Mart. & Zucc. (designated by Esser, Kew Bull. 53: 957 (1998)) [=*M. daphnoides* (Mart. & Zucc.) Müll. Arg.]

Elachocroton F. Muell., Hooker's J. Bot. Kew Gard. Misc. 9: 17 (1857). T.: *E. asperococcus* F. Muell. [=*M. chamaelea* (L.) Müll. Arg. (S.: *Sebastiania chamaelea* (L.) Müll. Arg.; B.: *Tragia chamaelea* L.)]

Stillingia sect. *Cnemidostachys* (Mart. & Zucc.) Baill., Étude Euphorb.: 515 (1858)

Tragiopsis H. Karst., Wochenschr. Gärtnerei Pflanzenk. 2: 5 (1859). T.: *T. fruticulosa* H. Karst. [=*M. corniculata* (Vahl) Griseb.]

Stillingia sect. *Microstachyopsis* Müll. Arg., Linnaea 32: 89 (1863); in DC., Prodr. 15(2): 1160 (1866). T.: *S. heterodoxa* Müll. Arg. [=*M. heterodoxa* (Müll. Arg.) Esser]

Sebastiania sect. *Microstachys* (A. Juss.) Müll. Arg. in DC., Prodr. 15(2): 1166 (1866); in Mart., Fl. Bras. 11(2): 545 (1874); Benth. in Benth. & Hook. f., Gen. Pl. 3: 336 (1880); Pax in Engl. & Prantl, Nat. Pflanzenfam. 3(5): 94 (1890); Pax & K. Hoffm. in Engl., Pflanzenr. 52: 91 (1912); in Engl. & Harms, Nat. Pflanzenfam. ed. 2, 19c: 192 (1931).

Sebastiania sect. *Elachocroton* (F. Muell.) Pax in Engl., Pflanzenr. 52: 114 (1912); Pax & K. Hoffm. in Engl., Pflanzenr. 52: 114 (1912); in Engl. & Harms, Nat. Pflanzenfam. ed. 2, 19c: 192 (1931); A. S. Oliveira, Arch. Jard. Bot. Rio de Janeiro 27: 3 (1983).

Sebastiania sect. *Microstachyopsis* (Müll. Arg.) Pax in Engl., Pflanzenr. 52: 118 (1912); in Engl. & Harms, Nat. Pflanzenfam. ed. 2, 19c: 193 (1931).

Sebastiania sect. *Cnemidostachys* (Mart. & Zucc.) G. L. Webster, J. Arnold Arbor. 48: 386 (1967).

Herbs or small shrubs, monoecious. Indumentum consisting of multicellular, branched or unbranched, colourless hairs. Leaves alternate, sessile or petiolate with short to moderately long, eglandular petioles, stipulate, simple or with small lobes at base, densely serrate with minute glandular teeth less than 0.5 mm apart or entire with a glandular-hyaline margin, penninerved, chartaceous to coriaceous, eglandular above, below with few glands at base or eglandular, lower surface smooth or papillate; stipules shortly triangular, undivided, often persistent. Inflorescences

Monoecious, glabrous shrubs to treelets. Leaves densely alternate, partly opposite, very shortly petiolate to subsessile, stipulate, simple, serrate, penninerved, chartaceous, above eglandular, below with few marginal glands, lower surface smooth; stipules conspicuously large and serrate, persisting. Inflorescences terminal, later sometimes opposite to leaves, solitary, bracts with one or several pairs of cup-shaped glands; 0–few basal ones with 1 female flower, numerous apical ones with 1–10 male flowers. Male flowers nearly sessile; calyx 3-lobed, usually partly fused; petals and disc 0; stamens 3, filaments distinct, free or fused (see note below); pistillode 0; receptacle flat. Female flowers sessile; calyx 3-lobed; petals and disc 0; ovary smooth, with 3 locules, ovules 1 per locule; style absent, stigmas 3, undivided, eglandular, conspicuously vertically compressed. Fruits subsessile, spheroidal, smooth, dehiscing septicidally into 3 mericarps; pericarp dry, woody; columella 3-quetrous, slightly alate, persistent. Seeds 3 per fruit, ovoid-globose, ecarunculate. Fig.: See Fig. 45, p. 367.

A hitherto insufficiently known genus with 3 or more species, some of them still undescribed, endemic to Madagascar.

This genus is currently under study by Valéry Malecot of Paris, who wishes to unite it with *Excoecaria*. In at least one (still undescribed) species, flowers of bisexual thyrses have a smaller calyx and fused filaments, whereas flowers of unisexual thyrses have large sepals and free stamens.

294. **Grimmeodendron** *Urb.*, Symb. Antill. 5: 397 (1908); Pax & K. Hoffm. in Engl., Pflanzenr. 52: 258 (1912); Fawc. & Rendle, Fl. Jamaica 4: 326 (1920); Britton & Millsp., Bahama Fl.: 232 (1920); Pax & K. Hoffm. in Engl. & Harms, Nat. Pflanzenfam. ed. 2, 19c: 204 (1931); Alain, Fl. Cuba 3: 119 (1953); G. L. Webster, Ann. Missouri Bot. Gard. 81: 123 (1994). LT.: *G. jamaicense* Urb. (designated by L. C. Wheeler, Taxon 24: 536 (1975)).

Monoecious, glabrous shrubs or trees. Latex white. Leaves alternate but often apically crowded, with short eglandular petioles, stipulate, simple, serrate, penninerved, coriaceous, above eglandular or with a pair of basal glands, with strictly marginal glands below, lower surface smooth; stipules small, eglandular, ciliate. Inflorescences terminal, solitary, bracts eglandular, 1 or 2 basal ones with 1 female flower, numerous apical ones with 2–4 male flowers; bracteoles present, cilia-like. Floral buds inclinate and initially covered by their large bracts, later on erect. Male flowers sessile to shortly pedicellate; calyx 3-lobed, partly fused at base; petals and disc 0; stamens 3, filaments distinct, largely fused, anthers basifixed, extrorse, 2-thecate, longitudinally dehiscent; pollen 3-colporate, distinctly 3-lobate in polar view, colpus transversalis narrow elongated, costae+, colpi narrow, tectate-perforate, columellae distinct; pistillode 0; receptacle flat. Female flowers subsessile; calyx 3-lobed, lobes free?, eglandular; petals and disc 0; ovary smooth, with 3 locules, ovules 1 per locule; style short, stigmas 3, undivided, eglandular. Fruits subsessile, spheroidal, smooth, dehiscing septicidally into 3 mericarps; pericarp dry, woody; septa of mericarps with 1 often divided vascular strand; columella 3-quetrous, apically distinctly alate, persistent. Seeds 3 per fruit, ovoid-globose, smooth, brownish striate, ecarunculate. Fig.: See Symb. Antill. 5: 398 (1908); Pflanzenr. 52: 259, t. 50 (1912); Fl. Baham. Archip.: 819, t. 338 (1982).

A ditypic genus of Jamaica, the Bahamas and Cuba.

295. **Bonania** *A. Rich.* in Sagra, Hist. Fis. Cuba, Bot. 11: 201 (1850); Benth. in Benth. & Hook. f., Gen. pl. 3(1): 255, 335 (1880); Pax in Engl. & Prantl, Nat. Pflanzenfam. 3(5): 98 (1890); Pax & K. Hoffm. in Engl., Pflanzenr. 52: 259 (1912); in Engl. & Harms, Nat. Pflanzenfam. ed. 2, 19c: 204 (1931); Alain, Fl. Cuba 3: 115 (1953); Correll & Correll, Fl. Baham. Archip.: 784 (1982); Alain, Fl. Española 4: 88 (1986); G. L. Webster, Ann. Missouri Bot. Gard. 81: 123 (1994); Esser, Syst. Hipp.: 36 (1994). T.: *B. cubana* A. Rich.

Excoecaria sect. *Bonania* (A. Rich.) Müll. Arg. in DC., Prodr. 15(2): 1212 (1866).
Hypocoton Urb., Symb. Antill. 7: 263 (1912). T.: *H. domingensis* Urb. [=*B. domingensis* (Urb.) Urb.]

Monoecious, glabrous shrubs, often with thorns. Leaves alternate, with short eglandular petioles, stipulate, simple, entire to serrate, penninerved, completely eglandular, lower surface smooth; stipules small, undivided, eglandular. Inflorescences on axillary short shoots, solitary, bracts eglandular, 1 basal one with 1 female flower, numerous apical ones with 1–4 male flowers; bracteoles absent. Male flowers sessile in bud, very shortly pedicellate when flowering; calyx nearly completely fused, without distinct lobes; petals and disc 0; stamens 2 or 3, filaments short, free to basally fused, anthers subdorsifixed, extrorse, 2-thecate, longitudinally dehiscent; pollen 3-colporate, distinctly 3-lobate in polar view, colpus transversalis+, costae+, colpi narrow, tectate-perforate, columellae small; pistillode 0; receptacle flat. Female flowers subsessile to shortly pedicellate; calyx 3-lobed, lobes free?, petals and disc 0, ovary smooth, with 3 locules, ovules 1 per locule; style absent, stigmas 3, undivided, eglandular. Fruits short-pedicellate; dehiscing septicidally into 3 mericarps; columella 3-quetrous, alate, persistent. Seeds 3 per fruit, globose, smooth, greyish-yellowish dotted, ecarunculate. Fig.: See Hist. Fis. Cuba, Bot. 11: 201, t. 68 (1850), n.v.; Ic. Fl. Cuba Pl. Vasc.: t. 68 (1863), n.v.; Fl. Baham. Archip.: 785, t. 328 (1982).

A genus of 8–10 species, occuring in the Bahamas, Haiti, Cuba and perhaps Venezuela.

296. **Adenopeltis** *Bertero ex A. Juss.*, Ann. Sci. Nat. (Paris) 25: 24 (1832); Baill., Étude Euphorb.: 532 (1858); Müll. Arg. in DC., Prodr. 15(2): 1164 (1866); Benth. in Benth. & Hook. f., Gen. Pl. 3(1): 255, 338 (1880); Pax in Engl. & Prantl, Nat. Pflanzenfam. 3(5): 100 (1890); Pax & K. Hoffm. in Engl., Pflanzenr. 52: 264 (1912); in Engl. & Harms, Nat. Pflanzenfam. ed. 2, 19c: 204 (1931); G. L. Webster, Ann. Missouri Bot. Gard. 81: 123 (1994). T.: *A. serrata* (Ait.) G. L. Webster (B.: *Excoecaria serrata* Ait.).

Monoecious glabrous shrubs, with distinctly scabrate-papillate bark. Leaves alternate, with short eglandular petioles, stipulate, simple, serrate with distinctly enlarged but distant glandular teeth, penninerved, membranaceous to chartaceous, eglandular above, eglandular below, lower surface smooth; stipules small, divided into several ciliae, eglandular. Inflorescences terminal and axillary, solitary, bracts with a pair of stipitate, disc-shaped glands, 0–2 basal ones with 1 female flower, numerous apical ones with 1–3 male flowers; bracteoles absent. Floral buds inclinate and initially covered by their large bracts, later on erect. Male flowers sessile; calyx absent; petals and disc 0; stamens 2, filaments long, partly fused, anthers basifixed, extrorse, 2-thecate, longitudinally dehiscent; pollen 3-colporate, distinctly 3-lobate in polar view, colpus transversalis+, costae+, colpi narrow, tectate-perforate, columellae distinct; pistillode 0; receptacle flat. Female flowers sessile; calyx nearly absent, ?2-lobed, partly fused; petals and disc 0; ovary smooth but with slightly thickened base, with 3 locules, ovules 1 per locule; style absent, stigmas 3, undivided, eglandular. Fruits sessile, spheroidal, smooth, dehiscing septicidally into 3 mericarps; ?pericarp dry, moderately thick; columella with a large, 3-cornute, persistent carpidiophore at base. Seeds 3 per fruit, subglobose, smooth, ecarunculate. Fig.: See Étude Euphorb. Atlas: t. 7 fig. 15–19 (1858); Hist. Pl. 5: 134, tt. 212 and 213 (1874) (as *Excoecaria*); Pflanzenr. 52: 60, t. 10 fig. f and 264, t. 52 (1912); Nat. Pflanzenfam. ed. 2, 19c: 191, t. 101 fig. f (1931).

A monotypic genus of Chile and Peru.

297. **Stillingia** *Garden ex L.*, Syst. Nat. ed. 12, 2: 637 (1767); Mant. Pl.: 19, 126 (1767); A. Juss., Euphorb. Gen.: 49 (1824); Benth. in Benth. & Hook. f., Gen. Pl. 3(1): 255, 334 (1880); Pax in Engl. & Prantl, Nat. Pflanzenfam. 3(5): 96 (1890); Pax & K. Hoffm. in Engl., Pflanzenr. 52: 180 (1912); in Engl. & Harms, Nat. Pflanzenfam. ed. 2, 19c: 198 (1931); D. J. Rogers, Ann. Missouri Bot. Gard. 38: 207 (1951); G. L. Webster, J. Arnold Arbor. 48: 388 (1967); Coode, Fl. Masc. 160: 86 (1982); G. L. Webster, Ann. Missouri Bot. Gard. 81: 122 (1994); Esser, Blumea 44: 190 (1999). T.: *S. sylvatica* Garden ex L.

Stillingia sect. *Eustillingia* Klotzsch ex Baill., Étude Euphorb.: 510 (1858), nom. inval.; Müll. Arg. in DC., Prodr. 15(2): 1155 (1866); in Mart., Fl. Bras. 11(2): 538 (1874).

Gymnostillingia Müll. Arg., Linnaea 32: 89 (1863); Baill., Adansonia 5: 338 (1865); Müll. Arg. in DC., Prodr. 15(2): 1163 (1866); Baill., Adansonia 11: 121 (1873); Müll. Arg. in Mart., Fl. Bras. 11(2): 540 (1874); Benth. in Benth. & Hook. f., Gen. Pl. 3(1): 334 (1880). LT.: *G. acutifolia* (Benth.) Müll. Arg. (B.: *Sapium acutifolium* Benth.) (designated by Baillon, Adansonia 5: 339 (1865)) [=*Stillingia acutifolia* (Benth.) Hemsl.]

Monoecious perennial herbs, shrubs or trees, sometimes with succulent stems, glabrous. Latex white. Leaves opposite, or alternate, sometimes apically crowded to nearly verticillate, with short to nearly absent eglandular petiole (but compare lamina base), stipulate, simple, serrate to entire, penninerved, chartaceous to coriaceous, sometimes succulent and fleshy, above eglandular or with a pair of glands at base on the junction with the petiole, below with a row of strictly marginal, sometimes even stipitate glands, lower surface smooth; stipules small, filiform, basally often glandular, often divided into several ciliae. Inflorescences terminal and axillary, solitary, bracts with a pair of cup- to disc-shaped glands, 0–c. 15 basal ones with 1 female flower, numerous apical ones with (1)3–13 male flowers; bracteoles filiform. Floral buds inclinate and initially covered by their large bracts, later on erect. Male flowers sessile, shortly pedicellate only when flowering; calyx 2-lobed, largely fused; petals and disc 0; stamens 2, filaments slightly longer than the anthers, free or basally fused, anthers basifixed, extrorse, 2-thecate, longitudinally dehiscent; pollen spheroidal to prolate, 3-colporate or with 1 lateral pore, nearly circular in polar view, if 3-colporate colpus transversalis+, costae+, colpi narrow, tectate-perforate, columellae distinct; pistillode 0; receptacle flat. Female flowers sessile to subsessile; calyx 3-lobed, nearly free or more rarely united into an annulus, eglandular, rarely absent at all; petals and disc 0; ovary smooth but with slightly enlarged and separated base, with 2 or 3 locules, ovules 1 per locule; styles short, stigmas 2 or 3, undivided, eglandular. Fruits nearly sessile, spheroidal to elliptic, smooth but with slightly thickened base, upper part dehiscing septicidally into 2 or 3 mericarps but lower part remaining unopened on the plant; pericarp dry, woody, thin to moderately thick; columella 2- or 3-quetrous apically, forming a large, 2- or 3-cornute, woody carpidiophore at base. Seeds 2 or 3 per fruit, elliptic to subglobose, smooth, brownish and often maculate, carunculate or ecarunculate. Fig.: See Fl. Bras. 11(2): t. 70 fig. 2, 3 (1874); Nat. Pflanzenfam. 3(5): 96, tt. 61 and 62 (1890); Pflanzenr. 52: 185–192, tt. 34–36 (1912); Nat. Pflanzenfam. ed. 2, 19c: 199, t. 105 (1931); Ann. Missouri Bot. Gard. 38: 214, tt. 2 and 3 (1951); Illustr. Fl. Pacific States 3: 33, tt. 3029–3031 (1960); J. Arnold Arbor. 48: 390, t. 5 (1967); Phytologia 14: 455, t. 3 (1967); Fl. Baja California: 133, t. 84 (1980); Fl. Vitiens. Nov. 2: 566–568, tt. 151, 152 (1981); Fl. Mascar. 160: 89, t. 17 (1982); Ann. Missouri Bot. Gard. 75: 1667, t. 1 (1988); Contr. Univ. Michigan Herb. 20: 211, t. 5 (1995); Fieldiana, Bot. n.s. 36: 36, t. 23 p.p. (1995); Blumea 44: 196, fig. 6 (1999).

A genus of c. 27 species, mostly in the New World ranging from the USA (Kansas) throughout S America to Argentina, but a few taxa also on Mauritius and Réunion, in Malesia in a few scattered localities, and on Fiji.

298. **Spegazziniophytum** *Esser*, **genus novum** patagonicum a genere *Colliguaja* frutice habitu sucido spinis instructo, foliis tantummodo in innovationibus inventis, deinde mox delapsis, in statu aphyllo florenti, cymulis masculis trifloris, sepalis 2 masculis majoribus, staminibus 2 liberis differens; praeterea a genere *Stillingia* fructu sicco totaliter dehiscenti, gynobase carenti, seminibus siccis, caruncula minuta distinguitur. Typus generis: *S. patagonicum* (Speg.) Esser **comb. nov.** (B.: *Colliguaja patagonica* Speg., Revista Fac. Agron. Univ. Nac. La Plata 3: 592 (1897))

Esser, Syst. Hipp.: 62 (1994), sine nom.

Monoecious, succulent shrubs, glabrous, branches transformed into thorns. Leaves present only on young branches, but usually leafless. Leaves alternate, subsessile, simple, entire, chartaceous, basally glandular. Inflorescences axillary, solitary, catkin-like, bracts with a pairs of cup- to disc-shaped glands; female flowers 1 per bract, male flowers 2 or 3 per bract. Male flowers sessile; calyx 2-lobed, fused; petals and disc 0; stamens 2, with distinct, free filaments; pistillode 0. Female flowers subsessile; calyx 3-lobed, nearly free; ovary smooth; style absent, stigmas undivided, glandless. Fruits sessile, 2-locular, regularly dehiscent; pericarp very thin (thickness much less than $^1/_{10}$ of fruit length); columella alate, but without a carpidiophore. Seeds 2 per fruit, subglobose, with a very characteristic, minute apical caruncle.

A monotypic genus endemic to Patagonia (Argentina).

Most similar to *Stillingia*, but because of the absence of a carpidiophore not to be included in it.

299. **Falconeria** *Royle*, Ill. Bot. Himal. Mts.: 354 (1839); Wight, Icon. Pl. Ind. Orient. 5(2): 20 (1853); Baill., Étude Euphorb.: 526 (1858); Müll. Arg., Linnaea 32: 83 (1863); Benth., J. Linn. Soc., Bot. 17: 242 (1878); in Benth. & Hook. f., Gen. Pl. 3(1): 335 (1880) (under *Sapium*); Esser *et al.*, Syst. Bot. 22: 619 (1998); Esser, Blumea 44: 160 (1999). LT.: *F. insignis* Royle (designated by Pfeiffer, Nomencl. Bot. 1, 2: 1334 (1874)) (S.: *Sapium insigne* (Royle) Benth.)

Excoecaria sect. *Falconeria* (Royle) Müll. Arg. in DC., Prodr. 15(2): 1211 (1866).
Sapium sect. *Falconeria* (Royle) Hook. f., Fl. Brit. India 5: 471 (1888); Pax in Engl. & Prantl, Nat. Pflanzenfam. 3(5): 98 (1890); Pax & K. Hoffm. in Engl., Pflanzenr. 52: 241 (1912); in Engl. & Harms, Nat. Pflanzenfam. ed. 2, 19c: 202 (1931).

Monoecious glabrous trees with slightly succulent branches. Latex white. Leaves deciduous, often flowering and fruiting when leafless. Leaves alternate but apically crowded on otherwise often leafless branches, with short petiole, stipulate, simple, serrate, penninerved, chartaceous, above with a pair of cup- to disc-shaped glands on the junction with the petiole, below with marginal glands, lower surface smooth and shiny; stipules small, divided into few ciliae, eglandular. Inflorescences terminal on distinctly thicker branches, solitary, but male and female flowers in separate, many-flowered thyrses; bracts with a pair of flattened to disc-shaped glands, in female thyrses with 1 female flower each, in male thyrses with 9–15 male flowers each; bracteoles divided into few ciliae. Male flowers nearly sessile in bud, with short but distinct pedicel during flowering; calyx 2-lobed, partly fused; petals and disc 0; stamens 2, filament slightly longer than anthers, free, anthers basifixed, extrorse, 2-thecate, longitudinally dehiscent; pistillode 0; receptacle flat. Female flowers nearly sessile; calyx 3-lobed, partly fused, eglandular; petals and disc 0; ovary smooth, with 2 or 3 locules, ovules 1 per locule; style short to nearly absent, stigmas 2 or 3, undivided, eglandular. Fruits very shortly pedicellate, globose, smooth, dehiscing only irregularly and not into mericarps; pericarp fleshy when young, papery when mature, very thin and fragile; columella 2- or 3-quetrous, alate but membranaceous and caducous. Seeds 2 or 3 per fruit, globose, smooth, with pale, thin arillus, ecarunculate. Fig.: See Ill. Bot. Himal. Mts.: t. 84 (1839) (as *Sapium*); Icon. Pl. Ind. Orient. 5(2): t. 1866 (1853) (as *Sapium*); Pflanzenr. 52: 242, t. 45 (1912) (as *Sapium*); Fl. Bhutan 1(3): 806, t. 50 fig. p–s (1987) (as *Sapium*); Esser, Blumea 44: 164 fig. 2 (1999).

A monotypic genus, ranging from the Himalaya region to SE Asia up to the Malay Peninsula.

300. **Sapium** *Jacq.*, Enum. Syst.: 9 (1760), nom. cons [non *Sapium* P. Browne, Civ. Nat. Hist. Jamaica 2: 338 (1756)]; A. Juss., Euphorb. Gen.: 49 (1824); Benth. in Benth. & Hook. f., Gen. pl. 3(1): 255, 334 (1880), pro parte; Hurus., J. Fac. Sci. Univ. Tokyo, Sect. 3, Bot. 6: 315 (1954); G. L. Webster, J. Arnold Arbor. 48: 391 (1967), pro parte; Jabl., Phytologia 16: 393 (1968); A. H. Gentry, Woody Pl. NW. S. Amer.: 415 (1993); G. L. Webster, Ann. Missouri Bot. Gard. 81: 123 (1994) pro parte; W. C. Burger & Huft, Fieldiana, Bot. n.s. 36: 151 (1995); Kruijt, Biblioth. Bot. 146: 7, 27 (1996); Gillespie in Acev.-Rodr., ed., Fl. St. John: 220 (1996). T.: *S. aucuparium* Jacq., nom. illeg. [see Kruijt & Zijlstra, Taxon 38: 320 (1989) = *S. glandulosum* (L.) Morong (B.: *Hippomane glandulosa* L.)]

Sapiopsis Müll. Arg., Linnaea 32: 84 (1863). T.: *S. cremostachys* A. St.-Hil. ex Müll. Arg. [= *S. glandulosum* (L.) Morong (B.: *Hippomane glandulosa* L.)]
Excoecaria sect. *Sapium* (Jacq.) Müll. Arg. in DC., Prodr. 15(2): 1202 (1866); in Mart., Fl. Bras. 11(2): 612 (1874).
Sapium sect. *Eusapium* Pax in Engl. & Prantl, Nat. Pflanzenfam. 3(5): 97 (1890), nom. inval.
Sapium subg. *Eusapium* Pax & K. Hoffm. sect. *Americana* Pax & K. Hoffm. in Engl., Pflanzenr. 52: 200 (1912), nom. inval.; in Engl. & Harms, Nat. Pflanzenfam. ed. 2, 19c: 199 (1931)

Monoecious (few species doubtfully dioecious) shrubs or trees, glabrous. Latex white. Leaves alternate, with distinct but short, eglandular but more often apically biglandular petioles, stipulate, simple, entire to serrate to ciliate, penninerved, chartaceous to coriaceous, above eglandular or sometimes with a pair of glands at base, below with strictly marginal (sometimes even stipitate) glands, lower surface smooth or whitish-farinose; stipules small, undivided, eglandular, often persistent and because of very similar axillary scales often seemingly 4. Inflorescences terminal and axillary, solitary, bracts with a pair of disc-shaped to flattened-elliptical glands, 0–10 basal ones with 1 female flower, numerous apical ones with (3)7–18 male flowers; bracteoles small, divided to ciliate. Floral buds inclinate and initially covered by their large bracts, later on erect. Male flowers sessile in bud, shortly pedicellate when flowering; calyx 2-lobed, largely fused; petals and disc 0; stamens 2, filaments usually slightly longer than anthers, free or nearly free, anthers basifixed, extrorse, 2-thecate, longitudinally dehiscent; pollen 3-colporate, hardly to distinctly 3-lobate in polar view, colpus transversalis sometimes elongated, costae almost equatorial, with equatorial ring, colpi narrow, tectate-perforate, columellae distinct; pistillode 0; receptacle flat. Female flowers sessile to subsessile; calyx 2- or 3-lobed, usually partly fused; petals and disc 0; ovary smooth, with 1–3 locules, ovules 1 per locule; styles short, stigmas 1–3, undivided, eglandular. Fruits sessile to subsessile, spheroidal, smooth, dehiscing septicidally into 2 or 3 mericarps or irregularly; pericarp dry or sometimes leathery-fleshy outside, woody, thin; septa of mericarps with usually 1 vascular strand; columella erect and 2- or 3-quetrous or falcate, alate, persistent. Seeds 1–3 per fruit, irregularly spheroidal, verrucate to nearly smooth but complelety covered by a red aril, ecarunculate. Fig.: See Euphorb. Gen.: t. 15, f. 51.1–6 (1824); Hooker's Icon. Pl. 27: tt. 2647–2650, 2677–2684 (1901), 28: t. 2757 (1903), 29: tt. 2878–2900 (1909), and 31: 3025 (1915); Contr. U.S. Natl. Herb. 12: tt. 10–17 (1908); Pflanzenr. 52: 207–227, tt. 38–43 (1912); Fl. Jamaica 4: 324, t. 109 (1920); Euphorb. Surinam: t. 5 (1931); Sellowia 11: t. 30 fig. a–g (1959); Phytologia 14: 454, t. 2 (1967); Ann. Missouri Bot. Gard. 54: 325, t. 22 (1967); Phytologia 16: 426, 427 (1968); Fl. Española 4: 212, t. 116-33 (1986); Fl. Ilustr. Catar. EUFO: 321, t. 41 fig. a–g (1988); Ann. Missouri Bot. Gard. 75: 1136, t. 5 (1988); Árvores Brasileiras: 110 (1992), and 2: 97 (1998); Woody Pl. NW. S. Amer.: 421, t. 121.4 (1993); Fieldiana, Bot. n.s. 36: 35, t. 22 and 36, t. 23 (1995); Euf. Reg. Ararac.: 150, t. 44 (1995); Biblioth. Bot. 146: 35–87, tt. 10–31 and 33–35 (1996); Fl. St. John: 221, t. 100H–M (1996).

A genus of c. 25 species, strictly Neotropical and ranging from the USA through Central America and the West Indies to temperate S America up to Paraguay and Argentina.

301. **Hippomane** *L.*, Sp. Pl.: 1191 (1753); Gen. Pl.: 499 (1754); A. Juss., Euphorb. Gen.: 51 (1824); Baill., Étude Euphorb.: 539 (1858); Müll. Arg. in DC., Prodr. 15(2): 1199 (1866); Baill., Hist. Pl. 5: 228 (1874); Benth. in Benth. & Hook. f., Gen. Pl. 3(1): 255, 333 (1880); Pax in Engl. & Prantl, Nat. Pflanzenfam. 3(5): 98 (1890); Pax & K. Hoffm. in Engl., Pflanzenr. 52: 261 (1912); Fawc. & Rendle, Fl. Jamaica 4: 327 (1920); Pax & K. Hoffm. in Engl. & Harms, Nat. Pflanzenfam. ed. 2, 19c: 203 (1931); Standley & Steyerm., Fieldiana, Bot. 24(6): 122 (1949); Alain, Fl. Cuba 3: 117 (1953); G. L. Webster, J. Arnold Arbor. 48: 393 (1967); Ann. Missouri Bot. Gard. 54: 328 (1967); Proctor, Kew Bull. Addit. Ser. 11: 546 (1984); Alain, Fl. Española 4: 158 (1986); R. A. Howard, Fl. Lesser Antilles 5: 54 (1989); A. H. Gentry, Woody Pl. NW. S. Amer.: 420 (1993); G. L. Webster, Ann. Missouri Bot. Gard. 81: 124 (1994); W. C. Burger & Huft, Fieldiana, Bot. n.s. 36: 124 (1995); Gillespie in Acev.-Rodr., ed., Fl. St. John: 213 (1996). LT.: *H. mancinella* L. (designated by M. L. Green, Prop. Brit. Bot.: 195 (1929)).

Mancanilla Mill., Gard. Dict. abr. ed. 4 (1754).

Monoecious, glabrous shrubs or trees. Latex white. Leaves alternate and apically slightly crowded, with distinct petioles shorter than the lamina and with one disc-shaped gland apically on the junction with the lamina, stipulate, simple, distantly and shallowly serrate to crenate, penninerved, chartaceous, above with one undivided gland at base (see petiole), eglandular below, lower surface paler but smooth; stipules small, entire, eglandular. Inflorescences terminal, solitary, floral bracts with a pair of flattened to disc-shaped glands, 1 or 2 basal ones with 1 female flower, numerous apical ones with c. 10–30 male flowers; bracteoles absent. Male flowers nearly sessile in bud, shortly pedicellate when flowering; calyx 2-lobed, partly fused; petals and disc 0; stamens 2, filaments nearly as long as anthers, slightly to mostly fused, anthers basifixed, extrorse, 2-thecate, longitudinally dehiscent; pollen prolate-spheroidal, 3-colporate, distinctly 3-lobate in polar view, colpus transversalis+, costae+, colpi narrow, tectate-perforate, columellae distinct; pistillode 0; receptacle flat. Female flowers nearly sessile; calyx 3- or 4-lobed, free, eglandular; petals and disc 0; ovary smooth, with 6–9 locules, ovules 1 per locule; style short but present, stigmas 6–9, undivided, eglandular. Fruits short-pedicellate, broadly transversely elliptic, smooth, indehiscent, drupaceous, apple-like; exocarp fleshy, thick; endocarp irregularly rugulose. Seeds 6–9 per fruit, compressed elliptical, dark, ecarunculate. Fig.: See Euphorb. Gen.: t. 16, f. 54.1–10 (1824); Étude Euphorb.: t. 6, f. 12–20 (1858); Gartenflora 15: t. 510 (1866); Nat. Pflanzenfam. 3(5): 99, t. 64 (1890); Silva N. Amer. 7: t. 310 (1895), n. v.; N. Amer. Trees: 603, t. 554 (1908); Pflanzenr. 52: 262, t. 51 (1912); Fl. Jamaica 4: 328, t. 110 (1920); Man. Trees N. Amer. 2: 653, t. 592 (1922); Nat. Pflanzenfam. ed. 2, 19c: 204, t. 109 (1931); Ann. Missouri Bot. Gard. 54: 329, t. 23 (1967); Fl. Galapagos Isl.: 590, t. 156, colour pl. 73 (1971); Fl. illustr. phan. Guadel. Martin.: 1556, t. 732 (1978); Fl. Baham. Archip.: 820, t. 339 (1982); Kew Bull. Addit. Ser. 11: 546, t. 167 (1984); Fl. Española 4: 160, t. 116-19 (1986); Fl. Lesser Antilles 5: 56, t. 20 (1989); Woody Pl. NW. S. Amer.: 421, t. 121.1 (1993); Fieldiana, Bot. n.s. 36: 45, t. 32 p.p. (1995); Fl. St. John: 215, t. 97A–G (1996).

A genus of 2 or 3 species, ranging from the USA (Florida) through Central America and the Carribean Islands to Venezuela and the Galapagos Islands.

302. **Dendrothrix** *Esser*, Novon 3: 245 (1993); G. L. Webster, Ann. Missouri Bot. Gard. 81: 131 (1994). T.: *D. yutajensis* (Jabl.) Esser (B.: *Sapium yutajense* Jabl.).

Monoecious shrubs or trees. Hairs multicellular, dendritic, usually coloured. Latex white. Leaves alternate but often apically crowded, with short to moderately long eglandular petioles, stipulate or exstipulate, simple, entire, penninerved but sometimes distinctly triplinerved, coriaceous, eglandular above, below with a pair of basimarginal glands, sometimes with additional irregularly dispersed laminar glands, lower surface minutely papillate; stipules (if present) small, undivided, eglandular. Inflorescences terminal, compound with bracts of side branches small, eglandular, coriaceous, floral bracts with one or several pairs of cup- to disc-shaped glands, 0–2 basal ones with 1 female flower, numerous apical ones with 8–15 male flowers; bracteoles absent. Floral buds inclinate and initially covered by their large bracts, later on erect. Male flowers nearly sessile in bud, with short but distinct pedicel during flowering; calyx 2-lobed, partly fused; petals and disc 0; stamens 2, filaments slightly longer than anthers when flowering, fused, anthers basifixed, fused, extrorse, 2-thecate, longitudinally dehiscent; pollen spheroidal, 3-colporate, perforate-tectate, psilate; pistillode 0; receptacle flat. Female flowers nearly sessile; calyx 3-lobed, partly fused, eglandular or rarely with marginal, cup-shaped glands; petals and disc 0; ovary smooth, densely pubescent, with 3 locules, ovules 1 per locule; style short but evident, stigmas 3, undivided, eglandular. Fruits very short-pedicellate, spheroidal, smooth, indistinctly pubescent to glabrous, dehiscing septicidally into 3 mericarps; pericarp dry, woody, moderately thick; septa of mericarps with 1 or 2 vascular strands; columella 3-quetrous, alate, persistent. Seeds 3 per fruit, elliptic, smooth, brown, with small caruncle (only known for two species). Fig.: See Mem. New York Bot. Gard. 17: 185, t. 24 (1967) (as *Sapium*); Novon 3: 247–250, tt. 1–3 (1993).

A genus of three species, confined to Venezuela and Brazil.

303. **Senefelderopsis** *Steyerm.*, Bot. Mus. Leafl. 15: 45 (1951); Jabl., Mem. New York. Bot. Gard. 12: 174 (1965); G. L. Webster, Ann. Missouri Bot. Gard. 81: 119 (1994); Esser, Mitt. Inst. Allg. Bot. Hamburg 25: 121 (1995); Murillo-Aldana & Franco-Rosselli, Euf. Reg. Ararac.: 153 (1995). T.: *S. croizatii* Steyerm.

Monoecious shrubs or trees. Hairs multicellular, uniseriate, usually coloured. Latex white. Leaves alternate but often apically crowded, with distinct eglandular petioles shorter than the lamina, stipulate, simple, entire, penninerved, chartaceous to coriaceous and sometimes very fragile, above with a pair of disc-shaped glands on base, below with a row of submarginal glands, lower surface minutely papillate or whitish-farinose; stipules small, undivided, eglandular. Inflorescences terminal, compound with bracts of side branches small, eglandular, coriaceous, floral bracts with a pair of elliptic glands, 0–2 basal ones with 1 female flower, numerous apical ones with 6–12 male flowers; bracteoles quite similar to bracts, undivided. Floral buds inclinate and initially covered by their large bracts, later on erect. Male flowers nearly sessile in bud, with short but distinct pedicel during flowering; calyx 3-lobed, partly fused; petals and disc 0; stamens 3(5), filaments slightly longer than anthers when flowering, free, anthers basifixed, extrorse, 2-thecate, longitudinally dehiscent; pollen prolate to subprolate, 3-colporate, perforate-tectate, psilate; pistillode 0; receptacle flat. Female flowers shortly but distinctly pedicellate, calyx 3(6)-lobed, partly fused, eglandular; petals and disc 0; ovary smooth, densely pubescent, with 3 locules, ovules 1 per locule; style short but evident, stigmas 3, undivided, eglandular. Fruits short-pedicellate, 3-locular, smooth, indistinctly pubescent and glabrescent, dehiscing septicidally into 3 mericarps; exocarp leathery-fleshy; mesocarp woody, moderately thick; septa of mericarps with 1 vascular strand; columella 3-quetrous, alate, persistent. Seeds 3 per fruit, elliptic, smooth, brown, with small caruncle or ecarunculate. Fig.: See Caldasia 3: 123 (1944) (as *Senefeldera*); Bot. Mus. Leafl. 15: 48, t. 16 (1951); Mitt. Inst. Allg. Bot. Hamburg 25: 124, t. 1 (1995); Euf. Reg. Ararac.: 154, t. 46 (1995).

A genus of two species, confined to the Guayana Highland of Colombia, Venezuela, and Guyana.

304. **Pleradenophora** *Esser*, **genus novum** mesoamericanum a congeneribus plurimis (*Stillingia, Sapio* et cetera) glandulis bractearum plerumque plurimis (dissectis), foliis supra basi saepe bi- usque pluriglandulosis, cymulis masculis plurifloris, sepalis masculis majoribus connatis, staminibus plurimis, fructibus siccis dehiscentibus pericarpio crasso, seminibus ovoideo-globosis manifeste ecarunculatis; praeterea a *Stillingia* ipsa basi fructus non ut carpidiophorum manenti, porro a *Sapio* ipso seminibus cinereo-brunneis exarillatis differt. Typus generis: *P. longicuspis* (Standl.) Esser **comb. nov.** (B.: *Sebastiana longicuspis* Standl., Field Mus. Nat. Hist., Bot. Ser. 11: 134 (1932)).

Monoecious, glabrous shrubs or trees. Latex white. Leaves alternate, shortly petiolate, often with a pair of disc-shaped petiolar glands above on the junction with the lamina, stipulate, simple, serrate, penninerved, chartaceous, above often glandular at base (see petiole), below with marginal glands or eglandular, lower surface smooth; stipules small, entire, eglandular. Inflorescences terminal or laterally displaced, solitary, floral bracts often (always?) with several pairs of disc- to cup-shaped glands, 1–3 basal ones with 1 female flower, numerous apical ones with c. 5–10 male flowers; bracteoles inconspicuous (absent?). Male flowers nearly sessile in bud, shortly pedicellate when flowering; calyx 3-lobed, partly fused; petals and disc 0; stamens 3(5) in largest flowers, 2 in smaller ones, filaments distinct, free, anthers basifixed, extrorse, 2-thecate, longitudinally dehiscent; pistillode 0; receptacle flat. Female flowers (sub)sessile; calyx 3-lobed, free, eglandular; petals and disc 0; ovary smooth, with 3 locules, ovules 1 per locule; style short to absent, stigmas 3, undivided, eglandular. Fruits (sub)sessile, subglobose, smooth, dehiscing septicidally into 3 mericarps; pericarp dry, woody, moderately thick; septa of mericarps usually with one bifurcated vascular strand; columella 3-quetrous, alate, persistent. Seeds 3 per fruit, ovoid-globose, smooth and apically often characteristically foveolate, greyish-brown and often maculate, ecarunculate. Fig.: See Contr. Univ. Michigan Herb. 20: 206, t. 4 (1995) (as *Sebastiania*).

A genus of 3, probably even more species, distributed from Mexico through Guatemala to Belize. *Sapium biloculare* S. Watson from the USA (Arizona) might also belong here.

Pleradenophora has usually in the past been treated as a part of *Sebastiania*, even though it comes closer to *Sapium* and *Stillingia* (leaf glands, many-flowered staminate cymules with fused sepals etc.). The distinctive characters given in the diagnosis above will also serve to distinguish them.

305. **Sclerocroton** *Hochst. in C. Krauss*, Flora 28: 85 (1845); Benth. in Benth. & Hook. f., Gen. Pl. 3(1): 337 (1880) (under *Excoecaria*); Kruijt & Roebers, Biblioth. Bot. 146: 7, 16 (1996). LT.: *S. integerrimus* Hochst. (designated by Kruijt & Roebers, Biblioth. Bot. 146: 17 (1996)).

Excoecaria sect. *Sclerocroton* (Hochst.) Müll. Arg. in DC. Prodr. 15(2): 1213 (1866).
Sapium sect. *Sclerocroton* (Hochst.) T. Post & Kuntze, Lex. Gen. Phan.: 498 (1903).
Sapium subg. *Sclerocroton* sect. *Armata* Pax & K. Hoffm. in Engl., Pflanzenr. 52: 243 (1912); in Engl. & Harms, Nat. Pflanzenfam. ed. 2, 19c: 202 (1931).

Monoecious shrubs or trees. Glabrous. Leaves alternate, with very short eglandular petioles, stipulate, simple, shallowly but distinctly serrate to entire, penninerved, chartaceous, eglandular above, below with a row of submarginal to laminar glands, lower surface often whitish-farinose; stipules small, undivided, eglandular. Inflorescences terminal and axillary, solitary, bracts with one to several pairs of disc- to cup-shaped glands, (0)1(2) basal ones with 1 female flower, numerous apical ones with (3)5–8 male flowers; bracteoles present, undivided. Male flowers subsessile in bud, distinctly pedicellate when flowering, calyx 3-lobed, partly fused; petals and disc 0; stamens 2 or 3, filaments shorter or longer than anthers, free, anthers basifixed, extrorse, 2-thecate, longitudinally dehiscent; pistillode 0; receptacle flat. Female flowers distinctly pedicellate; calyx 3-lobed, free or rarely partly fused, often glandular; petals and disc 0; ovary with three pairs of distinct appendages, with 3 locules, ovules 1 per locule; style distinct, stigmas 3, undivided, eglandular. Fruits pedicellate, spheroidal, with three pairs of appendages, dehiscing septicidally into 3 mericarps; exocarp dry and thin or thick and fleshy and often separating from the mesocarp; mesocarp moderately to very thick (thickness $^1/_{10}$ of fruit length or more); septa of mericarps with 1 vascular strand; columella 3-quetrous, alate, persistent. Seeds 3 per fruit, elliptic to subglobose, smooth, brown or whitish-brownish-grey mottled, with small caruncle or ecarunculate. Fig.: See Pflanzenr. 52: 245, t. 46 and 248, t. 47 (1912) (as *Sapium*); Nat. Pflanzenfam. ed. 2, 19c: 202, t. 107 (1931) (as *Sapium*); Biblioth. Bot. 146: 19–26, tt. 4–9 (1996).

A genus of six species, distributed in Continental Africa (from Guinea through Cameroon to Mozambique and South Africa) and Madagascar.

306. **Balakata** *Esser*, Blumea 44: 154 (1999). T.: *B. luzonica* (S. Vidal) Esser (B.: *Myrica luzonica* S. Vidal).

Sapium sect. *Pleurostachya* Pax & K. Hoffm. in Engl., Pflanzenr. 52: 243 (1912); in Engl. & Harms, Nat. Pflanzenfam. ed. 2, 19c: 202 (1931).

Monoecious glabrous shrubs or trees. Latex white. Leaves alternate, with very short to moderately long eglandular petioles, stipulate, simple, entire, penninerved, chartaceous to coriaceous, eglandular above, below with a row of marginal to submarginal glands, lower surface minutely papillate or smooth; stipules small, undivided, eglandular. Inflorescences terminal and axillary, compound with bracts of side branches small and eglandular, floral bracts with a pair of elliptical, irregularly flattened and sometimes fragmented glands, (0)3–13 basal ones with 1 female flower, numerous apical ones with 5–9 male flowers; bracteoles undivided. Male flowers distinctly pedicellate; calyx 2-lobed, partly fused; petals and disc 0; stamens 2, filaments slightly longer than anthers, free, anthers basifixed, extrorse, 2-thecate, longitudinally dehiscent; pistillode 0; receptacle flat. Female flowers very shortly to distinctly pedicellate; calyx 2-lobed, partly fused, eglandular; petals and disc 0; ovary smooth, with 2 locules, ovules 1 per locule; style short but evident, stigmas 2, undivided, eglandular. Fruits distinctly pedicellate, spheroidal, smooth, indehiscent and berry-like; pericarp fleshy, sometimes with more or less woody endocarp, moderately thick. Seeds 1 or 2 per fruit, depressed-globose, smooth, with a thin sarcotesta and a stony seed coat, ecarunculate. Fig.: See Icon. Pl. Ind. Orient. 5(2): t. 1950 fig. 2 (1853) (as *Sapium*); Sin. Gen. Pl. Leños. Filip. Atlas: t. 90 fig. B (1883) (as *Myrica*); Trees W. Pacif. Region: 185, t. 64 (1951) (as *Sapium*); Blumea 44: 158 fig. 1 (1999).

A genus of two species, ranging from NE India to China and Malesia up to the Philippines and New Guinea.

307. **Triadica** *Lour.*, Fl. Cochinch. ed. 1, 2: 598, 610 (1790), ed. 2, 2: 735, 748 (1793); A. Juss., Euphorb. Gen.: 50 (1824); Hurus., J. Fac. Sci. Univ. Tokyo, Sect. 3, Bot. 6: 315 (1954); G. L. Webster, Ann. Missouri Bot. Gard. 81: 123 (1994); Kruijt, Biblioth. Bot. 146: 7 (1996); Esser *et al.*, Syst. Bot. 22: 620 (1998); Esser, Blumea 44: 197 (1999). LT.: *T. sinensis* Lour. (designated by G. L. Webster, Ann. Missouri Bot. Gard. 81: 123 (1994)) [=*T. sebifera* (L.) Small (S.: *Sapium sebiferum* (L.) Roxb.; B.: *Croton sebiferus* L.)]

Stillingfleetia Bojer, Hortus Maurit.: 284 (1837). T.: *S. sebifera* (L.) Bojer. (B.: *Croton sebiferus* L.) [=*T. sebifera* (L.) Small]

Seborium Raf., Sylva Tellur.: 63 (1838). T.: *S. chinense* Raf. [=*T. sebifera* (L.) Small (B.: *Croton sebiferus* L.)]

Stillingia sect. *Triadica* (Lour.) Baill., Étude Euphorb.: 511 (1858).

Sapium sect. *Triadica* (Lour.) Müll. Arg., Linnaea 32: 121 (1863); Hook. f., Fl. Brit. India 5: 469 (1888); Pax in Engl. & Prantl, Nat. Pflanzenfam. 3(5): 98 (1890); Pax & K. Hoffm. in Engl., Pflanzenr. 52: 237 (1912); in Engl. & Harms, Nat. Pflanzenfam. ed. 2, 19c: 201 (1931); G. L. Webster, J. Arnold Arbor. 48: 392 (1967); Chakrab. & M. G. Gangop., J. Econ. Taxon. Bot. 14: 183 (1990).

Excoecaria sect. *Triadica* (Lour.) Müll. Arg. in DC., Prodr. 15(2): 1210 (1866); and in Mart., Fl. Bras. 11(2): 625 (1874).

Monoecious glabrous trees. Latex white. Leaves alternate, with moderately long eglandular petioles, stipulate, simple, entire, penninerved but triplinerved with basal pair of side veins forming the basal leaf margin and arising at a shallower angle, chartaceous, above with a distinct pair of spheroidal glands on the junction with the petiole apex, below with a submarginal row of glands, lower surface distinctly whitish to papillate; stipules small, undivided, eglandular. Inflorescences terminal and axillary, solitary; bracts with a pair of spheroidal glands, few basal ones with 1 female flower, numerous apical ones with 3–8 male flowers; bracteoles present, similar to bracts. Male flowers distinctly pedicellate; calyx with 3(6) lobes, partly fused; petals and disc 0; stamens 2 or 3, with filaments longer than anthers, free, anthers basifixed, extrorse, 2-thecate, longitudinally dehiscent; pollen 3-colporate, distinctly 3-lobate in polar view, colpus transversalis+, costae+, colpi narrow, tectate-perforate, columellae distinct; pistillode 0; receptacle flat. Female flowers distinctly pedicellate, calyx with 3(6) lobes, partly fused, eglandular or with few spheroidal glands at margin; petals and disc 0; ovary smooth, with 3 locules, ovules 1 per locule; style distinct, stigmas undivided, eglandular. Fruits distinctly pedicellate, subglobose, smooth, dehiscing nearly simultaneously septicidally and loculicidally into 6 valves but without mericarps that enclose the seeds; pericarp dry, woody, moderately thick; septa of mericarps very thin and early caducous, with 1 or 2 vascular strands; columella persistent, triquetrous, conspicuously alate, with attached seeds. Seeds attached to the columella for a considerable time, 3 (very rarely 2) per fruit, depressed-globose, completely covered by a whitish sarcotesta, ecarunculate. Fig.: See Euphorb. Gen.: t. 16, f. 52.1–10 (1824) (as *Stillingia*); Étude Euphorb.: t. 7, f. 26–30 (1858) (as *Stillingia*); Fl. Bras. 11(2): t. 84 (1874) (as *Excoecaria*); Nat. Pflanzenfam. 3(5): 97, t. 63 (1890) (as *Sapium*); N. Amer. Trees: 601, t. 552 (1908) (as *Sapium*); Pflanzenr. 52: 238, t. 44 (1912) (as *Sapium*); Nat. Pflanzenfam. ed. 2, 19c: 201, t. 106 (1931) (as *Sapium*); Fl. Malay Penins. 3: 316, t. 155 (1924) (as *Sapium*); Formos. Trees: 358, t. 313 and 359, t. 314 (1936) (as *Sapium*); J. Fac. Sci. Univ. Tokyo, Sect. 3, Bot. 6: 314, t. 45 (1954); Acta Phytotax. Sin. 5: pl. 22 (1956) (as *Sapium*); Fl. Pakistan 172: 73, t. 15 fig. d–h (1986) (as *Sapium*); Blumea 44: 203 fig. 7 (1999).

A genus of 3 or 4 species, ranging from India to W Malesia and China.

308. **Shirakiopsis** *Esser*, Blumea 44: 184 (1999). T.: *S. indica* (Willd.) Esser (B.: *Sapium indicum* Willd.).

Excoecaria sect. *Parasapium* Müll. Arg., Linnaea 32: 123 (1863), pro parte excl. typ.
Sapium sect. *Parasapium* (Müll. Arg.) Hook. f., Fl. Brit. India 5: 471 (1888), pro parte excl. typ.; Pax & K. Hoffm. in Engl., Pflanzenr. 52: 249 (1912), pro parte excl. typ.
Shirakia Hurus., J. Fac. Sci. Univ. Tokyo, Sect. 3, Bot. 6: 317 (1954), nom. illeg. non *Shirakia* S. Kawas., Bull. Geol. Survey Chosen 6(4): 98 (1934), pro parte excl. typ.; Kruijt, Biblioth. Bot. 146: 91 (1996), pro parte excl. typ.

Monoecious shrubs or trees. Hairs multicellular, unbranched, usually coloured. Latex white. Leaves alternate, with short eglandular petioles, stipulate, simple, shallowly but distinctly serrate, penninerved, chartaceous, eglandular above, below with a row of strictly marginal glands, lower surface paler but smooth; stipules small, undivided, eglandular. Inflorescences terminal, solitary, bracts with a pair of spheroidal-elliptical glands, (0)1–3 basal ones with 1 female flower, numerous apical ones with (3)5–7 male flowers; bracteoles + ? Floral buds inclinate and initially covered by their large bracts, later on erect. Male flowers pedicellate; calyx 2- or 3-lobed, partly fused; petals and disc 0; stamens 2 (African taxa) or 3 (Malesian taxa), filaments nearly as long as anthers, free, anthers basifixed, extrorse, 2-thecate, longitudinally dehiscent; pollen 3-colporate, distinctly 3-lobate in polar view, colpus transversalis+, costae+, colpi narrow, tectate-perforate, columellae distinct; pistillode 0; receptacle flat. Female flowers pedicellate; calyx 2- or 3-lobed, partly fused, eglandular; petals and disc 0; ovary smooth, glabrous, with 2 or 3 locules, ovules 1 per locule; style present, stigmas 2 or 3, undivided, eglandular. Fruits pedicellate, elliptic to transversely elliptic, smooth, glabrous, dehiscing septicidally into 2 or 3 mericarps or hardly or not dehiscent (drupaceous) to a varying degree; exocarp dry or leathery-fleshy; mesocarp woody, moderately thick to very thick (pericarp-thickness $^1/_{10}$ of fruit length or more); septa of mericarps with 1 vascular strand; columella 2- or 3-quetrous, alate. Seeds 2 or 3 per fruit, elliptic to subglobose, smooth, brownish, caruncle very inconspicuous to absent. Fig.: See Icon. Pl. Ind. Orient. 5(2): t. 1866 fig. 1 (1853) (as *Sapium*); Pflanzenr. 52: 250, t. 48 and 253, t. 49 (1912) (as *Sapium*); Nat. Pflanzenfam. ed. 2, 19c: 203, t. 108 (1931) (as *Sapium*); Woody Pl. Ghana: 252, t. 57 (1961) (as *Sapium*); Fl. Pl. Lign. Rwanda: 268, t. 92.1 (1982) (as *Sapium*); Fl. Rwanda 2: 240, t. 73.1 (1983) (as *Sapium*); Fl. Ethiopia and Erithrea 2(2): 330, t. 85.40 (1997) (as *Sapium*); Blumea 44: 188, fig. 5 (1999).

A genus of 6 species, 3 of them in Asia from India and Sri Lanka to Cambodia and throughout Malesia to the Solomon Islands, 3 of them in Africa from Ivory Coast to Ethiopia, Tanzania and South Africa.

309. **Mabea** *Aubl.*, Hist. Pl. Guiane 2: 867 (1775); A. Juss., Euphorb. Gen.: 40 (1824); Benth., Hooker's J. Bot. Kew Gard. Misc. 6: 363 (1854); Müll. Arg. in DC., Prodr. 15(2): 1148 (1866); in Mart., Fl. Bras. 11(2): 515 (1874); Benth. in Benth. & Hook. f., Gen. Pl. 3(1): 254, 331 (1880); Pax in Engl. & Prantl, Nat. Pflanzenfam. 3(5): 92 (1890); Pax & K. Hoffm. in Engl., Pflanzenr. 52: 26 (1912); in Engl. & Harms, Nat. Pflanzenfam. ed. 2, 19c: 187 (1931); Jabl., Mem. New York Bot. Gard. 17: 164 (1967); Hollander & C. C. Berg, Proc. Kon. Ned. Akad. Wetensch. C 89: 147 (1986); A. H. Gentry, Woody Pl. NW. S. Amer.: 420 (1993); Esser, Novon 3: 341 (1993); G. L. Webster, Ann. Missouri Bot. Gard. 81: 119 (1994); Esser, Syst. Hippom.: 76 (1994); Murillo-Aldana & Franco-Rosselli, Euf. Reg. Ararac.: 90 (1995). LT.: *M. piriri* Aubl. (designated by Pfeiffer, Nomencl. Bot. 2(1): 191 (1874)).

Monoecious shrubs or trees, sometimes climbing, or even vines. Hairs multicellular and dendritic, very rarely uniseriate, usually coloured. Latex white. Leaves alternate, with very short to short eglandular petioles, stipulate, simple, serrate to entire, penninerved but sometimes triplinerved, chartaceous to coriaceous, eglandular above, below with a row of marginal to submarginal glands or rarely eglandular, lower surface minutely papillate, whitish-farinose, or smooth; stipules small to moderately large, undivided, often serrate and glandular at base. Inflorescences terminal and axillary, solitary or compound with bracts of side branches small or leaf-like to varying degrees, eglandular or glandular, floral bracts sometimes elevated by a peduncle, with a pair of elliptic glands, rarely eglandular, 0–15 basal ones with one female flower, numerous apical ones with 1–5(8) male flowers; bracteoles absent. Floral buds inclinate and initially covered by their large bracts, later on erect. Male flowers distinctly pedicellate; calyx (4)5(6)-lobed, partly fused; petals and disc 0; stamens 3–c.100, filaments nearly absent to distinct, free, anthers basifixed, extrorse, 2-thecate, longitudinally dehiscent; pollen suboblate to prolate, 3-colporate, hardly to distinctly 3-lobate in polar view, colpus transversalis+, costae+, colpi narrow, tectate-perforate to semitectate, columellae present, psilate to striate; pistillode 0; receptacle elevated. Female flowers distinctly pedicellate; calyx (3)6(9)-lobed, partly fused, eglandular or with few to numerous marginal, elliptical glands; petals and disc 0; ovary smooth or with 3 pairs of appendages, densely pubescent, with 3 locules, ovules 1 per locule; style short to very long, stigmas 3, undivided, eglandular. Fruits distinctly pedicellate, spheroidal to elliptic, smooth or with 3 pairs of appendages, densely pubescent, dehiscing septicidally into 3 mericarps; pericarp dry, woody, moderately thick; septa of mericarps with 1 vascular strand; columella 3-quetrous, alate, persistent. Seeds 3 per fruit, elliptic to depressed-globose, smooth, brown or rarely greyish-brownish mottled, with small caruncle or ecarunculate. Fig.: See Hist. Pl. Guiane 4: t. 334 (1775); Euphorb. Gen.: t. 13, f. 39.1–9 (1824); Étude Euphorb. Atlas: t. 13 fig. 19–28 (1858); Fl. Bras. 11(2): tt. 73, 74 (1874); Nat. Pflanzenfam. 3(5): 93, t. 58 (1890); Trans. Linn. Soc. London, Bot. 2, 4: t. 30 fig. 1–5 (1905); Pflanzenr. 52: 29, t. 4 and 33, t. 5 and 41, t. 6 (1912); Dicc. Pl. Uteis Brasil: 499 (1926); Nat. Pflanzenfam. ed. 2, 19c: 188, t. 99 (1931); Ann. Missouri Bot. Gard. 54: 322, t. 21 (1967); Amer. J. Bot. 56: 754, t. 14 (1969); Bol. Mus. Nac. Rio de Janeiro, Bot. 62: t. 2 following page 4 (1981); Proc. Kon. Ned. Akad. Wetensch. C 89: 147–153, tt. 1 and 2 (1986); Bradea 4(47): 374, t. 1 and 375, t. 2 (1987); Ann. Missouri Bot. Gard. 75: 1126, t. 4 (1988); Boissiera 44: 70, t. 25 and 71, t. 26 (1990); Árvores Brasileiras: 106 (1992); Woody Pl. NW. S. Amer.: 421, t. 121.5 (1993); Novon 3: 342–350, tt. 1–5 (1993); Euf. Reg. Ararac.: 94–100, tt. 21–24 (1995); Fieldiana, Bot. n.s. 36: 44, t. 31 p.p. (1995).

A genus of 39 species, ranging from Mexico to Bolivia and Brazil, but absent from the Antilles.

310. **Gymnanthes** *Sw.*, Prodr.: 95 (1788); Klotzsch, Arch. Naturgesch. 7(1): 182 (1841); Baill., Étude Euphorb.: 530 (1858); Benth. in Benth. & Hook. f., Gen. Pl. 3(1): 255, 337 (1880); Pax in Engl. & Prantl, Nat. Pflanzenfam. 3(5): 101 (1890); Pax & K. Hoffm. in Engl., Pflanzenr. 52: 81 (1912); Fawc. & Rendle, Fl. Jamaica 4: 329 (1920); Pax & K. Hoffm. in Engl. & Harms, Nat. Pflanzenfam. ed. 2, 19c: 191 (1931); Alain, Fl. Cuba 3: 121 (1953); G. L. Webster, J. Arnold Arbor. 48: 387 (1967); Taxon 32: 304 (1983); G. L. Webster

& Huft, Ann. Missouri Bot. Gard. 75: 1129 (1988), pro parte excl. *Actinostemon*; G. L. Webster, Ann. Missouri Bot. Gard. 81: 122 (1994), pro parte excl. *Actinostemon*; Esser, Syst. Hipp.: 45 (1994); Gillespie in Acev.-Rodr., ed., Fl. St. John: 213 (1996); Esser *et al.*, Syst. Bot. 22: 619, 624 (1998); Esser, Blumea 44: 165 (1999). LT.: *G. elliptica* Sw. (designated by Britton & Shafer, N. Amer. Trees: 600 (1908)).

? *Sapium* P. Browne, Civ. Nat. Hist. Jamaica 2: 338 (1756). NT.: *Excoecaria glandulosa* Sw. (designated by Kruijt & Zijlstra, Taxon 38: 321 (1989)) [=*G.* sp.?]

? *Ateramnus* P. Browne, Civ. Nat. Hist. Jamaica 2: 339 (1756); Hallier f., Meded. Rijks-Herb. 36: 4 (1918); Pax & K. Hoffm. in Engl., Pflanzenr. 85: 204 (1924); in Engl. & Harms, Nat. Pflanzenfam. ed. 2, 19c: 207 (1931); Rothm., Feddes Repert. Spec. Nov. Regni Veg. 53: 5 (1944); C. D. Adams, Fl. Pl. Jamaica: 425 (1972); G. L. Webster, Taxon 32: 304 (1983); Proctor, Kew Bull. Addit. Ser. 11: 544 (1984); Oe, Rev. Ateramnus: 2 (1988); Kruijt & Zijsltra, Taxon 38: 322 (1989); G. L. Webster, Ann. Missouri Bot. Gard. 81: 122 (1994); Esser, Blumea 44: 165 (1999). LT.: *A. lucidus* (Sw.) Rothm. (B.: *Gymnanthes lucida* Sw.) (designated by Rothm., Feddes Repert. Spec. Nov. Regni Veg. 53: 5 (1944)) (NT.: *A. glandulosus* (Sw.) C. D. Adams (B.: *Excoecaria glandulosa* Sw.), designated by Kruijt & Zijlstra, Taxon 38: 322 (1989), is refused here; see Esser, Blumea 44: 165 (1999)).

Adenogyne Klotzsch, Arch. Naturgesch. 7(1): 183 (1841); Esser, Syst. Hipp.: 46 (1994). T.: *A. pachystachys* Klotzsch [=*G. pachystachys* (Klotzsch) Müll. Arg.]

Sarothrostachys Klotzsch, Arch. Naturgesch. 7(1): 185 (1841); Esser, Syst. Hipp.: 46 (1994); Welzen & Esser, Blumea 42: 256 (1997); Esser, Blumea 44: 165 (1999). LT.: *S. multiramea* Klotzsch ex Wawra (designated by L. C. Wheeler, Taxon 24: 537 (1975)) [=*G. multiramea* (Klotzsch ex Wawra) Müll. Arg.]

Sebastiania subsect. *Sarothrostachys* (Klotzsch) Müll. Arg. in DC., Prodr. 15(2): 1175 (1866).

Sebastiania sect. *Adenogyne* (Klotzsch) Benth. in Benth. & Hook. f., Gen. Pl. 3: 336 (1880); Pax in Engl. & Prantl, Nat. Pflanzenfam. 3(5): 94 (1890); Pax & K. Hoffm. in Engl., Pflanzenr. 52: 124 (1912); in Engl. & Harms, Nat. Pflanzenfam. ed. 2, 19c: 194 (1931).

Sebastiania sect. *Sarothrostachys* (Klotzsch) Benth. in Benth. & Hook. f., Gen. Pl. 3: 336 (1880); Pax in Engl. & Prantl, Nat. Pflanzenfam. 3(5): 94 (1890); Pax & K. Hoffm. in Engl., Pflanzenr. 52: 118 (1912); in Engl. & Harms, Nat. Pflanzenfam. ed. 2, 19c: 193 (1931).

Duvigneaudia J. Léonard, Bull. Jard. Bot. État 29: 15 (1959); Fl. Cong. Rwa.-Bur. 8(1): 139 (1962); G. L. Webster, Ann. Missouri Bot. Gard. 81: 123 (1994); Kruijt & Roebers in Kruijt, Biblioth. Bot. 146: 12 (1996); Esser, Syst. Hipp.: 47 (1994); Blumea 44: 165 (1999). T.: *D. inopinata* (Prain) J. Léonard (B.: *Sebastiania inopinata* Prain, Bull. Misc. Inform., Kew 1910: 128 (1910)) [=*G. inopinata* (Prain) Esser **comb. nov.**]

Monoecious or dioecious shrubs or trees, lateral short shoots sometimes transformed into thorns. Indumentum absent or consisting of multicellular, unbranched, coloured hairs. Latex white. Leaves alternate, sometimes apically crowded, with short eglandular petioles, stipulate, simple, entire or serrate, penninerved, chartaceous to coriaceous, eglandular above, below with a row of strictly marginal to submarginal glands (laminar glands in *G. hypoleuca* Benth.), lower surface smooth, sometimes whitish-farinose or minutely papillate; stipules small, undivided, eglandular or glandular at base. Inflorescences terminal and axillary, solitary or compound near base with bracts of side branches small and undivided, floral bracts sometimes elevated by a peduncle, with a pair of elliptical, sometimes stipitate glands, rarely eglandular, 0–3 basal ones with 1 female flower, numerous apical ones with 1–3(5) male flowers; bracteoles undivided. Floral buds inclinate and initially covered by their large bracts, later on erect. Male flowers distinctly pedicellate (subsessile only when bracts pedunculate); calyx 3-lobed, largely fused, or minute and free or even completely absent, rarely unilateral, inclinate; petals and disc 0; stamens (2)3–12, filaments nearly as long as or longer than anthers, free or basally fused, anthers basifixed, extrorse, 2-thecate, longitudinally dehiscent; pollen 3-colporate, hardly to distinctly 3-lobate in polar view, colpus transversalis+, costae+, colpi narrow, margo present or absent, tectate-perforate, columellae present; pistillode 0; receptacle flat. Female flowers pedicellate, pedicel often conspicuously elongating after flowering; calyx 3-lobed, partly fused or free; petals and disc 0; gynophore present in one species; ovary smooth or with 3 pairs of appendages, glabrous or pubescent, with 3 locules, ovules 1 per locule; styles short but distinct, stigmas 3, undivided, eglandular. Fruits distinctly, often very long pedicellate, spheroidal to elliptic, smooth or with 3 pairs of appendages, glabrous or pubescent, dehiscing septicidally into 3 mereicarps or rarely indehiscent; pericarp usually dry, woody, moderately thick; septa of mericarps with 1(2) vascular strands; columella 3-quetrous, alate, persistent. Seeds 3 per fruit, elliptic, smooth, brownish, with small caruncle or ecarunculate. Fig.: See Euphorb. Gen.: t. 16, f. 55.1–8 (1824) (as *Excoecaria*); Étude Euphorb.: t. 5, f. 19–20 (1858) (as *Adenogyne*); Fl. Bras. 11(2): t. 79 (1874) (as *Sebastiania*); Silva N. Amer. 7: t. 309 (1895), n. v.; N. Amer. Trees: 599, t. 551 (1908); Pflanzenr. 52: 83, t. 15 and 121–130, tt. 23–25 (1912) (partly as *Sebastiania*); Fl. Jamaica 4: 330, t. 111 (1920); Man. Trees N. Amer. 2: 65, t. 593 (1922); Nat. Pflanzenfam. ed. 2, 19c: 193, t. 102 (1931); Bull. Jard. Bot. État 29: 19, t. 1 (1959) (as *Duvigneaudia*); Fl. Cong. Rwa.-Bur. 8(1): 141, t. 10 (1962) (as *Duvigneaudia*); Hooker's Icon. Pl. 38(1): t. 3723 (1974) (as *Sebastiania*); Fl. Baham. Archip.: 783, t. 326 (1982); Kew Bull. Addit. Ser. 11: 545, t. 166 (1984) (as *Ateramnus*); Fl. Española 4: 157, t. 116-19 (1986); Tropenbos ser. 7: t. 103 (1994) (as *Sebastiania*); Fl. St. John: 214, t. 96A–G (1996); Biblioth. Bot. 146: 15, t. 3 (1996) (as *Duvigneaudia*); Blumea 44: 171 fig. 3 (1999).

A genus of c. 45 species in this expanded circumscription, most of them in the New World from the USA (Florida) through the Antilles to S America up to Paraguay, few species in Africa (Congo basin from Cameroun to Congo (Kinshasa)) and Asia (Malay Peninsula, Sumatra, Borneo).

311. **Ditrysinia** *Raf.*, Neogenyton: 2 (1825); Esser, Syst. Hipp.: 41 (1994). T.: *D. ligustrina* (Michx.) Raf. (B.: *Stillingia ligustrina* Michx.).

Gymnanthes sect. *Stillingiopsis* Müll. Arg., Linnaea 32: 96 (1863).
Gymnanthes sect. *Ditrysinia* Müll. Arg., Linnaea 34: 216 (1865).
Sebastiania sect. *Ditrysinia* (Raf.) Müll. Arg. in DC., Prodr. 15(2): 1165 (1866); Benth. in Benth. & Hook. f., Gen. Pl. 3(1): 336 (1880); Pax in Engl. & Prantl, Nat. Pflanzenfam. 3(5): 94 (1890); Pax & K. Hoffm. in Engl., Pflanzenr. 52: 117 (1912); in Engl. & Harms, Nat. Pflanzenfam. ed. 2, 19c: 192 (1931); Small, Manual SE Flora: 789 (1933); Correll & M. C. Johnst., Manual Vasc. Pl. Texas: 951 (1970).
Sebastiania sect. *Stillingiopsis* (Müll. Arg.) G. L. Webster, J. Arnold Arbor. 48: 386 (1967).

Monoecious shrubs. Hairs short and stiff to absent. Leaves alternate, with short eglandular petioles, stipulate, simple, entire, penninerved, chartaceous, eglandular above, below with scattered laminar glands, lower surface smooth; stipules small, undivided, eglandular. Inflorescences terminal, solitary, bracts with a pair of elongate kidney-shaped glands, 2–6 basal ones with 1 female flower, numerous apical ones with 1 male flower; bracteoles absent. Male flowers shortly but distinctly pedicellate; calyx 3-lobed, partly fused; petals and disc 0; stamens 3, filaments short, free, anthers basifixed, extrorse, 2-thecate, longitudinally dehiscent; pistillode 0; receptacle flat. Female flowers shortly to moderately pedicellate; calyx 3-lobed, nearly free, eglandular; petals and disc 0; ovary smooth, glabrous, with 3 locules, ovules 1 per locule; style short but evident, stigmas 3, undivided, eglandular. Fruits pedicellate, spheroidal, smooth, glabrous, dehiscing septicidally into 3 mericarps; pericarp dry, woody, moderately thick; septa of mericarps with 1 vascular strand; columella 3-quetrous, alate, persistent. Seeds 3 per fruit, elliptic, smooth, brown, with small caruncle. Fig.: See Manual SE Flora: 789 (1933) (as *Sebastiania*).

A monotypic genus from the Atlantic states of the USA.

312. **Anomostachys** *(Baill.) Hurus.*, J. Fac. Sci. Univ. Tokyo, Sect. 3, Bot. 6: 311 (1954); Kruijt, Biblioth. Bot. 146: 8 (1996). T.: *A. lastellei* (Müll. Arg.) Kruijt (B.: *Stillingia lastellei* Müll. Arg.)

Stillingia sect. *Anomostachys* Baill., Étude Euphorb.: 525 (1858).
Excoecaria sect. *Anomostachys* (Baill.) Müll. Arg. in DC., Prodr. 15(2): 1218 (1866); Pax & K. Hoffm. in Engl., Pflanzenr. 52: 159 (1912); in Engl. & Harms, Nat. Pflanzenfam. ed. 2, 19c: 196 (1931).

Monoecious glabrous shrubs. Latex whitish. Leaves alternate, with very short eglandular petioles, stipulate, simple, entire, penninerved, chartaceous, eglandular above, below with few marginal glands, shallowly serrate to entire, lower surface dull; stipules small, undivided, early deciduous. Inflorescences axillary, branched; bracts eglandular, few basal ones with 1 female flower, numerous apical ones with 2–8 male flowers; bracteoles + ?. Male flowers sessile; calyx with 3 lobes, partly fused; petals and disc 0; stamens 3, filaments short; pistillode 0; receptacle flat. Female flowers shortly but distinctly pedicellate; calyx with 3 free and often bifid lobes, eglandular; petals and disc 0; ovary with 4 or 6 appendages, with 2 or 3 locules, ovules 1 per locule; style short but present, stigmas undivided, eglandular. Fruits shortly but distinctly pedicellate, 1- or 2-locular, with appendages when young but smooth when mature, indehiscent; pericarp thin, dry; columella and septa unknown. Seeds 1 or 2 per fruit, elliptic to globose, smooth, brown, with a small caruncle. Fig.: See Biblioth. Bot. 146: 10, t. 2 (1996).

A possibly monotypic genus, endemic to Madagascar.

This genus is currently under study by Valéry Malecot. Two species of dubious affinity (for example, *Excoecaria perrieri* Leandri), both endemic to Madagascar, may belong here, if they do not represent an undescribed genus. They differ from typical *Anomostachys* in having the leaves glandular above, in having simple thyrses, and in having subsessile female flowers (Malecot, pers. comm.)

313. **Neoshirakia** *Esser*, Blumea 43: 129 (1998). T.: *N. japonica* (Siebold & Zucc.) Esser (B.: *Stillingia japonica* Siebold & Zucc.).

Excoecaria sect. *Parasapium* Müll. Arg., Linnaea 32: 123 (1863), pro parte quoad typus.
Sapium sect. *Parasapium* (Müll. Arg.) Hook. f., Fl. Brit. India 5: 471 (1888), pro parte quoad typus; Pax & K. Hoffm. in Engl., Pflanzenr. 52: 249 (1912), pro parte quoad typus; Ohwi, Fl. Japan: 591 (1965).
Shirakia Hurus., J. Fac. Sci. Univ. Tokyo, Sect. 3, Bot. 6: 317 (1954), pro parte quoad typus, nom. illeg. [non *Shirakia* S. Kawas., Bull. Geol. Survey Chosen 6(4): 98 (1934)]; Kruijt, Biblioth. Bot. 146: 91 (1996); Esser *et al.*, Syst. Bot. 22: 625 (1998). T.: *S. japonica* (Siebold & Zucc.) Hurus. (B.: *Stillingia japonica* Siebold & Zucc.).

Monoecious, glabrous shrubs or trees. Leaves alternate, shortly petiolate with the basilaminar glands sometimes displaced onto the petiole, stipulate, simple, entire to irregularly and shallowly serrate, penninerved, chartaceous, with a pair of disc-shaped, stipitate glands at the junction with the petiole, below with a row of submarginal to laminar glands, lower surface smooth, whitish; stipules distinct, more than 5 mm long, undivided, eglandular?. Inflorescences terminal and axillary, solitary, bracts with a pair of elliptical glands, 0–2 basal ones with 1 female flower, numerous apical ones with 3 male flowers; bracteoles + ?. Male flowers distinctly pedicellate, calyx 3-lobed, partly fused; petals and disc 0; stamens (2)3, filaments nearly as long as anthers, free, anthers basifixed, extrorse, 2-thecate, longitudinally dehiscent; pistillode 0; receptacle flat. Female flowers distinctly pedicellate; calyx 3-lobed, free or rarely partly fused, usually eglandular; petals and disc 0; ovary smooth, glabrous, with 3 locules, ovules 1 per locule; style distinct, stigmas 3, undivided, eglandular. Fruits distinctly pedicellate, spheroidal, smooth, glabrous, dehiscing septicidally into 3 mericarps; pericarp dry, woody, moderately thick; columella 3-quetrous, alate, persistent. Seeds 3 per fruit, globose, smooth, brown or whitish-brownish-grey mottled, ecarunculate. Fig.: See J. Fac. Sci. Univ. Tokyo, Sect. 3, Bot. 6: 312, t. 44 fig. a–d and 316, t. 46 (1954) (as *Shirakia*).

A monotypic genus ranging from C. China to Korea and Japan.

314. **Pseudosenefeldera** *Esser*, **genus novum** occidentali-amazonicum a congeneribus plurimis (*Anomostachyde*, *Gymnanthe* et cetera) foliis plerumque pseudoverticillatis supra eglandulosis, thyrsis compositis, glandulis paginarum inferiorum foliorum superficialibus neque marginalibus, ramulis lateralibus inflorescentiarum primum omnino bractea unica neque bracteis plurimis tectis, floribus masculis per anthesin valde zygomorphis, seminibus laevibus chalaza dorsali instructis differt. Typus generis: *P. inclinata* (Müll. Arg.) Esser **comb. nov.** (B.: *Senefeldera inclinata* Müll. Arg. in Mart., Fl. Bras. 11(2): 530 (1874)).

Gen. nov., Esser, Syst. Hipp.: 274 (1994). T.: *P. inclinata* (Müll. Arg.) Esser (B.: *Senefeldera inclinata* Müll. Arg.).
Senefeldera sect. *Inclinatae* Pax in Engl., Pflanzenr. 52: 25 (1912); in Engl. & Harms, Nat. Pflanzenfam. ed. 2, 19c: 187 (1931); Crepet & Daghlian, Amer. J. Bot. 69: 262 (1982).

Monoecious trees. Hairs multicellular, uniseriate, usually yellowish-brownish. Latex white. Leaves pseudo-verticillate, rarely crowded-alternate, with distinct eglandular petioles, stipulate, simple, entire, penninerved, chartaceous to coriaceous, eglandular above, below with laminar glands quite close to margin, lower surface shiny and smooth; stipules small, undivided, very soon caducous. Inflorescences axillary, yellowish, compound with bracts of side branches conspicuously large, membranaceous, eglandular or with laminar glands, each bract initially covering a whole lateral branch, floral bracts eglandular, 0 or 1 basal one with 1 female flower, numerous apical ones with 1 or 3 male flowers; bracteoles small, undivided. Floral buds inclinate and initially covered by their large bracts, later on erect. Male flowers distinctly pedicellate; calyx distinctly zygomorphic and unilateral, completely fused, without lobes; petals and disc 0; stamens 4–17, filaments hardly longer than anthers, free or fused at base, anthers basifixed, extrorse, 2-thecate, longitudinally dehiscent; pollen spheroidal, 3-colporate, distinctly 3-lobate in polar view, ?colpus transversalis+, ?costae+, ?colpi narrow, tectate-perforate, slightly striate, columellae distinct; pistillode 0; receptacle flat to slightly elevated. Female flowers distinctly pedicellate; calyx 3-lobed, partly fused, eglandular; petals and disc 0; ovary smooth, glabrous, with 3 locules, ovules 1 per locule; style distinct, stigmas 3, undivided, eglandular. Fruits short-pedicellate, spheroidal, smooth, glabrous, dehiscing nearly simultaneously septicidally and loculicidally into 6 valves that do not enclose the seeds; pericarp dry, woody, moderately thick; septa of mericarps very narrow, without any visible vascular strand; columella 3-quetrous, very distinctly alate, persistent. Seeds 3 per fruit, subglobose, with dorsal chalaza, smooth, brown, ecarunculate. Fig.: See Fl. Bras. 11(2): t. 75 fig. 2 (1874) (as *Senefeldera*); Mem. New York Bot. Gard. 12: 172, t. 27 fig. K–V (1965) (as *Senefeldera*); Amer. J. Bot. 69: 263, tt. 11–15 and 265, t. 21 (1982) (as *Senefeldera*); Syst. Hipp.: t. 36 (1994).

A monotypic genus of Amazonian South America from Venezuela to Peru.

315. **Actinostemon** *Mart. ex Klotzsch,* Arch. Naturgesch. 7(1): 184 (1841); Baill., Étude Euphorb.: 531 (1858); Müll. Arg., Linnaea 32: 108 (1863); in DC., Prodr. 15(2): 1192 (1866); and in Mart., Fl. Bras. 11(2): 591 (1974); Benth. in Benth. & Hook. f., Gen. Pl. 3(1): 255, 338 (1880); Pax in Engl. & Prantl, Nat. Pflanzenfam. 3(5): 99 (1890); Pax & K. Hoffm. in Engl., Pflanzenr. 52: 57 (1912); in Engl. & Harms, Nat. Pflanzenfam. ed. 2, 19c: 190 (1931); Jabl., Phytologia 18: 213 (1969). T.: *A. grandifolius* Klotzsch. [?=*A. concolor* (Spreng.) Müll. Arg. (B.: *Gussonia concolor* Spreng.)]

?*Gussonia* Spreng., Neue Entd. 2: 119 (1821); Klotzsch, Arch. Naturgesch. 7(1): 183 (1841); Kuntze, Revis. Gen. Pl. 2: 606 (1891); Esser, Syst. Hipp.: 34 (1994). LT.: *G. discolor* Spreng. (designated by G. L. Webster, Ann. Missouri Bot. Gard. 81: 121 (1994)).

Dactylostemon Klotzsch, Arch. Naturgesch. 7(1): 181 (1841); Müll. Arg. in DC., Prodr. 15(2): 1195 (1866); in Mart., Fl. Bras. 11(2): 599 (1874); Benth. in Benth. & Hook. f., Gen. Pl. 3(1): 338 (1880). T.: *D. glabrescens* Klotzsch [=*A. angustifolius* (Müll. Arg.) Pax (B.: *D. angustifolius* Müll. Arg.)]

Actinostemon subg. *Dactylostemon* (Klotzsch) Pax in Engl., Pflanzenr. 52: 59 (1912).

Monoecious shrubs or trees. Hairs multicellular, uniseriate, yellowish-brownish, sometimes absent. Apical buds often (always?) enclosed by the conspicuous large stipules. Leaves alternate to pseudoverticillate, with short eglandular petioles, stipulate, simple, entire, penninerved but sometimes distinctly triplinerved, membranaceous to chartaceous, eglandular above, below with laminar glands, lower surface smooth and shiny or whitish; stipules often large (at least 5 mm long), entire, eglandular. Inflorescence buds covered by large, usually conspicuously parallel-veined, sterile scales. Inflorescences terminal and/or axillary, solitary or basally compound with crowded branches, with bracts of side branches small and inconspicuous, eglandular, floral bracts irregular, small to cilia-like or nearly absent, eglandular or with a pair of elliptical glands, few basal ones with 1 female flower, numerous apical ones with 1–3(6) male flowers; bracteoles absent. Floral buds initially inclinate and protected by sterile scales. Male flowers pedicellate; calyx very small to completely absent, often 3-lobed, free; petals and disc 0; stamens c. 3–12, filaments longer than anthers, free or basally fused, anthers basifixed, extrorse, 2-thecate, longitudinally dehiscent; pollen 3-colporate, distinctly 3-lobate in polar view, colpus transversalis+, costae+, colpi narrow, tectate-perforate, columellae short; pistillode absent; receptacle nearly flat. Female flowers shortly distinctly pedicellate, pedicel sometimes conspicuously elongating after flowering; calyx absent to 3-lobed, eglandular, free; petals and disc 0, ovary smooth or with 3 pairs of appendages (echinate in one poorly known of doubtful affinity), glabrous or pubescent, with 3 locules, ovules 1 per locule; style absent to distinct, stigmas 3, undivided, eglandular. Fruits short- to very conspicuously pedicellate, elliptic to globose, smooth or with 3 pairs of appendages (echinate in one poorly known species), glabrous to pubescent, dehiscing septicidally into 3 mericarps; pericarp dry, woody, moderately thick; septa of mericarps with 1, sometimes divided, vascular strand; columella triquetrous, alate, persistent. Seeds 3 per fruit, elliptic, smooth, brown, with a small caruncle. Fig.: See Arch. Naturgesch. 7(1): t. 8 fig. A and D (1841) (partly as *Dactylostemon*); Étude Euphorb.: t. 5, f. 17–18 (1858); Fl. Bras. 11(2): tt. 82, 83 (1874) (partly as *Dactylostemon*); Pflanzenr. 52: 67–77, tt. 11–14 (1912); Nat. Pflanzenfam. ed. 2, 19c: 191, t. 101 fig. a, b (1931); Sellowia 11: t. 27 fig. h–m (1959); Phytologia 18: 233 and 235, tt. 2 and 3 (1969); Bradea 4(15): 97, t. 1 (1984); Fl. Ilustr. Catar. EUFO: 287, t. 37 fig. h–m (1988).

A genus of 13 species, ranging from Cuba, the Lesser Antilles and Venezuela to Uruguay and Argentina.

A. concolor (Spreng.) Müll. Arg. (B.: *Gussonia concolor* Spreng.), the species usually cited as lectotype of *Actinostemon,* was excluded from the latter by Klotzsch himself and cannot be a lectotype. *A. grandifolius,* however, is Klotzsch's only valid species (validated by an illustration) and therefore the correct type.

316. **Senefeldera** *Mart.*, Flora 24, Beibl. 2: 29 (1841); Klotzsch, Arch. Naturgesch. 7(1): 184 (1841); Baill., Étude Euphorb.: 535 (1858); Müll. Arg. in DC., Prodr. 15(2): 1153 (1866); in Mart., Fl. Bras. 11(2): 529 (1874), pro parte excl. *Pseudosenefeldera*; Baill., Hist. Pl. 5: 227 (1874); Benth. in Benth. & Hook. f., Gen. Pl. 3(1): 254, 332 (1880), pro parte; Pax in Engl. & Prantl, Nat. Pflanzenfam. 3(5): 93 (1890); Croizat, J. Wash. Acad. Sci. 33: 15 (1943), pro parte; Jabl., Mem. New York Bot. Gard. 12: 171 (1965), pro parte; G. L. Webster & Huft, Ann. Missouri Bot. Gard. 75: 1127 (1988); A. H. Gentry, Woody Pl. NW. S. Amer.: 423 (1993); G. L. Webster, Ann. Missouri Bot. Gard. 81: 119 (1994), pro parte; Esser, Syst. Hippom.: 245 (1994). T.: *S. multiflora* Mart.

Senefeldera sect. *Eusenefeldera* Pax in Engl., Pflanzenr. 52: 23 (1912), nom. inval.; in Engl. & Harms, Nat. Pflanzenfam. ed. 2, 19c: 187 (1931).

Monoecious shrubs or trees. Hairs multicellular, uniseriate, pale, or absent. Latex white. Leaves pseudoverticillate, rarely subalternate, with moderately long eglandular petioles, stipulate, simple, entire, penninerved, chartaceous, above with one or two disc-shaped glands on base, below with scattered laminar glands, lower surface shiny and smooth, neither papillate nor whitish; stipules small, undivided, very soon caducous. Inflorescences axillary, compound with bracts of side branches small, often glandular, coriaceous, floral bracts with a pair of elliptic glands, (1)2 or 3 basal ones with one female flower each, numerous apical ones with 1 or 3 male flowers; bracteoles quite similar to bracts, undivided. Male flowers nearly sessile to shortly pedicellate; calyx completely fused, without distinct lobes; petals and disc 0; stamens 6–13, filaments slightly shorter than anthers when flowering, free or partly fused at base, anthers basifixed, extrorse, 2-thecate, longitudinally dehiscent; pollen spheroidal, 3-colporate, perforate-tectate, psilate; pistillode 0; receptacle elevated. Female flowers shortly pedicellate; calyx 3-lobed, nearly free, eglandular or with small elliptical marginal glands; petals and disc 0; ovary smooth, glabrous, with 3 locules, ovules 1 per locule; style absent to short, stigmas 3, undivided, eglandular. Fruits short-pedicellate, spheroidal, smooth, glabrous, dehiscing first loculicidally and immediately afterwards septicidally, therefore never with mericarps that enclose the seeds; pericarp dry, woody, moderately thick; septa of mericarps with 1 sometimes divided vascular strand; columella 3-quetrous, alate, persistent, often splitting from top. Seeds 3 per fruit, elliptic, with patches of smooth and others of colliculate or foveolate cells, brown and yellowish mottled, with caruncle. Fig.: See Étude Euphorb. Atlas: t. 9 fig. 30, 31 (1858); Fl. Bras. 11(2): t. 75 fig. 1 (1874); Pflanzenr. 52: 24, t. 3 (1912); Woody Pl. NW. S. Amer.: 421, t. 121.7 (1993); Syst. Hipp.: t. 33 (1994).

A genus of 3 species, ranging from Panama to Brazil.

317. **Rhodothyrsus** *Esser*, Brittonia 51: 177 (1999). T.: *R. macrophyllus* (Ducke) Esser (B.: *Senefeldera macrophylla* Ducke).

Monoecious trees. Hairs unicellular and malphigiaceous-dibrachiate or multicellular and uniseriate, usually yellowish-brownish. Latex white. Leaves alternate, with long, apically pulvinate, eglandular petioles, stipulate, simple, entire, penninerved, chartaceous, eglandular above, below with laminar glands usually very close to or even touching the midrib, lower surface shiny and smooth; stipules small, undivided, often glandular at base. Inflorescences terminal, purple or orange, repeatedly compound with bracts of side branches small, scaly and usually glandular, floral bracts with small and inconspicuous, elliptic marginal glands, 0–2 basal ones with a single female flower each, numerous apical ones with 1 or 3 male flowers; bracteoles small, undivided. Floral buds inclinate and initially covered by their large bracts. Male flowers shortly pedicellate, inclinate for a long time and densely appressed to the central axis, only erect when flowering; calyx completely fused, without lobes; petals and disc 0; stamens 2–6, filaments short and fused only at base, anthers basifixed, extrorse, 2-thecate, longitudinally dehiscent; pollen oblate-spheroidal, 3-colporate, slightly 3-lobate in polar view, colpus transversalis+, costae+, colpi narrow, margo small to absent, semi-tectate reticulate, columellae large, psilate; pistillode 0; receptacle elevated. Female flowers shortly but distinctly pedicellate; calyx 3-lobed, partly fused, very often with marginal, disc-shaped glands; petals and disc 0; ovary smooth, glabrous or hirsute, with 3 locules, ovules 1 per locule; style short but distinct, stigmas 3, undivided, eglandular. Fruits short-pedicellate, spheroidal, smooth, glabrous or hirsute, dehiscing septicidally and loculicidally into 3 mericarps or 6 valves; pericarp dry, woody, moderately thick; septa of mericarps quite narrow, with 0 or 1 vascular strand; columella 3-quetrous, very distinctly alate, persistent. Seeds 3 per fruit, subglobose, with patches of smooth and brown and others of sculptured and paler cells, ecarunculate. Fig.: See Mem. New York Bot. Gard. 12: 172, t. 27 fig. A–J (1965) (as *Senefeldera*); Euf. Reg. Ararac.: titlepage and 152, t. 45 (1995) (as *Senefeldera*); Brittonia 51: 172–174, tt. 1–3 (1999).

A genus of two species, confined to NW Venezuela and to Amazonia from Brazil and the Guianas to Peru and Colombia.

318. **Dalembertia** *Baill.*, Étude Euphorb.: 545 (1858), and Ann. Sci. Nat. Bot., sér. 4, 9: 195 (1858); Müll. Arg. in DC., Prodr. 15(2): 1225 (1866); Baill., Adansonia 11: 124 (1873); Hist. Pl. 5: 232 (1874); Benth. in Benth. & Hook. f., Gen. Pl. 3(1): 255, 339 (1880); Pax in Engl. & Prantl, Nat. Pflanzenfam. 3(5): 100 (1890); Pax & K. Hoffm. in Engl., Pflanzenr. 52: 268 (1912); in Engl. & Harms, Nat. Pflanzenfam. ed. 2, 19c: 205 (1931); Miranda, Anales Inst. Biol. Univ. Nac. México 14: 34 (1943); Standl. & Steyerm., Fieldiana Bot. 24(6): 86 (1949); G. L. Webster, Ann. Missouri Bot. Gard. 81: 122 (1994). T.: *D. populifolia* Baill.

Alcoceria Fernald, Proc. Amer. Acad. Arts 36: 493 (1901). T.: *A. pringlei* Fernald [= *D. populifolia* Baill.]

Monoecious herbs or shrubs. Hairs multicellular and uniseriate or absent. Latex white. Leaves alternate, with long eglandular petioles, stipulate, usually palmately 3–11-lobed, entire or distantly serrate, penninerved and triplinerved, membranaceous to chartaceous, eglandular above and below, lower surface smooth; stipules small, undivided, eglandular. Inflorescences terminal, solitary, bracts with one pair of elliptical glands, 0–4 basal ones with 1 female flower, numerous apical ones with 1–3 male flowers; bracteoles absent. Male flowers pedicellate in bud, inclinate but covered by calyces; calyx completely fused without distinct lobes, strongly zygomorphic; petals and disc 0; stamen 1, enclosed in the calyx, filament short, anthers basifixed, 2-thecate, longitudinally dehiscent; pollen 3-colporate, distinctly 3-lobate in polar view, colpus transversalis+, costae+, colpi narrow, tectate-perforate; pistillode 0; receptacle flat. Female flowers long pedicellate, often bending downwards; calyx 3-lobed, free, eglandular or basally biglandular; petals and disc 0; ovary smooth, glabrous or pubescent, with 3 locules, ovules 1 per locule; style long, stigmas 3, undivided, eglandular. Fruits long-pedicellate, spheroidal, smooth, indistinctly pubescent to glabrous, dehiscing septicidally into 3 mericarps; pericarp dry, woody, moderately thick; septa of mericarps with 1 often divided vascular strand; columella 3-quetrous, persistent. Seeds 3 per fruit, subglobose, surface distinctly colliculate, reddish, ecarunculate. Fig.: See Étude Euphorb. Atlas: t. 5 fig. 11–15 (1858); Pflanzenr. 52: 269, t. 54 (1912); Nat. Pflanzenfam. ed. 2, 19c: 191, t. 101 fig. g, h (1931).

A genus of 4 or less species, endemic to Mexico and Guatemala.

319. **Maprounea** *Aubl.*, Hist. Pl. Guiane 2: 895 (1775); A. Juss., Euphorb. Gen.: 54 (1824); Müll. Arg. in DC., Prodr. 15(2): 1190 (1866); in Mart., Fl. Bras. 11(2): 541 (1874); Benth. in Benth. & Hook. f., Gen. Pl. 3(1): 255, 333 (1880); Pax in Engl. & Prantl, Nat. Pflanzenfam. 3(5): 98 (1890); Pax & K. Hoffm. in Engl., Pflanzenr. 52: 175 (1912); Prain, Fl. Trop. Afr. 6(1): 1002 (1913); Pax & K. Hoffm. in Engl. & Harms, Nat. Pflanzenfam. ed. 2, 19c: 197 (1931); J. Léonard, Fl. Cong. Rwa.-Bur. 8(1): 142 (1962); Allem, Acta Amazon. 6: 417 (1976); Senna, Rodriguesia 36(61): 51 (1984); Radcl.-Sm. in Fl. Trop. East Africa, Euph. 1: 395 (1987); L. B. Smith *et al.*, Fl. Ilustr. Catar. EUFO: 316 (1988); G. L. Webster & Huft, Ann. Missouri Bot. Gard. 75: 1131 (1988); A. H. Gentry, Woody Pl. NW. S. Amer.: 422 (1993); G. L. Webster, Ann. Missouri Bot. Gard. 81: 123 (1994); Radcl.-Sm.in Fl. Zamb. 9(4): 325 (1996); Esser, Novon 9: 32 (1999). T.: *M. guianensis* Aubl.

Aegopricum L., Pl. Surinam.: 15 (1775). T.: *A. betulinum* L. [=*M. guianensis* Aubl.]
Aegopricon L. f., Suppl. Pl.: 63 (1782)
Stillingia sect. *Maprounea* (Aubl.) Baill., Étude Euphorb.: 520 (1858)

Monoecious or dioecious, glabrous shrubs or trees. Leaves alternate, with moderately to very long eglandular petioles, stipulate, simple, entire, penninerved, membranaceous to chartaceous, eglandular above, below with laminar glands, rarely eglandular, lower surface paler to glaucous but not papillate; stipules small, undivided, eglandular. Inflorescences terminal on short lateral shoots, solitary, bracts with a pair of shortly stipitate, irregularly elliptical and rarely divided glands, (0)1–5 basal ones with 1 female flower, often distinctly separated from the male part or even on a different plant, numerous apical bracts with 1–3(5) male flowers, aggregated into a glomerulate capitulum; bracteoles absent. Floral buds inclinate and initially covered by their large bracts, later on erect. Male flowers sessile to subsessile; calyx 2- or 3-lobed, largely fused; petals and disc 0; stamens (1)2 or 3, filaments long and partly to completely fused, anthers basifixed, free, extrorse, 2-thecate, longitudinally dehiscent; pollen 3-colporate, distinctly 3-lobate in polar view, colpus transversalis+, costae+, colpi narrow, tectate-perforate, columellae present; pistillode 0; receptacle flat. Female flowers distinctly to very long pedicellate; calyx 3-lobed, partly fused, eglandular; petals and disc 0; ovary smooth, with 3 locules, ovules 1 per locule; style absent to distinct, stigmas 3, undivided, eglandular. Fruits long pedicellate, spheroidal, smooth, dehiscing septicidally into 3 mericarps; pericarp dry, woody, thin (thickness less than $^{1}/_{10}$ of fruit length); septa of mericarps very thin and fragile, with 0 or 1 vascular strand; columella basically 3-quetrous but wings very fragile and caducous, therefore often rod-shaped. Seeds 3 per fruit, subglobose, smooth or distinctly rugulose-foveolate, brownish, with a very large caruncle partly to largely covering the upper part of the seeds. Fig.: See Hist. Pl. Guiane 4: t. 342 (1775); Euphorb. Gen.: t. 17, f. 57.1–8 (1824); Pl. Usuel. Brasil: t. 65 (1824–1828); Étude Euphorb. Atlas: t. 7 fig. 20–25 (1858); Fl. Bras. 11(2): t. 81 (1874); Pflanzenr. 52: 176, t. 32 and 178,

t. 33 (1912); Fl. Cong. Rwa.-Bur. 8(1): 143, t. 9 (1962); Rodriguesia 36(61): 74–76, tt. 15–25 (1984); Fl. Trop. East Africa, Euph. 1: 396, t. 75 (1987); Boissiera 44: 73, t. 27 (1990); Woody Pl. NW. S. Amer.: 421, t. 121.6 (1993); Euf. Reg. Ararac.: 107, t. 27 (1995); Fl. Zamb. 9(4): 326, t. 70 (1996); Árvores Brasileiras 2: 95 (1998); Novon 9: 33 (1999).

A genus of 4 or 5 species, growing in Africa from Nigeria to Angola, and in the New World from Panama and Trinidad to Brazil and Bolivia.

Tribe 50. **PACHYSTROMATEAE** *(Pax & K. Hoffm.) Pax*, Bot. Jahrb. Syst. 59: 145 (1924); G. L. Webster, Ann. Missouri Bot. Gard. 81: 123 (1994). T.: *Pachystroma* Müll. Arg.

Acalypheae subtribe *Pachystromatinae* Pax & K. Hoffm. in Engl., Pflanzenr. 68: 3, 35 (1919).

Monoecious, glabrous shrubs or trees. Latex white. Leaves petiolate, stipulate, penninerved, spiny-serrate, eglandular. Inflorescences terminal, solitary, elongate thyrsal; bracts glandular, lower ones with 1 female flower, apical ones with 1–3 male flowers. Flowers erect in bud, subsessile, actinomorphic. Male flowers with 3–6-lobed calyx completely closed in bud; stamens 3, fused with their filaments forming a massive column. Female flowers with undivided, eglandular stigmas. Fruits with woody, persistent calyx, septicidally dehiscent, leaving a 3-cornute carpidiophiore. Seeds 3 per fruit, ecarunculate.

Because of the completely different floral buds with erect flowers and a completely closed, large calyx this monotypic tribe is sufficiently distinct from the *Hippomaneae*.

320. **Pachystroma** *Müll. Arg.*, Linnaea 34: 177 (July 1865); in DC., Prodr. 15(2): 713, 893 (1866); in Mart., Fl. Bras. 11(2): 387 (1874); Baill., Hist. Pl. 5: 228 (1874); Benth. in Benth. & Hook. f., Gen. Pl. 3(1): 307 (1880); Pax in Engl. & Prantl, Nat. Pflanzenfam. 3(5): 78 (1890); Pax & K. Hoffm. in Engl., Pflanzenr. 44: 99 (1910); Nat. Pflanzenfam. ed. 2, 19c: 151 (1931); Allem, Iheringia, Bot. 22: 29 (1977); Senna, Bradea 3(48): 421 (1983); L. B. Smith *et al.*, Fl. Ilustr. Catar. EUFO: 225 (1988); G. L. Webster, Ann. Missouri Bot. Gard. 81: 124 (1994). T.: *P. ilicifolium* Müll. Arg. [=*P. longifolium* (Nees) I. M. Johnst. (B.: *Ilex longifolia* Nees)]

Acantholoma Gaudich. ex Baill., Adansonia 6: 231 (Sept. 1865). T.: *A. spinosum* Gaudich. ex Baill. [=*P. longifolium* (Nees) I. M. Johnst. (B.: *Ilex longifolia* Nees)]

Monoecious, glabrous shrubs or trees. Latex white. Leaves alternate, with short eglandular petioles, stipulate, simple, distantly spiny-serrate to nearly entire, penninerved, chartaceous to coriaceous, eglandular above and below, lower surface smooth; stipules moderately large, undivided, eglandular, very soon caducous. Inflorescences terminal, solitary, bracts with a pair of disc-shaped glands, basally 0–2 with 1 female flower, numerous apical ones with 1 or 3 male flowers; bracteoles + ?. Male flowers erect in bud, buds with a completely closed and fused calyx, nearly sessile; calyx 3–6-lobate, valvate, largely fused; petals and disc 0; stamens 3, filaments completely fused forming a thick column, anthers basifixed, extrorse, longitudinally dehiscent; pollen spheroidal, 3-colporate, slightly 3-lobate in polar view, colpus transversalis elongated, costae thick, colpi short to long and narrow, tectate-perforate, psilate; pistillode 0; receptacle flat. Female flowers subsessile; calyx 3-lobed, partly fused, eglandular; petals and disc 0; ovary smooth, with 3 locules, ovules 1 per locule; style distinct, stigmas 3, undivided, eglandular. Fruits short-pedicellate with a woody persistent calyx, spheroidal-elliptic, smooth, dehiscing septicidally into 3 mericarps; pericarp dry, woody, moderately thick; columella triquetrous, basally with a large, 3-cornute receptacle. Seeds 3 per fruit, elliptic, smooth, brownish-maculate, ecarunculate. Fig.: See Adansonia 6: t. 1 (1866) (as *Acantholoma*); Fl. Bras. 11(2): t. 54 (1874); Pflanzenr. 44: 100, t. 35 (1910); Bradea 3(48): 423 and 424, tt. 1 and 2 (1983); Fl. Ilustr. Catar. EUFO: 230, t. 27 fig. a–d (1988); Árvores Brasileiras: 108 (1992).

A monotypic genus of SE Brazil.

Tribe 51. **HUREAE** *Dumort.*, Anal. Fam. Pl.: 45 (1829); Hutch., Amer. J. Bot. 56: 755 (1969); G. L. Webster, Taxon 24: 600 (1975); Ann. Missouri Bot. Gard. 81: 124 (1994). T.: *Hura* L.

Hippomaneae subtribe *Hureae* Müll. Arg., Linnaea 34: 203 (1865); in DC., Prodr. 15(2): 1036, 1228 (1866).

Hippomaneae subtribe *Hurinae* Pax in Engl., Pflanzenfam. 3(5): 101 (1890); Pax & K. Hoffm. in Engl., Pflanzenr 52: 271 (1912); Pax, Bot. Jahrb. Syst. 59: 150 (1924).

Monoecious shrubs or trees. Indumentum absent or uniseriate. Latex white. Leaves petiolate, stipulate, penninerved, often glandular at base or on petiole apex. Inflorescences terminal or axillary, solitary, elongate thyrsal; bracts eglandular, adnate to the axis and peltate or fused into a sheathing tunica, basal ones with 1 female flower, sometimes separated from the male part, apical ones with 1(12) male flowers. Flowers erect when breaking through the tunica, sessile to pedicellate, actinomorphic. Male flowers with 2–5-lobed or completely fused calyx; stamens 1–80, often fused. Female flowers with 3–6-lobed, free to fused calyx; styles nearly absent to very long, stigmas undivided, eglandular, in one genus united into an umbrella-shaped disc. Fruits septicidally or loculicidally dehiscent, often leaving a 3-cornute carpidiophiore. Seeds 3–30 per fruit, ecarunculate.

KEY TO THE GENERA

1. Styles united into a conspicuous column longer than the stigmas · · · · · 2
1. Styles short to distinct, but always shorter than the stigmas · · · · · · · · · · 3
2. Male flowers with 10–80 stamens; ovary 5–20-locular; stigmas forming an umbrella-shaped disc · 324. *Hura*
2. Male flowers with 1 stamen; ovary 3-locular; stigmas short but free · 323. *Ophthalmoblapton*
3. Petioles of leaves short (up to 12 mm long); flowers with 1–3 free stamens · 321. *Algernonia*
3. Petioles of leaves longer (10–90 mm); flowers with 2 or 3, partly to completely fused stamens · 322. *Tetraplandra*

321. **Algernonia** *Baill.*, Ann. Sci. Nat. Bot., sér. 4, 9: 198 (1858), and Étude Euphorb.: 546 (1858); Müll. Arg. in DC., Prodr. 15(2): 1230 (1866); in Mart., Fl. Bras. 11(2): 535 (1874); Baill., Hist. Pl. 5: 232 (1874); Benth. in Benth. & Hook. f., Gen. Pl. 3(1): 256, 339 (1880); Pax in Engl. & Prantl, Nat. Pflanzenfam. 3(5): 102 (1890); Pax & K. Hoffm. in Engl., Pflanzenr. 52: 276 (1912); in Engl. & Harms, Nat. Pflanzenfam. ed. 2, 19c: 206 (1931); Emmerich, Arq. Mus. Nac. Rio de Janeiro 56: 92 (1981); Bradea 3(20): 148 (1981); G. L. Webster, Ann. Missouri Bot. Gard. 81: 125 (1994). T.: *A. brasiliensis* Baill.

Monoecious, glabrous shrubs or trees. Leaves alternate but often apically crowded, with short eglandular petioles (up to 12 mm long), stipulate, simple, entire or shallowly and distantly serrate, penninerved, membranaceous to coriaceous, above with a pair of disc-shaped, sometimes stipitate glands on base, eglandular below, lower surface smooth and shiny; stipules small, undivided. Inflorescences terminal, solitary, floral bracts peltate and adnate to central axis, eglandular, 0–2 basal ones with 1 female flower, numerous apical ones with 1 male flower; bracteoles + ?. Floral buds inclinate, initially covered by the tunica-like sheath formed by the bracts, but soon erect. Male flowers nearly sessile; calyx 2–5-lobed, partly fused, or absent; petals and disc 0; stamens 1(3), filaments short, free, anthers basifixed, 2-thecate, longitudinally dehiscent; pollen 3-colporate, distinctly 3-lobate in polar view, colpus transversalis+, costae+, colpi narrow, tectate-perforate; pistillode 0; receptacle flat. Female flowers sessile; calyx 3(6)-lobed, often largely fused, eglandular; petals and disc 0; ovary smooth or with 3 pairs of appendages, with 3 locules, ovules 1 per locule; style distinct but shorter than the stigmas, stigmas 3, undivided, eglandular. Fruits depressed-subglobose, smooth or with 3 pairs of appendages, dehiscent; columella (sometimes?) with a basal 3-cornute, persistent carpidiophore, but unknown for most of the species. Seeds 3 per fruit, subglobose, smooth, ecarunculate. Fig.: See Étude Euphorb.: t. 2, f. 30–32 (1858); Fl. Bras. 11(2): t. 87 (1874); Pflanzenr. 52: 277, t. 56 fig. d–g and 278, t. 57 (1912); Nat. Pflanzenfam. ed. 2, 19c: 297, t. 111 fig. d–g (1931); Arq. Mus. Nac. Rio de Janeiro 56: 100–103, tt. 1–4 (1981); Bradea 3(20): 149 (1981).

A genus of five species, endemic to E Brazil.

322. **Tetraplandra** *Baill.*, Ann. Sci. Nat. Bot., sér. 4, 9: 200 (1858); Baill., Étude Euphorb.: 549 (1858); Müll. Arg. in DC., Prodr. 15(2): 1230 (1866); in Mart., Fl. Bras. 11(2): 533 (1874); Baill., Hist. Pl. 5: 231 (1874); Benth. in Benth. & Hook. f., Gen. Pl. 3(1): 340 (1880); Pax in Engl. & Prantl, Nat. Pflanzenfam. 3(5): 102 (1890); Pax & K. Hoffm. in Engl., Pflanzenr. 52: 274 (1912); in Engl. & Harms, Nat. Pflanzenfam. ed. 2, 19c: 206 (1931); Emmerich, Arq. Mus. Nac. Rio de Janeiro 56: 94 (1981); G. L. Webster, Ann. Missouri Bot. Gard. 81: 125 (1994). T.: *T. leandri* Baill.

Monoecious, glabrous shrubs or trees. Leaves alternate but often apically crowded, with moderately long eglandular petioles (10–90 mm long), stipulate, simple, entire or shallowly and distantly serrate, penninerved, membranaceous to coriaceous, above with distinctly glandular auricles at base that often form a pair of disc-shaped, sometimes even stipitate glands, eglandular below, lower surface smooth and shiny; stipules small, undivided, eglandular. Inflorescences terminal, solitary, floral bracts peltate and adnate to central axis, eglandular, 0–2 basal ones with 1 female flower, numerous apical ones with 1 male flower; bracteoles present. Floral buds inclinate, initially covered by the tunica-like sheath formed by the bracts, but soon erect. Male flowers nearly sessile; calyx 3–5-lobed, partly fused, or absent; petals and disc 0; stamens 2 or 3, filaments short, partly to completely fused, anthers basifixed, 2-thecate but often fused, longitudinally dehiscent; pollen 3-colporate, distinctly 3-lobate in polar view, colpus transversalis+, costae+, colpi narrow, tectate-perforate; pistillode 0; receptacle flat. Female flowers sessile; calyx 6-lobed, mostly free, eglandular or glandular; petals and disc 0; ovary smooth or with 3 pairs of appendages, with 3 locules, ovules 1 per locule; style short to absent, rarely distinct but shorter than the stigmas, stigmas 3, undivided, eglandular. Fruits subsessile, smooth or with 3 pairs of appendages, dehiscent into 3 mericarps; columella probably with a basal, 3-cornute, persistent carpidiophore, but unknown for most of the species. Seeds 3 per fruit, subglobose, smooth, brownish maculate, ecarunculate. Fig.: See Étude Euphorb. Atlas: t. 5 fig. 8–10 (1858); Fl. Bras. 11(2): t. 104 (1874); Pflanzenr. 52: 277, t. 56 fig. a–c (1912); Nat. Pflanzenfam. ed. 2, 19c: 207, t. 111 fig. a–c (1931); Arq. Mus. Nac. Rio de Janeiro 56: 104–110, tt. 5–11 (1981).

A genus of seven species, ranging from E Brazil to the Peruvian Amazon. Very similar to and perhaps identical with *Algernonia.*

323. **Ophthalmoblapton** *Allemão,* Pl. Novas Brasil: 4 (1849), Ann. Sci. Nat. Bot., sér. 3, 13: 119 (1849); Baill., Étude Euphorb.: 547 (1858); Müll. Arg. in DC., Prodr. 15(2): 1155 (1866); in Mart., Fl. Bras. 11(2): 531 (1874); Baill., Hist. Pl. 5: 231 (1874); Benth. in Benth. & Hook. f., Gen. Pl. 3(1): 255, 333 (1880); Pax in Engl. & Prantl, Nat. Pflanzenfam. 3(5): 99 (1890); Pax & K. Hoffm. in Engl., Pflanzenr. 52: 278 (1912); in Engl. & Harms, Nat. Pflanzenfam. ed. 2, 19c: 206 (1931); Emmerich, Bol. Mus. Nac. Rio de Janeiro,

Bot. 62: 1 (1981); L. B. Smith *et al.*, Fl. Ilustr. Catar. EUFO: 325 (1988); G. L. Webster, Ann. Missouri Bot. Gard. 81: 125 (1994). T.: *O. macrophyllum* Allemão.

Monoecious, glabrous shrubs or trees. Latex white. Leaves alternate but often apically crowded, with short to moderately long eglandular petiole, stipulate, simple, entire or shallowly and distantly serrate, penninerved, chartaceous to coriaceous, glandular above?, below sometimes with a pair of basal glands, lower surface smooth; stipules small, undivided, eglandular. Inflorescences terminal and axillary, solitary, floral bracts peltate and adnate to central axis, eglandular?, 0–2 basal ones with 1 female flower, numerous apical ones with 1–c. 12 male flowers. Floral buds inclinate, initially covered by the tunica-like sheath formed by the bracts, but soon erect. Male flowers subsessile; calyx 3–5-lobed, partly fused; petals and disc 0; stamen 1, filament short, anther basifixed, 4-thecate, longitudinally dehiscent; pollen 3-colporate, distinctly 3-lobate in polar view, colpus transversalis+, costae+, colpi narrow, tectate-perforate; pistillode 0; receptacle flat. Female flowers subsessile; calyx 6-lobed, mostly free, often glandular; petals and disc 0; ovary smooth, with 3 locules, ovules 1 per locule; style very long and thick and much longer than the stigmas, stigmas 3, very short, undivided, eglandular. Fruits smooth, dehiscent into 3 mericarps, but further details unknown. Seeds 3 per fruit, ovoid-globose, brown, ecarunculate. Fig.: See Pl. Novas Brasil: 4? (1849), n. v.; Fl. Bras. 11(2): t. 103 (1874); Pflanzenr. 52: 280, t. 58 (1912); Sellowia 11: t. 30 fig. h–n (1959); Bol. Mus. Nac. Rio de Janeiro, Bot. 62: t. 1 following page 4 (1981); Fl. Ilustr. Catar. EUFO: 321, t. 40 fig. h–n (1988).

A genus of four species, endemic to E Brazil.

324. **Hura** *L.*, Sp. Pl.: 1008 (1753); A. Juss., Euphorb. Gen.: 51 (1824); Baill., Étude Euphorb.: 541 (1858); Müll. Arg. in DC., Prodr. 15(2): 1229 (1866); in Mart., Fl. Bras. 11(2): 632 (1874); Baill., Hist. Pl. 5: 230 (1874); Benth. in Benth. & Hook. f., Gen. Pl. 3(1): 255, 339 (1880); Pax in Engl. & Prantl, Nat. Pflanzenfam. 3(5): 102 (1890); Pax & K. Hoffm. in Engl., Pflanzenr. 52: 271 (1912); Prain, Fl. Trop. Afr. 6(1): 1019 (1913); Fawc. & Rendle, Fl. Jamaica 4: 333 (1920); Pax & K. Hoffm. in Engl. & Harms, Nat. Pflanzenfam. ed. 2, 19c: 206 (1931); Standl. & Steyerm., Fieldiana, Bot. 24(6): 124 (1949); Alain, Fl. Cuba 3: 122 (1953); Backer & Bakh. f., Fl. Java 1: 500 (1963); D. G. Burch, Ann. Missouri Bot. Gard. 54: 330 (1967); Berhaut, Fl. Ill. Sénég. 3: 489, t. 488 (1975); Correll & Correll, Fl. Bahama Archip.: 821 (1982); Alain, Fl. Española 4: 161 (1986); Radcl.-Sm. in Fl. Trop. East Africa, Euph. 1: 397 (1987); R. A. Howard, Fl. Lesser Antilles 5: 57 (1989); A. H. Gentry, Woody Pl. NW. S. Amer.: 422 (1993); G. L. Webster, Ann. Missouri Bot. Gard. 81: 125 (1994); W. C. Burger & Huft, Fieldiana, Bot. n.s. 36: 126 (1995); Gillespie in Acev.-Rodr., ed., Fl. St. John: 214 (1996). T.: *H. crepitans* L.

Monoecious trees with spiny stem. Latex white. Hairs multicellular, uniseriate, yellowish-brown. Leaves alternate, long petiolate (nearly as long as lamina length) with a pair of petiolar glands above, stipulate, simple, distinctly serrate to nearly entire, penninerved, chartaceous, eglandular above, eglandular below, lower surface smooth; stipules large (at least 10 mm long), undivided, eglandular. Inflorescences terminal and axillary, solitary, bracts 1-flowered with usually 1 female flower in axis of or separate from male part, male part with long peduncle, bracts numerous, eglandular, peltate and united into a membranaceous, tunica-like sheath; bracteoles + ?. Floral buds initially covered by the tunica-like sheath formed by the bracts, erect when breaking through the tunica. Male flowers distinctly pedicellate, breaking though the sheath when flowering; calyx cupulate, completely fused; petals and disc 0; stamens numerous (10–80), filaments long and completely fused into a stout column, the anthers in 2 or more sessile whorls around the upper part of the column, extrorse, 2-thecate; pollen subprolate, 3-colporate, hardly 3-lobate in polar view, colpus transversalis broad, costae+, colpi narrow and long, margo absent, tectate-perforate, columellae distinct but slender, psilate; pistillode 0; receptacle flat. Female flowers distinctly pedicellate; calyx 5-lobed and mostly fused or cupular and completely fused; petals and disc 0; ovary smooth, glabrous, with 5–20 locules, ovules 1 per locule; styles united into a conspicuous column much longer than the stigmas, stigmas 5–20, undivided, eglandular, united into an umbrella-like, lobed disc. Fruits distinctly pedicellate, transversely elliptic and concave on both ends, smooth, dehiscing loculicidally and septicidally; pericarp dry, woody, moderately thick; each septum with 1 vascular strand; columella not seen. Seeds 5–20 per fruit, laterally compressed, smooth, brownish, ecarunculate. Fig.: See Étude Euphorb.: t. 6, f. 21–35 (1858); Fl. Bras. 11(2): t. 86 (1874); Hist. Pl. 5: 137, tt. 216–218 (1874); Pflanzenr. 52: 272, t. 55 (1912); Fl. Jamaica 4: 333, t. 112 (1920); Nat. Pflanzenfam. ed. 2, 19c: 205, t. 110 (1931); Handb. Syst. Bot. ed. 4: 678, t. 448 (1935); Ann. Missouri Bot. Gard. 54: 331, t. 24 (1967); Amer. J. Bot. 56: 756, t. 17 (1969); Illustr. Guide Trop. Pl.: 357 (1969); Fl. Ill. Sénég. 3: 489, t. 488 (1975); Fl. illustr. phan. Guad. Martin.: 159, t. 734 (1978); Fl. Bahama Archip.: 822, t. 340 (1982); Fl. Española 4: 163, t. 116-20 (1986); Fl. Lesser Antilles 5: 58, t. 21 (1989); Árvores Brasileiras: 103 (1992); Woody Pl. NW. S. Amer.: 421, t. 121.3 (1993); Fieldiana, Bot. n.s. 36: 45, t. 32 p.p. (1995); Fl. St. John: 215, t. 97H–N (1996).

A genus of two or three species, indigenous in the New World from Mexico through Central America and the West Indies to Peru and Brazil, and cultivated in most parts of the tropics.

Tribe 52. **EUPHORBIEAE** *Blume*, Bijdr. Fl. Ned. Ind.: 631 (1825); Pax & K. Hoffm., Nat. Pflanzenfam. ed. 2, 19C: 207 (1931); L. C. Wheeler, Amer. Midl. Naturalist 30: 477 (1943); G. L. Webster, Ann. Missouri Bot. Gard. 81: 125 (1994). T.: *Euphorbia* L.

Monoecious or less commonly dioecious trees, shrubs or herbs; latex milky, toxic or innocuous; indumentum simple or 0; leaves alternate, opposite or whorled, mostly elobate, penninerved and eglandular; stipules + or 0. Inflorescences pseudanthial, with connate bracts forming a glanduliferous cyathium, often with petaloid appendages, enclosing 1 terminal female flower and 4 or 5 lateral male mono- or dichasia; inner bracts eglandular; flowers erect in bud; male calyx minute or 0; stamen 1; pollen 3-colporate, colpae marginate, exine perforate-tectate; female calyx 3–6-lobate or 0; ovary (2)3(4)-locular; styles free or basally connate, mostly 2-fid, rarely entire. Fruit dehiscent, rarely drupaceous, 1–3-seeded. Seeds carunculate or ecarunculate; testa dry, smooth or ornamented.

A large tribe of c. 2375 species, divided here into 3 subtribes and 11 genera, following Webster (1975). A catalogue of all the species-names in the tribe was published by Oudejans (1990), with a supplement (1993).

Key to the subtribes

1. Male calyx +; involucre of 4 partially- or fully-fused bracts; cyathial glands commissural, not on rim · 52a. *Anthosteminae*
1. Male calyx 0; involucral bracts usually 5, connate into a cup or tube · · · 2
2. Petaloid appendages bracteal, opposite male monochasia; cyathial glands commissural; monochasial bracts large, imbricate, enclosing flowers; female sepals 5–6, large, imbricate · · · · · · · · · · 52b. *Neoguillauminiinae*
2. Petaloid appendages interbracteal, alternate with male monochasia, or 0; cyathial glands on rim of cyathium; monochasial bracts small or 0 · 52c. *Euphorbiinae*

Subtribe 52a. **Anthosteminae** *(Baill.) G. L. Webster*, Taxon 24: 600 (1975), and Ann. Missouri Bot. Gard. 81: 126 (1994). T.: *Anthostema* A. Juss.

Anthostemideae Baill., Ann. Sci. Nat. Bot., sér. 4, 9: 192 (1858).
Anthostemeae Klotzsch & Garcke, Monatsber. Königl. Preuss. Akad. Wiss. Berlin 1859: 247 (1859).

Monoecious or dioecious trees; leaves alternate, penninerved; stipules inconspicuous. Cyathia in terminal or axillary hyperdichasia; involucre of 4 outer bracts subtending or enclosing 4 subflorescences; cyathial glands projecting inward from infolded margins of outer bracts; female flower 1, central and terminal or apparently lateral; male flowers in dichasia ± enclosed by 4 secondary inner bracts; male and female calyx +, gamophyllous; ovary 3–4-locular; styles emarginate to 2-fid. Fruit dehiscent. Seeds carunculate or ecarunculate.

This palaeotropical bigeneric subtribe occurs in W. Africa and Madagascar. The cyathia are tetramerous and somewhat bilaterally symmetrical; tetramery also occurs in the next subtribe.

Key to the genera

1. Cyathia bisexual; outer bracts partially connate into an open 4-lobed involucre; cyathial glands at margins of inner bracts; hyperdichasia axillary; ovary 3-locular; seeds carunculate · · · · · · · · · 325. *Anthostema*

1. Cyathia mostly unisexual; outer bracts connate into a closed cup; cyathial glands fused by pairs into 4 lobes alternating with the involucral lobes; hyperdichasia terminal; ovary 4-locular; seeds ecarunculate · 326. *Dichostemma*

325. **Anthostema** *A. Juss.*, Euphorb. Gen.: 56 (1824); Baill., Ann. Sci. Nat. Bot. 9: 193 (1858), and Étude Euphorb.: 543 (1858); Boiss. in DC., Prodr. 15(2): 188 (1862); Benth. & Hook. f., Gen. Pl. 3: 261 (1880); Brown in Fl. Trop. Afr. 6(1): 607 (1912); Pax & K. Hoffm., Nat. Pflanzenfam. ed. 2, 19C: 207 (1931); Keay (ed.), Fl. W. Trop. Afr. ed. 2, 1: 416 (1958); Berhaut, Fl. Ill. Sénég. 3: 379 (1975); G. L. Webster, Ann. Missouri Bot. Gard. 81: 126 (1994). T.: *A. senegalense* A. Juss.

Monoecious glabrous laticiferous trees. Leaves alternate, petiolate, stipulate, simple, entire, shiny, penninerved, with numerous parallel nerves arising at a shallow angle from the midrib, eglandular; stipules small, soon deciduous. Inflorescences pseudocymose, hyperdichasial, axillary, cymes small, subsessile, dense; cyathia imperfect, involucre open on one side, composed of 4 ± connate outer bracts, each with 1–2 large, flattened rhomboid or narrowly scutiform commissural glands on their infolded margins, ± united laterally; male flowers numerous, crowded in uniparous dichasial clusters opposite the cyathial lobes, ± enclosed by 4 secondary inner bracts; females 1 per cyathium, somewhat excentric. Male flowers shortly pedicellate; calyx 3–4-toothed; petals 0; disc 0; stamen 1, filament short, articulating with the pedicel below the calyx, anthers basifixed, introrse with respect to the dichasia, 2-thecate, thecae parallel, contiguous, longitudinally dehiscent; pollen oblate-spheroidal, 3-colporate, circular to slightly 3-lobate in polar view, colpus transversalis +, costae +, colpi narrow, margo small or 0, intectate, reticulate, columellae large, exine thick; pistillode 0; receptacle 0. Female flowers stoutly pedicellate; calyx ± as in the male, but larger; petals 0; disc 0; ovary 3-locular, sessile, articulating with the stipe, ovules 1 per locule; styles 3, connate over $^1/_2$-way into a stout column, arms clavate or 2-lobate, stigmas canaliculate, minutely papillulose. Fruit 3-coccous, deeply 3-lobate, dehiscing septicidally and loculicidally into 3 2-valved cocci or 6 valves; exocarp thin, not separating; endocarp fairly thick, woody; columella clavate-capitate, 3-alate to 3-quetrous, persistent. Seeds large, ovoid, laterally compressed, carunculate; testa crustaceous, shiny, brownish, streaked and mottled buff; endosperm fleshy, copious; cotyledons broad, flat. Fig.: See Euphorb. Gen.: t. 18, f. 60.1–9 (1824); Étude Euphorb.: t. 5, f. 1–7 (1858); Fl. Ill. Sénég. 3: 378 (1975).

A tritypic genus, with 2 species in W. Africa and 1 in Madagascar.

326. **Dichostemma** *Pierre*, Bull. Mens. Soc. Linn. Paris 1(159): 1259 (1896); Hutch. in Fl. Trop. Afr. 6(1): 605 (1912); Pax & K. Hoffm., Nat. Pflanzenfam. ed. 2, 19C: 207 (1931); Keay (ed.), Fl. W. Trop. Afr., ed. 2, 1: 416 (1958); G. L. Webster, Ann. Missouri Bot. Gard. 81: 126 (1994). T.: *D. glaucescens* Pierre.

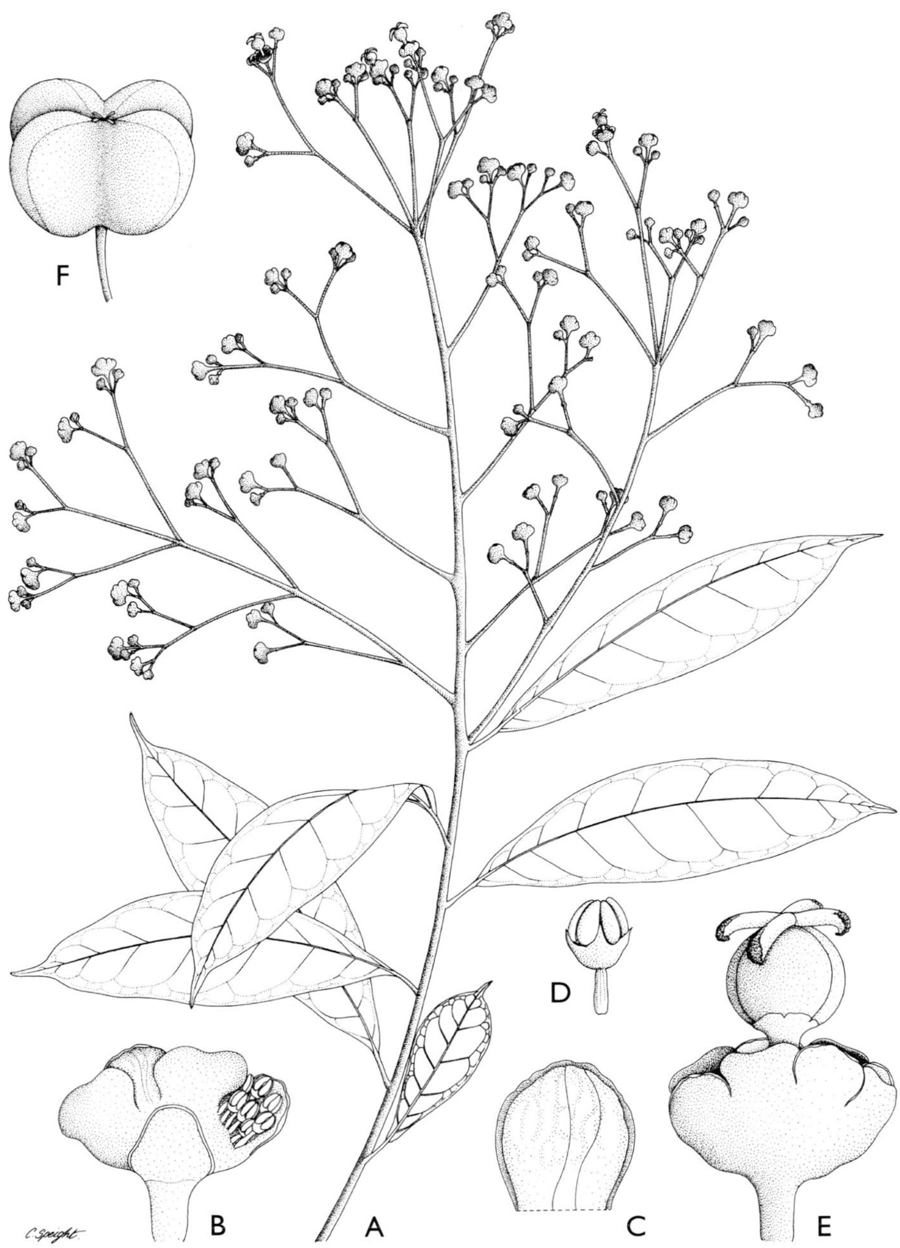

FIG. 46. *Dichostemma glaucescens* Pierre. **A** Habit × 1; **B** Male cyathium, vertical section to show male flowers × 6; **C** Involucral bract × 12; **D** Male flower × 12; **E** Female cyathium × 6; **F** Fruit × 1.5. **A–E** from *Tessmann* 610; **F** from *Thomas, Namata, Satabie & Nkongmenek* 7837A. Drawn by Camilla Speight.

Similar to 325. *Anthostema*, but often dioecious; indumentum pruinose, persistent, largely confined to the inflorescences; lateral nerves few, widely-spaced, brochidodromous; inflorescences terminal, pseudopaniculate, cyathia mostly unisexual, rarely bisexual, outer cyathial bracts connate into a closed cup-shaped involucre, glands fused by pairs into 4 lobes alternating with the involucral lobes; female flowers central, ovary 4-locular, styles 4; fruit 4-coccous, deeply 4-lobate, dehiscing into 4 2-valved cocci, endocarp thinly woody, columella 4-quetrous, splitting into 4 strands; seeds ecarunculate, 3-gonous, dorsally rounded, with 2 plane ventral sides at right angles to each other, brown, concolorous. Fig.: See Fig. 46, p. 401.

A monotypic W. African genus ranging from S. Nigeria to Congo (Kinshasa).

Subtribe 52b. **Neoguillauminiinae** *Croizat*, Philipp. J. Sci. 64: 408 (1938); G. L. Webster, Taxon 24: 601 (1975), and Ann. Missouri Bot. Gard. 81: 126 (1994). T.: *Neoguillauminia* Croizat.

Monoecious glabrous trees or shrubs; leaves alternate or opposite, exstipulate, entire. Cyathia in axillary pedunculate hypercymes; involucral bracts 4–6, dilated or petaloid, connate near the base; glands 4 or (6)8–10(12), in (3)4–5(6) pairs partitioning the male subflorescences; female flower central, male flowers in bracteate monochasia; male calyx 0; female sepals (4)5–6, imbricate; ovary 3-locular; styles connate at the base, elobate, dilatate or 2-fid. Fruit dehiscent. Seeds smooth, carunculate.

As defined by Webster (1975), this subtribe comprises 2 Australasian genera. The cyathia differ from those of *Euphorbia* in having the pseudopetals homologous with the involucral bracts.

Key to the genera

1. Cyathial glands (6)8–10(12), massive; pseudopetals large (c. 1 cm long); leaves spiral . . . . 327. *Neoguillauminia*
1. Cyathial glands 4, small; pseudopetals smaller (<1 cm long); leaves opposite . . . . 328. *Calycopeplus*

327. **Neoguillauminia** *Croizat*, Philipp. J. Sci. 64: 398 (1938), and Bull. Jard. Bot. Buitenzorg III. 17: 206 (1941); Guillaumin, Fl. Anal Synopt. Nouv.-Caléd.: 182 (1948); McPherson & Tirel, Fl. Nouv.-Caléd. 14(1): 22 (1987); G. L. Webster, Ann. Missouri Bot. Gard. 81: 127 (1994). T.: *N. cleopatra* (Baill.) Croizat (B.: *Euphorbia cleopatra* Baill.)

Monoecious glabrous tree or shrub. Latex white. Leaves alternate, spiral, aggregated at the extremities of the branches, petiolate, exstipulate, simple, entire, penninerved, with numerous parallel nerves arising at a shallow angle from the midrib, coriaceous, eglandular. Inflorescences axillary or subterminal, pseudocymose, pseudocymes lax; cyathia long-pedunculate;

involucral bracts 4–6, dilated or petaloid, large, connate near the base, obovate, later reflexed, reddish without, white within; glands (6)8–10(12), in (3)4–5(6) pairs, massive, fleshy, stipitate, partitioning the male subflorescences; subflorescences monochasial, bracteate; bracts broad, ± rectangular, erose; female flower, if present, solitary, central. Male flowers long-pedicellate; calyx 0; petals 0; disc 0; stamen 1, filament short, articulating with the pedicel, anthers basifixed, introrse with respect to the cyathia, 2-thecate, thecae parallel, contiguous, longitudinally dehiscent; pollen ± as in 325. *Anthostema*; pistillode 0; receptacle 0. Female flower shortly and stoutly pedicellate; sepals 5–6, imbricate; petals 0; disc 0; ovary 3-locular, ovules 1 per locule; styles connate at the base, shortly 2-fid, stigmas canaliculate, ± smooth. Fruit rounded-3-lobate, dehiscing into 3 2-valved cocci; exocarp thin, not separating; endocarp very thick, woody; columella capitate, 3-alate to 3-quetrous, persistent. Seeds ovoid, smooth, caruncolate, smaller than in 325. *Anthostema* and 326. *Dichostemma*; testa crustaceous, brown, concolorous; caruncle small. Fig.: See Philipp. J. Sci. 64: 412 opp., t. 1.1 (1938); Fl. Nouv.-Caléd. 14(1): 23, t. 3 (1987).

A monotypic genus endemic to New Caledonia.

Wood traded locally under the name "False Walnut".

328. **Calycopeplus** *Planch.*, Bull. Soc. Bot. France 8: 30 (1861); Boiss., Ic. Euph. t. 120 (1866); Benth., Fl. Austral. 6: 52 (1873), and Gen. Pl. 3: 261 (1880); Pax & K. Hoffm., Nat. Pflanzenfam. ed. 2, 19C: 221 (1931); Airy Shaw, Kew Bull. 35: 603 (1980); G. L. Webster, Ann. Missouri Bot. Gard. 81: 127 (1994); P. I. Forst., Austrobaileya 4(3): 418 (1995). T.: *C. ephedroides* Planch. [=*C. paucifolius* (Klotzsch) Baill. (B.: *Euphorbia paucifolia* Klotzsch)]

Euphorbia sect. *Calycopeplus* (Planch.) Boiss. in DC., Prodr. 15(2): 175 (1862). T.: *Eu. ephedroides* (Planch.) Boiss. (B.: *C. ephedroides* Planch.) [=*C. paucifolius* (Klotzsch) Baill. (B.: *Eu. paucifolia* Klotzsch)]

Monoecious treelets, shrubs or undershrubs. Latex white. Stems and branches virgate, terete, angulate or compressed, branches opposite or verticillate, usually exfoliate at anthesis. Leaves on young shoots opposite or verticillate, petiolate or sessile, sometimes stipulate, linear-lanceolate, simple, entire, penninerved, soon deciduous; stipules minute or 0. Cyathia small, axillary or terminal, solitary or binate, 2-bracteate, subsessile, campanulate or spreading, deeply 4-lobate, lobes small, petaloid, white or greenish; interlobate glands small or 0; male flowers in 4 oppositilobate 3–16-flowered bracteate clusters, outer 1–2 bracts enlarged, enclosing the cluster; female flower solitary, central. Male flowers pedicellate; calyx 0; petals 0; disc 0; stamen 1, filament articulating with the pedicel, anthers subdorsifixed, introrse with respect to the cyathia, 2-thecate, thecae parallel, contiguous, longitudinally dehiscent; pollen ± as in 325. *Anthostema*; pistillode 0; receptacle 0. Female flowers pedicellate; calyx 4–6-lobate, broadly cupular; petals 0; disc 0; ovary sessile or subsessile, 3-locular, ovules 1 per locule; styles 3, free or connate at the base, entire, 2-lobate or 2-fid. Fruit 3-coccous, 3-lobate, smooth, glabrous, dehiscing into 3 2-valved cocci; endocarp thinly crustaceous; columella slender, clavate-capitate. Seeds oblong-subglobose, carunculate, smaller than in 327. *Neoguillauminia*; testa crustaceous, smooth, shiny, dark brown; caruncle terminal, entire; endosperm fleshy; cotyledons broad, flat. Fig.: See Ic. Euph. t. 120 (1866); Philipp. J. Sci. 64: 412 opp., t. 1.2 (1938); Austrobaileya 4(3): 421, t. 1 and 425, t. 2 (1995).

A pentatypic genus endemic to Australia.

The xeromorphic vegetative morphology confers a very different *gestalt* on this genus from the other 3 genera in these first 2 subtribes, but the cyathial structure provides the unifying factor.

Subtribe 52c. **Euphorbiinae** *Hurus.*, J. Fac. Sci. Univ. Tokyo, Sect. 3, Bot. 6: 226 (1954); G. L. Webster, Ann. Missouri Bot. Gard. 81: 127 (1994). T.: *Euphorbia* L.

Pedilanthaceae Klotzsch & Garcke, Monatsber. Königl. Preuss. Akad. Wiss. Berlin 1859: 247, 253 (1859). T.: *Pedilanthus* Neck. ex Poit.
Subtribe *Pedilanthinae* (Klotzsch & Garcke) Hurus., J. Fac. Sci. Univ. Tokyo, Sect. 3, Bot. 6: 226 (1954).
Subtribe *Anisophyllae* Klotzsch & Garcke, Monatsber. Königl. Preuss. Akad. Wiss. Berlin 1859: 247 (1859). T.: *Anisophyllum* Haw. [=*Chamaesyce* S. F. Gray; = *Euphorbia* L.]
Subtribe *Tithymalinae* Klotzsch & Garcke, Monatsber. Königl. Preuss. Akad. Wiss. Berlin 1859: 247 (1859). T.: *Tithymalus* Scop. [= *Euphorbia* L.]

Monoecious or rarely dioecious trees, shrubs or herbs; latex milky, innocuous or toxic; indumentum simple or 0; leaves alternate, opposite or whorled, elobate or rarely lobate, sometimes inaequilateral at the base, usually penninerved; stipulate or exstipulate. Inflorescences terminal or axillary, cyathia often subtended by paired cyathophylls; cyathium distinctly cupular, radially or bilaterally symmetrical, usually with 4–5 (rarely only 1–2) interbracteal glands on the margin; glands usually discrete, with or without petaloid appendages; female flower central; male flowers in 4–5 monochasia, monochasial bracts minute; male calyx 0; female calyx 3–6-lobate, often minute, or 0; ovary 3-locular; styles distinct or connate, 2-fid, rarely entire. Fruit dehiscent, rarely drupaceous. Seeds carunculate or ecarunculate.

The subtribe is here considered to consist of 8 genera, as in addition to the 7 listed by Webster (1994), following Carter (1988) I have resurrected *Elaeophorbia*; other authors have subdivided *Euphorbia* into many other segregate genera, however. Further subdivision of the subtribe into smaller subtribes appears unnecessary. Typification of the taxa is reviewed by Wheeler (1943).

KEY TO THE GENERA

1. Cyathia distinctly bilaterally symmetrical; glands enclosed within the spur; styles connate into a long column · · · · · · · · · · · · · · · 336. *Pedilanthus*
1. Cyathia ± radially symmetrical; glands not enclosed within a spur; styles mostly free or almost so · 2
2. Involucral glands (1–2)4–5(8), free, interbracteal, on the cyathial margin · 3
2. Involucral glands not as above; cyathophylls often connate into a cup · · 5
3. Leaves opposite, inaequilateral at the base; veins chlorenchyma-sheathed; stipules +; main axis aborting above the cotyledons; seeds usually ecarunculate · 331. *Chamaesyce*
3. Leaves alternate, opposite or whorled, but if opposite then not inaequilateral at the base; leaf-veins not chlorenchyma-sheathed; stipules + or 0; main axis not aborting; seeds carunculate or ecarunculate · · · 4
4. Female perianth +; ovary not confluent with the pedicel; fruit not or rarely slightly fleshy, dehiscent · 329. *Euphorbia*
4. Female perianth 0; ovary confluent with the pedicel; fruit thick, fleshy, indehiscent, drupaceous · 330. *Elaeophorbia*
5. Female flower calyculate; involucral lobes fused; glands 2, fused into a shield; seeds ecarunculate; stems not succulent · · · · · · 332. *Cubanthus*
5. Female flower ecalyculate; involucral lobes and glands connate into a ring; seeds mostly carunculate; stems succulent · 6
6. Cyathia somewhat bilaterally symmetrical, the gland-ring open on one side · 333. *Monadenium*
6. Cyathia radially symmetrical, without a gap on one side · · · · · · · · · · · · 7
7. Involucral glands connate, conspicuous; ovary without angular crests · 334. *Synadenium*
7. Involucral glands distinct, inconspicuous; ovary with prominent double angular crests · 335. *Endadenium*

329. **Euphorbia** *L.* Sp. Pl.: 450 (1753), and Gen. Pl., ed. 5: 208 (1754); Roep., Enum. Euph. Germ.: 9 (1824); A. Juss., Euphorb. Gen.: 57 (1824); Baill., Étude Euphorb.: 281 (1858); Boiss. in DC., Prodr. 15(2): 7 (1862), and Ic. Euph.: 1 (1866); Müll. Arg. in Mart., Fl. Bras. 11(2): 666 (1874); Benth., Gen. Pl. 3: 258 (1880); Hook. f., Fl. Brit. India 5: 244 (1887); Norton, Annual Rep. Missouri Bot. Gard. 11: 85 (1899); Pax, Bot. Jahrb. Syst. 34: 61 (1905); Berger, Sukk. Euph.: 1 (1907); N. E. Brown in Fl. Trop. Afr. 6(1): 470 (1911), and in Fl. Cap. 5(2): 222 (1915); Fawc. & Rendle, Fl. Jamaica 4: 334 (1920); Denis, Euphorb. Îles Austr. Afr.: 23 (1921); Gagnep., Fl. Indo-Chine 5: 236 (1925); Pax & K. Hoffm., Nat. Pflanzenfam. ed. 2, 19C: 208 (1931); Prokh., Consp. Syst. Tith. As. Med.: 1 (1933); Croizat, Bull. Jard. Bot. Buitenzorg III.16: 351 (1940); Hurus., J. Jap. Bot. 16: 330 (1940); A. C. White, R. A. Dyer & B. Sloane, Succ. Euph. 1: 59 (1941); L. C. Wheeler, Amer. Midl. Naturalist 30: 478 (1943); Leandri, Notul. Syst. (Paris) 12: 64 (1945), 156 (1946); Robyns, Fl. Parc Nat. Alb. 1: 474 (1948); Prokh. in Fl. S.S.S.R. XIV: 304 (1949); Alain, Fl. Cuba 3: 128 (1953); Vindt, Monogr. Euph. Maroc: 23 (1953); Hurus., J. Fac. Sci. Univ. Tokyo, Sect. 3, Bot. 6: 228, 230 (1954); Ursch & Leandri, Mém. Inst. Sci.

Madagascar, Sér. B, Biol. Vég. 5: 111 (1954); Dyer, Bull. Jard. Bot. État 27: 487 (1957); Keay (ed.), Fl. W. Trop. Afr., ed. 2, 1: 417 (1958); Backer & Bakh. f., Fl. Java 1: 500 (1963); Rech. f. & Schiman-Czeika, Fl. Iran. 6: 8 (1964); Croizat, Webbia 20: 573 (1965); Jabl., Mem. New York Bot. Gard. 12: 186 (1967); G. L. Webster, J. Arnold Arbor. 48: 395 (1967), and Ann. Missouri Bot. Gard. 54: 332 (1968); Radcl.-Sm. & Tutin, Fl. Eur. 2: 213 (1968); Radcl.-Sm., Kew Bull. 26: 261 (1972), Hooker's Icon. Pl. 38: t. 3724, 3725 (1974), Kew Bull. Addit. Ser. 4: 110 (1975), 8: 81 (1980), Fl. Iraq 4 (1): 327 (1980), Kew Bull. 35: 629, (1980), 36: 294 (1981), 37: 18 (1982), Fl. Turk. 7: 571 (1982), Euph. Phil.: 24 (1983), Fl. Cyp. 2: 1433 (1985), and Fl. Pak. 172: 88 (1986); J.-G. Adam, Mém. Mus. Natl. Hist. Nat., B, Bot. 20: 478 (1971); Whitmore, Tree Fl. Mal. 2: 96 (1973); Berhaut, Fl. Ill. Sénég. 3: 441 (1975); Allem & Irgang, Fl. Ilustr. Rio Gr. do Sul 11: 15 (1975); M. C. Johnst., Wrightia 5: 120 (1975); L. C. Leach, Dinteria 12: 1 (1976); D. C. Hassall, Austral. J. Bot. 25: 430 (1977); Subils, Kurtziana 10: 83 (1977); Philcox, Fl. Trin. Tob. II(X): 686 (1979); S. Carter, Kew Bull.: 35: 413, 423 (1980), Hooker's Icon. Pl. 39: 5 (1982), Kew Bull.: 37: 73 (1982), 39: 643 (1984), 42: 369, 385, 673 (1987), Bot. J. Linn. Soc. 94: 67 (1987), Fl. Trop. E. Afr. Euph. 2: 409 (1988), and Kew Bull.: 45: 327, 653 (1990); LaFon *et al.*, eds., Euphorbia J. 1–10 (1980–1996); Correll, Fl. Bah. Arch.: 798 (1982); Huft, Ann. Missouri Bot. Gard. 71: 1021 (1984); Proctor, Kew Bull. Addit. Ser. 11: 547 (1984); Alain, Fl. Español. 4: 143 (1986); M. G. Gilbert, Kew Bull. 42: 231 (1987); Grierson & D. G. Long, Fl. Bhut. 1: 759 (1987); McPherson & Tirel, Fl. Nouv. Caléd. 14: 10 (1987); G. L. Webster & Huft, Ann. Missouri Bot. Gard. 75: 1137 (1988); Alain, Descr. Fl. Puerto Rico 2: 389 (1988); R. A. Howard, Fl. Lesser Antilles 5: 46 (1989); Oudejans, World Cat. Spec. Tr. Euph. (1990, Suppl. 1993); J.-P. Lebrun & Stork, Enum. Pl. Afr. Trop. 1: 213 (1991); Mayfield, Sida 14: 573 (1991); Oudejans & Molero, eds., Curr. Res. Tax. Euph. (Coll. Bot. 21): 5 (1992); G. L. Webster, Ann. Missouri Bot. Gard. 81: 128 (1994); R. Turner, *Euphorbias* (1995); Gillespie in Acev.-Rodr., ed., Fl. St. John: 210 (1996); Philcox in Dassan., ed., Fl. Ceyl. XI: 192 (1997). LT.: *E. antiquorum* L. (designated by Millspaugh, 1909).

Lathyris Trew, Herb. Blackwell. 1, C.II: t. 123 (1754). T.: (none designated; *E. lathyris* L. by default).

Euphorbium Hill, Fam. Herb., ed. 2: 136 (1755). LT.: *E. antiquorum* L. (chosen by Wheeler, 1943).

Athymalus Neck., Elem. Bot. 2: 353 (1790). T.: (none designated) [prob. =*E. anacantha* Ait.]

Keraselma Neck., Elem. Bot. 2: 353 (1790). LT.: *K. esula* (L.) Raf. [=*E. esula* L.] (chosen by Wheeler, 1943).

Tithymalus Gaertn., Fruct. 2: 115 (1790), nom. cons. LT.: *T. peplus* (L.) Gaertn. [=*E. peplus* L.] (chosen by Millspaugh, 1909).

Dactylanthes Haw., Syn. Pl. Succ.: 132 (1812). LT.: *D. anacantha* (Ait.) Haw. [=*E. anacantha* Ait.] (chosen by Wheeler, 1943).

Esula (Pers.) Haw., Syn. Pl. Succ.: 153 (1812), non Morandi (1761) (S.: *Euphorbia* subgen. *Esula* Pers.). LT.: *E. dalechampii* Haw. [=*Eu. esula* L.] (chosen by Wheeler, 1943).

Galarhoeus Haw., Syn. Pl. Succ.: 143 (1812); Hurus., J. Fac. Sci. Univ. Tokyo, Sect. 3, Bot. 6: 236 (1954). LT.: *G. helioscopia* (L.) Haw. [=*E. helioscopia* L.] (chosen by Wheeler, 1943).

Medusea Haw., Syn. Pl. Succ.: 133 (1812). LT.: *M. major* (Ait.) Haw. [=*E. caput-medusae* L. var. *major* Ait.] (chosen by Wheeler, 1943).

Treisia Haw., Syn. Pl. Succ.: 131 (1812). LT.: *T. clava* (Jacq.) Haw. [=*E. clava* Jacq.] (chosen by Wheeler, 1943).

Characias Gray, Nat. Arr. Brit. Pl. 2: 259 (1821). T.: *Ch. purpurea* (Lam.) Gray (B.: *Tithymalus purpureus* Lam.) [=*E. characias* L. subsp. *characias*]

Desmonema Raf., Atl. J. 1(6): 177 (1833). T.: *D. hirtum* Raf. [=*E.* ?]

Pleuradena Raf., Atl. J. 1(6): 182 (1833), non *Pleuradenia* Raf. (1825). T.: *Pl. coccinea* Raf. [=*E. pulcherrima* Willd.]

Poinsettia Graham, Edinburgh New Philos. J. 20: 412 (1836); Dressler, Ann. Missouri Bot. Gard. 48: 329 (1961); Proctor, Kew Bull. Addit. Ser. 11: 549 (1984). T.: *P. pulcherrima* (Willd.) Graham [=*E. pulcherrima* Willd.]

Lacanthis Raf., Fl. Tellur. 2: 94 (1837). T.: *L. splendens* (Bojer ex Hook.) Raf. (B.: *E. splendens* Bojer ex Hook.) [=*E. milii* Des Moul. var. *splendens* (Bojer ex Hook.) Ursch & Leandri]

Bojeria Raf., Fl. Tellur. 2: 95 (1837), nom. prov. T.: *B. splendens* (Bojer ex Hook.) Raf. (B.: *E. splendens* Bojer ex Hook.) [=*E. milii* Des Moul. var. *splendens* (Bojer ex Hook.) Ursch & Leandri]

Adenorima Raf., Fl. Tellur. 4: 112 (1838). T.: *A. punicea* (Sw.) Raf. [=*E. punicea* Sw.]

Aklema Raf., Fl. Tellur. 4: 114 (1838). T.: *A. nudiflora* (Jacq.) Raf. [=*E. nudiflora* Jacq.]

Agaloma Raf., Fl. Tellur. 4: 116 (1838). LT.: *A. corollata* (L.) Raf. [=*E. corollata* L.] (chosen by Rafinesque, 1838)

Allobia Raf., Fl. Tellur. 4: 116 (1838). T.: *A. portlandica* (L.) Raf. [=*E. portlandica* L.]

Cyathophora Raf., Fl. Tellur. 4: 117 (1838). LT.: *C. heterophylla* (L.) Raf. [=*E. heterophylla* L.] (chosen by Wheeler, 1943)

Kanopikon Raf., Fl. Tellur. 4: 114 (1838). T.: *K. atropurpurea* (Brouss. ex Willd.) Raf. [=*E. atropurpurea* Brouss. ex Willd.]

Lepadena Raf., Fl. Tellur. 4: 113 (1838). T.: *L. leucoloma* (Raf.) Raf. (B.: *E. leucoloma* Raf.) [=*E. marginata* Pursh]

Lophobios Raf., Fl. Tellur. 4: 116 (1838). LT.: *L. terracina* (L.) Raf. [=*E. terracina* L.] (chosen by Wheeler, 1943)

Murtekias Raf., Fl. Tellur. 4: 116 (1838). T.: *M. myrsinites* (L.) Raf. [=*E. myrsinites* L.]

Nisomenes Raf., Fl. Tellur. 4: 116 (1838). T.: *N. diffusa* (Jacq.) Raf. (B.: *E. diffusa* Jacq.) [=*E. exigua* L.]

Peccana Raf., Fl. Tellur. 4: 114 (1838). T.: *P. glauca* Raf. [=*E. cymosa* Poir., vel *E. graminea* Jacq.]

Tirucalia Raf., Fl. Tellur. 4: 112 (1838), p.p. T.: *T. indica* Raf. [=*E. tirucalli* L.]

Torfasadis Raf., Fl. Tellur. 4: 112 (1838). T.: *T. canariensis* (L.) Raf. [=*E. canariensis* L.]

Tumalis Raf., Fl. Tellur. 4: 114 (1838). T.: *T. bojeri* (Hook.) Raf. (B.: *E. bojeri* Hook.) [=*E. milii* Des Moul.]

Vallaris Raf., Fl. Tellur. 4: 114 (1838). LT.: *V. ipecacuana* (sic) (L.) Raf. [=*E. ipecacuanhae* L.] (chosen by Wheeler, 1943)

Zalitea Raf., New Fl. N. Amer. 4: 98 (1838). T.: *Z. linearis* Raf. [=*E. hexagona* Nutt. ex Spreng.]

Dematra Raf., Autik. Bot.: 96 (1840). T.: *D. sericea* Raf. [=*E. petiolata* Banks & Sol. ex Russ.]

Kobiosis Raf., Autik. Bot.: 94 (1840). T.: *K. mellifera* (Ait.) Raf. [=*E. mellifera* Ait.]

Alectoroctonum Schltdl., Linnaea 19: 252 (1847). LT.: *A. scotanum* (Schltdl.) Schltdl. [=*E. scotanum* Schltdl.] (chosen by Wheeler, 1943).

Anthacantha Lem., Ill. Hort. 4: Misc. 73 (1857). T.: *A. heptagona* (L.) Lem. [=*E. heptagona* L.]

Adenopetalum Klotzsch & Garcke, Monatsber. Königl. Preuss. Akad. Wiss. Berlin 1859: 250 (1859), non Turcz. (1858). LT.: *A. gramineum* (Jacq.) Klotzsch & Garcke [=*E. graminea* Jacq.] (chosen by Wheeler, 1939).

Arthrothamnus Klotzsch & Garcke, Monatsber. Königl. Preuss. Akad. Wiss. Berlin 1859: 251 (1859), non Rupr. (1848). LT.: *A. tirucalli* (L.) Klotzsch & Garcke [=*E. tirucalli* L.] (chosen by Millspaugh, 1909).

Dichrophyllum Klotzsch & Garcke, Monatsber. Königl. Preuss. Akad. Wiss. Berlin 1859: 249 (1859). LT.: *D. marginatum* (Humb., Bonpl. & Kunth) Klotzsch & Garcke (B.: *E. marginata* Humb., Bonpl. & Kunth, non Pursh) [=*E. bonplandii* Sweet] (chosen by Wheeler, 1943).

Eumecanthus Klotzsch & Garcke, Monatsber. Königl. Preuss. Akad. Wiss. Berlin 1859: 248 (1859). LT.: *Eum. ariensis* (Humb., Bonpl. & Kunth) Klotzsch & Garcke [=*Euph. ariensis* Humb., Bonpl. & Kunth] (chosen by Millspaugh, 1916).

Euphorbiastrum Klotzsch & Garcke, Monatsber. Königl. Preuss. Akad. Wiss. Berlin 1859: 252 (1859). LT.: *E. hoffmannianum* Klotzsch & Garcke [=*Euphorbia hoffmanniana* (Klotzsch & Garcke) Boiss.] (chosen by Wheeler, 1943).

Leptopus Klotzsch & Garcke, Monatsber. Königl. Preuss. Akad. Wiss. Berlin 1859: 249 (1859), non Decne. (1844). LT.: *L. adiantoides* (Lam.) Klotzsch & Garcke [=*E. adiantoides* Lam.] (chosen by Wheeler, 1943).

Sterigmanthe Klotzsch & Garcke, Monatsber. Königl. Preuss. Akad. Wiss. Berlin 1859: 252 (1859). LT.: *St. splendens* (Bojer ex Hook.) Klotzsch & Garcke (B.: *E. splendens* Bojer ex Hook.) [=*E. milii* Des Moul. var. *splendens* (Bojer ex Hook.) Ursch & Leandri] (chosen by Wheeler, 1943).

Tithymalopsis Klotzsch & Garcke, Monatsber. Königl. Preuss. Akad. Wiss. Berlin 1859: 249 (1859). LT.: *T. corollata* (L.) Klotzsch & Garcke [=*E. corollata* L.] (chosen by Small, 1913).

Trichosterigma Klotzsch & Garcke, Monatsber. Königl. Preuss. Akad. Wiss. Berlin 1859: 248 (1859) (sphalm. *Tricherostigma* - Boiss.). LT.: *Tr. fulgens* (Karw. ex Klotzsch) Klotzsch & Garcke [=*E. fulgens* Karw. ex Klotzsch] (chosen by Millspaugh, 1917).

Petaloma Raf. ex Baill., Adansonia I.1: 114 (1860). LT.: *P. leucoloma* Raf. ex Baill., pro syn. [=*E. marginata* Pursh] (chosen by Wheeler, 1943).

Petalandra F. Muell. ex Boiss. in DC., Prodr. 15(2): 27 (1862). T.: *P. euphorbioides* F. Muell. ex Boiss., pro syn. [=*E. micradenia* Boiss.]

Lyciopsis (Boiss.) Schweinf., Beitr. Fl. Aethiop. 1: 37 (1867), non Spach (1835). T.: *L. cuneata* (Vahl) Schweinf. [=*E. cuneata* Vahl]
Chylogala Fourr., Ann. Soc. Linn. Lyon, n.s. 17: 150 (1869). T.: *Ch. serrata* (L.) Fourr. [=*E. serrata* L.]
Epurga Fourr., Ann. Soc. Linn. Lyon, n.s. 17: 150 (1869). T.: *Ep. lathyris* (L.) Fourr. [=*Eu. lathyris* L.]
Euphorbion St.-Lag., Ann. Soc. Bot. Lyon, 7: 125 (1880). T.: (none designated; *E. antiquorum* L. by default).
Zygophyllidium (Boiss.) Small, Fl. SE. U.S.: 714, 1334 (1903). LT.: *Z. hexagonum* (Nutt. ex Spreng.) Small [=*E. hexagona* Nutt. ex Spreng.] (chosen by Small, 1913).
Diplocyathium H. Schmidt, Beih. Bot. Centralbl. 22(1): 40 (1907), nom. inval. T.: *D. capitulatum* (Rchb.) H. Schmidt [=*E. capitulata* Rchb.]
Euphorbiodendron Millsp., Publ. Field Columbian Mus., Bot. Ser. 2: 305 (1909). LT.: *E. latazi* (Humb., Bonpl. & Kunth) Millsp. [=*Euphorbia latazi* Humb., Bonpl. & Kunth; perhaps =*E. laurifolia* Lam.] (chosen by Webster, 1994).
Euphorbiopsis Lév., Repert. Spec. Nov. Regni Veg. 9: 446 (1911). T.: *E. lucidissima* (Lév. & Vaniot) Lév. [=*Euphorbia lucidissima* Lév. & Vaniot]
Dichylium Britton in Britton & P. Wils., Sci. Surv. P. Rico & Virg. Is. 5(4): 499 (1924). T.: *D. oerstedianum* (Klotzsch & Garcke) Britton (B.: *Poinsettia oerstediana* Klotzsch & Garcke) [=*E. oerstediana* (Klotzsch & Garcke) Boiss.]
Ctenadena Prokh., Consp. Syst. Tith. As. Med.: 28 (1933). T.: *Ct. lanata* (Sieb. ex Spreng.) Prokh. (B.: *E. lanata* Sieb. ex Spreng.) [=*E. petiolata* Banks & Sol. ex Russell]
Cystidospermum Prokh., Consp. Syst. Tith. As. Med.: 25 (1933). T.: *C. cheirolepis* (Fisch. & C. A. Mey. ap. Kar. ex Ledeb.) Prokh. [=*E. cheirolepis* Fisch. & C. A. Mey. ap. Kar. ex Ledeb.]
Sclerocyathium Prokh., Consp. Syst. Tith. As. Med.: 30 (1933). T.: *Scl. popovii* Prokh. [=*E. sclerocyathium* Korovin & Popov]

Monoecious or rarely dioecious, commonly glabrous, sometimes glaucous, trees, shrubs or perennial, biennial or annual herbs. Roots fibrous, or thick, fleshy and sometimes tuberous. Stems and branches succulent, semi-succulent or herbaceous, unarmed or spiny, terete or 2-many-angled, rarely flattened; angles obscure to deeply winged, straight, sinuate or toothed, often accentuated by horny spine-shields or podaria; podaria separate or contiguous in vertical or spiral series, each bearing a leaf-scar, 2 spines and sometimes 2 smaller stipular prickles; spines free, or occasionally partially to completely connate. Latex usually copious, milky, white or very rarely yellow, often strongly caustic. Indumentum, when present, simple; hairs usually unicellular, rarely multicellular. Leaves alternate, spiral, opposite or verticillate, sessile or shortly petiolate, stipulate or exstipulate, simple, entire or serrate, penninerved or palminerved, sometimes minute and soon deciduous; stipules, when present, minute, subulate, glandular or spiny. Inflorescences pseudopleiochasial, pseudodichasial or pseudumbellate, terminal or axillary; cyathia few to numerous, pseudanthial, each subtended by 2 variously-coloured free or connate modified leaves or cyathophylls; involucres cupuliform or infundibuliform, glanduliferous, lobate; glands (1–2)4–5(8), alternating with the lobes, erect or patent, free or contiguous, entire, bicornute, regularly or irregularly denticulate or pectinate, usually fleshy, variously-coloured, appendiculate or not; lobes variously-shaped, up to $^1/_2$ the length of the involucre, ciliate or glabrous. Male flowers few to many in 5 abbreviate cymose groups separated by usually ciliate membranes; each flower subtended by ciliate, fimbriate or lacerate bracteoles; pedicels persistent; calyx 0; petals 0; disc 0; stamen 1, filament articulating with the pedicel, anthers basifixed, introrse with respect to the cymes, 2-thecate, thecae subglobose, divergent, longitudinally dehiscent; pollen subprolate to oblate-spheroidal, 3-colporate, mostly 3-lobate in polar view, colpus transversalis +, costae +, colpi narrow, sometimes operculate, margo often broad, usually tectate, psilate, columellae distinct; pistillode 0; receptacle 0. Female flower solitary, central, subsessile or pedicellate; pedicel usually elongating and becoming reflexed in fruit, becoming erect again before dehiscence; calyx usually rim-like, sometimes obscurely 3-lobate or more conspicuous and irregularly toothed; petals 0; disc 0; ovary (2)3-locular, ovules 1 per locule; styles (2)3, free or connate at the base or for up to $^1/_2$ their length, 2-lobate to 2-fid, stigmas often thickened, rugose. Fruit 3-lobate to globose or conical, smooth, granulose, verrucose, cornute or rarely alate, dehiscing septicidally and loculicidally into (2)3 2-valved cocci; exocarp usually thin, separating or not; mesocarp, when present, sometimes fleshy, rarely spongy; endocarp woody, cartilaginous or crustaceous; columella various, persistent. Seeds ellipsoid, ovoid or subglobose, carunculate or ecarunculate; testa smooth, foveolate, reticulate, rugulose, verrucose or tuberculate, thinly crustaceous, brown or grey, usually concolorous; caruncle various; endosperm thick, fleshy; embryo straight; cotyledons flat. Fig.: See Euphorb. Gen.: t. 18, f. 61.1–16 (1824); Icon. Pl. 2: t. 182 (1837), 4: t. 346 (1841), and 7: t. 700 (1844); Étude Euphorb.: t.1; t. 2, f.1–16 (1858); Ic. Euph.: tt. 27–119 (1866); Mart., Fl. Bras. 11(2): tt. 92–97 (1874); Hooker's Icon. Pl. 14: t. 1305 (1880), 16: t. 1548 (1886), 24: t. 2347 (1894), and 26: tt. 2531, 2532 (1897); Sukk. Euph.: 4–124, tt. 1–33 (1907); Euph. Îles Austr. Afr.: 23–115, tt. 1–29 (1921); Fl. Indo-Chine 5: 241, t. 26 and 249, t. 27 (1925); Nat. Pflanzenfam. ed. 2, 19C: 209–219, tt. 112–118 (1931); Hooker's Icon. Pl. 32: t. 3193 (1933), 34: t. 3324 (1936), and 35: t. 3404 (1940); Succ. Euph.: pl. I–XXVI, tt. 2–1071 and 1091–1102 (1941); Fl. Parc Nat. Alb. 1: 477–480, tt. 23–25 (1948); Fl. S.S.S.R. XIV: 353–449, t. XVIII–XXIII (1949); Hooker's Icon. Pl. 35: tt. 3480, 3481 (1950); Fl. Cuba 3: 131, t. 47 (1953); Monogr. Euph. Maroc: 24–188, tt. 10–52 (1953); J. Fac. Sci. Univ. Tokyo, Sect. 3, Bot. 6: 227–274, t. 7–32 (1954) (mostly as *Galarhoeus*); Mém. Inst. Sci. Madagascar, Sér. B, Biol. Vég. 5: 115–184, tt. XXI–LVII (1954); Fl. W. Trop. Afr., ed. 2, 1: 420, t. 139 (1958); Ann. Missouri Bot. Gard. 48: 331–341, tt. 1–3 (1961) (as *Poinsettia*); Fl. Iran. 6: tt. 2–20 (1964); J. Arnold Arbor. 48: 396, t. 6a–m (1967); Mém. Mus. Natl. Hist. Nat., B, Bot. 20: 482, t. 173 (1971); Hooker's Icon. Pl. 37: t. 3696 (1971), and 38: tt. 3724, 3725 (1974); Fl. Ill. Sénég. 3: tt. 440-484 (1975); Fl. Iraq 4 (1): 329–357, tt. 61–64 (1980); Fl. Turk. 7: 601, t. 17 (1982);

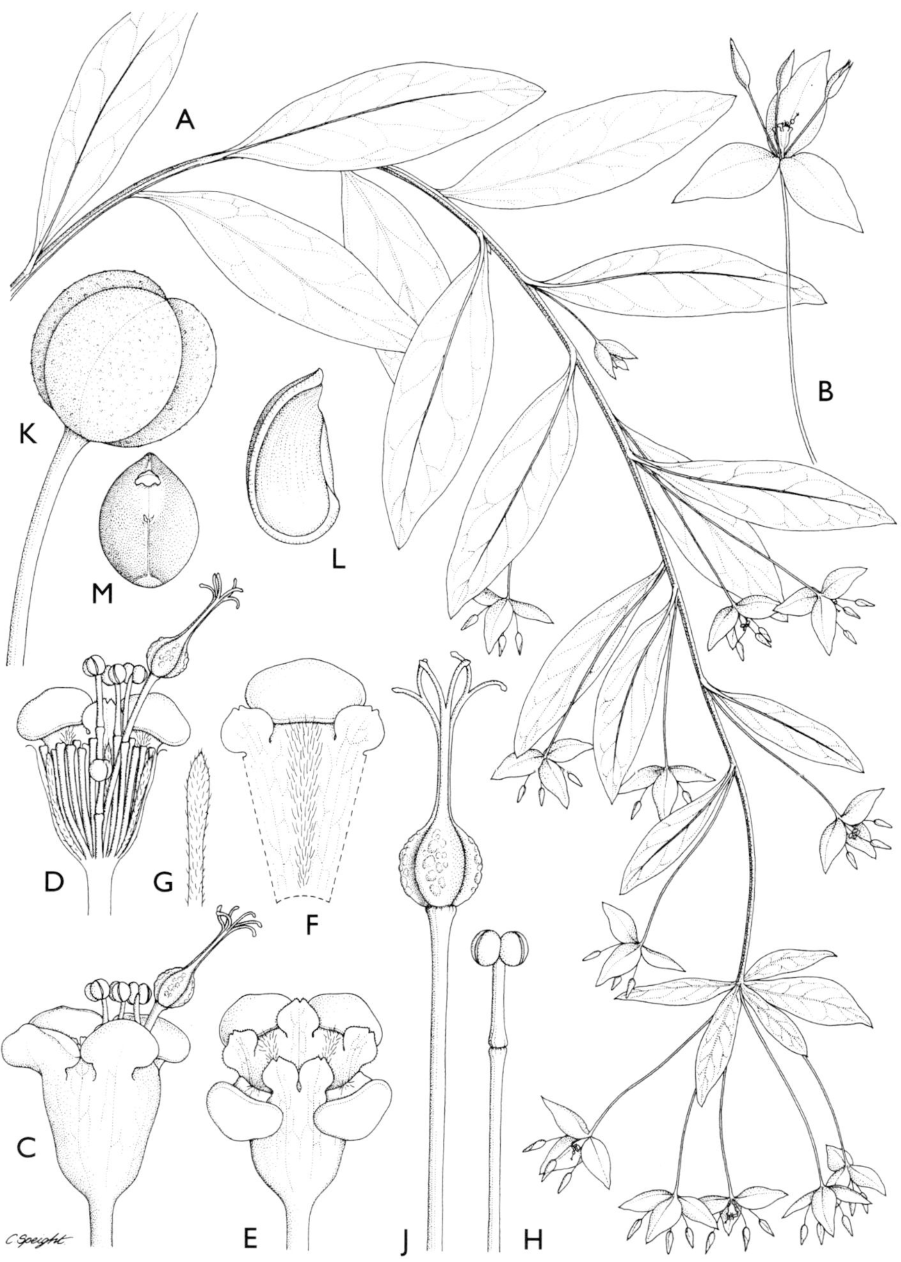

FIG. 47. *Euphorbia lancasteriana* Radcl.-Sm. **A** Habit × $^{1}/_{2}$; **B** Pseudumbel ray × 1; **C** Cyathium × 5; **D** Vertical section of cyathium showing male flowers at different stages of development × 5; **E** Cyathial involucre, flowers removed & 2 glands folded down to show disposition of involucral lobes × 5; **F** Portion of cyathial wall to show band of indumentum below gland within × 8; **G** Male bract × 8; **H** Male flower × 10; **J** Female flower × 10; **K** Fruit × 10; **L** Valve of fruit-coccus × 10; **M** Seed × 10. All from *Lancaster* 2023. Drawn by Camilla Speight.

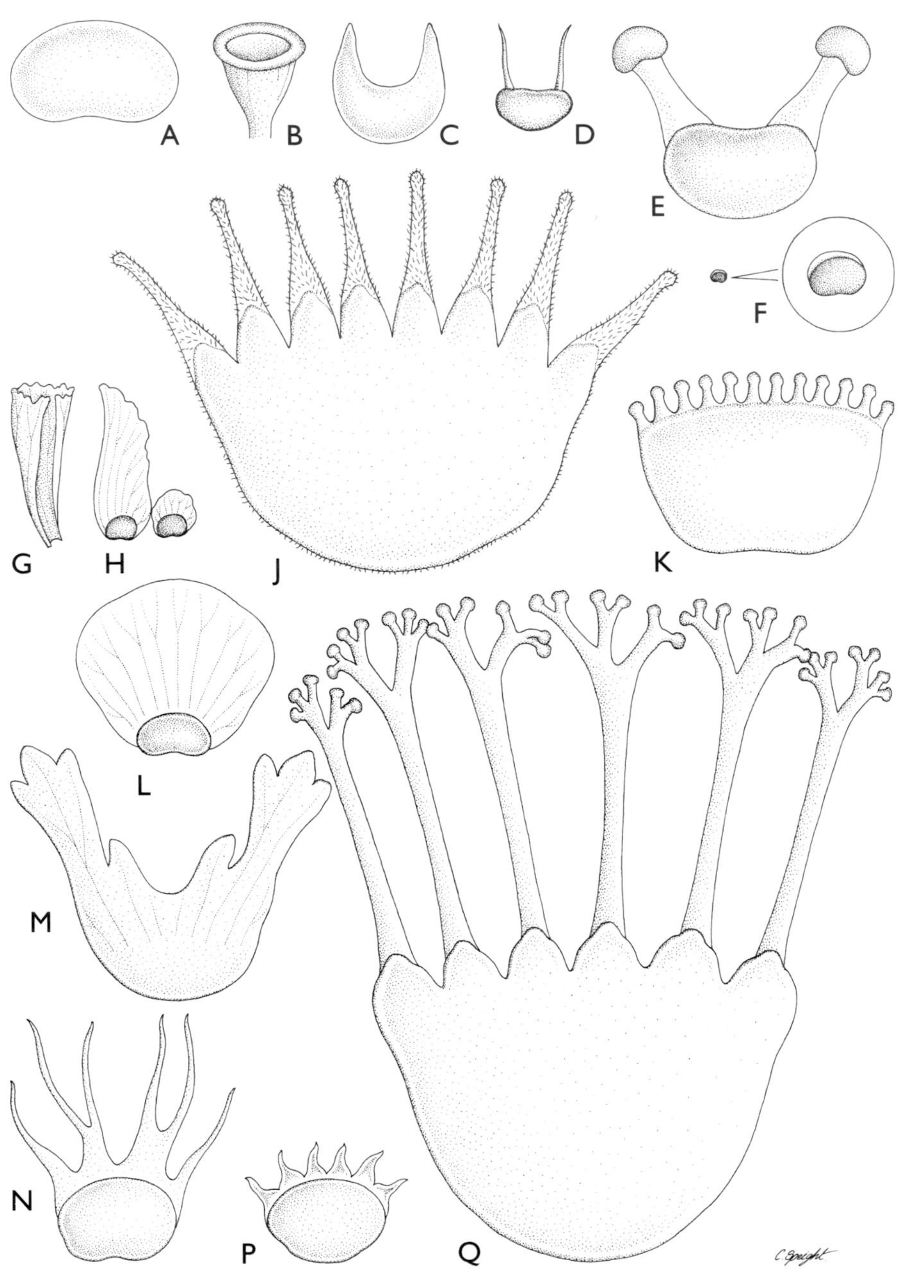

FIG. 48. *Euphorbia* L.: Cyathial glands of **A** *E. altissima* Boiss.; **B** *E. heterophylla* L.; **C** *E. robbiae* Turrill; **D** *E. schimperiana* Scheele; **E** *E. lathyris* L.; **F** *E. drummondii* Boiss.; **G** *E. longetuberculosa* Hochst. ex Boiss.; **H** *E. rosea* Retz.; **J** *E. noxia* Pax; **K** *E. denticulata* Lam.; **L** *E. marginata* Pursh; **M** *E. cheiradenia* Boiss. & Hohen.; **N** *E. goyazensis* Boiss.; **P** *E. schizolepis* F. Muell. ex Boiss.; **Q** *E. grantii* Oliv. All except **F** (enlargement) × 40. Drawn by Camilla Speight.

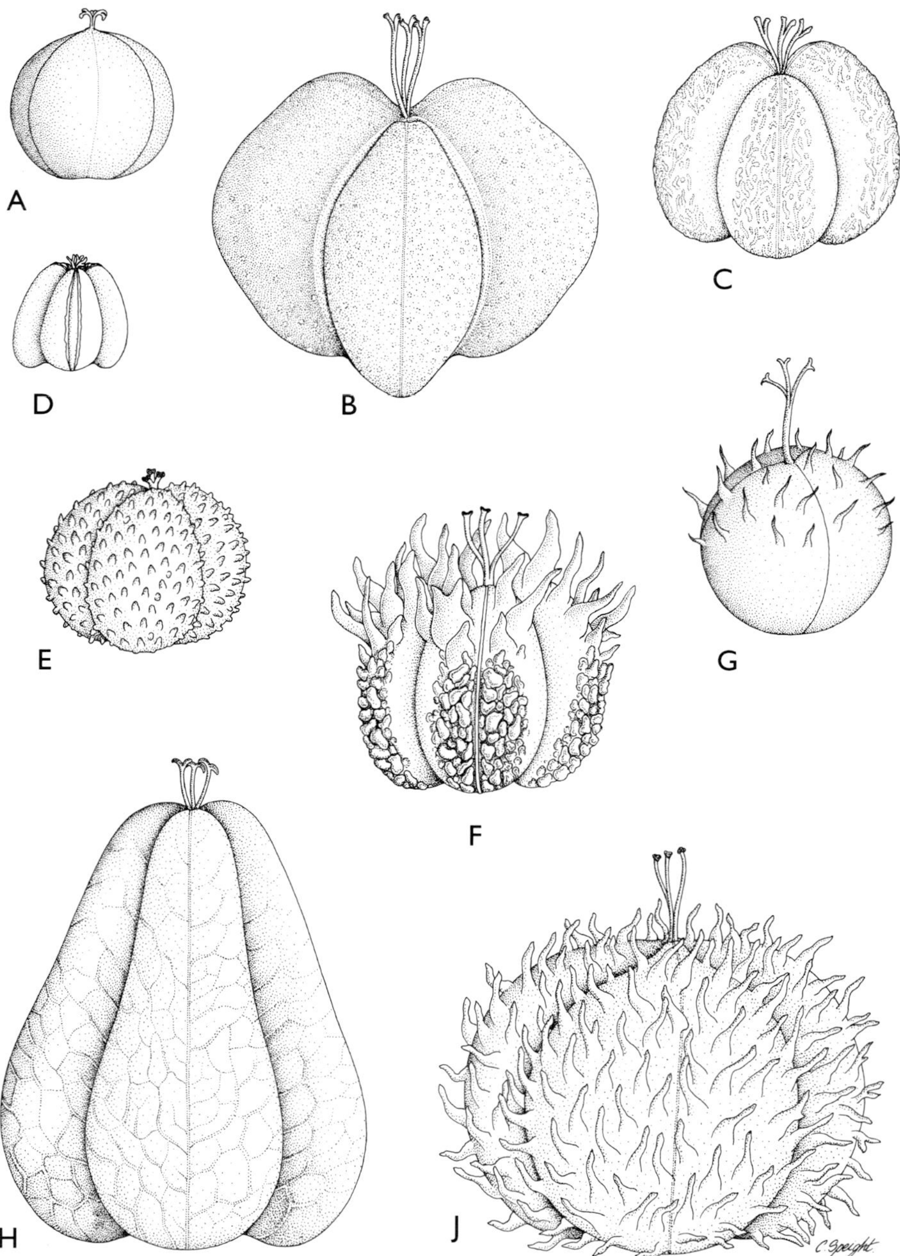

FIG. 49. *Euphorbia* L.: Fruits of **A** *E. microsphera* Boiss.; **B** *E. dendroides* L.; **C** *E. waldsteinii* (Soják) Radcl.-Sm.; **D** *E. peplus* L.; **E** *E. flavicoma* DC. (*E. brittingeri* Opiz ex Samp.); **F** *E. ceratocarpa* Ten.; **G** *E. valerianifolia* Lam. (*E. cybirensis* Boiss.); **H** *E. grossheimii* (Prokh.) Prokh.; **J** *E. macrocarpa* Boiss. & Buhse. All × 8. Drawn by Camilla Speight.

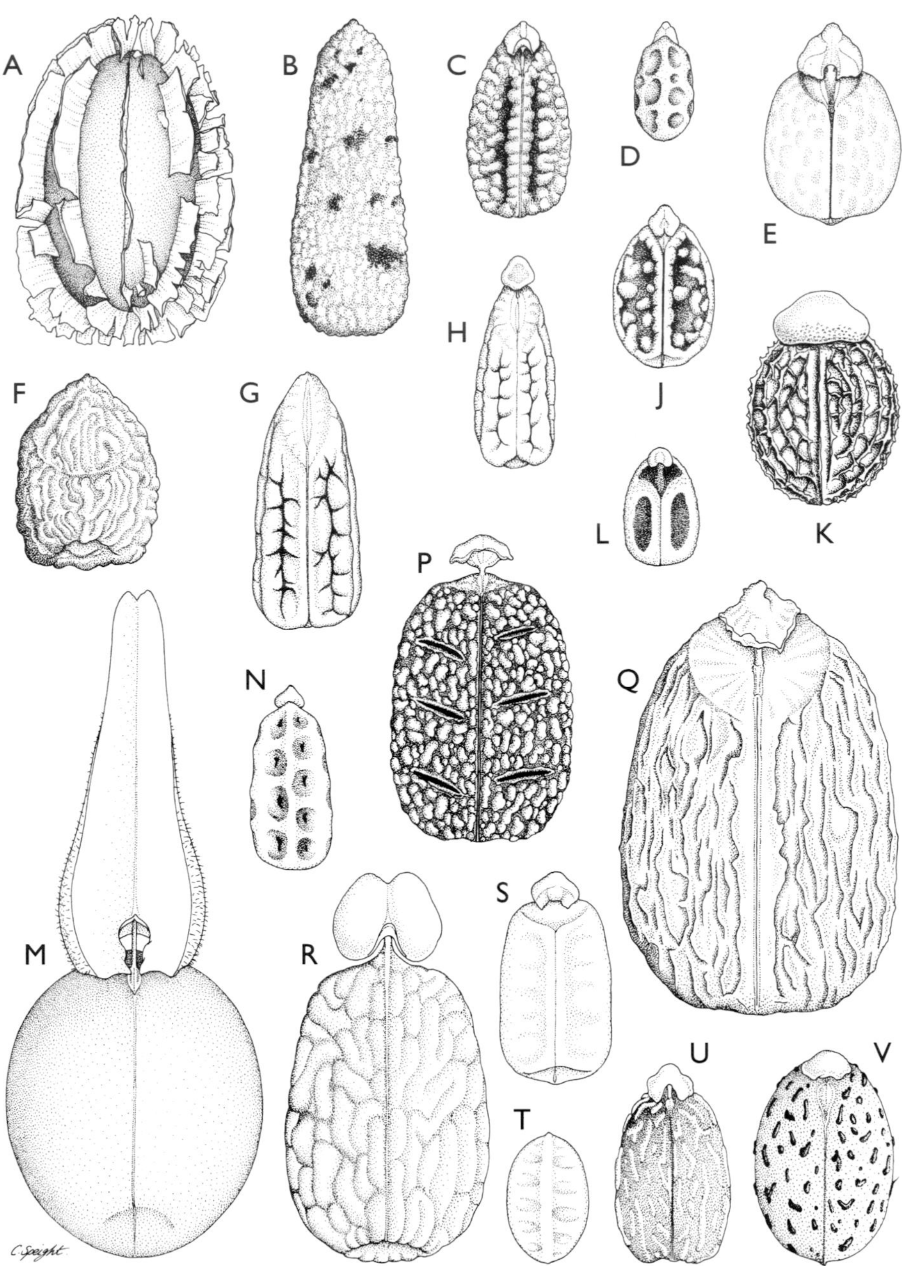

FIG. 50. *Euphorbia* L.: Seeds of **A** *E. guyoniana* Boiss. & Reut.; **B** *E. cheirolepis* Fisch. & C. A. Mey. ap. Kar. ex Ledeb.; **C** *E. francheti* B. Fedtsch.; **D** *E. peplus* L.; **E** *E. portlandica* L.; **F** *E. acalyphoides* Hochst. ex Boiss.; **G** *E. turczaninowii* Kar. & Kir.; **H** *E. inderiensis* Less. ex Kar. & Kir.; **J** *E. exigua* L.; **K** *E. helioscopia* L.; **L** *E. sulcata* Lens ex Loisel.; **M** *E. grossheimii* (Prokh.) Prokh.; **N** *E. chamaepeplus* Boiss. & Gaill.; **P** *E. phymatosperma* Boiss. & Gaill.; **Q** *E. lathyris* L.; **R** *E. myrsinites* L.; **S** *E. herniariifolia* Willd.; **T** *E. falcata* L.; **U** *E. medicaginea* Boiss.; **V** *E. hirsuta* L. (*E. pubescens* Vahl). All × 14. Drawn by Camilla Speight.

Euphorbia J. 1–10: tt. ∞ (1980–1996); Fl. Bah. Arch.: 803, t. 335 and 808, t. 336 (1982); Hooker's Icon. Pl. 39: tt. 3851–3875 (1982); Kew Bull. Addit. Ser. 11: 551, t. 168 (1984) (as *Poinsettia*); Fl. Cyp. 2: 1438, t. 83 (1985); Fl. Pak. 172: 109–163, tt. 20–34 (1986); Kew Bull.: 42: 232, t. 1 and 674, t. 1 (1987); Fl. Bhut. 1: 763, t. 47a,b (1987); Fl. Nouv. Caléd. 14: 15, t. 2 (1987); Fl. Trop. E. Afr. Euph. 2: 434–530, tt. 80–99 (1988); Descr. Fl. Puerto Rico 2: 394, t. 59-15 (1988); Fl. Lesser Antilles 5: 22, t. 12 (1989); *Euphorbias*: tt. pp. 24–125 (1995); Fl. St. John: 211, t. 95J–N (1996). See also Figs. 47–50, p. 411–414.

A vast subcosmopolitan genus of over 2,000 species, and one of the 5 largest genera of *Angiospermae*, the other 4 being *Astragalus, Carex, Senecio* and *Solanum*. The genus has been subdivided into many subgenera and sections, a number of which have also been treated as distinct genera at various times, as for example *Poinsettia*, accepted by Dressler (1961).

The literature, especially on the succulent taxa, is vast, and only a fraction has been cited here. Many references are given by Wheeler (1943), Vindt (1960), Webster (1967a), Carter (1988) and Lebrun & Stork (1991), and a large number of taxa are illustrated in the 10 volumes of *Euphorbia* Journal (1980–1996). Many species, both succulent and herbaceous, are in cultivation as ornamentals. Oudejans (1990, 1993) has catalogued all the binomials published in the genus up to that time. A large number of species are in cultivation, especially in the tropics (See Frontispiece).

330. **Elaeophorbia** *Stapf*, Hooker's Icon. Pl. 29: t. 2823 (1906), and in H. H. Johnst., Liberia 2: 646 (1906); N. E. Brown in Fl. Trop. Afr. 6(1): 604 (1912), p.p., and in Fl. Cap. 5(2): 221 (1915); Croizat, Bull. Jard. Bot. État. 15: 109 (1938); L. C. Wheeler, Amer. Midl. Naturalist 30: 465 (1943); Keay (ed.), Fl. W. Trop. Afr., ed. 2, 1: 423 (1958); Berhaut, Fl. Ill. Sénég. 3: 433 (1975); S. Carter, Fl. Trop. E. Afr. Euph. 2: 533 (1988). T.: *El. drupifera* (Schum.) Stapf [=*Euphorbia drupifera* Schum.]

Very like certain arborescent species of 329. *Euphorbia*, but pollen perforate-tectate; female perianth 0; ovary confluent with the pedicel; fruit thick, fleshy, indehiscent, drupaceous; endocarp osseous, 3-locular, each lobe dorsally grooved, 3-porate at apex and base; seeds sometimes only 1 or 2 per fruit by abortion; cotyledons elliptic, emarginate at the base, thick and fleshy. Fig.: See Hooker's Icon. Pl. 29: t. 2823 (1906); Fl. Ill. Sénég. 3: t. 432 (1975); Fl. Trop. E. Afr. Euph. 2: 532, t. 100 (1988).

A tetratypic tropical African genus ranging from Guinea west to Uganda and south to Angola.

331. **Chamaesyce** *Gray*, Nat. Arr. Brit. Pl. 2: 260 (1821); Croizat in O. & I. Deg., Fl. Hawaii (1936); L. C. Wheeler, Rhodora 43: 97 (1941), Contr. Gray Herb. 136: 97 (1941), and Amer. Midl. Naturalist 30: 461 (1943); Hurus., J. Fac. Sci. Univ. Tokyo, Sect. 3, Bot. 6: 271 (1954); Burch, Ann. Missouri Bot. Gard. 53: 90 (1966); G. L. Webster, J. Arnold Arbor. 48: 420 (1967); D. C. Hassall, Austral. J. Bot. 24: 633 (1976); G. A. Mulligan & D. R. Lindsay, Naturaliste

Canad. 105: 37 (1978); Proctor, Kew Bull. Addit. Ser. 11: 550 (1984); Koutnik, S. African J. Bot. 3: 262 (1984), and Allertonia 4(6): 331 (1987); R. A. Howard, Fl. Lesser Antilles 5: 21 (1989); G. L. Webster, Ann. Missouri Bot. Gard. 81: 129 (1994); Gillespie in Acev.-Rodr., ed., Fl. St. John: 205 (1996); Huegin, Feddes Repert. 109: 189–223 (1998). T.: *Ch. maritima* Gray, nom. illeg. [=*Ch. peplis* (L.) Prokh. (B.: *Euphorbia peplis* L.)]

Anisophyllum Haw., Syn. Pl. Succ.: 159 (1812), non Jacq. (1763). LT.: *A. peplis* (L.) Haw. (B.: *Euphorbia peplis* L.) [=*Ch. peplis* (L.) Prokh.] (chosen by Wheeler, 1941).
Ditritra Raf., Fl. Tellur. 4: 114 (1838). LT.: *D. hirta* (L.) Raf. [=*Ch. hirta* (L.) Millsp. (B.: *E. hirta* L.)] (chosen by Wheeler, 1943).
Endoisila Raf., Fl. Tellur. 4: 114 (1838). T.: *E. myrsinites* Raf., non *Euph. myrsinites* L. [?=*Ch. myrtillifolia* (L.) Millsp. (B.: *Euph. myrtillifolia* L.)]
Xamesike Raf., Fl. Tellur. 4: 115 (1838). LT.: *X. vulgaris* Raf. [=*Ch. canescens* (L.) Prokh. (B.: *E. canescens* L.) (S.: *E. chamaesyce* L.)] (chosen by Wheeler, 1941).
Aplarina Raf., New Fl. N. Amer. 4: 99 (1838). LT.: *A. prostrata* Raf., non *Ch. prostrata* (Ait.) Small (B.: *E. prostrata* Ait.) (chosen by Wheeler, 1941).
Euphorbia subgen. *Chamaesyce* (Gray) Caes. ex Rchb., Deut. Bot. Herb.-Buch.: 193 (1841).

Resembling certain herbaceous species of 329. *Euphorbia*, but differing primarily in the following vegetative characteristics: main axis aborting above the cotyledons; leaves opposite, inaequilateral at the base; veins chlorenchyma-sheathed; stipules +, interpetiolar. Also seeds usually ecarunculate. Fig.: See Ic. Euph.: tt. 1–26 (1866); Fl. Jamaica 4: 340, t. 113 (1920) (as *Euphorbia*); Rhodora 43: tt. 654–668 (1941); Contr. Gray Herb. 136: tt. 654–668 (1941); Fl. S.S.S.R. XIV: 477–489, tt. XXIV and XXV (1949) (as *Euphorbia*); J. Fac. Sci. Univ. Tokyo, Sect. 3, Bot. 6: 276–292, tt. 33–40 (1954); Mém. Mus. Natl. Hist. Nat., B, Bot. 20: 480, t. 172 (1971) (as *Euphorbia*); Kew Bull. 35: 415, t. 1 (1980), 39: 645, t. 1 (1984) (as *Euphorbia*); Kew Bull. Addit. Ser. 11: 553, t. 169 (1984); Allertonia 4(6): 335–383, tt. 1–24 (1987); Fl. Trop. E. Afr. Euph. 2: 410–427, tt. 76 *pro parte* –79 (1988) (as *Euphorbia*); Fl. Lesser Antilles 5: 22, t. 10 (1989); Kew Bull. 45: 329, t. 1 (1990) (as *Euphorbia*); Fl. St. John: 207, t. 94A–J (1996).

A subcosmopolitan genus of c. 250 species, mostly in tropical America and Africa, but with some specialized endemics in Hawaii. It is often treated as a subgenus of *Euphorbia*. Koutnik (1987) has strongly argued the case for generic recognition.

332. **Cubanthus** *(Boiss.) Millsp.*, Publ. Field Columbian Mus., Bot. Ser. 2: 371 (1913); Alain, Fl. Cuba 3: 127 (1953); and Fl. Español. 4: 133 (1986); G. L. Webster, Ann. Missouri Bot. Gard. 81: 129 (1994). LT.: *C. linearifolius* (Griseb.) Millsp. (B.: *Pedilanthus linearifolius* Griseb.) (chosen by Millspaugh, 1913).

Very like certain suffrutescent non-succulent species of 329. *Euphorbia*, but differing in having the involucral lobes fused, the 2 glands fused into a shield, and the female flower calyculate. The seeds are ecarunculate. Fig.: See Fl. Cuba 3: 127, t. 46 (1953).

A tritypic genus endemic to Cuba and Hispaniola.

333. **Monadenium** *Pax,* Bot. Jahrb. Syst. 19: 126 (1894); Stapf, Hooker's Icon. Pl. 27: t. 2666 (1900); N. E. Br. in Fl. Trop. Afr. 6(1): 450 (1911); Pax & K. Hoffm., Nat. Pflanzenfam. ed. 2, 19C: 222 (1931); A. C. White, R. A. Dyer & B. Sloane, Succ. Euph. 2: 941 (1941); L. C. Wheeler, Amer. Midl. Naturalist 30: 470 (1943); P. R. O. Bally, Genus *Monadenium*: 14 (1961); S. Carter, Kew Bull.: 42: 903 (1987), and Fl. Trop. E. Afr. Euph. 2: 540 (1988); M. G. Gilbert, Bradleya 8: 45 (1990); J.-P. Lebrun & Stork, Enum. Pl. Afr. Trop. 1: 230 (1991); S. Carter in LaFon *et al.*, eds., Euphorbia J. 8: 15 (1992); G. L. Webster, Ann. Missouri Bot. Gard. 81: 129 (1994). T.: *M. coccineum* Pax.

Lortia Rendle, J. Bot., 36: 29 (1898). T.: *L. erubescens* Rendle [=*M. erubescens* (Rendle) N. E. Br.]

Stenadenium Pax, Bot. Jahrb. Syst. 30: 343 (1901); N. E. Br. in Fl. Trop. Afr. 6(1): 448 (1911); Pax & K. Hoffm., Nat. Pflanzenfam. ed. 2, 19C: 222 (1931). T.: *St. spinescens* Pax [=*M. spinescens* (Pax) P. R. O. Bally]

Monoecious perennial herbs, shrubs or small trees, sometimes geophytic. Stems & branches ± fleshy or succulent. Indumentum, when present, simple. Latex caustic, milky. Roots thick, fleshy, sometimes tuberous. Leaves fleshy, mostly spirally arranged, shortly petiolate or subsessile, sometimes stipulate, simple, entire or rarely toothed. Stipules glandular, spinose or 0. Inflorescences axillary; cyathia sessile or shortly pedunculate, arranged in pseudodichasia, rarely solitary, somewhat bilaterally symmetrical; cyathophylls free or partially united along the upper side forming a cup enveloping the involucre; glands forming a rim open on the lower side; cyathial lobes 5, fimbriate. Male flowers in 5 groups, bracteolate, scarcely exserted, resembling those of 329. *Euphorbia*; pollen large, subprolate, 3-colporate, colpus transversalis broad, elongate, costate, colpi long, narrow, margo 0, tectate, psilate. Female flower pedicellate, pedicel becoming exserted through the gap in the gland-rim; calyx mostly rim-like, rarely shortly 3-lobate; petals 0; disc 0; ovary 3-locular, ovules 1 per locule; styles 3, connate at the base, stigmas 2-fid. Fruit 3-lobate, smooth, sometimes alate, dehiscing septicidally and loculicidally into 3 2-valved cocci. Seeds oblong, carunculate or rarely ecarunculate; caruncle pileiform. Fig.: See Hooker's Icon. Pl. 27: t. 2666 (1900); Nat. Pflanzenfam. ed. 2, 19C: 223, t. 120 (1931); Succ. Euph.: tt. 1072–1083 (1941); Genus *Monadenium*: tt. 12–95 (1961); Fl. Trop. E. Afr. Euph. 2: 543–558, tt. 102–104 (1988); Euphorbia J. 8: tt. 14–19 (1992).

A tropical African genus of 70 species, ranging from Somalia to the Transvaal.

334. **Synadenium** *Boiss.* in DC., Prodr. 15(2): 187 (1862); Benth., Gen. Pl. 3: 261 (1880); N. E. Br. in Fl. Trop. Afr. 6(1): 462 (1911), and in Fl. Cap. 5(2): 220 (1915); Pax & K. Hoffm., Nat. Pflanzenfam. ed. 2, 19C: 221 (1931); A. C. White,

R. A. Dyer & B. Sloane, Succ. Euph. 2: 951 (1941); L. C. Wheeler, Amer. Midl. Naturalist 30: 472 (1943); Robyns, Fl. Parc Nat. Alb. 1: 483 (1948); S. Carter, Kew Bull.: 42: 667 (1987), & Fl. Trop. E. Afr. Euph. 2: 534 (1988); G. L. Webster, Ann. Missouri Bot. Gard. 81: 129 (1994). LT.: *S. arborescens* Boiss. [=*S. cupulare* (Boiss.) L. C. Wheeler (B.: *Euphorbia cupularis* Boiss.)] (designated by White *et. al.*, 1941).

Monoecious shrubs or trees. Branches subfleshy, cylindric. Indumentum, when present, simple. Latex caustic, milky, copious. Leaves fleshy, mostly spirally arranged, shortly petiolate or subsessile, stipulate, simple, entire or rarely toothed, penninerved. Stipules small, glandular, conspicuous, dark brown. Inflorescences axillary; cyathia sessile or shortly pedunculate, arranged in pseudodichasia, these usually crowded together into subumbellate hyperflorescences at the branch-apices; cyathophylls paired, ± free, persistent. Involucral glands forming a conspicuous ± continuous rim; rim entire, rarely lobate, spreading, rugulose; cyathial lobes 5, fimbriate. Male flowers in 5 groups, shortly exserted, bracteolate, bracteoles included, flowers resembling those of 329. *Euphorbia*; pollen ± as in 333. *Monadenium*. Female flower shortly pedicellate; pedicel elongating slightly in fruit; calyx rim-like, 3-lobate; petals 0; disc 0; ovary 3-locular, smooth, ovules 1 per locule; styles 3, connate at the base, stigmas 2-fid. Fruit 3-lobate, smooth, sometimes pubescent, dehiscing septicidally and loculicidally into 3 2-valved cocci. Seeds ovoid, caruncuIate; caruncle sessile, often minute. Fig.: See Nat. Pflanzenfam. ed. 2, 19C: 222, t. 119A (1931); Succ. Euph.: tt. 1084–1090 (1941); Fl. Trop. E. Afr. Euph. 2: 536, t. 101 (1988).

An E. and S. tropical African genus of 19 closely-related species, ranging from Uganda to the Transvaal.

335. **Endadenium** *L. C. Leach*, Garcia de Orta, Sér. Bot. 1: 31 (1973); G. L. Webster, Ann. Missouri Bot. Gard. 81: 130 (1994). T.: *E. gossweileri* (N. E. Br.) L. C. Leach (B.: *Monadenium gossweileri* N. E. Br.)

Like 334. *Synadenium*, but cyathophylls large, enveloping the cyathia as in 333. *Monadenium*, nutant; involucral glands distinct, inconspicuous, located below the rim; ovary prominently 2-winged. Fig.: See Euphorbia J. 1: t. 128 (1992).

A monotypic genus from Angola.

336. **Pedilanthus** *Neck. ex Poit.*, Ann. Mus. Natl. Hist. Nat. 19: 388 (1812); A. Juss., Euphorb. Gen.: 59 (1824); Baill., Étude Euphorb.: 287 (1858); Klotzsch & Garcke, Monatsber. Königl. Preuss. Akad. Wiss. Berlin 1859: 253 (1859); Boiss. in DC., Prodr. 15(2): 4 (1862); Müll. Arg. in Mart., Fl. Bras. 11(2): 702 (1874); Millsp., Publ. Field Columbian Mus., Bot. Ser. 2: 353 (1913); Fawc. & Rendle, Fl. Jamaica 4: 346 (1920); Pax & K. Hoffm., Nat. Pflanzenfam. ed. 2, 19C: 223 (1931); L. C. Wheeler, Contr. Gray Herb. 124: 43 (1939), and Amer. Midl. Naturalist 30: 470 (1943); Alain, Fl. Cuba 3: 125 (1953); Dressler, Contr. Gray Herb. 182: 97 (1957); Backer & Bakh. f., Fl. Java 1: 503 (1963); Jabl., Mem. New

York Bot. Gard. 12: 189 (1967); G. L. Webster, J. Arnold Arbor. 48: 427 (1967), and Ann. Missouri Bot. Gard. 54: 346 (1968); Berhaut, Fl. Ill. Sénég. 3: 543 (1975); Philcox, Fl. Trin. Tob. II(X): 684 (1979); Correll, Fl. Bah. Arch.: 830 (1982); Proctor, Kew Bull. Addit. Ser. 11: 557 (1984); Koutnik, Euphorbia J. 3: 39 (1985); Alain, Fl. Español. 4: 181 (1986); Grierson & D. G. Long, Fl. Bhut. 1: 767 (1987); Alain, Descr. Fl. Puerto Rico 2: 410 (1988); R. A. Howard, Fl. Lesser Antilles 5: 69 (1989); G. L. Webster, Ann. Missouri Bot. Gard. 81: 130 (1994); Gillespie in Acev.-Rodr., ed., Fl. St. John: 219 (1996). LT.: *P. tithymaloides* (L.) Poit. (B.: *Euphorbia tithymaloides* L.) (chosen by Millspaugh, 1909).

Tithymalus Mill., Gard. Dict. Abr., ed. 4, 3 (1754), nom. rej. LT.: *T. myrtifolius* Mill. [=*P. tithymaloides* (L.) Poit. (B.: *Euphorbia tithymaloides* L.)] (chosen by Wheeler, 1939).

Tithymaloides Ortega, Tab. Bot.: 9 (1773), nom. rej. LT.: *T. myrtifolium* (L.) Kuntze (B.: *E. tithymaloides myrtifolia* L.) [=*P. tithymaloides* (L.) Poit. (B.: *Euphorbia tithymaloides* L.)] (chosen by Wheeler, 1939).

Ventenatia Tratt., Gen. Pl.: 86 (1802), non Cav. (1798). T.: *V. bracteata* (Jacq.) Tratt. (B.: *Euphorbia bracteata* Jacq.) [=*P. bracteatus* (Jacq.) Boiss.]

Crepidaria Haw., Syn. Pl. Succ.: 136 (1812). LT.: *Cr. myrtifolia* (Mill.) Haw. (B.: *Tithymalus myrtifolius* Mill.) [=*P. tithymaloides* (L.) Poit. (B.: *Euphorbia tithymaloides* L.)] (chosen by Wheeler, 1939).

Tirucalia Raf., Fl. Tellur. 4: 112 (1838), p.p. T.: (see also under *Euphorbia*).

Diadenaria Klotzsch & Garcke, Monatsber. Königl. Preuss. Akad. Wiss. Berlin 1859: 254 (1859). LT.: *D. pavonis* Klotzsch & Garcke [=*P. bracteatus* (Jacq.) Boiss. (B.: *Euphorbia bracteata* Jacq.)] (chosen by Wheeler, 1939).

Hexadenia Klotzsch & Garcke, Monatsber. Königl. Preuss. Akad. Wiss. Berlin 1859: 253 (1859). T.: *H. macrocarpa* (Benth.) Klotzsch & Garcke [=*P. macrocarpus* Benth.]

Monoecious woody or succulent trees or shrubs. Indumentum, when present, simple. Leaves alternate, distichous, shortly petiolate or subsessile, stipulate, simple, entire; stipules small, obtuse. Inflorescences terminal or axillary, pseudodichasial or pseudomonochasial by abortion, cyathophylls large or small. Cyathia strongly zygomorphic, bilaterally symmetrical; involucral lobes 5, partially connate, 3-typic: 2 ventral, sometimes dorsally glanduliferous (main lobes), 2 lateral, basally asymmetric, at least dorsally glanduliferous, ligulate, inconspicuous (lateral accessory lobes) and 1 dorsal, bilaterally symmetrical, laterally glanduliferous, ligulate, inconspicuous (median accessory lobe); glands 2, 4 or 6, appendiculate, appendages (spur lobes) partially connate with each other and adnate to the main involucral lobes, forming a ± elongate, solid, calcariform dorsal extension of the involucre, the lateral spur lobes completely adnate to it, the median spur lobes lying between the lateral lobes and anteriorly over the accessory lobes; bracteoles numerous, filamentous, or 0. Male flowers numerous, exserted, resembling those of 329. *Euphorbia*; pollen oblate-spheroidal to suboblate, slightly 3-lobate, 3-colpate, colpi broad, margo narrow, intectate, psilate, columellate. Female flower pedicellate; pedicel elongating and decurving or reflexing in fruit; calyx rim-like or 0; petals 0; disc 0; ovary 3-locular, smooth, ovules 1 per locule; styles 3, connate, stigmas 2-fid. Fruit 3-lobate, smooth, dehiscing septicidally and loculicidally into 3 2-valved cocci, or indehiscent. Seeds oblong-ovoid to subquadrate or ovoid-conic, carunculate, smooth or tuberculate. Fig.: See Étude Euphorb.: t. 3, f. 1–15 (1858); Fl. Bras. 11(2): t. 98 (1874); Fl. Jamaica 4: 347, t. 114 (1920); Nat. Pflanzenfam. ed. 2, 19C: 222, t. 119B (1931); Fl. Cuba 3: 126, t. 45 (1953); Contr. Gray Herb. 182: 29–30, tt. 12–14 and 33–182, pls. V–XXI (1957); Fl. Ill. Sénég. 3: 543, t. 542 (1975); Fl. Bah. Arch.: 833, t. 346 (1982); Euphorbia J. 3: tt. 38–42 (1985); Fl. Bhut. 1: 763, t. 47c,d (1987); Fl. Lesser Antilles 5: 68, t. 27 (1989); Fl. St. John: 218, t. 99H–K (1996).

A neotropical genus of c. 15 species, mostly from Mexico and C. America, *fide* Dressler (1957).

INCERTAE SEDIS FAMILIAE

337. **Afrotrewia** *Pax & K. Hoffm.*, Pflanzenr. 63: 14 (1914), and Nat. Pflanzenfam. ed. 2, 19C: 104 (1931). T.: *A. kamerunica* Pax & K. Hoffm.

This genus, known from flowering material from Kribi, between Ebolowa and Kampo in the Cameroon, was affined to *Crotonogynopsis* Pax and *Necepsia* Prain by Pax & Hoffmann, even though it is described as having a stellate indumentum, and the male flowers are simpler. The type-material has been destroyed, however, so no proper evaluation of it can be made until further material is collected.

338. **Chlamydojatropha** *Pax & K. Hoffm.*, Pflanzenr. 57: 125 (1912), and Nat. Pflanzenfam. ed. 2, 19C: 167 (1931). T.: *Chl. kamerunica* Pax & K. Hoffm.

This genus, known only from female material from the Cameroon, was affined to *Mildbraedia* Pax by Pax & Hoffmann, even though it is described as having a simple indumentum, but until male flowers are found, no proper evaluation of it can be made.

339. **Myladenia** *Airy Shaw*, Kew Bull. 32: 79 (1977). T.: *M. serrata* Airy Shaw.

This genus, known only from male material from E. Thailand, was considered by Airy Shaw to be related to *Austrobuxus* and *Dissiliaria*, but it differs from both in having a well-developed extrastaminal disc. However, the pollen, which is deeply 3-lobate in polar view, 3-colporate with elongate colpi, and tectate-perforate to reticulate, is non-spinous, which could rule *Myladenia* out of the *Oldfieldioideae*, although there are 3 genera with brevispinous or smooth pollen in the subtribe *Dissiliariinae*, but as no female flowers or fruits have ever been collected, the correct assignment of *Myladenia* must remain uncertain until this deficiency is rectified. The genus was completely overlooked by Webster (1994).

APPENDIX

List of new taxa, names and combinations appearing in this work

Synopsis of taxa

PHYLLANTHOIDEAE Asch.
Dispermae Zoll.
Phyllanthaceae Klotzsch & Garcke
WIELANDIEAE Baill. ex Hurus.
Wielandiiées Baill.
Wielandiinae Pax & K. Hoffm.
Heywoodia Sim
Savia Willd.
Maschalanthus Nutt.
Kleinodendron L. B. Sm. & Downs
Gonatogyne Klotzsch ex Müll. Arg.
Petalodiscus (Baill.) Pax
Blotia Leandri
Actephila Blume
Lithoxylon Endl.
Anomospermum Dalzell
Discocarpus Klotzsch
Lachnostylis Turcz.
Chonocentrum Pierre ex Pax & K. Hoffm.
Wielandia Baill.
AMANOEAE (Pax & K. Hoffm.) G. L. Webster
Amanoinae Pax & K. Hoffm.
Pentabrachion Müll. Arg.
Amanoa Aubl.
BRIDELIEAE Müll. Arg.
Bridelieae Thwaites
Bridelioideae Hurus.
Cleistanthus Hook. f. ex Planch.
Nanopetalum Hassk.
Lebidiera Baill.
Stenonia Baill.
Leiopyxis Miq.
Lebidieropsis Müll. Arg.
Kaluhaburunghos Kuntze
Schistostigma Lauterb.
Stenoniella Kuntze in T. Post & Kuntze
Zenkerodendron Gilg ex Jabl.
Paracleisthus Gagnep.
Godefroya Gagnep.
Bridelia Willd.
Candelabria Hochst.
Pentameria Klotzsch ex Baill.
Neogoetzea Pax
Gentilia Beille
Tzellemtinia Chiov.

PHYLLANTHEAE Dumort.
Phyllanthaceae Klotzsch & Garcke
Astrocasiinae G. L. Webster
Astrocasia B. L. Rob. & Millsp.
Leptopinae G. L. Webster
Leptopus Decne.
Lepidanthus Nutt.
Hexakistra Hook. f.
Thelypetalum Gagnep.
Chorisandrachne Airy Shaw
Archileptopus P. T. Li
Pseudolachnostylidinae Pax & K. Hoffm.
Chascotheca Urb.
Chaenotheca Urb.
Zimmermannia Pax
Zimmermanniopsis Radcl.-Sm.
Meineckia Baill.
Cluytiandra Müll. Arg.
Peltandra Wight
Neopeltandra Gamble
Pseudolachnostylis Pax
Keayodendron Leandri
Securineginae Müll. Arg.
Securinega Comm. ex Juss.
Andrachninae Müll. Arg.
Andrachne L.
Telephioides Ortega
Eraclissa Forssk.
Arachne Neck.
Phyllanthidea Didr.
Flueggeinae Müll. Arg.
Euphyllantheae Müll. Arg.
Sauropodeae Müll. Arg.
Phyllanthinae Pax
Flueggea Willd.
Acidoton P. Browne
Bessera Spreng.
Geblera Fisch. & C. A. Mey.
Colmeiroa Reut.
Pleiostemon Sond.
Neowawraea Rock
Richeriella Pax & K. Hoffm.
Margaritaria L. f.
Prosorus Dalzell
Zygospermum Thwaites ex Baill.
Wurtzia Baill.
Calococcus Kurz ex Teijsm. & Binn.
Phyllanthus L.
Niruri Adans.
Cicca L.
Xylophylla L.
Conami Aubl.
Meborea Aubl.
Genesiphylla L'Hér.
Cathetus Lour.
Nymphanthus Lour.
Emblica Gaertn.
Kirganelia Juss.
Tricarium Lour.
Epistylium Sw.
Geminaria Raf.
Menarda Comm. ex A. Juss.
Anisonema A. Juss.
Scepasma Blume
Synexemia Raf.
Hexadena Raf.
Moeroris Raf.
Nellica Raf.
Asterandra Klotzsch
Eriococcus Hassk.
Ceramanthus Hassk.
Macraea Wight
Reidia Wight
Chorisandra Wight
Dichelactina Hance
Staurothylax Griff.
Hemicicca Baill.
Williamia Baill.
Orbicularia Baill.
Phyllanthodendron Hemsl.
Aporosella Chodat
Flueggeopsis (Müll. Arg.) K. Schum.
Nymania K. Schum.
Uranthera Pax & K. Hoffm.
Dimorphocladium Britton
Ramsdenia Britton
Roigia Britton
Dendrophyllanthus S. Moore
Pseudoglochidion Gamble
Hexaspermum Domin
Arachnodes Gagnep.
Reverchonia A. Gray
Sauropus Blume
Agyneia Vent., non L.
Ceratogynum Wight
Diplomorpha Griff.
Synostemon F. Muell.
Breyniopsis Beille
Heterocalymnantha Domin
Breynia J. R. & G. Forst.

Foersteria Scop.
Melanthesa Blume
Melanthesopsis Müll. Arg.
Glochidion J. R. & G. Forst.
Agyneia L.
Bradleja Banks ex Gaertn.
Gynoon A. Juss.
Glochidionopsis Blume
Lobocarpus Wight & Arn.
Glochisandra Wight
Zarcoa Llanos
Coccoglochidion K. Schum.
Hemiglochidion (Müll. Arg.) K. Schum.
Tetraglochidion K. Schum.
DRYPETEAE (Griseb.) Hurus.
Drypeteae Griseb.
Putranjiveae Endl.
Putranjivaceae Endl. ex A. Meeuse
Cyclostemonées Baill.
Pierardieae Thwaites
Cyclostemoneae Baill. ex Müll. Arg.
Lingelsheimia Pax
Danguyodrypetes Leandri
Aerisilvaea Radcl.-Sm.
Drypetes Vahl
Koelera Willd.
Limacia F. Dietr.
Liparena Poit. ex Léman
Cyclostemon Blume
Hemicyclia Wight & Arn.
Astylis Wight
Sphragidia Thwaites
Dodecastemon Hassk.
Pycnosandra Blume
Cometia Thouars ex Baill.
Anaua Miq.
Laneasagum Bedd.
Stelechanteria Thouars ex Baill.
Freireodendron Müll. Arg.
Humblotia Baill.
Guya Frapp. ex Cordem.
Riseleya Hemsl.
Calyptosepalum S. Moore
Brexiopsis H. Perrier
Sibangea Oliv.
Putranjiva Wall.
Nageia Roxb.
Palenga Thwaites
Liodendron Keng
ANTIDESMEAE (Sweet) Hurus.
Stilagineae C. Agardh
Antidesmeae Sweet
Spondianthinae (G. L. Webster) G. L. Webster
Spondiantheae G. L. Webster
Spondianthus Engl.
Megabaria De Wild.
Uapacinae Müll. Arg.
Uapaceae (Müll. Arg.) Hutch.
Uapacaceae (Müll. Arg.) Airy Shaw
Uapaca Baill.
Scepinae (Lindl.) G. L. Webster
Scepaceae Lindl.
Aporuseae Lindl. ex Miq.
Protomegabaria Hutch.
Maesobotrya Benth.
Staphysora Pierre
Richeria Vahl
Guarania Wedd. ex Baill.
Jablonskia G. L. Webster
Baccaurea Lour.
Pierardia Roxb. ex Jack
Adenocrepis Blume
Calyptroon Miq.
Microsepala Miq.
Everettiodendron Merr.
Gatnaia Gagnep.
Apodiscus Hutch.
Ashtonia Airy Shaw
Aporosa Blume
Leiocarpus Blume
Lepidostachys Wall. ex Lindl.
Scepa Lindl.
Tetractinostigma Hassk.
Antidesminae Müll. Arg.
Stilaginae C. Agardh
Hieronymeae Müll. Arg.
Thecacoris A. Juss.
Cometia Thouars ex Baill.
Cyathogyne Müll. Arg.
Baccaureopsis Pax
Phyllanoa Croizat
Celianella Jabl.
Leptonema A. Juss.
Antidesma L.
Bestram Adans.
Stilago L.
Rhytis Lour.
Hieronyma Allemão
Stilaginella Tul.

Porantherinae (Müll. Arg.) Eg. Köhler
Porantthereae Müll. Arg.
Poranthera Rudge
Oreoporanthera (Grüning) Hutch.
HYMENOCARDIEAE (Müll. Arg.) Hutch.
Hymenocardiaceae Airy Shaw
Hymenocardia Wall. ex Lindl.
Samaropyxis Miq.
Didymocistus Kuhlm.
BISCHOFIEAE (Müll. Arg.) Hurus.
Bischoffieae Müll. Arg.
Bischofiaceae (Müll. Arg.) Airy Shaw
Bischofia Blume
Microelus Wight & Arn.
Stylodiscus Benn.
CENTROPLACEAE Radcl.-Sm., **trib. nov.**
Centroplacus Pierre
MARTRETIEAE Eg. Köhler ex J. Léonard
Martretia Beille
Tacarcuna Huft

OLDFIELDIOIDEAE Eg. Köhler & G. L. Webster
Petalostigmatinae Pax & K. Hoffm.
Toxicodendrinae Pax & K. Hoffm.
Dissiliariinae Pax & K. Hoffm.
Paivaeusinae Pax & K. Hoffm.
CROIZATEAE G. L. Webster
Croizatia Steyerm.
PODOCALYCEAE G. L. Webster
Podocalycinae G. L. Webster
Podocalyx Klotzsch
Tetracoccinae G. A. Levin ex G. L. Webster
Tetracoccus Engelm. ex Parry
Halliophytum I. M. Johnst.
Paradrypetinae G. A. Levin
Paradrypetes Kuhlm.
CALETIEAE Müll. Arg.
Hyaenanchinae Baill. ex Müll. Arg.
Hyaenancheae (Baill. ex Müll. Arg.) Hutch.
Toxicodendrinae Pax
Hyaenanche Lamb. & Vahl
Toxicodendrum Thunb.
Dissiliariinae Pax & K. Hoffm.
Austrobuxus Miq.
Buraeavia Baill.
Choriophyllum Benth.
Canaca Guillaumin
Dissiliaria F. Muell. ex Baill.
Sankowskia P. I. Forst.
Whyanbeelia Airy Shaw & B. Hyland
Choriceras Baill.
Longetia Baill. ex Müll. Arg.
Petalostigmatinae Pax & K. Hoffm.
Petalostigmateae (Pax & K. Hoffm.) G. L. Webster
Petalostigma F. Muell.
Pseudanthinae Müll. Arg.
Kairothamnus Airy Shaw
Scagea McPherson
Neoroepera Müll. Arg. & F. Muell.
Micrantheum Desf.
Caletia Baill.
Allenia Ewart
Pseudanthus Sieber ex Spreng.
Stachystemon Planch.
Chrysostemon Klotzsch
Chorizotheca Müll. Arg.
PICRODENDREAE (Small) G. L. Webster
Picrodendraceae Small
Picrodendrinae (Small) G. L. Webster
Piranhea Baill.
Celaenodendron Standl.
Parodiodendron Hunz.
Picrodendron Planch.
Paivaeusinae Pax & K. Hoffm.
Oldfieldia Benth. & Hook. f.
Paivaeusa Welw. ex Benth.
Cecchia Chiov.
Mischodontinae Müll. Arg.
Androstachydaceae Airy Shaw
Aristogeitonia Prain
Paragelonium Leandri
Mischodon Thwaites
Voatamalo Capuron ex Bosser
Androstachys Prain
Weihea Sim
Stachyandra J. F. Leroy ex Radcl.-Sm.

ACALYPHOIDEAE Asch.
Acalyphaceae Klotzsch & Garcke
CLUTIEAE (Müll. Arg.) Pax
Cluytieae Müll. Arg.
Cluytiinae Pax
Clutia L.
Altora Adans.
Cratochwilia Neck.
POGONOPHOREAE (Müll. Arg.) G. L. Webster
Pogonophora Miers ex Benth.

CHÆTOCARPEAE (Müll. Arg.) G. L. Webster
Trigonopleura Hook. f.
Chætocarpus Thwaites
Mettenia Griseb.
Regnaldia Baill.
Gaedawakka [L.] ex Kuntze
Neochevaliera Beille
PEREAE (Baill.) Pax & K. Hoffm.
Perideae Baill.
Peraceae (Baill.) Klotzsch & Garcke
Prosopidoclineae Klotzsch
Pera Mutis
Perula Schreb.
Spixia Leandro
Peridium Schott in Spreng.
Schismatopera Klotzsch
CHEILOSEAE (Müll. Arg.) Airy Shaw & G. L. Webster
Cheilosiformes Pax & K. Hoffm.
Cheilosa Blume
Neoscortechinia Hook. f. ex Pax
Scortechinia Hook. f.
Alcinaeanthus Merr.
ERISMANTHEAE G. L. Webster
Erismanthus Wall. ex Müll. Arg.
Moultonianthus Merr.
Syndyophyllum Lauterb. & K. Schum.
DICOELIEAE Hurus.
Dicoelia Benth.
GALEARIEAE Benth.
Galeariinae Pax
Bennettieaceae R. Br.
Bennettieae R. Br. ex Schnizl.
Pandaceae Pierre
Microdesmis Planch.
Galearia Zoll. & Moritzi
Cremostachys Tul.
Bennettia R. Br.
Panda Pierre
Porphyranthus Engl.
AMPEREAE Müll. Arg.
Monotaxis Brongn.
Hippocrepandra Müll. Arg.
Amperea A. Juss.
AGROSTISTACHYDEAE (Müll. Arg.) G. L. Webster
Agrostistachys Dalzell
Sarcoclinium Wight
Heterocalyx Gagnep.
Pseudagrostistachys Pax & K. Hoffm.
Cyttaranthus J. Léonard
Chondrostylis Boerl.
Kunstlerodendron Ridl.
SPHYRANTHEREAE Radcl.-Sm., **trib. nov.**
Sphyranthera Hook. f.
CHROZOPHOREAE (Müll. Arg.) Pax & K. Hoffm.
Regulares (Pax & K. Hoffm.) Pax & K. Hoffm.
Speranskiinae G. L. Webster
Speranskia Baill.
Ditaxinae Griseb.
Caperonieae Müll. Arg.
Caperonia A. St.-Hil.
Cavanilla Vell.
Lepidococca Turcz.
Philyra Klotzsch
Ditaxis Vahl ex A. Juss.
Aphora Nutt.
Serophyton Benth.
Stenonia Didr.
Paxia Herter
Paxiuscula Herter
Argythamnia P. Browne
Chiropetalum A. Juss.
Desfontaena Vell.
Chlorocaulon Klotzsch ex Endl.
Aonikena Speg.
Doryxylinae G. L. Webster
Doryxylon Zoll.
Sumbavia Baill.
Mercadoa Náves
Sumbaviopsis J. J. Sm.
Thyrsanthera Pierre ex Gagnep.
Melanolepis Rchb. f. & Zoll.
Chrozophorinae G. L. Webster
Chrozophora Neck. ex A. Juss.
Tournesol Adans.
Tournesol Scop.
Lepidocroton C. Presl
CARYODENDREAE G. L. Webster
Caryodendron H. Karst.
Centrodiscus Müll. Arg.
Discoglypremna Prain
Alchorneopsis Müll. Arg.
BERNARDIEAE G. L. Webster
Bernardiiformes Pax & K. Hoffm.
Bernardia Houst. ex Mill.
Bivonia Spreng.
Traganthus Klotzsch

Phaedra Klotzsch ex Endl.
Polyboea Klotzsch ex Endl.
Tyria Klotzsch ex Endl.
Alevia Baill.
Passaea Baill.
Necepsia Prain
Palissya Baill.
Neopalissya Pax
Amyrea Leandri
Paranecepsia Radcl.-Sm.
Discocleidion (Müll. Arg.) Pax & K. Hoffm.
Adenophaedra (Müll. Arg.) Müll. Arg.
PYCNOCOMEAE Hutch.
Pycnocominae G. L. Webster
Wetriariiformes Pax & K. Hoffm.
Pycnocoma Benth.
Droceloncia J. Léonard
Argomuellera Pax
Neopycnocoma Pax
Wetriaria (Müll. Arg.) Kuntze
Blumeodendrinae G. L. Webster
Blumeodendron (Müll. Arg.) Kurz
Podadenia Thwaites
Ptychopyxis Miq.
Clarorivinia Pax & K. Hoffm.
Botryophora Hook. f.
EPIPRINEAE (Müll. Arg.) Hurus.
Epiprineae Müll. Arg.
Epiprininae Müll. Arg.
Cephalocrotoneae Müll. Arg.
Cladogynifomes Pax & K. Hoffm.
Cleidiocarpinae Thin
Epiprinus Griff.
Symphyllia Baill.
Cleidiocarpon Airy Shaw
Sinopimelodendron Tsiang
Koilodepas Hassk.
Calpigyne Blume
Nephrostylus Gagnep.
Cladogynos Zipp. ex Span.
Adenogynum Rchb. f. & Zoll.
Chloradenia Baill.
Cephalocrotonopsis Pax
Cephalocroton Hochst.
Adenochlaena Boivin ex Baill.
Centrostylis Baill.
Niedenzua Pax
Cephalomappinae G. L. Webster
Cephalomappa Baill.
Muricococcum Chun & F. C. How
ADELIEAE G. L. Webster
Adeliiformes Pax & K. Hoffm.
Adelia L.
Ricinella Müll. Arg.
Crotonogynopsis Pax
Enriquebeltrania Rzed.
Beltrania Miranda
Lasiocroton Griseb.
Leucocroton Griseb.
ALCHORNIEAE (Hurus.) Hutch.
Alchorneinae Hurus.
Alchorneiformes Pax & K. Hoffm.
Orfilea Baill.
Diderotia Baill.
Laurembergia Baill.
Bossera Leandri
Alchornea Sw.
Cladodes Lour.
Hermesia Humb. & Bonpl. ex Willd.
Schousboea Schumach.
Stipellaria Benth.
Lepidoturus Bojer ex Baill.
Bleekeria Miq.
Caelebogyne J. Sm.
Aparisthmium Endl.
Conceveibum A. Rich ex A. Juss.
Bocquillonia Baill.
Ramelia Baill.
Conceveibinae G. L. Webster
Conceveiba Aubl.
Conceveibastrum (Müll. Arg.) Pax & K. Hoffm.
Veconcibea (Müll. Arg.) Pax & K. Hoffm.
Gavarretia Baill.
Polyandra Leal
ACALYPHEAE Dumort.
Ricininae Griseb.
Ricinus L.
Cataputia Ludw.
Adrianinae Benth.
Adrianeae (Benth.) Pax & K. Hoffm.
Adriana Gaudich.
Meialisa Raf.
Trachycaryon Klotzsch
Mercurialinae Pax
Mercurialiiformes Pax & K. Hoffm.
Mercurialis L.
Seidelia Baill.
Leidesia Müll. Arg.
Dysopsidinae Hurus.

Dysopsis Baill.
Molina Gay
Cleidiinae G. L. Webster
Wetria Baill.
Pseudotrewia Miq.
Cleidion Blume
Reidia Casar.
Psilostachys Turcz.
Lasiostyles C. Presl
Tetraglossa Bedd.
Sampantaea Airy Shaw
Macaranginae (Hutch.) G. L. Webster
Macarangeae Hutch.
Macaranga Thouars
Panopia Noronha ex Thouars
Mappa A. Juss.
Pachystemon Blume
Mecostylis Kurz ex Teijsm. & Binn.
Phocea Seem.
Tanarius Rumph.
Claoxylinae Hurus.
Claoxyliformes Pax & K. Hoffm.
Claoxyleae Hutch.
Erythrococca Benth.
Deflersia Schweinf. ex Penz.
Poggeophyton Pax
Chloropatane Engl.
Athroandra (Hook. f.) Pax & K. Hoffm.
Claoxylon A. Juss.
Erythrochilus Reinw. ex Blume
Quadrasia Elmer
Claoxylopsis Leandri
Discoclaoxylon (Müll. Arg.) Pax & K. Hoffm.
Micrococca Benth.
Lobaniliinae Radcl.-Sm.
Lobanilia Radcl.-Sm.
Mareyinae (Hutch.) Radcl.-Sm., **stat. nov.**
Mareyeae Hutch.
Mareya Baill.
Mareyopsis Pax & K. Hoffm.
Rottlerinae Meisn.
Trewiaceae Lindl.
Trewiiformes Pax & K. Hoffm.
Coelodisceae Müll. Arg.
Malloteae Hutch.
Mallotus Lour.
Echinus Lour.
Rottlera Roxb.
Adisca Blume
Lasipana Raf.
Plagianthera Rchb. f. & Zoll.
Hancea Seem.
Axenfeldia Baill.
Coelodiscus Baill.
Aconceveibum Miq.
Echinocroton F. Muell.
Diplochlamys Müll. Arg.
Deuteromallotus Pax & K. Hoffm.
Cordemoya Baill.
Boutonia Bojer ex Bouton
Coccoceras Miq.
Avellanita Phil.
Trewia L.
Neotrewia Pax & K. Hoffm.
Rockinghamia Airy Shaw
Octospermum Airy Shaw
Acalyphinae Griseb.
Acalypha L.
Mercuriastrum Heist. ex Fabr.
Cupameni Adans.
Caturus L.
Usteria Dennst.
Galurus Spreng.
Cupamenis Raf.
Linostachys Klotzsch ex Schltdl.
Odonteilema Turcz.
Calyptrospatha Klotzsch ex Baill.
Gymnalypha Griseb.
Corythea S. Watson
Schizogyne Ehrenb. ex Pax
Acalyphopsis Pax & K. Hoffm.
Lasiococcinae G. L. Webster
Lasiococca Hook. f.
Spathiostemon Blume
Polydragma Hook. f.
Clonostylis S. Moore
Homonoia Lour.
Lumanaja Blanco
PLUKENETIEAE (Benth.) Hutch.
Plukenetieae Benth.
Plukenetiinae Benth.
Haematostemon (Müll. Arg.) Pax & K. Hoffm.
Astrococcus Benth.
Angostylis Benth.
Romanoa Trevis.
Anabaena A. Juss.
Anabaenella Pax & K. Hoffm.
Eleutherostigma Pax & K. Hoffm.
Plukenetia L.

Pterococcus Hassk.
Ceratococcus Meisn.
Sajorium Endl.
Hedraiostylus Hassk.
Tetracarpidium Pax
Pseudotragia Pax
Angostylidium (Müll. Arg.) Pax & K. Hoffm.
Apodandra Pax & K. Hoffm.
Elaeophora Ducke
Vigia Vell.
Fragariopsis A. St.-Hil.
Accia A. St.-Hil.
Botryanthe Klotzsch
Tragiinae G. L. Webster
Cnesmone Blume
Cenesmon Gagnep.
Megistostigma Hook. f.
Clavistylus J. J. Sm.
Sphaerostylis Baill.
Tragiella Pax & K. Hoffm.
Tragia Plum. ex L.
Schorigeram Adans.
Bia Klotzsch
Leptorrhachis Klotzsch
Leucandra Klotzsch
Ctenomeria Harv.
Lassia Baill.
Leptobotrys Baill.
Zuckertia Baill.
Platygyne P. Mercier
Acanthocaulon Klotzsch ex Endl.
Acidoton Sw.
Gitara Pax & K. Hoffm.
Pachystylidium Pax & K. Hoffm.
Dalechampiinae (Müll. Arg.) G. L. Webster
Dalechampieae Müll. Arg.
Dalechampia Plum. ex L.
Cremophyllum Scheidw.
Rhopalostylis Klotzsch ex Baill.
Megalostylis S. Moore
OMPHALEAE (Pax & K. Hoffm.) G. L. Webster
Omphaleinae Pax & K. Hoffm.
Omphalea L.
Omphalandria P. Browne
Duchola Adans.
Ronnowia Buc'hoz
Hecatea Thouars
Hebecocca Beurl.
Neomphalea Pax & K. Hoffm.

CROTONOIDEAE Pax
MICRANDREAE (Müll. Arg.) G. L. Webster
Micrandreae Müll. Arg.
Micrandriformes Pax & K. Hoffm.
Micrandrinae Müll. Arg.
Micrandra Benth.
Pogonophyllum Didr.
Micrandropsis W. A. Rodrigues
Cunuria Baill.
Clusiophyllum Müll. Arg.
Heveinae Müll. Arg.
Hevea Aubl.
Siphonia Rich.
Caoutchoua J. F. Gmel.
Siphonanthus Schreb. ex Baill.
MANIHOTEAE (Müll. Arg.) Pax
Manihoteae Müll. Arg.
Manihot Mill.
Mandioca Link
Janipha Humb., Bonpl. & Kunth
Hotnima A. Chev.
Manihotoides D. J. Rogers & Appan
Cnidoscolus Pohl
Jussieuia Houst.
Bivonea Raf.
Victorinia León
ADENOCLINEAE (Müll. Arg.) G. L. Webster
Adenoclineae Müll. Arg.
Adenoclininae Müll. Arg.
Tetrorchidiinae Pax
Glycydendron Ducke
Klaineanthus Pierre ex Prain
Tetrorchidium Poepp. & Endl.
Hasskarlia Baill.
Tetrorchidiopsis Rauschert
Adenocline Turcz.
Diplostylis Sond.
Paradenocline Müll. Arg.
Ditta Griseb.
Endosperminae Pax & K. Hoffm.
Endospermum Benth.
Capellenia Teijsm. & Binn.
GELONIEAE (Müll. Arg.) Pax
Gelonieae Müll. Arg.
Suregada Roxb. ex Rottler
Gelonium Roxb. ex Willd.

Erythrocarpus Blume
Ceratophorus Sond.
Cladogelonium Leandri
ELATERIOSPERMEAE G. L. Webster
Elateriospermum Blume
JATROPHEAE (Meisn.) Pax
Jatropheae Meisn.
Jatrophinae Pax
Joannesieae (Müll. Arg.) Pax
Joannesieae Müll. Arg.
Jatropha L.
Curcas Adans.
Castigliona Ruiz & Pav.
Mozinna Ortega
Loureira Cav.
Adenoropium Pohl
Zimapania Engl. & Pax
Collenucia Chiov.
Vaupesia R. E. Schult.
Oligoceras Gagnep.
Deutzianthus Gagnep.
Loerzingia Airy Shaw
Joannesia Vell.
Anda A. Juss.
Andicus Vell.
Leeuwenbergia Letouzey & N. Hallé
Annesijoa Pax & K. Hoffm.
CODIAEAE (Pax) Hutch.
Codiaeinae Pax
Baloghia Endl.
Steigeria Müll. Arg.
Hylandia Airy Shaw
Ostodes Blume
Pausandra Radlk.
Dodecastigma Ducke
Pantadenia Gagnep.
Parapantadenia Capuron
Dimorphocalyx Thwaites
Fontainea Heckel
Codiaeum Rumph. ex A. Juss.
Phyllaurea Lour.
Synapisma Endl.
Junghuhnia Miq.
Acidocroton Griseb.
Ophellantha Standl.
Blachia Baill.
Bruxanel(l)ia Dennst. ex Kostel.
Strophioblachia Boerl.
Sagotia Baill.
Baliospermum Blume
TRIGONOSTEMONEAE G. L. Webster
Trigonostemon Blume
Enchidium Jack
Silvaea Hook. & Arn.
Athroisma Griff.
Telogyne Baill.
Tritaxis Baill.
Tylosepalum Kurz ex Teijsm. & Binn.
Nepenthandra S. Moore
Actephilopsis Ridl.
Prosartema Gagnep.
Poilaniella Gagnep.
Neotrigonostemon Pax & K. Hoffm.
Kurziodendron N. P. Balakr.
RICINOCARPEAE Müll. Arg.
Ricinocarpoideae (Müll. Arg.) Pax
Ricinocarpaceae (Müll. Arg.) Hurus.
Ricinocarpinae G. L. Webster
Ricinocarpos Desf.
Roeperia Spreng.
Echinosphaera Sieber ex Steud.
Alphandia Baill.
Beyeria Miq.
Calyptrostigma Klotzsch
Beyeriopsis Müll. Arg.
Bertyinae Müll. Arg.
Bertya Planch.
Myricanthe Airy Shaw
Cocconerion Baill.
Borneodendron Airy Shaw
CROTONEAE Dumort.
Mildbraedia Pax
Neojatropha Pax
Paracroton Miq.
Fahrenheitia Rchb. f. & Zoll.
Desmostemon Thwaites
Moacroton Croizat
Cubacroton Alain, **synon. nov.**
Croton L.
Cascarilla Adans.
Aroton Neck.
Cinogasum Neck.
Brunsvia Neck.
Tridesmis Lour.
Schradera Willd.
Lascadium Raf.
Decarinium Raf.
Heptallon Raf.
Drepadenium Raf.
Hendecandra Eschsch.

Halecus [Rumph. ex] Raf.
Kurkas Raf.
Semilta Raf.
Luntia Neck. ex Raf.
Penteca Raf.
Triplandra Raf.
Astrogyne Benth.
Merleta Raf.
Pleopadium Raf.
Medea Klotzsch
Podostachys Klotzsch
Astraea Klotzsch
Ocalia Klotzsch
Cleodora Klotzsch
Eutropia Klotzsch
Timandra Klotzsch
Engelmannia Klotzsch
Geiseleria Klotzsch
Pilinophytum Klotzsch
Lasiogyne Klotzsch
Tiglium Klotzsch
Brachystachys (L'Hér.) Klotzsch
Macrocroton Klotzsch
Angelandra Endl.
Barhamia Klotzsch
Cyclostigma Klotzsch
Gynamblosis Torr.
Crotonanthus Klotzsch ex Schltdl.
Calypteriopetalon Hassk.
Myriogomphos Didr.
Klotzschiphytum Baill.
Comatocroton H. Karst.
Monguia Chapel. ex Baill.
Anisophyllum Boivin ex Baill.
Argyra Noronha ex Baill.
Aubertia Chapel. ex Baill.
Argyrodendron Klotzsch
Leucadenia Klotzsch ex Baill.
Oxydectes [L. ex] Kuntze
Heterocroton S. Moore
Crotonopsis Michx.
Leptemon Raf.
Friesia Spreng.
Eremocarpus Benth.
Piscaria Piper
Julocroton Mart.
Cieca Adans.
Heterochlamys Turcz.
Centrandra H. Karst.
RICINODENDREAE (Pax) Hutch.
Ricinodendrinae Pax
Givotia Griff.
Ricinodendron Müll. Arg.
Schinziophyton Hutch. ex Radcl.-Sm.
ALEURITIDEAE Hurus.
Aleuritinae (Hurus.) G. L. Webster
Aleurites J. R. & G. Forst.
Camirium [Rumph. ex] Gaertn.
Telopea Sol. ex Baill.
Reutealis Airy Shaw
Vernicia Lour.
Dryandra Thunb.
Elaeococca A. Juss.
Garciinae Müll. Arg.
Garcia Vahl ex Rohr
Grosserinae G. L. Webster
Cavacoa J. Léonard
Grossera Pax
Tapoïdes Airy Shaw
Anomalocalyx Ducke
Sandwithia Lanj.
Tannodia Baill.
Tandonia Baill.
Neoholstia Rauschert
Holstia Pax
Domohinea Leandri
Crotonogyninae G. L. Webster
Cyrtogonone Prain
Crotonogyne Müll. Arg.
Neomanniophyton Pax & K. Hoffm.
Manniophyton Müll. Arg.
Neoboutoniinae (Hutch.) G. L. Webster
Neoboutoniiformes Pax & K. Hoffm.
Neoboutonieae Hutch.
Neoboutonia Müll. Arg.
Benoistiinae (Radcl.-Sm.) Radcl.-Sm., **stat. nov.**
Benoistieae Radcl.-Sm.
Benoistia H. Perrier & Leandri

EUPHORBIOIDEAE (Boiss.) G. L. Webster
Euphorbieae Boiss.
Sapioideae Hurus.
STOMATOCALYCEAE (Müll. Arg.) G. L. Webster
Stomatocalyceae Müll. Arg.
Stomatocalycinae Müll. Arg.
Plagiostyles Pierre
Pimelodendron Hassk.
Stomatocalyx Müll. Arg.

Hamilcoinae Pax & K. Hoffm.
Hamilcoa Prain
Nealchornea Huber
HIPPOMANEAE A. Juss. ex Bartl.
Stillingiidae Baill.
Excaecarieae Baill.
Hippomaninae Pax
Sapioideae Hurus.
Carumbiinae Müll. Arg.
Homalanthinae Pax & K. Hoffm.
Homalanthus A. Juss.
    Omalanthus A. Juss.
    Carumbium Reinw.
    Wartmannia Müll. Arg.
    Dibrachion Regel
Hippomaninae Griseb.
Euhippomaneae Müll. Arg.
Adenopeltinae Pax & K. Hoffm.
Excoecariinae Pax & K. Hoffm.
Gymnanthinae Pax & K. Hoffm.
Mabeinae Pax & K. Hoffm.
Stillingiinae Pax & K. Hoffm.
Dendrocousinsia Millsp.
Excoecaria L.
    Commia Lour.
    Glyphostylus Gagnep.
    Taeniosapium Müll. Arg.
Spirostachys Sond.
Sebastiania Spreng.
Colliguaja Molina
Microstachys A. Juss.
    Cnemidostachys Mart. & Zucc.
    Elachocroton F. Muell.
    Tragiopsis H. Karst.
Conosapium Müll. Arg.
Grimmeodendron Urb.
Bonania A. Rich.
    Hypocoton Urb.
Adenopeltis Bertero ex A. Juss.
Stillingia Garden ex L.
    Gymnostillingia Müll. Arg.
Spegazziniophytum Esser
Falconeria Royle
Sapium Jacq.
    Sapiopsis Müll. Arg.
Hippomane L.
    Mancanilla Mill.
Dendrothrix Esser
Senefelderopsis Steyerm.
Pleradenophora Esser
Sclerocroton Hochst. in C. Krauss
Balakata Esser
Triadica Lour.
    Stillingfleetia Bojer
    Seborium Raf.
Shirakiopsis Esser
    Shirakia Hurus., non S. Kawas. p.p.
Mabea Aubl.
Gymnanthes Sw.
    Sapium P. Browne?
    Ateramnus P. Browne?
    Adenogyne Klotzsch
    Sarothrostachys Klotzsch
    Duvigneaudia J. Léonard
Ditrysinia Raf.
Anomostachys (Baill.) Hurus.
Neoshirakia Esser
    Shirakia Hurus., non S. Kawas. p.p.
Pseudosenefeldera Esser
Actinostemon Mart. ex Klotzsch
    Gussonia Spreng.?
    Dactylostemon Klotzsch
Senefeldera Mart.
Rhodothyrsus Esser
Dalembertia Baill.
    Alcoceria Fernald
Maprounea Aubl.
    Aegopricum L.
    Aegopricon L. f.
PACHYSTROMATEAE (Pax & K. Hoffm.) Pax
Pachystroma Müll. Arg.
    Acantholoma Gaudich. ex Baill.
HUREAE Dumort.
Algernonia Baill.
Tetraplandra Baill.
Ophthalmoblapton Allemão
Hura L.
EUPHORBIEAE Blume
Anthosteminae (Baill.) G. L. Webster
Anthostemideae Baill.
Anthostemeae Klotzsch & Garcke
Anthostema A. Juss.
Dichostemma Pierre
Neoguillauminiinae Croizat
Neoguillauminia Croizat
Calycopeplus Planch.
Euphorbiinae Hurus.
Pedilanthaceae Klotzsch & Garcke
Pedilanthinae (Klotzsch & Garcke) Hurus.

Anisophyllae Klotzsch & Garcke
Tithymalinae Klotzsch & Garcke
Euphorbia L.
Lathyris Trew
Euphorbium Hill
Athymalus Neck.
Keraselma Neck.
Tithymalus Gaertn.
Dactylanthes Haw.
Esula (Pers.) Haw.
Galarhoeus Haw.
Medusea Haw.
Treisia Haw.
Characias Gray
Desmonema Raf.
Pleuradena Raf.
Poinsettia Graham
Lacanthis Raf.
Bojeria Raf.
Adenorima Raf.
Aklema Raf.
Agaloma Raf.
Allobia Raf.
Cyathophora Raf.
Kanopikon Raf.
Lepadena Raf.
Lophobios Raf.
Murtekias Raf.
Nisomenes Raf.
Peccana Raf.
Tirucalia Raf.
Torfasadis Raf.
Tumalis Raf.
Vallaris Raf.
Zalitea Raf.
Dematra Raf.
Kobiosis Raf.
Alectoroctonum Schltdl.
Anthacantha Lem.
Adenopetalum Klotzsch & Garcke
Arthrothamnus Klotzsch & Garcke
Dichrophyllum Klotzsch & Garcke
Eumecanthus Klotzsch & Garcke
Euphorbiastrum Klotzsch & Garcke
Leptopus Klotzsch & Garcke
Sterigmanthe Klotzsch & Garcke
Tithymalopsis Klotzsch & Garcke
Trichosterigma Klotzsch & Garcke
Petaloma Raf. ex Baill.
Petalandra F. Muell. ex Boiss.
Lyciopsis (Boiss.) Schweinf.
Chylogala Fourr.
Epurga Fourr.
Euphorbion St.-Lag.
Zygophyllidium (Boiss.) Small
Diplocyathium H. Schmidt
Euphorbiodendron Millsp.
Euphorbiopsis Lév.
Dichylium Britton
Ctenadena Prokh.
Cystidospermum Prokh.
Sclerocyathium Prokh.
Elaeophorbia Stapf
Chamaesyce Gray
Anisophyllum Haw.
Ditritra Raf.
Endoisila Raf.
Xamesike Raf.
Aplarina Raf.
Cubanthus (Boiss.) Millsp.
Monadenium Pax
Lortia Rendle
Stenadenium Pax
Synadenium Boiss.
Endadenium L. C. Leach
Pedilanthus Neck. ex Poit.
Tithymalus Mill.
Tithymaloides Ortega
Ventenatia Tratt.
Crepidaria Haw.
Tirucalia Raf.
Diadenaria Klotzsch & Garcke
Hexadenia Klotzsch & Garcke
Afrotrewia Pax & K. Hoffm.
Chlamydojatropha Pax & K. Hoffm.
Myladenia Airy Shaw

ACKNOWLEDGEMENTS

I would like to thank Dr Hans-Joachim Esser of Hamburg for his contribution to this treatment, especially for the exceptionally difficult Tribe HIPPOMANEAE A. Juss. ex Bartl., upon which he has done much new work, as well as for the small related tribes PACHYSTROMATEAE (Pax & K. Hoffm.) Pax & HUREAE Dumort. I would also like to thank Dr Lynn Gillespie of Ottawa for casting a very critical eye over my accounts of the groups in which she specializes, namely the tribes PLUKENETIEAE (Benth.) Hutch. & OMPHALEAE (Pax & K. Hoffm.) G. L. Webster, and Dr Stefan Dressler of Frankfurt am Main for doing likewise with the Tribe BRIDELIEAE Müll. Arg. I am also very grateful to Dr Wolfgang Stuppy, formerly of Kaiserslautern, for allowing me to see his thesis, 'Systematische Morphologie und Anatomie der Samen der Biovulaten Euphorbiaceen', Kaiserslautern (1995), and to quote from it here. I would also like to thank my former assistant, Alison Marriott, who provided the much-needed stimulus and encouragement for me to get started on the work, and who typed out the first subfamily; Veronica Marx who typed out the second; and Christine Barker and Melanie Thomas of the Section where I used to work, who gave me assistance in a variety of ways. My thanks also go to Suzy Dickerson of the Editorial Unit of Kew's Information Services Division for all her help in expediting the production of this publication, to Camilla Speight, whose 50 plates of excellent artwork grace its pages, to Christabel King for the colourful frontispiece and cover illustration, to Ruth Linklater for her extensive editorial work on the manuscript and to Christine Beard for her work on the page make-up.

REFERENCES

Agardh, C. A. (1825). *Stilaginaceae.* In Aphorismi Botanici, 14: 199–200.

Airy Shaw, H. K. (1960). Notes on Malaysian *Euphorbiaceae.* Kew Bull. 14: 353–397.

——— (1963). Notes on Malaysian and other Asiatic *Euphorbiaceae.* XXXII: *Coccoceras* Miq. reduced to *Mallotus* Lour. Kew Bull. 16: 349–352.

——— (1965). Diagnoses of new families, new names etc. for the seventh edition of Willis' "Dictionary". Kew Bull. 18: 249–272.

——— (rev.) (1966). A Dictionary of the Flowering Plants and Ferns (J. C. Willis), ed. 7, Addenda xxi–xxii, 1–1214. Cambridge University Press.

——— (1967). Notes on Malaysian and other Asiatic *Euphorbiaceae.* Kew Bull. 20: 379–414.

——— (1971). Two new species of *Micrococca* Benth. from Malaya. Kew Bull. 25: 524–526.

——— (1972). The *Euphorbiaceae* of Siam. Kew Bull. 26: 191–363.

——— (rev.) (1973). A Dictionary of the Flowering Plants and Ferns (J. C. Willis), ed. 8, 1–1245. Cambridge University Press.

——— (1974). Noteworthy *Euphorbiaceae* from Tropical Asia. Hooker's Icon. Pl. 38(1): tt. 3701–3725.

——— (1975). The *Euphorbiaceae* of Borneo. Kew Bull. Addit. Ser. IV: 1–245.

——— (1977). Additions and corrections to the *Euphorbiaceae* of Siam. Kew Bull. 32: 69–83.

——— (1978). Notes on Malesian and other Asiatic *Euphorbiaceae*. CC: *Clonostylis* S. Moore reduced to *Spathiostemon* Bl. Kew Bull. 32: 407–408.

——— (1980a). New *Euphorbiaceae* from New Guinea. Kew Bull. 34: 591–598.

——— (1980b). The *Euphorbiaceae* of New Guinea. Kew Bull. Addit. Ser. VIII: 1–243.

——— (1980c). A Partial Synopsis of the *Euphorbiaceae-Platylobeae* of Australia. Kew Bull. 35: 577–700.

——— (1981). The *Euphorbiaceae* of Sumatra. Kew Bull. 36: 239–374.

Baillon, H. (1858). Étude Générale du Groupe des *Euphorbiacées*, 1–669. V. Masson, Paris.

Balakrishnan, N. P. & Chakrabarty, T. (1993). The genus *Paracroton* (*Euphorbiaceae*) in the Indian subcontinent. Kew Bull. 48: 715–726.

Baldwin, J. & Schultes, R. E. (1947). A Conspectus of the genus *Cunuria*. Bot Mus. Leafl. 12: 325–351.

Bentham, G. (1867). *Burseraceae*. In Bentham, G. & Hooker, J. D., Genera Plantarum 1: 993. Lovell Reeve & Co., London.

——— (1880). *Euphorbiaceae*. In Bentham, G. & Hooker, J. D., Genera Plantarum 3: 239–340. Lovell Reeve & Co., London.

Bhatnagar, A. K. & Kapil, R. N. (1974). *Bischofia javanica* — its relationship with *Euphorbiaceae*. Phytomorphol. 23: 264–267.

Borhidi, A. (1972). La Taxonomia del Género *Platygyne* Merc. Ann. Hist.-Nat. Mus. Natl. Hung. 64: 89–94.

Bouchat, A. & Léonard, J. (1986). Révision du genre *Necepsia* Prain (*Euphorbiacée* africano-malgache). Bull. Jard. Bot. Belg. 56: 179–194.

Breteler, F. J. & Mennega, A. M. W. (1994). Novitates Gabonenses 17. *Conceveiba leptostachys*, a new *Euphorbiacea* from Gabon and Cameroun. Bull. Jard. Bot. Belg. 63: 209–217.

Britton, N. L. & Wilson, P. (1924). Botany of Porto Rico and the Virgin Islands 5: 471–506. New York Academy of Sciences.

Brunel, J. F. (1987). Sur le genre *Phyllanthus* L. et quelques genres voisins de la Tribu des *Phyllantheae* Dumort. (*Euphorbiaceae, Phyllantheae*) en Afrique Intertropicale et à Madagascar. Thèse présentée à l'Université L. Pasteur, Strasbourg.

Carter, S. (1988). *Euphorbiaceae*, Part 2. In Polhill, R. M., ed. Flora of Tropical East Africa, 409–564. A. A. Balkema; Rotterdam, Boston.

Chakrabarty, T. & Balakrishnan, N. P. (1985, publ. 1987). A note on the genus *Ostodes* (*Euphorbiaceae*). Bull. Bot. Surv. India 27: 259–260.

Coode, M. J. E. (1976). Typification of *Macaranga* Du Petit-Thouars. Taxon 25(1): 184.

Croizat, L. (1941). The Tribe *Plukenetiinae* of the *Euphorbiaceae* in Eastern Tropical Asia. J. Arnold Arbor. 22: 417–431.

——— (1942a). On certain *Euphorbiaceae* from the Tropical Far East. J. Arnold Arbor. 23: 29–54.

——— (1942b). New and Critical *Euphorbiaceae* chiefly from the Southeastern United States. Bull. Torrey Bot. Club 69: 445–460.

Cronquist, A. (1981). An Integrated System of Classification of Flowering Plants. Columbia Univ. Press, New York.

Daydon Jackson, B., ed. (1895). Index Kewensis Plantarum Phanerogamarum. 4 Vols. Clarendon Press, Oxford.

Dehgan, B. & Webster, G. L. (1979). Morphology and Infrageneric Relationships of the genus *Jatropha* (*Euphorbiaceae*). Univ. Calif. Publ. Bot. 74: 1–73.

Dressler, R. L. (1954). The genus *Tetracoccus* (*Euphorbiaceae*). Rhodora 56: 45–61.

——— (1957). The genus *Pedilanthus* (*Euphorbiaceae*). Contr. Gray Herb. CLXXXII: 1–188.

——— (1961). A Synopsis of *Poinsettia*. Ann. Missouri Bot. Gard. 48: 329–341.

Farr, E. R. *et al.*, eds. (1979). Index Nominum Genericorum (Plantarum). 3 vols. Utrecht.
Forman, L. L. (1966). The reinstatement of *Galearia* Zoll. & Mor. & *Microdesmis* Hook. f. (sic) in the *Pandaceae*. Kew Bull. 20: 309–321.
——— (1971). A Synopsis of *Galearia* Zoll. & Mor. (*Pandaceae*). Kew Bull. 26: 153–165.
Franco, R. P. (1990). The genus *Hyeronima* (sic) in South America. Bot. Jahrb. Syst. 111(3): 297–346.
Gilg, E. (1908). *Flacourtiaceae* Africanae. In Engler, A., ed., Beitr. Fl. Afr. 32. Bot. Jahrb. Syst. 40: 444–518.
Gillespie, L. J. (1988). Rev. Phyl. Anal. *Omphalea* (diss. ined.): 144.
——— (1993). A Synopsis of Neotropical *Plukenetia* (*Euphorbiaceae*) including two new species. Syst. Bot. 18: 575–592.
——— (1994). Pollen Morphology and Phylogeny of the Tribe *Plukenetieae* (*Euphorbiaceae*). Ann. Missouri Bot. Gard. 81: 317–348.
——— (1997). *Omphalea* (*Euphorbiaceae*) in Madagascar: A new species and a new combination. Novon 7: 127–136.
Grierson, A. J. C. & Long, D. G. (1987). Fl. Bhutan 1(3): 811. Royal Botanic Garden, Edinburgh.
Grüning, G. (1913). *Euphorbiaceae-Porantheroideae* et *Ricinocarpoideae*. In Engler, A., ed., Das Pflanzenreich, Heft 58: 1–97. Wilhelm Engelmann, Berlin.
Hallier, H. (1910). Über Phanerogamen von unsicherer oder unrichtiger Stellung. Meded. Rijks-Herb. 1: 1–41.
Hans, A. S. (1973). Chromosomal Conspectus of the *Euphorbiaceae*. Taxon 22(5/6): 591–636.
Hayden, W. J. (1977). Comparative Anatomy & Systematics of *Picrodendron*, genus incertae sedis. J. Arnold Arbor. 58: 257–279.
——— (1987). The identity of the genus *Neowawraea* (*Euphorbiaceae*). Brittonia 39(2): 268–277.
——— (1990). Notes on neotropical *Amanoa* (*Euphorbiaceae*). Brittonia 42 (4): 260–270.
——— (1994). Systematic Anatomy of *Euphorbiaceae* Subfamily *Oldfieldioideae*. 1. Overview. Ann. Missouri Bot. Gard. 81(1): 180–202.
———, Gillis, W., Stone, D. E., Broome, C. R. & Webster, G. L. (1984). Systematics & Palynology of *Picrodendron*. Further evidence for relationship with the *Oldfieldioideae* (*Euphorbiaceae*). J. Arnold Arbor. 65: 105–127.
Hemming, C. F. & Radcliffe-Smith, A. (1987). A Revision of the Somali species of *Jatropha* (*Euphorbiaceae*). Kew Bull. 42: 103–122.
Henderson, R. J. F. (1992). Studies in *Euphorbiaceae* A. L. Juss., sens. lat. I. A Revision of *Amperea* Adr. Juss. (*Acalyphoideae* Ascherson, *Ampereae* Müll. Arg.). Austral. Syst. Bot. 5: 1–27.
Hooker, J. D. (1887). *Euphorbiaceae*. Flora of British India, 5: 241–462. L. Reeve & Co., London.
Hooker, W. J. (1849). Niger Fl.: 503. Ballière, London.
Hurusawa, I. (1954). Eine nochmalige Durchsicht des herrkömmlichen Systems der *Euphorbiaceen* im weiteren Sinne. J. Fac. Sci. Univ. Tokyo, Sect. 3, Bot. 6: 209–342.
Hutchinson, J. (1969). Tribalism in the Family *Euphorbiaceae*. Amer. J. Bot. 57(7): 738–758.
——— (ined.). *Euphorbiaceae* for The Genera of Flowering Plants.
Ingram, J. (1967). A Revisional study of *Argythamnia* Subgenus *Argythamnia*. Gentes Herb. 10(1): 1–38.
——— (1980a). The generic limits of *Argythamnia* (*Euphorbiaceae*) defined. Gentes Herb. 11(7): 427–436.
——— (1980b). A Revision of *Argythamnia* Subgenus *Chiropetalum* (A. Juss.) Ingram. Gentes Herb. 11(7): 437–468.
Jablonski, E. (1967). *Euphorbiaceae*. In Maguire, B., ed., Botany of the Guyana Highland — Part VII. Mem. New York Bot. Gard. 17: 80–190.

Jussieu, A. de (1824). De *Euphorbiacearum* generibus medicisque earumdem viribus Tentamen, tabulis æneis 18 illustratum. 1–118. Didot Jr, Paris

Köhler, E. (1965). Die Pollenmorphologie der Biovulaten *Euphorbiaceae* und ihre Bedeutung für die Taxonomie. Grana Palynol. 6: 26–120.

Koutnik, D. L. (1987). A Taxonomic Revision of the Hawaiian species of the genus *Chamaesyce* (*Euphorbiaceae*). Allertonia 4(6): 331–388.

Kuntze, O. (1891). Revisio Generum Plantarum, 3 vols. Arthur Felix, Leipzig.

La Fon, R., ed. (1980–1996). *Euphorbia* Journal 1–10. Strawberry Press, Mill Valley, California.

Leandri, J. (1958). *Euphorbiacées*. Fl. Madagascar 111(1): 1–209. Typographie Firmin-Didot et Cie., Paris.

Lebrun, J.-P. & Stork, A. L. (1991). Énumération des Plantes à Fleurs d'Afrique Tropicale. I. Conservatoire et Jardin Botaniques de la Ville de Genève. Geneva.

Léonard, J. (1956). Notulae Systematicae XXI: Observations sur les Genres *Oldfieldia*, *Paivaeusa* et *Cecchia* (*Euphorbiaceae* Africanae). Bull. Jard. Bot. État 26: 335–343.

——— (1962). *Euphorbiaceae*. Flore du Congo et du Rwanda-Burundi, 8(1): 1–214. Publications de l'INÉAC, Bruxelles.

——— (1989). Révision du genre africain *Martretia* Beille. Bull. Jard. Bot. Belg. 59: 319–332.

——— (1995). Révision des espèces zaïroises des genres *Thecacoris* A. Juss. et *Cyathogyne* Müll. Arg. (*Euphorbiaceae*). Bull. Jard. Bot. Belg. 64: 13–52.

——— & Mosango, M. (1985). *Hymenocardiaceae*. In Fl. Afr. Cent.: 1–16. Jard. Bot. Nat. Belg., Bruxelles.

Levin, G. (1986a). Systematic foliar morphology of *Phyllanthoideae* (*Euphorbiaceae*) I. Conspectus. Ann. Missouri Bot. Gard. 73: 29–85.

——— (1986b). Systematic foliar morphology of *Phyllanthoideae* (*Euphorbiaceae*) II. Phenetic analysis. Ann. Missouri Bot. Gard. 73: 86–98.

——— (1986c). Systematic foliar morphology of *Phyllanthoideae* (*Euphorbiaceae*) III. Cladistic analysis. Syst. Bot. 11: 515–530.

——— & Simpson, M. G. (1994). Phylogenetic implications of pollen ultrastructure in the *Oldfieldioideae* (*Euphorbiaceae*). Ann. Missouri Bot. Gard. 81: 203–238.

Lindley, J. (1836). A Natural System of Botany, ed. 2. Longman & Co., London.

McPherson, G. (1985). *Scagea*, a new genus of *Euphorbiaceae* from New Caledonia. Bull. Mus. Natl. Hist. Nat., Sér. 4, B, Adansonia 7(3): 247–250.

——— & Tirel, C. (1987). *Euphorbiacées* I. Flore de la Nouvelle-Calédonie et Dépendances 14: 1–226. Muséum National d'Histoire Naturelle, Paris.

McVaugh, R. (1944). The genus *Cnidoscolus*: Generic limits and Intrageneric groups. Bull. Torrey Bot. Club 71: 457–474.

Meeuse, A. D. J. (1990). The *Euphorbiaceae* auct. plur.: an unnatural taxon. Eburon, Delft.

Mennega, A. M. W. (1987). Wood anatomy of the *Euphorbiaceae*, in particular of the Subfamily *Phyllanthoideae*. Bot. J. Linn. Soc. 94: 111–126.

Merrill, E. D. (1935). A Commentary on Loureiro's "Flora Cochinchinensis". Trans. Amer. Philos. Soc. 24: 1–49.

Metcalfe, C. R. & Chalk, L. (1950). Anatomy of the Dicotyledons. Clarendon Press, Oxford.

Miller, K. & Webster, G. L. (1962). Systematic Position of *Cnidoscolus* and *Jatropha*. Brittonia 14: 174–180.

Millspaugh, C. F. (1909). Praenunciae bahamenses II. Publ. Field Columbian Mus., Bot. Ser. 2: 289–322.

——— (1913). The genera *Pedilanthus* & *Cubanthus*, and other American *Euphorbiaceae*. Publ. Field Columbian Mus., Bot. Ser. 2: 353–377.

——— (1916). Contributions to North American *Euphorbiaceae.* VI. Publ. Field Columbian Mus., Bot. Ser. 2: 401–420.

——— (1917). *Trichosterigma benedictum.* Addisonia 2: 3–4, t. 42.

Milne-Redhead, E. W. B. H. (1949). Tropical African Plants XX. Kew Bull. 3: 456–457.

Müller Argoviensis, J. (1864). System der *Euphorbiaceen.* Bot. Zeitung (Berlin) 22: 324.

——— (1865). *Euphorbiaceae.* Vorläufige Mittheilungen aus dem für De Candolle's Prodromus bestimmten Manuscript über diese Familie. Linnaea 34: 1–224.

——— (1866). *Euphorbiaceae* (excluding *Euphorbieae*). In De Candolle, A., Prodromus Systematis Naturalis Regni Vegetabilis 15(2): 189–1260. V. Masson et fils, Paris.

——— (1874). *Euphorbiaceae,* part II. In Martius, C. F. P. von, Fl. Brasiliensis 11(2): 293–751.

Oudejans, R. C. H. M. (1990). World Catalogue of Species Names Published in the Tribe *Euphorbieae* (*Euphorbiaceae*) with their Geographical Distribution. Utrecht.

——— (1993). Ditto, Cumulative Supplement I. Scherpenzeel, The Netherlands.

Pax, F. (1890). *Euphorbiaceae.* In Engler, A. & Prantl, K., eds., Die Natürlichen Pflanzenfamilien, ed. 1, 3(5): 1–119. Wilhelm Engelmann, Leipzig.

——— (1910). *Euphorbiaceae-Jatropheae.* In Engler, A., ed., Das Pflanzenreich, Heft 42: 1–148. Wilhelm Engelmann, Berlin.

——— (1924). *Joannesieae.* In Die Phylogenie der *Euphorbiaceae.* Bot. Jahrb. Syst. 59: 129–182.

——— & Hoffmann, K. (1911). *Euphorbiaceae-Cluytieae.* In Engler, A., ed., Das Pflanzenreich, Heft 47: 1–124. Wilhelm Engelmann, Berlin.

——— (1912a). *Euphorbiaceaee-Gelonieae* & *-Hippomaneae.* In Engler, A., ed., Das Pflanzenreich, Heft 52: 1–41 & 1–319. Wilhelm Engelmann, Berlin.

——— (1912b). *Euphorbiaceae-Acalypheae-Chrozophorinae.* In Engler, A., ed., Das Pflanzenreich, Heft 57: 1–142. Wilhelm Engelmann, Berlin.

——— (1914). *Euphorbiaceae-Acalypheae-Mercurialinae* and Additamentum V. In Engler, A., ed., Das Pflanzenreich, Heft 63: 1–473. Wilhelm Engelmann, Berlin.

——— (1919). *Euphorbiaceaee-Acalypheaee-Plukenetiinae, -Epiprineae* & *-Ricininae, Dalechampieae* and Additamentum VI. In Engler, A., ed., Das Pflanzenreich, Heft 68: 1–134, 1–59 &1–63. Wilhelm Engelmann, Berlin.

——— (1922). *Euphorbiaceae-Phyllantheae.* In Engler, A., ed., Das Pflanzenreich, Heft 81: 1–349. Wilhelm Engelmann, Berlin.

——— (1931). *Euphorbiaceae.* In Engler, A. & Harms, H., eds., Die Natürlichen Pflanzenfamilien, ed. 2, 19C: 11–251. Wilhelm Engelmann, Leipzig.

Pfeiffer, L. (1874). Nomenclator Botanicus 2: 1128. T. Fischer, Cassell.

Prain, D. (1911). A review of the genera *Erythrococca* & *Micrococca.* Ann. Bot. 25: 575–638.

——— (1912). *Euphorbiaceae: Crotonogyne-Micrococca.* In Thiselton-Dyer, Sir W. T., ed., Fl. Trop. Afr. 6(1): 820–879. Lovell Reeve & Co., London.

——— (1920). *Euphorbiaceae: Croton-Sapium.* In Thiselton-Dyer, Sir W. T., ed., Flora Capensis 5(2): 410–516. L. Reeve & Co., London.

Punt, W. (1962). Pollen morphology of the *Euphorbiaceae* with special reference to taxonomy: 1–116. North-Holland Publishing Co., Amsterdam.

Radcliffe-Smith, A. (1968). An account of the genus *Givotia* Griff. (*Euphorbiaceae*). Kew Bull. 22: 493–505.

——— (1973). An account of the genus *Cephalocroton* Hochst. (*Euphorbiaceae*). Kew Bull. 28: 123–132.

——— (1981). Notes on African *Euphorbiaceae* IX. Kew Bull. 35: 777.

——— (1984). Notes on African *Euphorbiaceae* XIV. Kew Bull. 39: 794.

——— (1987). *Euphorbiaceae,* Part 1. In Polhill, R. M., ed., Flora of Tropical East Africa, 1–407. A. A. Balkema; Rotterdam, Boston.

——— (1988). Notes on Madagascan *Euphorbiaceae* I: On the Identity of *Paragelonium* and on the Affinities of *Benoistia* & *Claoxylopsis* (*Euphorbiaceae*). Kew Bull. 43: 625–647.

——— (1990). Notes on Madagascan *Euphorbiaceae* III : *Stachyandra.* Kew Bull. 45: 561–568.

——— (1991). Notes on African *Euphorbiaceae* XXV: *Jatropha* (VI). Kew Bull. 46: 141–157.

——— (1992). Notes on African *Euphorbiaceae* XXVII: *Clutia.* Kew Bull. 47: 111–119.

——— (1993). Notes on Australian *Euphorbiaceae* II: *Pseudanthus* and *Stachystemon.* Kew Bull. 48: 165–168.

——— (1996a). A second species of *Aristogeitonia* (*Euphorbiaceae*) for Tanzania. Kew Bull. 51: 799–801.

——— (1996b). *Euphorbiaceae,* Part 1. In Pope, G. V., ed., Flora Zambesiaca 9(4): 1–337. Royal Botanic Gardens, Kew.

——— (1997). Notes on African & Madagascan *Euphorbiaceae.* Kew Bull. 52: 171–176.

——— (1998a). Notes on Madagascan *Euphorbiaceae* VI: A synopsis of *Tannodia* Baill. (*Crotonoïdeae-Aleuritideae-Grosserineae*) with especial reference to Madagascar. Kew Bull. 53: 173–186.

——— (1998b). Notes on Madagascan *Euphorbiaceae* VII: A Synopsis of the Genus *Amyrea* Leandri (*Euphorbiaceae-Acalyphoideae*). Kew Bull. 53: 437–451.

——— (1998c). Notes on Madagascan *Euphorbiaceae* VIII: A third species of *Aristogeitonia* (*Euphorbiaceae*) for Madagascar. Kew Bull. 53: 977–980.

——— & Govaerts, R. (1997). New names & new combinations in the *Euphorbiaceae-Acalyphoideae.* Kew Bull. 52: 477–481.

——— & Ratter, J. A. (1996). A new *Piranhea* from Brazil, & the subsumption of the genus *Celaenodendron* (*Euphorbiaceae-Oldfieldioideae*). Kew Bull. 51: 543–548.

Rogers, D. J. & Appan, S. G. (1973). *Manihot, Manihotoides* (*Euphorbiaceae*). Fl. Neotrop. Monogr. 13: 1–272.

Schot, A. M. (1995). A Synopsis of Taxonomic Changes in *Aporosa* Bl. (*Euphorbiaceae*). Blumea 40: 449–460.

Schultes, R. E. (1952). Studies in the genus *Micrandra* I: The relationship of the genus *Cunuria* to *Micrandra.* Bot. Mus. Leafl. 15: 201–222.

——— (1970). The History of Taxonomic Studies in *Hevea.* Bot. Rev. (Lancaster) 36: 197–276.

——— (1990). A brief Taxonomic View of the genus *Hevea.* Malaysian Rubber Research & Development Board Monograph 14: 1–57.

Secco, R. (1985). Notas sobre o novo conceito de *Sagotia racemosa* em relação ás suas variedades. Acta Amazon. 15(1–2, suppl.): 81–85.

——— (1988). Dialissepalia do gênero *Sandwithia* Lanj.: Uma novidade botânica do alto Rio Negro e da Venezuela. Bol. Mus. Paraense Emilio Goeldi, N.S., Bot. 4: 177–185.

——— (1990). Revisão dos gêneros *Anomalocalyx* Ducke, *Dodecastigma* Ducke, *Pausandra* Radlk., *Pogonophora* Miers ex Benth. e *Sagotia* Baill. (*Euphorbiaceae-Crotonoideae*) para a América do Sul. Museu Paraense Emilio Goeldi, Belem.

Secco, R. & Webster, G. L. (1990). Materiales para a Flora Amazonica IX: Ensaio sobre a sistemática do gênero *Richeria* Vahl. (Euphorbiaceae). Bol. Mus. Paraense Emilio Goeldi, N.S., Bot. 6: 141–158.

Small, J. K. (1913). *Euphorbiaceae.* In Britton, N. L. & Brown, A., eds., An Illustrated Flora of the Northern United States, ed. 2, 2: 452–477. Charles Scribner's sons, New York.

——— (1917). The Jamaica Walnut. J. New York Bot. Gard. 18: 180–186.

Stuppy, W. (1995). Systematische Anatomie und Morphologie der Samen der Biovulaten *Euphorbiaceen.* Dissertation, Fachbereich Biologie, Universität Kaiserslautern, Germany. 1–364.

Susila Rani, S. R. M. & Balakrishnan, N. P. (1995). A Revision of the genus *Claoxylon* Adr. Jussieu (*Euphorbiaceae*) in India. Rheedea 5(2): 113–141.
Takhtajan, A. L. (1980). Outline of the Classification of Flowering Plants (*Magnoliophyta*). Bot. Rev. (Lancaster) 46: 225–359.
Thin, N. N. (1984). Tribus *Alchornieae* (*Euphorbiaceae*) of Vietnamese Flora. Tap Chi Sinh Hoc 6(3): 26–29.
——— (1988). Tribe *Epiprineae* (Müll. Arg.) Hurusawa (*Euphorbiaceae*) in Vietnam. Tap Chi Sinh Hoc 10(2): 30–33.
Thomas, D. W. (1990). *Conceveiba* Aublet (*Euphorbiaceae*) new to Africa. Ann. Missouri Bot. Gard. 77: 856–858.
van Welzen, P. C. (1994). Taxonomy, Phylogeny and Geography of *Neoscortechinia* Hook. f. ex Pax (*Euphorbiaceae*). Blumea 39: 301–320.
——— (1995). Taxonomy and Phylogeny of the *Euphorbiaceae* Tribe *Erismantheae* G. L. Webster. Blumea 40: 375–396.
———, Bulalacao, L. J. & van Ôn, T. (1995). A Taxonomic Revision of the Malesian genus *Trigonopleura* Hook. f. (*Euphorbiaceae*). Blumea 40: 363–374.
Verdcourt, B. (1954). Revision of the genus *Zimmermannia* Pax. Kew Bull. 9: 38–40.
Vindt, J. (1960). Monographie des *Euphorbiacées* du Maroc: Index Bibliographique. Trav. Inst. Sci. Chérifien, Sér. Bot. 19: 481–521.
Wallnöfer, B. (1991). Beschreibung der zweiten art in der Neotropischen gattung *Nealchornea* Huber (*Euphorbiaceae*). Linzer Biol. Beitr. 23(2): 775–785.
Webster, G. L. (1956). A monographic study of the West Indian species of *Phyllanthus*. J. Arnold Arbor. 37: 91–122.
——— (1965). A Revision of the genus *Meineckia* (*Euphorbiaceae*). Acta Bot. Neerl. 14: 323 –365.
——— (1967a). The genera of *Euphorbiaceae* in the Southeastern United States. J. Arnold Arbor. 48: 303–430.
——— (1967b). *Acidoton* (*Euphorbiaceae*) in Central America. Ann. Missouri Bot. Gard. 54: 191.
——— (1975). Conspectus of a new Classification of the *Euphorbiaceae*. Taxon 24(5/6): 593–601.
——— (1979). A Revision of *Margaritaria* (*Euphorbiaceae*). J. Arnold Arbor. 60: 403–444.
——— (1984a). A Revision of *Flueggea* (*Euphorbiaceae*). Allertonia 3(4): 259–312.
——— (1984b). *Jablonskia*, a new genus of *Euphorbiaceae* from South America. Syst. Bot. 9 (2): 229–235.
——— (1987). The Saga of the Spurges: of a Review of Classification and Relationships in the *Euphorbiales*. Bot. J. Linn. Soc. 94: 3–46.
——— (1993). A Provisional Synopsis of the Sections of the genus *Croton* (*Euphorbiaceae*). Taxon 42: 793–823.
——— (1994). Synopsis of the genera and suprageneric taxa of *Euphorbiaceae*. Ann. Missouri Bot. Gard. 81(1): 33–144.
——— & Armbruster, W. S. (1991). A Synopsis of the Neotropical species of *Dalechampia* (*Euphorbiaceae*). Bot. J. Linn. Soc. 105: 137–177.
——— & Burch, D. (1968). 97. *Euphorbiaceae*. In Woodson, R. E., Jr. & Schery, R. W., eds., Flora of Panama, Part IV. Ann. Missouri Bot. Gard. 54: 211–350.
———, Gillespie, L. & Steyermark, J. (1987). Systematics of *Croizatia* (*Euphorbiaceae*). Syst. Bot. 12(1): 1–8.
——— & Huft, M. J. (1988). Revised Synopsis of Panamanian *Euphorbiaceae*. Ann. Missouri Bot. Gard. 75: 1087–1144.
——— & Miller, K. I. (1963). The genus *Reverchonia* (*Euphorbiaceae*). Rhodora 65: 193–207.
Wheeler, L. C. (1939). A Miscellany of New World *Euphorbiaceae*. II. Contr. Gray Herb. 127: 48–78, tt. 3–4.

——— (1941). *Euphorbia* Subgenus *Chamaesyce* in Canada and the United States exclusive of Southern Florida. Rhodora 43: 97–154, 168–205, 223–286, tt. 654–668. (Reprinted as Contr. Gray Herb. 136).

——— (1943). The genera of living *Euphorbieae.* Amer. Midl. Naturalist 30: 456–503.

——— (1975). Euphorbiaceous genera lectotypified. Taxon 24(4): 534–538.

White, A., Dyer, R. A. & Sloane, B. L. (1941). The Succulent *Euphorbieae* (Southern Africa). 2 Vols. Abbey Garden Press, Pasadena, California.

Whitmore, T. C. (1973). *Euphorbiaceae.* Tree Flora of Malaya 2: 34–136. Longman, London.

INDEX TO TAXA

INDEX TO ILLUSTRATIONS

Euphorbia exigua L. – Seed, Fig. 50, p. 414
Euphorbia falcata L. – Seed, Fig. 50, p. 414
Euphorbia flavicoma DC. (*E. brittingeri* Opiz ex Samp.) – Fruit, Fig. 49, p. 413
Euphorbia francheti B. Fedtsch. – Seed, Fig. 50, p. 414
Euphorbia goyazensis Boiss. – Cyathial glands, Fig. 48, p. 412
Euphorbia grantii Oliv. – Cyathial glands, Fig. 48, p. 412
Euphorbia grossheimii (Prokh.) Prokh. – Fruit, Fig. 49, p. 413; Seed, Fig. 50, p. 414
Euphorbia guyoniana Boiss. & Reut. – Seed, Fig. 50, p. 414
Euphorbia helioscopia L. – Seed, Fig. 50, p. 414
Euphorbia herniariifolia Willd. – Seed, Fig. 50, p. 414
Euphorbia heterophylla L. – Cyathial glands, Fig. 48, p. 412
Euphorbia hirsuta L. (*E. pubescens* Vahl). – Seed, Fig. 50, p. 414
Euphorbia inderiensis Less. ex Kar. & Kir. – Seed, Fig. 50, p. 414
Euphorbia lancasteriana Radcl.-Sm – Fig. 47, p. 411
Euphorbia lathyris L. – Cyathial glands, Fig. 48, p. 412; Seed, Fig. 50, p. 414
Euphorbia longetuberculosa Hochst. ex Boiss. – Cyathial glands, Fig. 48, p. 412
Euphorbia macrocarpa Boiss. & Buhse – Fruit, Fig. 49, p. 413
Euphorbia marginata Pursh – Cover Design & Frontispiece; Cyathial glands, Fig. 48, p. 412
Euphorbia medicaginea Boiss. – Seed, Fig. 50, p. 414
Euphorbia microsphera Boiss. – Fruit, Fig. 49, p. 413
Euphorbia myrsinites L. – Seed, Fig. 50, p. 414
Euphorbia noxia Pax – Cyathial glands, Fig. 48, p. 412
Euphorbia peplus L. – Fruit, Fig. 49, p. 413; Seed, Fig 50, p. 414
Euphorbia phymatosperma Boiss. & Gaill. – Seed, Fig. 50, p. 414
Euphorbia portlandica L. – Seed, Fig. 50, p. 414
Euphorbia (Poinsettia) pulcherrima Willd. ex Klotzsch – Cover Design & Frontispiece.
Euphorbia robbiae Turrill – Cyathial glands, Fig. 48, p. 412
Euphorbia rosea Retz. – Cyathial glands, Fig. 48, p. 412
Euphorbia schimperiana Scheele – Cyathial glands, Fig. 48, p. 412
Euphorbia schizolepis F. Muell. ex Boiss. – Cyathial glands, Fig. 48, p. 412
Euphorbia sulcata Lens ex Loisel. – Seed, Fig. 50, p. 414
Euphorbia turczaninowii Kar. & Kir. – Seed, Fig. 50, p. 414
Euphorbia valerianifolia Lam. (*E. cybirensis* Boiss.) – Fruit, Fig. 49, p. 413
Euphorbia waldsteinii (Soják) Radcl.-Sm. – Fruit, Fig. 49, p. 413
Excoecaria cochinchinensis Lour. – Cover Design & Frontispiece.
Gavarretia terminalis Baill. – Fig. 24, p. 198
Gitara panamensis Croizat – Fig. 35, p. 259
Gonatogyne brasiliensis Müll. Arg. (S.: *Savia brasiliensis* (Müll. Arg.) Pax & K. Hoffm.) – Fig. 1, p. 8
Grossera vignei Hoyle – Fig. 42, p. 336
Haematostemon guianensis Sandwith – Fig. 31, p. 242
Jatropha integerrima Jacq. – Cover Design & Frontispiece.
Jatropha multifida L. – Cover Design & Frontispiece.
Keayodendron bridelioides (Mildbr. ex Hutch. & Dalziel) Leandri (B.: *Casearia bridelioides* Mildbr. ex Hutch. & Dalziel) – Fig. 4, p. 30
Koilodepas longifolium Hook. f. – Fig. 22, p. 181
Micrandropsis scleroxylon (W. A. Rodrigues) W. A. Rodrigues (B.: *Micrandra scleroxylon* W. A. Rodrigues) – Fig. 37, p. 269
Neotrewia cumingii (Müll. Arg.) Pax & K. Hoffm. (B.: *Mallotus cumingii* Müll. Arg.) – Fig. 30, p. 233